TEXTILE SCIENCE AND TECHNOLOGY 5

POLYPROPYLENE FIBERS – SCIENCE AND TECHNOLOGY

TEXTILE SCIENCE AND TECHNOLOGY

Volume 1 Open-end Spinning *by* V. Rohlena *et al.*
Volume 2 Processing of Polyester Fibres *by* O. Pajgrt *and* B. Reichstädter
Volume 3 Shuttleless Weaving Machines *by* O. Talavášek *and* V. Svatý
Volume 4 Fluorescent Brightening Agents *by* R. Williamson
Volume 5 Polypropylene Fibers — Science and Technology *by* M. Ahmed

TEXTILE SCIENCE AND TECHNOLOGY 5

POLYPROPYLENE FIBERS – SCIENCE AND TECHNOLOGY

M. AHMED

President,
International Fibers Inc., Reidsville, NC, U.S.A.

Sponsored by the
SOCIETY OF PLASTICS ENGINEERS, Inc.

ELSEVIER SCIENTIFIC PUBLISHING COMPANY
AMSTERDAM OXFORD NEW YORK 1982

ELSEVIER SCIENTIFIC PUBLISHING COMPANY
Molenwerf 1
P.O. Box 211, 1000 AE Amsterdam, The Netherlands

Distributors for the United States and Canada:

ELSEVIER SCIENCE PUBLISHING COMPANY INC.
52, Vanderbilt Avenue
New York, N.Y. 10017

Library of Congress Cataloging in Publication Data

Ahmed, M. (Mukhtar)
Polypropylene fibers, science and technology.

(Textile science and technology ; v. 5)
Includes bibliographies and index.
1. Polypropylene fibers. I. Society of Plastics
Engineers, inc. II. Title. III. Series.
TS1548.7.P6A36 677'.4745 82-7423
ISBN 0-444-42090-8 AACR2

ISBN 0-444-42090-8 (Vol. 5)
ISBN 0-444-41818-0 (Series)

SOCIETY OF PLASTICS ENGINEERS, INC.
EXECUTIVE OFFICE
14 FAIRFIELD DRIVE, BROOKFIELD CENTER, CONNECTICUT 06805
TELEPHONE 203-775-0471 — TELEX 643-712

ROBERT D. FORGER
Executive Director

FOREWORD

The Society of Plastics Engineers is pleased to sponsor and to endorse this new volume, Polypropylene Fibers - Science and Technology. This publication serves a long-standing need and represents yet another contribution by the publisher and the Society to the technology of plastics and polymers.

Through the medium of its Technical Volumes Committee, SPE has been involved in sponsoring books of this nature since 1956. During this period it has been the catalyst for publication of at least one book annually in some vital segment of plastics not previously covered in the literature. The committee has surveyed and determined needs for specific titles, located authors and editors, considered broad content, and, most importantly, reviewed final manuscripts to ensure accuracy of technical material.

It is this same technical competence that is the Society's hallmark in its other activities - education, programs, conferences, periodicals, and meetings. Its reservoir of 24,000 practicing plastics engineers is its greatest resource, a resource which has made it the largest organization of its type in plastics worldwide.

Robert D. Forger
Executive Director
Society of Plastics Engineers
January 15, 1982

Technical Volumes Committee
James L. Throne, Chairman
Lee. L. Blyler, Jr.
A. Wayne Cagle
Jerome L. Dunn
Thomas W. Haas
John L. Kardos
Richard R. Kraybill
Timothy Lim
Eldridge M. Mount,III
Nickolaos A. Peppas
B. J. Sexton

PREFACE

Polypropylene fiber is one of the newest fibers introduced in the last decade to the textile trade. In a relatively short developmental period, it has become an important member of the rapidly growing family of synthetic fibers. Early projections, based on ultra-optimistic forecasts of raw material costs, may not have been met as yet, but it seems that polypropylene fiber has its developmental period behind it and is ready to take off.

As this is a new fiber, its manufacturing technology and marketing prospects are not yet well known. This book is an attempt to consolidate such information from a unified point of view. It should not be considered as a manual of operating instructions for the practitioner, nor a discourse on theoretical aspects of polymer rheology and fiber structure for the academician. The volume is primarily intended for those who are engaged in fiber production, process and product development, and marketing activities. The emphasis is not on what is presently being produced from polypropylene but on what is possible. Thus, the book should be most useful to the process and product development engineers and marketing strategists who are charged with the task of taking care of tomorrow rather than of today.

In spite of the heavy emphasis on development work, the material covered in this volume is broad enough to be useful to such diverse groups as equipment designers, machinery manufacturers, polymer scientists, textile engineers, technical students, and the faculty. The volume should be of significant help also to those who are contemplating entering into polypropylene fiber manufacturing or diversifying their product lines. It is hoped that a happy balance of theory and practice has been maintained.

In view of the vast scope of the subject and the limited space available, some omissions are inevitable. In instances where a choice has to be made, the priority was given to information of practical importance. Where needed, the discussion has been frequently supplemented with extensive listings of patent references. An extensive bibliography has been included in each chapter to broaden the scope of the book and impart to it a comprehensive value.

I share the satisfaction of the completion of this project with my wife, Razia. This work would not have been possible without her active support and encouragement. I am indebted to Mrs. Dottie Meacher, whose typing services remained available to me throughout the course of this work. My thanks are also due to Hercules, Inc., whose excellent library facilities made this project possible. Needless to say, several of my colleagues had much to contribute to this work in one form or the other. I am indebted to them all. However, while improvements are theirs, the shortcomings are still mine.

Mukhtar Ahmed
President,

International Fibers Inc.
Reidsville, NC, U.S.A.

TABLE OF CONTENTS

Chapter I. THE POLYMER

I. INTRODUCTION

Polypropylene is an important member of the family of olefin plastics. It is the first synthetic stereoregular polymer to achieve industrial importance. The early attempts to initiate polymerization of propylene led to branched products of only a low degree of polymerization and resulted in the formation of non-crystalline products varying in viscosities from oils to semi-solids. These non-crystalline polymers were of little, if any, commercial importance. It was only after Ziegler and co-workers discovered a coordination catalyst system for the production of high-density polyethylene, that propylene could be polymerized in the form as we know it today in the industry.

In the 1950s Natta and co-workers at the Polytechnic Institute of Milan were the first to report the preparation of crystalline polypropylene. They used heterogeneous catalysts of the type discovered by Ziegler for the low-pressure polymerization of ethylene to high-density polyethylene. Subsequent work revealed that the hard, crystalline product consisted of stereoregular polymer chains. It represented not only a turning point in polymer chemistry but also opened up a whole vista of new technology of vinyl polymerization.

The properties of stereoregular polypropylene - more specifically, isotactic polypropylene - made it evident that this material offered vast possibilities in the thermoplastic industry as well as in film and fiber production. Basic patents were issued in 1954 and industrial production of isotactic polypropylene was started in 1957 by Montecatini in Italy. Subsequently several other manufacturers entered the field, largely as licensees of Montecatini, and today a number of major chemical companies are manufacturing polypropylene and its co-polymers all over the world.

The present world capacity of polypropylene is estimated as 2.5 to 3.0 million tons, divided mainly between the United States, Japan, West Germany, Italy, the United Kingdom, France, Belgium, and the Netherlands. This capacity is expected to rise sharply and may reach 6.5 million tons by 1980 (ref.1). The present United States capacity of polypropylene is estimated as 4.5 billion pounds (see Table I for a breakdown). This capacity is currently on a sharp rise. Based on a 10 to 15 percent growth rate, the United States production of polypropylene could be between 5.0 and 6.5 billion pounds by 1980.

The reason for rapid expansion in production capacity for polypropylene is the property advantages polypropylene has over polyethylene, its nearest competitor on the basis of cost. The expectation is that the raw material for polypropylene will be available considerably more cheaply than that for polyethylene. An economic edge in raw material cost and the newest developments in high efficiency catalysts would undoubtedly make polypropylene a very low-cost plastic material, a position which polyethylene and PVC have held so far.

TABLE I

U.S. Polypropylene Capacity (1979-80)

Companies	*Capacity (million pounds per year)*
Amoco Chemical	
Alvin, Tex.	250
New Castle, Del.	250
Atlantic Richfield	
La Port, Tex.	400
Eastman Chemical	
Longview, Tex.	I40
Exxon Chemical	
Baytown, Tex.	400
Gulf Oil Chemical	
Cedar Bayou, Tex.	400
Hercules Inc.	
Bayport, Tex.	400
Lake Charles, La.	700
Montreal, Que. Canada	200
Northern Petrochemical	
Morris, Ill.	200
Novamont	
Neal, W.Va.	200
La Port, Tex.	200
Phillips Petroleum	
Pasadena, Tex.	90
Rexene Polymer	
Odessa, Tex.	I40
Bayport, Tex.	I50
Shell Chemical	
Woodbury, N.J.	280
Norco, La.	300
TOTAL	4,700

Polypropylene is used as molding resin as well as in many extruded forms, including film, fiber, and slit film yarns. A breakdown of its uses in the United States, Western Europe and Japan is given in Table II. Currently about 30 percent of world production of polypropylene goes into various fiber applications. This use of polypropylene is expected to grow in proportion. Thus, it is anticipated that by 1980 the fiber usage of polypropylene will count for at least 40 percent of total production. It is estimated that by 1980, more than 800 million pounds of polypropylene will be consumed for fiber manufacturing in the United States alone. These optimistic forecasts are based on the continued penetration of natural fiber markets by polypropylene fiber, the entrance of several new polypropylene fiber manufacturers as a result of expiration of basic patents, the creation of new markets as a result of new fiber developments, and above all, on the relatively low price of the raw material.

TABLE II

Polypropylene Resin Usage in the Industrially Developed Countries

x 1000 tons

	Western Europe		*U.S.A.*		*Japan*	
	1974	*1980*	*1974*	*1980*	*1973*	*1980*
Split film	I90	350	83		44	III
Monofil	26	65	83		3	3
Spun bonded						
BCF	9	27	12	61	9	10
CF			30			
Staple and tow	45	78	26	45	41	42
Total fiber	270	520	234	748	97	166
Mouldings	270	450				
Film	60	160	730	1679	566	745
Extrusion	60	180				
Total (all uses)	660	1310	964	2427	663	911
Fiber % of Total	41	40	24	31	15	17

The use of polypropylene as a base material for fiber production was pioneered by Montecatini just after the commercialization of polypropylene in Italy and patents to that effect were issued as early as 1958-1960 (ref.3,4). Although it has not been too difficult to introduce monofilaments and coarse staple fibers to the market, only limited quantities of low denier multifilaments and fine dpf staple fibers were offered in early 1960s. It was not until the mid-sixties that polypropylene attracted much attention from fiber manufacturers and became established as one of the most potent raw material candidates for fiber production. Montecatini was, of course, the pioneer in Europe, but considerable efforts for commercialization of polypropylene fibers were expended by I.C.I. in the United Kingdom, by Canadian Celanese, Hercules, Reeves Brothers, and Vectra in North America.

The growth of polypropylene consumption for fiber manufacturing has been rather slow. This has been because of the well-established position of nylon, and later polyester, against which polypropylene fiber had little property advantage to offer. The price advantage in raw material was not realized until recently and patent barriers prevented newcomers from entering the field. Now, most of the basic patents for conversion of polypropylene into fibers have expired, more economical pigmentation processes have been developed, and the price advantage for raw material over nylon and polyester has finally become a reality, it is expected that many of small manufacturers will venture into the production of polypropylene fibers.

II. MANUFACTURE

II.1. Monomer

Propylene is the monomer of polypropylene. It is a liquifiable hydrocarbon gas boiling at -47°C. The production of propylene is initimately connected with petroleum refinery operation and the petrochemical industry in general.

Apart from a small production in Germany during the War years, 1939-45, all propylene since the beginning of the petrochemical industry has been made by the thermal decomposition of petroleum hydrocarbons in one or other of the following ways:

(a) As a co-product with ethylene (as well as higher hydrocarbons) in the high temperature pyrolysis of gaseous and liquid hydrocarbons obtained from petroleum distillation and natural gas fractionation.

(b) As a by-product in the manufacture of gasoline by the catalytic cracking of higher hydrocarbons and to some extent of thermal reforming of gasoline for higher octane values.

Propylene as a Co-product with Ethylene: Crude petroleum consists mainly of paraffin hydrocarbons (40-75 %), naphthenes (20-50 %) comprising mainly the derivatives of cyclopentene and cyclohexane, and aromatic hydrocarbons (6-22 %) with a high proportion of benzene. The composition varies widely from source to source. The carbon members of the distillation fractionations are given in Table III. The distillates contain little or no olefins, dienes, or acetylene. The gas fraction consists mainly of methane, ethane, propane, and butanes. If this fraction is not to be used on the site, it may be condensed and bottled and is known as LPG (liquified petroleum gases), consisting mainly of butane and propane.

Most of the distillation fractions are suitable feed-stocks for cracking operation, depending on economic considerations and the type of plant available. Typical yields of propylene from various feed-stocks are given in Table IV.

In the United States, catalytic cracking of heavy distillation fractions is widely used for upgrading the production of high octane gasoline. Large quantities of low-boiling saturated hydrocarbon gases are thereby obtained. Large volumes of ethane, propane and n-butane are also recovered from natural gas. These are widely used as the raw material for the production of ethylene and propylene by thermal cracking.

TABLE III

Approximate Distillation Ranges of Petroleum Fractions (ref.I4)

Product	*Distillation range (°C)*	*Carbon number*
Gas (LPG)	15 - 30	1 - 4
Light naphtha	20 - 100	4 - 7
Petrol	70 - 200	6 - 12
Paraffin (kerosene)	175 - 275	9 - 16
Diesel oil (gas oil)	200 - 400	15 - 25
Paraffin wax	230 - 300(50-75 mm Hg)	18 - 35
Lubricating oil, pitch	Residue	>30

Whereas the cracking of ethane gives very little propylene, propane cracking yields both ethylene and propylene in proportions depending upon the cracking severity. Typically, ethylene and propylene can be produced in a weight ratio of 1:0.3-0.6, a ratio of 0.4 parts by weight of propylene to one part of ethylene being about average.

TABLE IV

Typical Yield of Propylene From Various Feedstocks

Feedstock	*Propylene yield (% of feedstock)*
LPG	15 - 25
i-butane	80 - 85
Naphtha	15 - 20
Crude oil	6 - 12

n-Butane can possibly be recovered in large amounts in future from several sources.

(a) As more natural gas is processed, more n-butane can be recovered.

(b) More n-butane can be recovered from various streams in some refineries.

(c) n-Butane is blended in large amounts in gasoline and various liquified hydrcarbons. Stricter volatility requirements to reduce air pollution may restrict such use in future.

Iso-butane may be of considerable interest as a feed-stock in the future, especially if the demand for propylene and iso-butylene increases. Hydrocracking of heavy petroleum distillation fractions results in higher amounts of gasoline than catalytic cracking. But it also increases the amount of iso-butylene.

The co-production of ethylene and propylene (as well as higher unsaturated and saturated hydrocarbons) by the high temperature pyrolysis of naphtha and higher petroleum fractions has been known for long but it is only in the last three decades that this method has become of major industrial significance. The cracking of gas oil for the production of ethylene has been practiced in the USA on a limited scale for some time and the first cracking installation for this purpose in Europe, namely the plant of British Celanese commissioned at Spondon in 1942 was based on the pyrolysis of a paraffinic gas oil. However, it soon became apparent that the preferred raw material for gaseous olefin production in Europe (and also in Japan) was naphtha, the reasons being as follows:

(a) The establishment of a substantial petroleum refining industry in Europe, which took place in the 1950s, was, and stil is, based on Middle East or North African crudes with a relatively high content of light fractions.

(b) The demand for gasoline in Europe is low compared with the USA while fuel oil requirements are relatively high.

(c) It is thus possible in Europe to meet all motor gasoline requirements from straight run products, suitably upgraded by catalytic reforming and very little catalytic cracking of heavy fractions is necessary.

(d) By-product propylene supplies in Europe are limited by the relatively modest catalytic cracking capacity and also by the fact, unlike the USA, that no abundant and cheap supplies of propane and butane are available.

It thus came about that the co-production of ethylene and propylene (together with butenes and butadiene as well as prolysis gasoline) in Europe and Japan is based largely on naphtha cracking though limited amounts are made from higher fractions. Because of their high content of paraffin hydrocarbons, the naphtha fractions of the crude oils normally available in Europe and Japan are very suitable for conversion in the steam cracking process.

In the early naphtha crackers such as those installed by Montecatini in Italy and by ICI in England in 1950-1, most of the heat of cracking was supplied by addition to the prevaporized and preheated naphtha of a relatively large quantity of steam previously heated to about 925°C. Subsequent technological improvements have made it possible to operate steam cracking of naphtha at much lower steam ratios in tubular reactions, the heat of cracking being supplied by external heating. Nowadays, propylene yields may be 15 percent or more by weight on naphtha fed together with over 30 percent by weight of ethylene.

A further extension of olefin technology is the recently announced cracking reactors of Union Carbide and Monsanto. Feed-stock flexibility allows production from a wide range of petroleum feed-stocks including crude oil (Fig.1). These units can recover upto 70 percent of the contents of a barrel of crude oil as olefins. Conventional petroleum cracking operations give only 40 percent maximum chemical yield (ref.23).

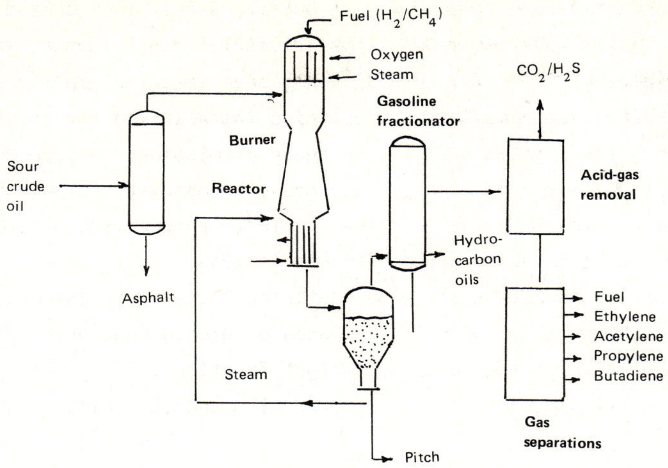

Fig. 1. Flow-sheet of Carbide Process for Producing Olefins Directly from Crude Oil (ref.23)

The yield of propylene and ethylene which can be obtained from the pyrolysis of hydrocarbons depends on the composition of the hydrocarbon feed-stock and on the cracking conditions employed. Table V gives typical yields obtainable from four widely different feed-stocks and Table VI shows the effect on the yield structure of varying the cracking severity for a naphtha feed-stock.

TABLE V

Yield of Major Products from Medium Severity Cracking of Hydrocarbons (ref.24)

	Feedstock			
	Ethane *% wt*	*Propane* *% wt*	*Naptha* *% wt*	*Gas oil* *% wt*
Hydrogen	3.5	1.6	1.0	0.5
Methane	3.6	20.6	14.3	8.8
Ethylene and acetylene	50.3	34.3	27.0	20.7
Ethane	38.2	2.9	3.6	4.2
Propylene and C_3 acetylenes	2.4	17.0	15.2	14.6
Propane	0.2	15.0	0.5	0.6
C_4 hydrocarbons	0.6	4.0	10.0	9.6
Gasoline	1.2	4.6	25.9	18.0
Fuel oil			2.5	23.0
	100.0	100.0	100.0	100.0

TABLE VI

Effect of Severity on Yields of Major Products from Naphtha Cracking (ref.24)

	Low % *wt*	*Medium* % *wt*	*High* % *wt*
Hydrogen	0.8	1.0	1.2
Methane	12.0	14.3	16.5
Ethylene and acetylene	24.0	27.0	29.5
Ethane	3.5	3.6	3.4
Propylene and C_3 acetylenes	17.3	15.2	13.0
Propane	0.8	0.5	0.3
C_4 hydrocarbons	12.0	10.0	8.0
Gasoline	28.1	25.9	23.1
Fuel Oil	1.5	2.5	5.0
Total	100.0	100.0	100.0

In areas where the demand for propylene is increasing, the trend is moving away from high severity cracking operations where the ratio of propylene to ethylene yield is in the region of 0.5, to less severe conditions where this ratio is nearer to 0.7. At the same time the cracking process should be highly selective in producing high yield of both propylene and ethylene so that feed-stock consumption is kept to a minimum. In this respect there are advantages in operating at low residence times in cracking furnaces. One example of a process using this principle is the Stone and Webster Ultra Selective Conversion cracking process, the other is the Kellogg's Millisecond Pyrolysis process.

Ethylene plants are now in operation and many more are being built to produce as much as 450,000 tons per year of ethylene and at the same time over 300,000 tons per year of propylene.

When high molecular weight feed-stocks are used, a much more complex product mixture is obtained than that obtained from gaseous feed-stocks. In the past, it was not always economically feasible to recover and purify all by-products, instead they were sometimes used only for their fuel value or were recycled. Current economics generally dictates the recovery of most by-products, requiring a more complicated plant. Capital costs are thus greater when high molecular weight feed-stocks are used. Plants using heavier feed-stocks should not as a rule be called ethylene-propylene plants. In a broad sense, they are basic petrochemical plants since they yield such a variety of products (Fig.2). To an increased extent, these plants are being designed to obtain many products that were previously considered rather undesirable. Butadiene, for example, will be produced in the future more frequently in pyrolysis plants than by the dehydrogenation of n-butane and n-butenes.

Polypropylene as a By-product of Petroleum Industry: Unlike ethylene, polypropylene is an important by-product of the cracking of petroleum fractions for gasoline manufacture as well as one of the key raw materials for the production of alkylate and polymer gasolines. The rapid growth of a vast cracking industry in the USA since 1913 was the result of an ever-increasing demand for gasoline arising from the development in that country of a tremendous automobile industry. Moreover the developments in automobile engines necessitated the production of gasolines of improved characteristics particularly higher octane number and this requirement very largely dictated the technological evolution of the cracking processes employed Outside the USA, major refineries were mostly located at the oilfields and a large cracking industry did not arise in Europe and Japan until after World War II.

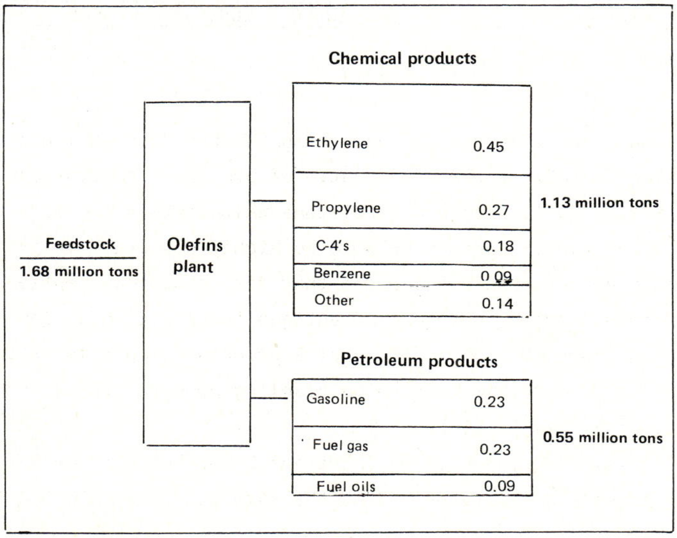

Fig. 2. The Product Mixture of a Typical Petrochemical Plant (ref.25)

Catalytic cracking is a major source of propylene and butenes, particularly in the USA, though very little ethylene is formed in this way. Total gas production (including C_4 hydrocarbons) is about 8-12 percent by weight of feed-stock (e.g. gas oil), the propylene yield being in the range of 3-5 percent by weight of feed-stock with much less propane. However, the use of molecular sieve-type catalysts in place of the older silica-alumina catalysts is now widespread and is likely to increase. With these newer catalysts, gasoline yield is increased but propylene production is reduced, possibly to below 3 percent by weight of feed-stock.

Another process which contributes substantially to by-product propylene availability is the thermal reforming process used to upgrade in anti-knock value of the heavy portion of naphtha formed by catalytic cracking. This reforming process, first introduced into the USA by Sun Oil Co. in 1955, yield propylene in amounts of 3-4 percent by weight of feed-stock together with similar quantities of propane and lesser amounts of ethylene. The corresponding catalytic reforming process, introduced into the USA in about 1950, gives a gaseous by-product almost free from olefins.

Finally mention should be made of the thermal coking process used to convert heavy, low-grade petroleum oils into gas oil and smaller quantities of gasoline, gas and coke. This process yields appreciable quantities of gas rich in propylene and its homologues.

A typical refinery gas might contain 9 percent by volume of propylene with about 7 percent of ethylene, the remainder being paraffin hydrocarbons and hydrogen. However, it should be pointed out that the refinery gas is a mixture of several different gas streams from various refinery operations and its composition will vary considerably, depending not only on the refining processes practiced and conditions of operation but also on the type of crude oil employed.

The future of catalytic cracking and its growth rate have recently become clouded by hydrocracking as an alternate means of producing light distillates from heavier crude oil components. It is generally acknowledged that high-pressure hydrocracking, requiring an outside hydrogen source, is more expensive than catalytic but it has feed flexibility over catalytic cracking on extra heavy stocks.

Although total US hydrocracking capacity is still small, it is expected to grow at a faster rate than catalytic cracking over the next few years. This trend was to be accelerated by the need to destroy heavy high-sulfur residual oils for anti-pollution purposes. Actually most US hydrocracking is installed on the West Coast where the first restrictions were placed on high-sulfur residual fuel oils.

Hydrocracking produces no olefins whatsoever. The total C_3 yield is propane. However, hydrocracking often produces 20 percent or more butane, most of which is iso-butane. Refinery alkylation units have historically been short of iso-butane, with sufficient butylenes and propylene available. Hydrocracking and the new zeolite catalysts are shifting the refineries from being short to being long on iso-butane. Therefore, the refineries have more incentive to utilize their propylene as alkylation feed as the value of iC_4 drops.

The preferred olefins for alkylation have always been the butylenes which yield a 2 to 3 higher octane-number alkylate than mixed propylene-butylene, and a 5 to 6 higher octane-number alkylate than propylene alone. However, when excess iso-butane is available, the refiner naturally turns to propylene to supplement alkylation feed.

Whereas in Europe and Japan, a major portion of the total propylene used is supplied by the steam cracking of naphtha, the position in the USA is quite different. Of the 16 billion lbs of propylene converted into chemicals and plastics in the USA in 1977, more than half was derived from refinery streams with propane cracking next in importance followed by naphtha and gas oil cracking; the cracking of the butanes contributed only a minor proportion. This situation is, however, rapidly changing for the USA. The future increase in propylene supply will be mainly from the steam cracking units rather than from gasoline production. The feed-stock situation for steam-cracking olefin plants is also changing. Only one of 13 recently announced olefin projects is based on NGL: others are based on naphtha and gas oil feed-stocks.

Separation and Recovery Processes (ref.I4,24): From the foregoing discussion of the methods of preparation of olefins, it can be seen that widely differing separation problems arise depending on the nature of the mixture. It is not proposed here to decribe the separation methods in detail as they all involve the standard techniques of fractionation, solvent-extraction, and absorption.

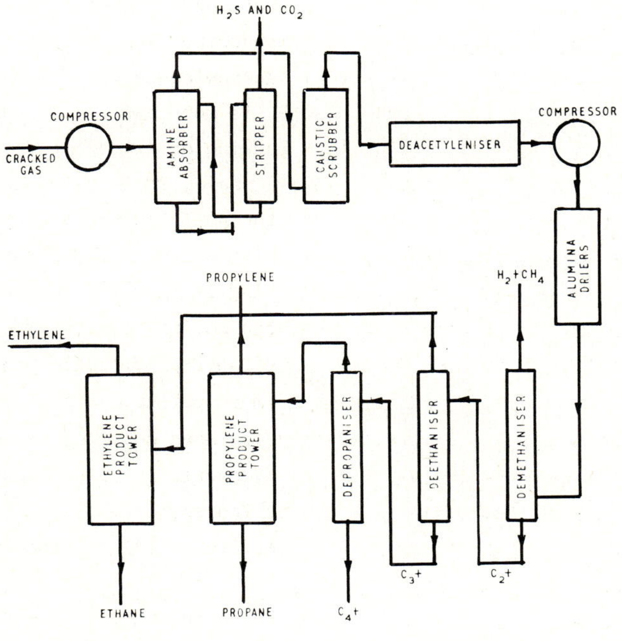

Fig. 3. Typical Scheme for Olefin Separation from Cracked Gas (ref.20)

A typical scheme for the treatment of cracked gas is shown in Fig. 3. After compression to 100-200 lb/in^2 the gas is scrubbed to remove carbon dioxide, carbon disulfide, and hydrogen sulfide. Acetylene removal may be either by scrubbing, for example, with dimethylformamide, or by selective hydrogenation over a platinum metal catalyst. Further compression may then take place to 400-600 lb/in^2, although in a low-pressure process the pressure may be kept at 100 lb/in^2 throughout. After drying, the gas is fed to the "demethanizer", where hydrogen and methane are separated from the other components. This may be achieved by fractionation in a 20-30 plate column with an overhead temperature of $-70^{o}C$ to $-95^{o}C$. Alternatively, selective solvent-extraction or "hypersorption" by means of activated carbon may be employed. Subsequent fractionating columns then remove the C_3 and C_4 fractions as bottoms, while ethylene and propylene are separated respectively as overheads. In some USA plants the ethane and propane are removed and fed to steam-cracking furnaces to furnish more olefins for recycling.

Selective solvents may also be used for separation of C_4 components, usually by the method of extractive distillation. For example, ammoniacal cuprous acetate or N-methylpyrrolidone may be used for extraction of butadiene, furfural, or acetonitrile for butadiene and n-butenes, and sulfuric acid for isobutene.

The recovery of propylene in a refinery is either carried out in the individual plants producing the material or in a centralised gas recovery plant, into which is fed the combined gas flows from all units. Although the composition of the gas varies, depending on the particular refinery arrangement, the processing scheme for the recovery of the propylene is similar.

Refinery gas recovering plants are usually based on oil absorption and are designed to recover between 70 and 90 percent of the propane-propylene in the gas stream. Processes have been developed based on adsorption using a moving bed of activated charcoal but these have not proved commercially attractive. Straightforward fractionation at low temperatures using refrigration could also be used and this route is the one normally followed in the recovery sections of olefin production plants, where a high recovery of ethylene and propylene is sought and separation of hydrogen and methane is also required. In refinery processes the concentration of valuable ethylene and propylene in the gas streams is much lower than in the olefin plants streams and the extra cost and complication of this scheme is not justified over the simpler oil absorption process. Whilst the gas recovery sections of refinery units and olefin plants can usually be considered quite separately, in special circumstances, where olefin plant forms part of an oil refinery complex, the recovery of ethylene, propylene and other light products is sometimes combined in one common unit.

A straightforward absorption process using a heavy hydrocarbon liquid stream to absorb the propylene and heavier components from the gas stream followed by fractionation of the rich oil to separate the propylene, provides the simplest recovery scheme.

Purification of Propylene (ref.24): Further treatment and purification of the propane-propylene stream from the depropanizer depends on the final product purity required.

Propylene is usually recovered from ethylene plants as "chemical grade" to provide feed-stock for the manufacture of derivatives such as cumene and acrylonitrile or as "polymer grade" for the production of polypropylene. The propane-propylene stream from the depropanizer would be suitable for sale as LPG without further treatment.

Typical specifications for the two product grades are given in Table VII. Both product specifications shown in the Table VII require the removal of methyl acetylene and propadiene down to a level of 40 ppm and sometimes even lower levels are required.

TABLE VII

Typical Polypropylene Product Specifications (ref.24)

		"Chemical" Grade	*"Polymer" Grade*
Polypropylene	% mol min.	93.00	99.5
Hydrogen	ppm mol max.	300	300
Methane	ppm mol max.	300	300
Ethylene	ppm mol max.	300	300
Ethane	ppm mol max.	1000	1000
Propane	% mol	balance	balance
Methyl acetylene	ppm mol max.	10	10
Propadiene	ppm mol max.	30	30
Butadiene	ppm mol max.	10	10
Sulphur	ppm mol max.	5	5
Water	ppm mol max.	5	5

For chemical grade propylene where no separation of propane from propylene is needed, the methyl acetylene and propadiene are usually removed by hydrogenation in either the vapour or liquid phase. In the vapour phase a high purity hydrogen stream produced within the ethylene plant is mixed with the propylene stream and the mixture passed over a palladium based catalyst under carefully controlled temperature conditions. The reaction is exothermic and it is common to use at least two catalyst beds with intercooling to avoid excessively high temperatures within the reactors. Under well-controlled conditions the hydrogenation process is selective, the methylacetylene and propadiene being hydrogenated mainly to propane, and losses of propylene due to hydrogenation should be insignificant.

Processes offered by Farbenfabriken Bayer AG and Institut Francaise du Petrole, hydrogenate the methylacetylene and propadiene in the propylene stream in the liquid phase. The Bayer process uses a tubular reactor cooled by refrigerant whilst the IFP process uses fixed bed reactors arranged in series with refrigerated pre-cooling. Higher selectivity in hydrogenation is claimed than for the vapour phase processes. Thus some propylene gain should result.

With the development of high severity cracking, the yield of methylacetylene and propadiene has increased and in some locations the recovery of these compounds become attractive for two reasons. Firstly, hydrogenation of acetylenes becomes more complicated due to the high concentrations in the propylene stream, and also when hydrogenated, the propylene concentration in the resulting propylene-propane stream can be lower than desirable due to the hydrogenation of the acetylenes to propane. Secondly, the methylacetylene-propadiene mixture can be marketed as a welding gas having properties in many ways superior to acetylene. Recovery of the methylacetylene and propadiene from the depropanizer overhead stream can be achieved either by fractionation or by absorption although care has to be exercised in keeping the concentration of the methylacetylene plus propadiene below about 60 percent to avoid the risk of detonation and explosion. If a chemical grade propylene product with a very low permissible content of methylacetylene and propadiene is required, a system involving two fractionators with intermediate hydrogenation may be necessary.

After hydrogenation, the propane-propylene stream contains small concentrations of hydrogen and methane orignating from the hydrogen added to the hydrogenation reactors. Also the stream is saturated with water since only part of the feed to the depropaniser has passed through the cracked gas driers. Both the light ends and the water must be removed and this can be normally done in a simple stripping tower. The hydrogen sulfide, mercaptan sulfur and carbonyl sulfide of the propylene stream must also be reduced to a low level in order to meet the sulfur specification of the product. Removal of hydrogen sulfide and mercaptans presents no problems since the processing sequence in the cracked gas compression section of the plant can generally be arranged so that all of the propylene and propane contained in the cracked gas passes through the caustic scrubber and hence is freed from these compounds. Where this is not possible a simple caustic wash system followed by water washing of the depropanizer overhead stream is sufficient.

To produce the high purity polymer grade propylene required for polypropylene manufacture it is necessary to separate the propane and propylene. This is a difficult separation since the relative volatility between propylene and propane is very low and the separation falls into the catagory of superfractionation. The average relative volatility between propylene and propane is 1.12 at 21 bars abs, 1.15 at 15.8 bars abs and 1.21 at 7 bars abs. Even though the higher relative volatility at the lower pressure would result in significant reduction in the flux ratio and the number of theoretical plates needed for the separation, the extra costs and complications at low pressures cannot usually be justified. Normally, therefore, the operating pressure of the fractionator depends on the temperature of the cooling water available for condensing the column overheads. With cooling water available at 25°C a column overhead vapour could be compressed and condensed in the reboiler using a heat pump principle to save operating costs. However, due to the very large quantities of vapour to be handled in a modern day ethylene plant this is not an economic solution. Also, although the cooling water load on the overhead condenser is very big, the cost of heating the boiler can usually be made very small since waste heat from the pyrolysis section of the plant can be used.

II.2. Stereospecific Polymerization

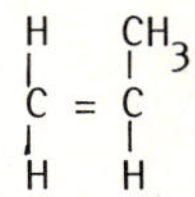
[I]

The chemical structure of propylene [I] shows that it can be polymerized with several different spatial arrangements of building blocks, i.e., propylene monomer units. The building blocks can be attached to each other in a linear but spatially random fashion or they can be made to string up in a regular spatial order. The two kinds of stereo arrangement of building blocks impart completely different physical properties to the polymer. The sterically random polymers are known as atactic; those with stereoregular configuration as tactic. In the case of atactic polypropylene, the side methylene groups are, sterically speaking, distributed randomly. This leads to poor molecular symmetry and little or no crystallinity. As a result, atactic polypropylenes are waxy or rubbery materials with few useful properties.

Among several theoretically possible stereoregular configurations, isotactic polypropylene is the most important commercially. In simple terms, isotactic polypropylene is one in which the building blocks are so arranged that if one passes along the chain from one block to the next, it will be found to be in the same relative position in space. Syndeotactic is the term used to describe the polymer where methyl groups alternate in a sterically regular fashion along the chain. Only imperfectly syndeotactic polypropylene has been made to date and has little practical significance so far. Fig.4 depicts the models of atactic, isotactic, and syndeotactic polypropylenes.

In addition to atactic, syndeotactic, and isotactic polypropylenes, it is also possible to make polymers that start off being isotactic and then reverse their position and continue in reverse steric order. Fig.4(d) depicts such an arrangement. These polymers are called stereoblock polymers. They differ from normal isotactic polypropylene in their physical and mechanical properties. In general these polymers are less crystalline than isotactic polypropylene.

Still further variation in stereospecificity of polypropylene is found in randeotactic polymers. By randeotactic is meant that the polymer chain is composed of random sections of atactic and isotactic arrangements of varying lengths. These polymers are in between atactic and isotactic polymers in crystallinity. Such polymers are produced from the rearrangement of isotactic polymers through simultaneous decomposition and polymerization (ref.5).

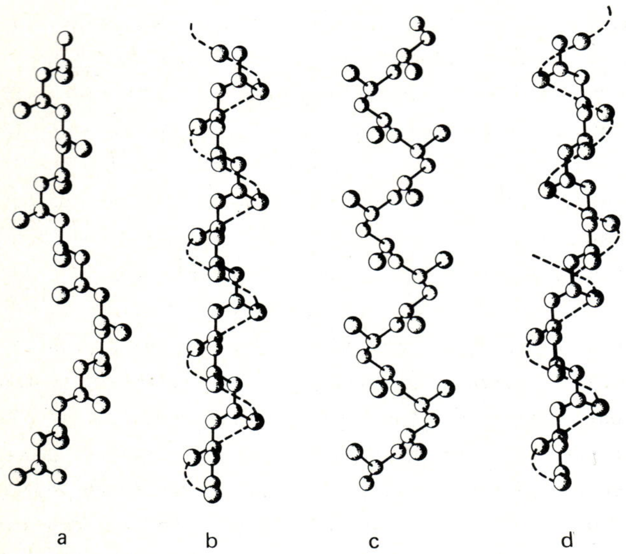

Fig. 4. Stereostructures of Polypropylene Chain
(a) atactic, (b) isotactic, (c) syndeotactic, (d) stereoblock

The formation of any one of these forms of polypropylene is matter of the polymerization catalyst and polymerization conditions used. Commercially the most important form of polypropylene is isotactic polypropylene. Our discussion in subsequent chapters will therefore be confined entirely to this form. The atactic form will be occasionally mentioned on the premise that commercial isotactic polypropylene is usually accompanied by small amounts of atactic by-product. The stereoblock and randeotactic polymers are only of significance in connection with the production of high-shrinkage fibers and conjugate filaments discussed in Chapters IX and XI, respectively.

The catalyst used for the polymerization of propylene is very important in determining the tacticity of the polymer. Manufacturers are, therefore, still very secretive about their particular catalyst composition and catalyst-preparing processesses. A wide variety of catalyst systems has been reported in patent literature. They are collectively called Ziegler-Natta catalysts, after their inventors. They are also known as anionic-coordinated catalysts. The term "anionic coordinated" makes some assumptions about the polymerization mechanisms which, although not yet proved, are becoming generally accepted. The developments in Ziegler-Natta catalysts and stereoregular polymerization have been reviewed by several authors.

A partial list of these references is given in the Bibliography of this chapter. From these reviews, further references may be cited for detailed informations.

Generally speaking, the Ziegler-Natta catalyst systems consist of a halide of transition metal of the fourth group of the Periodic Table (e.g., of titanium or vanadium), and an organometallic compound of a metal belonging to one of the three first groups of the Periodic Table (e.g., lithium, beryllium, aluminum). Some important examples are: [crystalline $TiCl_3$ with or without $AlCl_3$, plus AlR_3, AlR_2X, or BeR_2, where R is an alkyl group and X is a halogen], [$TiCl_4$ or $VoCl_3$ plus AlR_3, AlR_2X, BeR_2, LiR, ZnR_2] and [VCl_3 or β-$TiCl_3$ plus AlR_3, AlR_2X, BeR_2] . The transition metal chloride is often used in its highest valence state and reduced *in situ* to a lower state. However, the preformed crystalline $TiCl_3$ is a superior catalyst for production of stereoregular polypropylene and its availability has led to its widespread use.

While virtually hundreds of Ziegler-Natta catalyst combinations have been claimed only a few have found commercial acceptance. The two important criteria determining the commercial value of a Ziegler-Natta catalyst system are its stereospecificity and its overall catalytic activity.

It has often been suggested that stereoregularity depends on the organometallic compound and is practically independent of the type of transition metal compound. This view has, however, been recently challenged. Results have been obtained which show that the stereospecificity depends on the actual combination of the two components, and on the use of optimum conditions of polymerization which depend on the nature of the catalyst system. Similarly, the overall catalytic activity in polymerization depends on the nature of each component of the catalyst system. For example, δ-$TiCl_3$ is more active than α-$TiCl_3$ and, for a given type of $TiCl_3$, $Be(C_2H_5)_2$ is more active than $Al(C_2H_5)_3$ which, in turn, is more active than $Al(C_2H_5)_2Cl$ or $Al(C_2H_5)_2Br$.

Ziegler-Natta catalysts are frequently activated to increase their efficiency. Both mechanical and chemical methods are used to accomplish this. Dry milling of $TiCl_3$ has particularly proved to be advantageous, as has heating for prolonged periods. Among chemical compounds, secondary and tertiary amines and, more often, various Lewis acids can be mentioned as typical activators.

Ziegler-Natta catalysts are heterogeneous systems since the solid halides of the transition metals and the products obtained by reaction of the organometallic compound with liquid halides are insoluble in the hydrocarbon - a medium in which polymerization reaction takes place. The polymerization rate of propylene is first order with respect to both the concentration of the propylene and the amount of transition metal compound. The polymerization rate may be considered truly catalytic with respect to $TiCl_3$ and the polymerization with system $Al(C_2H_5)_3$-$TiCl_3$ at a constant partial pressure of propylene has practically the same value for several hours.

Several hypotheses have been put forward for the chemical mechanism of the polymerization of propylene and for the factors determining stereospecificity. It is generally believed that a catalyst complex is formed by alkylation of the transition metal halide with the organometallic compound. These complexes may be represented by $Cat\overset{\delta+}{\;}\!\!—\!\!\overset{\delta-}{R}$ where R is an alkyl group bound to the transition metal. The bond is polarized in such a way that a partial negative charge is located on the carbon atom. The monomeric units, i.e., propylene molecules, are inserted at a metal-carbon bond as shown by the equation below:

$$\overset{\delta+}{Cat}—\overset{\delta-}{R} + CH_2{=}\underset{CH_3}{\underset{|}{CH}} \longrightarrow \overset{\delta+}{Cat}—\overset{\delta-}{CH_2}—\underset{CH_3}{\underset{|}{CHR}}$$

Before entering into the chain, the propylene double bond coordinates with the transition metal of the catalyst complex. This type of polymerization has thus been termed anionic coordinate.

From a study of the kinetics of polymerization of propylene over $Et_3Al/\alpha\text{-}TiCl_3$ catalyst, using ^{14}C-labeled triethylaluminum and determining the M_v of the polymer, Natta has shown that:

$$\frac{1}{\bar{X}_n} = \frac{k_1C^* + k_2PC^* + k_3\left[Et_3Al\right]^{\frac{1}{2}}C^* + k_4\left[TiCl_3\right]^{\frac{1}{2}}PC^* + k_5\left[H_2\right]^{\frac{1}{2}}}{k_pPC^*}$$

where $\bar{X}_n$ = number average degree of polymerization
C^* = concentration of active sites
P = partial pressure of propylene
K_p = rate constant for propagation

The terms in the numerator correspond to transfer processes which have been observed to depend on the concentration of growing chains, the propylene, triethylaluminum, titanium trichloride, and molecular hydrogen (if added), respectively. The relative rates of the different transfer processes and the propagation rate and hence the molecular weight of the polymer depend on the Al:Ti ratio and temperature. The molecular weight varies inversely with temperature. It also increases with polymerization time asymtotically, the maximum values being reached in times varying from minutes to several hours.

The chain termination processes are essentially the following: (a) exchange of alkyl groups with organometallic compounds present in the solution; (b) chain transfer with monomers; and (c) spontaneous rupture of polymer chains. Among the organometallic compounds, dialkylzinc and dialkylcadmium are particularly effective in reducing the molecular weight of polypropylene during polymerization. Hydrogen is also effective as a chain-transfer agent and is used in industry to control the molecular weight of polypropylene.

There are various chemical and physical ways by which an active site may be destroyed. Oxygen, water, or carbon dioxide as impurities in the monomer stream will inhibit polymerization. Alcohol treatments are frequently used to destroy the catalyst. Mechanical blocking of the catalyst surface with polymer also has a retarding effect. The broad molecular weight distributions, which are normally obtained in heterogeneous polymerization, have been attributed either to the variation of rate constant with chain-length and/or to the nonuniformity of the activity of catalyst centers. On the other hand, amorphous polymers tend to be of much lower molecular weight than crystalline polymers, by a factor of about 7, which suggests either that the propagation rate is less at the nonstereospecific sites than at stereospecific sites or that the former are more succeptible to chain transfer.

II.3. Polymerization Processes

In as much as the details of the polymerization processes as they are actually practiced are usually closely guarded secrets, there is hardly any published information available on the production of polypropylene. Nevertheless, a typical polymerization process involves four major steps: catalyst preparation, polymerization, polymer purification, and polymer finishing.

Catalysts are now commercially available; however a number of companies prefer to prepare themselves the catalyst they need. In fact, many polymerization characteristics, polymer qualities and economic factors depend on the catalyst performance. Upto now, only titanium compounds are used, along with aluminum alkyls. Titanium trichlorides obtained from tetrachloride by reduction with hydrogen or with metal aluminum are preferred. The most common co-catalysts are triethylaluminum and diethylaluminum monochloride. Another kind of catalyst is titanium chloride obtained from tetrachloride by aluminum alkyl reduction. By careful control of shape and granulometric distribution of this last type of catalyst it is possible to produce polypropylene granules of narrow particle size distribution. This kind of polymer is commercially known as polypropylene flake and may replace standard pellets in extrusion processes.

Polymerization process consists in bringing monomer into contact with catalyst, at such a pressure, temperature and chemical environment, that polypropylene with the desired tacticity and molecular weight is obtained at the highest convenient polymerization rate and highest yield with respect to the catalyst.

Polymerization can be carried out in suspension in hydrocarbon diluent or in liquid monomer, or in the gas-phase.

The flow sheet in Fig.5 may be considered typical of a continuous suspension polymerization process. Choice of the process, design of the reactor, and operating conditions are largely based on engineering considerations, as heat removal, possibility of making the process continuous, possibility of simplifying subsequent operations (for instance, operating at such high yields with respect to the catalyst to avoid removal of catalyst residues), possibility of obtaining complete homogeneity within the reacting mass. Heat removal is obtained by jacket cooling, in conventional reactors, or by evaporation of part of hydrocarbon diluent or of monomer. Standard propylene polymerization enthalpy is -24.89 kcal/mol; standard propylene vaporization enthalpy is +3.59 kcal/mol.

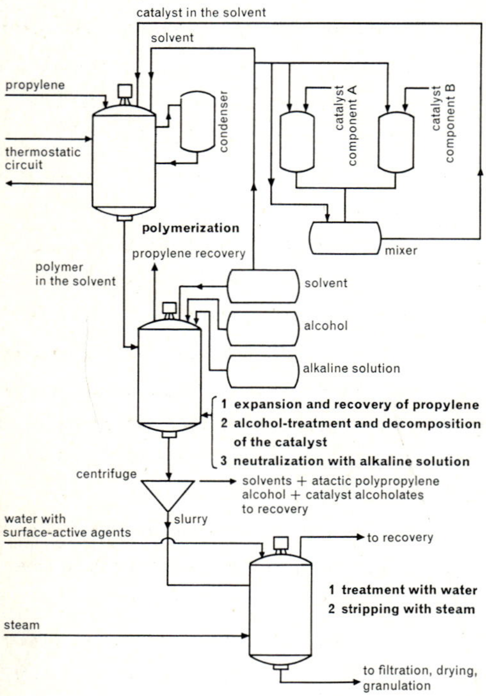

Fig. 5. Schematic Representation of Polypropylene Manufacture through Slurry Process

Currently, most of the polypropylene production is carried out in hydrocarbon diluents, such as hexane or n-heptane. Isotactic polypropylene is insoluble in n-heptane or hexane upto their boiling points at atmospheric pressure: for this reason the polymerization in hydrocarbon solvents is in fact a suspension polymerization if carried out at temperatures lower than around 100°C.

Suspension polymerization is particularly convenient because the viscosity of the polymerization medium is low even at high polymer concentration: ease of stirring and homogenisation is thus enhanced. Atactic polypropylene is soluble in hydrocarbons: thus for process reasons efforts are made to keep as low as possible the production of atactic polymer. Efforts are also made to regulate polymer molecular weight at the polymerization stage, even if methods are claimed for regulation of molecular weight by accurate thermomechanical degradation, for instance during granulation. Hydrogen is one of the most effective molecular weight regulators, while the use of certain metal alkyls, of hydrogen halides or of halogenated hydrocarbons is also claimed.

The catalyst is slurried in highly purified heptane and fed along with propylene and a diluent into the polymerization vessel. Diluent may be any of the paraffins and naphthenes of suitable boiling point; usually cyclohexane or n-heptane is used. Operating conditions commonly range between $50^{o}C$ and $80^{o}C$ temperature and between 2 and 4 atmospheric pressure. With a properly prepared catalyst system (much depends on the state of subdivision and the purity of the violet $TiCl_3$ used) crystalline polypropylene forms continuously and precipitates as a solid on the catalyst surface until the slurry becomes very thick (upto 40 percent solids).

The reactor slurry flows to a flash tank where unreacted propylene is vaporized for recirculation into the reactor. The polymer slurry from the flash tank is then centrifuged to remove diluent. Along with the diluent, most of the soluble atactic or low molecular weight polypropylene is also removed. The isotactic polypropylene from the centrifuge is treated with anhydrous isopropanol to remove catalyst residues (de-ashing process) and hydrocarbon diluent. It is then washed with methanol and dried. The dried polypropylene flake may be supplied directly to polymer processors or compounded with suitable stabilizers and pelletized. Most of the major fiber manufacturers employ virgin polymer and do their own compounding, while smaller plants purchase compounded polymer.

Rate of polymerization is adjusted by controlling the temperature, pressure, and catalyst content or, in some cases, by changing the method of catalyst preparation. The average molecular weight of polypropylene may be controlled by polymerization temperature; however, high temperatures tend to reduce the stereospecificity of the catalyst, leading to an increase in atactic content. As mentioned before, a common method of molecular weight control is utilization of hydrogen as chain terminator. It is fed into the reactor to create a constant partial pressure. The viscosity-average molecular weights of polypropylene obtained with bimetallic catalyst systems based on violet $TiCl_3$ generally range between 100,000 and one million, depending on the operating conditions. The index of polydispersity of molecular weights (M_w/M_n) of polypropylene obtained by continuous suspension polymerization generally ranges between 4 and 15.

The raw isotactic polypropylene always contain considerable amounts of atactic polymer and the catalytic residues. These must be removed because the former impairs the mechanical properties of the polymer and the latter impart discoloration to the finished products. It also contributes to the corrosion of the processing equipment. A great number of purification processes are described in the patent literature; almost the only common principle is that the two impurities are separated from the isotactic polymer through rendering them soluble with suitable chemical agents. As has already been said, hydrocarbon dissolves atactic polymer;

the simplest way of separating atactic from isotactic polymer is therefore to dissolve the atactic in hydrocarbon or halocarbon solvents and subsequent solid-liquid separation, e.g. by centrifugation. Centrifugation can be carried out quite simply on the slurry discharged from the polymerization reactor.

Catalyst components can be reacted with suitable chemicals, as for example, alcohols, and transformed into titanium and aluminum compounds soluble in hydrocarbons, or soluble in water or in alcohols. The polymer is then purified from catalyst residues either by centrifugation or by extraction with water or alcohol.

According to which processes are employed for the purification of the polymer from its atactic content and from catalyst residues, a solid polymer suspended in hydrocarbon diluent, or in water, or alcohol has to be handled. Conventional drying equipment and procedure may be employed to dry the polymer: steam stripping, flashing (in the case of low boiling hydrocarbons), heating in air or nitrogen tunnels.

Monomer and polymerization solvents must, of course, be carefully purified before recycling to the polymerization section. Purification is generally carried out by conventional physical methods: absorption, distillation, etc.

The largest single item which determines the production cost of polypropylene is the cost of the monomer. The second item is the capital cost of the plant (amortization, interest on investment, etc). Cost of monomer is outside the control of the polymer producer, as it depends on the cost of naphtha or other starting material, how the ethylene and other co-products from the cracking operation are valued and the cost of purifying the propylene to polymerization grade. On the other hand, the polymer producer can control the costs of his plant.

Major research efforts are therefore being carried out by a number of industrial companies to simplify production technologies and processes.. The recent development of "high milage" catalysts by Montecatini, BASF, Mitsubishi, Mitsui Toatsu and Solvay is very significant for the economy of polypropylene production. The usual catalyst produces 3,000 grams of polypropylene for each gram of titanium. The new catalysts require one gram of titanium to produce 80,000 to 250,000 grams of polypropylene. These developments result in less catalytyst residue in the polymer so that catalyst washing (de-ashing) is significantly reduced or entirely eliminated - a step that is expected to reduce the cost of production by as much as 20 percent.

Polypropylene can be polymerized also by gas phase processes and at least two such processes are being operated commercially. BASF is producing in excess of 25,000 metric tons/year at their Rheinische Olefinwerke in Wesseling, West Germany, and Norchem has started a sizable plant at Morris, Ill., USA under a liscence from BASF. The BASF polypropylene is quite different from conventional polypropylene. It has a relatively low level of crystallinity but its impact strength is much better than that of conventional polypropylene. This polymer is a suitable material for tape yarns manufacturing but not recommended for the production of multifilament yarns. The Norchem's polypropylene is reported to be identical to the conventional polypropylene in tacticity and ash content.

Figure 6 depicts a simplified flowsheet of a typical gas-phase polymerization of propylene. Feed and recycle propylenes are purified by distillation, and the distillate is dried with alumina. Other catalyst poisons are also removed by this technique. The dried liquid propylene is then continuously added to the bottom of the polymerization reactor, and a high-efficiency modified Ziegler-Natta catalyst is added at the top. A mixture of solid particles and upflowing propylene gas is present inside the reactor. The solid catalyst fragments as the propylene polymerises and as the size of the polypropylene powder particles grows.

Although BASF and Solvay do not report details of the reactor design or operation of the commercial units being used, a fluidized bed reactor or an agitated and expanded powder bed reactor is probably used. A reaction temperature of 90°C and pressures of 30 to 35 atm are reported. Temperature control in the reactor is accomplished by using liquid propylene, which partially evaporates in the reactor. About 15 percent of the propylene is in the reactor while the remaining 85 percent is being recycled. Unreacted propylene and powder polypropylene are removed from the reactor continuously. This propylene is recoved, compressed, repurified, and then recycled to the reactor. The polypropylene product contains some adsorbed propylene that has to be removed. The resulting polymer is then combined with suitable additives and stored or shippec.

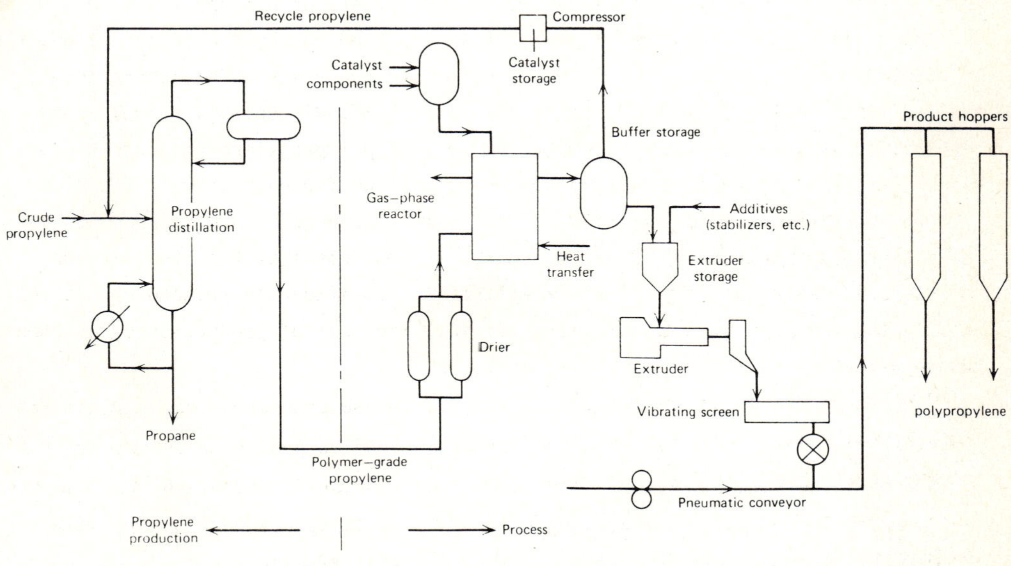

Fig. 6. A simplified Flowsheet of a Typical Gas-phase Polymerization Process

To be economically competitive, gas-phase process requires a catalyst for which essentially no separation steps are required after the polymerization reactor. Separation steps for catalyst residues or atactic polypropylene would increase the operating costs excessively. Very high-efficiency and extremely selective catalysts have recently been developed, as mentioned before. The use of these catalysts makes it possible to attain a gas-phase polymerized polypropylene with essentially the same stereoregularity and ash-content as that obtained in the conventional suspension-polymerized polypropylene. It is expected that most of the new polypropylene manufacturing plants will be based on gas-phase polymerization technology.

III. PROPERTIES AND CHARACTERIZATION

Since polypropylene was first offered commercially in 1958, the number of grades of the polymer for specific end uses has increased greatly. These different commercial grades vary in molecular weight, molecular weight distribution, isotactic content, stereoblock content, additive composition, and in recent years, in copolymerized ethylene content. For the manufacture of fibers, isotactic homopolymers of high tacticity are of particular interest. Copolymers, stereoblock, and randeotactic polymers may find some uses in speciality fibers such as conjugate fibers, elastic fibers, and high-shrinkage fibers.

Within the class of homopolymers, variations arise primarily on the basis of molecular weight, molecular weight distribution, atactic content, catalyst residues, and the degree of light and heat stability. Table VIII summarizes some of

TABLE VIII

Some of the Typical Properties of Polypropylene Homopolymers

Property	*ASTM method*	*(1)*	*(2)*	*(3)*
Flow rate at 230°C, g/I0 min	DI238-62T	0.5-I.0	2.0-8.0	8.0-20.0
Specific gravity	D792-64T	0.905	0.905	0.905
Tensile yield stength, psi	D638-64T	5000	5000	5000
Ultimate elongation, %	D638-64T	250	200	200
Tensile modulus, psi x 10^{-3}	D638-64T	155	160	160
Flextural modulus,psi x10^{-3}	D790-63	170	180	180
Notched Izod impact stength	D256-56	1.8	0.4	0.4
Heat deflection temp. °F	D648-56	220	230	230
Hardness, Shore D	D1706-61	72	72	72
Flamability rating	D635-63	slow	slow	slow
Thermal expansion x 10^{-5}/°C	D696-44	6.8	6.8	6.8
Thermal conductivity x 10^{-4}	C177	2.8	2.8	2.8
Specific heat(cal/g/°C)		0.46	0.46	0.46
Dielectric const(60 Hz-50mc)	D150-59T	2.28	2.28	2.28
Dielectric strength(v/mil)	D149-59	610	610	610
Volume resistivity,ohm.cm.	D257-58	1.0x10	1.0x10	1.0x10
Surface resistivity(ohm.cm)	D257-61	10	10	10

the important properties of two representative uncompounded homopolymers along with the methods of testing. Unidirectional orientation, as is present in fibers, may significantly alter some of these properties. Care should be taken, therefore, when applying these data to fibrous structures.

The technical data sheets prepared by the marketing departments of polypropylene manufacturing companies often list as many as 50 different grades of homopolymer polypropylene. They are generally divided into four or five broad catagories, reflecting their molecular weights and designated as very stiff, stiff, easy flow, extra easy flow, etc. Each catagory is then subdivided into several sub-groups reflecting the additive content and designed as flame retardant, light stabilized, heat stabilized, light and heat stabilized, extra light stabilized, extra heat stabilized, extra heat and light stabilized, carbon black filled, glass fiber filled, etc. For the production of fibers, generally the unfilled homopolymers of easy to extra easy flow grades come in question (melt flow indes 10 to 30). Large-scale fiber producers would prefer to purchase the uncompounded polymer and do their own compounding. Small-scale producers of fibers and film yarn may consider the heat and light stabilized grades of appropriate melt flow index.

The copolymers that are of interest to the fiber manufacturer are those having less than 5 percent ethylene, preferably copolymerized as block segments. In these polymers, the usual high-performance properties of polypropylene are preserved by isotactic polypropylene segments, while the less rigid regions of heterosegments improve the low-temperature performance of the polymer or impart to it high-shrinkage properties when exposed to heat. As mentioned earlier, copolymers and stereoblock polymers are somewhat less crystalline than pure isotactic polypropylene and their physical properties, such as heat shrinkage and tensile strength, are more sensitive to temperature. This fact makes some of these polymers suitable material for the production of high-shrinkage fibers, elastic yarns, or heterofil fibers with self-crimping properties. These polymers are, however, not suitable for the production of high-tenacity fibers and resilient products.

Polypropylene can be characterized by its molecular parameters such as molecular weight, molecular weight distribution, stereospecificity, etc., or by its bulk properties, such as melt-flow properties, crystallinity, mechanical, thermal, physical, and chemical properties of the polymer. In the following pages some of these properties will be discussed as far as they are of interest to fiber manufacturing, polymer specification, and fiber end-uses. The methods of characterization will be mentioned briefly in connection with the respective properties, where applicable.

III.1. Molecular Weight and Molecular Weight Distribution

Polypropylene, like other synthetic polymers, is polydisperse in molecular weight, i.e., it consists of a mixture of homologous molecules of different molecular weights. Molecular weight of polypropylene can therefore be spoken of only in terms of averages. This heterogeneity in molecular weight, although rarely affecting chemical properties, has an important influence on the physical, mechanical and rheological behavior of the polymer. Molecular weight and molecular weight distribution play a vital role in determining the melt-spinning characteristics of the polymer and the mechanical properties of the fiber spun from it. It is therefore instructive to discuss briefly of what is involved.

The most commonly used molecular weight averages are number average M_n, and weight average M_w, defined as

$$M_n = \frac{1}{\Sigma (w_i/M_i)} = \Sigma N_i M_i / \Sigma N_i$$; where $w_i = c_i/c$ is the weight fraction of species i in the polymer and N_i is the number of moles of species i.

$$M_w = \frac{\Sigma c_i M_i}{c} = \Sigma w_i M_i$$; where w_i is the weight fraction of species i and $c = \Sigma c_i$.

Weight average is always larger than number average. The former is particularly sensitive to the presence of larger species, whereas the latter is sensitive to the proportion by weight of small molecules. The ratio M_w/M_n is frequently used as a measure of the width of molecular weight distribution. For a completely random (Gaussian) distribution this ratio should be 2, as is the case with polycondensation polymers such as nylon and polyesters. The distribution of molecular weights in polypropylene is not Gaussian and therefore the ratio M_w/M_n deviates considerably from 2. In commercial polymers this ratio is in the range of 3 to 10.

Osmometry is the most common collegative method for the determination of M_n; the determination of M_w is commonly done by light scattering method. The description of experimental techniques and the discussion of the theoretical basis of these methods are readily available elsewhere (see Bibliography) and need not be repeated here. Both of these methods yield absolute values of respective molecular weight averages. It is seldom, however, that the determination of absolute molecular weight is required in the fiber industry. Generally a relative value is considered adequate and this can most conveniently be obtained by solution viscometry.

The viscosity of a polymer solution, among other factors, is a function of molecular weight and therefore can be used as an indirect method for molecular weight determination. The average molecular weight, so determined, is denoted as M_v (viscosity average). For practical purposes, M_v is related to M_w and M_n as $M_w > M_v > M_n$; M_w being approximately 20 percent higher than M_v for commercial polypropylene of M_w/M_n ratio of about 5 (ref.6).

The calculation of M_v by viscometry is based on Mark-Houwink equation: $[\eta] = K.M_v^a$; where $[\eta]$ is called intrinsic viscosity and is defined as : $[\eta] = \lim_{c\to o}\eta_{sp/c}$. In this equation c is the concentration of the polymer solution and η_{sp} (specific viscosity) is given by equation: $\eta_{sp} = \eta - \eta_o/\eta_o$; where η and η_o are the viscosities of a solution at a given concentration and that of pure solvent, respectively. In the commonly used viscometers, $[\eta]$ is expressed in deciliters/g. K and a in Mark-Houwink equation are constants. For polypropylene, a has a value of about 0.8 and K ranges from 0.8×10^{-4} to 1.1×10^{-4} depending on the temperature of measurement and solvent used. One of the following relations is commonly employed for routine testing, using decalin as solvent and making meacurements at 135°C:

$\eta = 1.0 \times 10^{-4} M_v^{0.80}$ (ref.7)

$\eta = 1.15 \times 10^{-4} M_v^{0.79}$ (ref.8)

$\eta = 0.90 \times 10^{-4} M_v^{0.80}$ (ref.9) using tetralin as solvent at 135°C.

Experimentally, intrinsic viscosity $[\eta]$ is determined by extrapolating to infinite dilution the reduced specific viscosity η_{sp}/c determined at different polymer concentrations. For practical purposes it may usually be sufficient to evaluate $[\eta]$ from a single determination of the reduced specific viscosity. In fact, this is the procedure most commonly used in industrial laboratories for routine testing and quality control purposes.

Numerous relationships have been proposed between $[\eta]$ of a polymer and its reduced specific viscosity, η_{sp}/c. Of these many relations, the Huggin's equation: $\eta_{sp}/c = [\eta] + K'[\eta]^2c$, and Martin's equation: $\log(\eta_{sp}/c) = \log[\eta] + K''[\eta]c$, have proved to be the most useful.

Elliot et al (ref.10) have recently published an excellent statistical analysis of dilute solution viscosity data for a wide range of polypropylene and polyethylene samples in Decalin at 135°C. Their analysis indicates that Huggin's equation fits the experimental data on polypropylene better than Martin's. This is especially true in the lower $[\eta]$ range commonly met for polypropylene grades meant for fiber spinning purposes. A grand average of 0.139 of K' is applicable to both polymers. Based on this, tables have been constructed permitting the ready determination of $[\eta]$ from a single relative viscosity measurement at a known concentration. Tables IX and X are two such listings suitable for $[\eta]$ of polypropylene in the range commonly met for fiber spinning.

The molecular weight distribution of a polydispersed polymer may be represented graphically by a histogram plot. Because the number of homologous species involved in a polymer is very large, a continuous curve, such as shown in Fig.6, is usually used to represent the distribution. As the diluent used in the polymerization process of polypropylene removes most of the very low molecular weight polymer,

TABLE IX

Relationship between η_{rel} and [η] for c = 0.100 g/dl, k = 0.139, for [η]= 0.0 to 10.9 (ref. 17)

[η]	0.0	0.1	0.2	0.3	0.4	0.5	0.6	0.7	0.8	0.9
0.0	1.000	1.010	1.020	1.030	1.040	1.051	1.061	1.072	1.082	1.093
1.0	1.103	1.114	1.125	1.136	1.146	1.157	1.168	1.180	1.191	1.202
2.0	1.213	1.225	1.236	1.248	1.259	1.271	1.283	1.294	1.306	1.318
3.0	1.330	1.342	1.355	1.367	1.379	1.392	1.404	1.417	1.429	1.442
4.0	1.455	1.468	1.480	1.494	1.507	1.520	1.533	1.546	1.560	1.573
5.0	1.587	1.600	1.614	1.628	1.642	1.656	1.670	1.684	1.698	1.713
6.0	1.727	1.742	1.756	1.771	1.786	1.800	1.815	1.830	1.845	1.861
7.0	1.876	1.891	1.907	1.922	1.938	1.954	1.969	1.985	2.001	2.017
8.0	2.034	2.050	2.066	2.830	2.099	2.116	2.133	2.149	2.166	2.183
9.0	2.201	2.218	2.235	2.253	2.270	2.288	2.305	2.327	2.341	2.359
10.0	2.377	2.396	2.414	2.432	2.451	2.470	2.488	2.507	2.526	2.545

Figures in columns are η_{rel} values

TABLE X

Relationship Between η_{rel} and [η] for C = 0.500 g/dl, k = 0.139, for [η]= 0.0 to 10.9 (ref. 17)

[η]	0.0	0.1	0.2	0.3	0.4	0.5	0.6	0.7	0.8	0.9
0.0	1.000	1.051	1.103	1.157	1.213	1.271	1.330	1.392	1.455	1.52[illegible]
1.0	1.587	1.656	1.727	1.800	1.976	1.954	2.034	2.116	2.201	2.28[illegible]
2.0	2.377	2.470	2.564	2.662	2.762	2.865	2.971	3.080	3.192	3.10[illegible]
3.0	3.425	3.546	3.670	3.798	3.929	4.064	4.203	4.345	4.491	4.64[illegible]
4.0	4.794	4.951	5.113	5.279	5.449	5.624	5.803	5.986	6.176	6.36[illegible]
5.0	6.565	6.768	6.976	7.189	7.408	7.632	7.861	8.097	8.383	8.58[illegible]
6.0	8.837	9.097	9.362	9.634	9.913	10.198	10.491	10.790	11.096	11.41[illegible]
7.0	11.731	12.060	12.397	12.742	13.094	13.456	13.825	14.205	14.591	14.98[illegible]
8.0	15.393	15.808	16.232	16.667	17.112	17.566	18.032	18.508	18.995	19.49[illegible]
9.0	20.002	20.524	21.057	21.602	22.159	22.729	23.312	23.909	24.518	25.14[illegible]
10.0	25.779	26.430	27.097	27.778	28.474	29.186	29.913	30.657	31.417	32.19[illegible]

Figures in columns are η_{rel} values

the resultant molecular weight distribution is markedly asymmetrical with a long "tail" on the large molecular weight side of the distribution maximum. The function represented by the curve in Fig.6 is often called the differential molecular-weight distribution function, W(M). Another often-used distribution function is the integral distribution function, I(M), which is defined by $I(M) = \int_0^M W(M)dM$. It is the weight fraction of all the sample having a molecular weight lower than M. In the industry, a more commonly used index of molecular weight distribution is the so-called Q-value, defined as M_w/M_n, where M_w and M_n are the weight average and number average molecular weights, respectively.

The determination of molecular weight distribution of polypropylene is rather an involved process and relevant literature should be consulted for experimental details (see Bibliography). The methods involve fractionation of the polymer into portions of different average molecular weights. Fractional precipitation technique is too tedious and time consuming to be given any serious practical consideration.

Elution fractionation in connection with intrinsic viscosity determination is probably the most commonly used method today. In recent years gel permeation chromatography has also come into wide spread use. The method is, however, not yet standardized and each laboratory has its own experience with this technique.

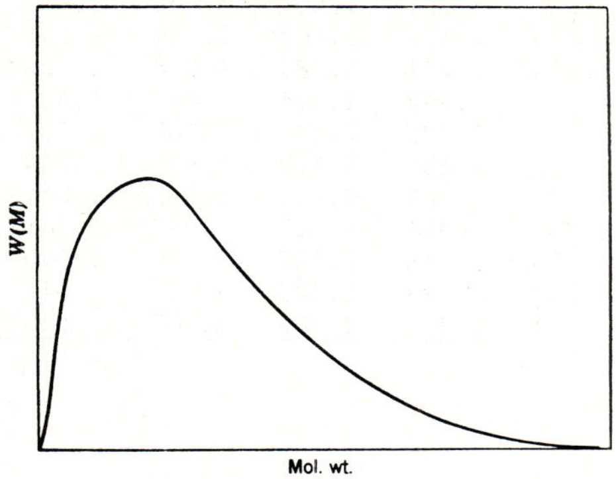

Fig. 6. A Typical Differential Molecular Weight Distribution Function for Polypropylene

Commercially, melt-flow rate (see chapter IV) is used to characterize a polymer for extrusion. It is a rough measure of molecular weight and molecular weight distribution. The ranges that might be expected are illustrated in Table XI for a series of Hercules' Pro-Fax samples.

TABLE XI

Melt-flow Rate, Intrinsic Viscosity, and Average Molecular Weight of a Series of Pro-Fax Polypropylene

Sample	*Melt-flow rate(dg/min)*	*Intrinsic viscosity(dl/g)*	$M_v \times 10^{-3}$	$M_w \times 10^{-3}$
61XX	60	1.3	120	156
62XX	30	1.5	140	180
63XX	15	1.7	170	220
64XX	6.5	2.05	210	270
65XX	4	2.25	250	325
66XX	2	2.6	290	380
67XX	0.8	3.1	360	470
68XX	0.4	3.5	420	550

Melt-flow rate by ASTM D I238-65T, 230°C; Viscosity in decahydronaphthalene at I35°C. Pro-Fax is a trade mark of Hercules Inc.

III.2. Tacticity

Commercial polypropylenes are highly isotactic and contain usually only a few percent of atactic material. The isotactic material is partially crystalline, the atactic portion is practically wholly amorphous. The determination of tacticity in polypropylene is generally done by extraction method. It is based on the selective solubility of atactic polypropylene fraction in standard solvents under standard conditions. Quoted figures for isotactic indices are usually in terms of "boiling Heptane insolubles" or "boiling Decalin insolubles". Due to its simplicity, this method has become widely used. Care should be taken, however, in the interpretation of the results, as the soluble fraction may include, in addition to atactic polypropylene, stereoblock polymer and very low molecular weight isotactic fractions. Also, the dissolution of various components greatly depends on the physical form of the sample.

Isotactic content of polypropylene is of great practical significance for physical and mechanical properties of the product. Commercial polypropylenes generally have about 0.95 or higher isotactic indices. High isotactic index is sought for, as it contributes to higher crstallization potential of the polymer and much improved mechanical properties of the products. Higher isotactic index of the polymer, for example, increases its yield stress, hardness, flexural elastic modulus, and brittle point. High isotactic index, however, has a somewhat negative effect on impact strength of the product. Very high isotactic index may also create some problems in cold-drawing of the fiber. The fibers produced from such polymers may have several desirable properties, such as high strength, creep-resistance, and resilience, but the manufacturing processes may be limited in production rates. Polymers with high atactic content are useful for the production of high-shrinkage fibers, conjugate fibers, and flexible yarns such as those used for baling twine.

III.3. Thermal and Thermodynamical Properties

The crystalline melting point of isotactic polypropylene is generally quoted as 167°C. As is the case with practically all other crystalline polymers, all melting does not occur at this temperature; a melting range is observed which starts as low as 130°C. The glass transition occurs at a temperature T_g at which the noncrystalline phase of a polymer will become glassy and brittle. Atactic polypropylene shows a marked glass transition temperature in the range of -12°C to -15°C. The agreement on T_g value of isotactic polypropylene derived from various experimental methods is not very good - it ranges from about -30°C to

$25^{\circ}C$. This is because different types of supermolecular structure bonds contribute differently to relative movement of the structural elements.

The temperature range within which polypropylene can be used is limited by the crystalline melting point, T_m, on the higher side, and by the glass transition temperature, T_g, on the lower side. Between these two extreme points, polypropylene undergoes considerable changes in some of its physical and mechanical properties. These changes and their relation with temperature or heat-input are important, from the viewpoint of both fiber properties and some aspects of fiber processing, such as cold-drawing, texturizing, embossing, dyeing, etc. Some of these temperature dependent changes are reversible, such as stiffness, but some changes are irreversible, such as those occuring in crystalline and supermolecular structures, discussed elsewhere in this book.

TABLE XII

Some Thermodynamical Properties of Polypropylene (ref.18)

Property at 230°C	
Heat content relative to 20°C, cal/g	129
Coefficient of thermal diffusion, cm^2/s	0.9×10^{-3}
$(\partial H/\partial T)_P$	0.55
$(\partial S/\partial T)_P$	1.10×10^{-3}
$(\partial H/\partial P)_T$	0.84
$(\partial S/\partial T)_T$	1.0×10^{-3}
Bulk modulus, dyn/cm^2	0.65×10^{10}
Density, g/cm^3	0.75
Density at 20°C, g/cm^3	0.90

H = enthalpy; S = entropy; P = pressure

Thermodynamic properties, such as specific heat, heat of fusion, thermal conductivity, etc., of a polymer are important from process design point of view. These are summarized in Table XII for isotactic polypropylene. The specific heat at constant pressure is about 0.5 cal/g.deg.C. This high heat content combines with low coefficient of thermal diffusion, $10^{-3} cm^2/s$, to give slow cooling and solidification rate of the melt in processing.

Polypropylene melts are highly compressible, having bulk moduli of the order of $10^{10} dyn/cm^2$. The density of the melt varies by about 0.1 percent per degree C, but changes of 25 percent or more can occur as crystallization takes place. From these measurements the basic thermodynamic parameters - the rate of change of entropy and enthalpy with temperature and pressure - may be derived. Classical thermodynamics may then be used to calculate, for instance, the rate of temperature rise during adiabatic compression or the rate of change of pressure with temperature at constant volume.

Considerable variation has been noticed in literature values for the heat of fusion of isotactic polypropylene (ref.11,12). These differences probably arise from experimental uncertainties as well as from variation in polymer crystallinity which is strongly dependent on the thermal history of the sample. A value of $\Delta H_{\mu} = 2.1 \pm 0.3$ kcal/mol is generally agreed upon.

In viscous flow, energy is dissipated as heat and the rate of temperature rise at constant pressure is equal to the product of shear stress times shear rate divided by the specific heat. This shear heating is an important feature of all large scale extruders and, in addition, local temperature rises of more than $50^{o}C$ may be encountered in particularly high rate of extrusion. High pressure spinning, discussed in Chapter VIII, is a special case taking advantage of shear heating of the melt.

III.4. Chemical and Biological Response

Being a hydrocarbon, the chemical resistance of isotactic polypropylene with respect to most chemicals is outstanding. Extensive information on chemical resistance of polypropylene, developed through exhaustive tests with hundreds of organic and inorganic chemicals, shows that polypropylene is very resistant to mineral acids, alkalies, aqueous solutions of inorganic salts, detergents, oil, and greases at room temperature. The substances that attack polypropylene are highly oxidative agents, such as hydroperoxides, fuming nitric acid, halogens, concentrated sulfuric acid, and chlorosulfonic acid. Other chemicals which adversely affect polypropylene are organic solvents. Most aliphatic, aromatic, and halogenated hydrocarbons cause swelling and dissolution of isotactic polypropylene at elevated temperatures. Table XIII gives data on the resistance of polypropylene to some chemicals.

At room temperature, polypropylene is resistant to the action of water and does not show any changes in dielectric or other physical properties in an atmosphere of high relative humidity or even on immersion in water. Water, alkali solutions, and some detergents, however, do leach out some of the heat and light stabilizers present in the formed polypropylene articles and may therefore affect its long-term properties. Polypropylene is oxidized at elevated temperatures or under the influence of UV light with initial formation of hydroperoxide groups in its chain. The increase in the oxidized content of polymer results in discoloration, odor, reduced molecular weight, and detereoration of mechanical properties of the product. Small amounts of stabilizers added to the polymer during processing prolong the resistance of polymer to atmospheric oxidation. This subject will be a topic of our discussion in the next chapter.

TABLE XIII

Chemical Resistance of Isotactic Polypropylene (ref.19)
(weight change in 30 days exposure)

Chemical	*Weight Change (%)*	
	23°C	*60°C*
Sulfuric acid 98 %	- 0.2	- 0.2
Nitric acid, fuming	- 0.1	-
Sodium hypochlorite 20%	0.1	- 2.1
Gasoline	4.8	6.6
Benzene	3.4	0.6
Xylene	7.0	0.3
Methylene chloride	5.5	1.6
Carbon tetrachloride	13.5	0.9
Turpentine	9.5	10.5
Transformer oil	0.4	14.9

The weight change as listed is due to the sum of the effects of swelling and dissolution.

Polypropylene is generally considered to be extremely resistant to fungi and bacteria. Some microorganisms, however, may grow even on small amounts of contaminants which may be present on the surface of the fiber. Such growth has no effect on the mechanical properties of the product. Pure polypropylene is well tolerated by living tissues but the stabilizers or other additives may complicate the situation to some degree in this respect. Polypropylene can not be digested by insects and related pests; polypropylene products are, therefore, not liable to attack by living organisms. This property is of high significance for polypropylene fibers as it renders them inherently "moth proof" without any additional finishing operation.

IV. REFERENCES

1. K.H. Roenitz, British Plastics, Feb.(1971)56.
2. Anon., Plastic World, Aug.(1972)15.
3. British Patent 828,023, Montecatini, 1959.
4. British Patent 828,791, Montecatini, 1960.
5. U.S. Patent 3,511,824, Johnson & Johnson, 1970.
6. R. Chiang, J. Polym. Sci., 36(1959)91.
7. R. Chiang, J. Polym. Sci., 28(1959)235.
8. J.B. Kinsinger, J. Phys. Chem., 63(1959)2002.
9. K. Yamaguchi, Macromol. Chem., 128(1969)19.
10. J.H. Elliott, J. Appl. Polym. Sci., 14(1970)2947.
11. W.R. Kirgbaum and I. Uematsu, J. Polym. Sci. Part A, 3(1965)76T.
12. V. Cappucio et. al., Chimica e Industria, 44(1962)463.
13. S.A. Mossmann, Man-made Textiles in India, Oct. 1976.
14. A.L.J. Raum in *Vinyl and Allied Polymers,* Ritchie, Ed., Chem. Pub. Co., Cleveland, Ohio, 1968.
15. *Encyclopedia of Polymer Science & Technology* Vol. II, 597, Wiley, New York, 1969.
16. H. Steiner, *Introduction to Petroleum Chemicals,* Pergamon Press, 1961.
17. J.H. Elliott, J. Polym. Sci., 14(1970)2947.
18. F.N. Cogswell, Plast. Polym., 38(1970)331.
19. H.P. Frank, *Polypropylene* Gordon and Breach, New York, 1969.
20. C.H. Davenport, Petr. Refiner., 39(1960)125.
21. G. Natta, J. Polym. Sci., 34(1959)21.
22. O.J. Sweeting in *Science and Technology of Polymer Films,* Vol. II, Wiley, New York, 1971.
23. C&EN, (ACS, Washington, D.C.), May 2, 1977.
24. Hancock, *Propylene,* Wiley, New York, 1973.
25. C&EN, (ACS, Washington, D.C.), April 4, 1977.
26. U.S. Patent 3,879,368, Union Carbide, 1975.

V. BIBLIOGRAPHY

V.I. Books and Monographs

1. T.O.J. Kesser, Polypropylene, Reinhold, New York, 1960.
2. M. Sittig, Polyolefin Resin Processes, Gulf Publ. Co., Houston, Texas, 1961.
3. J.R. Van Vazer, Viscosity and Flow Measurement - a Laboratory Handbook, Interscience, New York, 1963.
4. Raff and Doak, Ed., Crystalline Olefin Polymers, Part 1 and 2, Interscience, New York, 1964.
5. A.V. Galanti and C.L. Mantell, Polypropylene Fibers and Films, Plenum Press, New York, 1965.
6. L. Reich and A. Schneider, Polymerization by Organometallic Compounds, Interscience, New York, 1966.
7. R. Long, Ed., The Production of Polymer and Plastics Intermediates from Petroleum, Plenum Press, New York, 1967.
8. R.D. Ritchie, Ed., Vinyl and Allied Polymers, Vol. 1, CRC Press, Cleveland, Ohio, 1968.

8a. L.H.Tung, Fractionation of Synthetic Polymers, Marcel Dekker, New York, 1977.

9. Encyclopedia of Polymer Science and Technology, entry: Molecular Weight Determination of Polymers, Vol. 9, pp 182-93; entry: Polypropylene, Vol. 11, pp 597-619; Interscience, New York, 1968, 1969.
10. V.L. Erlich, Man-made Fibers - Science and Technology, Vol. 3, Chapter on Linear Polyolefins, pp 369-99, Interscience, New York, 1968.
11. H.P. Frank, Polypropylene, Gordon and Breach, New York, 1969.
12. O.J. Sweeting, Ed., The Science and Technology of Polymer Films, Vol. 2, Interscience, New York, 1971.
13. N.M. Bikales, Characterization of Polymers, Interscience, New York, 1971.
14. H.L. Peebles, Molecular Weight Distribution in Polymers, Interscience, New York, 1971.
15. T. Keii Kodansha, Kinetics of Ziegler-Natta Polymerization, Halstead Press, Div. Wiley, New York, 1972.
16. A.G. Sirota, Polyolefins, translated and published by Israel Program for Scientific Translations, Jerusalem, 1972.
17. Hancock, Ed., Propylene, Wiley, New York, 1973.
18. P.E.M. Allen, Kinetics and Mechanism of Polymerization Reactions, Wiley, New York, 1974.
19. J.C.W. Chien, Coordination Polymerization, Academic Press, New York, 1974.
20. L.F. Albright, Processes for Major Addition Type Plastics and Their Monomers, McGraw-Hill, New York, 1974.
21. Schildknecht, Ed., Allyl Compounds and their Polymers, Interscience, New York, 1974.
22. P.E. Slade, Ed., Polymer Molecular Weights, Part 1 and 2, Marcel Decker, New York, 1975.
23. Polyolefin Production Processes - Latest Developments, Noyes Data Corp., Park Ridge, N. J. 1977.
24. Technical Papers, SPE Reg. Tech. Conf., Polyolefins II, Houston, Tex. Feb. 1978.

V.2. Journal Literature

1. R.L. Miller, Characterization of Stereoregular Polymers, I. Theory, J. Polym. Sci., 46(1960)303-16.
2. S. Newman, The Characterization of Stereoregular Polymers, II. Polypropylene, J. Polym. Sci., 47(1960)111-37.
3. J.W.S. Hearle, The Polyolefins, Textile Manufacturer, 87(1961)439-41.
4. T.E. Davis and R.L. Tobias, "Fractionation and Molecular Weight Distribution of Polypropylene, J. Polym. Sci., 50(1961)227-43.
5. J. Van Schooten, Molecular Weight Distribution of Isotactic Polypropylene, Makromol. Chem., 43(1961)23-30.
6. T. Kawai, Analysis of Intrinsic Viscosity Data, J. Polym. Sci., 54(1961)343-62.
7. G.J. Howard, Molecular Weight Averages and the Tung Distribution Function, J. Polym. Sci., 59(1962)4-6.
8. M.L. Huggins, Report on Nomenclature Dealing with Steric Regularity in High Polymers, J. Polym. Sci., 56(1962)153-61.
9. R.A. Mendelson, Elution Fractionation of Crystalline Polypropylene, J. Polym. Sci., A-1(1963)2361-9.
10. H.W. Haines, Polypropylene, Ind. Eng. Chem., 55(1963)30-37.
11. A.M. Kotliar, Critical Evaluation of Mathematical Molecular Weight Distribution Models Proposed for Real Polymers. I. Effects of Low Molecular Weight Cut-off, J. Polym. Sci., Part A, 2(1964)4303-25.
12. D.A. Lewis, Polyolefins, Rept. Progr. Appl. Chem, 49(1964)601-21.
13. R. Chiang, Intrinsic Viscosities of Isotactic Polypropylene in Various Solvents, J. Phys. Chem., 70(1966)929-30.
14. B.C. See, Estimating Fractionation Conditions for Linear Polyethylene and Isotactic Polypropylene, J. Appl. Polym. Sci., 10(1966)1625-35.
15. D.S.Sanders, Propylene as a Chemical Feedstock, CEP, July, 1977, 40-45.

15. A.C. Ouans, The Molecular Weight Distribution of Polypropylene, J. Polym. Sci., Part C, 21(1967)309-15.
16. J.F. Johnson, Fractionation of Polymers, Encycl. Polym. Sci. Tech., 7(1967)231-60
17. G. Bobcock, Polyolefins, Rept. Progr. Appl. Chem., 52(1967)580-600.
18. R. Koningsveld, Polymer Fractionation - The Preparatory Problems, J. Polym. Sci., A-2, 6(1968)367-81.
19. H.K. Livingston, Molecular Weight Determination, Encycl. Polym. Sci. Tech., 9(1968)182.
20. R. Koningsveld, Polymer Fractionation - The Analytical Problem, J. Polym. Sci., A-2, 6(1968)383-406.
21. T. Sakai, Extrapolation Procedures of Intrinsic Viscosity and for Huggins Constant K, J. Polym. Sci., A-2, 6(1968)1659-72.
22. J. Hughes, Modified Ubelhode Viscometer for Measurement of Viscosity of Polymer Solutions, J. Sci. Instru., 12(1969)1134-5.
23. O.F. Solomon, Correct Determination of Intrinsic Viscosity, J. Appl. Polym. Sci., 12(1968)971-2.
24. M.D. Baijal, Molecular Weight Fractionation of Polypropylene, Macromolecules, 2(1969)679-81.
25. A. Rudin, Molecular Weight Distribution of Polymers, J. Chem. Educ., 46(1969)595-600.
26. J. Lanikova, Fractionation of Isotactic Polypropylene, Eur. Polym. J., 6(1970)25-28.
27. A. Yamamoto, Comparison of Various Methods of Determining Molecular Weight Distribution, Polym. J., 1(1970)304-11.
28. J.H. Elliot, One Point Intrinsic Viscosity Method for Polyethylene and Polypropylene, J. Appl. Polym. Sci., 14(1970)2947-63.
29. L.J. Housek, Fractionation of Isotactic Polypropylene, Eur. Polym. J., 6(1970)29.
30. K. Yamaguchi, Molecular Weight Distribution of Crystalline Polypropylene, Makromol. Chem., 132(1970)143-52.
31. H.P. Below, Molecular Weight Distribution of Polyolefins, Eur. Polym. J., 6(1970)29.
32. L. Ackerman, Single Point Determination of Limiting Viscosity Numbers, J. Polym. Sci. (Chem), 10(1972)2689-90.
33. I. Pasquan, Stereoregular Polymerization, MTM Int. Rev. Sci., Phy. Chem., Ser. one, 8(1972)159-98.
34. T. Ogawa, Fractionation of Polypropylene by Raising Temperature Method, J. Appl. Polym. Sci., 17(1973)2235-44.
35. P. Parrini, Molecular Weight Determination of Isotactic Polypropylene by High Sensitivity Ebulliometry, Makromol. Chem., 175(1974)935-44.
36. U.S. Patent 3,807,221 to Phillips Petroleum, Molecular Weight Determining Apparatus, 1974.
37. P.J.T. Tait, Ziegler-Natta Polymerization - Model, Mechanism, Kinetics, Chem. Tech., 5(1975)688-92.
38. H.G. Elias, Polymolecularity and Polydispersity in MW Determinations, Pure Appl. Chem., 43(1975)115-47.
39. A. Rudin and R.A. Wagner, New One Point Intrinsic Viscosity Method, J. Appl. Polym. Sci., 19(1975)3361-7.
40. C. Cozart, Stereospecific Polymerization of Polypropylene, Angewandte Chem., 44(1975)31-46.
41. G. Hinrichsen, Tasks and Objectives of Polymer Physics in Connection with the Production and Characterization of Synthetic Fibers, Melliand Textilber, 6(1977)I.
42. D.A. Blackadder, Dissolution of Polypropylene in Organic Solvents - Effect of Molecular Weight of Polymer, Polymer, 18(1977)547-550.
43. R.T.Magovern, Current Polyolefin Manufacturing Processes, SPE Reg. Tech. Conf. on Polyolefin, II, Houston Tex. Feb. 1978.
44. M.Klaus, Prospects of the Petrochemical Industry in the World, Chem.Econ.Engg.Review, April 1977, 1-16.

45. Matsuura, Rationalized Process for Polypropylene Production, Chem.Econ.Engg. Review, Oct. 1977, 28-31.
46. B.P. Ennis et al., Olefin Manufacture via Multisecond Pyrolysis, Chemtech, Nov. 1975, 693-699.
47. K.V.Erdman, The Outlook for Feedstock for Polyolefins, Symp. Polypropylene 20th Year in Textiles, Sept. 13-14, 1977, New York.
48. M.Johnson, Polyolefin Process Patents, SPE RECTEC "Polyolefins II", Feb.14-15, 1978, Houston, Texas.
49. G.Moraglio, [η] - M Relationship for Isotactic Polypropylene Obtained with High Mileage Catalysts, Polymer, 19(1978)1363-1364.

VI. DIRECTORY OF POLYPROPYLENE RESIN MANUFACTURERS

United States

Amoco Chemical Corp.
Atlantic Richfield
Eastman Chemicals
Exxon Chemicals
Gulf Oil Chemical Co.
Hercules Inc.
Northern Petrochemical Corp.
Rexene Polymer (Dart Industries)
Shell Chemical Co.
Novamont
Phillips Petroleum Co.

Canada

Hercules Inc.

Germany

Rheinische Olefinwerke
Societe Chimique des Charlonnages
Chemische Werke Hulls

France

Solvay
CDF Chimie

Holland

DSM

Belgium

Amoco Chemicals
Hercules Inc.

Sweden

Scandinaviska

South Africa

Shell Chemical Co.

England

Imperial Chemical Co.

Japan

Mitsubishi Petrochemical Co.
Mitsui Petrochemical Co.
Mitsui Toatsu/Nippon Oil
Showa Denka

Taiwan

Hercules Inc.

Italy

Montedison

Chapter II. STABILIZATION AGAINST HEAT AND LIGHT

I. INTRODUCTION

One of the shortcomings of polypropylene is its relatively low resistance to heat and light, which causes problems in fabrication and induces rapid changes in physical properties of the products in use. The problem of polymer degradation is especially serious in its use in film and fibers because of the relatively high temperatures of fabrication involved and because of the large surface area of the product exposed to weathering conditions. The task of improving the resistance of polypropylene to high temperatures and weathering conditions is, therefore, a major area of interest in fiber manufacturing. This chapter will be devoted to a discussion of the action of various degradative agencies on polypropylene and on investigation into measures to counter them.

Polypropylene is relatively stable to the action of heat and light alone. Its oxidation resistance is, however, poor. In fact it is almost impossible to process polypropylene under commercial operating conditions without the presence of anti-oxidants in the polymer. The molecular weight of the polymer, with accompanied effects on the viscosity of the melt and tensile properties of the fibers, is reduced with increasing degree of oxidation. Polypropylene is also sensitive to atmospheric oxidation under the action of light at normal temperatures of use. The effect of light leads polypropylene to the same kind of physical changes as those experienced under the effect of heat.

Attention must be paid to degradation of polypropylene in two phases. Firstly, the polymer must be protected from thermo-oxidative and pyrolytic degradation during fabrication, especially within the extruder, metering pump, spinning block, and the region just below the spinnerette, where the polymer is subjected to high temperatures in presence of atmospheric oxygen. Secondly, attention must be paid during subsequent processes such as drawing, texturizing, heat setting, scouring, dyeing, etc., and during the useful life of the product, when the fiber is subjected to mild but prolonged action of heat and light in the presence of atmospheric oxygen.

Through the use of certain chemical compounds the occurence of chemical changes leading to the deterioration of polymer properties can be slowed or their onset delayed to sufficient degree as to make the article of manufacture useful for a reasonably long period of time. Such additives are known as stabilizers; the knowledge of their composition and the mode of their action is important for the manufacture of commercially useful fibers.

The characteristic feature of the chemical substances commonly used for stabilization of polypropylene is the very small amount needed for effective protection. The concentration of stabilizers in the polymer is never more than 5 percent and generally lies between 0.1 and 2.0 percent of the polymer composition. The choice of suitable stabilizers for specific end-uses of the fiber or for the particular processing effect desired has remained, to date, largely an empirical art. However, on the basis of past experience in dealing with such formulations, combined with recently acquired insight into the mechanism of degradation and the mode of action of various stabilizers, it should be possible to make some intelligent predictions and set forth the direction for developmental work.

II. DEGRADATION UNDER HEAT AND LIGHT

In any specific industrial or commercial application of polypropylene, it is often difficult to distinguish between the effects of heat, heat-oxygen combination, or light on the polymer degradation. Most often the overall deterioration of polymer properties is the result of the composite process to which heat, light, and atmospheric conditions contribute. Nevertheless, their respective functions in overall deterioration process must be understood separately if the most effective possible steps are to be taken to counter the effect of these degradative agencies. The following discussion will provide an outline.

II.1. Pyrolytic Degradation

Polypropylene is relatively stable to heat in the absence of oxygen and quite high temperatures are necessary to cause any significant decomposition. It shows little or no reaction below 350°C. Of all the decomposition reactions, the degradation of this sort is the most difficult to inhibit when it occurs. However, the temperatures required for pyrolysis reactions of polypropylene are considerably higher than the processing temperatures generally employed. The pyrolytic degradation of polypropylene is, therefore, not a real problem in processing of the polymer.

In addition to heat, the shear of the melt also induces the breaking of polymer chains. The chain breaking under excessive shear is an important path of polymer decomposition, but in actual practice the localized overheating of the melt in areas of excessive shear may become the main cause of degradation even before chain breakage due to mechanical energy occurs. The areas of overheating caused by excessive shear rates are not entirely unknown in processing equipment, such as the metering zone of an extruder and the pumping areas of the machinery. As will be discussed in Chapter VII, in spinning operations an attempt is generally made to achieve a certain degree of shear to facilitate mixing and homogeneization of the melt. These areas of excessive shear and localized heating in these devices of polymer mixing are an undesirable feature of processing machinery. These defects can be largely remedied by a change in equipment and the choice of proper throughput rate.

II.2. Thermo-oxidative Degradation

In comparison with the relatively good thermal stability of polypropylene in the absence of oxygen, its resistance to heat in the presence of oxygen is very poor. The stability of polypropylene against oxidative degradation is, therefore, more crucial from the processing, as well as from the end-use, point of view than is the polymer decomposition due to heat alone.

Oxidative degradation of polypropylene is a chain reaction that can be triggered by heat or light. It can be accelerated by certain metal catalysts and subdued by certain chemical compounds known as antioxidants and UV stabilizers. The generally accepted scheme for simple oxidative degradation under the influence of heat is shown below. In the following outline of mechanism, R-H refers to the polypropylene molecule, R* to a free radical, ROO* to peroxide radicals, and ROOH to hydroperoxides.

Initiation	RH	$\xrightarrow{\text{Energy}}$	R* + H*
	RH	$\xrightarrow{\text{Energy/Oxygen}}$	R* + HOO*
Propagation	R* + O_2	$\longrightarrow$	ROO*
	ROO* + RH	$\longrightarrow$	ROOH + R*
	OH* + RH	$\longrightarrow$	H_2O + R*
	ROOH	$\longrightarrow$	RO* + OH*
Termination	R* + R*	$\longrightarrow$	R-R
	R* + ROO*	$\longrightarrow$	ROOR
	ROO* + ROO*	$\longrightarrow$	ROOR + O_2

The reaction of free radicals with molecular oxygen occurs almost instantaneously, with practically no energy of activation. Termination of the chain reaction is possible only by dimerization or disproportionating action of active free radicals.

However, due to their high reactivity, such events are effective only at relatively high concentrations of free radicals, when the polymer degradation is well advanced.

As the concentration of peroxides in the system increases, they begin to decompose and a new reaction cycle comes about:

$$2\ ROOH \xrightarrow{Energy} RO^* + ROO^* + H_2O$$
$$ROOH \xrightarrow{Energy} RO^* + OH^*$$

The branching, i.e., production of free radicals from peroxides, is a very slow process at room temperature but is greatly accelerated by heat, light, and metal impurities in the polymer. Thus, under the action of these agencies, new reaction cycles are started by R*, ROO* and OH* radicals which react with RH (polymer) and oxygen of the air. Oxidation now progresses *autocatalytically* by a chain radical mechanism in a typical "snowballing" manner.

The term *autocatalysis* is applied to describe the course of reaction where one of the reaction products functions as a catalyst. As a result of its accumulation, the rate of reaction increases with an increasing degree of conversion. The end result of such reactions is that the reaction rate increases exponentially with reaction time and after a certain period, called the *induction period*, reaches an extremely high value, causing catastrophic changes in material properties. The rate-determining step of autocatalytic oxidation of polypropylene is the relatively slow decomposition of peroxide molecules. Any factor that accelerates this decomposition also increases the rate of polymer oxidation.

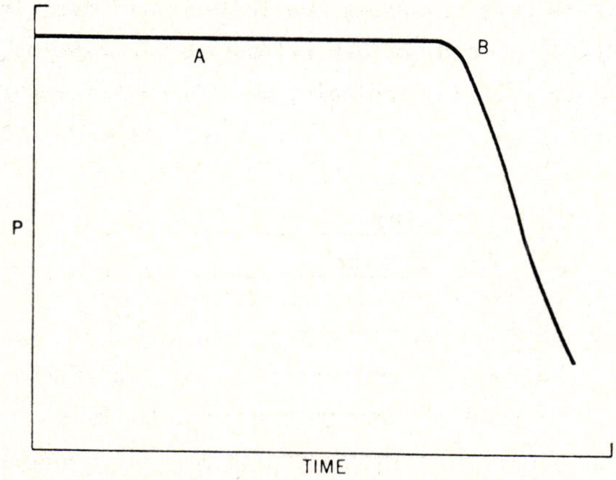

Fig. I. Schematic Representation of Aging of Polypropylene

The deterioration of polypropylene follows a course similar to that shown in Fig.1. When some property of practical importance, P, is plotted against time, no appreciable change in P occurs during the induction period, which ends near B. Then easily measurable changes begin and the polymer deteriorates with accelerating rapidity. The period of apparent inactivity, or the induction period, is often taken as a measure of stability of the product towards thermal and photooxidation.

As a result of oxidation of polypropylene, the molecular weight, melt viscosity, and intrinsic viscosity of the polymer decreases, the molecular weight distribution pattern alters, the carbonyl content increases, and elongation and tensile strength of the product is impaired.

II.3. Photo-oxidation

Unprotected polypropylene progressively loses its chemical and physical properties during weathering at ambient temperature. This is attributed to photochemical reactions initiated by absorbed UV light. The atmosphere absorbs most of the noxious radiation emitted by the sun; only about 5 percent of the energy reaching the surface of the earth occupies the UV region from 3000 to 4000 Å. However, the energy of this residue is still high enough as to initiate a large variety of chemical reactions which lead to the degradation of polymers. Direct photolytic cleavage of the backbone of the polypropylene molecule seems to be a minor cause of the photodegradation of the polymer: the photodegradation reactions of polypropylene are almost exclusively the autoxidation reactions triggered by UV radiation of 3000-3700 Å.

Polypropylene, when pure, does not usually absorb in the spectral region above 3000 Å and hence should not degrade in the ultraviolet light of the sun. The first law of photochemistry states that there cannot be a photochemical reaction if there is no absorption. However, processed polypropylene does in fact have a slight absorption which extends into the region above 3000 Å. The photochemistry of polypropylene depends, therefore, on low intensity, externally induced chromophores, the chemical nature of which is very hard to define. The best known photoactive species thermally produced in polypropylene during processing are carbonyl groups and hydroperoxides.

In any photochemical reaction, the ways in which the energy of an absorbed phototon is dissipated can be classified into photophysical processes and photochemical processes. The photophysical processes include fluorescence, which is simply re-emission at a slightly longer wavelength and slightly greater time interval; and energy transfer. The alternatives to these photophysical processes, the processes which compete with them for the energy of the quantum absorbed are photochemical reactions such as the formation of free radicals, intermolecular rearrangements, cyclization, photo-elimination, and photo-ionization. Nearly all of these chemical processes will ultimately cause a deterioration in the physical properties of a polymer. When stabilizing a polymer, it is therefore desirable to eliminate the photochemical reactions, i.e., to maximize the photophysical processes, particularly the conversion to thermal energy. If one can increase the efficiency of this process, a much more stable polymer will result.

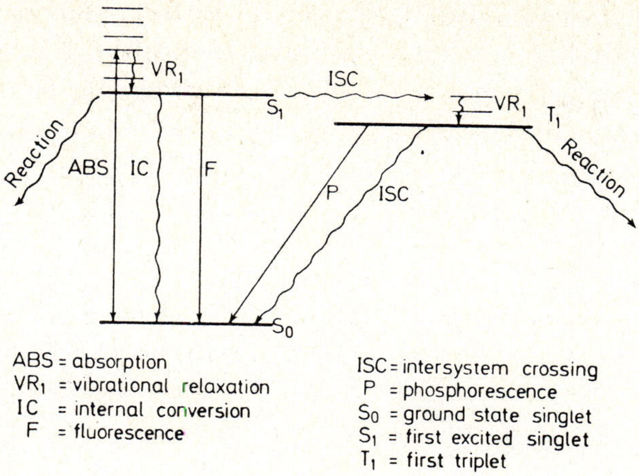

Fig. 2. Photo-physical Processes

Figure 2 shows an energy diagram which describes the kinds of photophysical processes that are possible. When a quantum of light is absorbed, a particular group is raised to some excited state, and since most groups will be in singlet ground state, they will be excited to a vibrationally excited singlet state. In the solid phase, there will be a rapid degradation of the vibrational energy down to the lowest vibrational level of the electronically excited singlet state, indicated as S_1. From this S_1 excited state all the processes that we are concerned with will occur. The simplest is fluorescence, that is, emitting by dropping down to the ground state singlet. The second is internal conversion, that is, conversion of the electronic energy into thermal energy. It is a spectroscopically forbidden process but there are various means by which it can occur; in many polymers it is the major process. The spin of the excited electron can also be inverted to give the triplet state, by intersystem crossing to vibrationally excited triplet and dropping down to the lowest vibrational state, T_1, of the triplet. Chemical reaction can occur from either the singlet or the triplet state. Similarly, by intersystem crossing one can convert the triplet energy to thermal energy,or a quantum of light can be emitted by phosphorescence.

Photochemical processes in the degradation of polypropylene are essentially the photochemical reactions of hydroperoxides and carbonyl groups, produced in the course of processing of the polymer at high temperatures or through long wavelength UV initiated photooxidation.

The alkyl hydroperoxides absorb a part of the UV light reaching the substrate and break into free radicals:

$$\text{ROOH} \xrightarrow{\text{Light}} \text{RO*} + \text{OH*}$$

The same applies to carbonyl groups, although the reactions of ketones after light absorption are more complex. There are two major reactions which are believed to occur: Type I, which gives two free radicals; and Type II, a molecular rearrangement which gives a methyl ketone and a double bond. In both cases the chain breaks. This ultimately leads to a lowering of molecular weight and, finally to degradation in the physical properties of the polymer.

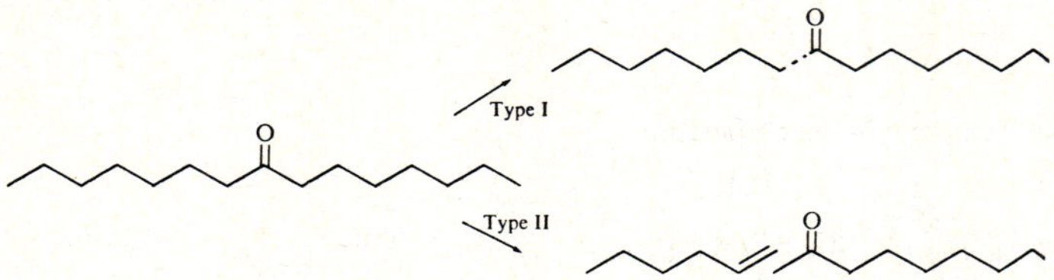

The free radicals produced in the above reactions may start a chain reaction, including chain propagation, oxidative processes, chain transfer, and chain termination. The net result is outlined below:

Propagation

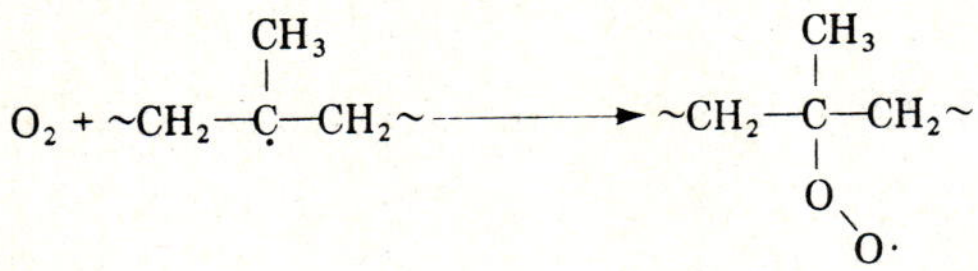

(a) Intermolecular

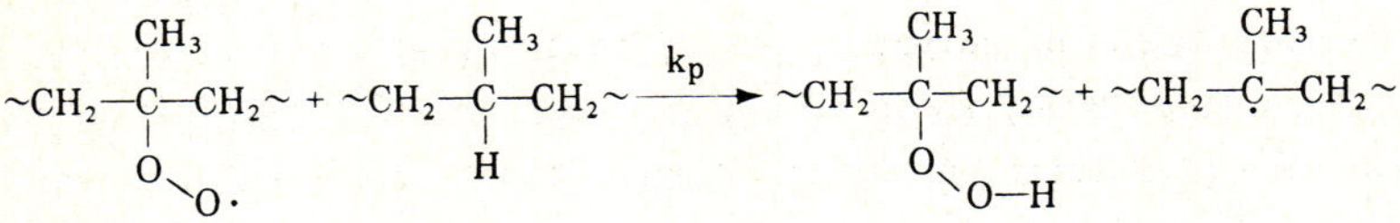

(b) Intramolecular

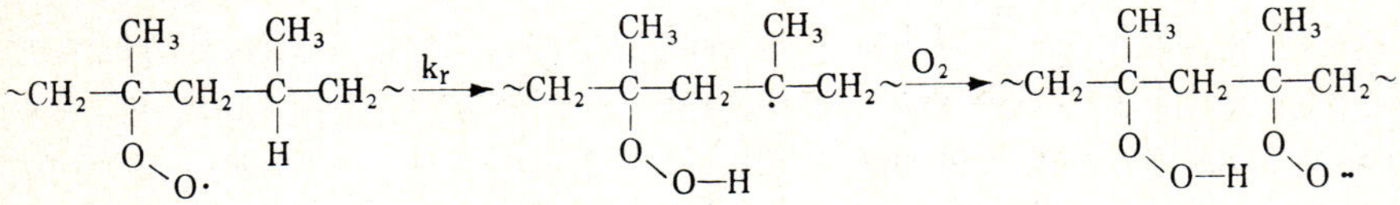

Tertiary Radical Termination

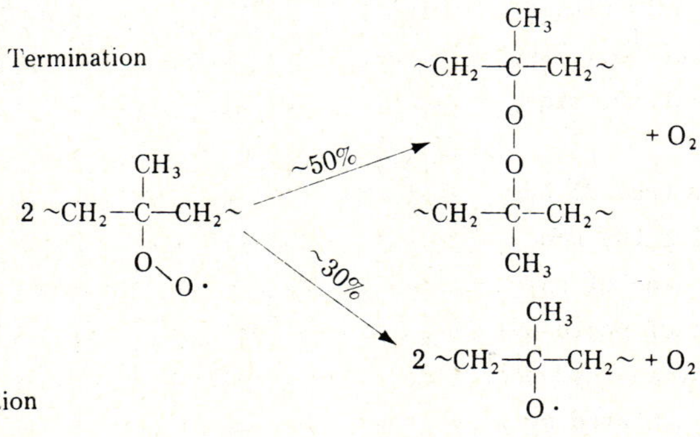

Alkoxy Abstraction

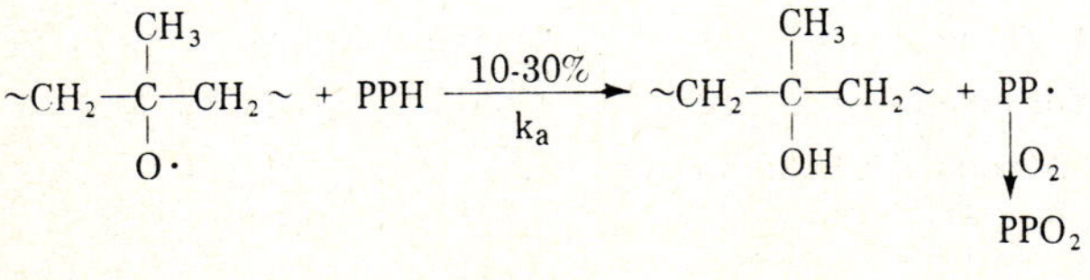

β-Scission

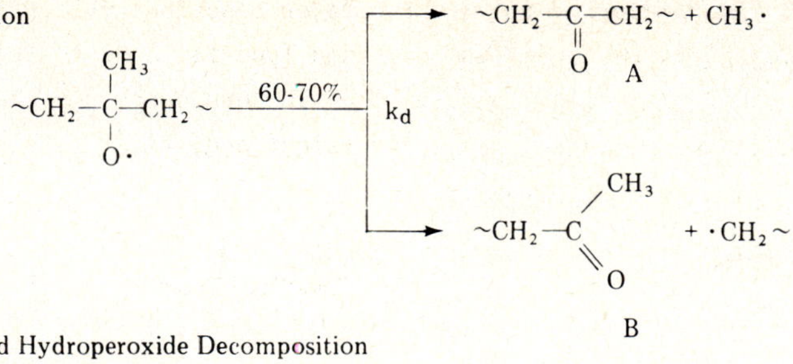

Induced Hydroperoxide Decomposition

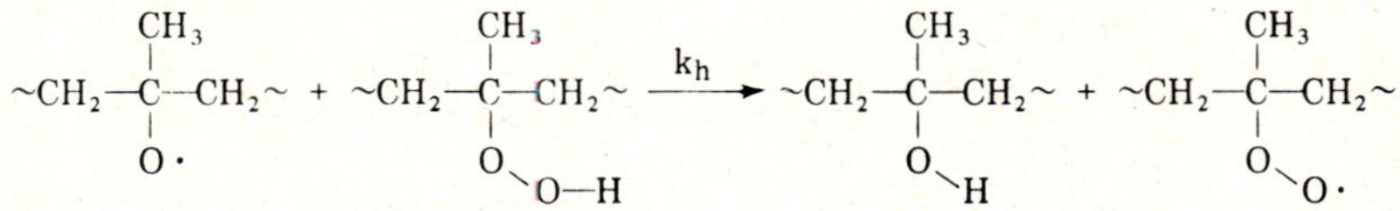

Cross Termination

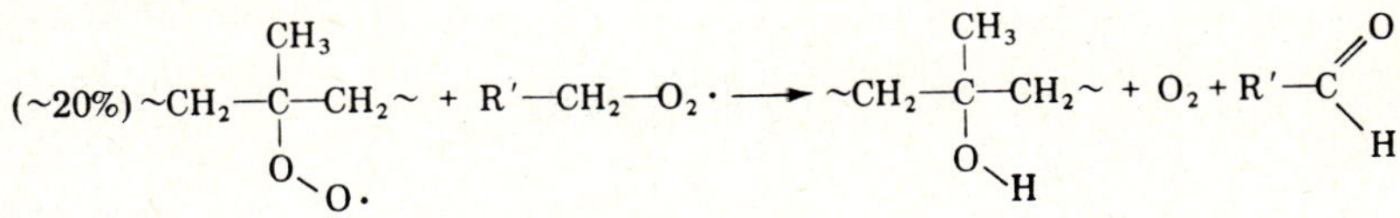

The UV degradation of polypropylene differs from thermal oxidation in the nature of initiation step. Here, the number of oxidation cycles per initiation step is small and rate of initiation high. Moreover, since more ketone groups are formed as by-products of degradation, the amount of energy absorbed tends to increase with time of irridiation.

Singlet oxygen ($^1\Delta_g$, $^1O_2$) has been reported to attack C=C unsaturation in rubbers to give hydroperoxides, and to attack either unsaturated or saturated sites in polyethylene to give hydroperoxides. It has been shown that many of the effective Ni(II) chelates commonly claimed to photostabilize polypropylene are extremely efficient quenchers of $^1O_2$. Taken together, these discoveries automatically lead to the suggestion that $^1O_2$ might well be an important factor in the photodegradation of polypropylene.

Carlsson and Wiles (ref.2) have shown that singlet oxygen is in fact a prominent factor in accelerating the photodegradation of polypropylene. By treating a polypropylene film with singlet oxygen (produced by microwaves) the usual induction period for the onset of photo-oxidation process was completely eliminated. The increase in polypropylene hydroperoxides and carbonyl products was the end result of photo-oxidation initiated by singlet oxygen.

The stationary concentration of $^1O_2$ in the atmosphere is likely to be very low in comparison with, for example, O_3 concentration, so that atmospheric $^1O_2$ can probably be ignored. A much serious source of $^1O_2$ comes from the quenching of UV excited species (S) within the polymer by ground state oxygen, $^3O_2$.

$$S \xrightarrow{\text{Light}} S^* \xrightarrow{\text{oxygen}} S + {}^1O_2$$

Known sensitizers for $^1O_2$ generation include many dyes, carbonyl groups, antioxidant degradation products, and polynuclear aromatic compounds. Ketonic chromophores might act both as $^1O_2$ source and also provide adjacent unsaturation sites as a result of the Norrish Type II cleavage:

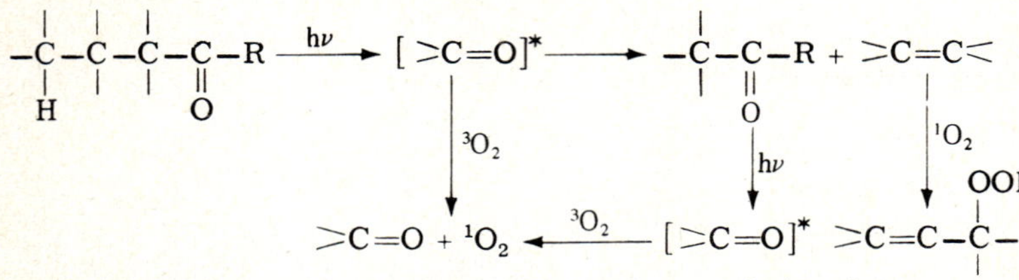

Oxygen quenching of ketone photoreactions and phosphorescence is now well documented in both liquids and solid polymers. In fact, only very low oxygen concentrations are apparently required to quench phosphoresence from the relatively long-lived macroketones in the solid polymers.

II.4. Catalytic Effects in Thermal- and Photo-oxidation

The stability of polypropylene to thermal and photo-degradation is strongly affected by the impurities present in the polymer. Most of the transition elements and their salts accelerate the oxidation processes of polypropylene through their catalytic action. They do so even in the presence of antioxidants. Some pigment formulations may contain traces of compounds of Cu, Mn, Fe, Co, etc., which are effective catalysts for thermal and photodegradation of polypropylene. Furthermore, polypropylene melt may get contaminated with traces of metals as a result of wear and corrosion of extruder, spin pump, filter blocks, etc.

The mechanism of metal catalysis can be visualised variously and probably differs from metal to metal. Three alternative mechanisms have been considered:
(a) activation of molecular oxygen, (b) activation of polymer, (c) catalytic decomposition of peroxides. The last-named mechanism is accepted in general and can be described as follows:

$$ROOH + M^{+2} \longrightarrow ROO^* + M^{+1} + H^+$$
$$ROOH + M^{+1} \longrightarrow RO^* + M^{+2} + OH^-$$

The decomposition of peroxides by metals leads to the chain-starting radicals, RO* and ROO*, and this causes the accelerated thermal and photo-oxidation of the polymer even at moderately low temperatures.

Conjugated unsaturation leads to intense UV absorptions in the near UV light. Although unlikely to be initially present in polypropylene, conjugated unsaturation might result from the attack of atmospheric SO_2 on adjacent hydroperoxide groups in the polymer:

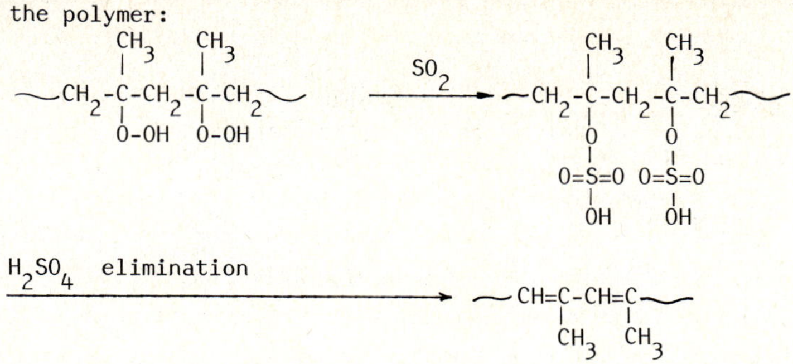

SO_2 exposure of PPOOH films and fibers gives a yellow-brown color, presumably due to conjugated unsaturation with no IR detectable sulphate groups in contrast to polyethylene hydroperoxides which apparently give stable sulphates. This conjugated unsaturation could act as an additional source of sensitization of polypropylene photodegradation.

Luminescence emission from polyolefins has frequently been reported. This emission was originally attributed to macroaldehyde or ketones but is now believed to result largely from PNA (polynuclear aromatics) impurities. These impurities are released into the atmosphere in large quantities from the combustion and pyrolysis of fuels and lubricants. For example, about 0.02g of PNA are released every hour by one car. PNA species are apparently absorbed by polyolefins upto equilibrium concentrations of about 1×10^{-3} M (100 ppm) naphthalene, about 1×10^{-4} M (10 ppm) phenanthrene, and about 1×10^{-5} M (1 ppm) anthracene. The observed PNA emission spectra indicate that alkyl derivatives of each compound are probably present rather than the parent PNA. These absorbed impurities may be removed from polypropylene film and fibers by a hexane wash, but build up to their equilibrium value within 10 days of storage in air (ref.2).

PNA compounds are efficient sources of singlet oxygen, $^1O_2$, by the reaction:

$$\text{PNA} \xrightarrow{\text{Light}} \text{PNA}^* \xrightarrow{\text{Oxygen}} \text{PNA} + {}^1O_2$$

The possibility that PNA impurities in polypropylene samples might act as *in situ* source of $^1O_2$ which can then either diffuse away to attack allylic $-CH_2-$ groups, or give photosensitive hydroperoxides, is supported by the effects of deliberately introduced PNA concentrations on the photostability of polypropylene fiber (ref.2).

III. STABILIZATION

The objectives of stabilization of polypropylene against heat and light are two-fold. Firstly, the polymer has to be stabilized so it can be processed at high temperatures of melting without excessive degradation and color alteration.

Secondly, the stabilization is to be imparted to the polymer to make the fabricated articles resistant to aging in the intended use.As processing conditions and end-use environments differ from product to product, it is evident that the stabilization requirements for each case will differ considerably.

The essence of stabilization of polypropylene against oxidative degradation is to delay the onset of severe degradation in the fabrication step and to slow the rate of chemical changes leading to deterioration of physical properties of the product. The accent is, therefore, to increase the induction period of oxidation.

The stabilizers used for polypropylene are commonly divided into heat and light stabilizers, depending on their predominant action in polymer stabilization against heat or light degradation, respectively. This distinction is, however, superficial because no stabilizer is strictly a heat or a light stabilizer. A more scientific classification would be:

(a) Chain terminators, inhibitors, or free radical scavengers
(b) Peroxide decomposers
(c) UV absorbers
(d) Energy quenchers
(e) HCl scavengers
(f) Metal deactivators

The antioxidants used for the purpose of stabilization against heat fall into two main catagories: those which react with free radicals to form more stable free radicals, less capable of continuing the oxidation chain, and those compounds which decompose the alkyl peroxides without the formation of free radicals. The stabilization of polypropylene against photo-oxidation may be enhanced with the help of UV absorbers and energy quenchers in conjunction with peroxide decomposers. UV absorbers are those additives which absorb most of the UV radiation and dissipate this energy in a harmless way in order to prevent light absorption by the carbonyl and other sensitizing groups in the polymer. Stabilizers acting through this mechanism are thought to behave essentially as UV filters. The quenchers have the function of transferring electronic energy from the excited states of the sensitizer molecule to the molecule of a convenient acceptor. These additives often function in conjunction with other stabilizers mentioned above.

In addition to these principal catagories of stabilizers, a means of preventing the metal catalysis of oxidation process may also be provided by utilizing suitable metal complexing agents. Furthermore, some provisions are to be made to protect the stabilizers from attack by acidic impurities present in the polymer such as polymerization catalyst residues. This can be done with the help of certain HCl-fixing compounds.

A stabilization formulation is often a complex recipe of several different stabilizer components, each providing one or several functions, discussed above.

In the following pages, these individual functions of different groups of stabilizers will be discussed and various compounds falling into the respective groups will be mentioned. It should, however, be noted that few compounds used as polypropylene stabilizers have only one function attributed to them. Most of the stabilizers function through more than one way, although one particular function may predominate. Another characteristic of stabilizer formulations is the concept of *synergism* . Often, the cumulative effect of different components in the stabilizer mixture is more than the additive effect of the individual components. This phenomenon is an important consideration in arriving at a suitable stabilization formulation for any particular use. The maximization of this objective function is the key-note of the proprietory formulations, so jealously guarded in the industry.

III.1. Free Radical Scavengers

Some organic compounds, known as free radical scavengers or inhibitors (also known as primary antioxidants or primary heat stabilizers) have the capacity of terminating the chain reaction of polymer oxidation by giving rise to relatively inert free radicals which are not sufficiently reactive toward molecular oxygen to continue the chain reaction. These inhibitors remove the active free radicals so rapidly that the main reactions are virtually eliminated as long as some effective amount of such inhibitors remain in the polymer matrix.

The general mechanism of oxidation inhibition or stabilization against oxidative degradation through chain termination consists of a fast reaction between the inhibitor AH and the free radical R*, producing an inactive group RH and a relatively stable free radical A*. The latter does not propagate the chain reaction but instead gets deactivated and regenerated to the original inhibitor molecule. Thus,

$$R^* + A \longrightarrow R + A^*$$

$$A^* \xrightarrow{\hspace{3cm}} \begin{array}{l}\text{Deactivation}\\ \text{Regeneration}\end{array}$$

Deactivation can happen through bimolecular recombination, internal rearrangement, or disproportioning. In addition to the above mechanism, there are also some other proposed mechanisms of radical inhibition: for example, by complex formation between the free radical and the inhibitor, or by surface reaction on inert surfaces such as those possible on pigment particles.

Apart from the reaction with free radicals leading to chain termination of oxidative reactions, inhibitor compounds may undergo several other side-reactions such as the chain starting, chain transfer, decomposition of peroxides to radical reaction products, direct oxidation to stable products, interaction with the polymerization catalyst residues and other additives, etc. An efficient inhibitor is that which favors chain transfer leading to relatively stable free radicals of the inhibitor molecules by combining with very reactive, short-lived free-radicals, produced in peroxide decomposition or otherwise, but is relatively immune to other side-reactions.

The respective rates of above mentioned side-reactions may differ at different temperatures and thereby cause a change in the efficiency or even the entire nature of the inhibiting reaction. This is why some of the commonly used inhibitors, such as Toponol CA, function as efficient antioxidants for polypropylene at moderate temperatures but act as prodegradants at high temperatures. Similarly, most of the inhibitors are good stabilizers at low concentrations but they act adversely at high concentrations.

Inhibitors cannot supress the oxidizability of polymer totally but their action is rather to retard the onset of oxidation reaction, i.e., to prolong the induction period to such an extent that the polymer gets processed without undue degradation and the product keeps its useful properties for the intended period of use. In general, while the inhibitors influence the induction period, they fail to influence the actual rate of oxidation thereafter (ref.3).

An important criterion for the kinetics of inhibited auto-oxidation is the relationship between the concentration of the antioxidant and the length of the induction period. In the case of polypropylene, a linear relationship, as expected from the kinetic considerations of chain-termination action, does not exist. A characteristic relationship between these two parameters is schematically represented in Fig.3. According to this schematic, the induction period, τ , increases very rapidly with concentration, but within a short range it reaches a plateau to a value τ_{max}. Above this limit, an increase in the inhibitor's concentration leads to a decrease in induction period. The concentrations corresponding to τ_k and τ_{max} depend on temperature, the nature of the inhibitor, degree of dispersion, and the polymerization catalyst residues or other impurities present in the polymer.

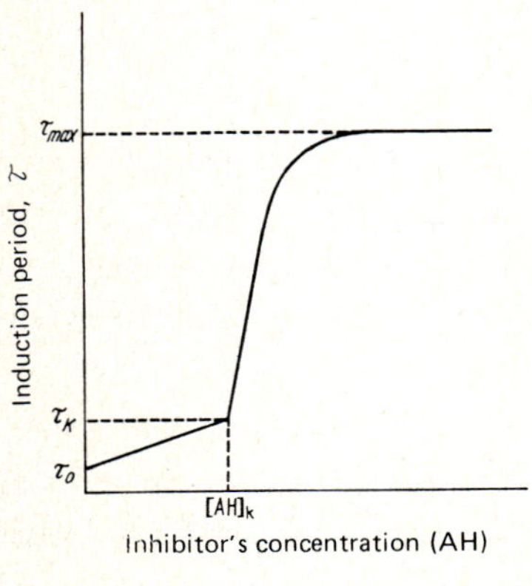

Fig. 3. Schematic Representation of the Dependence of Induction Period on Inhibitor's Concentration (ref.45)

Out of the large number of organic substances which may act as inhibitors, hindered phenolics have been found to be the most suitable for polypropylene. The most important commercial compounds falling into this class can be summarized under the following groups:

(a) Alkylated phenols, e.g., DBPC

(b) Alkylenebisphenols, e.g., Antioxidant 2246

(c) Thiobisphenols, e.g., Santonox

(d) Polyphenols, e.g., Ionox 330

(e) Hydroxyphenylalkanecarbonic esters, e.g., Irganox 1076

(f) Phenol substituted s-Triazins, e.g., Irganox 858

(g) Phenol substituted nitrogen heterocyclic compounds, e.g., Goodrite 3114

(h) Miscellaneous compounds, combining the functionality of hindered phenol with the functionality of tertiary phosphite, hydroxybenzophenone, triazole, and other such groups.

Simple di-alkylated phenols, such as DBPC [I] stil find some application in relatively non-critical uses, such as sacking yarns, fiberfill material, etc., but substituted polyphenols with improved sublimation and extraction resistance are generally preferred for fiber production. Of special importance are the bisphenols, such as 4,4'-butyllidenebis-(3-methyl-6-tert.-butylphenol) [II] and other condensation products of alkylphenols, aldehydes, or sulfur chlorides. Thiobisphenol, such as thiobis-(3-methyl-6-tert.-butylphenol) [III] are important as they, in addition to their inhibition reactions, also possess considerable activity for peroxide decomposition into non-radical components.

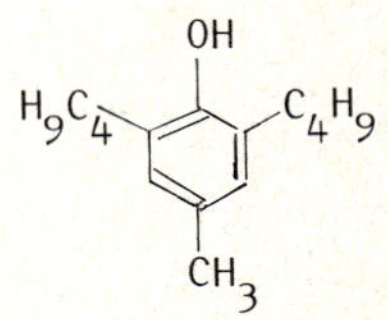

2,6-di-tert-butyl-p-cresol

[I]

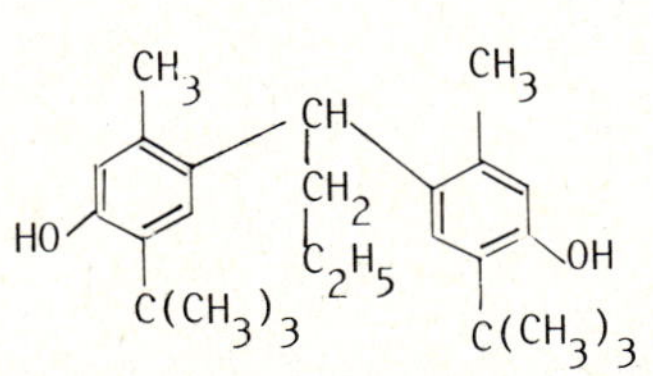

[II]

Butylidenbis-(3-methyl-6-tert.-butylphenol)

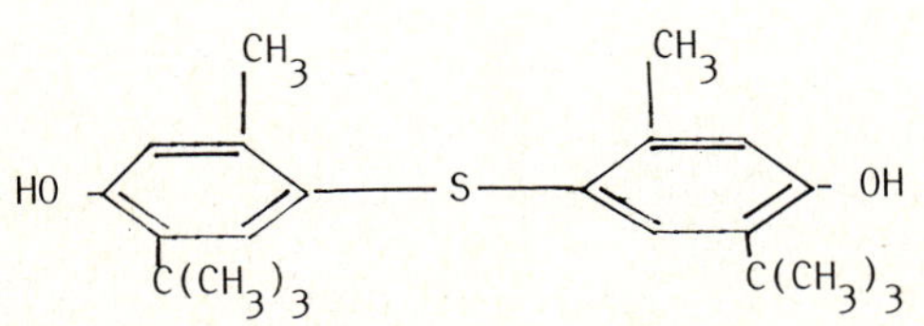

[III]

4,4'-thiobis-(3-methyl-6-tert.-butylphenol)

A special advantage in the use of newly developed phenolic antioxidants with three are more rings is their resistance to sublimation and water extraction. The former is important in processing, the latter is growing in importance with the penetration of textile field by polypropylene fiber where satisfactory washing performance is a prerequisite. A typical compound of this kind is Topanol CA |IV| or Antioxidant 80 |V|. The four-ring structures, such as Ionox 330 |VI|, Goodrite 3114 |VII|, Goodrite 3125, and other tris hindered phenols, are still better in this respect. They are specially suitable for uses such as sweater yarns, outdoor carpets, sand bags, etc., where the product is expected to be often exposed to the action of water.

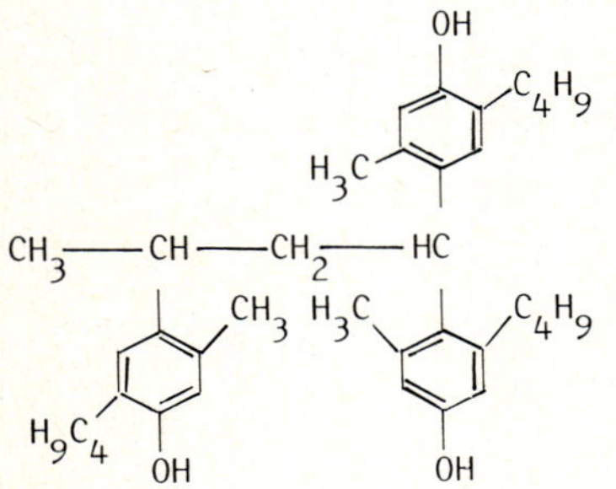

[IV]

Topanol CA

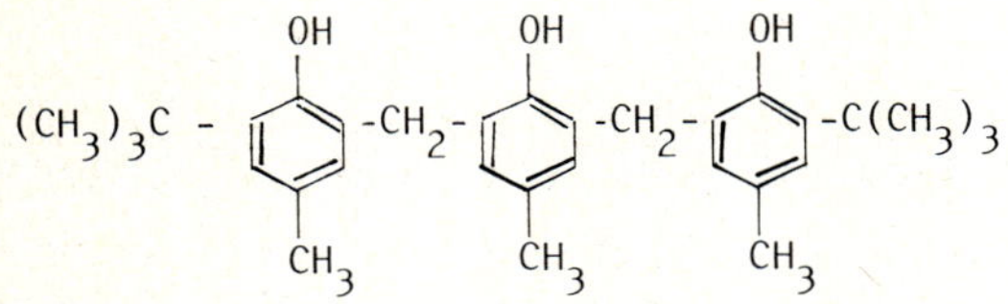

[V]

Antioxidant 80

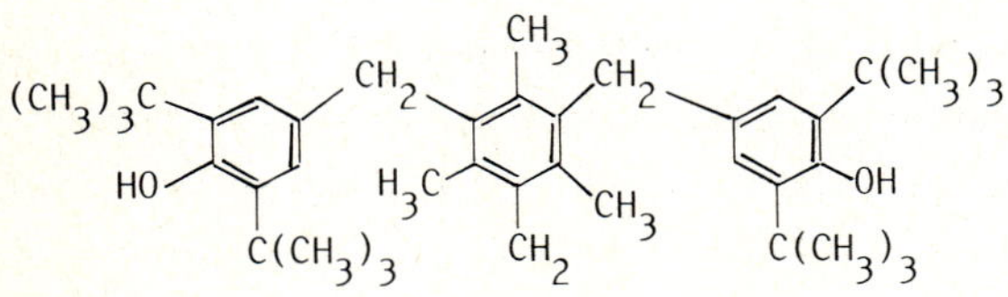

[VI]

Ionox 330

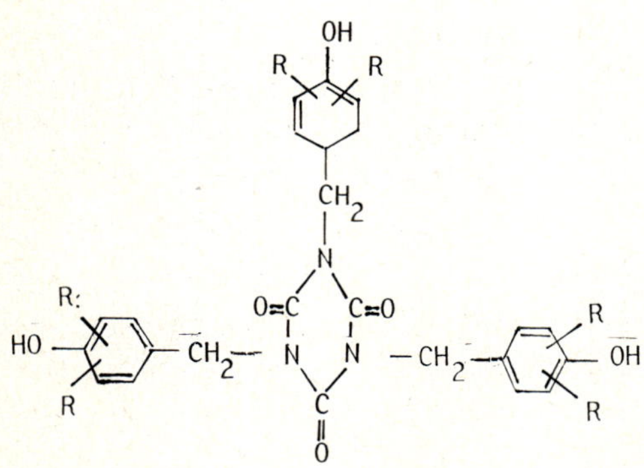

[VII]

Goodrite 3114

Other typical high molecular weight antioxidants are Irganox types [VIII], [IX], [X]. These products, which represent phenolic compounds of various sorts, are characterized by high sublimation resistance, extraction resistance, little or no tendency of discoloration, and high antioxidant efficiency.

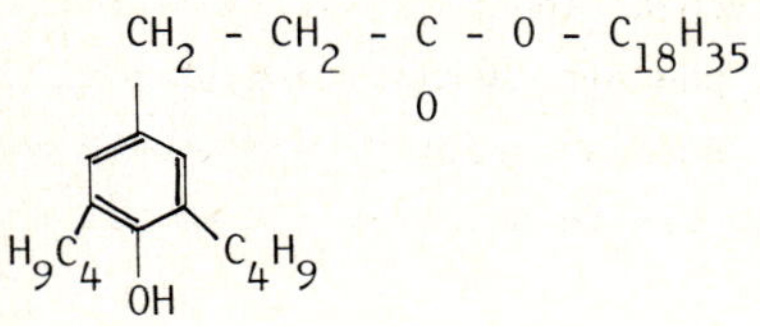

[VIII]

Irganox 1076

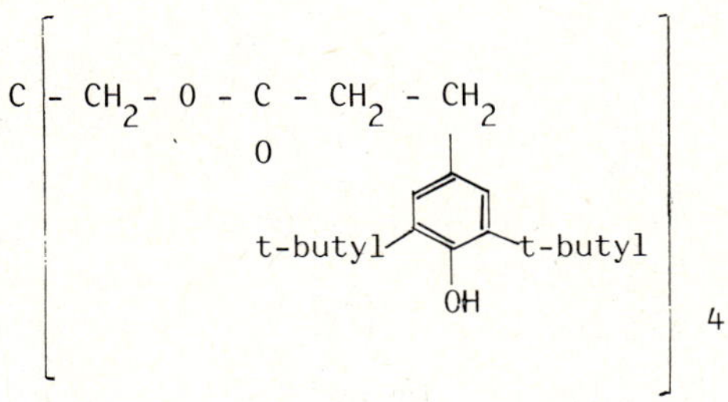

[IX]

Irganox 1010

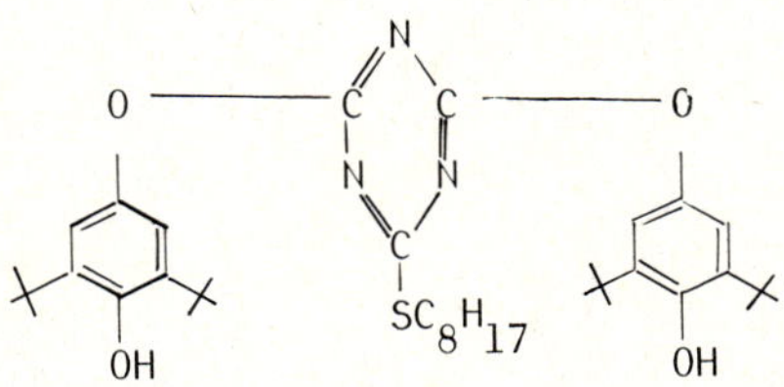

[X]

Irganox 858

The kinds of free radical scavengers discussed above are highly effective in the case of thermal oxidation, both in processing and in use. But their effectiveness against photo-oxidation is limited. This has often been explained on the basis of the kinetics of photo-oxidation process. In thermal oxidation reactions, the rate of chain starting is low and the kinetic chain length is long. In the oxidation reactions caused by light, the situation is reversed. The phenolic type inhibitors are supposed to be highly effective in the former case but ineffective in the latter case. Recent experiments do not lend credence to this explanation. The failure of phenolic inhibitors to stabilize polypropylene against photo-oxidation is now interpreted due to their rapid destruction or exudation during weathering, rather than to their inability to scavenge POO* radicals. Octadecyl-3-(3',5'-tert-butyl-phenyl-4'-hydroxyphenyl)propionates do stabilize polypropylene against light (ref.48), and since they are phenols, it is not inconceivable that much of their effectiveness stems from POO* radical scavenging. These phenols are relatively stable against UV light and so can persist long enough to scavenge POO* radicals.

Ni-chelates, discussed in section III.4., have also been found to be effective scavengers of OH* and PO* radicals at room temperatures (ref.49). In addition, these compounds are relatively stable under photo-oxidation conditions. The compounds of this group are therefore of high technological importance in the stabilization of polypropylene against heat and light. Their stabilization effect partly arises from their ability to scavenge free radicals.

Recently a great deal of development work has gone into the preparation and evaluation of substituted 1,2,2,6,6-pentamethyl piperidines and their corresponding nitroxides (see section III.4 for details). These piperidines are readily oxidized to the corresponding nitroxides which are very effective alkyl radical scavengers:

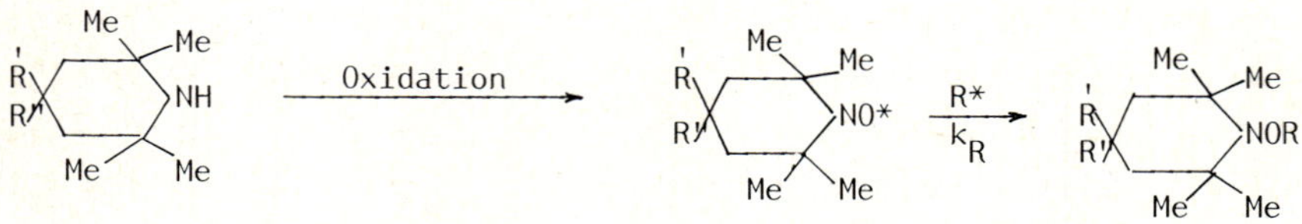

It has been shown that nitroxide radicals are regenerated during the thermal and photo-oxidation of polypropylene initially containing nitroxides. Each nitroxide radical can probably terminate upto 10 free radical chains by the above reaction follwed by:

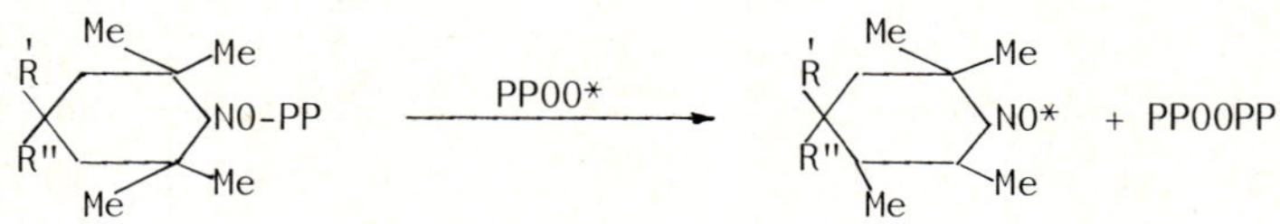

Both oximes and Ni(II) chelates of oximes are effective polypropylene stabiliaers. This effectiveness might again result from radical scavenging, as oximes are known to react with alkoxy radicals and OH* radicals. This reaction produces iminoxy or nitroxide radicals, which can in turn scavenge other alkoxy radicals:

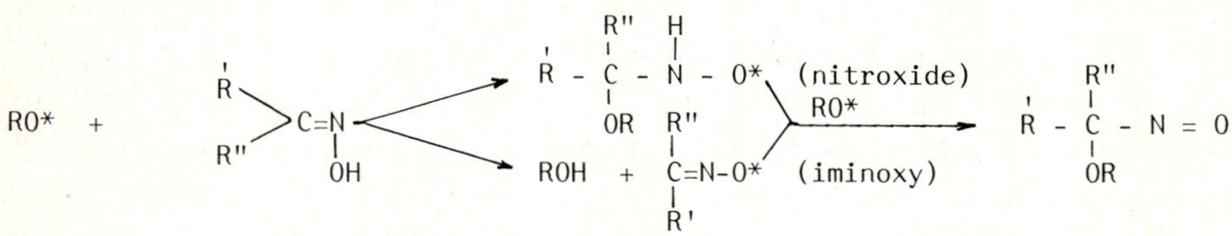

The nitroso product from the above reaction can again take part in radical scavenging processes:

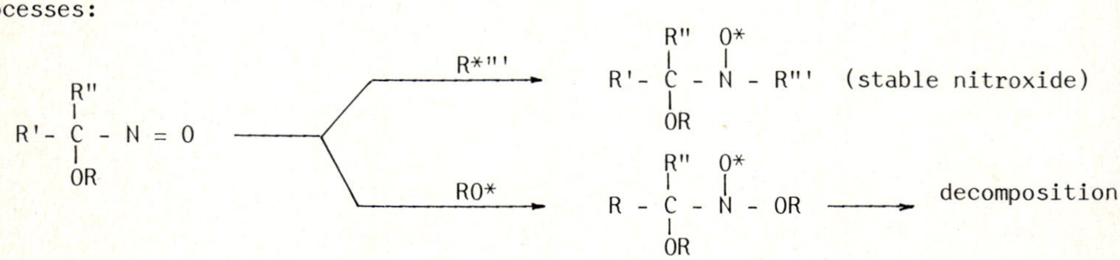

III.2. Peroxide Decomposers

As mentioned previously, the peroxides formed during the decomposition process of polypropylene decompose into free radicals on absorption of energy and initiate fresh cycles of oxidation reaction: $ROOH \xrightarrow{Energy} RO^{*} + OH^{*}$

The energy needed for the decomposition of peroxides is available in the form of heat and light. It should be possible to decrease the initiation rate of thermal and photo-oxidation by preferentially decomposing the alkyl peroxides into non-radical components before they get the opportunity to decompose into free radicals by the action of heat and light. Several organic substances are known to perform this function. They are called peroxide decomposers, secondary antioxidants, or synergists.

The commonly used peroxide decomposers are the organic compounds of sulfur, phosphorous, and nitrogen, especially thioesters, disulfides, tertiary phosphites, phosphonates, phosphorous-sulfur mixed organic compounds such as thiophosphites, secondary and tertiary amino compounds, hindered amines, Ni(II) chelates, or any combination of these structures. These substances are rarely used alone but commonly accompany the inhibitors in stabilization formulations. This is why they are known as secondary stabilizers.

In combination with a free radical inhibitor, the peroxide decomposers delay the consumption of the former by decomposing the peroxides to inactivate decomposition products. This corresponds to the reaction: $ROOH + AH \longrightarrow$ Stable products.

Among the sulfur compounds used as peroxide decomposers for polypropylene, esters of thiodipropionic acid have traditionally played the most important role. The common examples of thiopropionic acid esters are dilauryl thiodipropionate (DLTDP) [XI], distearyl thiodipropionate (DSTDP) [XII], and lauryl stearyl thiodipropionate (LSTDP) [XIII]. Although DLTDP is more popular, tests show that DSTDP is superior to DLTDP in stabilization of polypropylene against thermal oxidation at higher temperatures (ref.4). The volatility of the antioxidant probly becomes important so that the less volatile DSTDP shows better stabilization efficiency.

[XI] $(C_{12}H_{25}O\text{-}\overset{O}{\overset{\|}{C}}\text{-}CH_2\text{-}CH_2)_2S$

DLTDP

[XII] $(C_{18}H_{37}O\text{-}\overset{O}{\overset{\|}{C}}\text{-}CH_2\text{-}CH_2)_2S$

DSTDP

[XIII] $H_{25}C_{12}\text{-}O\text{-}\underset{\underset{O}{\|}}{C}\text{-}CH_2\text{-}CH_2\text{-}S\text{-}CH_2\text{-}CH_2\text{-}\underset{\underset{O}{\|}}{C}\text{-}O\text{-}C_{18}H_{37}$

LSTDP

The above mentioned thioesters are very effective peroxide decomposers at relatively high temperatures but their efficiency at ambient temperature is quite diminished. These compounds, therefore, function most effectively as process stabilizers, i.e., in the protection of polypropylene during extrusion, but do not contribute significantly to the stabilization of the fiber against UV degradation.

In recent years esters of phosphoric acid have invited considerable attention and a number of such compounds are now commercially available. Tertiary esters of the general formula [XIV] are almost exclusively used, where R', R'', and R''' are alkyl, aryl, or arylalkyl groups and A can be oxygen, sulfur, or nitrogen, or a combination of these. Triphenyl phosphite [XV] was probably the earliest and the most widely known compound of this family. Trialkyl phosphites are reported to be more effective in polypropylene than triphenyl phosphites (ref.5) and relatively higher molecular weight compounds such as [XVI] have the added advantage of inextractability. Some cyclic polyphosphites such as [XVII] have also been found effective peroxide decomposers. Other examples of high molecular weight compounds of this class are [XVIII] and [XIX].

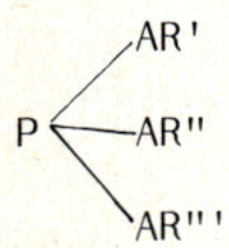

[XIV]

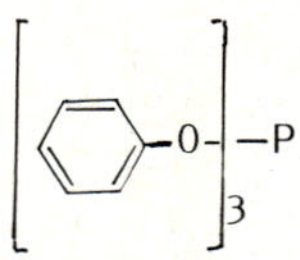

Triphenylphosphite

[XV]

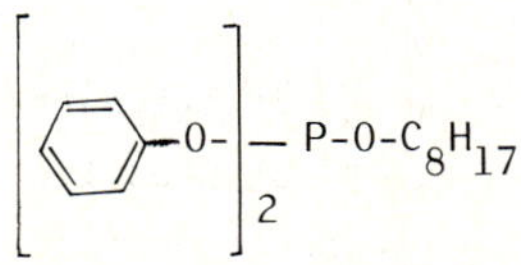

Diphenyl-isooctylphosphite

[XVI]

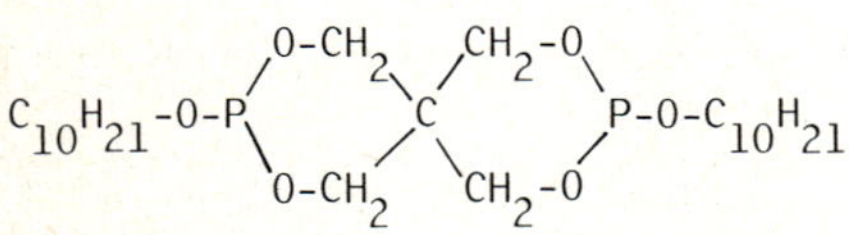

[XVII]

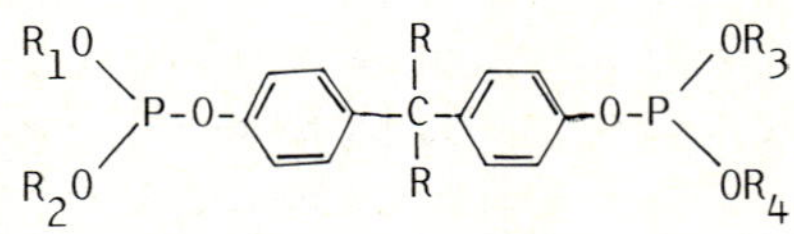

[XVIII]

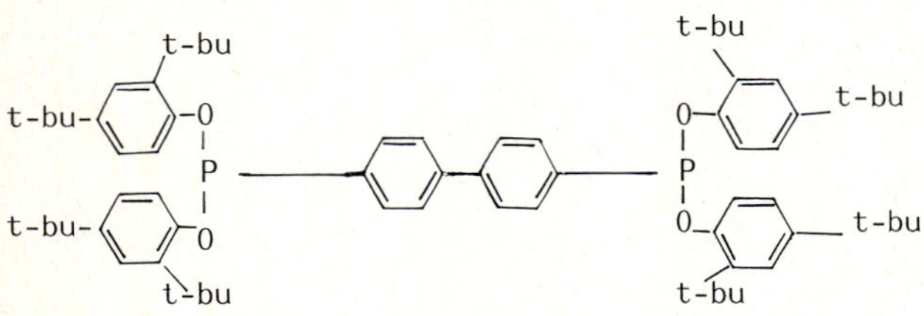

[XIX]

It has been demonstrated that at low temperatures organic trialkylphosphites are more effective peroxide decomposers than thioesters (ref.7). It has been reported, for example, (ref.8) that trialkylphosphites react with hydrogen peroxide even at $0^{o}C$. These compounds have, however, proved to be ineffective UV stabilizers for polypropylene, possibly owing their rapid photodestruction in the polymer. Nevertheless, they show some degree of synergism in combination with Ni(II)-chelates, mentioned previously and described in detail in section III.4. Figure 4 illustrates this effect. Even here, their effectiveness diminishes drastically in moist atmosphere, probably again owing to their hydrolysis under the action of light.

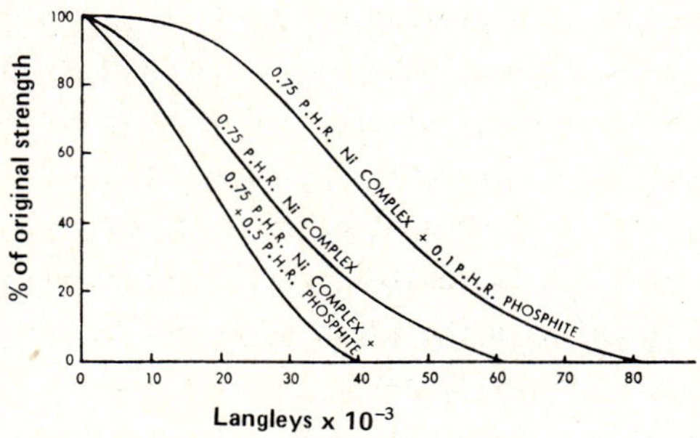

Fig. 4. Synergism of Phosphites with Nickel Complexes (ref.4I).

Inspite of the ineffectiveness of tertiary phosphites and related compounds in stabilization of polypropylene against UV degradation, this group of peroxide decomposers is generally preferred over DLTDP or DSTDP in stabilization formulations. This is due to the following three important reasons:

(1) The organic tertiary phosphites and related compounds are chemically more compatible with a number of free radical scavengers and energy quenchers than are the thioesters. They are also more compatible with polypropylene than thioesters. This results in less problems in melt-spinning operation.

(2) On weight-to-weight basis, the tertiary phosphites and phosphonites are considerably more effective synergists than thioesters in stabilization of polypropylene against thermal degradation at high processing temperatures (Fig.5).

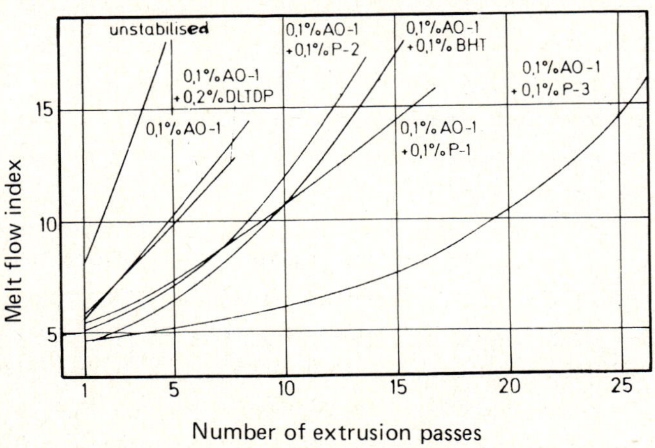

Fig. 5. Comparison of Synergist Compositions for Thermal Stabilization of Polypropylene (ref. 52).
AO-: Formula [I];
P-I:Formula [XVII]
P-2:Tris-(nonylphenyl)-phosphite
P-3:Formula [XIV]

(3) The phosphites and the related compounds are beneficial in supressing the discoloration of polypropylene in processing as well as in use. It appears that they react preferentially with oxidized residues of the phenolic antioxidants, thus discharging the color of the typical quinoid bodies.

The use of organic tertiary phosphites as synergists for high temperature oxidative stabilization is specially warranted in those cases where sulfur-bearing compounds cannot be tolerated because of their reaction with certain pigments or some other components of the stabilization system (see Chapter III for pigment-stabilizer interaction).

A comparatively recent development in stabilization of polypropylene is the use of thiophosphites, such as trilauryl-trithiophosphite [XX]. The presence of thio- as well as phosphite groupings in one single molecule makes these compounds extremely effective peroxide decomposers at high as well as at low temperatures. These compounds are relatively stable under photodegradative conditions. They are, therefore, useful not only for process stability of polypropylene but also for photo-oxidation stability of the fibrous product in use. They can be normally used without the presence of DLTDP or DSTDP in the formulation.

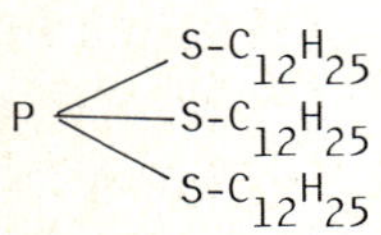

[XX]

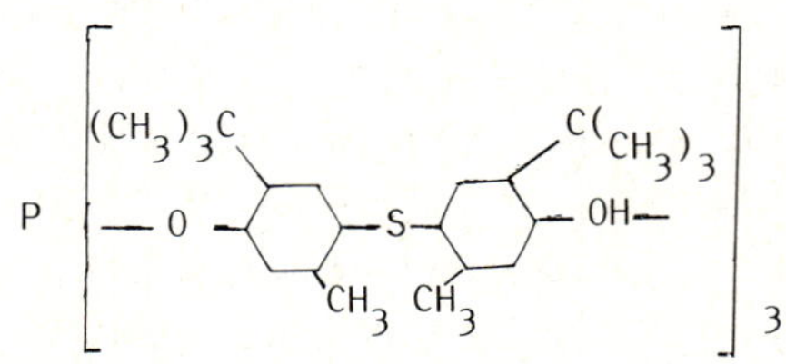

[XXI]

A still further improvement in tertiary phosphite stabilizers is based on the condensation product of Santonox with phosphorous trichloride in 3:1 mole proportions. The resulting molecule, such as [XXI], while retaining the essential characteristics of a tertiary phosphite, also has thiobisphenolic groups in the molecule. These types of compounds are expected to show not only the low and high temperature effectiveness as peroxide decomposers, but also function as free radical inhibitors. These compounds have been found quite effective as heat stabilizers but their effectiveness against UV degradation has not been proved.

Most of the phosphite compounds mentioned heretofore are viscous liquids or tacky solids. Consequently, they offer some difficulty in mixing with the polymer and tend to exude to the surface of the fiber in processing and in use. This shortcoming is remedied through the use of recently developed polymeric phosphites, especially the polyphenyl phosphites, which are the reaction products of substituted hydroquinones and phosphorous trichloride (ref.9). The effectiveness of polyphenyl phosphites is well demonstrated in aging studies. Heat aging tests, for example, indicated an exceptionally high degree of synergism between polyphenyl phosphites and thidipropionate esters, suggesting a hindered phenolic functionality. On the other hand, processing stability studies suggest that the compounds also function much like arylphosphites under conditions of high shear and low oxygen concentration, e.g., those encountered during extrusion operations (ref.9). Similarly, the protection against embrittlement by light obtained with polymeric polyphenyl phosphites is known to be more than double of that obtained with either the monotrisphosphite or the normal hindered phenols.

A recent development in the field of peroxide decomposers for the stabilization of polypropylene is the synthesis of compounds having the general formula [XXII] which have proven to be better peroxide decomposers than dialkylthiodipropionates, simple tertiary phosphites, or some of the trithiophosphites (ref 10).

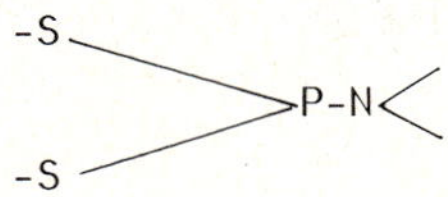

[XXII]

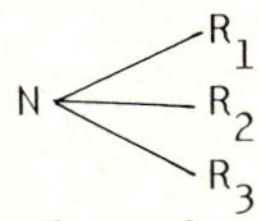

[XXIII]

Tertiary aliphatic amines of the general formula [XXIII] have also been recently reported as very efficient peroxide decomposers (re.11,12) and consequently good stabilizers for polypropylene. Tertiary amino compounds are effective at low as well as at high temperatures. These compounds surpass the sulfur- and phosphorous-bearing compounds in their effectiveness as processing stabilizers. The stabilization compounds containing tertiary amino groups are, therefore, particularly useful in those cases where a high degree of heat stabilization is required. These compounds have also been noticed for their light-stabilization effect (ref.12). Their color stability is, however, not very good. In fact, some of these compounds tend to impart a yellowish tinge to the extrudate. This may be due to their own reactions or, more often, to secondary amine contamination.

The use of tertiary amino compounds is not yet common for the stabilization of polypropylene, but considerable research efforts are being expended on this class of stabilizers. This contention is based on the fact that these structures are easy to prepare, have a very high cost/efficiency factor and, if added in sufficent quantity, can also function as dye-additives. Montecatini, for example, recently described a condensation product of epichlorohydrin and di-secondary diamines as an efficient polymeric stabilizer for polypropylene extrusion (ref.13). Such amino compounds are primarily used as dye-additives and have been the subject of Montecatini's numerous patent disclosures for the enhancement of dyeability of polypropylene fibers. Similarly, various copolyamides with tertiary amino groups, used primarily as dye-additives, have been mentioned as effective auxiliary antioxidants (ref.14).

Several of the Ni-chelates mentioned in section III.4 can also decompose -OOH groups in polypropylene fibers and film. This has been demonstrated by immersing prephoto-oxidized polypropylene film in solutions of various stabilizers. Film portions were periodically removed for hydroperoxide determination as the stabilizer diffused into the film. The result showed that most UV stabilizers failed to reduce the -OOH concentration whereas Ni(II) or Zn(II)bis(di-n-butyldithiocarbamate) and Ni(II)bis(di-isopropyledithiophosphate) were apparently very effective hydroperoxide decomposers for polypropylene.

Finally, some hindered amines, such as 1,2,2,6,6-pentamethyl piperidines, have been shown to be effective peroxide decomposers:

$$>NH \; + \; ROOH \longrightarrow \; >NOH \; + \; ROH$$

The >NOH can further combine with alkyl radical, generating a nitroxyl which is an effective quencher of singlet oxygen (see section III.4).

$$>NOH \; + \; R^{*} \longrightarrow \; >N^{*}O \; + \; RH$$

This class of compounds is particularly effective in the photo-stabilization of polypropylene.

In summary, thioesters such as DLTDP and DSTDP are most commonly used as peroxide decomposers in polypropylene stabilization. They are very effective at moderately high temperatures and serve as low-cost general-purpose process stabilizers in conjunction with other synergistic components. They are recommended as general purpose synergistic stabilizers against polymer degradation in extrusion.

The incorporation of tertiary phosphites and related compounds in stabilization formulations for process degradation has the advantage of better color stability than that achieved by thioesters or tertiary amino compounds. Thiophosphites are probably the most suitable peroxide decomposers for use against polymer degradation and discoloration during processing as well as in weathering. The compounds which intend to combine the functionality of phosphite with that of hindered phenol or that of thiobisphenol are better all-round single component process stabilizers. However, it is doubtful if they offer any distinct advantage in stabilization economy against a mixture of phosphites, thioesters, and hindered phenols.

The tertiary aliphatic amines and related compounds are especially suitable in high temperature melt stabilizer formulations. The use of tertiary aliphatic amines in synergistic compositions is recommended where high extrusion temperatures are involved or where the residence time of the melt in the spinning system is extraordinary long and where coloration is not a problem. Ni-chelates and hindered amines are particularly beneficial peroxide decomposers under photo-oxidative conditions.

III.3. U.V. Absorbers

One potential means of protecting polymer from oxidative degradation is to prevent the polymer from absorbing the initiation energy. There is no way of controlling the absorbed energy which is supplied to the polymer in the form of heat at any given temperature. However, in contrast to the initiation reaction in thermal degradation of polypropylene, the initiation reaction of photo-oxidation can be suppressed to a great extent by decreasing the photoenergy absorbed by the polymer. This can be done by utilizing certain organic compounds which strongly absorb UV radiation and thereby shield the polymer from its action. The technique of mixing the polymer with suitable UV absorbers has now become a common practice in stabilization of several polymers, including polypropylene.

One of the additives which has long been used to render polypropylene stable against light is carbon black. The absorption spectrum of carbon black is, however, too broad, including all visible light, thus producing only jet black fibers. It is well known that several other pigments, in addition to carbon black, are good stabilizers against light degradation in thick sections. Zinc oxide is the best inorganic UV absorber, highly efficient from 2400 Å upto 3655 Å. But above 3900 Å it is an opaque reflector of visible light, thus producing only dull or semi-dull fibers. The most important UV absorbers presently in use are the organic products that allow stabilization without pigmentation by virtue of their transparency to visible light and a strong absorption maximum in the range of near UV radiation.

The absorption spectrum of all useful UV absorbers is characterized by a strong extinction in UV range with practically no absorption above 4000 Å. An ideal UV aborber would show practically vertical decrease in extinction at 4000 Å. If this upper cut-off is sharp and the slope is steep enough, they combine strong protection with absence of added color. In practice, however, few, if any, UV absorbers show this behavior.

The most widely held view of the mechanism of light stabilization by UV absorbers reduces to the UV screening effect of the additive as contended in the previous paragraphs, i.e., it is suggested that the light stabilizers act as a filter, weakening the flux of UV radiation incident on the polymer, the filtering action being brought into play by undergoing temporarily tautomeric chelation and reversible formation of a six-membered hydrogen bonded ring, eventually releasing the excess energy as fluorescence, or more commonly as heat, according to the scheme described below:

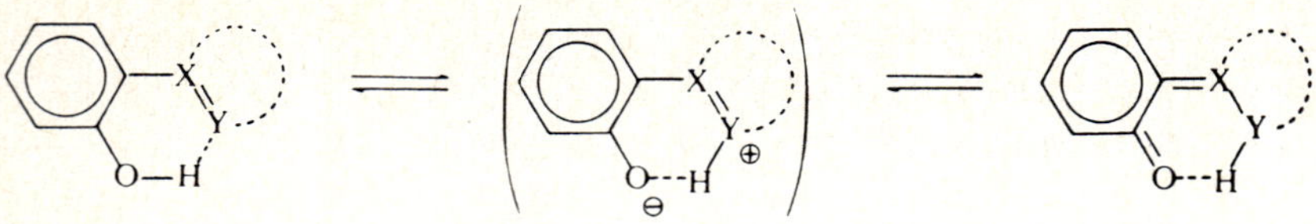

It is assumed that, in the excited state, the keto-form is more stable than the enol-form, while in the ground state the enol is more stable. Attempts are then made to establish a correlation between the absorption spectrum and the efficiency of the light stabilizer. If the absorption spectrum of the UV absorber is known, it is not difficult to determine, in accordance with Beer-Lambert's law, the energy absorbed by the polymer specimen without an additive, Δe_o, and the energy absorbed by the same specimen with added light absorber, Δe_s. Then the protective effect of the light absorber may be determined from the relation:

$$\text{Protective effect} = \Delta e_o / \Delta e_s .$$

It follows from the above equation that since Δe_s is always less than Δe_o, protective effect is always greater than unity, i.e., almost any substance absorbing UV light at a given wavelength should exhibit a stabilization effect in the range of that wavelength. Experimental results show, however, that this is not necessarily so (ref.17,18). In general, there is no direct correlation between UV spectra of UV ansorbers and their stabilization effect on the polymer. Conversely, there are frequent observations showing that the conventional UV absorbers are often more effective than can be expected on the basis of the light-filtering hypothesis. This justifies the assumption that the stabilization effect is based not only on the light-filtering capability of these compounds, but also on other effects which may even become the primary source of stabilization in some cases.

Guillory and Cook (ref.19) and Heller and Blatman (ref.20) have recently examined the mechanism of commonly used UV absorbers and have shown that they do not behave as light screeners only. They also act as energy quenchers for the excited state of polymer molecules, although their quenching effectiveness is considerably less than that of the Ni-chelates and hindered amine stabilizers, discussed in the next section.

For a UV absorber to function as a light stabilizer, it is not only necessary for it to strongly absorb UV light in the near UV region but also it must be able to dissipate the absorbed energy in a harmless manner. The absorbed photo-energy must be quickly transformed into heat or light of low enery level and be dissipated in the surrounding atmosphere. The absorbed energy in no case should be consumed in the formation of free radicals. In this case the UV absorbers would act as photo-sensitizers rather than a photo-stabilizer for the polymer.

The commonly used UV absorbers for polypropylene are the derivatives of hydroxybenzophenone, such as Cyasorb UV 531 |XXIV|, Eastman DOBP |XV|, and some phenyl benzoate esters. Derivatives of benzotriazole, such as Tinuvin 326, 327, 328 |XXVI|, and Tinuvin P, are another class of UV absorbers commonly used for polypropylene.

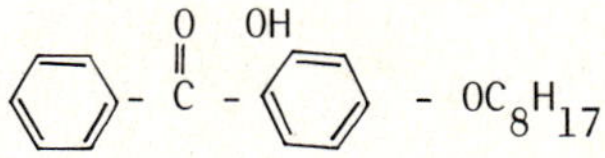

Cyasorb UV 531

|XXIV|

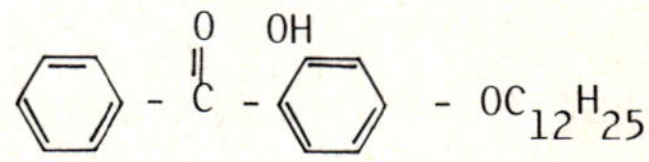

Eastman Inhibitor DOBP

|XXV|

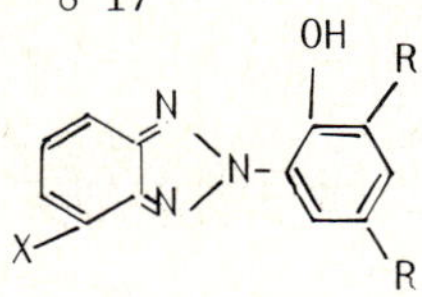

|XXVI|

Most of the above mentioned compounds are soluble in organic solvents and some of them also show sublimation tendency. To overcome these problems, high molecular weight UV absorbers, generally based on hydroxybenzophenone and benzotriazole structures, have been prepared. For example, one class of such polymeric or polymerizable UV absorbers can be prepared by reacting 4-hydroxybenzophenone or triazole molecules with suitable unsaturated acid and polymerizing:

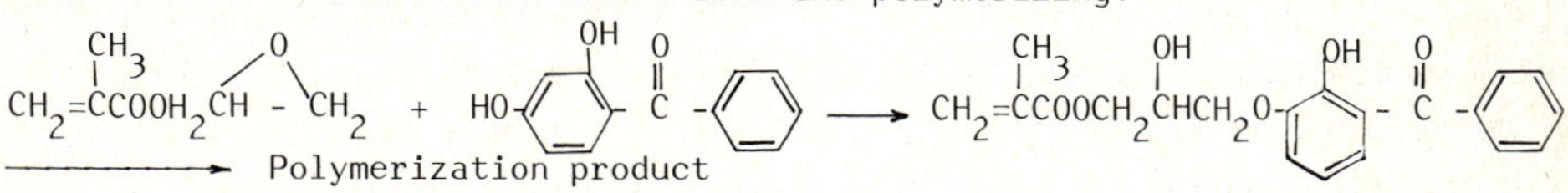

The interest in prepolymers or *in situ* polymerizable compounds is especially strong because of their easy miscibility with the polymer. Several different kinds of structures have been mentioned; some of them combine hindered phenol structures with UV absorbing hydroxybenzophenone or triazole structures.

The common UV absorbers can lead to a considerable improvement in the light stability of polypropylene when the product is relatively thick in cross-section. However, their efficiency decreases as the product thickness decreases. It is generally accepted that in thin cross-sections there is not a sufficient depth of UV absorber for efficient screening. Thus, in polypropylene multifilaments, where the fiber diameter is less than 50 microns, the effectiveness of conventional UV absorbers is substantially less than it is in monofilaments or molded plastics. Figure 6 compares the outdoor light-stability (Florida exposure) of 5 mil monofilaments and 210/35 multifilament yarns containing 0.0 , 0.5, and 1.0 phr UV absorber. The 0.5 and 1.0 phr UV absorber in 5 mil monofilament increases the sample life to 60 and 100 percent, respectively, whereas the same concentration of the absorber in 210/35 denier multifilament yarn leads to only a 33 and 70 percent increase in life, respectively.

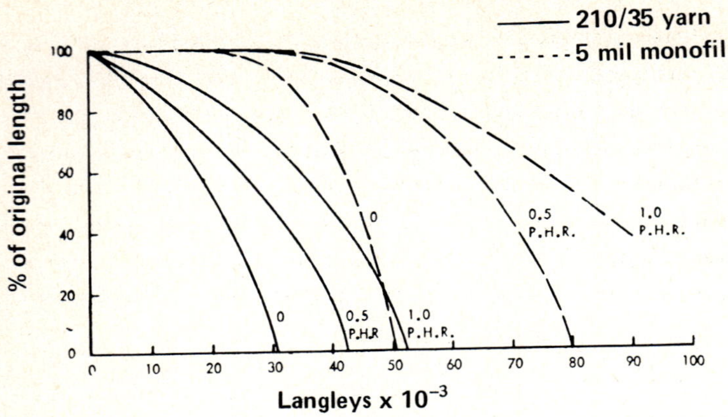

Fig. 6. Comparative Light Stability of UV-stabilized Polypropylene Filaments of Two Different Thicknesses (ref. 41).

III.4. Energy Quenchers

The excitation of chromophores in polypropylene (hydroperoxides, peroxides, carbonyl groups, catalyst residues, polynuclear aromatic impurities, etc) will produce excited states which can then take part in initiation reactions (cleavage, hydrogen abstraction) and so start oxidative degradation of the polymer. In general, the excited state of a chromophore (D) in polypropylene may react, fragment, return to the ground state by physical routes, or transfer its excess electronic energy to a stabilizer (Q):

$$D \underset{}{\overset{\text{Energy } (h\nu)}{\rightleftarrows}} D^* \xrightarrow{Q} D + Q^* \longrightarrow Q + h\nu'$$
$$D^* \rightarrow \text{decomposition, etc.}$$

The stabilizer must then harmlessly dissipate this energy. If energy transfer to the quencher Q can compete with decomposition, the chromophore is deactivated and the polymer stabilized. Efficient energy transfer can only occur if the energy level of the accetor is below that of the donor.

Energy transfer may occur by two important processes: long-range (or Foster) transfer, and collision (or exchange) transfer. Long-range transfer may occur between an excited state and a quencher separated by tens of ångstroms if appreciabl overlap exists between the emission spectrum of the donor D* and the absorption spectrum of the quencher Q. Collisional transfer requires physical contact between the excited species and the quencher, so that D* and Q must be close (within 10 to 15 Å) if deactivation is to occur within the short lifetime of D*. Efficient chromophore quenching by a collisional or contact transfer would require a quencher concentration of $\geqslant$10 weight percent if it were uniformly distributed. However, the concentration of oxidation products and the stabilizer in the amorphous zones of solid polymers will partly offset this high concentration, as will association by hydrogen bonding between photo-oxidation products and the quenchers.

The cleavage of UV excited carbonyl chromophores in polyolefins has been widely believed to be an important source of free radicals during the early stages of polyolefin photodegradation. The most probable process for energy exchange involves quenching of the excited singlet of the carbonyl group by the acceptor molecule which is postulated to have a triplet ground state. This can be written schematically in the following way:

$$R_2C{=}O^o \xrightarrow{h\nu} R_2C{=}O^* \quad \text{(singlet)}$$
$$R_2C{=}O^* + Q^o \longrightarrow R_2C{=}O^o + Q^* \quad \text{(triplet)}$$
$$Q^* \longrightarrow Q^o + \text{heat}$$

The quenching of excited carbonyl chromophores by long-range energy transfer to UV stabilizers such as Ni-chelates has been widely accepted as a desirable stabilization mechanism. Efficient quenching at 50 Å separation between chromophore and stabilizer would lead to protection by <0.1 weight percent of a uniformly dispersed stabilizer (mol.wt.$\approx$300). However, the limited published experimental support for long range energy transfer to stabilizers is open to other interpretations.

It has been recently concluded that sensitization of polypropylene photo-oxidation by carbonyl chromophores is of much less importance than previously suggested, and, by the same token, UV stabilization by the quenching of these chromophores is apparently restricted to collisional process rather than the elegant but unsubstantiated method of long-range energy transfer.

Peroxides and hydroperoxides photolise extremely readily to give free radicals even with long wavelength UV radiation. The absence of structure in the absorption spectra of peroxides and hydroperoxides indicates that transient excited states do not exist upon UV absorption; only instantaneous dissociation at the O-O bond occurs. This implies that quenching of this photolysis is impossible. Carlsson and Wiles (ref.49) have measured quantum yield for POOH loss in the presence of well known energy quenchers and confirmed that no reduction in POOH quantum yield results with any of these stabilizers.

The polynuclear aromatics, naphthalenes, phenanthrene, and anthracene (or derivatives of these) are often present in polypropylene articles as a result of absorption from the atmosphere. All absorb in the near-UV and can generate $^1O_2$ (singlet oxygen) within the polymer and in the case of anthracene yield an endoperoxide and anthraquinone, both of which photolise to give free radicals.

The generation of both $^1O_2$ and aromatic ketones such as anthraquinone from PNA impurities within polypropylene samples could be prevented by the deactivation of PNA excited states. It has been shown that some Ni(II) ketoximes and substituted benzotriazole stabilizers will in fact quench anthracene triplets in polypropylene fiber more efficiently than can $^3O_2$.

The potential importance of singlet oxygen in polypropylene photo-oxidation has already been well established. The deactivation of singlet oxygen to ground state oxygen has been demonstrated for many aliphatic and aromatic amines in the liquid phase. The aromatic amines such as N-isopropyl-N'-phenyl-p-phenylene diamine can quench at rates approaching the bimolecular encounter rate. These compounds are used extensively as antioxidants in rubbers, and may owe some of their protective action to $^1O_2$ quenching. They are seldom used in plastics largely because of staining problems (due to intensely colored oxidation products).

Considerable progress has been in recent years made in polymer stabilization through quenching of singlet oxygen. Several stabilizers of this type have been made commercially available. Heller and Blattman (ref.50) have ranked the following stabilizers according to singlet oxygen quenching efficiency:

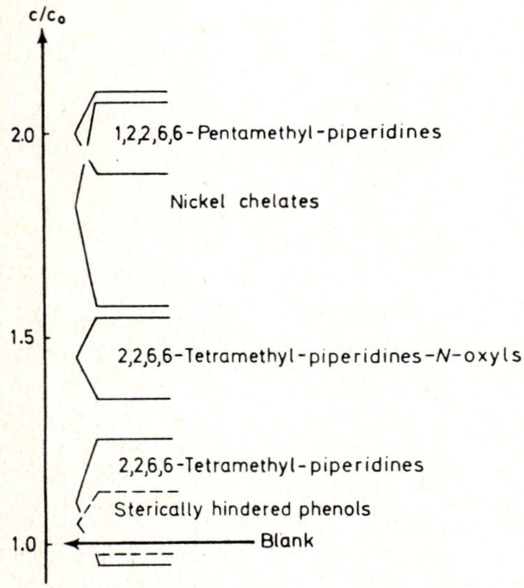

Similar results have been reported by other workers. The results indicate that of the tested compounds, the most efficient quencher of singlet oxygen is the N-methyl substituted piperidine derivatives, followed by Ni(II)-chelates and nitroxyl compounds of the hindered amines. The compounds of this type are often referred to as "reactive stabilizers".

Some of the useful Ni-chelate structures are listed below:

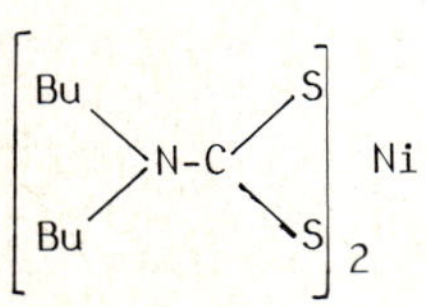

Nickel-dibutyldithiocarbamate (various sources)

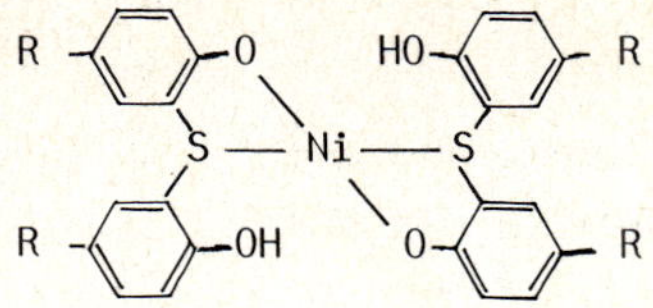

AM 101 and related structures
(Ferro Corp.)

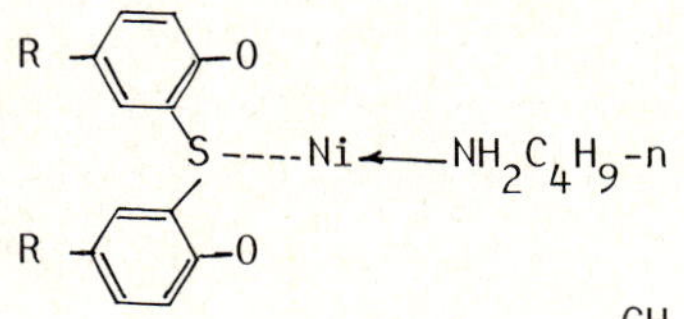

Cyasorb 1084 (American Cyanamide)

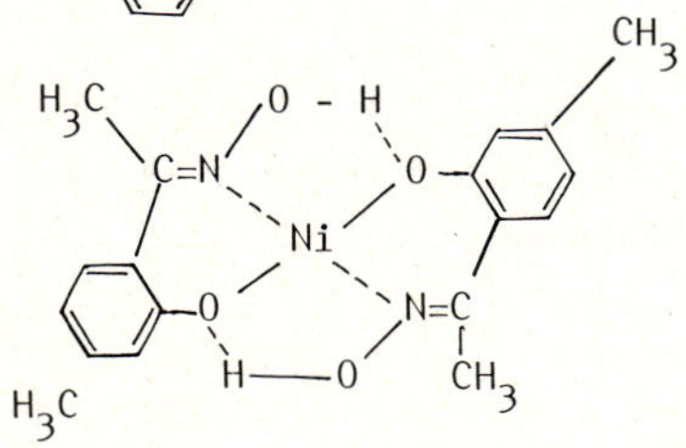

Hegapex A (ICI)

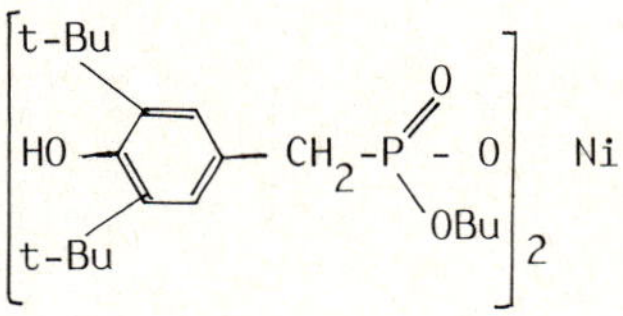

Irgastab 2002 (Ciba-Geigy)

The chromium chelates are also effective, and cobalt complex of the structure of Cyasorb 1084 has been found to surpass its nickel counterpart in light-stabilization of polypropylene.

The stabilizers that function in accordance with quenching mechanism can be expected to be less dependent on product thickness than are the UV absorbers. These systems are, therefore, of particular interest in the manufacturing of polypropylene fibers and film. The stabilization efficiency of these compounds is higher than that of the best UV absorbers. Figure 7 compares one common UV absorber with one Ni-complex stabilizer in their effect on the light stability of 210/35 denier multifilament yarn. The stability of the yarn containing the Ni-complex is clearly superior to that of the yarn containing UV absorber.

The nickel chelate stabilizers are generally of somewhat blue or green color, which may be inconvenient in certain uses where purity of fiber color is of utmost importance. Moreover, they have a tendency to decompose to black nickel sulfide at the high temperatures needed for processing of high-viscosity polymers. The necessary sulfur comes from the ligand itself or from the thioesters used as peroxide decomposers for heat stabilization. Nickel sulfide may also be produced by reaction with sulfides present in certain pigments as impurities or with pigments which are primarily based on metal sulfides (see Chapter III for details).

The nickel sulfide formation can be prevented to some extent by incorporating zinc stearate in stabilization formulation. Zinc sulfide would preferably be formed which is a colorless compound.

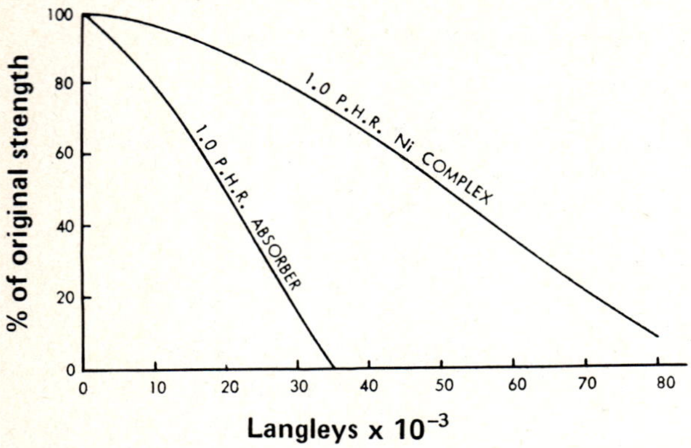

Fig. 7. Comparative Efficiency of a Common UV-absotber and a Ni-complex Stabilizer for the Light Stabilization of Polypropylene Multifilament Yarn (ref.4I)

In addition to the formation of nickel sulfide, some metal complex stabilizers are inherently prone to decompose into colored components at elevated temperatures of extrusion. Thus, UV 1084, for example, would tend to impart a bluish or purple tint to the extrudate at temperatures more than, say, 280°C. The intensity of discoloration is also a function of residence time for which the melt is held at elevated temperatures. The newer complexes are believed to be sufficiently stable to make them suitable for a broad range of processing temperatures. Irgastab 2002 is one such example.

A further complication in the use of the nickel chelate stabilizers arises where gas fading of the product is of any significant importance. The slight blue or green coloration imparted by these stabilizers to the fiber is fugitive to nitrogen oxides, which effectively change the original hue of the pigmented or dyed fiber. With extended exposure to nitrogen oxides, present in the industrial atmosphere, the white fibers ultimately change to yellowish coloration. It has been found that not all nickel stabilizers are equal in this respect. It has been claimed for example, that Irgastab 2002 is free from this shortcoming. Similar claims are made for some members of AM UV stabilizers (Ferro) and for some newer stabilizers from Cyanamide.

Active research still continues by leading fine chemical manufacturers to enhance the heat stability and gas-yellowing resistance of this class of compounds. Research efforts are also being expended to synthesise non-nickel organometallic compounds with possibly higher quenching efficiency, heat stability, and gas-yellowing resistance.

Recently, some very efficient light stabilizers of the hindered amine type, believed to function primarily through the quenching of singlet oxygen, have been introduced to the industry. This class of stabilizers is a breakthrough in the stabilization of polypropylene, especially in its use as fibers and film. Early publications have concentrated on N-oxyl free radicals, such as:

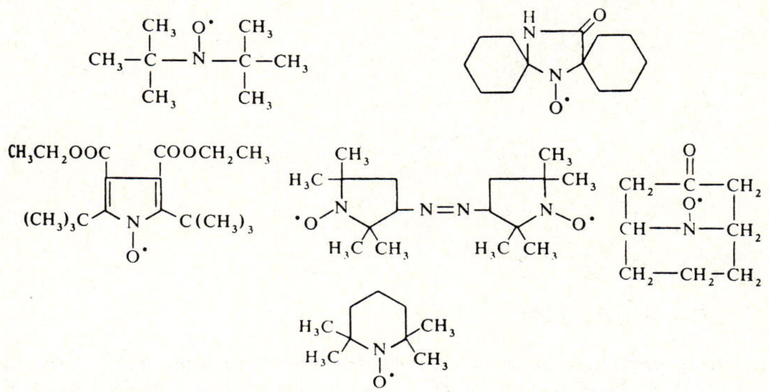

The application of nitroxyl radicals in the stabilization of polymers was originally considered on the basis of their ability to trap the free radicals essential in the degradation of polymers. Accordingly, the simple nitroxyls were thought to be good antioxidants. However, their efficiency in preventing therm-oxidative degradation was not sufficient to allow commercial usage for this purpose. Later on, the usefulness of these additives in polymer light stabilization, particularly polyolefins, was discovered. Their efficiency in suppressing photo-degradation of polyolefins can - depending upon the specific substitution - be quite remarkable. However, the color of nitroxyl radicals - yellow to red - effectively prevented their usage as commercial plastic additives in the concentration range normally utilized, e.g., around 0.5 percent based on the polymer.

A big step forward was the surprising finding by chemists of Sankyo company that not only the free radicals described , but also specific free amines, e.g., [XXVII] are effective stabilizers. Heller and Blattman (ref.20) have recently

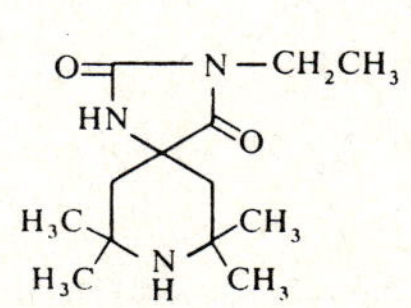

[XXVII]

examined the mechanism of their action as light stabilizer and have come to the conclusion that the effective members of this class of compounds, e.g., 2,2,6,6-tetramethyl-piperidine [XXVIII] function as efficient light stabilizers mainly through their ability to chemically consume the singlet oxygen. They further

TABLE I

Effectiveness of UV Stabilizers in .00I inch Thick Oriented Polypropylene Film (Base stabilizer 0.I % AO-I, exposed 45 South direct, Arizona) (ref.5I)

Light Stabilizer 0.3 %	*Natural*	*Kilolangley to 50 % Retention of Tenacity*			
		Pigmented (0.4%)			
		TiO	*Red BR*	*Phthalo.Blue*	*Yellow 3G*
Blank	13	23	15	16	22
LS-1	80	92	58	94	40
LS-4	28	27	22	23	23
LS-5	30	39	35	33	31
LS-6	36	NT	34	32	25
LS-7	32	33	34	35	29
LS-8	39	47	43	51	33

Structures:

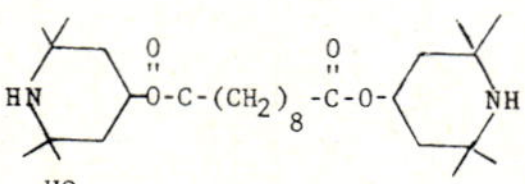

Bis(2,2,6,6-Tetramethyl-Piperidinyl-4) Sebacate

LS-1

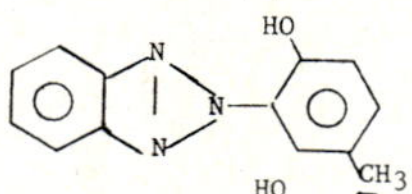

2-(2'-Hydroxy-5'-Methyl Phenyl) Benzotriazole

LS-2

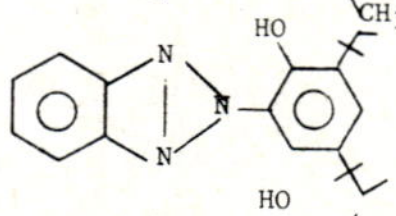

2-(3',5'-Di-t-Amyl-2'-Hydroxyphenyl) Benzotriazole

LS-3

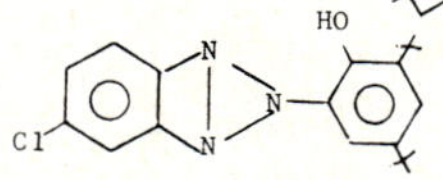

2-(3',5'-Di-t-Butyl-2'-Hydroxyphenyl)-5-Chloro Benzotriazole

LS-4

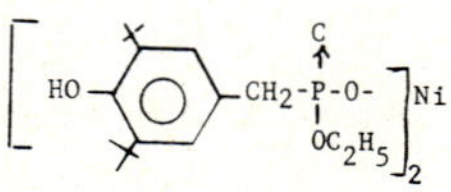

Nickel Bis[O-Ethyl(3,5 Di-t-Butyl-4-Hydroxybenzyl)]Phosphonate

LS-5

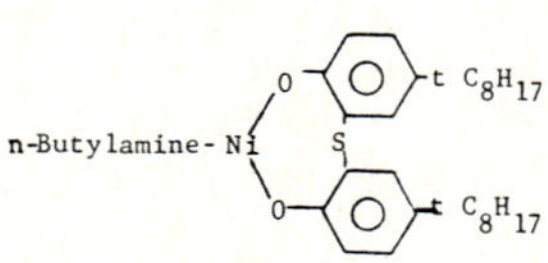

2-2'-Thiobis(4-t-Octylphenylato n-Butylamine Nickel III

LS-6

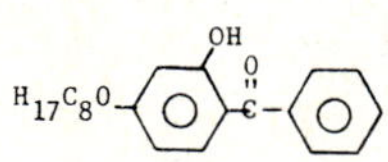

2-Hydroxy-4-n-Octoxybenzophenone

LS-7

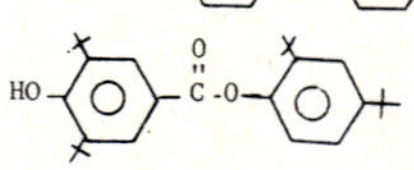

2,4-Di-t-Butylphenyl-3,5-Di-t-Butyl-4-Hydroxy Benzoate

LS-8

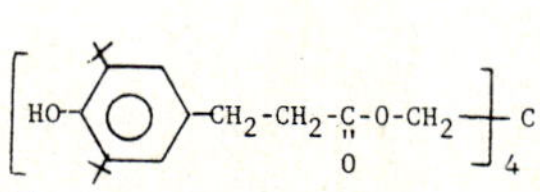

Tetrakis[Methylene(3,5-Di-t-Bu 4 Hydroxyhydrocinnamate)]Methane

AO-1

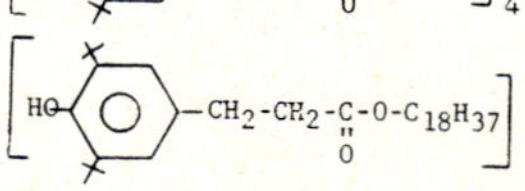

Octadecyl 3,5-Di-t-Butyl-4-Hydroxyhydrocinnamate

AO-2

2,2,6,6-Tetramethyl-piperidines:

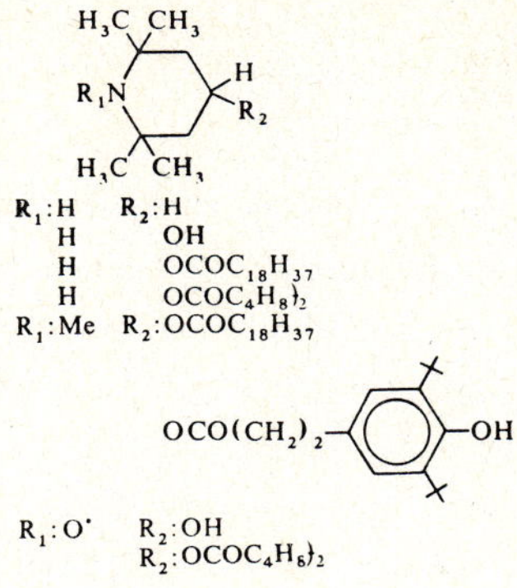

[XXVIII]

showed that 2,2,6,6-tetramethyl-piperidines are not unique in this respect. Other established light stabilizers, such as Ni-chelates, are also capable of quenching singlet oxygen, although to a lesser extent.

The light stabilizers of the type 2,2,6,6-tetramethyl-piperidine have been claimed to be as much as three to four times as effective as nickel chelates in thin cross-sectional products of polypropylene (Table I). They should, therefore be particularly valuable in the stabilization of polypropylene in its use as fibers and film. Ciba-Geigy is the forerunner in the commercialization of these stabilizers. The structure of the first member of this class of quenchers is given below [XXIX]:

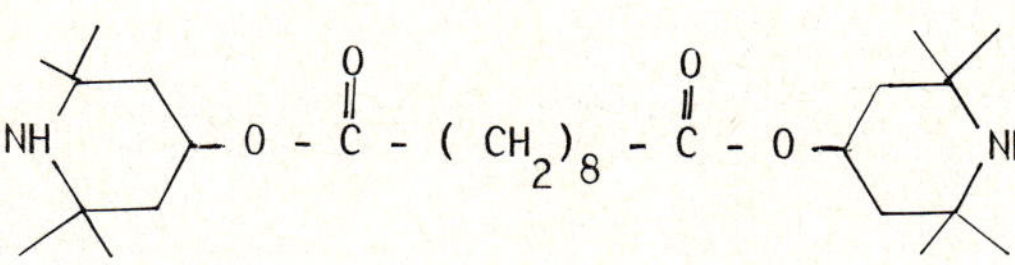

Tinuvin 770 (Ciba-Geigy)

[XXIX]

III.5. Metal Deactivators

The role of some trace metals and their compounds in catalytically accelerating the oxidation rate of polypropylene has been discussed previously. An increased amount of antioxidants would not suffice to inhibit the catalytic oxidation of the polymer. Some metal complexing organic compounds are, however, known to inhibit such catalytic acceleration of thermal and photo-oxidation by transforming the metal ions to harless organometallic compounds. They can, therefore, be used as auxiliary stabilizers. Among these metal complexing compounds, the so-called chelating agents are of special significance. The chelating agents bind the metal atom in two positions and form a stable ring structure as depicted in [XXX]

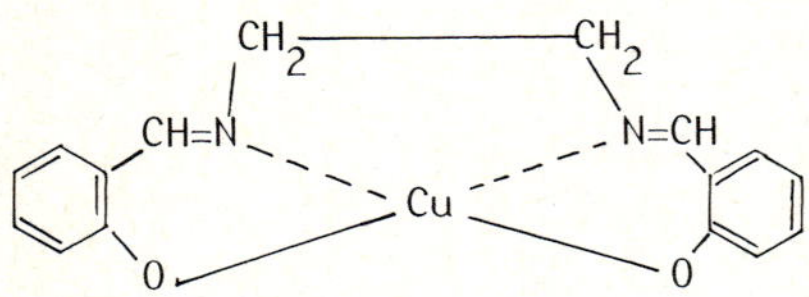

[XXX]

The effect of the chelating agents on the activity of metal ions is based first, on the inability to sterically shield the metal ions and block the formation of complexes of metals with peroxides, thereby preventing the homolytic decomposition of peroxides, and second, on their ability to change the redox potential of metal ions (ref.24) so that the catalytic reactions are prevented. However, the mere formation of metal complex or a change in redox potential is no guarantee of deactivation. In some cases, it might in fact accelerate the catalytic degradation of polypropylene.

Patent literature is rich in synergistic formulations using metal chelating agents as auxiliary antioxidants, but few systematic studies have been undertaken. Among a large number of possible metal complexing compounds (ref.25,26), oxanilide, oxalic acid dihydrazide, and oxalic acid di(2-phenylhydrazide) have been found particularly useful for polypropylene.

Various oximes have been shown to complex with titanium halides. The oximes are known to be quite effective UV stabilizers for polypropylene. Bisphenols have also been shown to complex with Ti(III) and Ti(IV) compounds, both at room temperature and at elevated temperatures. In addition, various other stabilizers, such as phosphites may be able to chelate trace metal initiators and so prevent metal-sensitized reactions. It has been found that the acceleration of polypropylene thermal oxidation by catalyst residues can be reduced by the inclusion of expoxides which probably deactivate the transition metal.

Polypropylene is generally free from excessive amounts of metal compounds and, therefore, does not need special stabilization against metal catalysis of the oxidation process. For practical purposes, the presence of tertiary phosphites used as peroxide decomposers, or epoxide compounds used as HCl-scavengers, or certain polyphenyl free radical inhibitors, is enough to deactivate the trace metal ions present inadvertantly in the polymer and no extra complexing agents are needed. However, if substantial quantities of metal residues or even a trace amount of copper or manganese are suspected in the polymer, or a danger of contamination exists, some suitable metal deactivators may be needed in the stabilization formulation. The use of metal complexing agents should be avoided when a strong metal complex stabilizer is present in the polymer formulation. The use of metal deactivators is strongly recommended in products which may come in prolonged contact with copper. One such example is the yarns and fabrics used for copper wire insulation.

III.6. HCl Scavengers

It is well known that efficiency of thermoxidative stabilization of polypropylene is strongly influenced by the presence of residues of Ziegler-Natta catalysts, which, if not removed by thorough extraction, cause a shortening of induction period of oxidation, discoloring, and deterioration of mechanical properties of the polymer. This effect is in addition to, and apart from, the metal catalysis discussed previously. Their deteriorating action on the stabilization of polypropylene is through a reduction of the efficiency of antioxidants present in the polymer (ref.27,28).

The reactions of catalyst residues with phenolic antioxidants involve first the formation of hydrogen chloride and colored titanate (ref.28). The liberated hydrogen chloride, owing to its diffusion mobility in the molten polymer, comes in contact with further molecules of the antioxidant, whose degradation and dealkylation reactions it catalyses.

HCl acceptors scavenge the free HCl as it is formed in processing of the polymer or during aging of the product, thereby protecting the phenolic antioxidant molecules. A large number of HCl scavengers is known. These are extensively used in the stabilization of PVC plastic. For polypropylene, five classes of compounds have proven useful:

(1) Polyvalent metal salts of organic acids
(2) Di-butyl tin compounds
(3) Epoxy compounds
(4) Organic basis, such as diphenylthiourea, (5) Organic triphosphites.

The reaction mechanism of these classes of compounds is represented:

$$(C_{11}H_{23}COO)_2Ca \ + \ 2\ HCl \longrightarrow CaCl_2 \ + \ 2\ C_{11}H_{23}COOH$$

$$(C_4H_9)_2Sn(C_{11}H_{23}COO)_2 \ + \ 2\ HCl \longrightarrow (C_4H_9)_2SnCl_2 \ + \ 2\ C_{11}H_{23}COOH$$

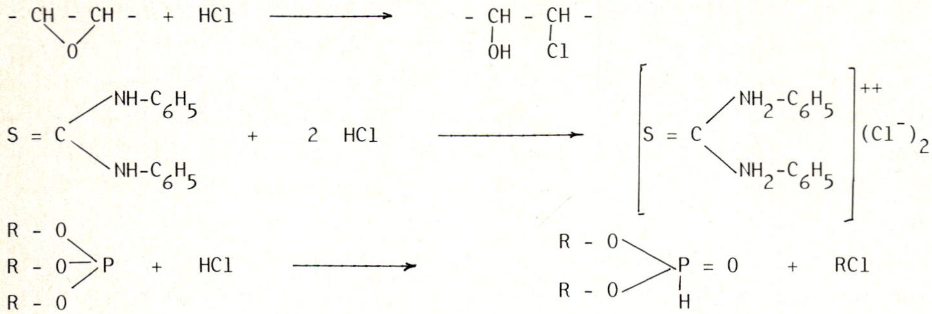

Among the polyvalent metal salts of organic acids, calcium stearate and zinc stearate are commonly used at 0.2 to 0.5 percent concentration. Fig.8, for example, depicts the benificial effect of Ca-stearate on the thermoxidative stability of a stabilized polypropylene composition. In addition to their HCl scavenging action, the stearates perform as processing aids by decreasing the effective melt viscosity in extrusion. Zinc stearate has better antacid properties than calcium stearate. Zinc stearate has the additional advantage of preventing discoloration of polypropylene fiber through sulfide formation when the stabilization formulation contains nickel compounds. Its action as processing aid is assumed, however, to be somewhat inferior to that of calcium stearate. Often it is advantageous to use a combination of both.

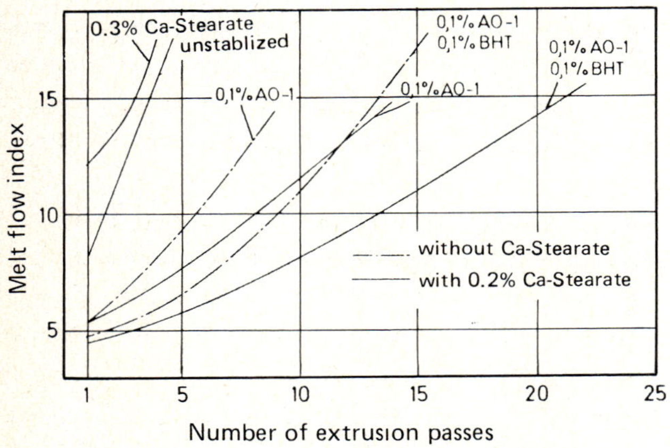

Fig. 8. Effect of Calcium Stearate on the Thermal Stability of Polypropylene (ref.52)

Epoxy compounds contain a three-membered heterocyclic oxirane ring

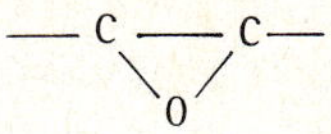

They have long been known as stabilizers for PVC and have been mentioned in patent literature as auxiliary stabilizers for polypropylene. The stabilization effect of these compounds is partly based on their ability to open up the ring structure and combine free HCl to chlorohydrins. Hydroquinone diglycidyl ether has commonly been mentioned as suitable for polypropylene. Higher molecular-weight epoxyethers, such as polyalkyl-glycidylethers and polycondensation products of epichlorohydrin and bisphenol A, glycerine-glycid-dylether, etc., and epoxyesters, such as triglyceride of epoxystearic acid, are usually favored because of their resistance to extraction.

In addition to their action as hydrochloride acceptors, epoxy compounds have the capacity to complex trace metals and thereby reduce the catalytic oxidation of the polymer. The effectiveness of some epoxy compounds to prevent the thermal and photo-oxidation of polypropylene is so high that it cannot be explained solely on the basis of their function as HCl acceptors and metal deactivators. Probably some free radical inhibition takes place wherein the chain structure of the epoxy compound takes part.

Epoxy compounds can potentially function as very effective low-price stabilizers for polypropylene fibers, but they generate an excessive filter pack pressure in spinning. This is probably in part due to self-polymerization at elevated temperatures and partly due to segregation and accumulation on filter media. They cannot, therefore, be used in large amounts where a fine degree of filteration is needed, such as spinning of fine-denier multifilament yarns. Their use for split film or monofilament is, however, feasible.

The polymer compositions containing substantial quantities of halogenated compounds as flame retardant additives, play havoc with the aromatic stabilizers through decomposition by HCl, evolved in the decomposition of the fire retardant additives. No effective means of stabilizing such compositions has yet been discovered although aforementioned HCl scavengers may offer some measure of relief. Compounds such as tetrakis-3-(3-di-tet-butyl-4-hydroxphenyl) propionyloxymethyl methane; 4,6-di-(4-hydroxy-3,5-di-tert-butyl-phenoxy)-2-octylthio-i,3,5 triazine (ref.29) and di-n-octylphthalate (ref.30) have also been specifically proposed.

III.7. Synergism in Stabilizers

The combination of two or more stabilizers in one stabilization formulation very often results in a higher degree of protection than the additive effects of the individual components. The phenomenon is known as "synergism". Although considerable information is available on the mode of action of the individual stabilizer components, the knowledge of synergism between different components is less precise and to a great extent still an art. The synergistic formulations for specific end products are often closely held secrets of polypropylene fiber manufacturing companies.

It is usual, in practice, to call one of the components of the stabilizer's formulation "primary antioxidant" and the others "secondary antioxidants", "synergists", or "auxiliary stabilizers". The latter components are generally the ones which alone will show very little antioxidant properties. In polypropylene stabilization, the phenolic stabilizers (for heat) and UV absorbers or energy quenchers (for light), are usually primary antioxidants and the peroxide decomposers, HCl acceptors and metal deactivators the synergists. The division is, however, completely empirical and, in view of the current knowledge about stabilization, quite impractical and erroneous. What is more important in this connection is the simple realization that synergism exists among some particular combinations of different classes of stabilizers.

For thermal stability, the synergism between a free radical inhibitor and a peroxide decomposer is by far the most important as compared with synergism in combination with other components, (Fig.5 and Fig.9). In combination with a radical inhibitor, the peroxide decomposer delays the consumption of the former by decomposing the peroxide to nonradical decomposition products. On the other hand, the radical inhibitors prevent the formation of long-reaction chains which lead to the formation of peroxides, thus protecting the peroxide decomposers from rapid depletion.

The absolute quantities and quantitative ratios of phenolic inhibitors and peroxide decomposers are important in any stabilization formulation. It has been variously shown that the thermoxidative resistance increases with increasing amounts of DLTDP, DSTDP, or tertiary phosphite at a given amount of phenolic antioxidant in the mixture, until the weight ratio of synergist to phenolic inhibitor exceeds 2.5 - 3.0. A further increase in the amount of the former does not increase the antioxidant efficiency of the mixture. On the other hand, the excess of phenolic antioxidant over thioester or phosphite contributes little to the stabilization effect. Above a certain concentration, phenolic antioxidants may even enhance the degradation rate of the polymer rather than inhibiting it. The stabilization effect of peroxide decomposers without the use of phenolic antioxidants is also negligible. This applies specially to the oxidation at high temperatures. Fig.9 typically shows this effect. The compounds, like thiobisphenols and phenolic phosphites, because of their dual function, act as "autosynergist" for thermal stabilization.

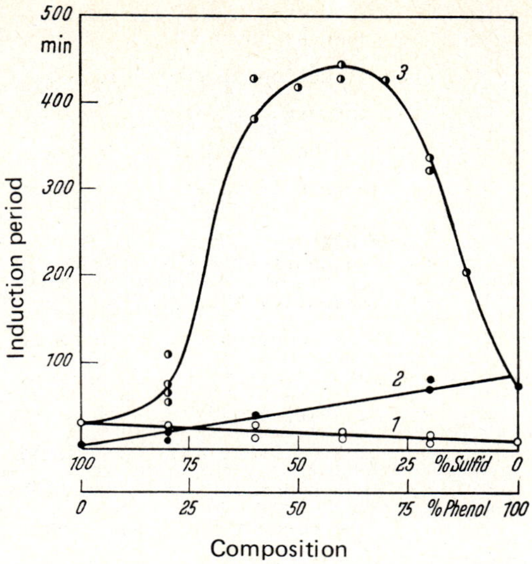

Fig. 9. Synergism Between a Peroxide Decomposer, didecyl sulphide, (1), and Free Radical Inhibitor, 2,4-di-(a,a-dimethylhexyl)-4- methylphenol, (2). Curves 1 and 2 Depict the Effect of Single Components and Curve 3 represents the effect of the two in mixture. 100% = 0.20 mol/kg. (ref.15).

Like thioesters and tertiary phosphites, tertiary aliphatic amino compounds also show significant synergistic effect for heat stability of polypropylene at high temperatures when used in combination with phenolic stabilizers, as shown in Table II. It is evident from these data that the combination of phenolic inhibitors with tertiary amine compounds yields better results than the cumulative effect of the two. The real significance of the synergism between these two groups of antioxidants becomes apparent at high temperatures. This synergistic effect is not impaired by the presence of other synergistic compounds in the mixtures, shown in Table II.

TABLE II

Synergism of Amino Compounds With Phenolic Antioxidants (ref.42)

Formulation	*Parts*					
Polypropylene	100	100	100	100	100	100
di-tert.-butyl-cresol			0.25	0.25		
iso-octyl-diphenyl phosphite					0.14	0.14
4,4-thiobix(2-tert-butyl-methyl phenol)					0.05	0.05
Zinc 2-ethylhexoate					0.06	0.06
DLTDP					0.25	0.25
Ethylene diamine tetraacitic acid		0.1		0.1		0.05
Initial melt index	1.0	1.0	1.0	1.0	2.5	2.5
Melt index after 3 min. at 375°F	8.6	8.2	1.0	1.0	3.0	3.0
Melt index after 30 min. at 600°F	48.0	30.1	8.0	3.0	30.3	11.4

Apart from the synergistic effect between a free radical inhibitor and a peroxide decomposer, there are two other kinds of combinations that lead to synergistically increased stabilization against thermal oxidation. These are: 1) metal deactivators combined with inhibitors and peroxide decomposers, 2) HCl acceptors combined with phenolic inhibitors and other stabilizers. The role of metal deactivators, as described previously, is the depression of catalytic oxidation of polypropylene. These compounds may not show any significant reaction when used alone but they synergistically reinforce the antioxidants present in the composition. The rate of decomposition of peroxides into free radicals caused by metal ions is considerably faster than that caused by the peroxide decomposers. In the case of strong metal catalysis, therefore, the peroxide decomposer present in the stabilization composition is unable to function and the synergistic action due to its presence is never realized. The presence of metal deactivators, in effect, enhances the synergism between the inhibitor and the peroxide decomposer.

Antacid stabilizers, discussed in Section III.6. do not show any antioxidant properties if used alone but they considerably enhance the effectiveness of other antioxidants present in the formulation. By preferentially removing the hydrochloride released in the reactions of titanium chlorides at high temperatures, the HCl fixing compounds protect the antioxidant compounds and keep them available for their respective function (Fig.8).

Like the stabilizer's formulations primarily directed against thermal oxidation, the photostabilization formulations are also based on synergistic effect of several stabilizers. This is because no single light stabilizer of any class is sufficiently effective in protecting polypropylene against actinic degradation so as to make it a universal stabilizer for all occasions. Moreover, by taking advantage of synergism between stabilizers of different classes, the total amount of the additives can be effectively decreased or some favorable economies achieved. Several types of synergistic effects for light stability have been reported and many more combinations are possible. The following is a summary of some important synergist combinations:

TABLE III

Synergism Between Free Radical Inhibitor and UV Absorber for Photostability of Polypropylene (ref.27)

Components		*Induction period (hrs)*
Tinuvin 326	Topancol CA	
0.6%	-	445
-	0.1%	210
0.6%	0.1%	855

1. Although common phenolic inhibitors have been found to exert no or very little photostabilization effect for the polymer, there has been ample evidence that such compounds do have some beneficial effects when synergistically combined with UV absorbers. One such example is shown in Table III. The photoactivity of inhibitors in the presence of absorbers may be interpreted on the basis that the light energy available for chain initiation is small in the presence of UV absorber, the reaction chains are still induced but their number is so small that the "snowballing" effect is checked by the inhibitor. Based on the synergistic results shown by these formulations, compounds have been prepared which combine the functionality of hindered phenol with UV absorption of hydroxybenzophenones or benzotriazoles, such as [XXXI].

[XXXI]

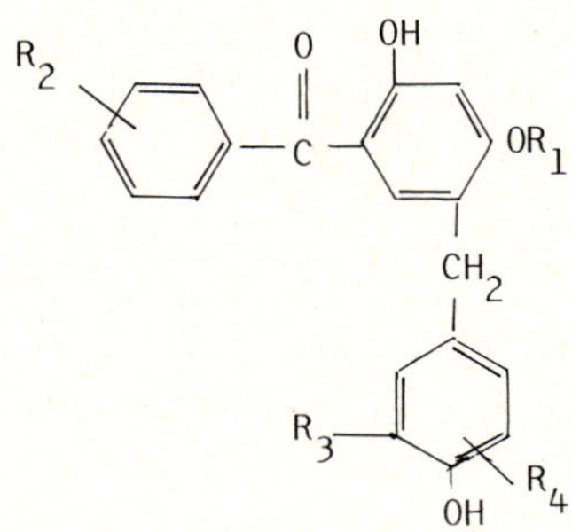

2. A similar synergism is shown by a combination of a peroxide decomposer with a UV absorber. Fig. 10 illustrates the interplay of a dithio-propionate-type peroxide decomposer and a triazole-type UV absorber. It is seen from these results that, while the antioxidant does not significantly contribute to the light stability of polypropylene when used alone, it does dramatically improve the efficiency of the UV absorber when present in the formulation as a second component. Tertiary phosphite-type peroxide decomposers are preferred because of their low-temperature activity. Compounds have been prepared which combine the functionality of phosphites with UV absorbancy of common UV absorbers (ref.33). These compounds show some degree of autosynergism.

3. Much more important for practical consideration is the synergism shown by antioxidant formulations (phenolic compounds in conjunction with peroxide decomposers) with UV absorbers than that between UV absorber and phenolic inhibitor or that between UV absorber and peroxide decomposer alone. Table IV exemplifies this effect.

4. A similarly strong synergism is exhibited by energy quenchers in combination with inhibitors and peroxide decomposers. Table V shows one such example.

5. A strong synergism is noticed between a UV absorber and a quencher (ref.34).

6. The synergistic effect between a quencher and a peroxide decomposer is of

particular importance (ref.35). As shown earlier, phosphites are better peroxide decomposers at ambient temperatures than thiodiproprionate compounds. They are, therefore, better synergists than the latter. Fig. 11 depicts this synergistic

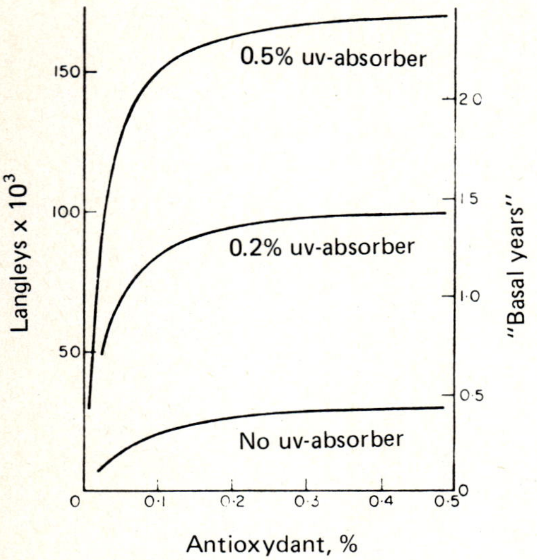

Fig. 10. Synergism Between an Antioxidant (propionate type) and a UV Absorber (triazole derivative) (ref.47).

TABLE IV

Synergism Between Antioxidants and UV Absorber for Light Stability (ref.43)

Antioxidants	*UV Absorber*	*Weather-o-meter stress cracking life (hrs)*
None	None	70
0.1% BHT + 0.3% DLTDP	None	100
None	1.0% Tinuvin 326	210
0.1% BHT + 0.3% DLTDP	1.0% Tinuvin 326	1733

effect. It is also apparent from Fig. 11 that there is an optimum level of tertiary phosphite concentration for use with nickel complex stabilizer. Above this optimum level, the phosphite leads to an overall reduction in the light stability of the formulation.

7. Synergism has also been shown in mixtures of various UV absorbers, such as that between the derivatives of hydroxybenzophenone and those of chlorobenzatriazole (ref.36). Recent trends in photostabilization formulations have been to use more than one light stabilizer along with auxiliary additives. This is

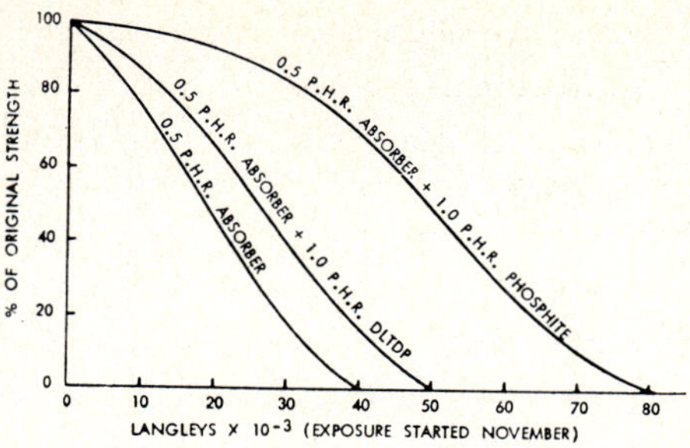

Fig. 11. Comparison of Synergistic Effect of Phosphite and DLTDP with UV Absorber (ref.41).

specially necessary in the case of multifilament polypropylene yarns because of the varied demands put on them in use. For example, light stabilization formulations for a high degree of UV stability may include 0.1 to 0.3 percent UV absorber or a mixture of UV absorbers, 0.1 to 0.5 percent energy quencher, 0.1 to 0.5 percent tertiary phosphite-type compound along with metal deactivator and HCL scavenger. This is obviously in addition to the stabilizer components employed primarily for heat stabilization as discussed previously. The exact composition, of course, would depend on the intended use of the product, the expected environmental conditions to be met in use, and economical constraints. Thus, for every specific end-use of the fiber, the stabilization formulation would be composed of several components delicately balanced for various effects.

TABLE V

Synergism Between Energy Quenchers and Antioxidants for Light Stability of Polypropylene (ref.43)

Free radical inhibitor	*Peroxide decomposer*	*Energy quencher*	*Fadeometer hours for ½ strength*
None	None	None	70
0.1% BHT	0.3% DLTDP	None	100
None	None	1.0% Ferro AM 101	320
0.1% BHT	0.3% DLTDP	1.0% Ferro AM 101	790

Experience in compounding and stabilization formulation would show that in a multiple-component system an ingredient cannot be taken out of the formulation or substituted by another component without disturbing the delicate balance between various components. It must, therefore, be clearly realized as the cardinal principle that the characteristics of a stabilization formulation is not a cumulative result of the functionalities of individual components plus a bonus factor of synergism. Each synergistic formulation has its own characteristics, hardly duplicated by other combinations. Because of the synergistic interaction of individual components, the efficiency of any particular component would naturally differ from formulation to formulation.

The secret of successful formulation is the optimization of results, both with respect to heat stability and light stability, with least expenses, keeping in mind the possible side effects which might appear in mixtures. The optimum formulations would not only depend on the type of products, denier per filament, nature of pigments present in the fiber, the severity of processing conditions, environment of use, etc., but also on such factors as whether the fiber is dyeable or pigmented, whether the product is expected to undergo frequent washing, and whether the retention of physical and mechanical properties of the fiber is more important than slight changes in the tint and hue of the fiber color in use. All these factors are important but they should be viewed realistically in line with the overall life of the product. Thus, in a carpet fiber, for example, there is no point of achieving light stability for an expected life span of 20 years where it is known that the product will be discarded long before for failure in other properties such as color or matting.

III.8. Stabilizers Formulation and Choice of Components

For a judicial choice of stabilizer components, extensive experience in stabilization formulation is necessary. Apart from the stabilization efficiency of individual components and their synergistic interaction with each other, numerous other considerations are taken into account. Some of these are discussed below.

Compatibility: Compatibility of stabilizers with the polymer is one of the primary requirements for the effectiveness of a stabilizer. Adequate compatibility is required for three reasons:

(a) Good compatibility of stabilizer components helps to distribute it intimately in the polymer, thus reducing the critical amount of stabilizers needed for stabilization.

(b) Good compatibility of stabilizers with the polymer is necessary to ensure trouble-free spinning operation, uniform filament thickness, good drawing performance, and uniform mechanical properties of the fiber.

(c) It has been found that a compatible stabilizer is generally resistant to sweating-out or exudation under the action of heat in processing or in use.

The equilibrium concentration of antioxidants and UV stabilizers in polyolefins at room temperature is extremely low. The equilibrium solubility of thiobisphenols in PE is only ~2×10^{-3} wt percent, as compared to ~0.1 wt percent normally included in the polymer. Excess antioxidant slowly exudes to the surface over several months at 25°C. Antioxidant effectiveness will depend not only on chemical behavior, but also on the solubility and diffusivity of the stabilizer; these same criteria apply also to UV stabilizers. The diffusion coefficient for a thiobisphenol in PE is ~1×10^{-9} cm^2 sec^{-1} at 25°C. Both solubility and diffusion coefficient decrease with increasing molecular weight.

The diffusion coefficient (D) of 2-hydroxy-4-C_nH_{2n+1}O-benzophenones in polypropylene is a maximum for n = 4 to 8 (D_{25^oC}~5×10^{-10} cm^2 sec^{-1}), whereas compatibility is independent of the size of the alkoxy substituent. However, compatibility (as measured by melting point depression) increases with increasing chain length. The $C_{10}H_{21}$O- to $C_{16}H_{33}$O-substituted 2-hydroxybenzophenones have been found most resistant to exudation from polyolefins as compared to both shorter and longer side-chain derivatives.

Although low compatibility and a high diffusion rate can obviously lead to the rapid loss of a stabilizer from a polymer sample, the importance of diffusion by the stabilizer to its photoprotective role has not yet been resolved. Oxygen has a diffusion coefficient of ~1×10^{-7} cm^2 sec^{-1} at 25°C in amorphous polypropylene, as compared to 10^{-9} to 10^{-10} cm^2 sec^{-1} for 2-hydroxybenzophenones and hindered phenols. This large difference in diffusion coefficient will be offset by the higher concentration of UV stabilizer (~0.5 wt percent, i.e.,~0.1 M, possibly much higher in the amorphous zones) as compared to O_2 (~0.001 M). Thus it is conceivable that stabilizer diffusion to a radical or excited chromophore could be an essential part of protection. However, it has been found that 2-hydroxy-4-alkoxybenzophenone and copolymerized 2-hydroxy-4-allyloxybenzophenone in polyethylene and polystyrene, respectively, imparted at least equal UV protection. In some cases the polymerized stabilizer is more effective and is much more resistant to leaching. Higher stabilizer loadings are also possible with the copolymer without massive exudation. A 2-hydroxybenzophenone-containing polymer was also found to be of similar effectiveness as the free stabilizer in protecting polypropylene against photodegradation, but much more resistant to extraction. Thus it would appear that diffusion of 2-hydroxybenzophenones does not play an essential part in their mechanism. Alternatively, reactive species may move to this stabilizer within their lifetimes. The concentration of stabilizer in amorphous zones and its exclusion from the crystalline zones of the polymer will appreciably increase the operational concentration of stabilizer in these amorphous, photo-oxidation sensitive zones and may minimize the need for molecular motion.

Carlsson and Wiles (ref.49) have shown that Ni(II) bis(dibutyldithiocarbamate) can diffuse into solid polypropylene film and destroy OOH groups. An optimum ease of diffusion is probably essential to the operation of a hydroperoxide scavenger.

Volatility: Resistance to volatilization of stabilizers at elevated temperatures is important for two reasons:

(a) Quite a substantial amount of stabilizer can be wasted during fiber extrusion, decreasing the effective concentration of that component in the product. An excessively volatile component may not only prove an economic waste but also a nuisance for spinning operation, requiring frequent cleaning of air quenching screens and the surrounding area.

(b) The volatile substances may further be lost in drawing, texturizing, and heat-setting operations, further decreasing its effective concentration in the fiber and fouling the processing equipment.

Several of the stabilizers offered in the early days of polypropylene development were considerably volatile or prone to sublimation at spinning temperatures, but most of the recently introduced stabilizers are fairly resistant to volatilization and sublimation at working temperatures.

Resistance to Heat and Light Degradation: Some of the commonly used phenolic and most of the nickel stabilizers are prone to degradation under high temperatures of spinning. The heat degradation of stabilizers not only decreases their effective concentration in the final product but may also cause undesirable coloration. The phenolic stabilizers, for example, would impart a yellow or brown coloration to the fiber while the nickel stabilizers would degrade to purple color compounds. Particular care should be exercised in the use of those stabilizers which are suspected of heat degradation. In such cases high-temperature spinning is generally not feasible.

Stabilizer components are also expected to be resistant to deteriorating action of light. Very little is known of this aspect but it is generally believed that most of the commercial stabilizers are fairly resistant to light and do not offer any significant problem. In a few low-cost phenolic stabilizers a yellow coloration is noticed on aging but it can be subdued to a large extent by using phosphite-type stabilizers as synergists.

Color: A strongly colored stabilizer component evidently cannot be used in formulations meant for fiber production. Although most of the stabilizer components used for polypropylene do not suffer from this shortcoming, a few UV stabilizers do. Of particular importance are some nickel-based stabilizers, which if incorporated in excessive amounts impart to the fiber an undesirable bluish tinge and make the final color of the yarn somewhat dull. A commonly used UV absorber, UV 531, imparts to the fiber a slightly yellowish tinge when used in the range of 0.5 percent concentration. Tinuvin-type UV absorbers are relatively free from self-coloration. While self-coloration is a problem in most of the textile uses of polypropylene fiber, in certain uses it might not be a significant factor. For example, the strongly greenish color imparted by some nickel or cobalt complexes should not be a problem in the manufacture of synthetic grass even at a high level of such compounds.

Chemical Interactions: At elevated temperatures of extrusion or under excessive rate of shear, some stabilizers or their degradation products would chemically react with other stabilizers or their degradation products. Such is the case, for example, with nickel-based stabilizers that tend to react with the degradation products of sulfur-bearing stabilizers to produce nickel sulfide, an intensely black-colored product that imparts an undesirable grey color to the fiber. This kind of interaction can also take place between a nickel stabilizer and a sulfur-bearing pigment, such as cadmium and mercury sulfides.

Another example is the reaction of TLTTP with cadmium sulfide yellow. At higher spinning temperatures some CdS yellows may be bleached and the full color value of the formulations may not be realized. Apart from the interaction of stabilizers with each other, or with pigments, they can also react with other additives, such as antistatic compounds, processing aids, flame retardant, etc., present in the formulation. The formation of colored compounds is not always the case. One important case of chemical interaction without accompanied coloration is met when halogen-containing fire retardant additives are present in the polymer formulations.

Physical Interaction with Pigments: One type of physical interaction considered to be of some significance in determining the efficiency of an antioxidant is absorption of stabilizer by pigments. The amount of absorption depends not only on the type of pigment but also on the structure of the antioxidant. Channel black is an order of magnitude more efficient as an absorbent compared to the other pigments.

Smell and Toxity: Although smell and toxity are important factors to be considered in stabilization formulation, the polymer processor is generally spared screening for these factors; the stabilizer manufacturers would normally not release a compound to the polymer processor which is either potentially toxic or strong-smelling. In development work, however, it is always prudent to keep these factors in mind while screening new chemical compounds for stabilization effect. Sometimes, the stabilizer compound itself would be odorless and non-toxic, but minor impurities may impart considerable smell to the product. Such cases are normally met in the case of sulfur-containing compounds.

Gas Yellowing: A major requirement for effective stabilization of polypropylene for fiber end-use is its resistance to gas yellowing. This is particularly important in home furnishing fabrics such as drapery, upholstery, and carpets, where the product is expected to be exposed to nitrogenous gases for much of its useful life. The end result of gas yellowing is the change in shade of the dyed or pigment product. Most of the phenolic antioxidants become discolored to varying degrees by nitrogen oxides, although some are less sensitive. Benzophenone-type UV stabilizers appear to react more with nitrogen oxides than the triazole-type UV stabilizers. Nickel stabilizers, such as UV 1084, are also sensitive to nitrogen oxides and gas-yellow noticeably. Thio groups in such molecules are believed to play a part in gas discoloration, as some nickel complex stabilizers which do not contain any sulfur in their structure are more resistant to gas yellowing than those which contain sulfur as part of their molecular structure. One such structure is recently introduced Irgastab 2002.

Launderability: An added criterion for the suitability of a stabilizer component for polypropylene fiber is its resistance to extraction in washing. This is specially important in apparel which is repeatedly subjected to warm or hot solutions of detergents in a washing machine and to hot blowing air in a dryer. Weimer and Conner (ref.37) have investigated the factors affecting the launderability of polypropylene multifilament yarns. They found that antioxidant is lost more rapidly during cyclic wash-dry laundering than during washing or drying alone. The data indicates that the loss mechanism is more complex than one gross extraction during washing and volatilization during drying, although both of these factors are involved. High molecular-weight stabilizers are obviously more resistant to laundering than the normal antioxidants. In addition to the molecular-weight, the loss of stabilizers in washing-drying cycles depends on several other factors, such as it:

(a) increases with increasing drying temperature
(b) decreases with increasing denier per filament
(c) increases with water flow rate
(d) appears to be proportional to the amount of the stabilizer in the fiber

To explain why antioxidants are lost faster in a cyclic wash-dry operation than total wash plus one total dry period, it has been suggested (ref.37) that the dominant mechanism for removal of stabilizer during cyclic wash-dry laundering involves diffusion of the stabilizer by thermal pumping in the drying cycle from within the polymer matrix to the fiber surface, which is then washed away in the washing cycle. Thus, the alternation of concentration profile results in a much more rapid loss of stabilizer than would occur from either washing or drying alone.

Economics: One basic rule of all stabilization efforts is that the maximum degree of stabilization is not always the goal and that the required degree of stabilization can be achieved by utilizing several alternative components in varying ratios. Almost invariably the minimum degree of stabilization for any particular end use is kept in mind and the stabilization formulation evolved to meet that basic requirement. It is evident that a high degree of UV stability would not be necessary for products which are to be used indoors, or good launderability is not required for such uses as carpets, fiber fill, or upholstery fabrics. This is why, for almost every end use, a separate stabilization formulation is involved.

The stabilizer formulations vary in price, a factor of major consideration in stabilization formulations. A prime advantage of polypropylene fiber over nylon and polyester is the low cost of its raw material. If high-priced stabilization formulations are employed across the board, the advantage of low-cost material is completely lost.

In arriving at the cost of stabilization for any particular use, the cost of stabilizer components as well as their needed amounts are important. Thus, it is not necessary that a low-cost stabilizer with low-stabilization efficiency would yield a lower cost formulation than an expensive but efficient stabilizer. As indicated earlier, the intensity of synergism between the various components of the formulation may also significantly change its cost structure.

IV. TESTING AND EVALUATING STABILIZERS

A variety of methods for evaluating stabilizers in polypropylene has been developed during the past several years. In the fiber industry, five sets of test methods are commonly applied to evaluate the relative effectiveness of various antioxidant systems in polypropylene:

1. Process stability tests
2. Weathering tests
3. Accelerated aging tests
4. Launderability tests
5. Miscellaneous observations

IV.1. Process Stability Tests

By process stability is meant the resistance of the polymer formulation to thermal degradation under processing conditions in the spinning system. The test methods commonly used to assess the stability of polypropylene include melt-flow drift measurements using melt indexer, retention of melt-flow after repeated extrusion, melt-viscosity retention measurements using a torque rheometer, differential thermal analysis or differential scanning calorimetry, and the measurement of intrinsic viscosity of the extrudate at definite conditions of extrusion.

There is no ideal simulated test method for assessing process stability: the farther the test conditions get from actual processing conditions, the less likely the performance of the stabilizers will correlate. The question of the best method is not only a question of best correlation but also one of expediency. Whatever the method chosen and whatever the test conditions selected, at best there is only a fair correlation expected between the simulated processing test methods and the actual processing results. The various test methods are, however, useful tools for differentiating polymer formulations whose stabilities vary considerably - although these methods may fail to differentiate small changes in stability.

The laboratory test methods, whether torque rheometry or melt indexing or DTA or any other, are regarded as screening tests only. For more reliable and more differentiating test results, actual spinning of the polymer formulation is invariably necessary. Three criteria can be used to measure relative stabilization to varying degrees of precision. They are:

(a) power requirement on the extruder under control conditions of temperature and back pressure at any fixed throughput rate,

(b) continuous measurement of melt viscosity by in-line rheometer,

(c) intrinsic viscosity measurement or melt-flow measurement of the polymer composition and the spun yarn, respectively.

The above indications are an approximate estimation of the degree of stabilization. The higher the stabilization efficiency, the higher the power requirement for a given throughput, the higher is the measured melt viscosity and the higher is the spin tension. These three methods necessitate the use of the same base polymer for comparative evaluation of stabilizers. For more exact evaluation and with somewhat more flexibility, although more time-consuming than the first three methods, is the actual determination of intrinsic viscosity or the melt-flow index of the extrudate and comparison of it with that of the blend.

IV.2. Weathering Tests

Weatherability is defined as the net effect of outdoor exposure on the characteristics of the material. Commonly, emphasis is placed on discernible changes in appearance and mechanical properties of the fiber and the fabric. Weathering resistance is obviously of special importance for polyproplene fibers destined for outdoor uses, such as outdoor-indoor carpets, synthetic grass, ropes, fishnets, sandbags, etc. It is also important for the products not meant for outdoor usage but expected to be exposed to direct or indirect sunlight for a prolonged period of time. Such uses include carpeting, wall coverings, upholstery fabrics, and related products.

Weathering tests are intended to measure the stability of the fiber product against light and environmental conditions. A true qualitative comparison of the effectiveness of different stabilizer formulations for their protection of polypropylene against light and environmental conditions is possible only under test conditions approaching closely the actual conditions of use. Atmospheric weathering tests are, therefore, employed in which the filaments are exposed to natural sunlight and the changes in their appearance and physical properties are periodically determined. For realistic and reliable test results the range of weathering variables at the location under consideration should be taken into account. Climatic conditions around the world are of great diversity, and optimum and economic product design for outdoor uses must reflect these climatic differences.

The rate of deterioration of physical properties of the fiber is dependent on the intensity of light, the degree of chemical pollution in the air, and the temperature of the environment. Thus, the aging rate should depend not only on the geographical situation of the testing station but also on the seasonal variations. As one would expect, solar radiation shows seasonal periodicity with the maximum intensity occurring during the summer months. Another important factor affecting solar energy received by a sample is the angle of incidence. This effect usually depends on the time of the year.

Because of these difficulties, the industry some time ago adopted Langley as an integral measure of the sunlight energy received by the test sample, whereby one Langley = 1 gram cal/cm^2 = 3.69 BTU/sq.ft.

The measurement of Langley value of sunlight is done by pyrheliometer. In this way the daily and seasonal variations are to a great extent eliminated. Sometimes an energy unit called "sun hour" is also used in reporting of sunlight exposure results. A sun hour is defined as that exposure hour wherein the radiation intensity exceeds 0.823 $gcal/cm^2/min$. The energy value of a sun-hour may fluctuate between 1 and 100 Langleys and, therefore, is only an approximate integral measure of sunlight energy. The weathering results are reported in Langleys or sun-hours for complete breakdown of the sample, or more often, the Langleys or sun-hours required for a certain percentage of the strength retained by the sample, usually 50 percent.

Statistical methods are quite useful in the analysis of weathering results. Some of these methods may be valuable in planning and conducting outdoor weathering experiments, particularly in reducing the disturbing effects of uncontrollable experimental variables, selecting test methods, and attaching limits of certainty to the results. Analysis of variance and correlation-regression strategy have proven specially useful in planning and analyzing the weathering results.

IV.3. Accelerated Aging Tests

Laboratory aging is a commonly used method for evaluating the degradative effect of UV radiation and heat on polypropylene fiber under controlled and accelerated conditions. The two main difficulties encountered in outdoor exposure, i.e., lack of repeatability and long duration of test, are, thereby, to some extent eliminated. Two distinct types of tests are conducted in the laboratory.

1. Oven aging tests, which put primary emphasis on heat aging and
2. Light exposure tests, which put primary emphasis on light stability.

Oven Aging Test: Oven aging is extensively employed as a laboratory test method to assess the effect of heat on the physical and mechanical properties of the fiber. Fiber samples, suitably mounted in frames or wound in skeins, are aged in a forced air circulation oven at elevated temperatures. Specimens are taken out at frequent intervals and their mechanical properties measured.

The relative thermal stability of test specimens is dependent on the temperature of testing and the thickness of the specimen. While it is not difficult to decide what denier of the test fiber to choose, the choice of aging temperature is a matter of considerable deliberations. Aging temperatures varying from 70°C to 135°C are being used, depending on the degree of stability of the fiber and the expediency of the test. A 50 percent loss in fiber strength and elongation or, preferably, in the toughness factor, $T.E^{\frac{1}{2}}$, is generally taken as the end of induction period and is considered as a relative measure of polymer stability at test temperature. Where the determination of $T.E^{\frac{1}{2}}$ is not practical, the aging is done to the point where the product begins to powder on gentle rubbing.

Despite its frequent use, the oven aging test is little understood and has only a fair precision. The variability of oven aging data in repeated tests is still a problem and often frustrates efforts to detect minor or moderate differences in stabilization formulations. A list of variables affecting oven aging data is given by Forsman (ref.38). Of these, temperature and air velocity are the most critical (ref.38). By suitable oven modifications and mechanically rotating the specimens simultaneously on two axis on a "ferris wheel" mechanism, Forsman (ref.38) was able to improve the precision in a conventional oven from ±40 percent (95 percent confidence limits) to ±5-10 percent.

Oven aging of fibers on thin filaments presents great difficulty in handling, mounting, and keeping track of failures as they occur. DiBattista (ref.39) has recently described an oven aging method for polypropylene filaments which makes use of a "tubular oven", and a novel technique for suspending the filaments in the oven. The data obtained with a tubular oven have been shown to be much more precise and repeatable than the data obtained with a conventional forced draft oven (ref.39).

Laboratory Light Exposure Test: The question of laboratory light exposure is mainly a question of selecting a suitable source of artificial light and its spectral characteristics. Carbon arc and xenon arc lamps are both widely used although the latter is becoming more popular because of the close resemblance of its spectrum with that of natural light.

Mounting techniques significantly affect the precision of test data. From the precision point of view, a single layer of filament mounting with nominal load to each filament bundle is probably the most desirable arrangement. A convenient rack to achieve this end has been described by Forsman (ref.40). This rack allows the fiber to contract under fixed load without danger of breakage.

Laboratory light exposure data are treated in the same manner as those from oven aging studies. Exposure hours, whether in xenotest or weather-o-meter, up to the end of induction period is commonly employed as a measure of UV stability under the test conditions. Here, too, statistical methods of analysis are useful and highly recommended for the planning of experiments and analysis of test data.

As is the case with the laboratory aging methods for the assessment of thermal stability (oven aging), the accelerated exposure methods are generally regarded as screening tests and not a substitute for outdoor aging. Nevertheless, frequent attempts have been made to correlate laboratory data to those obtained under the sun. Generally such attempts have not been successful because of the poor reproducibility obtained in outdoor weathering and because of the difficulty of reproducing the spectral quality of sunlight in the laboratory. However, the relative correlation generally seems to emphasize the usefulness of artificial light, specially that of the xenon arc lamp.

IV.4. Launderability Tests

As indicated earlier, the resistance of stabilizers to washing is of utmost importance in the apparel field. Also, washing resistance is not necessarily synonymous with resistance to water extraction of stabilizers. The results quoted elsewhere in this chapter show that stabilizers are lost much more rapidly during cyclic wash-dry laundering than during simple washing or simple drying alone. Testing may be done under simulated home and commercial laundering and drying conditions on fabrics knitted from suitable denier yarns. In most commonly used home laundering tests, washing is carried out for 25 minutes with any commercial non-ionic detergent plus a suitable quantity of hypochlorite bleach at 70°C. Drying is done in an electrically heated tumble dryer for 45 minutes at 110°C, the maximum operating temperature of most domestic dryers. The results of washing tests can be conveniently expressed by the term "launderability resistance" or "laundering cycle lifetime" which refers to the number of wash-dry cycles required to reduce the strength of a fabric to a point where it can be torn or punctured easily with the fingers. One laundering cycle is defined as one wash plus one drying phase.

Apart from the temperature and pH of the washing water and the temperature of drying, the laundering resistance of stabilized compositions depends on denier per filament. The construction of the test specimen should be kept constant.

There is no quick and effective method of testing the wash resistance of light stabilizer components of the formulation. A test method can, however, be evolved by coupling the wash-dry cycle test with xenon exposure test and the results expressed as percentage decrease in light stability (in xenon hours) after some fixed wash-dry cycles.

IV.5. Spontaneous Combustion

A serious problem encountered in the use of polypropylene fiber in its early days of development was a tendency of spontaneous combustion in the course of wash-drying of fabrics or prolonged storage in a confined space. Most of these instances have been closely related with the launderability of the product, but instances are also known where spontaneous combustion was not related to laundering but was caused by the absence of any effective heat stabilizer in the stabilization composition. There has been no such incident reported in recent years but it is always prudent to guard against any such eventuality. A convenient method to determine the tendency for spontaneous combustion is to wash-dry the sample for about 50 cycles under home laundering conditions, pack it into a glass cell fitted with a thermocouple, and age it at 120-150°C in an oven. A sharp exotherm would indicate a tendency for spontaneous combustion.

IV.6. Miscellaneous Tests and Observations

Apart from the stability considerations, a stabilizer formulation has to meet several other requirements to make it suitable for practical use. There are no set methods of laboratory evaluation of these technical requirements but observations are made with regard to these properties in actual spinning evaluation. The important factors are color, odor, filter plugging, and sublimation. In addition, some tests and observations may be required which are common in the textile industry.

V. REFERENCES

1. G.R. Cotton, W. Sacks, J. Polym. Sci., A-1(1963)1345.
2. D.J. Carlson, D.M. Wiles, J. Polym. Sci., Polym. Letter Ed., 11(1973)959.
3. J.E. Wilson, Ind. Chem. 47(1955)2201.
4. B.A. Marshal, ACS Symp. on Stabilization, Miami Beach, 1967.
5. B. Brown, SPE Trans., 2(1962)250.
6. U.S. Patent 3,639,513, Mitsubishi, 1972.
7. C.R. Buss, quoted in reference (41).
8. C. Walling, R. Rabinowitz, J. ACS., 81(1959)1243.
9. P.A. DePaolo, H.P. Smith, ACS Symp. on Stabilization, Miami Beach, 1967.
10. U.S. Patent 3,502,614, ICI, 1970.
11. Belgium Patent 634,640, Geigy.
12. U.S. Patent 3,347,875, Hercules, 1970.
13. Netherland Patent 131,625, Montecatini, 1971.
14. U.S. Patent 3,235,623, Soc. Rhodiaceta, 1966.
15. J. Voigt, Die Stabilisierung der Kunststoffe, Springer Verlag, Berlin, 1968.
16. J.H. Clemdet, SPE Trans., 1(1961)26.
17. B. Wright, Plastics, 28(1963)125.
18. R. Schimper, Plast. u. Kaut., 11(1964)466.
19. J.P. Guillory, C.F. Cook, J. Polym. Sci., A-1, 9(1971)1529.
20. J.J. Heller, H.R. Blattman, Pure Appl. Chem., 30(1972)145.
21. Y.A. Erschov et. al., Russ. Chem. Rev., 38(1969)147.
22. A.P. Pivovarov, et. al., Plast. Massy, 10(1966)7.
23. P.G. Briggs, J. Appl. Polym. Sci., 12(1968)1825.
24. A.J. Chalk, Trans. Faraday Soc., 53(1957)1214.
25. R.H. Hansen, Symp. on Polypropylene Fibers, Birmingham, Ala., 1964.
26. A.S. Kuzminskii, et. al., Vysokomol Soed., 4(1962)1682.
27. L. Balaban, et. al., Kunststoffe, 60(1971)341.
28. J. Kresta, J. Majer, J. Appl. Polym. Sci., 13(1969)1859.
29. U.S. Patent 3,652,494, Phillips, 1972.
30. U.S. Patent 3,624,028, Phillips, 1972.
31. U.S. Patent 3,632,858, Cyanamic, 1962.
32. U.S. Patent 3,538,048, ICI, 1970.
33. Japan Patent 70/04,623, ICI, 1970.
34. U.S. Patent 3,464,953, Eastman, 1969.
35. Brit. Patents 840,648; 888,673; 945,441; U.S. Patent 3,003,814.
36. U.S. Patent 3,502,613, Shell, 1970.
37. R.P. Weimer, W.P. Conner, Text. Res. J., 39(1969)1150.
38. J.P. Forsman, SPE J., 20(1964)729.
39. DiBattista, Polym. Eng. Sci., 10(1970)332.
40. J.P. Forsman, J. Appl. Polym. Sci., 9(1965)2511.
41. J. Bonkowski, Text. Res. J., 34(1969)245.
42. U.S. Patent 3,414,532, Argus, 1968.
43. Canadian Patent 834,281, Eastman, 1970.
44. J.W. Tamblyn, J. Appl. Polym. Sci., 9(1965)2251.
45. D. Sysvy, Vysokomol Soed., 3(1953)464.
46. B. Felder, quoted by H.J. Heller, Pure Appl. Chem. (London), 36(1973)141.
47. H.J. Heller, Europ. Polym. J., Suppl., (1969)105.
48. D.J. Carlson, D.M. Wiles, J. Maeromol Sci. Rev. Maeromol Chem., C14(1976)155.
49. D.J. Carlson, D.M. Wiles, J. Polym. Sci. Polym. Chem. Ed., 12(1974)12217.
50. J.J. Heller, H.R. Blattman, Pure Appl. Chem., 36(1973)141.
51. J.J. Usilton, ACS Preprints 393, 1976.
52. F. Mitterhoffer, Kunststoffe, 67(1977)151.

VI. BIBLIOGRAPHY

VI.1. Books

1. W.O. Lundberg, Autoxidation and Antioxidants, Vol. 1, Interscience, New York, 1961.
2. N.K. Baramboim, Mechanschimistry of Polymers, Maclaren & Sons, London, 1964.
3. G.G. Yakuulevich, Stabilization of Synthetic High Polymers, translated from the Russian, D. Davey, New York, 1964.
4. A.S. Kuzminskii, Crystalline Olefin Polymers, Consultants Bureau, New York, 1965.
5. M.B. Neiman, Aging and Stabilization of Polymers, translated from the Russian, Consultants Bureau, New York, 1965.
6. S.H. Pinner, Weathering and Degradation of Plastics, Columbia Press, London, 1966.
7. J. Voigt, Die Stabilisierung der Kunststoffe gegen Licht und Waerme, Springer Verlag, Berlin, 1966.
8. ACS Symp., Stabilization, Miami Beach, April 19, 1967, ACS Washington, D.C., 1967.
9. R.F. Gould, Ed., Stabilization of Polymers and Stabilization Processes, ACS, Washington, 1968.
10. R.T. Couley, Varcal Deckker, Thermal Stability of Polymers, Vol. I and II, New York, 1970.
11. Band I u. II. K. Thinius, Stabilisierung und Alterung Von Plastwerkstoffen, Verlag Chemie, Weinheim, Ger. 1971.
12. A.S. Kuzminskii, The Aging and Stabilization of Polymers, Elsevier, N.Y., 1971.
13. W.C. Hawkins, Polymer Stabilization, Interscience, New York, 1972.
14. B. Ranby, Photodegradation, Photo-oxidation and Photostabilization of Polymers, Interscience, New York, 1975.
15. J.J.C., Antioxidants: Synthesis and Applications, Noyes Data Corp., Park Ridge, N. J., 1975.
16. C.H. Bamford, Ed., Comprehensive Chemical Kinetics, Vol. 14., Degradation of Polymers, Elsevier, New York, 1975.
17. G. Geuskens, Ed., Degradation and Stabilization of Polymers, Appl. Sci. Publ. Barking, England, 1975.
18. W.Ranney, Antioxidants, Noyes Data Corp., Park Ridge, N.J., 1979.

VI.2. Journal Literature

1. K. Wisseroth, Thermal Decomposition of Polyethylene and Polypropylene, Angew. Chem. 72(1960)866-7.
2. F. Bueche, Mechanical Degradation of High Polymers, J. Appl. Polym. Sci., 4(1960)101-6.
3. C.A. Russell, J.V. Pascale, The Early Stages of Isotactic Polypropylene Oxidation, ACS, Div. Polym. Chem. Prep. 2(1961)19-23.
4. J.H. Chaudet, Mechanisms of UV Stabilization of Plastics, SPE Trans. 1(1961)26-30.
5. J. Lunenschloss, Weathering Fastness of Polypropylene, Textil-Praxis, 16(1961)875-77.
6. F.R. Eirich, H.F. Mark, Principles of Polymer Stability at Elevated Temperatures, Soc. Chem. Ind. Monograph 13(1961)43-59.
7. J. van Schooten, P.W.O. Wijga, Degradation of Polypropylene, Soc. Chem. Ind. Monograph No. 13(1961)432-52.
8. A.F. Strobel, S.C. Catino, Relationship Between UV Absorber, Structural Types and Photostability of Plastics, Ind. Eng. Chem. Prod. Res. DCV. 1(1962)241-48.

9. C.W. Taylor, The Resistance to Light of Ulstron Polypropylene Yarns, Textil-Praxis, 17(1962)989-90.
10. V.S. Nanda, A Theoretical Study of the Changes in the Statistical Character of a Polydisperse Polymer System Undergoing Random Degradation, Roy. Soc. Proc., Ser. A-270(1962)14-26.
11. J.A. Melchore, Weathering and Stabilization of Polyolefins, Ind. Eng. Chem. Prod. Res. Devel., 1(1962)232-235.
12. N.Z. Searle, Bibliography on UV Degradation and Stabilization of Plastics SPE Trans., 2(1962)32-54.
13. T.E. Davis, et al, Thermal Degradation of Polypropylene, J. Polym. Sci., 56 (1962)485-500.
14. D.E. Brown and N.P. Navreiter, Stabilization of Polypropylene, ACS, Div. Polym. Schm. Prep., 4(1963)310-316.
15. C. Booth, The Mechanical Degradation of Polymers, Polymer, 4(1963)471-478.
16. F.H. Winslow, et al, Effect of Morphology on Oxidative Crystallization of Polyolefins, ACS Div. Polym. Chem., Prep., 4(1963)706-716.
17. N.T. Notley, Thermal Oxidation of Polyolefins, Trans. Fared. Soc., 60 (1964)88-92.
18. J.P. Forsman, Oven Aging of Polypropylene, SPEJ, 20(1964)7-9.
19. R.H. Hanse, The Effect of Morphology, Antioxidation, and Copper on the Oxidation of Polypropylene, Proc. Symp. Polypropylene Fibers, Birmingham, Alabama, (1964)137-182.
20. G.C. Massen, Antioxidant, Encyl. Polym. Sci. Tech., 2(1964)171-197.
21. J. Bailey, Mechanism of the Thermal Decomposition of Polypropylene, Proc. Symp. Polypropylene Fibers, Birmingham, Alabama, (1964)121-136.
22. F. Geleji, Synergistic Stabilization Systems for the Protection of Polypropylene from Thermal Orientation, J. Polym. Sci., 7(1965)3695-3707.
23. H.H.G. Jellineck, Some Aspects of Polymer Degradation Processes, Am. Ass. Test. Mat. Spec. Tech. Publ. No. 382(1965)3-12.
24. R.H. Hansen, Stabilization of Polypropylene, Polym. Eng. Sci., 5(1965)223-226.
25. R.D. Deann, Stabilization of Plastics, SPEJ, 22(1966)12-17.
26. G. Kust, Weathering Trials with Threads and Cords from Polyolefin Fibers, Textil-Praxis, 20(1965)4-7.
27. C.A. Russell, Effects of Stearate Processing Aids upon Polypropylene Stability, SPE Trans., 5(1965)84-89.
28. R.H. Hansen, Determining Effectiveness of Antioxidants in Polyolefins, Mod. Plast., 42(1965)134-140.
29. J.W. Tamblyn and G.C. Newland, Induction Period in the Aging of Polypropylene, J. Appl. Polym. Sci., 9(1965)2251-2260.
30. J.P. Forsman, Effect of Mounting Technique on the Laboratory Weathering of Polypropylene Monofilaments, J. Appl. Polym. Sci., 9(1965)2511-2525.
31. H.J. Oswald, Deterioration of Polypropylene by Oxidative Degradation, Polym. Eng. Sci., 5(1965)152-158.
32. G.C. Newland, Metal-Organic Stabilizer and Antistabilizers for Polyolefin Plastics, J. Appl. Polym. Sci., 9(1965)1947-1953.
33. J.P. Forsman, Effect of Mounting Techniques on the Laboratory Weathering of Polypropylene Monofilaments, J. Appl. Polym. Sci., 9(1965)2511.
34. B. Schmitt and O. Brehm, The Heat Stability of Polyolefins Determined According to DIN 53381, Sheet 3, Kunststoffe, 55(1965)660-661.
35. M.R. Kamal, Effect of Variables in Artificial Weathering of Selected Plastics, Polym. Eng. Sci., 6(1966)333-340.
36. S.L. Fitton and W. Taylor, Thermal Stability of Polypropylene, Plastics 31(1966)1139-1142.
37. J.M. Heaps and A. Austin, Degradation of Polyolefins, Weathering Degradation Plast. Symp., (1966)104-118.
38. P.G. Kellehar, Thermal Oxidation of Thermoplastics, J. Appl. Polym. Sci., 10(1966)843-857.
39. J.P. Forsman, Causes of Polypropylene Oven-Aging Variability, Am. Soc. Test. Mater. Spec. Tech. Pub. 375(1964)73-80.

40. R.J. Martinovich, Weatherability of Plastics: Practical Approach to the Study of Polyolefin Weatherability, Mod. Plast. 44(1967)155.
41. J.C.W. Chien and C.R. Buss, Polymer Reactions V. Kinetics of Auto-oxidation of Polypropylene, J. Polymer Sci., 5(1967)309-310.
42. G.C. Newland, Weatherability of Plastics. Actinometry of Sunlight, Modern Plastics, 44(1967)145.
43. F.H. McTigue, M. Blumberg, Factors Affecting Light Resistance of Polypropylene, Appl. Polym. Symp., (1967)175-188.
44. F.H. McTigue, Weatherability of Plastics. Light Resistance of Polypropylene, Modern Plastics, 44(1967)145.
45. J.C.W. Chien, Polymer Reactions, VI. Inhibilet Auto-oxidation of Polypropylene, J. Polym. Sci., 5(1967)1683-1697.
46. J.P. Forsman, Measurement of Polypropylene Oxidative Stability, Mod. Plast., 44(1967)955-958, 244,246.
47. D.A. Gordon and E.C. Rothstein, Stabilization of Polypropylene, Polym. Eng. Sci., 6(1966)231-239.
48. L. Balahan, Photooxidative Degradation of Polypropylene, J. Polym. Sci., 22(1967)1059-1071.
49. P.J. Brigg, Mechanism of Photostabilization of Polypropylene by Nickel Oxime Chelates, J. Appl. Polym. Sci., 12(1968)1825-1833.
50. M. Heskins and J.E. Guillet, Mechanism of UV Stabilization of Polymers, Macromol., 1(1968)97-98.
51. A.C. Morris, Environmental Resistance of Polypropylene, Brit. Plast. 41(1968) 85-86, 89-91.
52. R.W. Moncrieff, Improving the Thermal Stability of Polypropylene, Textile Mfg., 94(1968)33-34.
53. J. Kresta and J. Majer, Effect of Titanium Compounds on Thermal Oxidation Properties of Stabilized Polypropylene, J. Appl. Polym. Sci., 13(1969)1859-1872.
54. J. Bonkowski, Light Stabilization of Polypropylene Multifilaments, Text. Res. J., 39(1969)243-247.
55. R.P. Weimer and W.P. Conner, Launderability of PP Multifilaments, Text. Res. J., 39(1969)1150.
56. E. Hemingway, UV Absorbers, Rep. Prog. Appl. Chem., 54(1969)159-163.
57. T.C. Virgil, Patent Aspects of Plastic Stabilizers, Advanc. Chem. Ser. No. 96(1969)76-78.
58. Y. Kato, Photo-oxidation of Polypropylene: Some Effects of Molecular Order, J. Appl. Polym. Sci., 13(1969)1447-1458.
59. A. DiBattista, A New Method of Determining Oven Aging of Stability of Polypropylene Monofilaments, Polym. Eng. Sci., 10(1970)332.
60. F.J. Golemba, UV Stabilization of Plastics, SPE J, 26(1970)88-91.
61. D. Rysany, Oxidative Thermal Stability of Isotactic Polypropylene as a Function of the Composition of the Stabilizer Mixture and the Temperature of Observation, Kunststoffe, 60(1970)118-120.
62. J.E. Goudrich, Torque Rheometer Evaluation of Polypropylene Process Stability, Polym. Eng. Sci., 10(1970)215.
63. R.D. Deanim, et al, Mechanism of UV Stabilization of Plastics, Polym. Eng. Sci., 10(1970)228-234.
64. C.W. Uzeimeier, Effect of Pigments on the Heat and Light Stability of Polypropylene, SPE Annu. Tech. Conf. Pap., 28(1970)29-33.
65. L.I. Nass, Stabilization, Encyl. Polym. Sci. Tech., Vol. 12(1970)725-768.
66. W. Kloepffer, Recent Theories on the Mechanism of UV Stabilization of Plastics, Kunststoffe, 60(1970)385-388.
67. D.A. Gordon, UV Absorbers, Kirk-Othmer Encycl. Chem. Tech. 2nd Ed., 21(1970) 115-122.
68. G.R. Lappin, UV Radiation Absorbers, Encyl. Polym. Sci., Tech., 14(1971)125-148.
69. G. Scott, New Concepts in Polymer Stabilization, Brit. Polym. J., 3(1971)24-27.
70. J.P. Guillory and C.F. Cook, Mechanism of Stabilization of Polypropylene by UV Absorbers, J. Polym. Sci., 9(1971)1529-1536.
71. L. Reich, et al, Kinetics of Thermal Oxidation of Polyolefins. Review, Polym. Eng. Sci., 11(1971)265-273.

72. G. Scott, Mechanism of Polymer Stabilization, Pure Appl. Chem., 30(1972) 267-289.
73. F.R. Mayo, Oxidation of Polypropylene, J. Polym. Sci. Polym. Lett. Ed. 10(1972)921-923.
74. J.C.W. Chien, Autoxidation of Polypropylene, J. Polym. Sci. Polym. Lett. Ed., 10(1972)915-919.
75. G. Scott, New Concepts in Polymer Stabilization, Brit. Polym. J., 3(1971) 24-27.
76. L. Goldfarb, Polypropylene Oxidation Kinetics, J. Polym. Sci., 10(1972) 3289-3294.
77. G. Scott, Mechanisms of Polymer Stabilization, Pure Appl. Chem. 30(1972) 267-289.
78. B. Miller, Autoignition of Polymers, J. Appl. Polym. Sci., 17(1973)628-642.
79. C. Savides, Guide to UV Stabilizers, SPE J, 29(1973)38-43.
80. J. Pospisil, Transformation of Phenolic Antioxidants During the Inhibited Oxidation of Polymers, Pure Appl. Chem. 36(1973)207-232.
81. P. Vink, Changes in Concentration of Some Stabilizers During the Photo-oxidation of Polypropylene Films, J. Polym. Sci. Symp., 40(1973)169-173.
82. O. Cicchetti, Titanium-Catalyzed-Inhibited Autoxidation of Polypropylene and its Models, Europ. Polym. J. 9(1973)1205-1229.
83. D.C. Mellor, et al, Effect of Processing Conditions on the UV Stability of Polyolefins, Eur. Polym. J., 9(1973)219-225.
84. H. Heller and H.R. Blattman, Some Aspects of Stabilization of Polymers Against Light, Pure Appl. Chem., 36(1973)141-161.
85. D.J. Carlsson and D.M. Wiles, The Involvement of Singlet Oxygen in the Photodegradation of Polypropylene Speculation on the Origin of $^1O_2$ in the Polymer, J. Polym. Sci. Polym. Lett. Ed., 11(1973)959-965.
86. K.J. Humphris and G. Scott, Mechanisms of Antioxidant Action. Phosphite Esters, Pure Appl. Chem., 36(1973)103-176.
87. S.E. Guillet, The Role of Energy Transfer in the Stabilization of Polymers, Pure Appl. Chem., 36(1973)127-140.
88. D.J. Carlsson, Photostabilization of Polypropylene-III, Stabilizers and Macroketones, Macromolecules, 7(1974)259-262.
89. J.S. Zannucci, et al, Mechanism of Stabilization of Some Polypropylene UV Stabilizers, Macromolecules, 7(1974)393-394.
90. B. Baum, The Weathering Degradation of Polyolefins, Polym. Eng. Sci., 14(1974) 206-211.
91. A. Tuzzi, Light Stabilizers for Polyolefins, Polym. Age, 5(1974)272.
92. R. Roe, Physical Factors Controlling Retention of Additives in Polyolefins, ACS Coat. Plast. 34(1974)132.
93. R.H. Hansen, et al, Novel Antioxidants for Polyolefins vs. Conventional Antioxidants: Principles and Practice, ACS Org. Coat. Plast., 34(1974)90.
94. A. Adamszyk, F. Wilkinson, Triplet-state Quenching by Photostabilizers of Polyolefins, J. Appl. Polym. Sci., 18(1974)1225.
95. B. Ranby and J.F. Rabek, Some Problems of the Photoxidative Degradation of Polymers by Singlet Oxygen, ACS Org. Coat. Plast., 35(1975)148-153.
96. A. Zweig, Singlett Oxygen and Polymer Photoxidations, I. Sensitizers, Quenchers, Reactants, J. Polym. Sci., Chem. Ed., 13(1975)717-736.
97. D.A. Plank, Polycarbonates. New Concept in Stabilization for Polypropylene, SPE Pap. Reg. Tech. Conf. Baytown, Texas, 1975.
98. W.C. Ray, Phosphites in Polyolefin Process Stabilization, Polym. Eng. Sci., 15(1975)703-707.
99. J.F. Rabek, Role of Singlet Oxygen in Photoxidation Degradation and Photo-stabilization of Polymers, Polym. Eng. Sci., 15(1975)40-43.
100. J.F. Westlake, Diffusion of Stabilizers in Polymers I, in Polyolefins, J. Appl. Polym. Sci., 19(1975)319-334.
101. P.A. Kirpichikov, Kinetic Aspects of the Stabilization of Polymers Using Phosphites, Plastic Kautsch., 22(1975)400-403.
102. J. Durmis, Loss of Light Stabilizers from Polypropylene, Eur. Polym. J., 11(1975)219-222.

103. N.S. Allen, Photoxidation and Stabilization of Commercial Polyolefins, Che. Soc. Rev. 4, 4(1975)533-547.

104. D.J. Carlsson, Photoxidation of Polypropylene Monofilaments Part I. Chemical Changes and Mechanical Deterioration Textil Res. J., 46(1976)590-599; Part II. Physical Changes and Microstructure, Tex. Res. J., 46(1976)641-648.

105. D.J. Carlsson and D.M. Wiles, The Photoxidation of Polypropylene Part I, J. Macromol. Sci. Rev. Macromol. Chem. C-14, 1(1976)65-106; Part II, C-14, 2 (1976)155-192.

106. F. Mitterhofer, Process-stability of Polypropylene, Kunststoffe, 67(1977) 151-153.

107. D.J. Carlsson, et al, Photoxidation of PP Films. VI. Possible UV Stabilization Mechanism, J. Appl. Polym. Sci., 16(1977)615.

108. J.Holcik, Oxidation Thermal Degradation of Polypropylene and the Influence of Transition Metal Chelates, Europ.Polymer J., 14(1978)769-772.

109. N.S.Allen, Photostabilization of Commercial Polypropylene by a Hindered Amine Stabilizer, J.Appl.Polym.Sci., 22(1978)3277-3282.

110. P.Maltese, Light Stability of Polypropylene Fibers and Carpets: Correlation between Artificial and Natural Weathering, Plast. Rubber Inst., London, Conf. Polyp. Fibers & Textiles, York, Sept. 26-28, 1979.

APPENDIX I

Suppliers of Antioxidants and UV Stabilizers in the United States

American Cyanamide Co., Bondbrook, N. J., 08805
Argus Chemical Corp., Div. Witco Chemicals, 633 Court St., Brooklyn, N. Y. 11231
Ashland Chemical Co., P. O. Box 2219, Columbus, Ohio 43216
Borg-Wagner Chemical Corp., Parkersburg, W. Va. 26101
Ciba-Geigy, Saw Mill River Road, Ardsley, N. Y. 10502
Cincinnati Milacron Chemical Co., West Street, Cincinnati, Ohio 45215
Eastman Chemical Products, Kingsport, Tenn. 37662
Evans Chemical Inc., 90 Tokanake Rd., Darien, Ct. 06820
GAF Corp., 140 W. 51 St., New York, N. Y. 10020
B. F. Goodrich Chemical Co., 6100 Oakitru Blvd., Cleveland, Ohio 44131
ICI United States Inc., New Murphy Road, Wilmington, Del. 19897
Sandoz Color & Chemicals, 10 E, Hanover, N. J. 07936
Uniroyal Chemical Div., Spencer St., Naugatauck, Ct. 06770
R. T. Vanderbilt Co., Winfield St., Norwalk, Ct. 06855
Ferro Chemicals, One Erie Plaza, Cleveland, Ohio 44114

Chapter III PIGMENTATION

I. INTRODUCTION

Convenient, inexpensive methods for providing a variety of stable colors are essential for large-volume usage of any textile fiber. Dyeing methods useful for polypropylene fiber have been investigated by many laboratories throughout the world but, because of the nonpolar nature of the polypropylene molecule, satisfactory results still challenge the ingenuity of the polymer chemist and dyeing technologist. In spite of the recent introduction of some dyeable polypropylene fibers by Hercules, Montecatini, Toyobo, Phillips, and others, most coloring of commercial polypropylene fiber is carried out *en masse*, known in the industry as "dope dyeing", "mass coloration", or "pigmentation". It involves the intermixing of colorants with the polymer prior to fiber extrusion. The admixing of colors with the polymer prior to fiber formation is not unique to polypropylene; the same technique is applied to large-scale coloration of viscose and cellulose acetate fibers (hence the term "dope dyeing"). This method has the disadvantages in complicating production and requiring an inventory of fibers in multitudes of colors. However, the light, wash, and dry-cleaning fastness is outstanding. The cost of coloring fibers by this method has also been found to be quite reasonable.

With increasing experience in production of melt-colored fibers and the availability of an increasing variety of colorants, the feeling that dyeing methods are absolutely essential for all large-volume applications is not so strong as it once used to be. In fact, the success of pigment coloration of polypropylene has been such that nylon and polyester fiber producers are now also turning to this technique to achieve fast-colored fibers for certain large-volume usages. These "producer dyed" or "bulk dyed" fibers are slowly making inroads in difficult colors, such as black.

The rapid growth of polypropylene fiber in tufted and nonwoven carpeting and upholstery fabrics, which are still the principal areas of application for colored fibers, has been dependent upon the development of pigmentation techniques. Pigment colors have economically provided the aesthetics which have made possible utilization of the desirable technical characteristics of polypropylene in these applications. Pigmentation provides color permanence not attainable any other way. The light-stability and color-fastness properties in washing and dry cleaning generally exceed the useful life of the product. Since the pigments are uniformly distributed throughout the fiber, crocking or change of colors with wear is generally no problem. There is also no adverse effect on the inherent chemical and biological resistance of polypropylene fiber.

The addition of colors by the fiber producer gives him the responsibility to provide the colors desired by the textile trade, and to provide them with uniformity and reproducibility. This means added cost not only for the pigments, but also for the color matching and for the complications resulting from production and stocking of fibers in many colors. Some of these problems, such as coloring cost, are merely a transfer of expenses from the dyer to the fiber manufacturer and by no means represent an overall extra cost to the finished goods. The problem of inventorying a multiple color stock is a real one, although it is not so serious as it initially appears to be. In the home furnishing field, in which most of the colored polypropylene fiber is currently being consumed, only ten to twenty colors are in fashion each year. The inventorying of this limited number of colors is, therefore, not a big complication. For large numbers of customers, the inconvenience of purchasing the stock-colored fibers is balanced by elimination of the cost of carrying out fiber or fabric dyeing. The problems of frequent color matching and inventory will become serious, however, as polypropylene fiber gains acceptance in the apparel field. Here the frequency of fashion changes in color is quite high and it will not be possible to satisfy the customer's demand with only a score of colors.

Pigmentation of polypropylene for the production of fibers is connected with the dry coloration of plastics on one hand and coloration of fibers in a conventional sense on the other hand. It uses the technology and know-how of pigmentation of plastics but keeps in view the particular requirements of the textile industry. Every fiber manufacturer has his own way of doing things in this respect. Different procedural and material preferences have emerged which are the outcome of long compounding experience of the respective companies. The key ingredient of any successful pigment formulation is experience and the keynote of pigmentation technology is the shear routine work of formulation, evaluation, and reformulation. The basic principles of color perception, pigment selection, polymer-pigment interaction, pigment-stabilizer compatability, etc., however, help to make the progress faster. These principles will be discussed in brief in this chapter. In spite of the various terms commonly employed in the industry to describe the process of coloring of polymer prior to fiber formation, the term "pigmentation" will be used in this discussion. Pigmentation of polypropylene for fiber production includes not only the process of imparting color by admixing dry colorants with the polymer, but also the incorporation of dulling agents or optical brighteners.

II. COLORANTS

Theoretically, two basic families of colorants can be considered for the mass coloration of polypropylene: dyes, and pigments. Dyes are synthetic organic chemicals soluble in most common solvents. They are characterized by good transparancy, high tinctorial strength and low specific gravity. But their inherent disadvantages specially their low heat resistance and their tendency to migrate and to sublime, limit their use in mass coloration of polypropylene. The colorants which are best suited for the melt coloration of polypropylene, and used most in practice, are pigments. Pigments are intensely colored solids, insoluble in the medium in which they are used, which produce color by existing as discrete particles in the medium, selectively absorbing and reflecting specific areas of light energy from wavebands which constitute white light. Pigments, because of their insolubility, must be uniformly and finely dispersed in the polymer by mechanical means to provide optimum properties.

As is well known, pigments can be divided into two main chemical classes: inorganic pigments and organic pigments. While the two groups differ widely in their properties from each other, the individual pigments have much in common within a particular group of chemical constitution. The inorganic pigments are insoluble in common solvents while the majority of the organic pigments exhibit some degree of solubility in a few organic solvents. Inorganic pigments are also more temperature-resistant than the organic pigments. Color intensity, on the other hand, is generally better in organic pigments than that in inorganic pigments. Added to these groups of pigments, which are used to impart color to the fiber, there is another group of compounds which are used much like a pigment but are devoid of any color. They are called "optical brighteners" or "optical brightening agents" and are used to brighten the whiteness of the fiber. They are exclusively organic compounds.

The behavior of pigments as colorants is related to two fundamental factors: light absorption and light scattering. The coloring material absorbs certain wavelengths of visible light and reflects or transmits the other wavelengths. The wavelengths absorbed is a function of the chemical constitution, the crystal structure, and even the particle size of the pigment. Pigments also scatter light. This scattering varies with wavelength but is primarily a function of the refractive index difference between the pigment and the polymer, and the particle size and particle shape of the pigment. A pigment which provides a large refractive index difference between the pigment and the polymer, and which has a particle size of about half the wavelength of light, will provide the most efficient scattering of light and, therefore, will have maximum opacity. Pigments which do not selectively absorb light energy but have a strong scattering power are used as delustering or dulling agents. The classical example of a pigment with these properties is Rutile TiO_2.

In general, the chemical constitution of a colorant is the prime determinant of its color. However, the particle size plays a great role in influencing the hue, color intensity, and transparency or opacity. The particle size also determines the tendency of the pigment to flocculate and aggregate to form lumps and rocks. Apart from the average particle size of the colorant, the particle size uniformity contributes to the purity of the color.

II.1. Inorganic Pigments

Inorganic pigments are natural or synthetic metallic oxides, sulfides and other salts. They are insoluble in common solvents as well as in polypropylene. They are in general superior to organic pigments in heat and light stability, weathering and migration resistance, and they are characterized by high specific gravity and good chemical resistance. This family of colorants produces more opaque colors than the other type. Its tendency to plug the filter pack of the spinnerette assembly is considerably higher than that of organic pigments. This is because of their nonfusible nature.

Inorganic pigments are relatively easy to disperse. When used in high concentrations, they yield considerably bright colors, although these may not be as bright as those possible with organic pigments. Tinctorially, the inorganics are weaker than the organics. Table I lists some of the important inorganic pigments which are suitable to varying degree for polypropylene coloration. The most important inorganic pigments are rutile-grade titanium oxide, zinc oxide, (used as delustering agents), cadmium yellows, chrome yellows, and titanates of barium, zinc, colbalt, and nickel (for yellows and oranges), cadmium and mercadium reds, molybdate red (for reds and oranges), iron oxides (for brown and black), and carbon black (for olives, greys, and blacks).

II.2. Organic Pigments

Several types of organic pigments have been developed recently and quite a few of them are well suited to polypropylene coloration. Organic pigments vary greatly in chemical structure, thereby accounting to a large degree for their wide range of colors and other properties. Products of similar composition can be expected

TABLE I

Inorganic Pigments Suitable For Polypropylene

<u>White</u>

Titanium Dioxide, Rutile
Zinc Oxide
Antimony Oxide

<u>Black</u>

Channel Black
Furnace Black

<u>Yellows and Orange</u>

Cadmium Sulfo-Selenide
Chrome Yellow
Chrome Orange
Molybdate Orange
Strontium Yellow
Zinc Chromates
Ni-titanium Yellow
Iron Oxide
Neaple Yellow
Primrose Cadmium Yellow
Mercadium Yellow, Orange
Zinc Titanate Yellow
(Zinc-Iron) and Titanium Antimony Yellows

<u>Violet, Maroons, Reds</u>

Mercadium Reds
Cadmium Selenide
Strontium Rubine Red
Iron Oxide

Remarks:

1) *TiO_2 pigments should be used in low concentrations for delustering only. Zinc oxide can be used at high concentration. Antimony oxide comes in question mainly for flame retardant compositions.*
2) *Furnace blacks have generally replaced channel black in the USA. Beware of the quality of dispersion!*
3) *Cadmium and mercadium yellows and oranges are excellent at all pigment levels but are not recommended for outdoor uses.*
4) *Lead chromate yellows and molybdate oranges have poor heat stability limiting their use to low extrusion temperatures.*
5) *Strontium rubines are limited to deep shades and low extrusion temperatures. However, the color is inexpensive and should not be overlooked for special applications. Iron oxide can be used for rust and gold colors but care should be taken with regard to stabilization.*

to behave alike; hence classification according to chemical type, irrespective of color, is useful in predicting performance. The commercially available organic pigments which can be considered for the coloration of polypropylene are classified below in approximate order of their date of introduction.

Phthalocyanine
Vat (Anthraquinone, Thioindigo, Perinone and Perylene)
Dioxazine
Azo
Quinacridone
Isoindolinone

Compared to inorganics, organic pigments are generally lower in specific gravity, higher in surface area, more transparent, stronger in color, higher in price (although not necessarily more expensive to use), and clearer in shades. However, there are many exceptions to these generalizations and the formulator would do well to consider at least the basic chemical types within each group for more reliable prediction of performance.

Organic pigments suffer from relatively low resistance to thermal decomposition, sublimation, solubility in organic solvents, and low light fastness. The pigment chemist, by preparing pigments of very much larger molecular size, is overcoming some of these problems. This search for larger molecules stems from the elucidation some 30 years ago of the structure of the phthalocyanines and from the realization that their excellent properties were due in part to their large size. Gaertner (ref.1) has reviewed the development of the modern yellow, orange, red, and purple organic pigments having very similar properties to those of blue and green phthalocyanines. Reeve (ref.2) has recently described the chemical structures of some commercial organic pigments, suitable for polymeric materials. At present, about all of these modern high-grade organic pigments are expensive and have not automatically displaced inorganic pigments, although for some uses they may be cheaper in the long run, or their use may be indispensable irrespective of relative costs.

Organic pigments are finding increasing use in the plastics industry but due to their relatively low heat stability their use in the coloration of polypropylene is limited. The most suitable and widely used organic pigments for polypropylene are blues, violets, and greens based on phthalocyanine, quinacridone and anthraquinone, and one or two reds and oranges based on perylene and perinone. Table II lists a few of the organic pigments which are suitable for coloring of polypropylene.

TABLE II

Organic Pigments Suitable For Coloration of Polypropylene

Violet, Maroons, Reds	Orange, Yellow
Quinacridone Red	Benzidine Yellow Analine
Quinacridone Violet	Benzidine Yellow HR
B.O.N. Red (2B.Ba salt)	Benzidine Orange
B.O.N. Red (2B.Ca salt)	Dis-Azo Yellow, Orange, Red, Brown
Helio Bordeau Maroon	Pyranthrone Orange
Thioindigo Red	GR Parinone
Thioindigo Violet	Isoindilinone Yellow, Orange
Dioxazine Violet	Flavanthrone Yellow
Red Lake R	Anthrapyrimidine Yellow
Anthraquinone Red	
Isoindolinone Red	**Blue, Green**
Pyrylene Reds	
Isoviolanthrone Violet	Phthalocyanine Blues
Perinone Red	Phthalocyanine Greens
	Indanthrone Blue

Remarks:

1) *Perylene reds and dis-azo reds are excellent in deep shades. Perylene reds are excellent in most concentrations for tints and beiges.*
2) *Organic yellows are best for deep, high-luster colors but should be avoided for tints and beiges.*
3) *Quinacridone and carbazole violet pigments have excellent pigmenting properties in full shades and in tints.*
4) *Phthalocyanine blues and greens are excellent for all turquoise and aqua, and for full shades or tints of blue.*

II.3. Optical Brighteners

In addition to the pigments used for coloring and delustering of polypropylene, there are some organic pigments available which are primarily used as optical brighteners for fiber. The optical brighteners are a special class of essentially colorless fluorescent organic substances which absorb UV radiation of about 3000 Å and emit this as visible radiation below 5500 Å. Brighteners are useful in correcting discoloration produced by some stabilizers and other additives in polypropylene, to enhance the whiteness of the fiber, and to brighten the color of the dyed or pigmented fiber. Several types of optical brighteners are available commercially or in developmental quantities. Some of them are quite stable at high temperatures and can be used for polypropylene. The structures most suitable for polypropylene are the so-called nonreactive, as opposed to those structures which are used for optical brightening of nylon, either in mass or from dye bath.

The efficiency of whitening agents is dependent on several factors, the most important being the concentration of the brightening agent and the presence of UV absorbers in the polymer formulation. The concentration at which an optical brightening agent should be applied depends on its chemical structure; the normal concentration level is between 0.02 to 0.2 percent. Since optical brighteners are only activated in the presence of UV light, their effectiveness is greatly reduced in the presence of strong UV absorbers and TiO_2. Fig. 1 shows UV absorption over the range of 300 to 390 mm for two UV stabilizers widely used in the stabilization of polypropylene and that of a bis (benzoxazolyl) derivative optical brightener. When the absorption ranges of UV stabilizer and optical brightener overlap, the latter is often deprived of UV light with the result that the degree of whiteness attained is lower than that which would be possible in formulations not containing any UV absorbers. Because of this limitation, the use of optical brightening agents is usually restricted to only those formulations which do not contain any UV absorbers in their stabilization recipes. Optical brightening agents are best compounded in conjunction with nickel or hindered amine stabilizers where the use of UV absorbers is only optional.

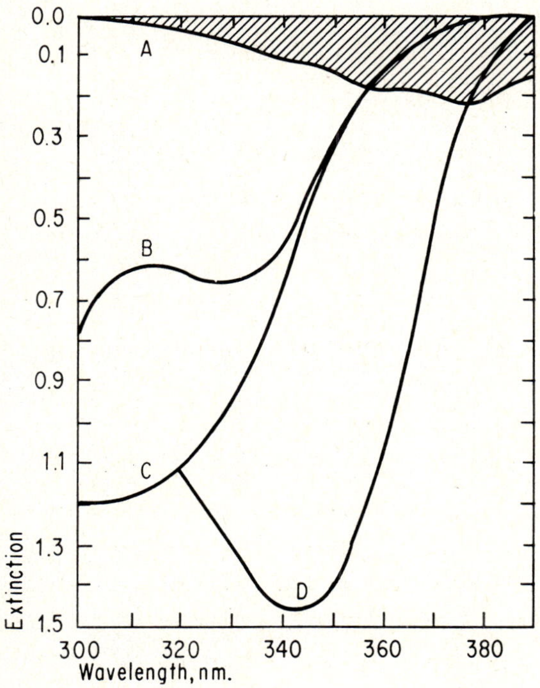

Fig. 1. UV Absorption of OBA and UV Absorbers (ref.34).
A) 0.0002% OBA
B) 0.002% UV stabilizer, benjophenone type
C) 0.002% oxalic acid-diarylamide
D) 0.002% UV stabilizer, benzotriazole type

III. CHOICE OF PIGMENTS AND COLOR MATCHING

In selecting a specific pigment for use as colorant for polypropylene fiber, consideration has to be given to many factors pertinent to processing conditions and end-use environments of the product. A selection based solely on hue, brightness, tinctorial strength, transparency, etc., would in many cases lead to disasterous affects. Thus, once the tinctorial properties of a pigment have been established, consideration must be given to its fastness properties, its resistance to heat, light, moisture, household chemicals, etc., and its interactions with the polymer or with the stabilization system. These requirements are briefly discussed in this section.

III.1. Color Characteristics

The color characteristics of a pigment include a description of its opacity, relative tinctorial strength, and optical effects (e.g., irridescence, metamerism) in addition to its spectrophotometric response. Tinting or tinctorial strength of a pigment is defined as the degree to which the colorant imparts color to a standard white pigment. It corresponds to the color yield of a pigment. The manufacturers of pigments generally furnish information on tinctorial strength of pigments in the form of their tinting strength ratings. These are obtained by grinding a weighed amount of colorant into a lithographic varnish for a specified length of time. This procedure, however, provides no tangible information on the dispersing properties of the pigment in polypropylene and, therefore, is not a reliable index of tinting strength obtained in the coloration of polypropylene fibers.

TABLE III

Refractive Index of Some White Pigments

Pigment	Refractive Index	Delustring Power
Calcium sulphate	1.59	Increases ↓
Barium sulphate	1.64	
Zinc oxide	2.00	
Lthiphone	2.01	
Antimonioxide	2.19	
Zinc sulphide	2.37	
Titanium dioxide (Anatas)	2.55	
Titanium oxide (Rutile)	2.70	

Although the degree of dispersion has a profound effect on the tinting strength and hue of the pigment, the primary basis for determining its tinctorial characteristics is the chemical and crystalline structure of the colorant. Some pigments are inherently weak in tinctorial strength; ultramarine blue, cadmium sulphoselenide, cadmium mercury pigments, iron oxides, molybdates, cobalt aluminate, Ba, St, Zn chromates, etc., can be cited as examples.

Closely related to tinctorial strength is the hiding power of a pigment. This property is of primary importance in the paint industry; in coloration of fibers, the hiding power is important mainly in the case of white pigments which are used as delustering agents. The hiding, covering, or delustering power of these pigments is a function of their refractive index. As Table III shows, rutile-grade TiO_2 leads other pigments in this respect.

III.2. Weatherability

Weatherability refers to the ability of a pigment to withstand all forms of weathering, including visible and UV light, moisture, temperature, and chemicals in whatever combination they may exist during the life of the product. Weatherability traits of importance include color fastness, resistance to chalking, and retention of physical properties. Weatherability is often a function of the light stability and oxidation resistance of a pigment since these elements are encountered in the course of weathering.

No matter how good its other properties, a pigment is of little value if it will not withstand the effects of exposure to light and weathering. This is specially true for fibers, because the very small cross section of the filament makes the pigment more vulnerable to the action of light and other weathering agents than the thick cross-sectional forms generally encountered in the plastic industry. A large proportion of polypropylene fiber is currently being used in home furnishing and outdoor uses which require particularly prolonged exposures to weathering conditions. In fact it was the exceptional light fastness of polypropylene products which was offered to the consumer as one of the primary strong points of polypropylene over the other manmade fibers to gain a sizable proportion of the home furnishing market.

The light stability ratings of colorants are often expressed as a 1-8 scale: 1 being the poorest and 8 being the highest. Most of the pigments offered commercially are known to command high light-fastness ratings. The quoted ratings for light fastness are, however, not wholly reliable, as light stability of most of these pigments in polypropylene does not correspond with the relative ratings quoted by the pigment manufacturers in other polymeric systems. This is because of a strong interaction of many pigments with polypropylene substrate, the presence of various stabilizers and antistatic additives, and the obviously thin cross section of the fiber. In general the light-fastness ratings of pigments are lower in polypropylene than those in other substrates. A careful testing is, therefore, needed before a pigment can be recommended for production use. Inorganic pigments generally have excellent stability to light. The organic pigments range in quality from poor to excellent according to their structure. Furthermore, a pigment of poor stability to light will appear to be affected more significantly in a formulation containing a white pigment, such as TiO_2.

Fading may be due either to a loss of color of the pigments, to fiber becoming white through surface crazing, or to fading or discoloring of stabilizers present in the fiber. The color change due to fading or discoloration of stabilizers is important only in the case of some nickel-containing stabilizers and to some extent in the case of simple phenolic antioxidants. The problem is, however, not serious in most of the cases.

Some pigments show little change in color on exposure to light alone but they fail rapidly when exposed to outdoor conditions. Cadmium-based pigments can be cited as examples. When used in interior applications they exhibit excellent stability against weathering, both in mass tone and tint applications, to both sunlight and artificial light. Outdoors, in tints or pastels, fading occurs rather rapidly. These changes - fading, darkening, or change in hue - are usually due to the presence of reactive groups on the pigment which are affected by moisture and atmospheric chemicals.

Light fading is often dependent upon the concentration of the pigment employed. As is true in the case of colored fibers of other origins, pastel shades exhibit lower fastness ratings than the heavy shades. The degree of dispersion has also a profound effect on the light stability of the pigment. Highly dispersed pigment would fade more rapidly than poorly dispersed pigment. The inclusion of UV absorbers in pigment formulation to improve light fastness usually gives disappointing results, probably because the amount of absorber in the thin layer of polymer between the surface of the fiber and the first pigment particles is too low to be effective.

III.3. Heat Stability

Heat stability is defined as resistance to change in color or performance as a result of heat encountered during processing. These changes may result from chemical decomposition, crystal change, or interaction with other ingredients in compounded polymer. Heat stability is a function of temperature and time and may affect performance without affecting color.

The temperature and time of compounding during melt spinning of pigmented polypropylene resin into fibers may vary from a few minutes at 250°C to 30 minutes at 300°C. Moreover, the environment is seldom chemically inert; acid, oxidative, or reductive conditions may exist or develop owing to the thermal breakdown of the system. The combined effect of time, temperature, and reactive medium frequently leads to marked color changes in the coloring matter, or even to its complete breakdown. The degree to which a colorant withstands these conditions is termed its stability to processing or more simply, but less accurately, its heat stability.

Because of the high processing temperatures employed in melt spinning, excellent resistance to heat is a prerequisite for a pigment to make it useful for coloration of polypropylene fibers. Most organic pigments can withstand a few minutes at high temperatures, in the range of 250° to 300°C, although prolonged exposure leads to decomposition, darkening, or sublimation. In contrast to the generally poor heat resistance of organic pigments, inorganic pigments exhibit excellent resistance to heat. High spinning temperatures eliminate most organic pigments . The palette of useful pigments for polypropylene, therefore, comprises of mostly inorganic pigments. Of the useful organic pigments stable at the processing temperatures of polypropylene, phthalocyanine green and blue and carbazole violet possess the greatest overall utility, followed by quinacridones, isoindolinones, and perinones. The greatest deficiency is noted in the red and yellow regions of organic pigments. This is certainly an area for future innovative organic pigment technology.

III.4. Chemical Resistance

Chemical resistance refers to inertness and compatability with other ingredients present within the compounded polymer as well as resistance to external environment. It is often associated with heat stability since reaction may take place during high-temperature processing that would not otherwise occur.

Some of the commonly used pigments or the impurities present in them may react with such other additives as stabilizers, antistatic agents, and processing aids present in the compounded polypropylene. The reaction of sulfur-bearing pigments or impurities with nickel stabilizers and the reaction of lead impurities with sulfur-bearing stabilizers have been discussed in Chapter 2. These reactions normally result in severe discoloration of the melt or, in minor cases, in a dull shade. Chrome pigments are normally contaminated with lead compound. They cannot be used, therefore, in combination with those stabilizers which may produce colored sulfides as a result of their interaction. The same is the case with neaple yellow (lead antimonate). Cadmium reds and oranges, as well as mercadium reds, are not suitable in mass tones in conjunction with Ni-complex stabilizers, since the formers are based on sulfides.

There have been reports of chemical interaction of certain pigments with a few antioxidants giving rise to yellow-colored products (ref.3). For example, it has been shown by extraction and chromatographic studies that a portion of 4,4-methylene-bis(2,6-di-t-butylphenol) was converted into hydrogalvinoxyl structure when repeatedly brought in contact with channel black in refluxing hexane solution. It is believed that other phenolic antioxidants are oxidized to quinoid-type structures in the presence of certain pigments (ref.3). Ultramarines are sensitive to hydrochloric acid. If the concentration of polymerization catalyst residue is significant in the polymer, the HCl liberated from $TiCl_3$ of the catalyst residue may interact with the pigment to change its color, hue, or intensity.

Apart from the Ni-complex stabilizers, sulfur-bearing compounds, and catalyst residues, there is also the possibility of interaction of pigments with some other additives or even with other pigments. Common antistatic agents used in polypropylene formulation, for example, behave as reducing agents at high temperature, discoloring certain organic pigments that are normally stable at the processing temperatures. Some stabilized polypropylenes will "base-exchange" with alkaline earth metal toners, changing their color from bluish red to orange. Metal free phthalocyanines will complex with metals (normally lead) from other pigments present in the formulation.

III.5. Catalytic Action on Polypropylene Stability

In addition to thermal- and photodecomposition of colorants to colorless or off-color compounds, some of the colorants are in a position to sensitize the decomposition of polypropylene under the action of light and heat (ref.5,6). There are also some colorants which act favorably in this respect and retard the decomposition of the polymer (ref.7,10). Inorganic pigments are particularly noticeable for their catalytic action on polypropylene stability while some organic pigments are also known to accelerate the decomposition of polypropylene by supplying free radicals to polymer decomposition chain. Some pigments, particularly carbon black and zinc oxide, protect the fiber from photodecomposition by virtue of their high degree of screening power; other organic pigments may help to retard the oxidative degradation of polypropylene by virtue of their ability to scavenge the free radicals produced in the process of oxidative decomposition. Careful attention is, therefore, necessary in the selection of pigments with regard to their degradative and stabilizing action on the polymer during processing and during the useful life of the product.

The effect of pigments on thermal- and photodegradation of unstabilized polypropylene has been extensively studied by Takahashi and Suzaki (ref.4), and

Uzelmeier (ref.3,5). Uzelmeier has concucted detailed studies on the effect of pigments on the heat and light stability of stabilized polypropylene. Table IV summarizes the changes in heat stability brought about by various types and amounts of some common pigments relative to an unpigmented stabilized control sample. The data show that none of the pigments improve the heat stability of polypropylene in the presence of an effective phenolic antioxidant. In fact, serious deleterious effects are noticed in several cases. The pigments, listed in order of decreasing antagonism in Table IV may be roughly divided into three classes - those that are severely prodegradant, including channel black, chromophthal red 3B, quinacridone magenta, phthalocyanine blue, and iron oxide tan; those that are moderately prodegradant, including phthalocyanine green, mineral and furnace blacks, ultramarine blue, and chromium oxide; and those that are slightly prodegradant, including cadmium yellow, mercadium red, and rutile titanium oxide. Of special interest is the channel black which acts as a mild antioxidant in the absence of phenolic antioxidant but causes a 50 percent reduction in heat stability of stabilized polypropylene with as little as 0.10 percent addition.

Uzelmeier (ref.3,5) determined the relative rates of hydroperoxide decomposition for various pigments and compared them with their relative prodegradant activity in stabilized polypropylene. No correlation was found. It thus appears that the reduced stability observed with many pigments in stabilized polypropylene is caused by interactions of pigments and stabilizers rather than the peroxide decomposition or chain starting power of the former. One type of interaction that has been considered is the absorption of stabilizers by pigments (ref. 11,12).

Because of their light-screening power, pigments are expected to contribute favorably toward light stabilization of polypropylene. Table V shows the effect of pigments on the light stability of stabilized polypropylene. It is shown that the light-protective action of pigments is generally realized at a high level of pigment loading but it is generally negligible in thin cross sections or at a low level of pigment loading. Takahashi and Suzuki (ref.4) showed that some pigments, instead of protecting the polymer from actinic degradation, accelerate its decomposition.

It follows from the above discussion that pigments should be carefully examined for their interference with the stability of the polymer. The use of those pigments which excessively accelerate the heat and light degradation of the polymer should be avoided. When a catalysing pigment has to be used for economic or tinctorial reasons, a better degree of stabilization would be in order. It follows further that, to meet the quality control standards of thermal- and photostabilization of the fiber equally with each color, the stabilization formulation must be adjusted to take into account the deleterious effects of the pigment formulation on polymer stability. This adjustment is recommended not only when the

TABLE IV

The Effect of Pigments on the Heat Stability of Stabilized Polypropylene

Pigment	phr	Decrease in Heat Stability, %[1)]	Pigment	phr	Decrease in Heat Stability, %
Channel black	.01	17	Mineral black	.10	46
Channel black	.05	37	Mineral black	.50	25
Channel black	.10	56	Mineral black	2.0	49
Channel black	.25	76	Furnace black	.10	15
Channel black	.50	93	Furnace black	.50	29
Channel black	1.0	95	Furnace black	2.0	51
Cromophtal red 3B	.10	29	Ultramarine blue	.10	15
Cromophtal red 3B	.50	78	Ultramarine blue	.50	15
Cromophtal red 3B	2.0	95	Ultramarine blue	2.0	49
Quinacridone magenta	.10	29	Cr_2O_3 green	.10	5
Quinacridone magenta	.50	51	Cr_2O_3 green	.50	17
Quinacridone magenta	2.0	85	Cr_2O_3 green	2.0	49
Phthalocyanine blue	.10	22	Cadmium yellow	.10	12
Phthalocyanine blue	.50	54	Cadmium yellow	.50	27
Phthalocyanine blue	2.0	85	Cadmium yellow	2.0	10
Fe_2O_3 tan	.10	39	Mercadium red	.10	15
Fe_2O_3 tan	.50	63	Mercadium red	.50	15
Fe_2O_3 tan	2.0	85	Mercadium red	2.0	15
Phthalocyanine green	.10	29	Rutile TiO_2	.10	10
Phthalocyanine green	.50	44	Rutile TiO_2	.50	15
Phthalocyanine green	2.0	56	Rutile TiO_2	2.0	12

1) Relative to unpigmented, heat-stabilized control; 10-mil films, days to embrittlement in a circulating air oven at 150°C.

nature of the pigment changes in the formulation but also when the concentration of the same pigment changes to a considerable degree.

TABLE V

The Effect of Pigments on the Light Stability of Stabilized Polypropylene

Pigment	%	Increase in UV Stability, %[1)]	Pigment	%	Increase in UV Stability, %[1)]
Channel black	.01	70	Rutile TiO_2	.10	60
Channel black	.05	220	Rutile TiO_2	.50	290
Channel black	.10	370	Rutile TiO_2	2.0	600
Channel Black	.25	890	Fe_2O_3 tan	.10	30
Channel black	.50	900	Fe_2O_3 tan	.50	160
Channel black	1.0	1300	Fe_2O_3 tan	2.0	610
Phthalocyanine blue	.10	170	Cr_2O_3 green	.10	40
Phthalocyanine blue	.50	600	Cr_2O_3 green	.50	140
Phthalocyanine blue	2.0	900	Cr_2O_3 green	2.0	510
Mercadium red	.10	70	Quinacridone magenta	.10	30
Mercadium red	.50	390	Quinacridone magenta	.50	80
Mercadium red	2.0	890	Quinacridone magenta	2.0	160
Cadmium yellow	.10	70	Cromophtal red 3B	.10	20
Cadmium yellow	.50	380	Cromophtal red 3B	.50	80
Cadmium yellow	2.0	890	Cromophtal red 3B	2.0	160
Phthalocyanine green	.10	80	Mineral black	.10	50
Phthalocyanine green	.50	290	Mineral black	.50	40
Phthalocyanine green	2.0	900	Mineral black	2.0	70
Furnace black	.10	20	Ultramarine blue	.10	0
Furnace black	.50	190	Ultramarine blue	.50	0
Furnace black	2.0	890	Ultramarine blue	2.0	0

1) Relative to an unpigmented, stabilized control; 10-mil films, Carbon Arc Weather-Ometer hours to Brittle flex failure.

III.6. Effect on Tensile Properties

Pigments can have significant effect on the mechanical properties of the fiber. Most of the pigments tend to impair the drawability and tensile properties of polypropylene fibers (ref.13). Similarly it has been shown that increasing amounts of pigments increasingly impair tensile properties. The deleterious effect of pigments on tensile properties of the fiber may have several reasons. The pigment particles and agglomerates can serve as weak points in the structure, the physical properties can be impaired through the induced decomposition of the polymer in the presence of pigment, or the fiber properties may be affected through morphological changes in the fiber. In the latter case, the drawability may not always decrease and the tensile strength may even experience a slight gain.

III.7. Rheological Properties

Flow properties of polypropylene melt are sometimes profoundly affected by the presence of certain pigments in the formulation. This is mainly due to the stabilizer/pigment interaction or to catalytic decomposition of the polymer, as described previously. The change of rheological properties through internal lubricating action of pigments is not known. Extensional flow properties of polypropylene are generally affected more by the presence of pigments than by the shear flow properties. An enhanced "rain effect" (nonuniform draw-down of the molten filament) is a common phenomenon. The cause of this kind of flow disturbance is not fully understood but it is theorized that the uneven drawability of the molten jet may be due to microscopic inhomogeneity in polymer viscosity. These microscopic inhomogeneities in polymer viscosity may be caused by localized catalytic degradation of polypropylene surrounding the functional pigment particle. Another obvious possibility is the poor compatability of the pigment or the pigment carrier with the polymer.

III.8. Morphological Properties

Some pigments, by virtue of their nucleating power, may influence the morphological structure of polypropylene fiber (ref.14,15). The fibers, composed of a large number of relatively small crystallites, are expected to show pronounced shrinkage in boiling water and the tape yarns of such morphological structure should have a greater tendency of fibrillation. Phthalocyanine blues and greens are known to produce these effects. On the other hand, quinacridone reds are known to considerably increase the rigidity and decrease the impact strength of polypropylene articles (ref.15).

III.9. Resistance to Migration

There are three types of colorant migration which are important for polypropylene fiber: a) bleeding: the migration of the colorant from the fiber to the material with which it is in contact. Poor wash and dry cleaning fastness is the obvious result of bleeding. Pigments generally do not suffer from this defect. b) crocking: migration of the colorant to the surface of the fiber so that it can be rubbed off. Crocking may be caused by pigment migration, but may also result from poor dispersion, high pigment loading or abrasion or deterioration of the fiber surface, such as through weathering. Poor rubbing fastness, either dry or wet, is the result of crocking. This property is specially important in the case of polypropylene fiber because quite a substantial portion of polypropylene fiber is being used for carpeting and upholstery fabrics. Inorganic pigments are generally not of much concern in this respect but some organic pigments may lack an adequate crock fastness. c) Blooming, chalking, bronzing: a chalky or metallic effect that is produced when a pigment migrates to the surface of the fiber, either in high temperature processing (bronzing) or cold drawing (chalking). This phenomenon is almost specifically associated with organic pigments which are not compatible with polypropylene because of their hydrophylic groups. The blooming of pigments is of special concern in those uses where the product is going to be subjected to some sort of heat treatment, such as autoclaving and latexing. During cold drawing, pigments may bloom or chalk and be deposited on draw rolls and yarn guides.

III.10. Dispersibility

Pigment manufacturing processes are generally designed to produce a particle size that will yield optimum spectral characteristics. This so-called primary particle size ranges from a few hundredths of a micron for most organic pigments to one micron or more for some inorganics (ref.16). However, for various reasons, these primary particles tend to aggregate or agglomerate during manufacture to form clusters of particles several times their intended size. The pigment powder thus consists of aggregate ranging in size from a few microns up to perhaps 200 microns (ref.17). If the user is to obtain full value from the pigment in terms of strength development, transparency, depth of color, brightness, and opacity, sufficient work must be done on the pigment-binder system to disperse those agglomerates to something approaching the particle size of single crystals formed in the original precipitation or calcination processes.

Dispersibility describes the ease with which pigment aggregates can be reduced to an acceptable particle size. The process of dispersion involves the production of discrete particles as near as practical to their primary particle size and maintaining them in that condition. The practical aspects lie in recognizing that the dispersion process is often costly and that reducing the pigment agglomerates to their absolute primary particle size is not essential or possible in practice. The optimum particle size for each pigment will vary, and pigments susceptible to overgrind are known. In most cases the optimum particle size of pigment dispersion for good optical properties generally lies in the range of 0.3 to 0.8 μ (ref.18).

To achieve even coloration, two criteria must be satisfied: the color must be evenly distributed throughout the medium and colorant must be adequately dispersed. Dispersion in this context refers to methods for wetting out and reducing in size the agglomerates and aggregates of dry pigments and not to means of uniformly distributing it in the system after dispersion. It is normally not difficult to control even distribution but problems often arise in dispersion.

Colorists and pigment technologists are aware that the degree of pigment dispersion in a medium has a direct effect on visual properties. Transparency increases with reduction in aggregate size and approaches a maximum as the degree of dispersion reaches the optimum. Tinctorial strength, and along with it, the economy of shade, is affected in the same fashion. What is perhaps not as fully appreciated is the fact that changes in hue also are directly related to dispersion. Blues become cleaner and greener as the dispersion improves, greens and reds become cleaner and yellower. Obviously, in any situation involving close color control from lot to lot, it is of paramount importance that the pigment dispersion be of consistent quality.

In addition to the problems of color intensity and hue, the degree of dispersion has a pronounced effect on the filtration characteristics of the melt. A poorly dispersed pigment would lead to rapid clogging of the filter pack and necessitate frequent interruptions in spinning for filter change. Additional direct effects of inadequate dispersion include high cost, increased color control problems, poor mechanical properties of the fiber, tendency to chalking and crocking, and increased number of thick-and-thin filament sections.

Pigments differ in their ease of dispersion. Pigments such as TiO_2 and cadmium or chrome pigments normally are among the easiest to disperse; others, such as carbon blacks and iron oxides, are the most difficult. Practically any pigment which has a particle size of less than about 0.2 micron will show appreciable aggregation during processing and will present some problem in dispersion. The severity of aggregation is also dependent on the type of pigment. Hydrophilic pigments aggregate severely when dried from aqueous media; organophilic pigments have less tendency toward aggregation as the particle size is reduced.

III.11. Filtration Characteristics

Excessive pressure rise in melt filtration during the spinning of pigmented polypropylene fibers is an ever-present technical problem, necessitating a frequent change of filter and causing a large amount of waste, high labor costs, and increased periods of equipment shutdowns. Filtration characteristics of a pigmented polypropylene blend is a function of the degree of pigment dispersion and, probably more importantly, is dependent on the uniformity of dispersion.

A material with a few, very large agglomerates may not perform in practice as well as one containing a large number of smaller agglomerates. All those factors which influence the degree and uniformity of pigment dispersion also affect the filtration characteristics of the blend.

III.12. Knife Blunting

In the production of polypropylene staple fibers the hardness of the pigment particles results in greatly increased wear on the cutter's knives, resulting in blunting of cutting roller edges in pacific converters. This gives rise to fused and hooked staple ends, causing difficulties in staple processing into spun yarns. Similarly, in tape yarn processes the hardness of pigments is reflected in the increased wear of film slitting blades and fibrillating pins. The effect of blunt slitting blades is seen as ragged edges on the tapes which cause tape snaps during orientation or weaving.

Pigments differ greatly in their knife-blunting effect, depending on their hardness and the quality of dispersion. Some pigment formulations would give a knife life of several days while the others would necessitate a blade change within a few hours. Knife life is, therefore, similar to filter pack life. All pigments increase knife blunting but organic pigments are less damaging than the inorganics. Among inorganic pigments, TiO_2, iron oxides, and chromium oxides are the hardest and the most damaging with respect to knife wear. Carbon black, cadmium sulfide, chrome yellows, and zinc sulfide are less damaging.

III.13. Cost Effectiveness

One of the most important considerations to be kept in view in color formulation is the economy of coloration. Sometimes alternatives do not exist because of the particular demands on processing or tinctorial and color fastness properties of the formulation. But most often there is the alternative of choosing one out of two or more possible pigment candidates for a specific end use and particular, although not the best, processing and color fastness properties. The overall efforts in these cases is to minimize the cost of coloration while meeting the minimum color fastness and processing specifications. Pigments differ widely in their prices. The price tag is influenced by two main factors: a) the chemical nature of the colorant, b) the physical state of the colorant, e.i., whether it is sold as dry color, treated dry color, resin dispersion, color concentrate, solvent dispersion, or aqueous dispersion.

A distinction should be made between the price of the pigment and the price of coloration. A low-price pigment may not necessarily result in a low price of coloration. For example, although the dispersions are comparatively more expensive than the corresponding dry colors, the use of the former may end up in a low pigment concentration needed for a comparable shade because of the higher color yield of well-dispersed colorants. Very often the higher price paid for a well-dispersed pigment can be offset by improved spinning continuity and low waste. Similarly, the color concentrates may appear expensive in comparison with color dispersions or dry colors, but the savings in blending procedures may offset the additional cost paid for color concentrates.

There is still no hard and fast rule to determine whether inorganic or organic pigments should be used if offered the choice. The former are relatively inexpensive and have excellent resistance to heat, light, and migration. However, they have lower tinctorial strength than the organics. At present, almost all of the modern high-performance organic pigments are rather expensive and have not automatically replaced inorganics in comparative formulations. But by virtue of their high tinctorial power, organic pigments may be cheaper in the ultimate analysis for some cases.

IV. COLOR MATCHING

Useful as chemical classification of pigments may be for understanding and predicting performance, the usual problem facing the formulator is that the color he must match has been predetermined. He is faced with selection of pigments that will produce that shade and offer the properties desired at economical costs. This exercise is called color matching.

Simply stated, color matching implies : Color A = Color B
although in actual practice Appearance A = Appearance B is more true.
The problem in color matching is: given Color A (or more correctly, Appearance A), how are the pigment variables (type of pigments, degree of dispersion, quantity, ratios, etc.), adjusted to reproduce this visual respons? This is a very complicated problem for which no universally accepted solution can be given.

The selection of pigments will depend on the following factors (ref.15):

1) The sort of color to be matched. Is it opaque, transparent, or translucent?

2) If it is transparent, can requirements of heat and light stability be met with available organic pigments?

3) If it is opaque, are there pigments in stock which are bright enough to achieve the required color, or will an organic pigment in combination with a delustering agent have to be used?

4) If a translucent color is needed, what additional requirements are there? Will the color have to be similar when viewed in reflected and in transmitted light?

5) How metameric will the matching be? Will the metamerism be an important consideration in this instance, and can anything be done about it anyway?
6) How accurate must the match be and how consistent must successive production runs be?

It is not easy for the colorist to obtain reliable answers to all these questions initially, but they emerge as he attempts to obtain approval for the matching.

A color formulator can arrive at the proper choice of a pigment recipe to match a given color by two routes. One is the time-tested method of experience; the other is by using color measuring instruments and computers. The coloration of polypropylene for fiber production is currently done mainly through the first approach although the second approach is gradually replacing the first.

Regardless of which method is followed, the pigment formulator always starts with a sample of the color the marketing department wishes him to reproduce. This is commonly in the form of a piece of fabric or a package of yarn. An experienced formulator usually starts with a reasonably good idea of those pigments he needs to match the color. He also knows that he will probably have to undergo considerable trial and error before he finds the exact combination of pigments for his purpose. Frequently, numerous pilot plant batches must be made before the right combination of pigments is established. The task becomes more difficult when a spectral match is required. When the pilot plant batch is satisfactory for color and tested for all other requirements, a production-batch formula is written.

Visual color matching, while fairly simple, is extremely time consuming. Hence, more efficient methods are required. One step toward more efficient color matching was taken some years ago when the role of the spectrophotometer was explored. By comparing spectral curves of in-stock pigments with the spectral curves of the standard, it should be theoretically possible to select the appropriate pigments from the available ones and to calculate their respective amounts in order to achieve a spectral match of the standard. Indeed, computer programs have been written for determining the best formulae to match a given shade. The basic data for these computations are provided in the form of reflectance curves of the individual colorants and the reflectance curve of the standard match. These methods permit the entire process of color formulation to be carried out in 15 to 20 minutes. The first formulation is generally a trial formulation. Further refinements are carried out by visual inspection of the samples.

A given color can be reproduced by using several formulae consisting of different pigments. These formulae may differ in their performance characteristics as well as in cost. The approximate formulae, derived from the color-measuring instruments and computer's calculations, are useful to determine whether, for example, the small amount of the expensive organic pigments or the large amount of the cheaper inorganic pigments is more economical.

A discussion of the theories of color, techniques of color difference measurement, and various practices of instrumental color matching, is not within the scope of this book. Excellent reference works and review papers are available (ref.19-32) and this literature is rapidly growing. Here we will dwell briefly on a few points which are of particular importance to the pigmentation of fibers.

A necessary preliminary step toward successful computer formulation is the development of a pigment library. This consists essentially of data derived from a series of spectrographic curves on each of the pigments normally available to the formulator. It is from these data that the computer selects pigments needed to match a given color. The pigments are dispersed in the polymer in 10 to 15 concentrations, ranging from 0.05 to 2 percent, and spun into fibers. As fiber cross section markedly affects the color strength of the product, a separate color library is required for each type of cross section. Also, as the filament thickness (denier per filament) affects the depth of color, a correlation between filament denier and depth of color is needed for each cross section. These data are referred to as pigment data.

For the measurement of color characteristics of the reference pigment concentrations, the fibers are chopped into small lengths, randomized, and pressed between two optically clear glass plates. In some cases, e.g., in the case of textured yarns, it should be advantageous to randomize the continuous filaments and press them into reference plates. The reference plaque should be thick enough to be opaque. This is necessary because the Kubelker-Munk law, the basic set of equations used for color calculations, assumes an opaque reflecting material. The pigment data must be qualified with the type of stabilization formulation used and the presence or absence of any delustrant in the formulation. The first qualification is needed as some stabilizers have their own color which they impart to the fiber. This correction should be taken into account in calculations. The second qualification is necessary because, unlike a human eye, the color instruments fail to distinguish between translucent and opaque color . In a spectrophotometric sense, white is not a color.

Unfortunately there are numerous limitations which constitute a difficult barrier to realizing the goal of color matching solely by reference to instrumental measurements and computations. Some common causes of difficulty are outlined below.

For computer formulation to be successful, uniform color pigments are necessary. The incoming shipments of pigment dispersions or color concentrates may be within the specifications which have been set forth for grain size, dust percentage, and moisture content. They may however, differ in the average particle size of the dispersed pigment and particle size distribution. The quality of pigment dispersions is difficult to measure, the tolerance limits are of necessity very broad and the control is, therefore, not effective.

Failure to obtain a reliable measure of reflectance of the standard is another difficulty. It is caused by a lack of correspondence of instrument geometry with visual viewing conditions, lack of sensitivity or accuracy of the instrument, and the nature of samples with respect to size or texture.

Variations in the blending process cause other problems. For a number of reasons, such as concentration errors, intensity of blending, or variation in blending time, variations in the extruder's screw temperature and back pressures, and other such factors influence the blending and dispersion of the colorants in the polymer melt.

The most serious problems arise through metamerism, although it is no more a problem to the polymer colorist than to the dyer or to the paint manufacturer. It is the greatest single cause of disagreement between customer and supplier regarding the quality of the match. The color of an object is a cumulative result of the abstraction of certain wavelengths of incident light by the colorant molecules. A true color match is possible, therefore, only if spectroscopically similar colorants are used for the match or those present in the standard. Often it is possible, however, to obtain closely sinilar reflectances without resorting to identical colorants. This type of match is called metameric match. As the spectral characters of individual colorants constituting the metameric match differ from the spectral characters of the individual colorants constituting the standard, the two compounded colors will not appear the same when compared under different lights. The difficulty a color matcher encounters in attempting to produce a metameric match that satisfies the customer in several light sources is evident. Winey and Longley (ref.33) have offered some practical solutions to problem of metamerism. Failure to prepare a match acceptable to the customer also arises when the single sample submitted for matching is near one of the limits rather than the center of the acceptable range.

In spite of the shortcomings of the instrumental methods of color matching, the future in this area is well defined. If color can be measured or quantified, its control becomes a matter of applying the already well-developed methods of statistical quality control. The ability to record, transmit, and statistically evaluate the control data is a major advantage of instrumental methods over visual methods. The greatest economic incentive for applying scientific color measurement methods comes from the day-to-day control of production. In this environment, time is essential and can be equated directly to costs and potential profits.

It must be emphasized that any color formulation corresponding to a particular match is valid only for a particular denier per filament of the yarn or staple.

If the thickness of the filament changes, the color depth of the yarn changes as well. Thicker filaments (high dpf) need lower pigment concentrations than the thinner filaments for equivalent depth of shade. Generally, the color depth is proportional to the cube root ($\sqrt[3]{dpf}$) of the filament denier. Thus, if A g of a pigment are needed to color a yarn of 27 dpf to a particular depth of shade then 3/2 A g of the pigment will be needed to color a yarn of 8 dpf to the same depth of shade. This relationship is, however, an approximation, especially when dealing with those pigment formulations wherein one or two components are present in very minute quantities.

V. DRY COLORS, PREDISPERSED PIGMENTS AND COLOR CONCENTRATES

Pigments are available in three basic forms:

1) Pure or surface treated pigments
2) Color dispersions or predispersed pigments
3) Color concentrates

Coloring with free or surface-treated pigments has been used in the early stages of polypropylene fiber development and is having a revival in high-volume production of pigmented thermoplastics. However, these attempts have not yet met with any degree of success in the production of high-quality pigmented melt which is necessary for a trouble-free spinning operation. For coloring of polymers with free pigments, there must be dispersion in the polymer in the form of prime particles. The pigment agglomerates have to be broken up by high shear forces and it is essential that a polymer film be formed around the dispersed color particles in order to prevent the formation of new agglomerates.

Using an extrusion or screw-type melting device, it is not possible to develop the amount of shear that is required to adequately disperse the pigment in the polymer without causing its excessive degradation. In order to optimize equipment utilization and attain a high degree of pigment dispersion without degrading the polymer, the utilization of a high-shear, slave extruder has been proposed (ref.34). This arrangement offers processing advantages based on specific energy requirements for dispersion and for adequate physical properties of the product. Only a small amount of the total pigmented product (masterbatch) is subjected to the high shear forces necessary to obtain optimum dispersion in the side extruder. In most advanced systems, the pigment concentrate is produced in a side compounder and metered into the main polymer stream for distribution (let down). Because of the economical factors and quality of pigment dispersion attainable, the dry coloration technique is limited in its use for polypropylene fiber coloration.

It is not always appreciated that while the quality of dispersion required in the fiber coloration is as demanding in terms of particle size as it is in the paint industry, the machinery and process generally used in the fiber industry are so much less efficient in doing work on the pigment to break the agglomerates than those used in the paint industry. Pressure has, therefore, been brought to bear on the pigment manufacturers to produce pigments that do not form aggregate during packaging, transport, storage, and blending. And if they do form agglomerates, these can be easily broken down by machinery used in polymer extrusion. This has been achieved by predispersing pigments in suitable media by the pigment manufacturer or pigment blender. These pigments are known in the plastic industry as pigment dispersions or predispersed pigments.

A wide palette of predispersed pigments is available commercially. The general principle of their preparation is to envelop discrete particles of the pigment in a "carrier" or "matrix". It is hoped that the enclosure of the individual particles will prevent them from forming agglomerates during the subsequent stages of processing. The carrier will readily disperse or dissolve in the polymer and thereby release the discrete pigment particles at the best possible moment of the mixing process.

It is particularly important that a suitable carrier be chosen. The required properties of the carrier are manifold. One of the primary prerequisites is its ability to "wet" the pigments. The wetting ability of a carrier depends to a large extent on its affinity for or compatability with the pigment. Polypropylene is a hydrocarbon and completely devoid of affinity for any substances except those of hydrocarbon nature. Polypropylene cannot, therefore, be used as an efficient carrier for dispersion of pigments. The color dispersions used for polypropylene are generally based on emulsifiable low molecular-weight polypropylenes, polyethylenes, or their copolymers. Highly polar materials or the materials widely differing from polypropylene in their thermal characteristics cannot be used as carriers for dispersions. Although the choice of carrier is important, the method of applying the carrier so that each particle of the pigment is coated by the carrier is the secret of success or failure. The pigment industry has recently made available excellent predispersed colors for polypropylene. In appearance most of them are powders or small granulates, although a number of liquid dispersions are also being offered. These are solvent-based dispersions to be metered directly into the extruder.

The use of predispersed colors, as pointed out in the foregoing discussion, is a good starting point to achieve a satisfactory pigment dispersion in polypropylene melt. Blending of polypropylene with pigment dispersions offers no special problem when the polymer and the dispersions are available in comparable particle sizes or when the blending equipment is sophisticated enough to ensure start-to-end and batch-to-batch uniformity. However, when the polymer is available only in pellet form and elaborate pigment dozing equipment is not available, satisfactory blending of pigment dispersion with the polymer becomes problematic. In such cases, a "color concentrate" is often used.

By definition, a color concentrate is a high loading of highly dispersed pigment in a polymer. Several color dispersions and pastes can theoretically be considered as color concentrates. In the plastic industry, however, a color concentrate is known to be a dispersion of pigment or a mixture of pigments in a polymeric medium to which it will eventually be blended in lower concentrations. A large number of pigment concentrates in polypropylene is commercially available. In addition, there are numerous polyethylene-based pigments concentrates available which can be used adequately in the coloration of polypropylene for noncritical uses, such as fibrillated yarns. The color concentrates are generally marketed in granular form. The advantages of color concentrates over pigment dispersions are their ability to blend easily with the bulk polymer, less sensitivity to color matching errors, and completely dust-free operation. On the negative side, the color concentrates are more limited in number and choice than the pigment dispersions, the problem of matching the rheological characteristics of the vehicle with the virgin polymer to be used exists, and above all, the cost is somewhat high.

VI. TESTING AND EVALUATION

The evaluation of a pigment formulation can be grouped into three classes of tests:

a) Process evaluation (heat stability, filter plugging rate, change of color with processing temperature, interaction with stabilizers and other additives, catalytic degradation of polymer, spin continuity, sublimation, chalking on drawing, etc.)

b) Weathering evaluation (effect on light stability of polymer formulation, change in color in polluted atmosphere, etc.)

c) Common textile testing (dry cleaning, wash fastness, light fastness, gas fading, crocking, bleeding, etc.)

The actual details of process evaluation procedures practiced by various fiber manufacturers differ, and it is largely proprietory information. The essence of all these evaluation procedures is the actual production of the fiber and yarn on laboratory or pilot plant equipment under varied processing conditions, from blending to extrusion to texturizing or staple cutting, then observation of the various steps for any anticipated shortcomings in the pigment formulation.

A $1\frac{1}{2}$ inch diameter extruder, fitted with a convenient filter pack and spinnerette assembly and equipped with spin-draw arrangement, is a convenient tool for routine testing for process evaluation. For heat stability evaluation, a series of melt temperatures and throughputs are employed and the subtle changes in color of the fiber and drastic changes in melt-flow index or intrinsic viscosity of the extrudate noted. The same applies to sublimation and chalking resistance, although it may take long spinning runs to detect these deficiencies. The filtration characteristics are normally determined by measuring the rate of pressure rise in the spinnerette pack using a small area of multiple layers of progressively finer screens (up to B.S.N. 400 mesh). The base polymer (stabilized) is taken as the comparison point.

Measuring and counting pigment particles in the extrudate by such techniques as the Quantiment particle-size analyzer (ref.6,23) gives a good approximation of performance but it suffers from the fact that the sample examined is very small. The number of agglomerates causing spinning troubles in a reasonably good material is generally so small that they can easily be missed in a small sample. Break index (counts of threadline breaks per 100 lbs of yarn) or minor break index (the counts of filament breaks without causing a break in threadline) is another indication of the degree of dispersion and compatability of the pigment with the polymer. A change in tensile factor ($TE^{\frac{1}{2}}$) caused by addition of a pigment can also be used as an indication of degree of dispersion, compatability, effect on polymer stability, rain-effect, and morphological effects of a pigment.

The weathering tests for the evaluation of pigments for their effect on light stability are common with those employed for the testing of stabilizers and are generally undertaken concurrently with the latter. In these tests, the changes in color as well as tensile properties are observed. Some pigments are very

TABLE VI

Common Textile Test Methods for Pigment Evaluation

Test	AATCC Test Method
Acid & Alkali fastness	6 - 1957
Bleaching with chlorine	3 - 1962
Crocking fastness	8 - 1968
	116 - 1966
Dry cleaning fastness	85 - 1968
Light fastness	16 - 1964
Phototropism	32 - 1952
Gas fading	75 - 1956
Perspiration fastness	15 - 1962
Wash fastness	61 - 1968
Water & light fastness	125 - 1967

stable to the action of light alone but they are poor under the simultaneous action of light and moisture. Similarly, some pigments are sensitive to sulfur pollution of the atmosphere. This makes the free weathering test mandatory.

The common textile tests for evaluation of light fastness, gas fading, etc., are well known and are listed in Table VI.

VII. REFERENCES

1. H. Gaertner, J. Oil, Col. Chemists Ass., 46(1963)13.
2. T.B. Reeve, J. Macromol. Sci., Rev., Polym. Tech., DI(2)(1971)217.
3. C.W. Uzelmeier, SPE Conf. on "Pigmentation", N.Y., 1970.
4. T. Takahashi, K. Suzuki, Kobunshi Kagaku, 21(1964)487,497.
5. C.W. Uzelmeier, SPE J., 26(1970)69.
6. W.L. Hakins, et. al., Rubber Age, 88(1960)279.
7. E.M. Bevilacqua in *Autoxidation and Antioxidants* Vol. II, Chapter 18, Interscience, N.Y., 1962.
8. W.L.Hawkins, F.H. Wilson, in *Chemical Reactions of Polymers*, Chapter XIII, Interscience, N.Y., 1964.
9. J.Voigt, *Stabilisierung der Kunststoffe gegen Licht und Warme*, Springer Verlag, Berlin, 1966.
10. W.L. Hawkins, et. al., J. Appl. Polym. Sci., I(1959)37.
11. W.L. Hawkins, M.A. Worthington, J. Polym. Sci., 62(1962)106.
12. F.G. Geleiji, Faserforschung u. Textiltech., 16(1965)395.
13. H. Schonborn, J.P. Luogo, Macromol, 2(1969)364.
14. E.J. Bailey, J.S.D.C., 87(1971)571.
15. T.B. Reeve, W.L. Dills, SPE Conf. on "Pigmentation", N.Y., 1970.
16. W.G. Huckle et. al., 20th ATC SPE, Atlantic City, N.J., Jan., 1964.
17. W. Lutzkendorf, Farbe u. Lacke, 59(1953)7.
18. F.W. Billmeyer, Jr., M. Salzman, *Principles of Color Technology*, Interscience, N.Y., 1966.
19. R.M. Johnston, M. Saltzman, Symp. "Industrial Color Technology", Am. Chem. Soc., Washington, D.C., 1971.
20. G. Wyszecki, W.S. Stiles, *Color Science - Concepts and Methods, Quantitative Data and Formulae*, John Wiley, New York, 1967.
21. F.W. Billmeyer, Jr., Apply. Optics, 8(1969)737.
22. F.W. Billmeyer, Jr. Appl. Optics, 8(1969)775.
23. F.W. Billmeyer, Jr., R.T. Marcus, Appl. Optics, 8(1969)763.
24. F.W. Billmeyer, Jr., R.L. Abrams, SPE ANTER, Chicago, May 1972.
25. R.M. Johnson, J. Color Appearance, 1 (2)(1971)27.
26. F.T. Simon, SPE ANTEC, N. Y. May, 1970.
27. H.R. Davidson, H. Hemmendinger, SPE J., 19(1963)104.
28. J.L. Saunderson, J.J. Opt. Soc. Am. 32(1962)727.
29. H. Hemmendinger, J. Paint Technol. 42(1970)132.
30. H.R. Davidson, J. Paint Technol. 42(1970)140.
31. R.M. Johnston, J. Paint Technol. 42(1970)145.
32. R.K. Winey, W.V. Longley, SPE ANTEC, Detroit, 1967.
33. R.L. Adams, SPE Conf. on "Pigmentation", N.Y., 1970.
34. E. Preininger, Mod. Plast. 127, June 1970.

VIII. BIBLIOGRAPHY

VIII.1. Books

1. Dieter Hayer, Einfarben von Kunststoffen, Hansen Verlag, Munchen, 1962.
2. D.B.Judd, Color in Business, Science and Industry, John Wiley, New York, 1963.

3. R.W.Burnham et al., A Guide to Basic Facts and Concepts of Color, John Wiley, New York, 1963.
4. Bilmeyer, Salzman, Principles of Color Technology, Interscience, New York, 1966.
5. Salzman, Ed., Industrial Color Technology, ACS, Washington, D. C., 1971.
6. M. Ahmed, Coloring of Plastics, Van Nostrand Reinhold, New York, 1979.

VIII.2. Journal Literature

1. R.A. Charvat, Dry Coloring of Polypropylene, Modern Plast. Nov. 1959.
2. W.D. Mohr, Mixing and Dispersing in E.C. Bernherdt, Ed. Processing of Thermoplastics Materials, Reinhardt Publishing Co., New York, (1959)117-152.
3. Max Saltzman, Color Measurement with the Eye and Other Instruments, Color Eng., 1(1963)12-18.
4. F. Geleji, Pigmentation of Polypropylene Fibers, Faserforsch u. Textiltech. 16(1965)395-400.
5. N.M. Bikales, Compounding, Encyclopedia of Polymer Science and Technology, Interscience, 4(1966)118-129.
6. W.G. Huckle and G.F. Jones, Use of Pigments in Plastics, SPE ANTEC XX, Atlantic City, 1966.
7. Processor's Guide to Mixing, Blending and Compounding Equipment, A State-of-the-Art Report, Plastics Technology, July, 1967.
8. A.L. Baseman, Coloring Plastics, Plast. Tech., 13(1967)31-38.
9. P.J. Papillo, Man Coloring of Plastics, Mod. Plast., 45(1967)131-146.
10. Pigments, Encyclopedia of Polym. Sci. and Tech., 10(1968)157.
11. H.F. Scharz, Phthclicyamine Blue Pigments for Plastics, SPE ANTEC, New York, 1968.
12. L.C. Komar, Titanium Dioxide in Polymer, SPE ANTEC XIV, New York, 1968.
13. D.P. Miller, Pigment Dispersions, Color Engineering, July-August, 1968.
14. E.J.G. Balley, The Coloration of Plastics, J.S.D.C., 85(1969)571-578.
15. W.L. Dills and T.B. Reeve, Effects of T_iO_2 on Weathering of Polyolefins, SPE Retee, Cleveland, 1969.
16. E.J. Balley, Colorants of Plastics, J.S.D.C., (1969)571-578.
17. B.G. Murray, Automatic In-Plant Dry Coloring -- The State of Art, 28th SPE ANTEC, New York, 1970.
18. B.G. Murray, Automatic In-Plant Coloring -- The State of the Art, SPE ANTEC, New York, 1970.
19. C.W. Uzelmeier, How Heat and Light Affect Pigment Polypropylene, SPE J. 26(1970)69-74.
20. G. Kaufman, Behavior of Organic Pigments in Polymers, Angew. Macromel. Chem., 10(1970)117.
21. G.F. Sonn, Colorants for High Temperature Polymers, SPE ANTEC, 28(1970)445-449.
22. R.M. Johnston, Applications of Color Difference Formulas, J. Paint Tech. 42(1970)145-150.
23. D.C.C. Thornley, Pigments, Plates and Progress, J.C.D.C., 1970.
24. M.Z. Schneider, Color Concentrates for Thermoplastics - a Different View, SPE ANTEC, New York, 1970.
25. R.M. Johnston, Color Measuring Instruments: A Guide to Their Selection, J. Col. App., 1(1971)27-38.
26. R. Deverell-Smith, Easily Dispersed Pigments Polymer Age, 2(1971)267.
27. C. Eckardt and H. Hefti, Comparison of Fluorescent Brighteners in the Production of Synthetic Polymer Fibers and Textile Industry, J.S.D.C. 87(1971)365.
28. T.J. Gravy, Daylight Fluorescent Pigments in Plastics, SPE Rey. Tech. Conf., Cleveland, 1971.
29. W.H. Mess and M.D. Garrett, The Dispension of Carbon Black, J.O.C.C.A., 54(1971)24-83.
30. R.L. Adams, Producing Pigmented Thermoplastics in Volume, SPE J., 27 (1971)54-57.
31. T.B. Reeve and W.L. Dills, Principles of Pigments Dispersion in Plastics, J. Color Appearance, 1(1971)25-29.
32. R. Deverell-Smith, E - Dispersed Pigments, Polymer Age, 2(1971)267-271.

33. T.B. Reeve, Principles of Pigment Dispersion in Plastics, J. Color App., 1 (1971)25-29.
34. T.B. Reeve, Organic Colorants for Polymers, J. Macromel Sci. Rev. Polym. Tech. D1, 2(1971)217-235.
35. C. Eckhardt and H. Hefti, Comparison of Fluorescent Brightening in the Production of Synthetic Polymer Fibers and in the Textile Industry, J.S.D.C., 87(1971)365-370.
36. C.E. McColough, New Approach to Coloring, Aust. Plast. Rubber, 23(1972) 25-29.
37. G. Martin, Coloring of Plastics, Kunststoffe Tech., 11 (10), 262-266, (11), 294-296, 12(1972)329-335.
38. M.J. Smith, Measurement of Pigment Dispersion Rate, Powder Tech., 5(1972) 229-361.
39. R.J. Ryan, Daylight Fluorescent Pigments and Colors, Pigment and Resin Tech., Sept., 1972.
40. P.J. Barnes, Proposed Classification for Pigments, Pig. Res. Tech., (1972) 13-20.
41. M. Berenson, Dispersion Problems Magnified, SPE ANTEC 30th, Chicago, 1972.
42. McDowell, Migration of Organic Pigments in Plastic Materials, J.S.D.C., (1972)212.
43. M. Berenson, Dispersion Problems - Magnified, SPE ANTEC, 30th, Chicago, 1972.
44. W. McDowell, Migration of Organic Pigments in Plastic Materials, J.C.D.C., June, 1972.
45. G. Robbins, The Utilization of Color Concentrates, 30th SPE ANTEC, Chicago, 1972.
46. G.F. Sonn, Colorants for High Temperature Polymers, 30th SPE ANTEC, Chicago, 1972.
47. M.Z. Schneider, Color Concentrates for Thermoplastics - A Different View, 30th SPE ANTEC, Chicago, 1972.
48. J. Richter, Daylight Fluorescent Pigments and Colors, Pig. Res. Tech., Sept., 1972.
49. H.G. Volz, Advanced Methods of Testing the Optical Properties of Pigments, Prog. in Organic Coating, 1(1973)1-34.
50. H.U. Hackney, Liquid Colors for Thermoplastics, SPE, Reg. Tech. Conf., Sept., 1973.
51. J. Richter, Fluorescent Dyes and Pigments, Europlast. 46, 65, Oct., 1973.
52. E.T. Conner, The Promise of Improved Measurement and Application, SPE Reg. Tech. Conf. 1973, Cherry Hill, N. J., Sept., 1973.
53. J.A. McCormack, Dispersions for the Plastics Industry, SPE Reg. Tech. Conf. 1973, Cherry Hill, N. J., Sept., 1973.
54. H.U. Hackney, Liquid Colors for Thermoplastics, SPE Reg. Tech. Conf., Cherry Hill, N. J., Sept., 1973.
55. M.J. Smith, Some Physical Properties of Pigments - II, J.O.C.C.A., (1973) 156.
56. M.J. Smith, Some Physical Properties of Pigments - III, J.O.C.C.A., 56(1973) 165.
57. S.L. Hauden, Pigment Coloring of Melt Spun Synthetic Filament and Fiber Products, SPE, Tech. Pap. 19(1973)400-402.
58. R.B. Doherty, In-Plant Coloring of Plastics: Procedure, Pitfalls and Payouts, SPE REG TEC CON, Cherry Hill, N. J., Sept., 1973.
59. A. Turturro, et al, Crystallization Kinetics and Morphology of High Density Polyethylene Containing Organic and Inorganic Pigments, Br. Polym. J. 5 (1973)129-139.
60. H.W. Hackney, Reconsider Liquid Coloring for Plastics, Plastics Eng., (1974) 61-63.
61. M.J. Smith, Some Physical Properties of Pigments and Their Importance in Relation to Dispersion in Plastics, J. Oil Col. Chem. Ass., 57(1974)36-43.
62. P. Ackroyd, The Mass Coloration of Man-Made Fibers, Rev. Prog. Coloration, 5(1974)86-96.
63. H. Schneider, Melt Dyeing of Fibers, Lenzinger Berichte, 40(1976)219-226, (also in Chemiefasern 26/78, E60-E61, (1976)307-310.

IX. APPENDIX I

Suppliers of pigments, pigment dispersions, concentrates and coloring systems in the United States

Allied Color Industries, 4758 Warner Road, Cleveland, Ohio 44125
American Cyanamide Corp., Bondbrook, N.J. 08805
American Hoechst Corp., Rt. 202-206, North Somerville, N.J. 08876
Americhem Inc., 2038 Main Street, Cuyahoga Falls, Ohio 44221
Amtek, P.O. Box 9, Nesquehoning, Pa. 18240
Ampacet Corp., 250-s Terrace Ave., Mt. Vernon, N.Y. 10550
Avecor Inc., 13596 Vaughn Str., San Fernando, Ca. 91330
BASF-Wyandotte, Wyandotte, Mo. 48142
Bee Chemical Co., 2700 E 170 Str., Lansing, Ill. 60438
Carolina Color Corp., 1000 Long, Salisbury, N. C. 28144
Chemetron Corp., P. O. Box 66251, Chicago, Ill. 60666
Chroma Corp., McHenry, Ill. 60050
Ciba-Geigy Corp., Saw Mill River Road, Ardsley, N.Y. 10502
Color Chip Corp., 124 South Ave., Garwood, N.J. 07027
Colorcon Inc., Moyer Blvd., West Point, Pa. 19486
Coz Chemical Co., Div. Allied Product, Providence Road, North Bridge, Ma 01534
Crompton & Knowels Corp., Campus Drive, Sommersette, N.J., 08873.
Day Glo Color Corp., 4732 St. Clair Ave., Cleveland, Ohio 44103
DuPont de Nemoures Co., 1002 Market St., Wilmington, Del. 19898
Eastman Chemical Products, Kingsport, Tenn. 37662
Ferro Corp., One Drive View Plaza, Cleveland, Ohio 44114
General Color & Chemical Co., P.O. Box 7, Valley Street, Minerva, Ohio 44657
Hammond Plastics Co., 1001 S. Bridge Street, Worcester, Ma. 01610
Harshaw Chemical Co., 1945 E. 97th St., Cleveland, Ohio 44106
Hercules Inc., 910 Market St., Wilmington, Del. 19899
Hilton Davis Chemical Co., 2235 Landow Farm Road, Cincinnati, Ohio 45237
ICI America Inc., New Murphy Road, Wilmington, Del. 19897
Indol Chemical Co., P.O. Box 537, Cliffwood, N.J. 07721
Inmont Corp., 150 Wagaraw Road, Hawthorne, N.J. 07507
Mearl Corp., 41 E. 42nd St., New York, N.Y. 10017
Nortech Div. Northern Petrochemical Co., 830 Main St., Clinton, Ma. 01510
Pigment Dispersion Inc., P.O. Box 412, Edison, N.J. 08817
Polychrome Dispersions Inc., 13429 S. Western Ave., Los Angeles, Cal. 90249
Reed Industries, 12333 Wilkins Ave., Rockville, Md. 20852
Sandoz Color & Chemical Co., Rt. 10 E. Hanover, N.J. 07936
Service Color Corp., 11524 W. Addison St., Franklin Park, Ill. 60131
Sun Chemical Co., 4226 Chickering Ave., Cincinnati, Ohio 45232
Teknor Apex Co., 505 Central Ave., Pawtucket, R.I. 02862
US Industrial Chemical Co., 99 Park Ave., New York, N.Y. 10016
Verona Dyestuffs, P.O. Box 385 Union Metro Park, Union, N.J. 07083
Standridge Color Corporation, 10200 Frontage Rd., Covington, Ga. 30209

Chapter IV RHEOLOGY OF THE MELT

I. INTRODUCTION

Rheology is the study of the deformation and flow of materials. The knowledge of deformation and flow of polymer material is essential not only for the polymer manufacturer who must tailor his polymer to meet the specific requirements of his customer, but also for the process engineer who must choose the polymer material for specific processing conditions or develop proper processing conditions for a specific polymer. Because polymer melt rheology is related to the molecular structure on one hand and product properties on the other, its study is of extreme importance to both the polymer physicist and the fiber technologist. An understanding of the melt flow is essential in design work - for extruder's screw configuration, melt distribution channels, metering pumps, filter packs, spinnerettes, etc., which are the critical components of processing machinery for the production of fibers - as well as in development of the optimum process conditions for the desired product. It is equally important to have experimental means of measuring the rheological properties of a given polymer and predicting its flow behavior under anticipated processing conditions on the basis of laboratory data.

Usually the rheological characteristics are considered as being determined by two groups of factors: the factors connected with the conditions of flow and those connected with the material itself and its state, i.e., molecular weight, molecular-weight distribution, additives, temperature, pressure, etc. Such a division is rather arbitrary but it is useful in simplifying any discussion of melt flow. We will follow this scheme in the forthcoming treatment of the subject. First we will discuss the relation between stress and deformation of the melt and then examine how this relation is affected by material parameters and its conditions of state. An attempt will be made here to expose the reader to the rudiments of melt rheology of polymers in general and those of polypropylene in particular. No attempt will be made to develop the phenomenological mathematics of viscoelastic flow or to catalog in any complete sense the existing and rapidly growing melt flow data for polypropylene. An extensive bibliography is provided at the end of this chapter for this purpose.

The rheological behavior of polymer melts can adequately be considered in terms of deformation under simple shear and deformation under simple extension, respectively. Both kinds of deformation play an important role in fiber spinning operations. The first kind of deformation is important in the flow of the polymer melt up to the spinnerette hole and the second kind becomes important in the subspinnerette area. The deformation itself can be divided into viscous and elastic components. The distinction between elastic and viscous deformation is that elastic is recoverable and viscous is not.

II. DEFORMATION UNDER SIMPLE SHEAR

The rheological properties of polymer melts have characteristics of both elastic solids and viscous fluids: hence they are classified as viscoelastic fluids. For an ideal elastic solid (Hookean element) we may write the relationship between stress and strain:

$$\delta = G.\gamma. \neq \int(t)$$

that is, the stress δ is proportional to strain γ and independent of time, t. The proportionality constant, G, is called modulus of elasticity and is a material constant. At the other extreme from the perfectly elastic solid is the perfect viscous fluid (Newtonian fluid) characterized by proportionality between stress and rate of strain:

$$\delta = \eta \frac{d\gamma}{dt} = \eta\dot{\gamma}$$

where the constant of proportionality, η is the coefficient of viscosity for the material.

The Hookean solid and Newtonian fluid represent two extremes in the rheological characteristics of materials. Viscoelastic materials lie between these two extremes. Most generally a viscoelastic substance can be described by the equation:

$$\delta = \int(\gamma, t)$$

that is, stress δ is a function of strain γ and time t. This nonproportionality of stress to strain, although of definite practical importance, is not amenable to simple treatment. The much more amenable equation:

$$\delta = \gamma \int(t)$$

in which stress δ is proportional to product of strain γ and a function of the time t may be taken as the defining equation of linear viscoelasticity. This is the fundamental description by which the mechanical and rheological properties of polymers may be understood. We will first consider the first part of this equation and later on revert to a brief discussion of time-dependent deformations.

For *Newtonian* fluids stress is proportional to rate of strain and the proportionally constant is assumed to be constant. The equation of flow under simple shear can be written as:

$$\delta = \eta.\dot{\gamma}$$

where δ is the shear stress and $\dot{\gamma}$ is the shear rate; η is viscosity coefficient, often simply called viscosity. When shear rate $\dot{\gamma}$ is plotted against shear stress, δ

these materials yield a straight line with a slope of η as depicted by curve A in Fig. 1.

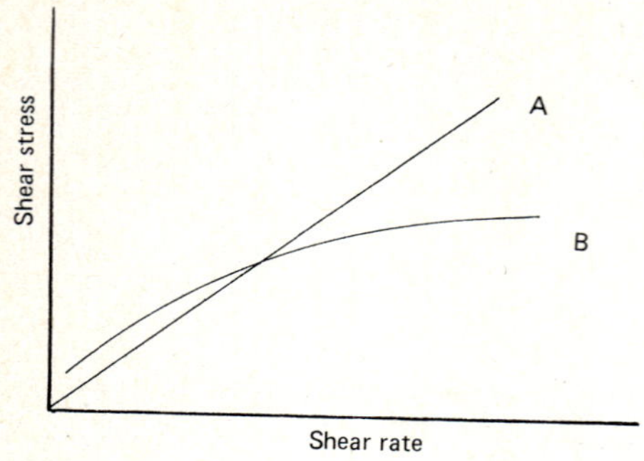

Fig. 1. Forms of Viscous Flow
A. Newtonian Flow
B. Shear Thinning

Polypropylene melt, like several other polymeric materials, deviates considerably from the ideal behavior and when sheared does not render a straight line as shown by curve A in Fig.1. Two types of deviations from Newtonian flow are commonly observed. One is shear thinning, a reversible decrease in viscosity with increasing shear rate. The second deviation from Newtonian flow is the exhibition of a yield value, a critical stress below which no flow occurs. At low shear stress, the δ-$\dot{\gamma}$ curve approximates a straight line of constant slope, generally called the limiting viscosity, η_o. At higher shear rates the line becomes curved, concave to the shear rate axis, i.e., an incremental change in shear stress results in an increasing change in shear rate or flow rate as depicted by curve B of Fig.1. The viscosity coefficient, η , is therefore not constant as in equation $\delta = \eta\dot{\gamma}$ but is itself a function of the shear rate or shear stress. However, at each value of shear rate it takes on an instantaneous value. This quantity is known as apparent viscosity η_α and is defined as:

$$\eta_\alpha = \frac{\text{shear stress}}{\text{shear rate}} = \frac{\delta}{\dot{\gamma}}$$

Several equations have been proposed (ref.1.) to describe pseudoplastic behavior of melts, but the only equation which has found any real use is the power law equation: $\delta = K\,(\dot{\gamma})^n$ or $\log \delta = \log K + n \log \dot{\gamma}$.
This means that log-log plots of δ versus $\dot{\gamma}$ will yield a straight line. The materials which follow this relationship are called "power law fluids". It is found that for polypropylene melts reasonably straight lines can be drawn over one to two decades of shear rate, but over a wider range the curve begins to deviate considerably from a straight line.

At this stage it is useful to look at the relationship between apparent viscosity η_α and the power law constants K and n. Since, $\eta_\alpha = \frac{\delta}{\dot{\gamma}}$ and $\delta = K\ (\dot{\gamma})^n$, combining the two equations we obtain:

$$\eta_\alpha = K\,(\dot\gamma)^{n-1}$$

It is also possible to relate the apparent viscosity, η_α, to shear rate, $\dot\gamma$, and n in terms which do not involve K. If η is the apparent viscosity at some shear rate, $\dot\gamma$, and η_r is the apparent viscosity at some arbitrary shear rate, then it immediately follows that:

$$\frac{\eta}{\eta_r} = \left|\frac{\dot\gamma}{\gamma_r}\right|^{n-1}$$

When γ_r is put equal to $1s^{-1}$, this simplifies to:

$$\eta = \eta_1 |\dot\gamma|^{n-1}$$

where η_1 is the apparent viscosity at a shear rate of 1 sec^{-1}, which for all practical purposes is equal to η_0.

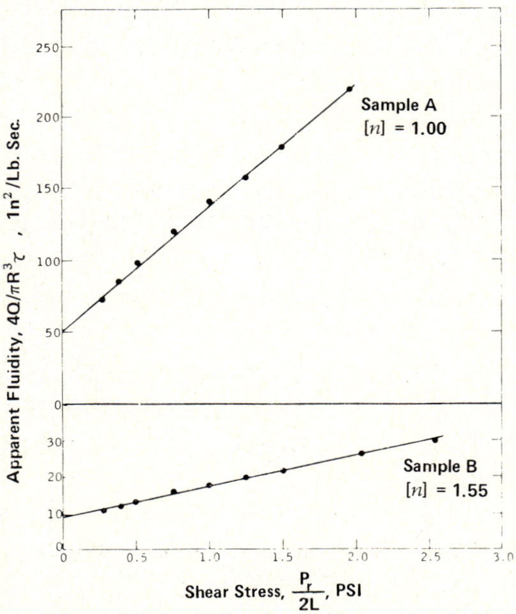

Fig. 2. Apparent Fluidity vs. Shear Stress at 400°F for Two Polypropylene Samples (ref.40)

Low-shear Newtonian melt viscosity η_0 is an important material characteristic and is often used in rheological work. It is determined by extrapolation of fluidity - shear stress curve to zero shear stress, as depicted in Fig. 2. The zero shear viscosity of polymers has been shown to be proportional to their molecular weights or intrinsic viscosities. For polypropylene, it is proportional to 3.6th power of the weight-average molecular weight. This topic will be discussed further in Section V of this chapter.

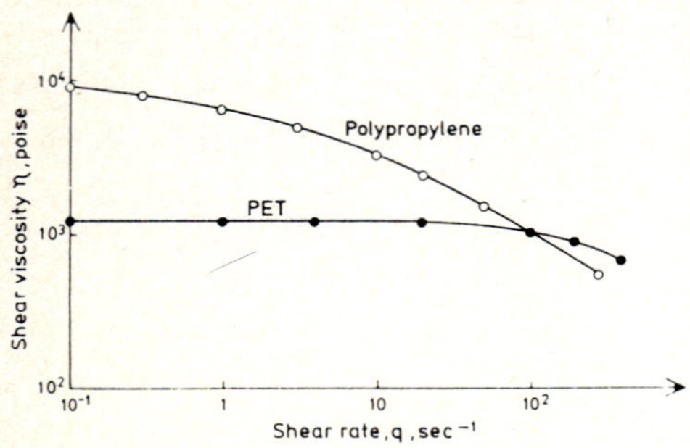

Fig. 3. Melt-Flow Behavior of Polypropylene in Comparison with PET (ref.46)
1. Polypropylene, [η] = 1.56, 250°C
2. Polyethylene terephthalate, [η] = 0.65, 285°C

The apparent viscosity of polypropylene melt decreases rapidly with increasing shear rate. This is illustrated by curve 1 in Fig. 3, which represents the value η_α as a function of $\dot{\gamma}$ for a polypropylene sample of 1.65 intrinsic viscosity. This pseudoplastic behavior is one of the characteristics differentiating molten polypropylene from many other spinnable polymers, such as polyethyleneterephthalate, whose flow properties are nearly Newtonian within reasonable shear rates (see curve 2 of Fig. 3). The deviation of polypropylene melt from Newtonian behavior is strongly dependent on its molecular weight and molecular-weight distribution. The higher the weight average-molecular weight of the polymer, the more it deviates from Newtonian behavior. Similarly, the polymers of broad molecular-weight distribution deviate more from Newtonian behavior than those with narrow molecular-weight distribution (Fig. 4).

Figure 5 shows the typical behavior of a pseudoplastic material, including polypropylene, over a wide range of shear rate. In this schematic representation it may be seen that the lower shear rate region of the curve in Fig. 5(a) corresponds to curve B of Fig. 1 with its low shear limiting viscosity η_0 and its curved region concave to the shear rate axis. At higher shear rates the curve passes through an inflection point and a high shear limiting Newtonian viscosity η_∞ is achieved. This behavior may be seen more clearly in Fig. 5(b), which shows the same curve represented as the logrithm of the apparent viscosity versus the logarithm of shear rate. Regions A and C represent constant low shear and high shear limiting Newtonian viscosities, respectively; region B represents the range through which η_α is a changing function. Section C is not very important from the process point of view in fiber spinning although some patent publications do mention the processing of polypropylene melt at very high shear rates, presumably in the region of η_∞.

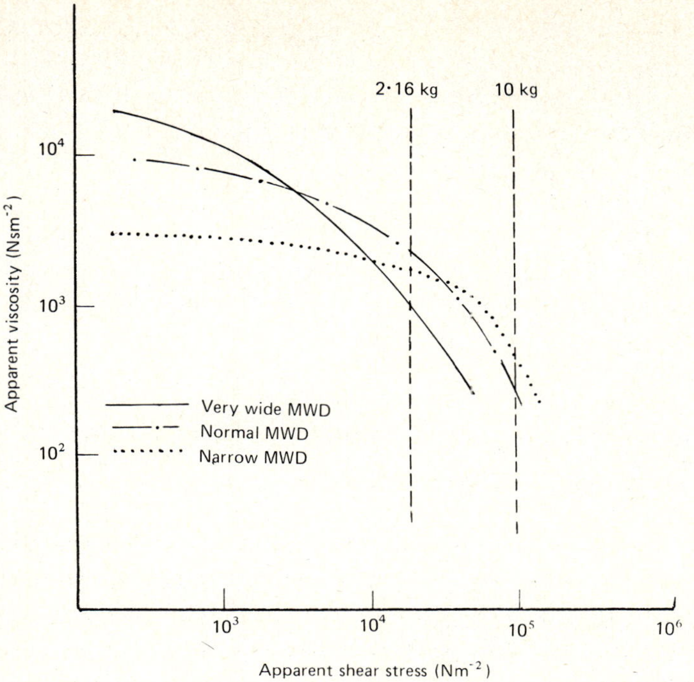

Fig. 4. Effect of Molecular-Weight Distribution (MWD) on Shear Sensitivity of Melt Viscosity (ref.31)

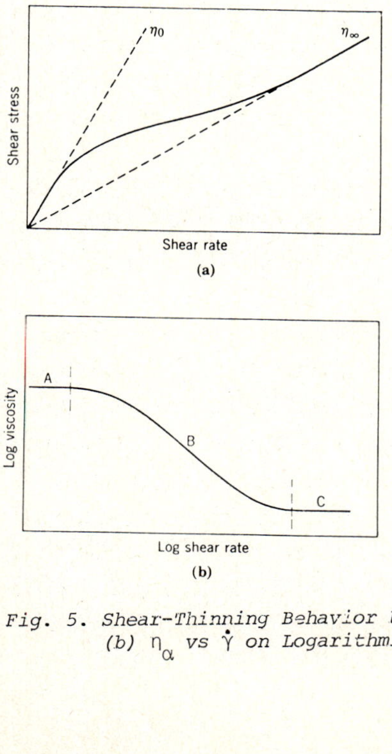

Fig. 5. Shear-Thinning Behavior Represented by (a) δ vs $\dot{\gamma}$ on Linear Coordinates; (b) η_α vs $\dot{\gamma}$ on Logarithmic Coordinates (ref.39)

The rheological behavior of a Newtonian fluid can be conveniently described as that exhibited by a dash-pot (Fig. 6 b) while the deformation of an ideal elastic element can be pictured as that experienced by a spring (Fig. 6 a). Maxwell model of a viscoelastic fluid combines the two models in a series (Fig. 6 c). The rheological equation for a Maxwell fluid is written as:

$$\gamma = \gamma_e + \gamma_f \quad \text{or} \quad \dot{\gamma} = \dot{\gamma}_e + \dot{\gamma}_f$$

where $\dot{\gamma}_e$ is the rate of pure elastic deformation and $\dot{\gamma}_f$ is the rate of pure viscous deformation.

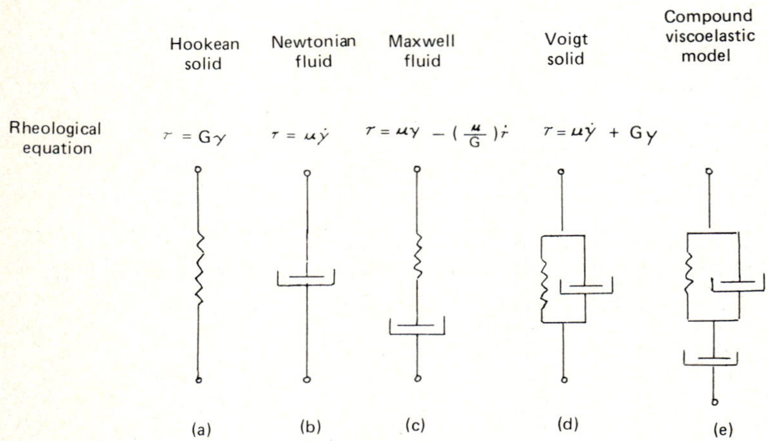

Fig. 6. Mechanical Models of Rheological Equations

Fig. 7 graphically shows a pure viscoelastic response representative of a Maxwell model. Here a constant applied stress renders an instantaneous elastic strain, ab, followed by a viscous strain, bc, for the duration of the stress. When the stress is removed, an instantaneous elastic recovery, cd, occurs so that the length cd is equivalent to that of ab.

The elastic response of real viscoelastic materials is never instantaneous, as described by the solid line in Fig. 7. Actually, the spring of the model is retarded as it is set up in parallel with a dash-pot, shown in Fig. 6d. The time response of this system is shown by the dashed line in Fig. 7. The time factor is called the relaxation time or retardation time, τ, and is equal to $\eta/_G$, where η is the viscosity of the dash-pot and G is the modulus of the spring.

The existence of a discrete relaxation spectrum for a polymer melt has been variously demonstrated. When the shear rate involves time of the order of the spectrum of retardation times of the system, then the response will be viscoelastic and nonlinear. However, if the shear rate involves time outside this range,

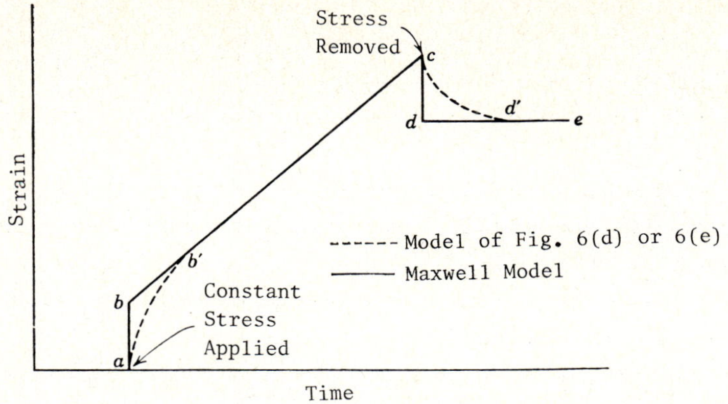

Fig. 7. Strain Curves for Viscoelastic Flow Models (ref.43)

then the system will approach Newtonian behavior. Thus, at very low rates, the response will be primarily viscous, and since the energy is dissipated the apparent viscosity will be high, but fairly constant. At very high shear rates the response will be largely elastic. The energy will be stored and thus the apparent viscosity will be low but fairly constant. This provides an explanation of the existence of η_o and η_∞ respectively, mentioned earlier.

It is evident from the above discussion that the shorter the time scale of deformation the more important the elastic component becomes. As the spinning of fibers is not a thermodynamically equilibrium process, the relaxation time of the melt is a very important factor in determining the speed of spinning, the dimensions of the spinnerette holes, pressure drop across the spinnerette and filter pack, and metering efficiency of the extruder screw or the metering pump. The ever-increasing trend to faster processing speeds and the need for more exact predictions of polymer behavior in the extrusion system demand more attention to the understanding of this property than has hithertofore been given.

The elastic properties of polymers are sometimes expressed in terms of its compliance, J_e, (reciprocal of shear modulus, G) which is a measure of how polymer molecules comply to a given stress and, as expected, is a function of molecular weight, molecular-weight distribution, and temperature (ref.2). Another frequently used measure of melt elasticity is normal stress differences which has been shown (ref.3) to be related to the exit pressure. Still another measure of elastic properties of the melt is the recoverable strain, γ_e, which is defined as:

$$\delta = \gamma_e/J_e$$

where δ is the shear stress and J_e is the steady-state compliance. The die-swell phenomenon encountered in the extrusion of viscoelastic material is believed to result from the elastic deformation incurred while the material flows through the extrusion orifice (ref.4,5). Although this may be an oversimplification (ref.6,7), die swell is still considered to a great extent a measure of the elastic component of the melt flow.

Since there are different ways of characterizing the elastic properties of the melt, the experimental results from various investigators are considerably at variance or even in contradiction with each other. Because of this uncertainty of definition and measure, the elastic properties of polypropylene melts have not been thoroughly investigated and reported.

The molecular picture explaining the observed dependence of polymer melt viscosity on shear rate is rather involved. For simplicity's sake we may assume that polymer melt consists of a partially entangled large molecule. Application of small stress changes the structure very little: molecules have time to flow without disentanglement and flow is thus Newtonian at high viscosity range. At very high shear stress, disentanglement has progressed as far as possible so that viscous resistance is again Newtonian but at a lower level of apparent viscosity. Other reasons causing the observed shear thinning is believed to be the anisotropy of molecules, aggregates, or flow units. This results from the tendency of the applied force to disturb the long chains from their preferred equilibrium in the direction of shear. At low shear rate, the chain molecules have enough time to continually revert to their equilibrium positions and behave as though nothing has changed in their configuration. At relatively high shear rate the chain molecules do not get enough time to orient into their statistically favored configuration and thus flow as elongated bodies.

III. DEFORMATION UNDER TENSION

The dominating type of deformation in the major part of the spin line is uniaxial extension. Viscoelastic behavior of polymer melt under extensional stress largely determines the speed of spinning, stability of the threadline and the super-molecular structure of spun filaments. It is, however, unfortunate that not enough attention has been paid to this aspect of rheological studies with the result that our understanding of polymer deformation under tension is only sketchy.

There are two types of extensional flow. Examples of obvious extensional deformations include casting of film (a low-stress phenomenon) and melt drawing of polymer jet into fibers (a high-stress phenomenon). The second, or concealed, class is the extensional component of convergent or divergent flows, which are common in practical deformations of melt in the entry region of spinnerette capillary in the fiber manufacturing process. Here, we will be concerned mainly with uniaxial deformation which is involved in the stretching of molten jets into thin filaments.

Uniaxial elongational flow occurs when a uniaxial tensile stress is applied to a fluid. By considering the fluid threadline of the spinning process as being subjected to a uniaxial stress (consisting primarily of gravitational and externally applied stresses), then the resulting deformation is that of uniaxial elongational flow.

Just as shear viscosity, η, is defined as the ratio of the shear stress, δ, to the shear rate, $\dot{\gamma}$, ($\eta = \frac{\delta}{\dot{\gamma}}$), the uniaxial elongational viscosity, β, is defined as the ratio of the uniaxial tensile stress, σ_1, to the uniaxial elongation rate, ε, i.e.:

$$\beta_1 = \frac{\sigma_1 (t)}{\varepsilon} \quad \text{or} \quad \frac{\sigma_1}{\varepsilon(t)}$$

in experiments where either the tensile stress or the elongation rate is constant in time. In flows where neither the stress nor strain rate is constant in time, one can define an apparent viscosity coefficient, β, as:

$$\beta = \frac{\sigma\ (t)}{\varepsilon\ (t)}$$

It should be pointed out here that the apparent uniaxial elongational viscosity, β, given by the above equation, may not be a unique function of the instantaneous elongation rate but may be dependent also on the strain history of the material.

Before proceeding into a discussion of elongational flow in the fiber spinning process, credit should be given to Trouton who, in 1906, initiated the research interest in elongational flow. Trouton's investigations into the elongational flow of very viscous materials has formed the basis for much of the current research into the elongational flow in fiber spinning. Trouton defined the coefficient of viscous traction (or elongational viscosity) as:

$$\beta = \frac{\sigma}{\varepsilon} = \frac{F/A}{dv/dx}$$

where F = applied axial force (force of traction)

A = cross-sectional area

dv/dx = velocity gradient in the axial direction

This definition has formed the basis for determining elongational viscosity in the fiber spinning process.

By resolution of the axial stress into components of shear and dilation, Trouton correctly determined that the elongational viscosity is three times the shear viscosity, i.e.:

$$\beta = 3\eta$$

It has since been shown by Middleman (ref.47) and Reiner (ref.48) that the above relation (known as Trouton's equation) holds good for Newtonian fluids. This relation is also valid for non-Newtonian fluids as the strain rate approaches a zero value, i.e., at relatively low elongational rates. Figure 8 shows this relation for polypropylene of various molecular weights. Under these conditions the material may be described as Troutonian.

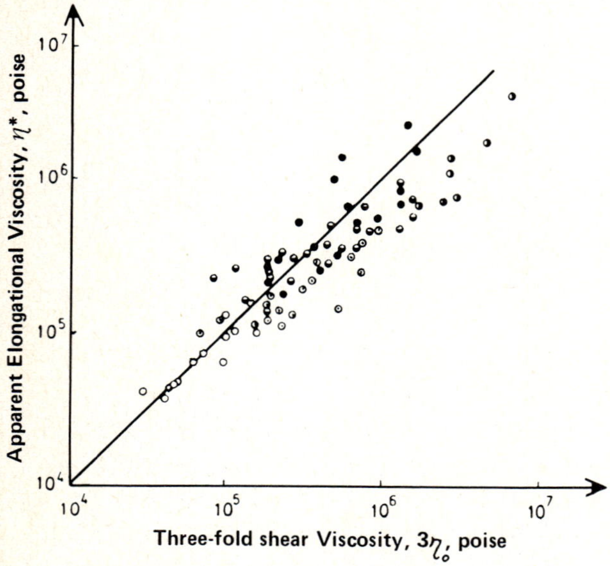

Fig. 8. Apparent Elongational Viscosity of Polypropylene as Measured in a Melt-Spinning Experiment Vs. Three-fold Newtonian Viscosity from Rheometric Data. Individual Points Correspond to Different Temperatures and Various Molecular Weights (ref.42).

The dependence of apparent viscosity on stress or rate of extensional deformation is an important rheological characteristic. However, the knowledge of this dependence is presently scanty at best. The only experimental data available are due to Han (ref.8) and Cogswell (ref.9). Figure 9 and Figure 10 show these results. Both of these authors showed that the elongational viscosity decreases with an increase of deformation rate. Han proposed a generalized, empirical equation to account for the elongation rate dependence of elongational viscosity at constant temperature, T_o. This equation is given by:

$$\beta = 3\eta_o\left[a + b\,(\varepsilon)^{q-1}\right]$$

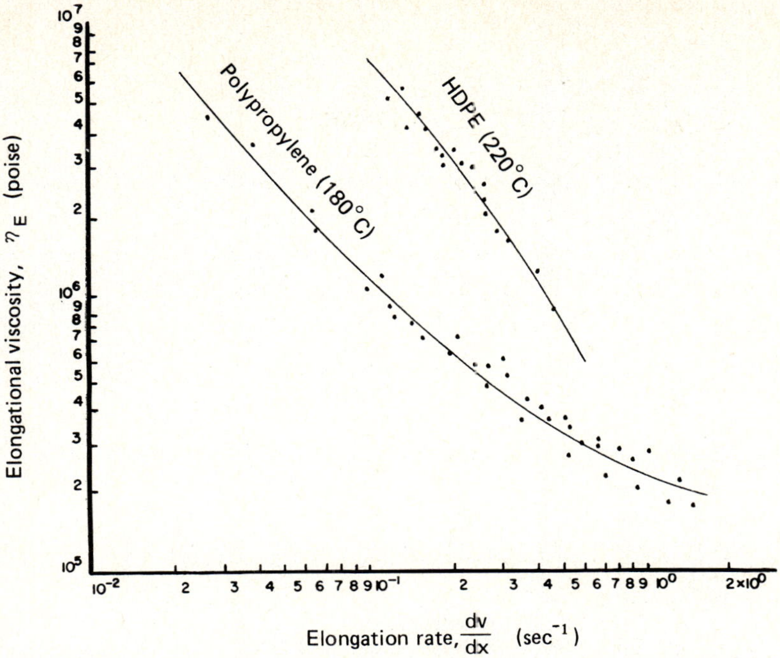

Fig. 9. Plots of Elongational Viscosity Vs. Elongation Rate of Polymer Melts Under Isothermal Spinning Conditions (ref.8)

where η_0 denotes zero-shear rate viscosity, and a, b, and γ are constants, characteristic of the material at temperature T_0.

The dependence of elongational viscosity on the rate of elongation is commonly thought to be bearing upon the spinnability of a polymer melt. In practical cases, however, it is the dependence on stress or the "resistance to deformation", an additive function of viscosity and modulus, which dictates the stability of a deformation process such as a spinning threadline. The modulus under tension is equal to three times the modulus under simple shear (ref.9) and increases with increasing tension. Although the dependence of modulus on stress is the same or similar under tension and under simple shear, the relaxation time spectrum or ratio of viscosity to modulus is quite different owing to the great difference in viscosities. Thus, elasticity plays a larger part in extensional flows than in simple shear flows of the same magnitude. Figure 10 shows the dependence of viscosity and elasticity (modulus) under simple shear and simple extensional stress for an extrusion-grade polypropylene.

IV. EFFECT OF TEMPERATURE

It is well known that temperature has an important effect on the rheological

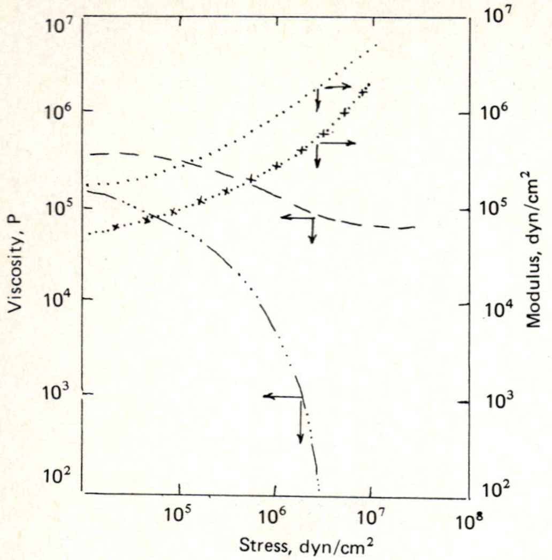

Fig. 10. Rheological Properties of a Polypropylene Sample (ref.9)

——···– *viscosity under simple shear*
– –– *viscosity under simple tension*
x··x··x··x *modulus under simple shear*
.... *modulus under simple tension*

properties of polymer melts. In general, melt viscosities decrease and relaxation times shorten with increasing temperature.

The most commonly used expression relating viscosity of a Newtonian fluid to temperature is the Arrhenius equation:

$$\eta = Ae^{E/RT}$$

where R is the gas constant, E the energy of activation for flow, and A a coefficient depending upon the nature of the fluid. At very low shear rates most non-Newtonian fluids exhibit Newtonian behavior. In this region the flow behavior can be characterized by a single constant, the limiting viscosity, η_0. Thus, for low shear rates, the dependence of melt viscosity on temperature can be expressed as:

$$\eta_0 = Ae^{E/RT}$$

The non-Newtonian viscosity η is a function of the state of shear as well as a function of temperature. Therefore, when the variation of η with temperature is investigated, it is necessary to specify whether the variation is to occur at constant shear rate or at constant shear stress. Each type of variation gives rise to a different energy of activation. It has been shown (ref.10) theoretically that for pseudoplastic fluids:

$$(\partial\eta/\partial T)_{\delta} > (\partial\eta/\partial T)_{\dot{\gamma}}$$

Assuming that the temperature dependence of η can be expressed through an Arrhenius-type equation, we have:

$$\eta_a = A\,e^{E_s/RT} \qquad \text{or} \qquad A\,e^{E_{\dot{\gamma}}/RT}$$

The first equation applies to changes occurring at constant shear stress and the second to changes occurring at constant shear rate.

The activation energies of flow for polypropylene at low shear stresses and low shear rates are the same - generally in the range of 18-20 kcal/g. E_{γ} decreases with increasing shear rate and may reach as low as 5 kcal/g in 10^3 - 10^5 sec^{-1} range. E_s, on the other hand, increases with increasing shear stress. Activation energies are generally reduced at higher temperature range. They are also strongly affected by molecular weight of the polymer: the viscosities of lower molecular-weight polymers being less sensitive to variation in temperature than those of higher molecular-weight polymers (ref.11).

Parallel with the changes of shear viscosity, tensile viscosity is also strongly dependent on temperature. Because of obvious experimental difficulties in controlling the temperature in the measurement of extensional viscosity, the reported data on polymer melts are scanty.

Kase and Matsuo (ref.12) appear to be the first to make Arrhenius-type plots of elongational viscosity and derive from them an activation energy of elongational flow. They deduced the values of $\beta(T)$ for polypropylene from experimentally measuring the cross-sectional area, $A(x)$, and the temperature, T, of running filaments along a thread line, and expressed the dependence of β on temperature as:

$$\beta = 0.040e^{3500/T}$$

The exponent in the above relation shows the activation energy of elongational flow to be about 7.0 kcal/mol. Similar experiments were carried out by Sano and Yamade (ref.49) with polypropylene samples of different intrinsic viscosities. The plots of log of elongational viscosity versus inverse temperature were divided into two linear regions: one above 130°C and one below 130°C. Apparent activation energies E_a, were evaluated and, for temperatures greater than 130°C, E_a was 4.5 to 7.5 kcal/mol., depending on the intrinsic viscosity. These values compare with the activation energy of shear viscosity of 8.7 kcal/mol. Below $130^{o}C$, E_a was 12 to 17 kcal/mol. The increase in E_a below $130^{o}C$ was attributed to orientation and crystallization. Ishibashi, et al (ref.13) and Hamana, Matsui and Kato (ref.50) have reported similar values for nylon 6 and PET:

$\beta = 0.37e^{5300/T}$ (ref.50); $\beta = 0.034e^{3250/T}$ (ref.13) for Nylon 6
$\beta = 0.73e^{5300/T}$ (ref.48) for PET

The exponents in these equations indicate an activation energy of approximately 10 kcal/mol. for both polymers. Han and Lamote (ref.8) have extended this work and developed the following empirical equation:

$$\beta = 3\alpha e^{\delta/T}\left[k_1 + k_2\varepsilon^{q-1}\right]$$

where $\alpha = \eta_o(T_o)e^{-E/RT_o}$

$\delta = E/R$

$E = E_s + E_e$

$k_1 = a(T_o)$

$k_2 = b(T_o)$

Here, E is the activation energy, determined experimentally, and consisting of two parts; one, the shear flow activation energy E_s, which comes from the zero shear viscosity; and the other, the elongational flow activation energy E_e; k_1 and k_2 are the values of a and b evaluated at some reference temperature T_o.

V. EFFECT OF MOLECULAR WEIGHT AND POLYDISPERSITY

The effect of molecular weight of polypropylene on its melt viscosity has been discussed by several authors (ref.14-16). As the molecular weight increases, the number of chain segments whose motion must be coordinated obviously increases. From a purely statistical standpoint, then, the probability of all segments moving properly is greatly reduced; there is more wasted motion, and the viscosity is therefore higher. The relation between η_o and molecular weight is schematically represented by Fig. 11. Above a critical molecular weight, M_c, (approximately 5,000 for polypropylene) η_o is proportional to molecular weight $\overline{M}_w$, so that $\eta_o = K'\,(M)^c$ or $\log\ _o = A' + c \log \overline{M}_w$.

The constant A' is a function of temperature, while the value of c for polypropylene lies around 3.6, in agreement with other polymers. Frank (ref.15) reports the equation:

$\log \eta_o = 3.69 \log M_w - 11.90$ is applicable to polypropylene melt at 190°C.

Figure 12 depicts the relationship between η_o measured at 190°C and $[\eta]$ measured in tetralin at 135°C. Figure 13 is another set of measurements taken at 205°C. Thus, for most grades of commercial polypropylene, the melt viscosity increases with about 5 power of $[\eta]$ or about 3.6 power of average molecular weight, M_w.

According to the molecular picture of polymer flow, one would expect the elasticity of the melt to be enhanced if the average molecular weight of the sample increased or if the molecular-weight distribution broadened. These expectations seem in general to be confirmed experimentally. The relaxation spectrum is extended to longer times as molecular weight of the polymer increases and it shifts to shorter times as the molecular weight of the polymer decreases (ref.17). The shape of relaxation spectrum is changed by a change in molecular-weight distribution (ref.18). Narrowing it narrows the spectrum and widening it widens the spectrum.

When polypropylene melt is exposed to high temperature and high shear during processing significant degradation often results, causing major changes in molecular weight and molecular-weight distribution. The melt viscosity decreases accordingly and melt elasticity may also experience considerable change. The extent of the change in molecular weight and molecular-weight distribution of the extrudate is not uniform for all polymers but depends on the molecular weight and molecular-weight distribution of the initial polymer. This is illustrated in Fig. 14 which shows the molecular weight of the extrudates produced from three

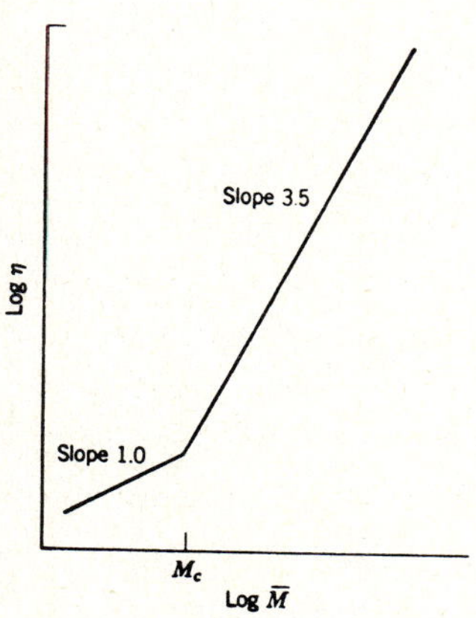

Fig. 11. Dependence of Melt Viscosity of Polypropylene on its Molecular Weight (Schematic)

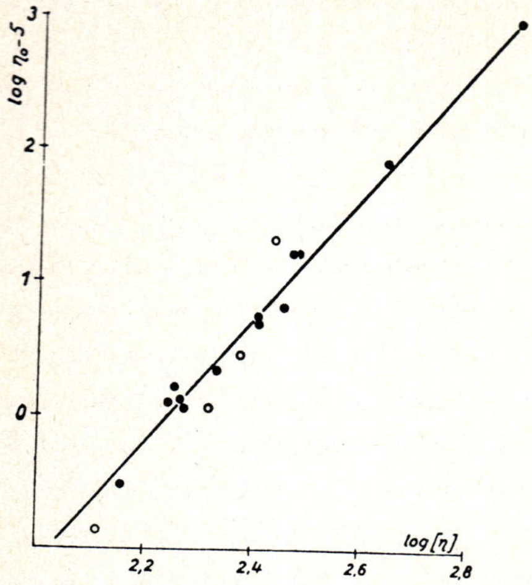

Fig. 12. Relationship between [η] and Low Shear Melt Viscosity η_o (at 190°C) of Polypropylene (ref.15)

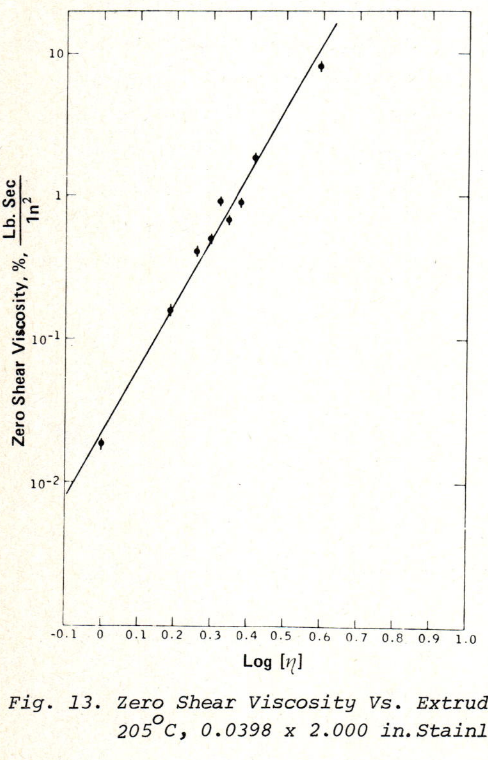

Fig. 13. Zero Shear Viscosity Vs. Extrudate Intrinsic Viscosity: 205°C, 0.0398 x 2.000 in. Stainless Steel Capillary (ref.40)

polymers of different initial molecular weights but with equivalent molecular-weight distribution when extruded under identical conditions.

Thermal degradation of polypropylene narrows its molecular-weight distribution. The narrowing of the molecular-weight distribution is related to the degree of degradation which can be described in terms of a ratio of the final, M_w, to the initial polymer molecular weight M_{wo}, respectively. This index, M_w/M_{wo}, is often used as an indicator of the change in molecular-weight distribution on degradation. A large value indicates a broad distribution of the extrudate and, conversely, a low value , a narrow distribution. This relation is based on the presumption that thermal degradation only slightly changes the value of M_n for isotactic polypropylene (ref.19) over a wide range of degradation. The factor M_w/M_{wo} is therefore essentially a ratio of the weight-average to number-average molecular weight, M_w/M_n.

Using a Monte Carlo sampling technique, Kotliar (ref.20) evaluated the distributions of molecular weight sizes and molecular weight averages resulting from random chain scissions from broad distributions. Figure 15 illustrates the effect of random degradation on the resulting distribution width, M_w/M_n, starting from polymers of three varied distribution widths. It may be seen from this figure that distribution width changes vary rapidly with the degree of degradation and that the broader the initial distribution, the greater the change.

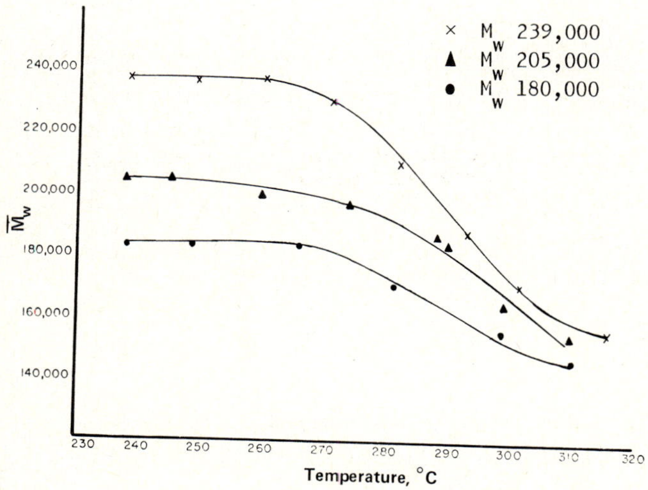

Fig. 14. Molecular Weight of Extrudate as a Function of Extrusion Temperature (ref.44)

Kotliar (ref.20) also evaluated the ratio of intrinsic viscosities of the degraded to the original polymer for different initial molecular-weight distribution and for different degrees of degradation. The results show that the broader

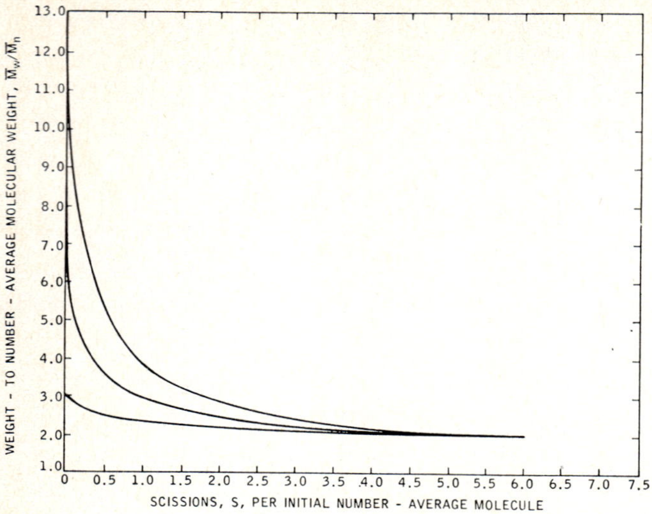

Fig. 15. Effect of Degradation on Molecular-Weight Distribution (ref.20)

the molecular-weight distribution of the original polymer, the higher the change in $[\eta]$ with an equivalent degree of degradation. Such induced distribution variations on polymer degradation may be used to explain the flow differences observed for the polymers of equivalent $[\eta]$ but different molecular-weight distributions.

The pronounced changes in molecular-weight distribution upon degradation of polypropylene has been used as a technical method to produce polymers of narrow molecular-weight distribution from polymers of broad molecular weight distribution. These types of polymers are particularly useful for the production of high-tenacity fibers and show a generally beneficial effect in spinning and drawing. The change in width of the molecular weight distribution curve can conveniently be calculated as (ref.19):

$$Q' = M_w/M_n = Q[\eta] \;/\; [\eta_o]\; I/a$$

where Q denotes the breadth of molecular-weight distribution of the original polymer, Q' the breadth of molecular-weight distribution of degraded polymer, $[\eta_o]$ its intrinsic viscosity, and $[\eta]$ the intrinsic viscosity of the extrudate; a is 0.80 from Chiang's equation (ref.21). The above relation is based on the simplifying assumption that M_n of the polymer is the same before and after extrusion, as reported by Davis and co-workers (ref.19).

Elongational viscosity is also known to depend on the molecular weight of the polymer. Sano and Yamade (ref.49) studied the elongational viscosity of isotactic polypropylene of three groups of intrinsic viscosities (average value range: 1.36 to 2.19). An increase of elongational viscosity with increase in intrinsic viscosity was observed over all temperatures considered. The activation energy of elongational flow was also observed to be dependent on intrinsic viscosity. Similar results have been reported by Ishibashi (ref.13) for Nylon 6 and by Hill and Cuculo (ref.51) for PET.

VI. EFFECT OF PROCESSING AIDS ON MELT RHEOLOGY

Polypropylene melts are noted for their high viscosity. This imposes various limitations on productivity and equipment design. Attempts are often made to decrease the melt viscosity of polymer by adding so-called processing aids to the polymer mass. Other terms commonly used to describe these agents include internal release agent, friction reducer, internal lubricant, viscosity depressant, and flow promoter.

The most commonly used melt lubricants or flow promoters for polypropylene are stearates of calcium, magnesium, zinc and sodium, several kinds of microcrystalline waxes, many esters of higher fatty acids, and paraffins. A marked decrease in melt viscosity has also been noticed by adding several organosilicon liquids of low viscosity. Andrianova and Kargin (ref.22), for example, showed that a very small addition of commercially available polyethylensiloxane greases produced a more than ten-fold drop in viscosity.

A characteristic of internal lubricants is that they are effective in relatively low quantities, in the order of 0.1 to 2.0 percent. To be useful as an additive, an internal lubricant must meet several requirements: it must be compatible with the polymer during processing and in the final product; it should not cause a haze or bloom in the product; it should not "plate out" of the resin on the processing machinery; it should not work as an efficient plasticiser; and it should not impart any coloration to the product. These additives are thought to be positioned between the flow elements of the melt or between the polymer mass and the metal surface of the conduit, with the effect of lubricating the unidirectional flow of the melt. An essential qualification of these additives is that although they are compatible with polypropylene, they are not completely soluble in it. Some processing aids may be preferable to others because of desirable side effects. For example, zinc stearate, in addition to its action as internal lubricant (ref.23), also serves as a color stabilizer in the presence of nickel-containing stabilizers. Similarly, nickel stearate may be used as much for its function as internal lubricant as a dye site for nickel-dyeable polypropylene fiber.

A decrease in viscosity of polypropylene melt with small quantities of organosilicons, metal soaps, and microcrystalline waxes cannot be explained on the basis of theories of plasticization, which calls for the interaction of the additive and the polymer on molecular scale, nor can it be explained by the wetting out of the metallic surfaces of the polymer flow channels. As these substances are not soluble in polypropylene (a necessary condition for plasticization) but are miscible with it at the microscopic level, it is more probable that a large decrease in melt viscosity occurs through the sliding of simple structural units (the so-called melt-flow elements) on each other.

A decrease in apparent melt viscosity of polymer melt helps to increase the extrusion capacity of the equipment, decrease the power requirement, and often improves uniformity of the extruded product. Some of the internal lubricants exude to the surface of the fiber in subsequent processing or on aging and may act in lieu of external lubricants or finishes commonly applied to the fiber. Similarly, some internal lubricants or processing aids may also contribute to the luster of the filament by causing the spun yarn to draw without the formation of any microcracks or voids.

The yarns containing internal lubricants are often easier to cold draw. This contributes to the economy of the process as well as to the quality of the product. Of course, the presence of such modifiers also affects the crystallization behavior of polymer and, if added in large quantities, may affect the mechanical properties of the product. More often than not, the presence of processing aids also exerts significant effect on the heat and light stability of the polymer. This is specially the case with soaps of fatty acids. This type of lubricant, such as calcium stearate, on one hand acts as HCl fixer, performing as a mild auxiliary stabilizer and anti-corrosion agent; on the other hand they are reported to enhance the degradation of polypropylene in the presence of trace metals (ref.24). Calcium stearate processing aid also tends to impair the color of the polymer in the presence of trace metal impurities. Other processing aids may not affect the stabilization and color characteristics of the polymer but they may impair the mechanical properties of the product. The cost benefits should therefore be weighed carefully before deciding to use any processing aid.

VII. MEASUREMENT OF FLOW PROPERTIES OF POLYMER MELT

Numerous techniques have been developed for the measurement of rheological parameters of polymer melt. These include the determination of shear stress associated with the corresponding shear rate to assess viscous flow behavior, die-swell ratios and high-speed shearing for the assessment of elastic properties, and simulated melt spinning techniques to assess the tensile properties of the melt. Thus, no single rheometer can adequately describe all parameters; a variety of types have evolved.

Most of the rheological work on polypropylene has been carried out by the use of extrusion-type rheometers (Fig.16). With these rheometers the molten mass is forced through a capillary of known dimension by either mechanical or pneumatic means. The melt is maintained at isothermal conditions by the usual electric or circulating liquid temperature control techniques. Either the extrusion pressure or the volumetric flow rate may be controlled as the independent variable; the other becomes the measured dependent variable. This type of rheometer is specially suitable for measurement required in fiber industry because the flow process in the rheometer capillary is very similar to that in a commercial spinnerette hole. The measurements are fairly easy to carry out and apparent viscosities and elastic content of the melt can be measured at very high shear rates, commonly met in commercial spinning processes. Moreover, since each increment is sheared for only a short time, heat dissipation is less critical than with other methods of measurement.

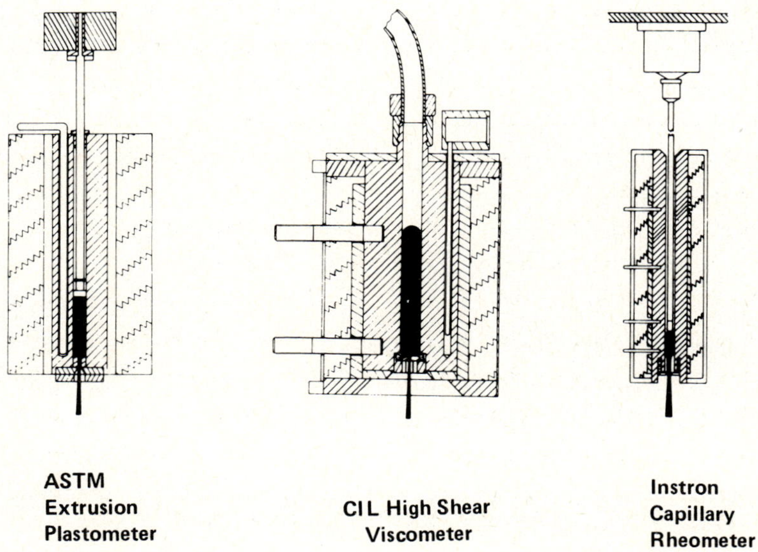

Fig. 16. Three Different Types of Extrusion Rheometers

The most widely used extrusion rheometer is the ASTM extrusion rheometer. Pressure is applied to the melt in the reservoir by means of a piston which is loaded with a standard weight. The flow rate through a capillary of known dimensions is determined by weighing the amount of resin extruded during a measured period of time. A visual observation of the extrudate gives indication of the phenomena attached with flow instabilities. This is specially true for the recognition of the onset of melt fracture or extrudate roughness at excessive shear rates.

An improvement on the simple extrusion rheometer is the C.I.L. high shear viscometer. It also consists of a thermostated barrel and capillary. However, in this case the polymer is driven through the capillary by applied gas pressure at various selected pressures. Another capillary rheometer - considerably more versatile, commercially available, and very popular in industrial laboratories - is the Instron capillary rheometer (ref.25). In addition to measuring the flow rate, this instrument also provides means of measuring melt densities and pressure-volume relationships. Moreover, the occurrence of melt-flow irregularities, such as melt fracture or slip-stick phenomena, may be observed by an irregular trace on the stress recorder.

The shear rate $\dot{\gamma}$ and shear stress δ for Newtonian flow at the wall of a capillary are given by Hagenlack and Poiseulle equations:

$$\dot{\gamma} = 4Q/\pi r^3; \ \delta = \Delta P.r/2l$$

respectively, where Q is flow rate in millimeters per second, r and l are capillary radius and length in centimeters, respectively, and ΔP is the pressure drop through the capillary in dynes per square centimeter. For non-Newtonian flow, $\dot{\gamma}$ is corrected as follows:

$$\dot{\gamma} = (3n+1/4n)\ 4Q/\pi r^3$$

where n is an exponent in Ostwald de Waele's equation $\delta = K.\dot{\gamma}^n$ and can be calculated from the "apparent viscosity curve" (log δ vs log $\dot{\gamma}$). The quantity (3n+1)/4n is commonly referred to as Rabinowitsch correction for non-Newtonian behavior of the fluid. At low shear rates this correction is negligible. Similarly, a correction should be applied to shear-stress equation to compensate the entry factor as follows:

$$S = \Delta P/2(^{l}/r+e)$$

where e is calculated from a plot ΔP versus l/r at constant shear rate.

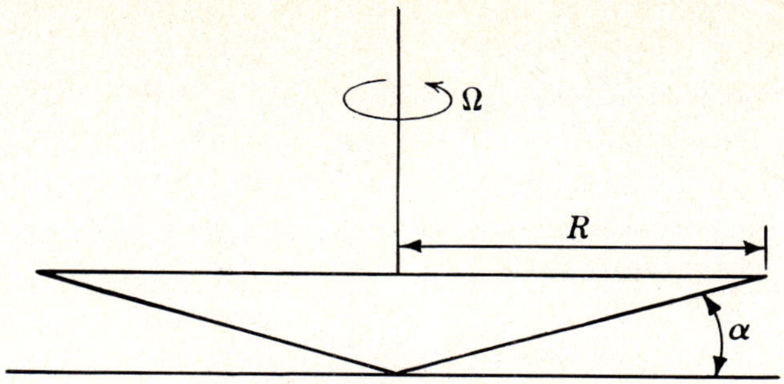

Fig. 17. Schematic Illustration of the Cone-and-Plate Rheometer

Rotational methods of rheometry involve shearing of a sample between two instrument surfaces which are in rotational motion relative to each other. As is the case with capillary instruments, there are numerous examples of rotational rheometers. Perhaps the most common of these is the cone-and-plate rheometer shown schematically in Fig. 17. The sample is placed in the horizontal gap between a rotating conical surface and a stationary flat surface. The viscous resistance of the sample to the motion of the platen is transmitted as a torsional moment M to a torsion bar whose small deformation is measured electronically. The shear stress throughout the sample at any given shear rate is given by:

$$\delta = 3M/2\pi R^3$$

where R is the radius of the cone. As a consequence of the increasing gap width with increasing distance from the axis of rotation, the shear rate is constant over the entire sample and is given (for small cone angles) at each angular velocity Ω by the equation:

$$\dot{\gamma} = \Omega/\alpha \quad (\alpha \text{ is cone angle}).$$

This constancy of shear rate throughout the sample is a major advantage of the cone-and-plate type of rotational rheometers. An additional feature of plate-and-cone rheometers such as Weissenberg Rheogoniometer, is that it is designed to allow direct quantitative measurement of the normal stress tending to separate the platens, which when combined with the shear stress allows the calculation of the shear modulus and elastic recoverable strain. The rotational rheometer, sometimes also called a torque rheometer, is a useful laboratory instrument since it requires a small amount of sample and offers a variety of test variables. The torque rheometer data can be correlated with a number of fundamental properties of the polymer.

The single most important rheological parameter commonly used in practice is the so-called Melt-Flow Index (MFI). In essence, MFI is measured (as grams extruded in ten minutes) by forcing molten polymer through a standard cylindrical die in a flat entry extrusion rheometer under standard conditions of 2.16 kg/cm^2 load and 230° temperature (ref.26). MFI is thus inversely proportional to melt viscosity and is used to define this parameter.

Precise measurement of MFI calls for very careful control of all possible variables and the test is by no means as simple as it appears at first sight. In fact none of the standard methods (ref.26) really defines test procedures accurately enough for polypropylene. According to ASTM D1238, the accuracy of the value of MFI obtainable from a single measurement, as carried out by different operators at different locations, is found to range from $\pm$ 9 percent to $\pm$ 15 percent (95 percent confidence limits) depending upon the magnitude of the MFI. Similar or larger variations arise with the British Standard and ISO test methods. To make matters worse, it is not infrequently found that test equipment is in a poor condition and that operators have adopted techniques of their own which are not always compatible with those of other operators. In practice, therefore, variations on individual measurements of up to $\pm$ 40 percent have been observed on normal MFI levels.

Analysis of the effects of altering the major variables possible in the internationally adopted methods of MFI measurement of polypropylene shows the major part played by the condition of the equipment itself and the consistency of the methods of operation. Table I ranks the variables in order of their importance.

All of the variables can be related, directly or indirectly, to three principal sources.

1) Fundamental polymer variables arising from the viscoelastic nature of the melt.

2) Variations in the extent of oxidative degradation taking place during the test.

3) Failure to maintain constant test conditions and procedure.

The variables attributable to the polymer itself are those relating to its viscoelasticity. The melt viscosity of polymers is not constant but decreases progressively with increasing rate of shear.

Establishment of steady state conditions in the region above the die takes between one and one-and-one-half minutes, so MFIs should not be measured immediately after applying or changing the load. This is particularly important if a quantity of polymer has been forced from the barrel under a load applied by hand, as is sometimes done to bring the piston starting height down to the specification level of 50 to 20 mm. The deviation attributable to this factor can be seen from Fig. 18. The maltreated material gives rise to an initial reduction in MFI which takes about one and a half minutes to return to the equilibrium level.

At 220°C and above, oxidative degradation is a major source of variability in MFI measurements on polypropylene. Exposure of the melt to entrapped oxygen gives rise to a rapid degradation that stops when the oxygen has been used up. The magnitude of the degradation is dependent on the amount of oxygen entrapped; hence large variations in MFI can result from differences in degree of compression of the granules after charging the barrel. The degradation can be eliminated if the test is carried out under a nitrogen atmosphere instead of in air. MFIs measured in air are always higher than those measured under a nitrogen blanket but the differences can be minimized if the charging time is kept as short and as consistent as possible. Degradation does not occur if the test is

TABLE I

Sources of Variability and the magnitude of Their Effects (ref.31)

Factor	Error in extreme instances
1. Poor packing during charging (resulting in oxidation)	+150% or higher
2. Dirty barrel	+100% (very variable)
3. Partially blocked die	-100% (very variable)
4. Piston height	
(a) in excess of standard 50 mm	-50%
(b) within standard 50 mm	-15%
5. Delays between charging and packing (oxidation)	+10%
6. Extended test time (oxidation)	+5% (more if polymer poorly stabilised)
7. Worn die	+10% for 0.5 mm wear in diam.
8. Worn piston (Automatic recording of piston speed only)	+10% for 0.13 mm wear
9. Test temperature	$\pm$1% for $\pm$1°C
10. Variations in load	pro rata

carried out at 190°C.

The importance of maintaining equipment in good condition cannot be overemphasized, as we have seen from consideration of Table I. Most of the test variables can be reduced to an acceptable level if the chosen test method is followed closely. However, it is the author's experience that familiarity with the equipment, or, alternatively, inadequate instruction, frequently causes standards to fall below those set down in the specifications. Examples of all the above variables have been observed in practice. It is not always made clear to operators that MFI measurement is a precise test which requires great care in operation and in maintenance of equipment. Use of suitable automatic measuring devices can overcome some operator errors.

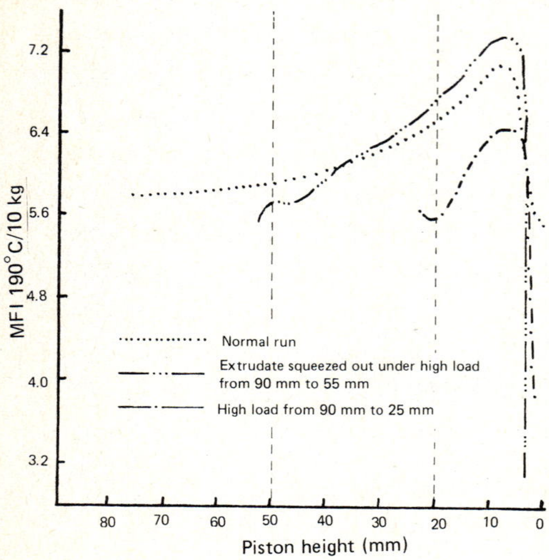

Fig. 18. Effect of Piston Height on Melt Flow Index for Samples with Different Pre-treatments (ref.31)

When the foregoing variables have been controlled there still remains the errors which arise because of variation within the limits of the specification. The most important of these are the results of oxidative degradation, especially in tests at 220°C and above. Variations in packing of the granules during charging of the barrel can give rise to MFIs which are more than two and half time the "true" value, through allowing the polymer to warm up and melt in the presence of entrapped air. The ISO recommendation leaves this problem particularly open and the ASTM method is not entirely satisfactory either, since it merely specifies that charging should be completed within one minute. The preferred technique for MFIs at 220°C and above on polypropylene is as follows:
1) Weigh out the correct amount of granules so that the test can be run at piston heights between 50 and 20 mm without forcing the polymer out by hand. This amount can be predetermined and should never exceed that required to fill the barrel without packing down.
2) Pour the sample into the heated barrel as smoothly and quickly as possible.
3) Ignoring any spilled granules, compress those in the barrel, applying

reasonable force with the "charging tool" to exclude air and produce a clear melt. This should be done quickly in one stage and preferably completed within 10 seconds of commencing charging, although consistency is as important as speed in this operation.

4) Fit the piston into the barrel and carry out the test as defined in the test method using the specified warming-up period.

Oxidation effects are most noticeable with polymers of low MFI because of the greater difficulty in squeezing out the entrapped air during the compression stage. In addition, the amount of polymer charged should be judged accurately enough to ensure measurement in the specified range of piston heights.

Currently, the internationally agreed test conditions for polypropylene are 230°C with a load of 2.16 kg. These conditions are unsuitable for powders where a temperature of 190°C is preferred to minimize degradation. Low loadings (e.g. 2.16 kg) at 190°C give very low flow rates with high molecular weight polymers. Consequently, use of the 190°C/10 kg test is currently recommended for all polypropylene powders. The same argument can be applied to the MFI test with granular polymers, although conditions are less critical. On balance, testing at 190°C/5 kg or 190°C/10 kg is less likely to give rise to errors than testing at 230°C/2.16 kg, especially on samples with low MFI.

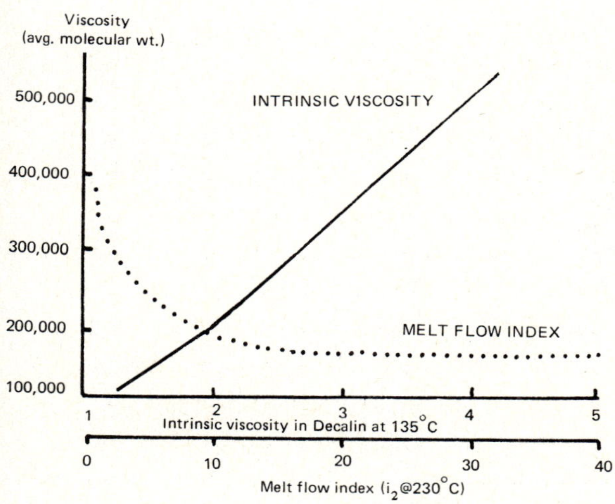

Fig. 19. Relationship Between MW, |η|, and Melt-Flow Index of Polypropylene Over a Broad Range of Molecular Weight (ref.45)

Even when all of the foregoing variables have been brought under control, it appears that significant variations will still exist with single measurements made at different sites using the same sample of polymer. Variations in measurements obtained by the polymer manufacturers should in theory be less than those for single samples (for example as measured by customers) because each of the manufacturer's measurements has the build-in check of measurement made on samples manufactured immediately before and after the material tested. On single measurements, therefore, the acceptance limits adopted by customers should be wider than those of the manufacturing specification.

MFI is a convenient indicator of melt viscosity and molecular weight of the polymer. A typical relationship between melt-flow index and intrinsic viscosity with corresponding molecular weights is shown in Fig. 19. Similar relations have been published by Baijal and Strum (ref.29), Grant and Dieckman (ref.27), and Frank (ref.28). The data shows that melt-flow index is more sensitive to high molecular weight while $[\eta]$ is more sensitive to low molecular weight.

An approximate relationship between intrinsic viscosity and melt index can be represented as:

$$\log [\eta] = A + B \log MFI$$

where A and B are constants depending on molecular weight and molecular-weight distribution of the polymer. For a limited range of MFI, the value of B can be

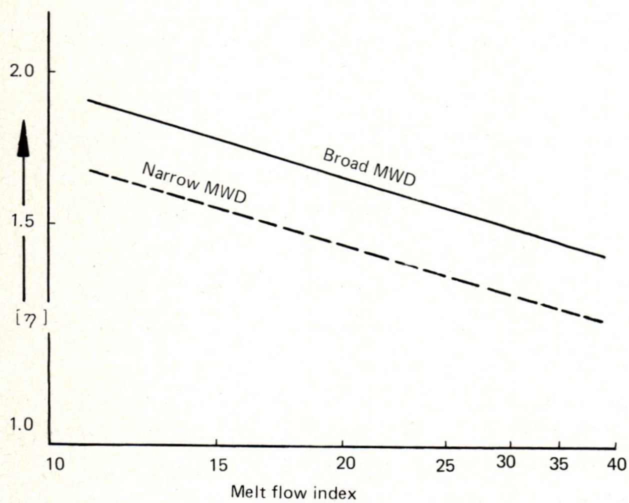

Fig. 20. Typical Relationship Between Intrinsic Viscosity and Melt Flow Index of Polypropylene for Two Different Molecular Weight Distributions.

regarded as constant: the constant A then gives an indication of the molecular-weight distribution of the polymer. Figure 20 depicts this relationship for polymers of two different polydispersities. As indicated, for a given $[\eta]$, the degraded polymers (narrow molecular-weight distribution) the MFI is higher than that for broad molecular-weight distribution, and vice versa. For polymers of broad molecular-weight (undegraded) the following equation may be considered as a good working relationship between the MFI and the $[\eta]$:

$$\log [\eta] = 0.50 - 0.197 \times \log [\mathrm{MFI}]$$

Table II lists some experimental results on Hercules' Pro-Fax[R] polypropylenes.

TABLE II

Melt Flow Index (MFI) and Intrinsic Viscosity $[\eta]$ of Uncompounded Pro-Fax[R] Polypropylene

MFI	$[\eta]$
0.3	4.0
0.6	3.5
3.3	2.5
10.2	2.0
44	1.5

[R] Pro-Fax is a registered trade mark of Hercules, Inc. for their polypropylene resins.

Since melt index is only a one-point determination, its value is rather limited in works other than routine checking. Because flow rates of different molecular-weight distributions respond differently to different shear stresses, the one-point melt index fails to give any information about the flow characteristics of a particular sample at different shear rates. A two-point measurement at two different loads has been suggested (ref.30,31) which is a definite improvement over the one-point method.

Studies of the elastic behavior of polymer melts have generally been carried out by expensive rotational viscometers involving large deformation and recovery periods. The time scale of deformation and recovery and the absolute magnitude of deformation are important factors in determining the elastic properties of viscoelastic melts. Often these factors vary widely in different tests and therefore the results cannot be compared from laboratory to laboratory. A more convenient experimental approach is through capillary rheometry. This approach is specially suitable in connection with fiber research and development as the melt-spinning process closely resembles the experimental extrusion of polymer through a capillary rheometer.

The estimate of elastic properties of melt through extrusion rheometry is, however, an approximate measure. It may be obtained by determining the melt recovery index, MRI, or die-swell ratio, as

$$\text{MRI or Die-Swell Ratio} = (d_e/d_c - 1)\ 100$$

where d_e is the diameter of the extrudate and d_c the diameter of the capillary. It should be obvious that the validity of the results would strongly depend on the reliability of the die-swell determination. A survey of the existing techniques for determination of die-swell indicates that considerable ambiguity exists in these measurements. The following techniques have been tried in order to obtain melt die-swell data to a reliable degree:

a) Vertical extrusion of a short length of polymer which is frozen and subsequently measured with a micrometer (ref.32),
b) Photographing the polymer stream as it emerges from a capillary,
c) A refinement of method a) in which the extrudate is held at the extrusion temperature for a period of time in order to permit complete relaxation (ref.33),
d) A refinement of method b) in which the material is extruded into a heated chamber held at the extrusion temperature until a melt stream of about 3 inches in length has emerged. At this time the flow is stopped by stopping the extrusion, pictures are taken of the lower end of the suspended stream at various time intervals through the glass window until the melt has fully relaxed. The extrudate diameter is then determined by comparison of the melt image to that of a standard (ref.34). This method is expected to yield the most consistent results, although the method is somewhat time consuming,
e) A modification of method d) whereby the extrudate stream is supported on a slowly moving screen (ref.35). By adjusting the screen speed and the shroud's temperature, and by continuously measuring the maximum swell of the molten extrudate, a quick estimate of the polymer's elastic properties can be obtained.

The techniques for evaluation of viscosity and modulus under tension, i.e., the characterization of extensional flow, are few and far between. For materials which undergo viscous flow, it is generally quite difficult to generate steady extensional flows. An apparatus for studying these parameters has been described in detail by Cogswell (ref.36). Meissner (ref.37) has described a technique employing a uniform steady strain rate in a horizontal sample generated by means of two sets of gripping wheels.

Laboratory equipment for the study of extensional flows is now available commercially in Germany through Karl Frank GmbH, Mannheim-Rheinau (ref.38). One device is essentially an attachment for that firm's melt indexer, which provides for drawing off the extrudate at a controlled rate and measurement of the resulting tensile force. This is a simulation of the melt-spinning process, where the rate of extension is not uniform throughout the sample. A second device is an extensional rheometer based on the design of Meissner (ref.37). The firm Gottfert-Feinwerk-Technik GmbH of Buchen/Odenwald reportedly manufactures a device called "Rheotens" which is a controllable draw-off attachment for use with that firm's "Extrusionmeter" laboratory extruder (ref. 38).

A number of investigators have in recent years described results, variously called drawing force, melt strength or draw-down tension, of a simple and apparently useful tests for characterization of some aspects of polymer melt spinning behavior using the extensional rheometers described above. However, at this time it is not clear precisely what material property or combination of properties is being measured by the melt tension.

It has been shown that the force measured in the melt tension test is a function not only of the viscosity of a polymer, but also of its activation energy and cooling rate, and of the extrusion conditions which affect the extrudate die swell. It is unlikely, therefore, that the test can be used to determine differences in elongational viscosity behavior of polymers, without at least making independent measurements of other properties such as low shear rate viscosity and die swell. Melt tension may be a useful tool for correlating process behavior or for specific quality control test purposes. However, it suffers from the disadvantage of all such tests which are influenced by a number of properties; the correlations that exist over a limited range of variables often break down when extended to new polymers or to a wider range of operating conditions.

VIII. REFERENCES

1. A.H.P. Skelland, *Non-Newtonian Flow and Heat Transfer*, John Wiley, New York, 1967.
2. J.D. Ferry, *Viscoelastic Properties of Polymers*, Interscience, New York, 1961.
3. C.D. Han, Polym. Eng. Sci., 11(1971)205.
4. A.B. Metzner, et al, Modern Plastics, 133, Nov. 1960.
5. E.B. Bagley, et al, J. Appl. Polym. Sci., 7(1963)1661.
6. S. Richardson, Rheol. Acta, 9(1970)193.
7. Y. Mori, J. Appl. Polym. Symp., 20(1973)209.
8. C.D. Han, R.R. Lamotte, J. Appl. Polym. Sci., 16(1972)3286.
9. F.N. Cogswell, P. Lamb, Plast. Polym. 38(1965)331.
10. A.B. Bestul, H.V. Belcher, J. Appl. Polym. Phys., 24(1953)696.
11. G.V. Vinogradev, Rheol. Acta, 3(1964)156.
12. S. Kase, T. Matsuo, J. Appl. Polym. Sci., 11(1967)251.
13. T. Ishibashi, et al, J. Appl. Polym. Sci., 14(1970)1597.
14. H.P. Schreiber, J. Appl. Polym. Sci., 9(1965)2101.
15. H.P. Frank, Rheol. Acta, 5(1966)89.
16. G. Natta, M. Compostella, Z. Ges. Textilind., 69(1967)827.
17. H.J.M.A. Mieras, J. Appl. Polym. Sci., 13(1969)309.
18. A.V. Tobolsky, *Principles and Structure of Polymers*, Interscience, New York, 1960.
19. T.E. Davis, et al, J. Polym. Sci., 56(1962)485.
20. A.M. Kotliar, J. Appl. Polym. Sci., 9(1965)507.
21. R. Chiang, J. Polym. Sci., 36(1959)91.
22. G.P. Andrianova, V.A. Kargin, Vysokomol. Soed., A 13(1971)1564.
23. *Encycl. Polym. Sci. Tech.*, Interscience, New York, 11(1969)603.
24. G.A. Russel, SPE Reg. Tech. Conf. June, 1964, Washington.
25. *Encycl. Polym. Sci. Tech.*, Interscience, New York, 8(1968)614.
26. ASTM - D 1238 - 70; ISO Recommendations R1133 Plastics; British Standard 2782: Part I, 1970.
27. D.E. Grant, J. Appl. Polym. Sci., 9(1975)3237.
28. H.P. Frank, *Polypropylene*, Gordon and Breach, New York, 1969.
29. M.D. Baigal, C.L. Strum, J. Appl. Polym. Sci., 14(1970)1651.
30. J.E. Guillet, J. Appl. Polym. Sci., 9(1965)757.
31. D.E. Hansen, Plast. Today (ICI), Nov. 1973.
32. E.B. Bagley, J. Appl. Polym. Sci., 7(1963)1661.
33. K. Kamide, et al, Kabunshi Kagaku, 22(1965)244.
34. C.D. Han, Trans. Soc. Rheol. 14(1970)213.
35. U.S. Patent 3,526,126, Esso Res., 1970.
36. F.N. Cogswell, Plast. Polym. 36(1968)109; Trans. Soc. Rheol. 16(1972)385.
37. J.M. Meissner, Rheol. Acta, 8(1969)78.
38. J.M. Dealy, Polym. Eng. Sci., 11(1971)433.
39. *Encyclopedia of Polymer Science and Technology*, Interscience, New York, 8(1968)591.
40. D. Poller, M.M. Kotliar, J. Polym. Sci., B1(1963)381.
41. V. Cupuccio, Chim.Ind. (Milan), 44(1962)463.
42. Y. Sano, Sen-i Gakkaishi, 24(1968)147.
43. L.D. Moore in *Crystalline Olefin Polymers*, I., Interscience, New York, 1965.
44. S.E. Ross, J. Appl. Polym. Sci., 9(1965)2729.
45. R.W. Ivett, Am. Dye. Rept. 43, March 1965.
46. V. Cuppuccio, Chim.Ind. (Milan), 44(1962)463.
47. S.M. Middleman, *The Flow of High Polymers* Wiley - Interscience, New York, 1968.
48. M. Reiner, *Deformation, Strain and Flow*, Lewis, 1969.

49. Y. Sano, et al, Sen-i Gakkaishi, 24(1968)146.
50. I. Hamana, et al, Melliand Textilber 50(1969)382.
51. T.W. Hill and J.A. Cuculo, J. Appl. Polym. Sci., 18(1974)2569.

IX. BIBLIOGRAPHY

IX.1. Books

1. J.M. McKelvey, Polymer Processing, John Wiley, New York, 1962.
2. E.T. Severs, Rheology of Polymers, Rheinhold Co., New York, 1962.
3. A.G. Fredrickson, Principles and Applications of Rheology, Prentice Hall, Englewood Cliffs. USA. 1964.
4. I.Raff, Ed., Crystalline Olefin Polymers, Interscience, New York, 1965.
5. J.R.A. Pearson, Mechanical Principles of Polymer Melt Processing, Pergamon, 1966.
6. V.F.R. Eirich, Ed., Rheology, Academic Press, New York, 1969.
7. J.A. Brydson, Flow Properties of Polymer Melts, Eliffe, London, 1970.
8. J.D. Ferry, Viscoelastic Properties of Polymers, John Wiley, New York, 1970.
9. R. Lenk, Rheologie der Kunststoffe, Hansen, Munich, 1971.
10. A. Ziabicki, Fundamentals of Fiber Formation, John Wiley & Sons, New York, 1976.
11. C.D. Han, Rheology in Polymer Processing, Academic Press, New York, 1976.
12. S. Middleman, Fundamentals of Polymer Processing, McGraw-Hill, New York, 1977.
13. L.E. Neilsen, Polymer Rheology, Marcel Dekker, New York, 1977.

IX.2. Journal Literature

1. F. Bueche and F.N. Kelley, Shear Viscosity - Molecular Weight Relations for Polymers, J. Polym. Sci., 45(1960)267-269.
2. A.J. DeVries, J. Tuchow, The Non-Linear Viscoelastic Behavior of Molten Polymers, Cahier Groupe Franc, Etudes. Rheology, 5, (1,2,3,4)(1960)173-189, (French).
3. S.J. Kinner, Polymer Rheology, J. Appl. Polym. Sci., 5(1961)55.
4. J.F. Carley, Effect of Static Pressure on Polymer Melt Viscosities, Modern Plastics, 39(1961)123-125, 127-128, 130, 198, 202, 204.
5. J.A. Faucher, Extension of the Eyring-Ree Theory of Non-Newtonian Flow, J. Appl. Phys., 32(1961)2336-2338.
6. R.A. McCord and B. Maxwell, New Light on Melt Elasticity, Mod. Plast., 39(1962)116-730, 232-240.
7. S. Strella, Relation Between Steady Flow and Dynamic Viscosity, J. Polym. Sci., 60(1962)9-11.
8. R.V. Carley, Apparent Increase in Melt Index During Measurement, J. Appl. Polym. Sci., 6(1962)19.
9. D.I. Marshall, Reply to Skinner; Polymer Rheology, J. Appl. Polym. Sci., 6(1962)46-48.
10. R.E. Price, Time Effects in Polymer Melt Viscometers, J. Appl. Polym. Sci., 7(1963)43-45.
11. H. Schott, Viscous Flow and Degradation of Molten Polypropylene, SPE Trans., 3(1963)145-151.
12. R. Sabia, Characterization of Non-Newtonian Flow of Polymers, J. Appl. Polym. Sci., 7(1963)347-355.
13. D. Poller and A.M. Kotliar, Low Shear Viscosity of Polypropylene Melts, J.Polym.Sci., B-1(1963)381-384.
14. J.G. Brodnyan, The Correlation of Non-Newtonian Flow of Polymer Melts with the Theories of Bueche, SPE Trans., 4(1964)277-281.
15. A.K. Van der Vegt, Flow Properties of Molten Polypropylene, Plast. Inst. (London), Trans. J., 32(1964)165-169.

16. G.V. Vinogradov, Rheology of Polymers. Viscous Properties of Polypropylene, Rheol. Acta, 3(1964)156-163.
17. J. Ui, Flow Properties of a Polymer Melt Studied with a Screw Extruder, SPE Trans., 4(1964)294-305.
18. R. Sabia, On the Characterization of Non-Newtonian Flow - Part II, J. Appl. Polym. Sci., 8(1964)1053-1061.
19. D. Poller, O.L. Reedy, Designing Dies for Monofilaments, Mod. Plastics, 41(1964)133-138.
20. T.W. Huseby, Structure-Property Studies in Polymer - Melt Rheology, Polym. Eng. Sci., 3(1965)130-134.
21. R. Longworth, Melt Index as a Measure of Newtonian Melt Viscosity, J. Polym. Sci., B3(1965)221-226.
22. H.P. Schreiber, Separation of Elastic and Viscous Effects in Polymer Melt Extrusion, J. Appl. Polym. Sci., 9(1965)887-892
23. D. Poller, A.M. Kotliar, Effects of Degradation on Polypropylene Melt Rheology, J. Appl. Polym. Sci., 9(1965)501-512.
24. D.E. Grant, S.F. Dieckmann, Melt Flow Properties of Polypropylene, J. Appl. Polym. Sci., 9(1965)3231-3244.
25. A.P. Metzger, J.D. Matlack, Extrudate Swelling at High Shear Rates, J. Polym. Sci., B3(1965)921-922.
26. H. Kobayashi, Viscous Flow of Molten Polypropylene, Bull, Inst. Chem. Res. Kyoto Univ., 43(1965)179-192.
27. H. Schouhorn, Surface Tension of Molten Polypropylene, J. Polym. Sci., B3(1965)235-237.
28. W. Pillippoff, R.A. Stratton, Correlation of the Weissenberg Rheogoniometer with Other Methods, Trans. Soc. Rheol., 10(1966)467-487.
29. J.A. Boydson, Some Uses and Abuses of Rheology in Polymer Processing, Plast. Inst. (London) Trans. J., 34(110)(1966)65-71.
30. A.Y. Malkin, C.V. Vinogradov, Universality of the Temperature-Independent Viscosity Characteristics of Polymer Melts and Some of its Consequences, J. Appl. Polym. Sci., 10(1966)767-771.
31. G.V. Vinogradov, Rheological Properties of Polymer Melts, J. Polym. Sci., A2,4(1966)135-154.
32. H.P. Frank, Rheological Measurements on Melts of Polypropylene. I. Relationship Between Melt Viscosity Molecular Weight and Distribution of Molecular Weight, Rheol. Acta, 5(1966)89-93.
33. J.F. Carley, Fundamentals of Melt Rheology, Heat Generation and Heat Transfer as Applied to Polymer Processing, Polym. Eng. Sci., 6(1966)158-168.
34. R.A. Mendelson, The Temperature Dependence of the Melt Viscosity of Crystalline Polypropylene, J. Polym. Sci., B5(1967)295-299.
35. I.J. Duvdevani, Analysis of Polymer Melt Flow in Capillaries Including Pressure Effects, SPE J, 23(1967)41-45.
36. J. Zimmerman, Some Quantitative Aspects of Bulge Formation at the Exit of a Spinnerette Capillary, Appl. Polym. Symp. No. 6, (1967)19-26.
37. W. Weber, Rheology (a review), Fortsch. Verfahrentech. 1966-67. 8(1968)34-55.
38. A. Ziabicki. Rheological Aspects of Fiber Spinning, Ind. Chem. Belge., 33(1968)46-50.
39. J.W. Lyons, Rheology, in Kirk-Othmer Encycl. of Chem. Tech., 2nd Ed., (1968)423-444.
40. G. Petraglia, Rheology of Polypropylene, Annu. Tech. Conf. SPE, Pap. 26(1968)168-171.
41. H.P. Frank, Rheological Measurements of Polypropylene Melts III. Effect of Molecular Weight and Molecular Weight Distribution on the Flow Curves of Polypropylene, Rheol. Acta, 7(-968)344-348.
42. V. Semjonow, Melt Viscosities of Polymers, Forsch. Hochpoly. Forsch., 5(1968)387-450.
43. R.A. Mendelson, Prediction of Melt Viscosity Flow Curves at Various Temperatures for Some Olefin Polymers and Copolymers, Polym. Eng. Sci., 8(1968)235-240.

44. C.L. Sieglaff, Rheological Phase Transition in Polypropylene, J. Macromol Sci. Phys., 2(1968)793-797.
45. H. Mieras, Elastic Behavior of Some Polymer Melts, Nature, 218 (S144)(1968)865-866.
46. F.N. Cogswell, Rheology of Polymer Melts Under Tension, Plast. Polym. 36(1968)109-111.
47. R.C. Kowalski, Introduction to Melt Rheology and Application to Extrusion, SPE J, 24(1968)47-51.
48. G.M. Fehn, Steady-State Drawing of Polymer Melts, J. Polym. Sci., A-1, 6 (1968)247-251.
49. J. Meissner, Rheometer for the Study of Mechanical Properties of Deformation of Plastic Melts Under Definite Tensile Stress, Rheol. Acta, 8(1969)78-88.
50. M.A. Matovich, J.R.A. Pearson, Spinning of Molten Threadline: Steady State Isothermal Viscous Flows, Ind. Eng. Chem., 8(1969)512-520.
51. M. Zidan, Rheology of Spinning Process, Rheol. Acta, 8(1969)89-123.
52. R.L. Combs, Effects of Molecular Weight Distribution and Branching on Rheological Properties of Polyolefin Melts, J. Appl. Polym. Sci., 13(1969) 519-534.
53. R.M. Gwynne, Reliability of Capillary Rheometer Measurement of Polymer Melts, J. Appl. Polym. Sci., 13(1969)1069-1072.
54. F.N. Cogswell, Tensile Deformation in Molten Polymers, Rheol. Acta, 8(1969)187-194.
55. H.P. Frank, Rheological Measurements of Polypropylene Melts - IV. Effect of Low Concentrations of Low MW Organic Compounds, Angew. Makromol. Chem., 9(1969)106-110.
56. C.D. Han, Correlation of Exit Pressure with Die Swell Ratio, Al Chem. E. J., 16(1970)1087-1088.
57. W. Graessley, Die Swell in Molten Polymers, Trans. Soc. Rheol. 14(1970)519-544.
58. R.I. Tanner, Theory of Die-Swell, J. Polym. Sci., A2,8(1970)2067-2078.
59. D.P. Thomas, Influence of Molecular Weight on the Melt Rheology of Polypropylene, SPE Annu. Tech. Conf. 20(1970)5-12.
60. G. Petraglia, Rheology of Polypropylene - Influence of Molecular Weight Distribution on Melt Elasticity in Capillary Flow, Polym. Eng. Sci., 10(1970)148-153.
61. C.D. Han, Method of Determining the Melt Elasticity in Capillary Flow, Polym. Eng. Sci., 10(1970)148-153.
62. C.D. Han, Determination of Polymer Melt Die Swell, Trans. Soc. Rheol., 14(1970)213-218.
63. M.D. Baijal, Melt Flow Rate - Intrinsic Viscosity Correlation for Polypropylene, J. Appl. Polym. Sci., 14(1970)1651-1653.
64. N. Siskovic, Viscoelastic Behavior of Molten Polymers in Porous Media, A 1 Ch E J, 17(1971)281-285.
65. C.D. Han, Effect of Temperature on the Elastic Properties of Polymer Melts, Polym. Eng. Sci., 11(1971)205-210.
66. C.D. Han, Rheology of Shaped Fiber Formation, J. Appl. Polym. Sci., 15(1971) 1091-1097.
67. J.M. Dealy, Extensional Flow of Non-Newtonian Fluids - a Review, Polym. Eng. Sci., 11(1971)433-445.
68. C.P. Sharma, Stratified Flow of Polymers, Text. Res. J., 41(1971)552.
69. A. Peterlin, Rheology of Polymer Systems, J. Elastoplast, 4(1972)112-136.
70. F.N. Cogswell, Converging Flow of Polymer Melts in Extrusion Dies, Polym. Eng. Sci., 12(1972)64-73.
71. C.D. Han, Melt Spinning I. Effect of Molecular Structure and Molecular Weight Distribution on Elongational Viscosity, Trans. Soc. Rheol., 16(1972)447-472.
72. F.N. Cogswell, Measuring the Extensional Rheology of Polymer Melts, Trans. Soc. Rheol., 16(1972)383-403.
73. Y. Chen, Die Swell, Advanc. Polym. Sci. Eng. Proc. Symp. 239-247 1972, Plenum, New York, 1972.

74. B.P. Levitskii, Influence of the Spinnerette Hole Dimensions on the Filament Diameter, Fiber Chem. 2(1972)491-494.
75. O.M. Yakhno, Polymer Flow through Spinnerette Holes of Dissimilar Shapes, Fiber Chem. 4(1972)34-35.
76. Y. Mori, Die Swell in a Molten Polymer, Appl. Polym. Symp. No. 20., (1972)209-220.
77. D.E. Hanson, Measurement of Melt Flow Index of Polypropylene, Plast. Today, 46(1973)7-9.
78. K.F. Wissbrun, Interpretation of Melt Strength Test, Polym. Eng. Sci., 13(1973)342-345.
79. R. Longworth, Measurement of Melt Flow and Melt Stability, Polym. Eng. Sci., 14(1974)98-100.
80. R.C. Kowalski, Melt Rheology and Its Application to Extrusion, SPE Reg. Tech. Conf. Newark, 1974.
81. K. Maciejewski, Rheological Flow Behavior of Polymer Blends, SPE Tech. Pap. 1974, 20(1974)23-27.
82. A.T. Serkov, Rheological Force, Fiber Chem., 6(1974)142-144.
83. J. Meissner, Development of Experimental Methods for the Study of Rheological Behavior of Polymer Melts, Colloid Polym. Sci., 252(1974)760-765.
84. K. Maciejewski, Rheological Flow Behavior of Polymer Blends, SPE Tech. Pap., 20(1974)23-27.
85. G. Marrucci, Limiting Concepts in Extensional Flow, Polym. Eng. Sci., 15(1975)229-33.
86. V.G. Kulichikhin, General Principles and Methods of Investigating the Rheology in the Uniaxial Extension of Polymeric Systems, Fiber Chem., 7(1975)20-30.
87. F.N. Cogswell, Polymer Melt Rheology During Elongational Flow, Appl. Polym. Symp. 27, 1975.
88. M. Matsui, Non-Isothermal Rheological Response in Melt Spinning and Idealized Elongation Flow, Polymer Eng. and Sci., 16(1976)735.
89. A.E. Everage, Extensional Deformation of Polymer Melts, J. Appl. Polym. Sci., 21(1977)841.
90. J.L. White, End Pressure Losses in Extrusion of Polymer Melts Through Dies, J. Appl. Polym. Sci., 21(1977)2289.
91. J.W. Hill, Elongational Flow Behavior of Polymeric Fluids, J. Macromol. Sci. Rev. Macromol. Chem., 14(1976)107-153.

Chapter V. FUNDAMENTALS OF MELT SPINNING

I. INTRODUCTION

The melt spinning process received wide attention from scientists and engineers since the early days of the synthetic fiber industry. Studies dealing with polypropylene fibers form only a small part of the published literature, but additional insight into the process can be gained from results on other fiber-forming polymers such as the polyamides and polyesters.

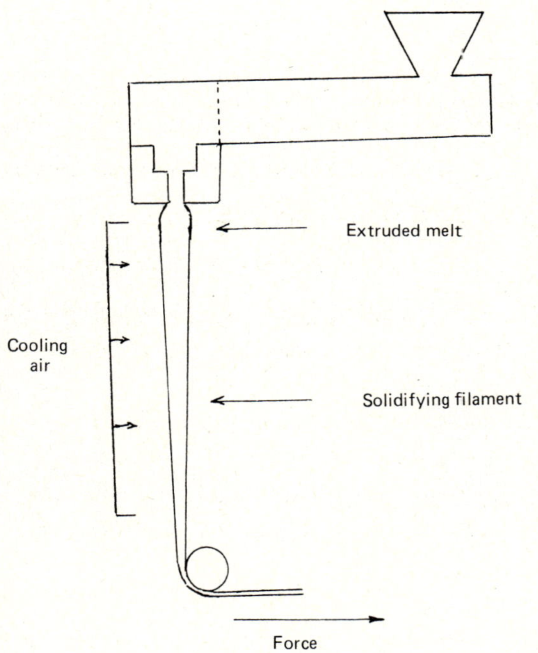

Fig. 1. Schematic Illustration of Melt Spinning.

In the process of melt spinning, a polymer melt is formed into filaments by extrusion through a spinnerette and simultaneously applying an extensional force to the extrudate. As the extrudate thins down, it solidifies on cooling below its melt temperature and transforms into a filament. This process is outlined schematically in Fig. 1. A spinnerette can be visualized as a plate with many holes. For textile and industrial yarns, spinnerettes with 10 to 150 holes are used, while spinnerettes with as many as 2,000 holes are used to spin staple fibers.

The physical phenomena which are important in melt spinning may usefully be grouped into three areas: (1) the rheological behavior of the melt as a jet is formed in the spinnerette capillaries; (2) the attenuation and cooling of the liquid jet; and (3) the crystallization of the jet and development of fiber structure. The studies in melt spinning may also deal with the optimization of fiber properties for any particular use and, by implication, with the optimization of process for the desired properties of the fiber. In this chapter, we will confine our discussion to the fundamental principles which are involved in the melt-spinning process in general and in the melt-spinning of polypropylene in particular. The practical utilization of these principles will be indicated wherever possible but no attempt will be made to discuss specific products or specific processes. Such a discussion will be postponed to subsequent chapters.

II. FLOW THROUGH SPINNERETTE CAPILLARIES

The flow pattern of molten polymer through the holes of a spinnerette is of great importance in the process of melt spinning. The mode of flow through spinnerette holes determines the technical parameters affecting the spinning behavior of polymer under imposed process conditions. Flow within the capillary is also closely associated with phenomenon appearing in the exit zone. A somewhat idealized view of such flow is shown in Fig. 2a and the corresponding stress and strain rates in Fig. 2b.

The capillary approach or "counter bore" section of the spinnerette hole is several times larger in diameter than the capillary section. Flow in Zone A of this section is characterized by small shear rates and relatively high viscosities. The behavior of melt flow in this zone is predictable from the melt index or intrinsic viscosity of the polymer used.

The flow behavior in Zone B, just above the entrance into the capillary, is quite complex. The flow in this entry zone, which can strongly affect the distortion of the extrudate in the exit region, has been investigated using color motion pictures (ref.1). The flow patterns of polypropylene consist of Newtonian-like streamlines, oriented radially inward, as shown in Fig. 3. The behavior persists over the entire range of stable laminar flow. It has also been shown (ref.1) that the cone angle of the entrance flow pattern decreases as the flow rate increases. At some critical flow rate, the tip of the cone begins to oscillate and whip around the entrance of the capillary. This critical flow rate coincides with the onset of flow instability, called "melt fracture".

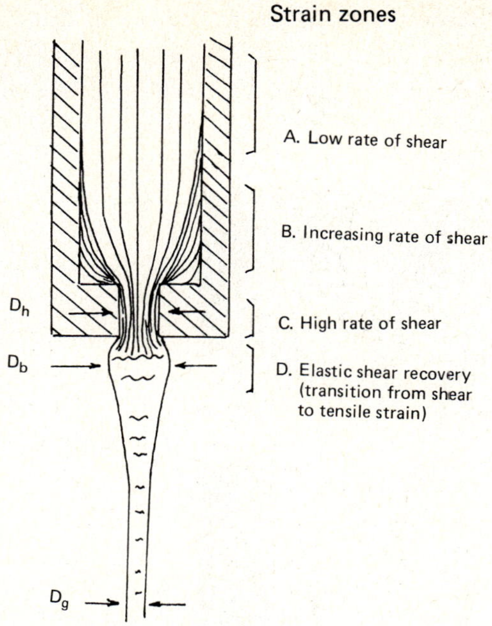

Fig. 2a. Rheological Analysis of Flow Through a Capillary Showing Various Strain Zones (ref.57)

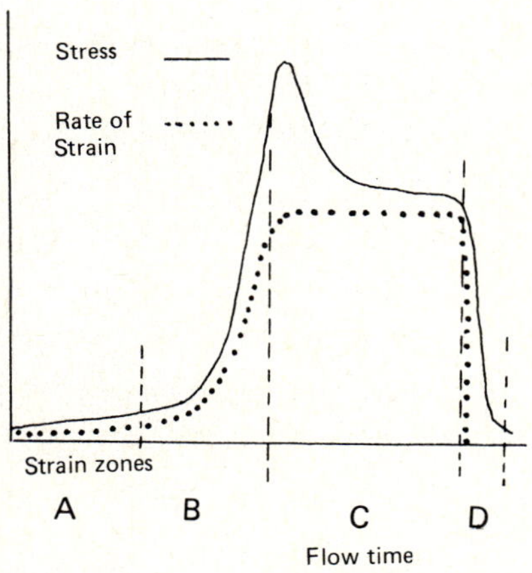

Fig. 2b. Stress and Rate of Strain in the Various Strain Zones of Capillary Flow and Drawdown (ref.57)

(See Chapter VI for a discussion of this phenomenon.)

In the case of a flat entrance to the capillary, such as that shown in Fig. 2, the polymer may reside (stagnate) for an extended period of time in the corners of the entry zone and be subjected to unwanted thermal degradation. A portion of the degraded material can periodically enter the main stream of flow through the capillary and produce instabilities in the spinning process. The stagnant region may be reduced by steamlining the entrance to the capillary as shown schematically in Fig. 4. The optimal taper angle depends on the rheological characteristics of the melt; high-viscosity melts such as polypropylene requiring smaller angle than low-viscosity polymers such as nylon. A taper of 60° to 90° is generally employed commercially.

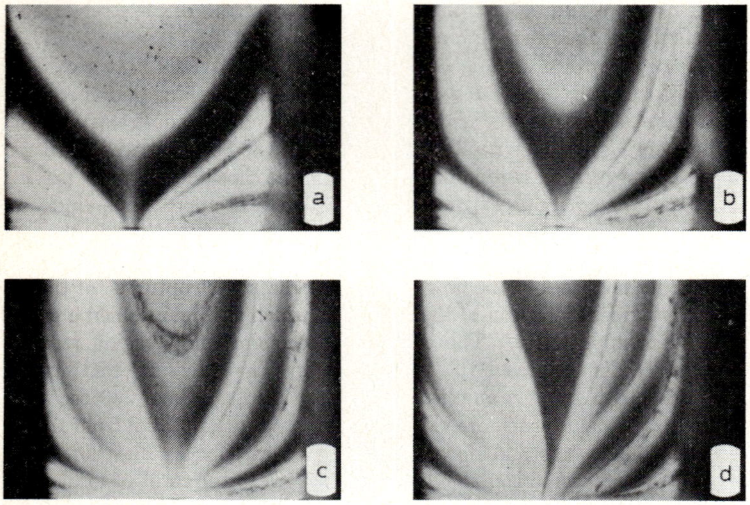

Fig. 3. Entrance Flow Patterns for Isotactic Polypropylene for Increasing Capillary Shear Rates. Temperature is 180°C. (a) Shear rate = 2.1 sec^{-1}; (b) shear rate = 22.1 sec^{-1}; (c) shear rate = 45.1 sec^{-1}; (d) shear rate = 236 sec^{-1} (spiralling instability) (ref.1)

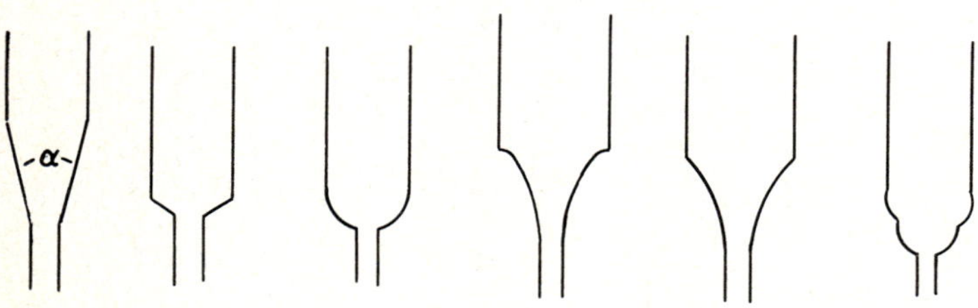

Fig. 4. Some Examples of Tapered Capillary Approach

The flow of polymer in Zone B is characterized by a high rate of stress rise, a rapidly decreasing viscosity, and the buildup of recoverable (viscoelastic) shear strain. Since polypropylene melts are highly non-Newtonian, and since the rheological behavior depends strongly on the molecular weight, molecular-weight distribution, and concentration of additives in the polymer, the spinning behaviors of different polypropylenes and different polymer formulations are markedly differentiated in Zone B.

The flow within the capillary, i.e., in Zone C, is characterized by a nearly constant shear rate and by a shear stress which decreases in the first portion of the zone and then remains constant. The recoverable strain decreases in this zone due to the decrease in stress and to relaxation of the polymer melt. The dimensions of the capillary determine the pressure drop, ΔP, required for obtaining a given volume output, Q, of polymer. The capillary dimensions also determine the extent of relaxation of elastic strains bearing upon die swell and specify the conditions for the onset of instabilities in flow ("melt fracture").

For non-Newtonian fluids whose shear-dependent viscosity can be approximated by Power Law (see Chapter IV):

$$\eta = \eta_o[\dot{\gamma}]^{n-1}$$

radial velocity distribution within the capillary reads:

$$V_x(r) = V_x(o)[1-(r/R_o)^{\frac{n+1}{n}}]$$

where x and r are cylindrical coordinates. By further manipulations of this equation, one obtains the total output Q and velocity gradient γ_ω on the capillary wall:

$$\dot{\gamma}_\omega = (\partial V_x/\partial_r)|_{r=R_o} = -(\Delta PR_o/2\eta\ell_o)^{\frac{1}{n}}$$

At n = 1 (Newtonian fluids) the above equations reduce to the classical Poiseuille theory with parabolic velocity distribution and output::

$$Q = \Pi R_o^4\ \Delta P/_{8\eta_o\ell_o}$$

Polypropylene melts usually exhibit shear-decreasing viscosities which correspond to exponent n in Power Law equation being less than unity. For such systems the velocity profile is more flat than parabolic, and output Q is greater than that predicted for Newtonian fluids. Under fiber-spinning conditions effective viscosity of polypropylene melt can be many times less than its initial value, η_0, and so also is the pressure difference ΔP required for a given output value Q.

All that has been said above concerns steady-state flow in very long capillaries. In fiber spinning we have to deal with polymer extrusion through channels whose length-to-radius ratio (L_o /R_o) lies within 1 to 4 and whose so-called transit times t* ($t^* = \Pi R_0^2\ell_0/Q$) are comparable with relaxation times of the melt.

In such cases the entrance and relaxation effects become important. The effect of finite capillary length, L_o, is often characterized by the ratio of the effective pressure difference, ΔP_{eff}, i.e., that responsible for the true output rate Q, to the applied pressure difference, ΔP_{appl}, as follows:

$$\Delta P_{eff}/\Delta P_{app} = \frac{1 - m\ell\, V_o^2/\Delta P_{app}}{1 + (R_p/L_o)\,(k + S_R/2) + Z(t^*/\tau)}$$

Here V_o is the average velocity in the capillary, as $V_o = Q/\ R_o^2$,k and m are constants, ℓ is the density of the polymer, R_o and L_o are the radius and length of the capillary, S_R is the recoverable shear strain and Z is the ratio t^*/τ where t^* is the average residence time or "transition time" of the melt in the capillary, given by $t^* = R^2 L_o/Q$, and τ is the relaxation time. The transition times for fiber spinning are of the order of 0.1 to 100 m sec., while the relaxation times for high-viscosity materials are as long as 100 to 700 m sec. Such polymer melts would therefore not relax much within the capillary. It is evident from the above equation that $\Delta P_{app}/\Delta P_{eff}$ should decrease as the ratios L_o/R_o and τ/t^* increase.

When the polymer emerges from the capillary (Zone D), the character of the flow changes rapidly. It is characterized by a rapid decrease in shear rate and shear stress, partial cooling of the melt, an increase in apparent viscosity, and the release of elastic energy stored in the melt. Viscoelastic polymer jets, such as that of polypropylene melt, swell after leaving the capillary in a manner depicted in Zone D of Fig. 2a. This enlargement phenomenon, found in all polymers to a varying degree, is particularly evident in the case of polypropylene. The "die swell", sometimes also known as 'Barus effect' and quantitatively defined as $x = D_{max}/D_o$, usually amounts to 2 to 3 in commercial spinning of polypropylene, although under extreme conditions values far larger have been observed. Here D_o is the diameter of the capillary and D_{max} is the maximum diameter attained by the extrudate.

The swelling of the emerging jet can cause difficulties in the elongation to which the molten filament is subjected during spinning. It can have a profound effect on the speed of spinning as well as the general quality of the fibers produced; and for fibers with profiled cross sections, significant distortions in shape can result. Because of its importance in commercial operations, the phenomenon of die swell has been extensively investigated and a variety of empirical approaches to the problem has been adopted.

Most experimental studies of die swell have been concerned with flow through long capillaries. Commercial fiber-spinning operations employ relatively short capillaries in which the liquid jet does not attain a steady-state velocity profile. The actual situation of bulge formation under fiber spinning conditions is further complicated by the fact that the liquid jet is subjected to a positive pull by the take-up godet. Despite these differences, the functional dependences noted in studies of flow in long capillaries have provided useful insight into the die-swell phenomenon in the course of spinning.

Some typical results for the swelling of polypropylene melt emerging from relatively long capillaries are shown in Fig. 5. Fig. 5a shows that the die swell increases as the throughput rate increases. An extension of the curve in Fig. 5a would show that as the fluid motion becomes slow enough for the fluid to be Newtonian, the die swell approaches a value of about 1.1. At high throughput rates, "melt fracture" occurs (see Chapter VI) and the die swell is not uniquely defined. Fig. 5b shows the effect of capillary length and Fig. 5c shows the effect of diameter on die swell. Die swell rapidly decreases with increasing capillary length and increasing capillary diameter. A limit is reached, however, beyond which die swell practically does not change.

Die-swell phenomenon is conventionally explained on the basis of the storage of elastic energy on entering the capillary and its release on leaving the capillary. Polymer flow down the axial region of the capillary is characterized by a rapid, irrotational deformation on entering the capillary (a cylindrical element becomes longer and thinner). As the material flows through the capillary, some relaxation of the orientation associated with this deformation takes place; and as it leaves the capillary a significant and rapid recovery takes place (the volume element shortens and thickens). The observed decrease in die swell with increasing L/D_o ratio of the capillary (Fig. 5b) reflects the relaxation of the stored elastic energy as the liquid passes through the capillary. The observed decrease in die swell with increasing diameter of the capillary (Fig.5c) or with decreasing throughput rate (Fig. 5a) reflects the smaller pressures - and hence smaller amounts of stored energy for a given die configuration.

On the molecular level, the swelling of the extrudate at the capillary exit is attributed to a strain recovery process of the entanglement network, and to a transition from an approximately parabolic velocity distribution in the die to a constant distribution in the extrudate. A useful working hypothesis is to assume that the molecules are oriented in the die and on emergence into the atmosphere recoiling takes place, with a contraction in the direction of flow being offset by lateral expansion.

The stored energy, and hence the die swell, is expected to increase with η, the apparent viscosity of the polymer melt. Since η decreases with increasing temperature and with decreasing molecular weight, the die swell is expected to show similar variations. These are illustrated by the data in Table I for two polypropylene resins of different intrinsic viscosities at three extrusion

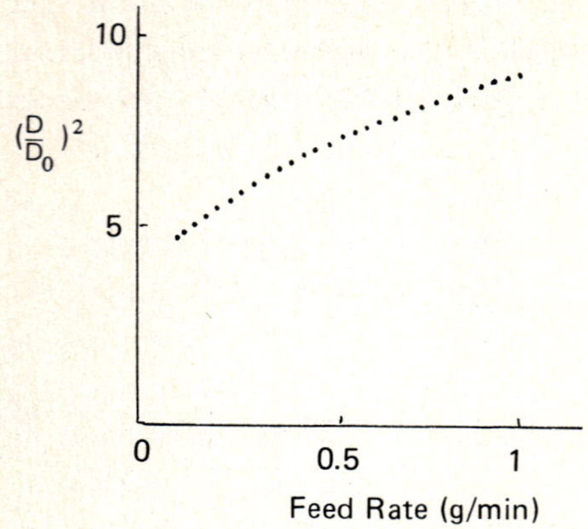

Fig. 5a. Effect of Throughput on Die-Swell Ratio (ref.58)
T = 190 C; L = 8 mm;
D = 1 mm; [η] = 1.9

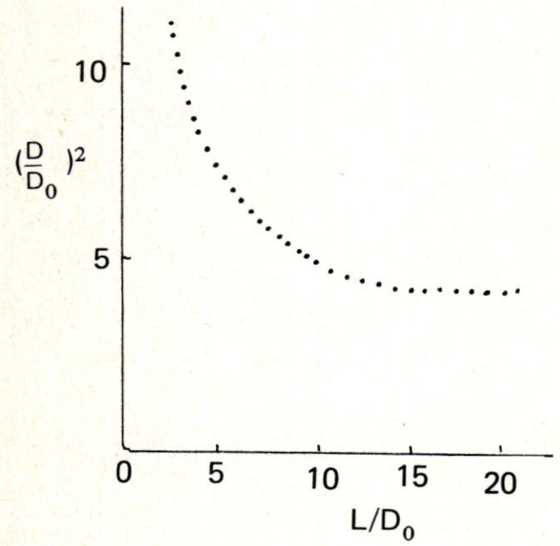

Fig. 5b. Effect of L/R on Die-Swell Ratio (ref.58)
T = 190 C; feed rate = 0.3 g/min;
D = 0.8 mm; [η] =1.9

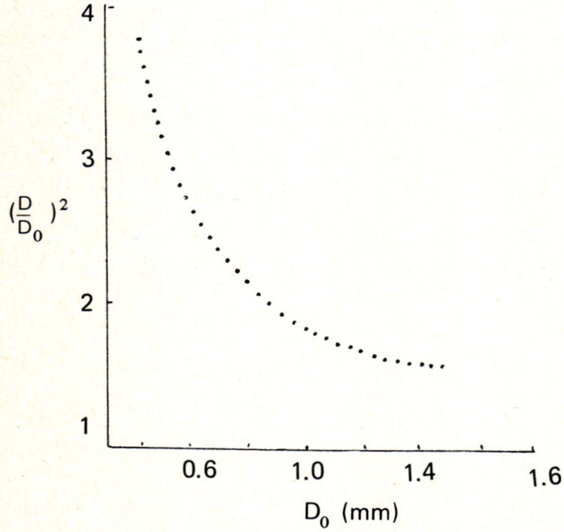

Fig. 5c. Effect of Capillary Diameter on Die-Swell Ratio (ref.58)
T = 280 C; feed rate = 0.3 g/min; L/D = 20
[η] = 1.9

temperatures. Polymers with broad molecular-weight distribution generally display greater die swell than those with narrow distribution for equivalent melt viscosity, very likely reflecting a sensitivity to the high molecular-weight portion of the distributions.

TABLE I

Effect of Melt Temperature and Viscosity on Die Swell For Two Polypropylenes (ref.58)

Temperature (^{o}C)	*Die Swell*	
	High Viscosity	*Low Viscosity*
190	4.2	2.8
230	2.8	1.5
280	2.1	1.3

Q = 0.5 g min^{-1}; capillary length = 16 mm; capillary diameter = 0.8 mm

The shape or entry angle of the capillary also has a significant effect on die swell. Flat (180°) entry holes usually produce more swelling than tapered entries to the capillaries. This is expected from the fact that sudden changes in flow pattern produce larger elastic strains in the polymer. Data obtained (ref.4) on polypropylene extruded through capillaries with length-to-diameter ratios of 8-1 at a shear rate of 1000 sec^{-1} indicate a small decrease in die swell with changes in entrance angle from 180° to 90° but little change for variations in entrance angle from 90° to 20°. Similar data have been reported (ref.5) for capillaries with length-to-diameter ratios of 20-1.

The occurrence of die-swell limits the minimum size of filament which can be obtained from any combination of capillary-polymer-throughput-temperature. The curves in Fig. 6a illustrate the dependence of the minimum size of polypropylene filaments on throughput, capillary diameter and length-to-diameter ratio. As seen in Fig. 6a, finer filaments are obtained under those conditions that produce smaller die swell (small throughput, large length-to-diameter ratio). As seen in Fig. 6b, the minimum filament size depends not only on the diameter of the extruded material (in which case its variation would follow the broken line) but also on the extent of the enlargement.

III. ATTENUATION AND COOLING OF THE FILAMENTS

The attenuation of the enlarged (die-swollen) filament takes place concurrently with its cooling and with changes in orientation, and after some period, with the

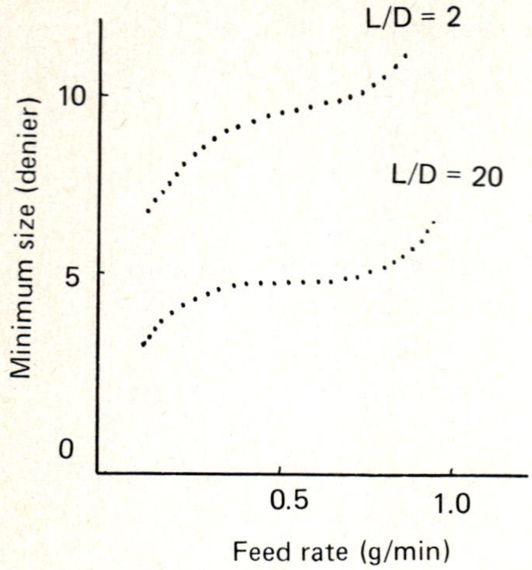

Fig. 6a. Effect of Feed Rate and L/R Ratio on Minimum Filament Size (ref.58)

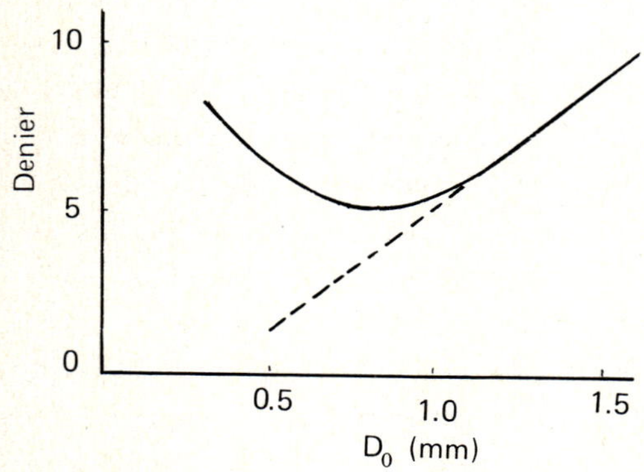

Fig. 6b. Effect of Capillary Diameter on Minimum Filament Size (ref.58)

development of crystallinity and the characteristic fiber structure. The development of microstructure in turn determines the processibility of the spun fiber and its ultimate physical properties.

The dynamics of filament attenuation in steady-state melt spinning prior to the onset of crystallization can be described (ref.6) by the following approximate relations:

$$v(\partial T/\partial x) + (\partial T/\partial t) = -(2\sqrt{\pi}h/\sqrt{A}\rho C_p)(T-T^*)$$

$$(\partial v/\partial x) = (F/\beta A)$$

$$v(\partial A/\partial x) + (\partial A/\partial t) = -A(\partial v/\partial x)$$

Here x is the distance from the spinnerette; t is time; T is the temperature of the filament and T* is the temperature of the surrounding air; v is the velocity of the filament, A its cross-sectional area, ρ its density, C_p its heat capacity and β its tensile viscosity; h is the heat transfer coefficient at the surface of the filament; and F is the filament tension.

It is assumed in the above equations that: (1) the filaments are circular in cross section, (2) the temperature T is uniform across the whole cross section, (3) heat conduction within the filament in the x direction is negligible, (4) die swell can be neglected, (5) the tensile viscosity of the polymer is dependent only on temperature, and (6) the tension on the filament depends on time but is independent of position.

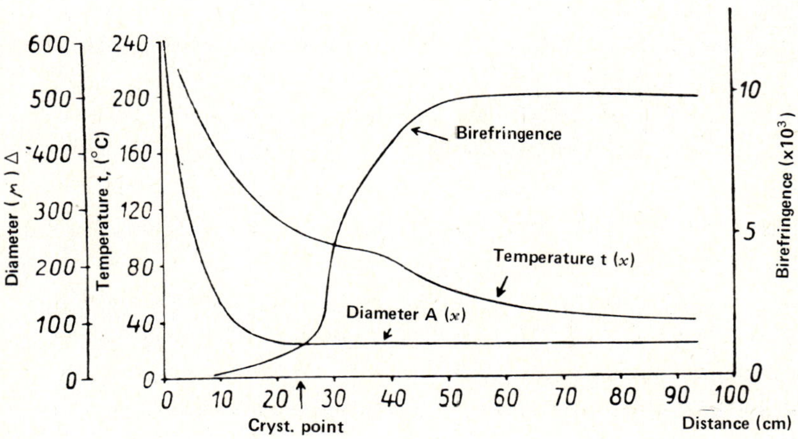

Fig.7. Changes in Filament Diameter, Birefringence, and Surface Temperature Along the Spinning Path for Polypropylene (ref.52).

These assumptions are rather drastic. As we have seen, polypropylene melts exhibit large die swell. Further, their tensile viscosities are dependent on elongation rate as well as on temperature (ref.7), and the force F is a function of distance from the spinnerette (ref.8). Despite these simplifications, a good correlation has been shown between the solutions of these equations and experimental data (ref.6,9). The principal findings can be summarized with reference to Fig. 7 as follows (ref.9):

(i). Increasing the diameter of the spinnerette hole with no change in take-up speed or take-up denier moves the diameter, A(x), and temperature, T(x), curves in the x direction and increases the filament tension F.

(ii). Increasing the take-up speed while maintaining a constant take-up denier moves both the A(x) and T(x) curves in the direction of x but changes F very little.

(iii). Placing an insulated cylinder below the spinnerette delays the thinning and cooling of the polymer stream and F is thereby decreased.

(iv). Decreasing the temperature of the cooling air produces more rapid thinning of the filament, and the filament tension increases.

(v). Increasing the speed of the cooling air has the same effect as lowering its temperature.

(vi). Raising the spinnerette temperature raises T(x) and decreases the filament tension, but has little effect on A(x).

(vii). Increasing the output at a constant take-up speed greatly increases the distance required for the filament to solidify, but has little effect on the filament tension.

(viii). Changing the intrinsic viscosity $[\eta]$ of the polymer, at least for materials like polypropylene with small activation energies for extensional flow, does not greatly affect A(x) or T(x).

More refined treatments of the thermal and rheological aspects of fiber spinning have been developed (see Bibliography). Die swell has been included in these calculations. Direct measurements of the filament thickness and temperature at various locations down the spin line have been carried out. Close agreement is found between the measured temperature and that predicted theoretically for all locations down the spin line prior to the onset of crystallization. Once crystallization is initiated, the latent heat of fusion must be included in the thermal analysis. By its inclusion, informations have been obtained about the rate of crystallization in the spinning operation.

IV. SOLIDIFICATION AND STRUCTURE FORMATION

The transformation of an attenuating melt into a solid polymer is one of the most important elements of fiber formation. Solidification involves irreversible changes of many structural characteristics of the material, such as molecular orientation and crystallinity. The rate and the extent of these changes depend on several variables which include common thermodynamic parameters of state (e.g. temperature) as well as instantaneous, non-equilibrium value of structural characteristics, such as degree of crystallinity, molecular orientation, degree of crystallite aggregation, density of intermolecular bonds, etc. One part of the problem consists of the relation between the instantaneous state of the system and its macroscopic behavior. The other part, associated with time variation of the individual variables, is the kinetics of elementary structural and thermodynamic processes.

The dynamic, kinematic and structural variables which determine all the macroscopic characteristics of the spinning system are the velocity vector, the stress tensor and the variables of state. All of these variables are functions of the position in the spinning line. A rigorous treatment of the subject is therefore fraught with numerous difficulties. No comprehensive treatment of the course and the mechanism of structure formation has yet emerged although several excellent studies on the subject have been reported lately. A list of selected references is given in the Bibliography of this chapter.

IV.1. Crystallization

Polypropylene fiber, like many other polymeric materials, exhibits those properties normally associated with crystals (e.g. diffraction of x-rays and evolution of latent heat on solidification). These same samples, however, exhibit concurrently those properties associated with non-crystalline materials (e.g. diffuse x-ray scattering and diffuse melting point). They are therefore called partially crystalline materials. It has proved impossible to separate mechanically the crystals from the non-crystalline phase. This implies that the partially crystalline polymeric materials exist in the form of a structure which at some points has three-dimensional periodicity, characteristics of a crystal, and at some other points is completely amorphous. In fact, a whole range of "order" may exist, ranging from a three-dimensional crystal to two-dimensional periodicity of chain segments to complete randomness of molecular chains.

Since the various mechanical and physical properties of polypropylene fibers are attributed to the presence of crystalline order in the material, there are practical reasons for understanding these domains and the mechanism by which these domains develop in the course of fiber formation.

Numerous investigations, both theoretical and experimental have been made of crystallization of polymeric materials, including polypropylene. Main attention has been focused on the thermodynamics and kinetics of crystallization under isothermal conditions. These studies have led to a better understanding of fundamental features of polymer crystallization, but do not provide adequate information for solidification of the polymer under melt spinning conditions. Crystallization during melt spinning occurs non-isothermally. Complicating the matter further is the fact that the crystallization during melt spinning occurs under molecular orientation. It is but natural, therefore, that the process of crystallization during the course of fiber formation is not yet well understood. In fact, it is only recently that the subject is being given any substantial attention. In the following pages, we will review these studies with special references to polypropylene. However, before we take up the subject of crystallization during melt spinning, a review of some background information on the crystal and morphological structure of polypropylene is in order.

Crystal structure of polymers deals with the organization of long chain molecules into some sort of sub-microscopic array. If the array, on the average, is highly ordered in three dimensions, the material is said to have a high crystallinity. If the packing of the chains, averaged over the sample, is not very regular, the crystallinity is low. Polypropylene undrawn fibers are rated as highly crystalline while the as-spun fibers of PET show little or no crystallinity.

Morphology deals with structure on the microscopic level. The simplest concept of morphology treats the molecular aggregation as two distinct phases - a nearly perfect crystalline phase embedded in an amorphous entanglement of chains with many chains joining the two phases to provide strength. These joining segments of chains are then called "tie chains". This classical concept is useful for providing simple quantitative description of morphology from which a "percent crystallinity" can be derived from x-ray diffraction, infrared, or density data. For the most part, physical property data have been correlated with percent crystallinity and orientation. However, the morphology of polypropylene is far more complicated than this simple model. The molecule, fold to form thin ribbons or layer shaped crystals termed "lamellae" and these in turn often aggregate into spherical or cylindrical clusters called "spherulites" or "cylindrites". Unfortunately, the knowledge of these morphological types is limited, identification and analysis of the structure are complex.

The modern view of superstructure or morphology in polymers comes from their ability to form single crystals in dilute solutions. In some ways the single crystals are more definitive structures for study than the large semi-crystalline aggregates in which we encounter crystallizing polymers commercially, and although polymer crystals are not wholly representative of large-scale crystallization behavior, they have yielded much important information on the morphology of crystallized polymers in the fibrous form.

When a polymer of sufficiently regular structure, such as stereoregular

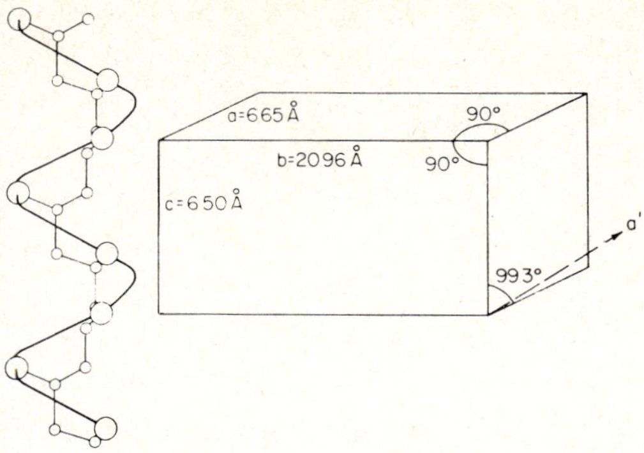

Fig. 8. Unit Cell of Monoclinic Polypropylene

polypropylene, is crystallized from a dilute solution, single crystals can be obtained which are in the form of regular lamellae of uniform thickness. As first shown by Natta (ref.17), polypropylene crystallizes into a monoclinic unit cell. The dimensions are: a = 6.65 Å, b = 20.96 Å, and c = 6.50 Å; α = 99 20'. (see Fig.8). It contains 12 monomer units and it has a calculated density of 0.936 g/cm^3. There are four chains passing through the unit cell, each of which has a 3_1 helical conformation with the a-axis running parallel to the axis of the helix.

Other crystalline forms of isotactic polypropylene exist. The molecule has the same conformation in all forms. The existence of hexagonal form has been amply documented by several investigators. Rapidly quenched polypropylene is obtained in a triclinic (smectic) or paracrystalline form which converts to the monoclinic form when heated above 60°C. The x-ray diffraction pattern of smectic material consist of two broad peaks instead of the well known four major peaks of the monoclinic form.

The surprising feature of lamellae formation is that (a) the chains are more or less perpendicular to the basal plane of the lamellae, and (b) the lamellae thickness - of the order of 100 Å - is independent of the length of the molecule which itself will be of the order of 10^5 Å long. This immediately leads to the conclusion that the molecules must be folded backwards and forwards on themselves to produce a structure which is represented in diagramatic form in Fig. 9. The fold length is determined by the crystallization temperature or, more exactly, by the degree of supercooling. The higher the crystallization temperature, the longer the fold length. The fold length thus becomes an important structural parameter in polymer science. The discovery of chain folding raises in turn a number of important problems. These include the origin of chain folding and the nature of the fold surface, i.e., to what extent is this surface disordered

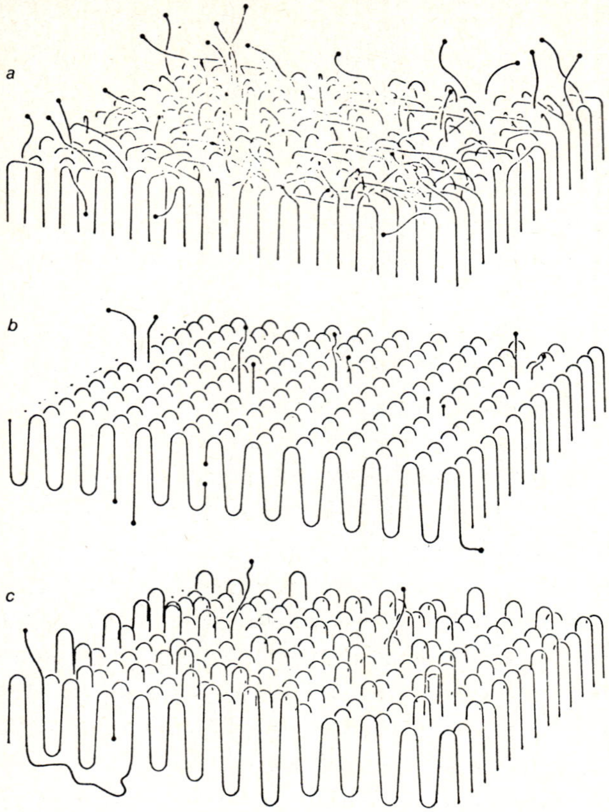

Fig. 9. Three Possibilities of Chain Entry in a Lamella
In (a), sometimes referred to as the "switchboard" model, the chains re-enter at random. In the other configurations, both "adjacent re-entry" models, surfaces where the chains fold are either smooth (b) or rough (c).

as the result of nonre-entrant adjacent folds.

The tie molecules in samples crystallized from melt are sections of long chain molecules which cannot crystallie because the molecules have started independently to crystallie in two different crystals so that the intervening section cannot be included in any one of them. Occasionally, the strained tie molecules may give rise to filament crystallization which provides very strong fibrillar links between the two crystals connected. In cold-drawn material the extremely large number of taut tie molecules originates from partially unfolded chains during micronecking which tears folded chain blocks from the lamellae of the spherulitic material and incorporates them into the microfibrils. They firmly connect subsequent crystal blocks and thus make the long and narrow microfibril a strong basic element of the fibrous structure. This topic will be further discussed in Chapter VII.

IV.2. Degree of Crystallinity

A numerical measure of "crystallinity" is often used as a measure of the degree of crystalline order in a semi-crystalline polymer. As mentioned before, the term implies the presence of a two-phase system of crystalline and amorphous regions. This was a reasonable interpretation of the diffraction patterns of polypropylene. Diffraction from polypropylene resembles the superposition of a crystalline diffraction pattern on an amorphous pattern, which in turn appears to be an extrapolation of the diffraction pattern of the melt. It is now generally considered that this interpretation is a gross oversimplification, but the resolution of such pattern into amorphous and crystalline "areas" and the derivation of a degree of crystallinity from such an operation can still be empirically useful. The problem in crystallinity determination is two-fold: (a) specification of experimental techniques capable of responding to differences in "crystallinity" and (b) identification of an experimentally determined "crystallinity" with the physical composition of the sample studied. The solution of the first problem is straight-forward: in principle, the properties to be expected of the pure crystalline or amorphous material may be measured or estimated. From a measured property, an experimental crystallinity is determined as if the sample consisted solely of a mixture of amorphous and crystalline regions. The second problem is not resolved as easily. One must be able to define the properties associated with the "crystalline" and "amorphous" regions actually present and the response of each region of intermediate order to the measuring technique employed. Differences in experimentally determined crystallinities are to be expected because of differences in the partition of intermediate regions between amorphous and crystalline.

I.R. Spectroscopy, NMR, and differential thermal analysis. The first two methods, i.e., density measurement and x-ray diffraction are probably the simplest and oft-quoted methods of evaluation of crystallinity. The agreement between the two methods is fairly good in most of the cases but the two methods give different estimates in the case of cold drawn polypropylene fibers.

The crystallinities of undrawn polypropylene filaments are most commonly calculated from the fiber density by the following equation:

$$1/d = x/d_c + (I - x)/d_a$$

where d is the measured density of the sample, d_a represents the density of wholly non-crystalline fraction (0.850 g/cc and d_c the theoretical density of wholly crystalline polymer (0.936 g/cc.). However, the determination of the degree of crystallinity by this method is subject to certain reservations; there are no two completely defined amorphous and crystalline phases. Moreover, different transitional states or "degree of order" exist. The density method does not distinguish between the true three dimensional crystalline state and the two dimensional para-crystalline order.

X-ray techniques give a deeper insight into the structure of crystalline polymers. These methods permit determination not only of the quantitative crystalline/amorphous ratio but also of the size of the crystallites and the form and dimensions of the unit cell. The intensity ratio of the X-rays scattered by the crystalline and non-crystalline regions is taken as a measure of the degree of crystallinity.

IV.3. Kinetics of Isothermal Crystallization

The overall kinetics of isothermal crystallization in polymers is generally described in terms of the Avrami relation:

$$V_c/V = 1\text{-}\exp(Kt^n)$$

where V_c/V is the volume fraction crystallized in time t, K is a constant which depends on the rates of nucleation and growth of crystals (and hence on temperature) as well as their geometry: and the exponent n depends on the time dependence of the nucleation rate and growth rate.

From the form of the above equation, a plot of fraction crystallized vs time should be a sigmoid curve such as that shown in Fig. 10. Curves of this form are commonly observed in studies of polymer crystallization. The data are most often obtained using dilatometry — i.e., measuring the change in density as a function of time. Such measurements can be made with considerable accuracy since the density of the crystalline regions in polymers exceeds that of the amorphous material by as much as 10 percent or more.

The overall rate of crystallization, i.e., $\frac{d}{dt}\left[\frac{Vc}{V}\right]$, is expected to increase with increasing undercooling below the melting point, pass through a maximum, and then decrease with further increases in undercooling. The maximum reflects a competition between the driving force for crystallization, which increases with increasing undercooling, and the molecular mobility, which decreases with increasing undercooling. For polypropylene, the temperature of maximum crystallization rate has been suggested to occur in the range of 100° to $110^{o}C$. The addition of some effective nucleating agents to the material can raise this temperature to about $140^{o}C$.

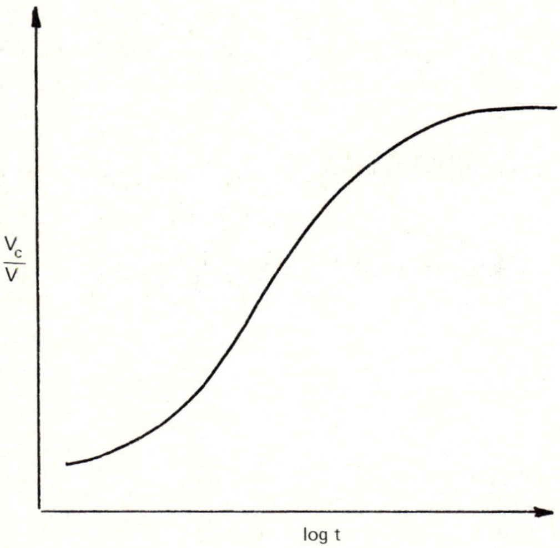

Fig. 10. Schematic Presentation of the Kinetics of Crystallization
V_c/V = fraction crystallized
t = time

Changes in material and process parameters can have a significant effect on the nucleation rate and growth rate, and hence on the overall crystallization rate and the final degree of crystallinity achieved in a sample. The following three material characteristics may be considered:

1) Molecular weight and molecular weight distribution of the polymer
2) Degree of tacticity
3) Presence of additives which may act as nuclei for crystallization.

The dependence of crystallization velocity on molecular weight has been demonstrated for several polymers. The effect of molecular weight and molecular weight distribution on crystallinity of polypropylene has been studied by several investigators (ref.28-32). These results can be summarized as follows:

1) The degree of crystallinity increases as the molecular weight decreases.
2) The rate of number of spherulites formation (nucleation rate) increases as the molecular weight of the polymer decreases (ref.31).
3) Contrary to the expectation, Kamide (ref.31) showed that the crystallization rate of polypropylene increases as the molecular weight increases, while other researchers (ref.29,32) state that the crystallization rate decreases with

increasing molecular weight. Kamide's data get support from Donald et al, (ref.30), who measured crystallization rate by differential scanning calorimetry and concluded that the sample of 6.06×10^5 molecular weight had higher crystallization rate than that of 1.96×10^5. These apparently conflicting data can be rationalized on the basis of the degree of supercooling, which is different with different molecular weights (ref.31).

4) The crystallization rate increases as the molecular weight distributions, M_w/M_n, increases (ref.31).

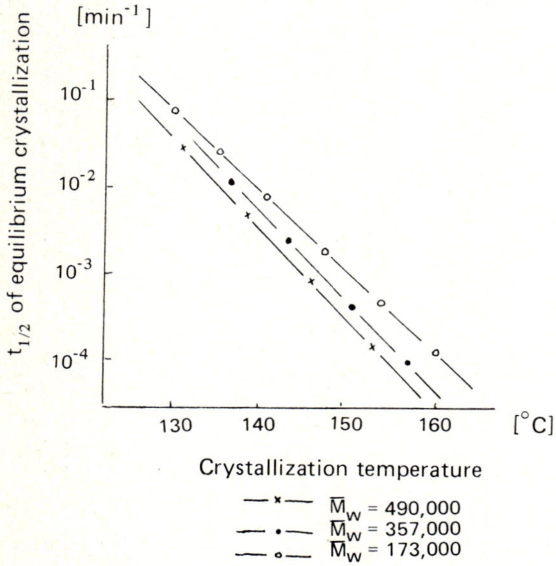

Fig. 11. Crystallization Rate of Polypropylene as it Depends on Crystallization Temperature (ref.32)

Figure 11 shows the dependence of crystallization rate on the crystallization temperature of three polypropylene samples. As is evident from this figure, the rate of crystallization increases rapidly upon lowering the crystallization temperature. Because of the too rapid crystallization rate of polypropylene, it has not been possible to investigate the crystallization rate at still lower temperatures. Qualitative information obtained through other sources, however, permit the assumption that the crystallization rate would reach a maximum and then drop off, in accordance with the schematics of Fig. 12.

The process of nuclei formation in polypropylene is very rapid and is influenced, apart from temperature, by the presence of nucleating agents. The presence of catalytic residues, some stabilizers, pigments, processing aids and deliberately added "nucleating agents", such as benzoic acid, increases the nuclei concentration and thereby accelerates the overall rate of crystallization. The increase in nuclei concentration, however, leads to finer and more numerous crystallites. The presence of nucleating agents also affects the temperature of maximum crystallization, T*. In the case of polypropylene, T* can be raised from 110°C without nucleating agents up to 140°C with most effective nucleating agents available.

The thermal history of the melt prior to the start of crystallization is also very important in determining the rate of crystallization. It is well known that any pre-orientation and quasi-crystalline order in the melt (sometimes also called 'melt structure') would facilitate the formation of nuclei in the process of crystallization. This melt structure is destroyed by prolonged heating at elevated temperature. The rate of nuclei formation is thereby significantly reduced. Since the linear rate of growth of a given nucleus is essentially independent of these variations, the overall crystallization rate is reduced. This is exemplified in Fig. 13

TABLE II

Effect of Stereoregularity on Crystallinity of Polypropylene (ref.34)

Fraction	Soluble in	Insoluble in	Crystallinity, %	Tm, °C
I		trichloroethylene	75-85	176
II		n-octane	64-68	174-176
III	n-octane	2-ethylhexane	60-66	174-175
IV	2-ethylhexane	n-heptane	52-64	168-170
V	n-heptane	n-hexane	41-54	147-159
VI	n-hexane	n-pentane	25-37	110-135
VII	n-pentane	ethyl ether	15-27	106-114

Polypropylene owes its ability to crystallize to the stereoregular character of its molecule. As several mechanical and physical properties of polypropylene are connected with the degree of crystallinity, an effort is always made to produce polymers of high isotactic content. Although there can be no direct relationship between tacticity and the degree of crystallinity of a particular sample, Natta (ref.34) and Miller (ref.35) have demonstrated the profound effect that tacticity of a sample has on its equilibrium crystallinity. This is evident from Table II. As the solubility in heptane decreases (a measure of atactic content), the equilibrium degree of crystallinity increases. Tacticity of polypropylene, through its effect on crystallinity of the product, is evidently expected to influence the mechanical properties of the product. Table III summarizes the various effect of tacticity on the mechanical properties of the polymer.

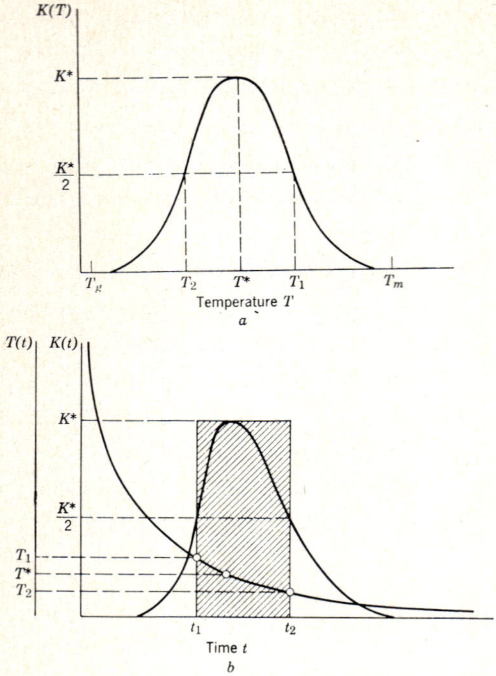

Fig. 12. Crystallization Rate Characteristics (Schematic)
(a) Temperature Dependent Crystallization Rate K(T); (b) Cooling Curve T (t) and Resulting Time-Dependent Crystallization Rate K(t) (ref.33)

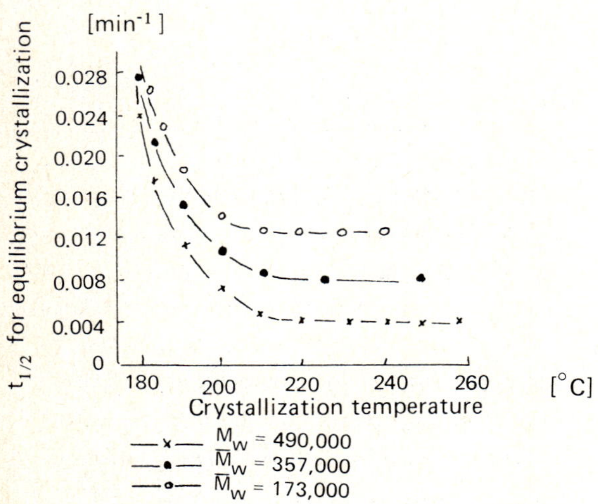

Fig. 13. The Rate of Crystallization of Polypropylene at 140°C as it Depends on the Temperature at Which Melt Was Held For One Hour Prior to Crystallization (ref.32)

The effect of nucleating agents on nucleus density, rate of crystallization and the size of crystallites has been briefly mentioned previously. A large number of nucleating agents for polypropylene has been discovered (ref.37-43). These substances make good control of the nucleation density and its homogeneity possible. This leads in turn to a change in several product properties, such as gloss, drawability, void formation on cold stretching, elasticity, skin-core structure of the fiber, etc. Nucleation agents for polypropylene are insoluble in polymer melt and are always crystalline substances.

TABLE III

Properties of Polypropylene Depending on Isotacticity and Molecular Weight (ref.36)

Property	With increasing Isotacticity	With increasing Molecular weight
Stiffness	Increases	Decreases
Impact strength	Decreases	Increases
Tensile yield stress	Increases	
Tensile strength at break	No change-Increase	Increases
Tensile modulus	Increases	
Tensile elongation at break	Decreases	Increases
Hardness	Increases	Decreases
Brittleness	Increases	
Low temperature brittleness point	Increases	Increases
Softening point	Increases	
Mold shrinkage	Increases	
Abrasion resistance	No change	Increases

The basic idea underlying most models of heterogeneous nucleation is that the total interfacial free energy of the crystalline embryo is reduced by a foreign phase which already provides part of the interface to be formed; this brings about a reduction in the activation energy for nucleation and thereby a decrease in the degree of supercooling required for nucleation. The amount of reduction in interfacial free energy is dependent on both the geometrical form of the foreign phase and its chemical nature.

The most effective nucleating agents for polypropylene are found among the alkali, Ca, Ba, Mg, and Aluminum salts of aromatic and cyclo-aliphatic mono- or di- carboxilic acids, aliphatic mono-carboxilic acids bearing aromatic substituents, aromatic phosphonic acids, phosphoric and phosphorous acids, some aromatic esters, and various pigments such as phthalocyanines, anthraquinone, pigments of condensed aromatic structures such as flavoathrone and quinacridons. Some polymeric compounds, such as poly - 3 methyl butane, have also been mentioned as effective nucleating agents for polypropylene. Inorganic salts and oxides (e.g. silica), contrary to what has been claimed in several patents, do not show any activity as nucleating agents (ref.40).

IV.4. Crystallization Under Non-Isothermal Conditions

So far we confined our discussion mainly to the crystallization of polymers under isothermal conditions. However, melt spinning is a process during which crystallization occurs non-isothermally under molecular orientation. The kinetics of non-isothermal crystallization are not well-understood, even more so is the influence of molecular orientation in non-isothermal crystallization. Nakamura (ref.44) and Ziabicki (ref.45) have done extensive work in this area of research; the following discussion is based on their treatments.

The kinetics of non-isothermal crystallization can be expressed by the following equation, which resembles Avarami equation mentioned before:

$$\delta_{(t)} = 1\text{-exp}\left\{-\left(\int_0^t K\left[T\right]_t dt\right)^n\right\}$$

Where n is the Avarami index determined from the data of the isothermal crystallization: $K\left[T\right]_t$ is related to $K\left[T\right]$ in Avarami equation through the following relation:

$$K\left[T\right]_t = k(T)^{\frac{1}{n}}$$

The $\left[T\right]_t$ in the above relation means the temperature T at time t of a solidifying system.

The Avarami-like equation for non-isothermal crystallization can be expanded as:

$$\delta_{(t)} = \int_0^t Kdt - \tfrac{1}{2}\left(\int_0^t Kdt\right)^2 + \frac{1}{6}\left(\int_0^t Kdt\right)^3 - - - - - -$$

For small degrees of transformation one can take only the first term in the above equation, and integration over time can be replaced by that over temperature:

$$\delta_{(t)} \approx \int_0^t Kdt = \int_{T(t=0)}^{T(t)} K(T)\xi^{-1}\, dT, \quad \text{with} \quad \xi = \frac{dT}{dt}$$

It is evident from the above equation that if the temperature dependence of K (T) and the course of cooling of the crystallizing system are known, the degree of crystallinity $\delta_{(t)}$ at any moment t can be calculated. The function K(T) is a material characteristic and can be found directly from experiments.

The typical shape of the experimentally found K(T) is shown in Fig. 13. The crystallization rate constant K exhibits a maximum value K* at some temperature T* and falls to zero just below the melting temperature T and at some inferior limit usually believed to lie near the glass transition temperature T_g. The glass transition temperature of polypropylene is below 0^oC. However, because of the very highly crystalline nature of polypropylene, crystallization process practically stops above room temperature. T_1 and T_2 are temperatures corresponding to the K value, half of K*, $D = (T_1 - T_2)$ is the half-width of the crystallization curve, and the product G = (k* D) is the approximate value of the area under this curve.

The product G = (K* D) has been termed by Ziabicki as "kinetic crystallizability". The "kinetic crystallizability" G characterizes and for small δ* is equal to the degree of transformation obtained over the entire crystallization region with unit cooling rate. Parameter G thus serves as a comparative measure of crystallinity of various materials when kinetic factors are important. The higher G is, the larger is the degree of crystallinity δ and the higher is the cooling rate required to undercool a liquid to an amorphous state without appreciable crystallization in the course of cooling. The calculated values of G for polypropylene is 33, compared with 0.5-1.0 for polyester, 6.4 for nylon 6 and 133 for nylon 66 (ref.45).

To calculate the degree of crystallinity of as-spun fiber, δ_{as}, we must find the integral appearing in equation:

$$\delta_{(t)} = \int_0^t K dt$$

For any particular cooling function T(t) one can transform the characteristic K(T) into the time function K(t) = K[T(t)] (see Fig. 13b). The approximate value of the integral will be obtained through multiplication of the maximum value K* by the half-width of Δt of the K(t) function:

$$\int_{t=0}^{t=\text{winding}} K(t)dt = \int_{t(Tm)}^{t(\text{solidification})} K[T(t)]dt \approx K^* \Delta t$$

where $\Delta t = t(T_2) - t(T_1)$ is the time difference corresponding to the passing by the spinning line two characteristic temperatures, T_1 and T_2, and is determined by spinning variables. More time a solidifying thread takes to pass between these two temperatures, higher degree of crystallinity it acquires.

The length of time an attenuating jet spends in a particular temperature range can be quantitatively expressed by the length of time the polymer requires to cool 1^oC, as given by the following equation:

$$\frac{D\tau}{DT} = \left(\frac{\ell A}{Q}\right)\left(\frac{dx}{dT}\right)_{T^*}$$

Where τ is the time, ℓ is the density, A is the cross-sectional area of the filament, and Q is the rate of flow of the polymer. The $\left(\frac{dx}{dT}\right)_{T^*}$ term depicts the temperature change along the length of the fiber in the vicinity of T* (the temperature of maximum crystallization rate). If we consider that T* is a material constant, the degree of crystallinity, as given by the density of the fiber, ℓ, can be theoretically calculated. Some of these results can be summarized as follows: i) Higher spinnerette temperature, an increase in cooling rate, an increase in take-up denier and a decrease in take-up speed at constant denier decreases the cooling time DT/Dt and consequently decreases ℓ or the degree of crystallization. ii) An insulating cylinder placed below the spinnerette reduces DT/Dt and experimentally measured ℓ. Temporarily delaying the cooling of polymer jet and achieving less crystallinity of the fiber sounds somewhat paradoxical. This apparent paradox comes from the fact that the cooling time DT/Dt is equal to the ratio of dt/dx over the yarn speed V. Since the insulating cylinder delays cooling, it allows the filament to thin and gain a higher speed V by the time the filament is eventually cooled to 100°-120°C. This makes the cooling time small and consequently the yarns are less crystalline.

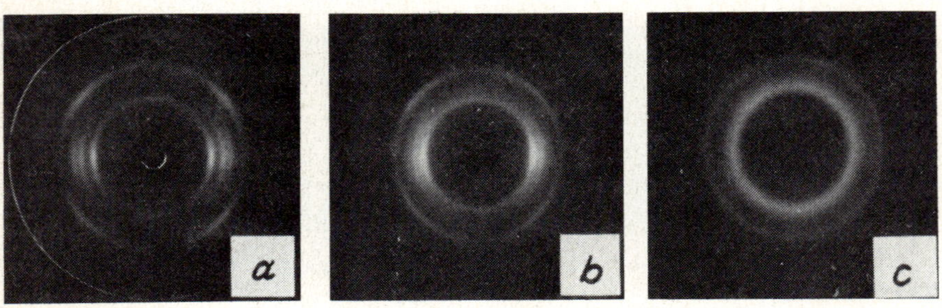

(a) 290°C (b) 300°C (c) 310°C

Fig. 14. X-ray Diagrams of Polypropylene Fibers Spun at Different Temperatures (ref.46)

Apart from the degree of crystallinity, the cooling rate may also influence the morphology of the fiber structure. The existence of two forms of crystalline order in polypropylene, i.e., smectic and monoclinic, has been amply demonstrated. In melt spinning of polypropylene one form or the other is predominantly formed depending on the melt temperature and the rate of quenching and orientation. Figure 14 represents the x-ray diagrams of three samples of polypropylene fiber prepared at different melt temperatures. For a given degree of super-cooling, relatively low melt temperatures favor the formation of a more stable, monoclinic structure in spun fibers (Fig.14). On the other hand, a high melt temperature leads to a less stable paracrystalline structure (Fig.14).

In addition to the temperature of the melt, the cooling rate has a marked effect on the crystalline structure of the fiber. Figure 15 shows how crystalline structure of the fiber is affected by the cooling rate. At high quenching rate (Fig.15.a.) the polymer solidifies practically completely to a paracrystalline, thermodynamically less stable structure while at low quenching rate (Fig.15.c.) the fiber structure develops into monoclinic order. This change in relative order is also reflected in the fiber density which increases as the quenching rate decreases (ref.46).

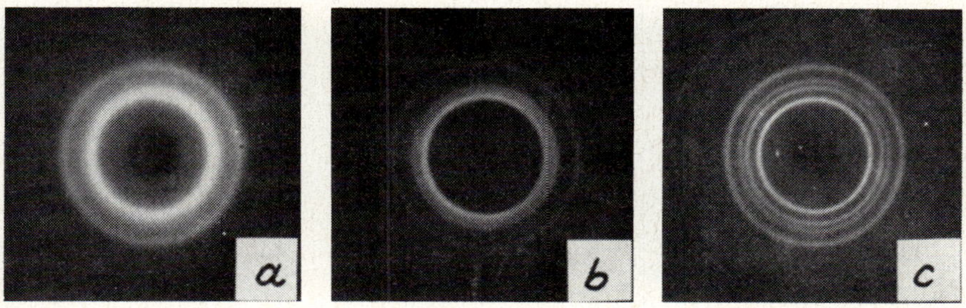

Fig. 15. X-Ray Diagrams of Polypropylene Fibers Quenched at Different Rates (ref.46)
(a) Dry Ice and Acetone (-65°C); (b) Water (10°C); (c) Air (22°C)

IV.5. Crystallization Under Stress

The preceding text has dealt with the prediction of crystallinity under non-isothermal conditions, taking no account of molecular orientation which is an essential part of melt spinning process. It has been mentioned previously and discussed in detail in the next section that the molecular orientation in the spinning fiber is essentially a function of tensile stress at the crystallization point of the solidifying filament. Thus, the crystallization kinetics of a melt spinning filament are reduced to the crystallization kinetics of a strained polymer system under non-isothermal conditions. As both the rate of cooling and the tensile stress at crystallization point change in a very complex manner with changes in spinning conditions and material parameters, a complete solution of the problem is quite complex. Some qualitative answers, however, have been developed intuitively or experimentally. These are briefly reviewed in this section.

Consider a polymer melt which is subjected to elongational deformation under constant tension. The cross-sectional area of the specimen is ever-decreasing and any equilibrium state cannot be imagined. It is assumed, however, that the conformation of molecules reaches a pseudo-equilibrium conformation corresponding to some recoverable macroscopic strain-induced crystallization for cross-linked polymers. The thermodynamic theory of strain-induced crystallization of cross-linked systems can, therefore, be extended to non-crosslinked polymers, such as polypropylene. The thermodynamic function G in the theory is defined by the following expression:

$$G = H - TS - FL$$

where H, S, F, and L are the enthalpy, entropy, tension, and length of a specimen, respectively. The difference between G before and after crystallization is the driving force to crystallization:

$$\Delta G = \Delta H - T\Delta S - F\Delta L$$

The decrease in ΔS with molecular orientation causes an increase in crystallization rate.

Spruiell and coworkers (ref.48,49,59) have recently conducted x-ray diffraction studies along the spinline of polypropylene and polyethylene filaments. Their results bear out the above expectations, at least qualitatively. The crystallization temperatures of polypropylene, for example, cooled under quiescent conditions were shown much lower than the crystallization temperature under several melt spinning conditions. The temperature for the onset of crystallization increased with increasing spinline tension (Fig.16). The relatively high crystallization temperatures in melt spinning indicate that molecular orientation resulting from elongational flow and draw-down forces in the filament greatly increase crystallization rate. It was shown that relatively modest spinning conditions are required to provide a sizeable effect on the rate of crystallization.

At any take up velocity and mass flow rate, the temperature at which crystallization begins is the result of a balance between the effect of super-cooling effect provided by the high cooling rate and the increased crystallization rate caused by the molecular orientation in the filament. It is believed that the effect of molecular orientation on crystallization rate reaches a saturation value at rather modest draw-down ratios and that beyond this level the crystallization rate is determined by the cooling conditions of the filament. Conversely, even at modest draw-down ratios, an excessively large number of nuclei

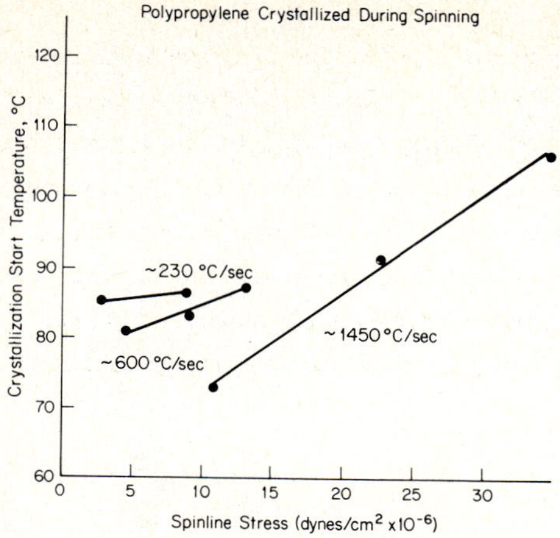

Fig. 16. Crystallization Start Temperature Versus Spinline Stress for Isotactic Polypropylene (ref.48)

is produced in the form of axially oriented molecules. As the density of nuclei is already high under these conditions, the addition of nucleating agents does not materially affect the rate of crystallization or the size of the crystallites. Thus, the effect of nucleating agents in the spinning of polypropylene fibers is only marginal. They may have some beneficial effect, however, in the formation of monofilaments and bristles, where the orientation effects are minimal.

Figure 17 shows WAXS patterns of polypropylene filaments ($\overline{M}_w = 2.77X10^5$) spun either into ambient air or into water at 9^oC. The effect of increasing take-up velocity (and spinline stress) is also shown. It can be seen from Fig.17.a-c that the samples are highly crystalline and the C-axis orientation is increasing with increasing take-up velocity. For the smectic form in Fig.17.d-f, there is a transformation in crystal structure from an unoriented paracrystalline to highly oriented partially monoclinic type with increasing take-up velocity (or spinline stress).

The effect of spinning temperature and take-up velocity on WAXS pattern is shown in Fig.18 for a polymer of 6.6 MFI ($\overline{M}_w = 2.77X10^5$). Increasing the spinning temperature, with other variables unchanged, decreases the viscosity in the upper part of the threadline and results in lower orientation in the spun filament. A combination of high take-up velocity and low spinning temperature (Fig.18.b) produces a highly crystalline filament with pronounced degree of crystalline orientation.

The effect of molecular weight on WAXS pattern is illustrated in Fig.19. Comparing the WAXS patterns in 19.a-c at a melt extrusion rate of 2.1 g/min and take-up velocity 200 m/min, there seems to be a noticeable decrease in the 110 and 130 arc widths with increasing molecular weight indicating an increase in crystalline orientation.

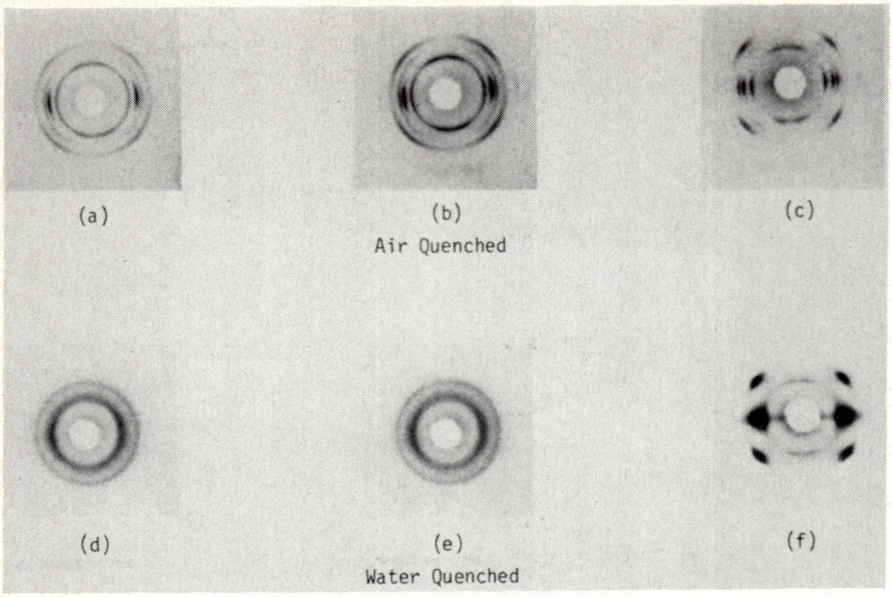

Fig. 17. Effect of Take-Up Velocity and Cooling Medium on WAXS Patterns of Polypropylene Fibers ($\overline{M}_w$ = 2.77X10^5). Extrusion Temperature = 230°C. Air Quenched: (a) 50 m/min; (b) 200 m/min; (c) 550 m/min. Water Quenched = (d) 25 m/min; (e) 100 m/min; (f) 300 m/min. (ref.59)

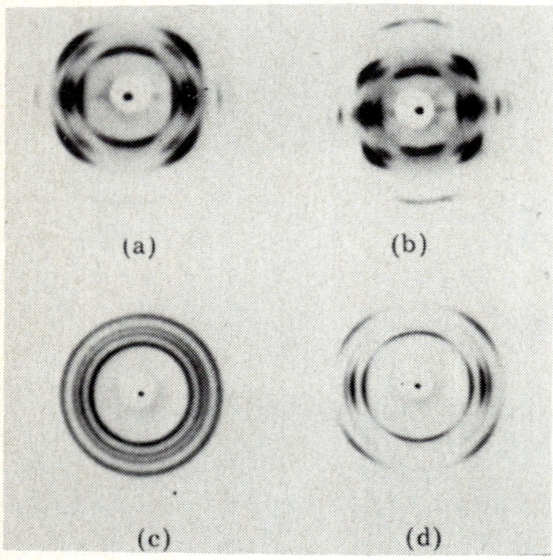

Fig. 18. Variation of WAXS Patterns For Polypropylene ($\overline{M}_w$ = 2.77X10^5) With Extrusion Temperature. T_E and Take-up Velocity V_L = (a) T_E = 200°C; V_L = 50 m/min; (b) T_E = 200°C, V_L = 550 m/min; (c) T_E = 260° C, V_L = 50 m/min; (d) T_E = 260° C, V_L = 550 m/min. (ref.59)

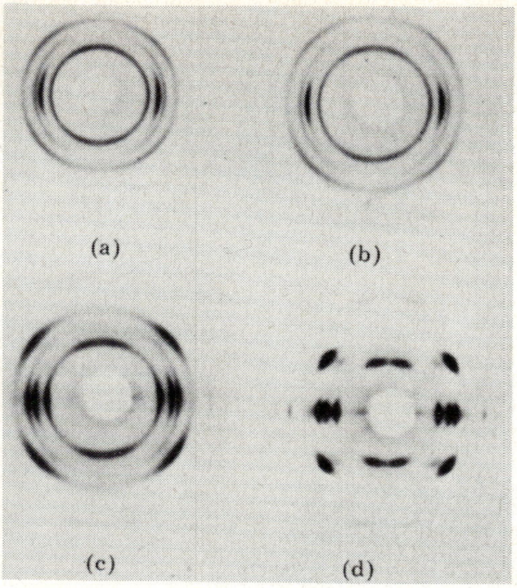

Fig. 19. Effect of Molecular Weight and Rate of Extrusion on the WAXS Patterns of Samples Spun at 200 m/min. Extrusion Temperature is 230°C. Extrusion Rates for (a), (b), and (c) is 2.1 g/min: (a) Hercules Profax 6323, MFI = 12.0, [η] = 1.9; (b) Eastman Tenite 4241, MFI = 9.0, [η] = 2.01; (c) Eastman Tenite 4221, MFI = 2.55, [η] = 2.71; (d) Hercules Profax 6823, MFI = 0.42, [η] = 3.21. (ref.59)

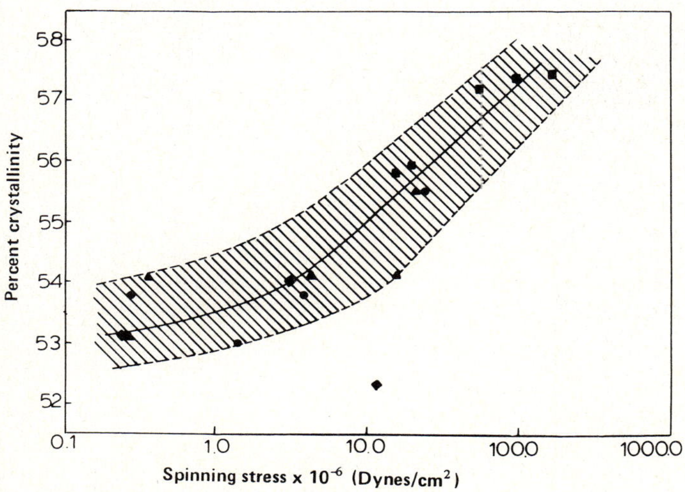

Fig. 20. Crystallinity Level of Melt-Spun Fibers as a Function of Spinline Stress (ref.59).

Figure 20 depicts the percent crystallinity data for monoclinic filaments obtained by spinning in air. These measurements indicate crystallinities of the order of 55 ± 3% for all samples. There is, however, a trend toward increased crystallinity with increasing spinline stress.

Figure 21 shows the SAXS patterns of various polypropylene filaments spun under a variety of conditions. Under conditions which give rise to low spinline stress such as low take-up velocity and high extrusion temperature, the SAXS pattern exhibits a continuous ring of nearly uniform intensity (Fig.21.a). These are the same conditions which lead to low orientation in the WAXS patterns (Fig.18.c). As the spinning conditions are changed so as to increase the spinline stress, the patterns evolve to definite "two-point" patterns. Spruiell and co-workers (ref.48,60) have interpreted these features to indicate spherulitic morphology at very low spinning stresses which undergoes a transition to a row-nucleated or cylindrilic morphology as the spinning stress increases.

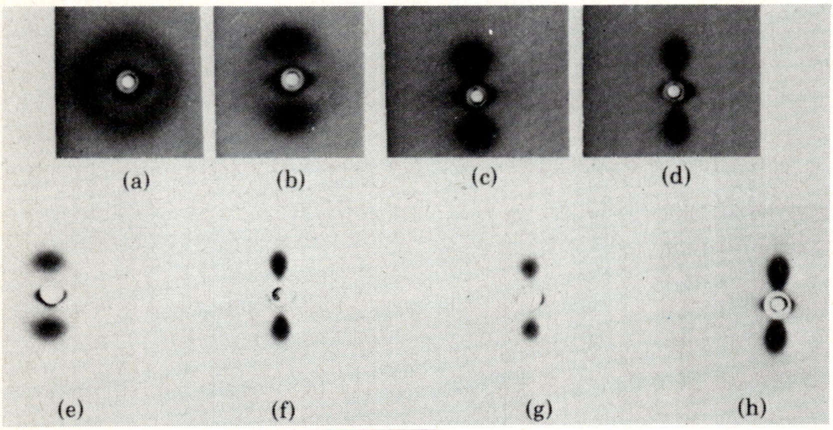

Fig. 21. SAXS Patterns For Melt-Spun Polypropylene Filaments: (a) $[\eta] = 2.26$, $T_E = 260^oC$, $V_L = 50$ *m/min; (b)* $[\eta] = 2.26$, $T_E = 260^oC$, $V_L = 300$ *m/min; (c)* $[\eta] = 2.26$, $T_E = 230^oC$, $V_L = 300$ *m/min; (d)* $[\eta] = 2.26$, $T_E = 230^oC$, $V_L = 550$ *m/min; (e)* $[\eta] = 2.71$, $T_E = 230^oC$, $V_L = 50$ *m/min; (f)* $[\eta] = 2.71$, $T_E = 230^oC$, $V_L = 200$ *m/min; (g)* $[\eta] = 3.21$, $T_E = 230^oC$, $V_L = 50$ *m/min; (h)* $[\eta] = 3.21$, $T_E = 230^oC$, $V_L = 200$ *m/min. (ref.59)*

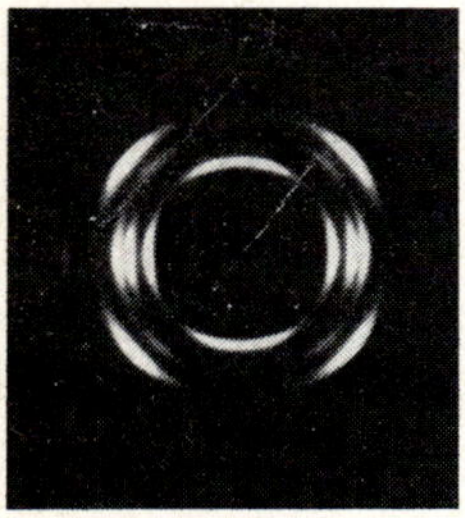

Fig. 22. X-Ray Diagram of a "Cross-Oriented" Polypropylene Fiber

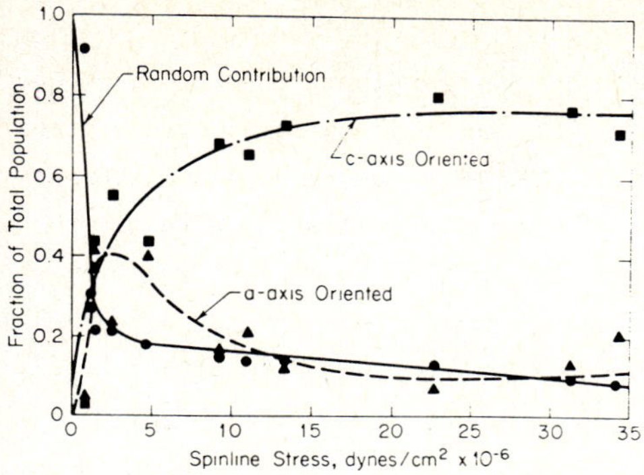

Fig. 23. Fraction a- and c-Axis-Oriented Crystallites Versus Take-Up Stress For Polypropylene (ref.49)

Polypropylene is distinctive and perhaps unique in that when crystallized from the melt under conditions of extensional flow, the x-ray diffraction pattern shows a type of molecular orientation which can be interpreted only in terms of two different, and apparently independent, orientations of the unit cells, characterized by one component with the c- (chains) axis parallel to the direction of flow and a second component with the c-axis approximately perpendicular to the direction of flow. The former component is often referred to as the c-axis-oriented population, while the latter is referred to as a-axis-oriented population. This bimodal orientation is shown by what appears to be "extra spots" on the first layer line of x-ray diffraction pattern (Fig.22).

Figure 23 shows the relative proportions of the c-axis and a-axis-orientation as a function of spinline stress. It is clear from these data that at high spinline stresses the largest proportion of the sample is in a c-axis-oriented population, with only about 10-20 percent of the sample belonging to the a-axis-oriented population. The presence of the bimodal orientation is thus not likely to invalidate the broad qualitative features of the morphologic structure described in the preceding pages.

The precise mechanism for bimodal orientation and its relation to the fiber morphology is not as yet fully understood. The following picture, however, evolves on considering the results of various workers (ref.25,32,47,51,61).

The crystallization from strained melt, as for instance in the melt-spinning operation, produces a row-nucleated structure. Linear nuclei are formed parallel to the strain direction (Fig.24). They contain more or less extended polymer chains. Secondary epitaxial nucleation on the surface of such linear row nuclei produces folded chain lamellae which are oriented perpendicular to the strain (Fig.25). In such a case the sample exhibits a high uniaxial orientation of chain axes in the strain direction with random orientation of the a- and b-axes

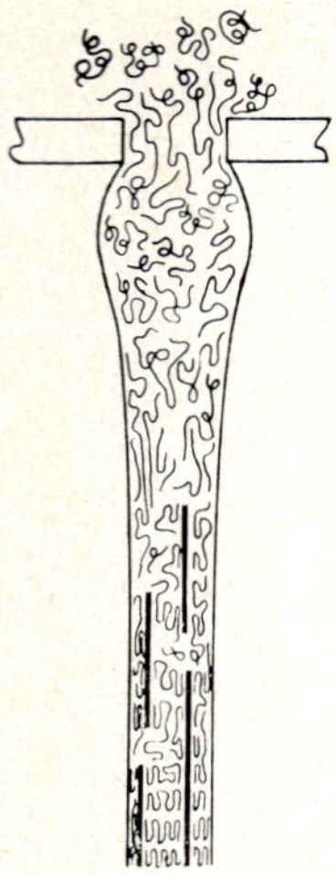

Fig. 24. A Model For Row Nucleation During Melt Spinning (ref.47)

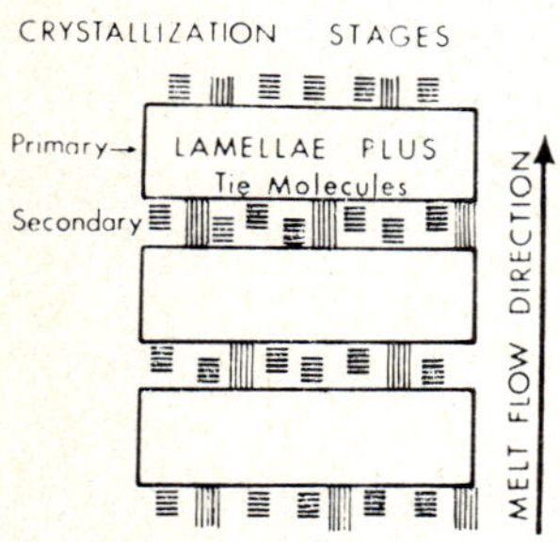

Fig. 25. Morphological Model For Bimodal Orientation of Polypropylene Proposed by Clark and Spruiell (ref.51)

perpendicular to it. If the growing lamallae exhibit a helical twist, the chain orientation in the strain direction is very soon replaced by orientation of the axis of maximum growth rate perpendicular to the strain direction and a more random orientation of the remaining two axes with a maximum in the strain direction. Such a row-nucleated structure has parallel cylindrical spherulites as its basic supercrystalline element. Polypropylene melt-spun fibers are particularly characterized by this kind of structure.

Clark and co-workers (ref.51) have proposed a morphological model for flow-crystallized polypropylene shown schematically in Fig.25. In the development of this model it has been assumed that crystallization proceeds in two stages. In the first stage, a row structure of planar lamellae develops from some sort of fibril nuclei. These lamellae grow with the c-axis perpendicular to the broad faces of the lamellae with the growth axis of the lamella parallel to the a-axis of the unit cell. During this primary crystallization, some of the less tactic molecules are rejected from the growing lamellae into regions between them. These less tactic molecules crystallize a short time later in a secondary stage of crystallization. The orientation of the crystals formed is presumably controlled by an epitaxial matching between the fold surface of the primary crystals and the polypropylene molecule. This matching requires that the chain of the secondary component be perpendicular to the chain axes of the primary component. The proposed interpretation of the bimodal orientation thus consists of a row structure of planar lamellae interconnected by the tie molecules, which give rise to the c-axis oriented contribution, together with small interlamellar crystals in the a'-axis orientation as a result of their epitaxial relationship with the primary row structure.

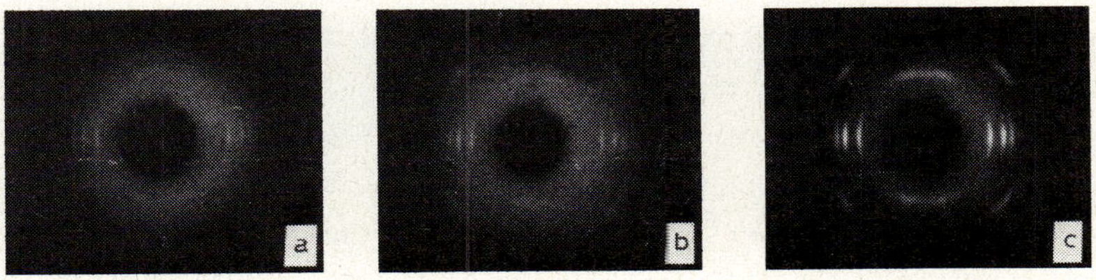

Fig. 26. X-Ray Photographs of the Running Thread of Polypropylene at Various Distance From the Spinnerette (a) 24 cm, (b) 30 cm and (c) 100 cm (ref.52)

A model such as that shown in Fig.25 would apply to melt spun polypropylene in the range of spinning stresses where a well developed two-point SAXS pattern is observed.

X-ray diffraction photographs reproduced in Fig.26 depict the structure formation of polypropylene fiber during melt spinning. It is seen from this figure that at the first stage of crystallization, only c-axis orientation appears. In the next stage an a-axis orientation develops and then both parts increase with distance from the spinnerette.

IV.6. Orientation

Flow of polymer jet during spinning leads to elongation and orientation of molecules along the axis of the filament. This parallelization and extension of macromolecules and other structural units (crystallites, fibrils, etc.,) along the fiber axis affects many physical properties of manmade fibers such as tensile strength, modulus, sorption of solvents, optical and electric behavior. On the other hand, orientation affects polymer crystallization, kinetics of solidification, dynamics of deformation, etc., and thus plays a very important role in fiber formation process.

The orientation in polymeric systems is reflected in marked anisotrophy in many physical properties, whose study can provide useful information about the orientation in fiber structure. Among techniques which are widely used in this regard, measurements of birefringence and sonic velocity provide information about the combined orientation of amorphous and crystalline regions, while X-ray diffraction studies define primarily the orientation of crystalline regions.

The orientation developed in fibers during melt spinning is controlled by two opposing molecular processes: (1) orientation and deformation of structural elements subjected to the action of the velocity field; and (2) disorientation due to thermal motions of the molecules. Molecular orientation can be produced by either: (1) streaming deformation developed in elongational flow of the liquid, or (2) deformation during and after crystallization.

Streaming orientation results from flow and disappears when the flow is interrupted. It can produce kinetically stable orientation in fibers provided that subsequent solidification occurs and that the rate of solidification is greater than the rate of disorientation (structural relaxation). The contribution of streaming within the capillary to the overall orientation is insignificant. This is because of the small velocity gradient on emerging from the capillary as well as the high temperature and hence small viscosity of the melt. During the subsequent elongational flow, orientation-controlling velocity parallel to the fiber axis is high and, because of the large increase in viscosity with falling temperature, the time for molecular relaxation becomes long. In contrast to behavior just below the spinnerette, molecular orientation produced in the solidification zone freezes in before it has the opportunity to relax and disorient. Structural characteristics measured along the spin line indicate that the molecular orientation increases very quickly as the polymer approaches the end of thinning and begins to solidify (see Fig.7).

In practical spinning operations the product $q.\tau$ is the most important parameter determining the orientation of the fiber. The larger the value of this product, the higher will be the overall orientation. Here q is the velocity gradient and τ is the relaxation time of the polymer. To obtain more detailed insight into the expected behavior, the theory of rubber elasticity can be used to relate the tensile stress, $(F/A)_c$, acting on the filament at the point of solidification to the elongation ratio, λ, of molecules as follows:

$$\left(\frac{F}{A}\right)c = K.T.\mu.(\lambda^2 - \frac{1}{\lambda})$$

where μ is the effective number of chains per unit volume and K is the Boltzmann's constant. The extension ratio is in turn related to the birefringence, Δn, as:

$$\Delta n = M.\mu.(\lambda^2 - \frac{1}{\lambda})$$

where M is a constant determined by the optical properties of the molecule. It follows (ref.9) that:

$$\Delta n = \frac{M}{kT}\left[F/A\right]_c$$

The birefringence of the yarn is therefore expected to be directly proportional to the tensile stress $(F/A)_c$ at the point of solidification. This stress is in turn approximately equal to the tensile stress (F/A) at the take-up roll. Hence the factors which contribute to an increase in spinline tension should also increase birefringence and orientation of the fiber. This conclusion is supported by experimental results on polypropylene.

The changes in filament orientation with variations in spinning conditions may be briefly summarized with reference to Fig.27. As seen from these results, the yarn birefringence Δn of polypropylene increases with (a) increasing viscosity of the polymer, (b) decreasing temperature of spinning, (c) increasing throughput at a given denier, (d) increasing rate of air quenching, (e) decreasing denier of the filament, and (f) removing an insulating cylinder placed below the spinnerette.

IV.7. Skin-Core Structure

In the discussion so far it has been tacitly assumed that the structure of the filament is uniform throughout the filament. In practice however, the skin of the fiber has somewhat different structure than the core. There are several potential sources of non-uniform, radially differentiated structure of undrawn, melt spun fibers. The primary factor is heat transfer from the surface of the melt-spinning fiber leading to radial temperature gradients. The temperature gradient itself can be the cause of radially-oriented polymer crystallization and indirect effects of temperature distribution are associated with variation of polymer viscosity, relaxation times and stress in fiber cross-section. A

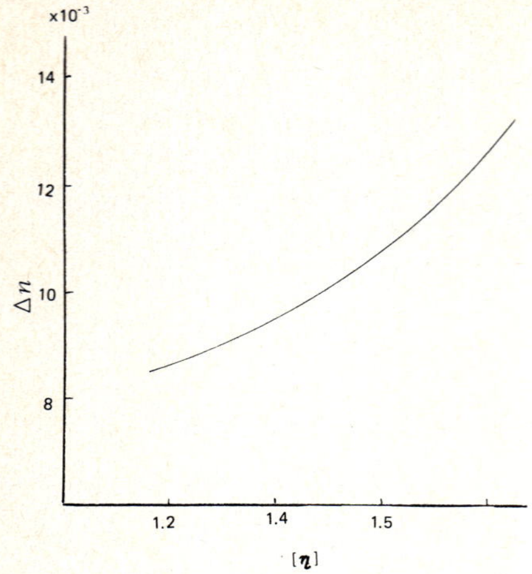

Fig. 27a. Dependence of Birefringence on Yarn Viscosity (ref.9)

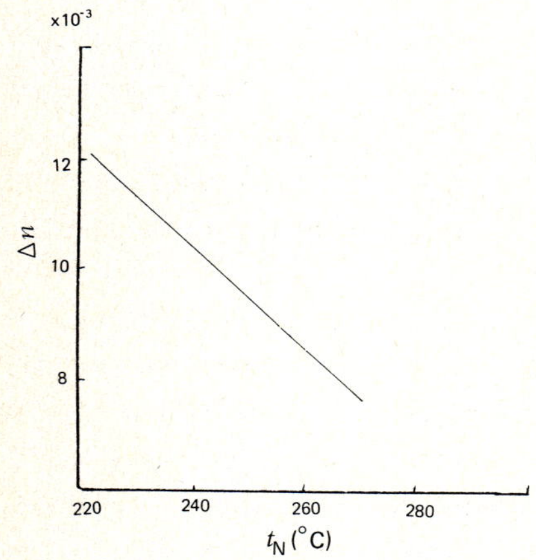

Fig. 27b. Changes in Birefringence With Melt Temperature t_N (ref.9)

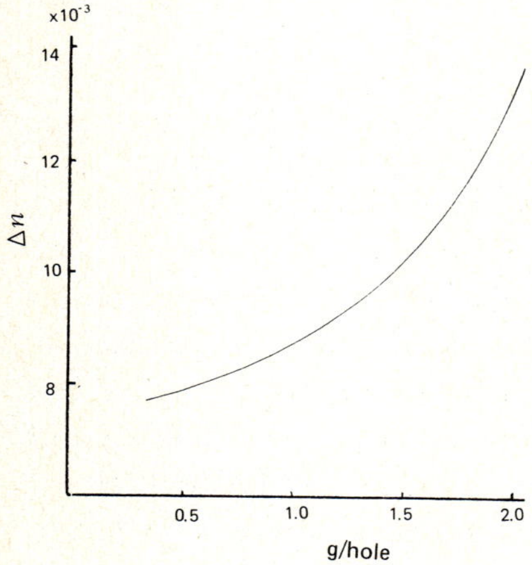

Fig. 27c. Changes in Yarn Birefringence With Throughput at Constant Denier (ref.9)

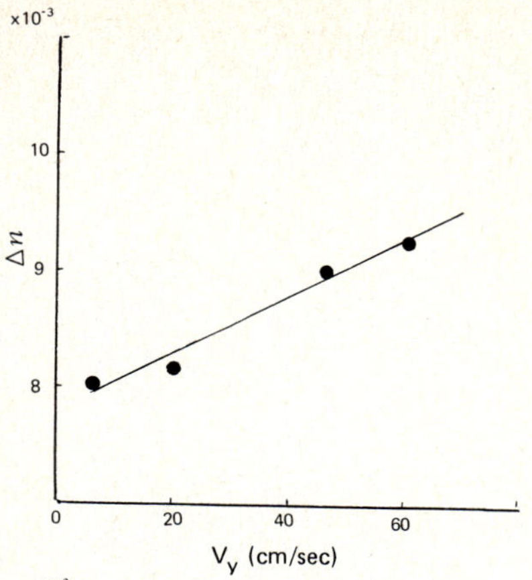

Fig. 27d. Changes in Yarn Birefringence With Change in Quench Air Velocity v_Y (ref.9)

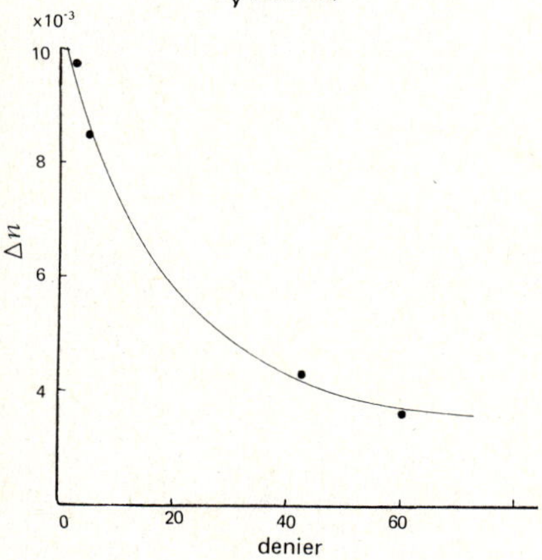

Fig. 27e. Changes in Birefringence With Filament Denier (ref.9)

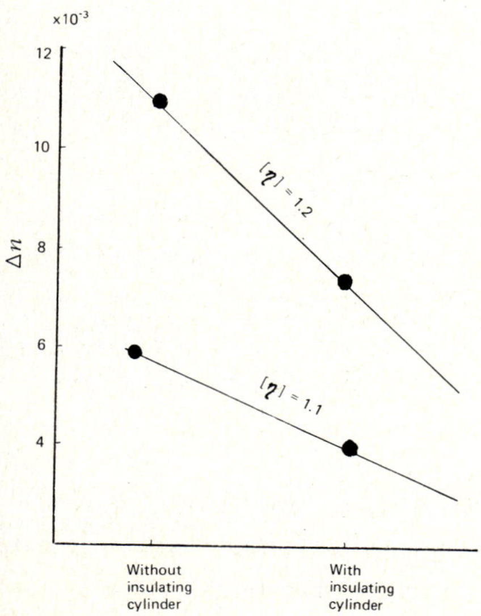

Fig. 27f. Effect of Insulating Cylinder on Yarn Birefringence (ref.9)

simple one-phase model of a cooling filament predicts maximum radial temperature gradients in melt spun filaments of the order of 10^3 - 10^4 deg C/Cm.

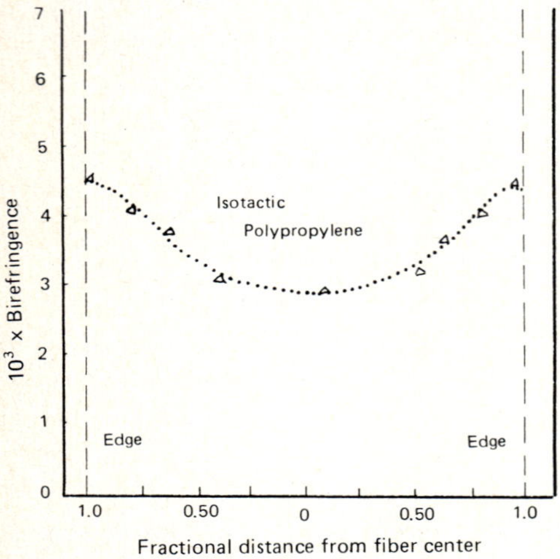

Fig. 28. Profile of Birefringence Measured on a 10-Micron Thick Section from a 70-Micron Diameter Polypropylene Filament (ref.53)

One consequence of radial viscosity and stress gradient should be non-uniform orientation, higher in the outer layer and lower in the core. Figure 28, for example, depicts the birefringence profile of an undrawn polypropylene monofilament cooled in a static environment. Interesting results were obtained by studying polypropylene fibers melt-spun with cross air blow. As would be expected from theoretical considerations, the filaments exhibited non-uniform orientation, the observed birefringence being higher on the windward side and decreasing leeward (Fig.29). The other effect of radial temperature gradient is due to crystallization. It was found by several authors that crystallization of polypropylene in a temperature field brings about crystallites oriented in the direction of the temperature gradient. Katayama, et. al. (ref.52) demonstrated this effect under conditions which resembled radial temperature distribution in melt-spinning. It is worth noting that orientations obtained in a temperature gradient differ from those observed under conditions of stress or molecular orientation. In melt-spun polypropylene fibers one would expect some perpendicularly-oriented crystals to be superimposed on those oriented in a velocity gradient.

The skin-core structural differences, discussed above, are a source of nuisance when the skin is not of uniform thickness around the core. Such a situation often arises in spinning of polypropylene under high tension (high molecular weight polymer, high rate of cooling, high take-up velocity, low spinning temperatures, etc.) with cross air blow. The filament tends to curl when relaxed. This tendency is further augmented in the process of cold drawing. If the asymmetric skin-core structure is very pronounced, it can be made of use for the production of "self-crimping" fibers. Indeed, several patents to that effect have been issued lately (see Chapter XI for details).

The development of asymmetric skin-core structure is an undesirable feature in the production of continuous multifilament yarns. The curls or filament liveliness create problems in downstream yarn processing train; the products prepared from such yarns are fuzzy in appearance. This tendency is often remedied by cold-drawing at somewhat higher temperatures than that normally required for symmetrically quenched yarns.

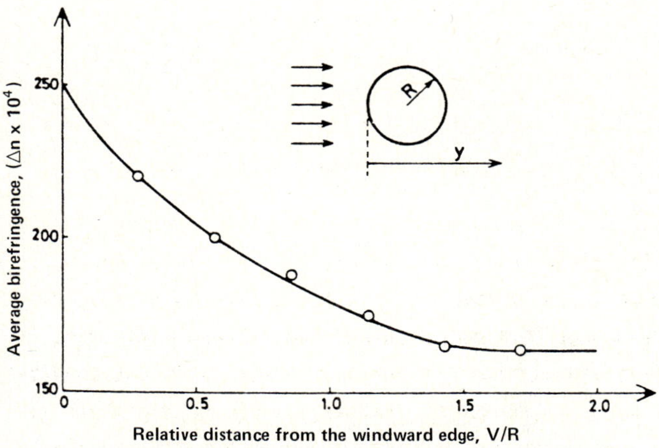

Fig. 29. Average Birefringence of an Undrawn Polypropylene Fiber Spun With Cross Air Blow (ref.56)

IV.8. Physical Properties and Fiber Structure

The large difference in morphology which occurs in polypropylene filaments is also reflected by the mechanical properties. Figure 30 compares the stress-elongation curves for filaments spun at low and high stress levels. The filament spun at low stress exhibits a yield point and necks in the usual way. The filament spun at high stress exhibits no yield point nor any noticeable neck. It does have a higher modulus and tensile strength and a lower elongation to break. A similar effect is shown in Fig. 31 which presents stress-strain curves of undrawn polypropylene fibers spun at velocities ranging from 8.4 to 258 m/min.

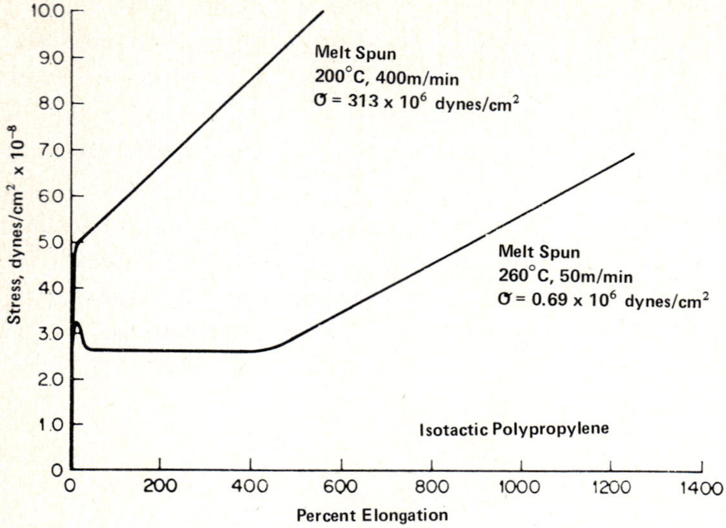

Fig. 30. Stress Versus Elongation Curves for Melt-Spun Polypropylene Fiber Spun at Low Stress Level (Upper Curve) and High Stress Level (Lower Curve) (ref.49)

The observed increase in stress and modulus and the reduction of elongation is due to increasing orientation with increasing spinning speed.

Two mechanical characteristics determine the course of drawing — maximum draw ratio R_{max} and yield stress, σ^*. Both these characteristics are very sensitive to material structure; usually the maximum draw ratio decreases, while the yield stress increases with the degree of orientation and crystallinity. Figure 32 presents maximum draw ratios of undrawn polypropylene fibers plotted versus the orientation factor for crystalline regions. It is worth noting that R_{max} also depends on the crystal structure and different correlations are obtained with prevailing hexagonal smectic and monoclinic modification.

The effect of undrawn fiber structure on the structure and properties of drawn and finished material is due to the fact that the original structure is not completely destroyed in the course of drawing. The more pronounced the crystalline structure in undrawn fibers, the stronger the effect on properties of drawn fibers. Figure 33 illustrates the tenacity of drawn polypropylene fibers as a function of the crystalline orientation factor, f_{cr}, of undrawn material (ref.32). Two separate curves correspond to hexagonal-smectic, and monoclinic crystal forms. It is evident that not only the crystal form (and apparently also the degree of crystallinity) affects tenacity of the drawn material, but there is also some effect of the orientation of undrawn fibers. Figure 34 shows the tenacity of drawn and annealed polypropylene fibers plotted

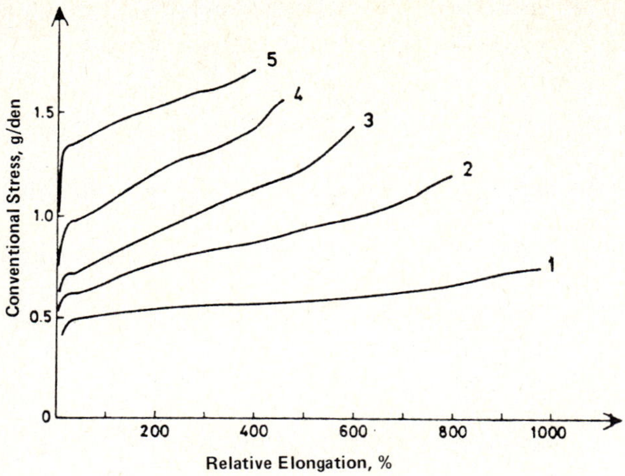

Fig. 31. Stress-Strain Characteristics for Undrawn Polypropylene Fibers Spun at Various Take-Up Velocities V : (1) 8.4 m/min; (2) 36 m/min; (3) 116 m/min; (4) 170 m/min; (5) 258 m/min (ref.54)

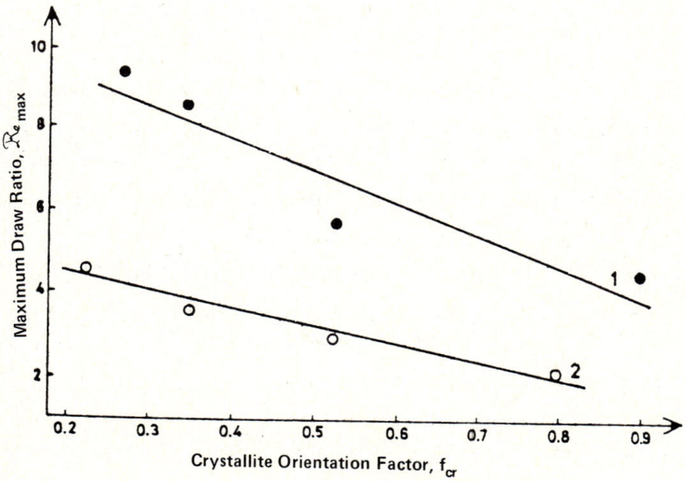

Fig. 32. Maximum Draw Ratio of Melt-Spun Polypropylene Fibers Versus Orientation Factor of the Crystalline Phase. (1) Fibers With Prevailing Hexagonal Modification; (2) Fibers With Prevailing Monoclinic Modification (ref.32)

versus the orientation angle of drawn material. Quite different correlations were obtained for the material which was only slightly oriented before drawing (Curve 1) and for material obtained by drawing of strongly spin-oriented fibers (Curve 2).

All of the above indicates that the texture of undrawn fibers and their structural uniformity plays an important role in fiber formation as determining the conditions of further technical operations (conditioning, drawing, texturizing) and affecting the structure and textile properties of the finished product.

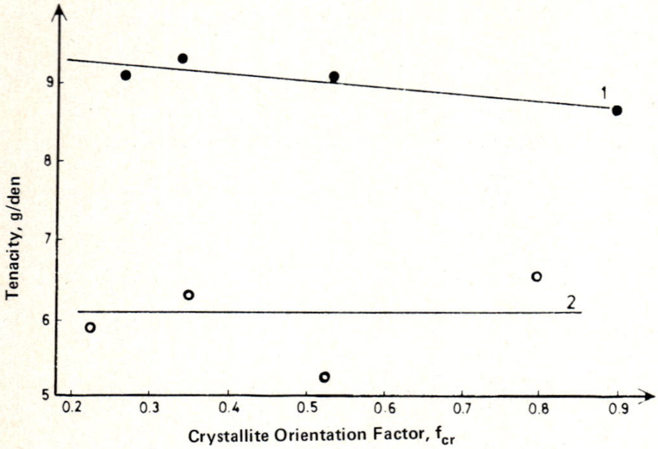

Fig. 33. Tenacity of Drawn Polypropylene Fibers as a Function of Crystallite Orientation in the Undrawn Material. (1) Fibers With Prevailing Hexagonal Modification; (2) Fibers With Prevailing Monoclinic Modification (ref.32)

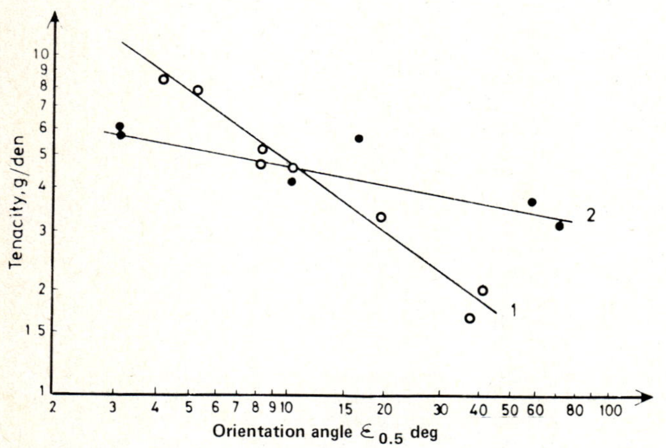

Fig. 34. Tenacity of Drawn Polypropylene Fibers as a Function of Crystallite Orientation in the Drawn Material. (1) Unoriented-Spun Filaments; (2) Oriented-Spun Filaments (ref.55)

V. REFERENCES

1. F. Ballinger and J. White, J. Appl. Polym. Sci., 15(1971)1949.
2. A. Ziabicki, in *Man-Made Fibers,* Vol. 1, Wiley, New York, 1968.
3. J.A. Brydson, *Flow Properties of Polymer Melts,* Van Nostrand Reinhold, New York, 1970.
4. D. Poller and O.L. Ready, Modern Plastics, March 1964.
5. K. Kamide et.al., Sen-i Gakkaishi, 21(1960)505.
6. K. Kase and T. Matsuo, J. Polym. Sci., 3(1965)2541.
7. R.R. Lamonte and C.D. Ham, J. Appl. Polym. Sci., 16(1972)3285.
8. T. Ishibashi, J. Appl. Polym. Sci., 14(1970)1597.
9. S. Kase and T. Matsuo, J. Appl. Polym. Sci., 11(1967)251.
10. M. Jaffe, Celanses Research Company, Private Communication.
11. F.J. Padden and H.D. Keith, J. Appl. Phys., 30(1959)1479.
12. F.L. Binsbergen, U.S. Patents 3,326,880 and 3,327,021, (1963).
13. F.L. Binsbergen, Polymer, 11(1970)253.
14. J.W. Christian, *Theory of Transitions in Metals and Alloys,* (Pergamon Press), New York, 1965.
15. A. Keller and F.M. Willmouth, J. Macromol Sci., Phys., 6(1972)493.
16. R.B. Williamson and W.F. Busse, J. Appl. Phys., 38(1967)4187.
17. G. Natta and P. Corradini, Nuovo Cirento, 15(1960)40.
18. K. Katayama, Kolloid z.z. Polymer, 226(1968)125.
19. S.E. Spruiell, and J. White, Appl. Polym. Symp., 27(1975)121-157.
20. A. Ziabicki, Appl. Polym. Symp., 6(1967)1-18.
21. L.E. Abbot and J.L. White, Appl. Polym. Sci. Symp., 20(1973)247.
22. J.L. White and E.S. Clark, J. Appl. Polym. Sci., 18(1974)2539.
23. V.S. Ishizuka, et.al., J. Chem. Soc., Japan Ind. Chem., 65(1962)603.
24. E.S. Clark, Bul. Amer. Phys. Soc., March 1970.
25. P.G. Anderson and S.H. Carr, J. Mat. Sci., 10(1975)870.
26. C.H. Garben and E.S. Clark, J. Macromol. Sci., 4(1970)499.
27. R.G. Quinn and H. Brody, J. Macromol. Sci., 5(1971)721.
28. J.H. Griffith, J. Polym. Sci., 38(1959)107.
29. P. Parrini, Macromol. Chem., 62(1963)83.
30. H.J. Donald, J. Polym. Sci., 6(1964)93.
31. K. Kamide, Kabunshi Kagaku, 22(1965)596.
32. M. Compostella, Angew. Chem., 74(1962)618.
33. A. Ziabicki, *Man-Made Fibers,* Vol. 1, Interscience, New York, 1968.
34. G. Natta, J. Polym. Sci., 34(1959)531.
35. R.L. Miller, J. Polym. Sci., 57(1962)975.
36. R. Raff, *Crystalline Olefin Polymers,* Vol.I. , Interscience, New York, 1965.
37. P.W. Wijga, U.S. Patents 3,207,735; 3,207,738; 3,299,029 (1960).
38. M. Wales, U.S. Patents 3,207,739 (1961); 3,207,737 (1962).
39. F.L. Bingsbergen, U.S. Patents 3,326,880; 3,327,021 (1963).
40. F.L. Bingsbergen, Polymer, 11(1970)253.
41. Donng, German Patent 1,118,279 (1963).
42. G.C. Vonk, Kolloid Z, 206(1965)121.
43. U.S. Patent 3,509,013 (to Hercules) 1970.
44. K. Nakamura, J. Appl. Polym. Sci., 16(1972)1077; 17(1973)1031; 18(1974)615.
45. A. Ziabicki, Appl. Polym. Symp., 6(1967)1-8; in *Man-Made Fibers,* Vol. 1, Interscience, New York, 1968; in *Fundamentals of Fiber Formation,* Interscience, New York, 1975.
46. M. Jambrich, Faserforsch, U. Textil Tech., 18(1967)103.
47. K. Katayama, Kolloid z. 226(1968)125; Appl. Polym. Symp., 20(1973)237-246.
48. J.R. Dees and S.E. Spruiell, J. Polym. Sci., 18(1974)1053.

49. S.E. Spruiell and J.L. White, Appl. Polym. Symp., 27(1975)121.
50. A. Keller and M.J. Machin, J. Macromol. Sci., 1(1967)41.
51. E.S. Clark, Polymer Preprints (ACS), 16(1975)343.
52. K. Katayama, Kolloid z., 226(1968)125.
53. P. Fung, Polym. Eng. Sci., 13(1973)295.
54. S. Oya, Sen-i Gakkaishi, 17(1961)1088.
55. W.C. Sheehaw, J. Appl. Polym. Sci., 8(1964)2359.
56. T. Matsuo, Sen-i Gakkaishi, 24(1968)512.
57. L.D. Moor, in *Crystalline Olefin Polymers*, Vol. 1, Interscience, New York, 1965.
58. V. Cappuccio, et al, Chim Industria, 44(1962)463.
59. H. Nadella, et al, J. Appl. Polym. Sci., 21(1977)3003.
60. J.E. Spruiell and J.L. White, Polym. Eng. Sci., 15(1975)660.
61. E.S. Clark, Polym. Eng. Sci., 16(1976)176.

VI. BIBLIOGRAPHY

VI.1. Books

1. H. Mark, Ed., Man-Made Fibers-Science and Technology, Vol. 1, Interscience, New York, 1968.
2. A. Ziabicki, Fundamentals of Fiber Formation, John Wiley & Sons, New York, 1976.
3. Z.K. Walczak, Formation of Synthetic Fibers, Gordon & Breach, New York, 1977.

VI.2. Journal Literature

1. G. Farrow, The Measurement of Crystallinity in Polypropylene Fibers by X-Ray Diffraction, Polymer, 2(1961)409-417.
2. M. Compstella, et al, Fibers and Films of Isotactic Polypropylene,Preparation and Properties With Regard to its Crystalline Structure, Angew. Chem., 74(1962)618-624.
3. A. Selwood, Axial Air Drag on Monofilaments, J. Text Inst. 53(1962)T576.
4. G. Farrow, Crystallinity, Crystallite Size and Melting Point of Polypropylene, Polymer, 4(1963)191-197.
5. J. Majer, Secondary Crystallization of Polypropylene, Kunststoffe-Rundschau 10(1963)9-11.
6. P. Parrini, Influence of Molecular-Weight Distribution on the Primary Crystallization Kinetics of Isotactic Polypropylene, Makromol Chem., 62(1963)83-96.
7. A. Weidlyer, X-Ray Small Angle Maxima in Oriented Polypropylene Fibers, Makromol Chem., 62(1963)139.
8. M. Dole, The Glass Transition in Isotactic Polypropylene, ACS, Div. Polym. Chem., 4(1963)631-636.
9. V. Groebe and H. Versaumer, On Fiber Formation in Melt Spinning, I, Faserforsch U. Textiltech 14(1963)249-256.
10. V. Groebe and H. Versaumer, On Fiber Formation in Melt Spinning, II, Faserforsch U. Textiltech 14(1963)288-294.
11. J.W.S. Hearle, The Fine Structure of Fibers and Crystalline Polymers, II, The Growth of Crystalline Regions in Fibers, J. Appl. Polym. Sci., 7(1963)1193-1206.
12. T. Manabe, Some Problems in Melt Spinning Processes, Part I, Methods for Estimating the Tension Acting on a Fiber, Textil Res. J., 33(1963)221-224.
13. A. Ziabicki, Studies on the Orientation Phenomenon by Fiber Formation From Polymer Melts, IV, J. Appl. Polym. Sci., 6(1962)361-367.

14. P.H. Lindenmeyer, Crystallization Habit and Fiber Formation, Textil. Res. J. 34(1964)825-830.
15. O.L. Reedy, Designing Dies for Monofilaments, Modern Plastics, 41(1964)133-138.
16. G. Bodon, Determination of the Crystallinity of Polypropylene, Faserforsch u. Textiltech., 15(1964)527-532.
17. G. Farrow, Measurement of the Smectic Content in Undrawn Polypropylene Filaments, J. Appl. Polym. Sci., 9(1965)1227-1232.
18. A.P.Metzger, Extrudate Swelling at High Shear Rates, J. Polym. Sci., 3(1965)921-922.
19. T. Manabe, Some Problems in Melt Spinning Processes, Part II, Molecular Orientation in the Course of Melt Spinning, J. Appl. Polym. Sci., 8(1964)1097-1111.
20. C.G. Vonk, The Induction of Low Orientation in Polyethylene by Foreign Particles, Kolloid Z.Z. Polymere, 206(1965)121-122.
21. F. Geleji, Fiber Formation From Polypropylene and Modified Polypropylene, Abhandl. Deut. Akad. Wiss. Berlin, Kl. Chem. Geol. Biol., 3(1965)29-35.
22. S. Kase, T. Matsuo, Melt Spinning I, Fundamental Equations on the Dynamics of Melt Spinning, J. Polym. Sci., 3(1965)2541-2554.
23. S.E. Ross, Some Observations Concerning the Behavior of Polypropylene Polymers and Fibers, J. Appl. Polym. Sci., 9(1965)2729-2748.
24. M. Gordon, The Bulk Crystallization Kinetics of Polypropylene and Polybutane, Polymer, 6(1965)213-219.
25. F. Rybnikar, Effect of Melting Conditions on the Isothermal Crystallization of Polypropylene, J. Polym. Sci., 16(1966)129-140.
26. C.W. Hock, Morphology of Polypropylene Crystallized From the Melt, J. Polym. Sci., 4(1966)227-242.
27. T. Manabe, The Effect of Spinning and Drawing Conditions on Mechanical Properties and Breaking Frequencies in Spinning and Drawing, ACS, Div. Polym. Chem. Preprints, Appl. Polym. Symp., Interscience, 7(1966)876-879.
28. A. Ziabicki, The Kinetics of Polymer Crystallization and Molecular Orientation in the Course of Melt Spinning, ACS, Div. Polym. Chem. Preprints, Appl. Polym. Symp., 7(1967)1-18.
29. M. Copley and N.H. Chamberlain, Filament Attenuation in Melt Spinning and its Effect on Axial Temperature Gradient, ACS. Div. Polym. Chem. Preprints, Appl. Polym. Symp., 7(1966)679-690.
30. G. Wilhelm, Cooling in the Spinning Compartment of a Polymeric Fiber Spun From Melt, Kolloid, zz, Polymere, 208(1966)97-123.
31. G. Wilhelm, Cooling of Melt Spinning Polymer Fibers, Kolloid, Z.Z.Polymere, 208(1966)97-123.
32. A. Ziabicki, Kinetics of Polymer Crystallization and Molecular Orientation in the Course of Melt Spinning, Appl. Polym. Symp, 6(1967)1-18.
33. Van der Vegt, Crystallization Phenomena in Flowing Polymers, Soc. Chem. Ind. (London) 26(1967)313-323.
34. S. Kase and T. Matsuo, Studies on Melt Spinning II, Steady State and Transient Solutions of Fundamental Equations Compared With Experimental Results, J. Appl. Polym. Sci., 2(1967)251-287.
35. R. Harrington, Flow Birefringence of Polymers, Encyl. Polym. Sci. Tech., 7(1967)100-179.
36. M. Jambrich and I. Diacik, The Molecular Structure of Polymer Systems at the Transition From the Melt to the Solid Phase, Faserforsch. u. Textiltech., 18(1967)103-107.
37. H.N. Beck, Heterogeneons Nucleating Agents for Polypropylene Crystallization, J. Appl. Polym. Sci., 11(1967)673-685.
38. F. Rybnikar, The Determination of Crystallinity of Isotactic Polypropylene, Kunststoffe, 57(1967)199-203.
39. P.G. Stern, On the Structure of Polypropylene Fibers, Polymer, 9(1968)471-477.
40. R. Zannetti, The Order Zones in Isotactic Polypropylene III, General Interpretation of the Small Angle X-Ray Diffraction Spectra, Eur.Polym.J., 4(1968)399-411.

41. F.L. Binsbergen, Morphology of Polypropylene Crystallized From the Melt, Polymer, 9(1968)23-40.
42. I.H. Hillier, Crystallization Kinetics and Bulk Polymer Morphology II, Polypropylene, Polymer, 9(1968)19-22.
43. D.M. Gezovich and P.H. Geril, Morphology of Quenched Polypropylene, Polym. Eng. Sci., 8(1968)209-219.
44. J.A. Turner, The β-Crystalline Form of Isotactic Polypropylene, J. Polym. Sci., 6(1968)539-546.
45. T. Kitao, Kinetic Interpretation of Melt Spinning, Proc. Int. Cong. Rheol., 4(1968)409-420.
46. U. Fujiwara, Spherulite Structure of Isotactic Polypropylene Obtained by Oriented Crystallization from the Melt, Kolloid Z.Z.Polymere, 226(1968)135-138.
47. K. Katayama, et al, Structure Formation During Melt Spinning Process, Kolloid Z.Z.Polymere, 226(1968)125-134.
48. J. Zimmerman, Some Quantitative Aspects of Bulge Formation at the Exit of a Spinnerette Capillary, Appl. Polym. Symp., 6(1967)19-26.
49. H. Mark, Recent Investigations into the extrusion and Drawing of Filaments, Chemicefasern, 18(1968)118-121.
50. M.J. Hill and A. Keller, Direct Evidence for Distinctive Stress Induced Nucleus Crystals in the Crystallization of Oriented Polymer Melts, J. Macromol, Sci. Phys., 3(1969)153-169.
51. D.R. Morrow, Polymorphism in Isotactic Polypropylene, J. Macromol. Sci. Phys., 3(1969)53-65.
52. F. Rybnikar, Efficiency of Nucleating Additives in Polypropylene, J. Appl. Polym. Sci., 13(1969)827-833.
53. I. Hamana, et al, Filament Formation in Melt Spinning, Melliand Textilber, 50(1969)382-388 and 499-503.
54. F.N. Binsbergen, Heterogeneous Nucleation in the Crystallization of Polyolefins II, Kinetics of Crystallization of Nucleated Polypropylene, Polymer, 11(1970) 309-326.
55. T. Kitoa, Kinetics in Interpretation of Melt Spinning, Proceeding of the Intern, Congress of Rheology, 5(1970)409-420.
56. V.G. Baranov, et al, Super Molecular Organization of As-Spin Fibers from Melts of Crystalline Polyolefins, J. Polym. Sci., 30(1970)305-310.
57. G.S. Fashyan, Super Molecular Organization of Fibers Extruded from the Melts of Crystalline Polymers, Fiber Chem., 2(1970)342-345.
58. S. Richardson, The Die Swell Phenomenon, Rheol. Acta., 9(1970)193-199.
59. W. Conti, On the Cooling of Spinning Fibers, Faserforsch u. Textiltech. 21(1970)509-513.
60. C.D.Han, Correlation of Exit Pressure with Die Swell Ratio, Aiche J., 16(1970)1087-1088.
61. W. Graissley, Die Swell in Molten Polymers, Trans. Soc. Rheol., 14(1970) 519-544.
62. R.I. Tanner, Theory of Die Swell, J. Polym. Sci., 8(1970)2067-2078.
63. C.D.Han, Determination of Polymer Melt Die Swell, Trans. Soc. Rheol., 14(1970)213-218.
64. J. Farukawa, et al, Orientation of Polymer Molecules During Melt Spinning, I, Estimation of Degree of Orientation by Retraction Above Tm, J. Polym. Sci., 9(1971)299-310.
65. T. Nagasawa, Film Forming Process of Crystalline Polymers, II, Microstructure, Appl. Polym. Symp. 20(1972)295-313.
66. K. Katayama, Problems in Structural Formation During Melt Spinning, Appl. Polym. Symp., 20(1972)237-246.
67. L.E. Abbott and J.L. White, Melt Spinning of High Density and Low Density Polyethylene Fibers. Development of Orientation and Crystallinity and Mechanical Properties of Spin Fibers, Appl. Polym. Symp., 14(1972)247-268.
68. R.R. Lamonte and C.D.Han , Melt Spinning. II. Analysis of the Deformation and Heat Transfer Processes, J. Appl. Polym. Sci., 16(1972)3285-3306.

69. K. Nakamura, Some Aspects of Non-isothermal Crystallization of Polymers. I. Relationship Between Crystallization Temperature, Crystallinity, and Cooling Conditions, J. Appl. Polym. Sci., 16(1972)1077-1091.
70. J.L. White, Critique on Flow Patterns in Polymer Fluids at the Entrance of a Die and Instabilities Leading to Extrudate Distortion, Appl. Polym. Symp., 20(1972)155-174.
71. Y.Chen, Die Swell, Advan. Polym. Sci. Eng. Proc. Symp., Plenum, New York, (1972)239-247.
72. B.P. Levitskii, Influence of the Spinnerette Hole Dimensions on the Filament Diameter, Fiber Chem., 2(1972)491-494.
73. A.T. Serkov, Forming of the Jet in Man-Made Fiber Extrusion, Feber Chem., 4(1972)147-150.
74. C.D.Han, Study of Shaped Feber Formation, J. Appl. Polym. Sci., 17(1973)187-200.
75. F.A. Seyer, Analysis of Flow Behavior of Polymer Molecules. Viscoelastic Functions and Crystallization, Kolloid zz, Polymer, 251(1973)108-116.
76. P.Y.E. Fung and S.H. Carr, Development of Stress Crystallized Morphology During Melt Spinning of Polypropylene, Polym. Eng. Sci., 13(1973)295-299.
77. K. Nakamura, Some Aspects of Non-isothermal Crystallization of Polymers. II. Consideration of the Isokinetic Conditions, J. Appl. Polym. Sci., 17(1973)1031-1041.
78. J.R.A. Pearson and M.A. Matovich, Spinning of Molten Threadline, Ind. Eng. Chem. Fund., 8(1973)605-609.
79. Y. Mori, On Die Swell in Molten Polymers, Appl. Polym. Symp., 20(1973)209-220.
80. H. Lewgering and G. Kirsch, Effect of Crystallization from Oriented Melts on Crystal Structure of Isotactic Polypropylene, Angew. Macromol. Chem., 33(1973)17-23.
81. T. Kitao, Orientation of Polymer Molecules During Melt Spinning. II. Orientation of Crystals in As-Spin Polyolefin Fibers, J. Polym. Sci., Polym. Phys., 11(1973)1091-1109.
82. E. Cernia and W. Conti, Dynamics of Melt Spinning, J. Appl. Polym. Sci., 17(1973)3639-3652.
83. H. Mark, Quantitative Observations on Melt Spinning, Leuzinger Ber., 35(1973)5-9.
84. N. Kawasaki, Capillary Length Dependence of Barrus Effect of Polymer Melts, Japan Polym. Sci. Tech. Eng. Ed., 2(1973)5157.
85. M. Fleissner, Macroscopic and Molecular Orientation in Polymers, Kolloid Z.Z. Polymere, 251(1973)1006-1014.
86. P. Tucker and G. Waller, Microfibers Formed *ab initio* in extensional melt flows, Textile Res.J.,44(1974)56-70.
87. A. Keller, Chain Orientation and Crystallization, Pure Appl. Chem., 39(1974) 195-224.
88. K. Nakamura, Non-isothermal Crystallization of Polymers. III. Crystallization During Melt Spinning, J. Appl. Polym. Sci., 18(1974)615-623.
89. Y.I. Surov, Temperature Field of Monofilaments During Melt Spinning, Fiber Chem., 6(1974)22-26.
90. J.R. Dees and J.E. Spruiell, Structure Development During Melt Spinning of Linear Polyethylene Fibers, J. Appl. Polym. Sci., 18(1974)1053-1078.
91. S. Kase, Melt Spinning III. Velocity Field Within the Thread, J. Appl. Polym. Sci., 18(1974)3267-3278.
92. C.D.Han, Flow of Viscoelastic Polymeric Melts Through Converging Ducts, I and II, Rheol. Acta, 14(1975)173-191.
93. H. Gehrke, Theoretical Calculation of Diameter and Temperature of a Melt Spinning Fiber, Chem. Eng. Sci., 30(1975)859-869.
94. H.P. Nadella, et al, Melt Spinning of Polypropylene Structure Development and its Relation to Properties, ACS, Div. Polym. Chem., 16(1975)339-342.
95. A. Prastaro and P. Parrini, Meltermatical Model for Spinning Molten Polymers and Conditions of Spinning, Text. Res. J., 45(1975)118-127.
96. H. Barcia, Flow Dynamics and Fiber Spinning, Fiber Producer, 3(1975)59.

97. J.R. Gaylord, Orientation of Crystallites Formed in Stretched Polymer Networks, J. Polym. Sci., Polym. Letters, Ed. 13(1975)337-340.
98. D.J. Fraser, Spherulite Content and Distribution in Polypropylene Filaments, The Microscope, 23(1975)145-155.
99. E.S. Clark and J.E. Spruiell, Bimodal Orientation in Flow Crystallized Polypropylene, ACS, Div. Polym. Chem., 16(1975)343-348.
100. L.A. Utracki, A Method for the Measurement of the True Die Swell of Polymer Melts, J. Appl. Polym. Sci., 19(1975)481-502.
101. J.E. Spruiell and J. White, Structure Development During Polymer Processing. Melt Spinning of Polyethylene and Polypropylene Fibers, Polym. Eng. Sci., 15(1975)660-667.
102. R. Hoseman, Paracrystalline State of Synthetic Polymers, J. Polym. Sci. Polym. Symp., 27(1975)265-281.
103. J.E. Spruiell and J.L. White, Structure Development During Melt Spinning of Fibers, Appl. Polym. Symp., 27(1975)121-157.
104. M.R. Mackley and A. Keller, Flow Induced Polymer Chain Extension and its Relation to Fiberous Crystallization, Philos. Trans. Roy. Soc., 278(1975) 29-66.
105. R.W. Dunlap, Effect of Melt History, Crystallization Pressure and Crystallization Temperature on the Crystallization and Resultant Structure of Bulk Isotactic Polypropylene, J. Appl. Polym. Sci., 19(1975)1037-1060.
106. A. Prastaro, Mathematical Model for Spinning Molten Polymers and Conditions of Spinning, Text. Res. J., 45(1975)118-127.
107. A.M. Chetterjee, Heterogeneous Nucleation of Crystallization of High Polymers from Melt I, II, III, J. Polymer Sci. Polym. Phys., 13(1975)2369-2400.
108. A. Ya Malkin, Rheology of Polymer Systems in Uniaxial Extension, A Review, Fiber Chem., 7(1975)114-123.
109. H. Ishihara and S. Kase, Studies in Melt Spinning VI, J. Appl. Polym. Sci., 20(1976)169-191.
110. D.J. Fraser, Spherulite Content and Distribution in Polypropylene Filaments, The Microscope, 23(1975)145.
111. J.L. White, Extrudate Swell During the Melt Spinning of Fibers - Influence of Rheological Properties and Take-Up Force, J. Appl. Polym. Sci., 20(1976)1005-1023.
112. T. Matsuo and S. Kase, Studies on Melt Spinning. VII. Temperature Profile Within the Filament, J. Appl. Polym. Sci., 20(1976)367-376.
113. Y.W. Kim, Studies on Melt Spinning. VII. Elongational Viscosity and Fiber Morphology of Multiphase Polymer Systems, J. Appl. Polym. Sci., 21(1977)515.
114. G. Hinrichsen, Tasks and Objects of Polymer Physics in Connection with the Production and Characterization of Synthetic Fibers. I., Melliand Textilber Intern, 6(1977)1-4.
115. H.P. Nadella, Melt Spinning of Isotactic Polypropylene: Structure Development and Relationship to Mechanical Properties, J. Appl. Polym. Sci., 21(1977)3003.
116. V.G. Baravov, Orientation Crystallization of Polymers, Fibre Chem., 9(1977)223
117. K.E.Prepelkin, Time Similitude in Man-made Fire Spinning and its Use in Technological Analysis, Fibre Chem., 9(1977)546-555.

Chapter VI INSTABILITIES IN MELT-SPINNING

I. INTRODUCTION

A steady-state spinline may be defined as one exhibiting cylindrical symmetry and whose surface coordinates at any point in space are invariable with time, or

$$\frac{d}{dx}(\ell vA) = 0$$

where ℓ is the density of the polymer, v is the speed of threadline, A is the cross-sectional area of the filament, and x is the distance from the spinnerette. Consider the idealized steady-state spinline shown in Fig.1. The polymer melt is extruded at a constant volumetric rate through an orifice of diameter D_n, whereupon a swelling occurs to form a bulge of diameter D_b immediately below the orifice exit. The filament is attenuated below the region of the bulge, solidifying *en route* and reaching a solidified diameter of D_g. It leaves the system at a constant velocity V_g. In this steady-state condition, the following continuity equation applies:

$$D\ell/D_t + \ell(\nabla.v) = 0$$

Here ℓ is the density of the polymer, t is time and v is velocity.

In actual practice of fiber spinning the conditions of steady-state are not strictly met: some fluctuations in mass flow along the spinline are always present. These result in nonuniformity of fiber thickness and variation of structure along the length of the filament. Under some conditions of spinning, these mass fluctuations can become so gross that filament break occurs and the spinning process comes to a halt. "Rain effect" is a visual description of thick and thin sections of the filaments passing through a cross section of multifilament threadline. In some cases the visual effect is not pronounced, but the filaments may still contain nonuniformities of shape and structure along their length.

The variation in fiber cross-section along the filament length is detrimental to the quality of the product in several ways.

(1) The yarns of nonuniform filament cross-section show poor tensile properties because of the various weak links in the individual filaments.

(2) The thin sections of the filament have different degrees of molecular

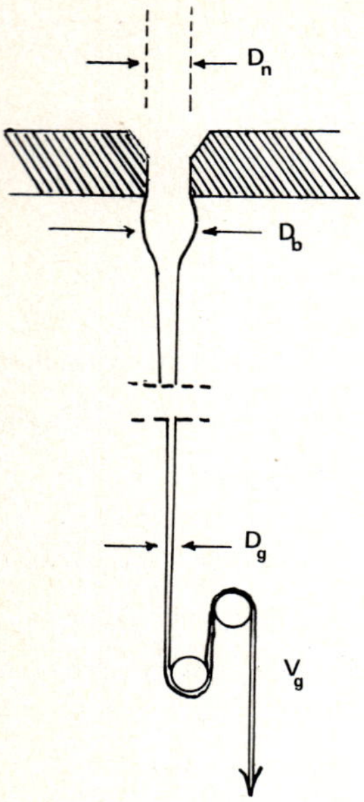

Fig. 1. Schematics of Melt-Spinning Process

orientation and different crystalline orders, with the result that their cold-drawing properties differ from those of the relatively thick sections of the filament. This produces numerous broken filaments at high draw ratios and severe nonuniformity in filament cross-sections at low draw ratios.

(3) The drawn sections from thin and thick filaments have greatly different shrinkage properties. These filaments, when subjected to heat treatment in texturizing or finishing, further accentuate the nonuniformities produced in the cold-drawing process.

(4) One common problem in polypropylene spinning of multifilament yarns at commercial spinning speeds is the occurrence of "loops" in the filament bundle. By "loops" is meant here frequent sections of filaments, from a few inches to a few feet in length, which stand out as loose and relaxed filaments even when the filament bundle is wound on a package under considerably high winding tension. Or, conversely, the occurrence of numerous filament sections of a few inches to a few feet in length which appear to be somewhat shorter than the rest of the filaments in any particular section of the filament bundle. Although the average length of all filaments in the bundle remains the same on a long-term basis, the lengths of some filaments differ from others over short distances. These "loops" therefore differ from those commonly encountered in the textile industry. The presence of loops in multifilament spun yarn is an indication of poor quality, even if

the tensile strength is not of primary interest for the eventual use of the product. The phenomenon of loop formation is intimately related to thick and thin filament sections along the filament.

(5) Apart from the quality of the finished product, there is always the question of production economics. The instabilities in melt spinning obviously lead to frequent threadline breaks and thereby to higher production cost, due to increased waste and unnecessary down-time of the equipment in spinning as well as in processing departments. The cost of quality control also increases and at some stage the whole concept of quality control becomes meaningless.

An obvious indication of instable spinning is the high "break-index" in production, i.e., the breaks occurring in the spinline per 100 lb of yarn, or the number of spinning breaks per threadline per shift.

For purposes of investigation, two alternative methods of testing and reporting are generally employed.

(1) Direct measurement of mass variation along the fiber or yarn length, via a capacitative technique, such as in the Uster Unevenness Tester. Results are reported in terms of percent CV (coefficient of variation), defined as:

$$\% \text{ CV} = \frac{\theta}{\overline{X}} \cdot 100\%$$

where $\overline{X}$ is the mean and θ the standard deviation.

(2) Cross-sectional area measurement of individual filaments across the filament bundle. Generally the results are reported in terms of a cross-section ratio, X, defined as:

$$X = \frac{\text{area of smallest filament observed}}{\text{area of largest filament observed}}$$

A more accurate, but more tedious, procedure is to measure individual areas of fiber cross-section in filament bundle and compute a percent CV. Even then, proper sampling requirements demand several such measurements at points along the yarn.

During a yarn Uster Unevenness Test, one determines essentially the mean value of the linear density of fiber (denier) across a bundle of filaments and computes the coefficient of variation of this mean from point to point along the yarn. In this situation, the coefficient of variation of the constituent filament is given by:

$$\% \text{ CV filaments} = \% \text{ CV yarn} \cdot \sqrt{n}$$

where n is the number of filaments in the yarn.

The above equation emphasizes that yarn CVs, which typically range from 2 to 5 percent for textile products, are reassuringly deceptive. For example, an as-spun 40-filament yarn with a percent CV of 4.5 actually has a constituent filament CV of 28.5% and in consequence the same percent CV across the yarn bundle. This is a far more frightening number than the percent CV number for the total yarn bundle would suggest. The larger the percent CV of the yarn bundle, the greater is the probability that individual filaments will vary in size along their lengths. One may expect it to increase exponentially with yarn percent CV.

The various factors causing unstable spinning and nonuniformities of filament cross-section can be broadly classified into four groups:

(1) Those connected with mechanical factors, such as fluctuations in extruder pressure, variation in pumping action, nonuniform dimensions of filament orifices, eccentricity of spin-finish applicator, etc. The mechanical aspects of the production of fibers from polymer melts are of considerable technical significance, and hence undoubtedly receive a great deal more attention than the published literature may indicate.

(2) Those connected with rheological properties of the melt, especially the unsteady flow of rheologically nonhomogeneous melt and the unsteady flow of homogeneous melt under excessive shear rate or under excessive rate of extention.

(3) Those connected with heat transfer, such as sudden changes in cooling air temperature or speed.

In this chapter we will examine some of these factors and see how spinline uniformity responds to various processing conditions. However, before we proceed with the discussion of instabilities in spinning, it is instructive to examine the subject of spinnability of the melt itself. Limited spinnability is one of the many sources of instability in technical fiber spinning. An understanding of the general principles of spinnability may help the fiber engineer choose spinning conditions that are safe from instabilities. Variations of the technical parameters of the spinning process as well as the variation of spinning melt characteristics makes possible considerable extension of the range of stable spinning.

II. SPINNABILITY OF POLYPROPYLENE MELTS

Polymer melts are called spinnable if they are capable of yielding long, continuous fluid threads when extended, are easily transferable into solid polymers, and possess useful properties when solidified. The spinnability of a melt is then defined by the maximum length of the filament that can be drawn from the melt. In the industrial process of fiber formation, the spinnability of a melt can be defined as the maximum draw-down that the melt is capable of yielding. The nominal draw-down is the ratio of the cross-sectional area of the spinnerette hole divided by the cross-sectional area of the solidified filament. Both of these definitions are, however, vague and erroneous if taken quantitatively. The maximum obtainable filament length or the maximum draw-down ratio depends not only on the material characteristics, but also on temperature and the rate of fiber formation. In spite of the imprecision of definition and a lack of quantitative criteria of spinnability, the concept of spinnability is quite useful in qualitative terms.

The problem of spinnability has been a subject of interest to fiber engineers, biologists, colloid and physical chemists, and polymer scientists for over 50 years. Several attempts have been made to explain spinnability in terms of macroscopic properties of a liquid. Nitschman and Schrade (ref.1), for example, based their theory on tensile viscosity as a prime macroscopic physical property of the liquid determining its spinnability, and Hirai (ref.2,3) considered the surface tension as the critical factor for spinnability of a polymer. The importance of shear viscosity for spinnability has been known, but it was repeatedly shown not to be the sole determining factor.

Probably the most successful approach to the problem to date is the qualitative treatment of Ziabicki and Taskerman-Krozer (ref.4,5) who based their analysis on the physical process that can lead to fracture of liquid thread. The following discussion is based largely on their analysis of the problem. A more rigorous treatment of the subject has been recently published by Lee and Rubin (ref.6).

According to Ziabicki, there are two main modes of fracture of fluid thread: (1) Cohesive or brittle fracture, operating at higher range of viscosities, and (2) Capillary breakup of fluid thread under the action of surface tension.

In any viscoelastic fluid with a definite relaxation time, τ, some fraction of deformation energy is stored during deformation. There is some critical level of stored elastic energy above which the material undergoes fracture (ref.7). Ziabicki and Taskerman-Krozer (ref.4,5) derived the approximate expression for the maximum thread length x^*_{coh} or spinnability of a polymer as:

$$x^*_{coh} \approx \tfrac{1}{2}[\ln(2K/E - 2\ln(v_0 \tau \mathcal{E})]\mathcal{E}$$

where V_o is the initial velocity of the jet, $\mathcal{E}$ is the axial deformation gradient, K is the cohesive energy density, and τ is the relaxation time. According to this equation:

(i) the spinnability increases with an increase in cohesive energy density. Conversely, any additive present in the polymer and causing a decrease in cohesive energy density of the melt would decrease the spinnability of the polymer.
(ii) the spinnability of the polymer melt decreases with increasing spinning speed, V_0.
(iii) the spinnability decreases with increasing deformation gradient, $\dot{\varepsilon}$, i.e., high deformation rates of spinline are limited by cohesive fracture.
(iv) the spinnability decreases with increasing relaxation time, τ.
(v) since the relaxation time, τ, can be theoretically equated to $\tau = 3\eta/E$ (where η is the viscosity and E is the Young's modulus of the spinning material), the spinnability is expected to decrease with increasing η and decreasing E, meaning that highly viscous liquids are limited in their spinnability due to cohesive fracture.

The second process limiting thread length or spinnability is connected with surface tension of the melt. Here the break is due to a spontaneous development of surface waves on the free surface of the liquid thread, a result of the fluid's surface tension. A cylindrical surface is unstable under surface tension and will tend to break into droplets when perturbed.

Ziabicki and Taskerman-Krozer (ref.4,5) developed an approximate relation for this effect as:

$$x^*_{cap} \approx 2\varepsilon^{-1} \left[\ln(R_o/S_o) - \frac{\alpha}{3\eta V_o \varepsilon R_o} \right]$$

where S_o denotes axially symmetrical distortion of the jet radius R_o, and α is the surface tension. According to this relation, it is evident that:
(i) the tendency to capillary breakup x^*_{cap} decreases with increasing viscosity η, deformation rate ε, radius of the jet R_o, and the spinning velocity V_o.
(ii) the x^*_{cap} decreases with surface tension.

The dependence of spinnability on the product of viscosity, spinning velocity, and extension rate, rather than on these factors separately, is predicted by both the above-mentioned fracture mechanisms and provides an explanation for the experimental data reported variously in several empirical studies and observed in commercial practice.

Theoretically, the two breaking mechanisms described above can act independently. Under a given set of spinning conditions, the mechanism that causes earlier breakage is expected to determine the spinnability of a polymer. It should be apparent from the above discussion that the two breaking mechanisms run in opposite directions with their own characteristic spinnability paths. In Figure 2 the maximum thread length $\overline{x}$ calculated for both breaking mechanisms is plotted versus (ηV_o). It is evident that $\overline{x}$ increases with (ηV_o) for smaller values of the parameter when the capillary breakup mechanism prevails and decreases in the higher (ηV_o) region, corresponding to cohesive fracture. These predictions are consistent with the experimental results and explain the practical experience in melt spinning.

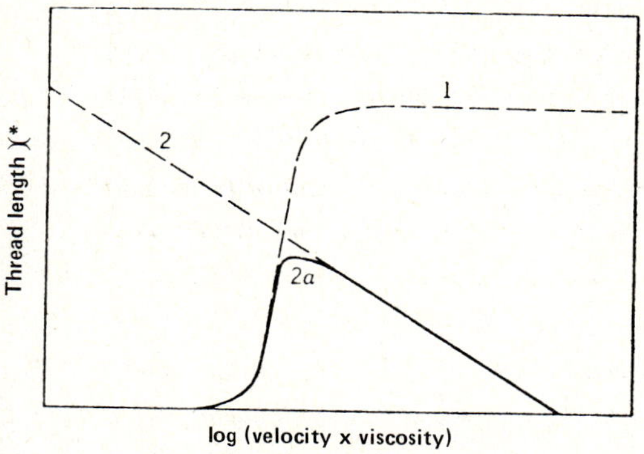

Fig.2. Liquid Thread Length x^* vs. Product (Velocity X Viscosity) (Theoretical). Dotted Lines: Independent Mechanisms - Curve 1, Capillary Breakup: Curve 2, Cohesive Fracture. Solid Line 2a, Cohesive Fracture in the Presence of Capillary Waves (ref.5).

Thus, for steady spinning operation, a high spinning speed and high deformation rate of the fluid jet is required in low η regions; and a slow spinning speed and low deformation rate in high η regions. Figure 2 does not take into account the effect of solidification on spinnability and the changes in η and α with a change in filament temperature. The effect of these factors is complex and depends on the conditions of flow and cooling temperatures.

For the spinning of polypropylene the region of high ηV_o values is more important than that characterized by low values of ηV_o. In technical practice, the instability due to cohesive fracture determines the upper limit of take-up velocity and melt draw-down ratio above which no spinning is possible. According to the theory of cohesive fracture, spinnability is controlled by the relaxation time of the material, τ, its tensile strength, and the deformation conditions. Cohesive fracture hardly affects melt-spinning of polycondensates, such as nylon and polyester, (τ of the order of 10 - 10^2 msec) which can be spun at take-up velocities as high as 4000 m/min and melt draw-down ratios up to 1000. Such behavior is due not only to the short relaxation times of spinning melts but also to relatively long solidification distances and rather high tensile strength of the melts. Melt-spinning of polypropylene, whose average molecular weight is much higher than that of polycondensates and whose relaxation times reach as high as 1 sec, is on the other hand associated with serious limitations due to cohesive instability.

Relaxation times are very sensitive to temperature and molecular-weight distribution of the polymer. Maximum take-up velocity for polypropylene depends on the extrusion (melt temperature, spinnerette hole diameter, extrusion rate through the spinnerette hole, etc.) and solidification conditions (temperature and velocity of cooling air, filament diameter, etc.).

Kamide and Inamoto (ref.8) investigated the effect of molecular weight, molecular-weight distribution and spinning temperature on the spinnability of polypropylene melts. Their data, coupled with experience in the industry, confirm the broad conclusions drawn from the above-mentioned considerations.

For every polymer melt there is a certain range of temperature in which spinning is practically possible. At the lower end of this temperature range the limit of spinnability is reached because of brittle fracture of the polymer jet. At the higher end of temperature range the spinnability becomes limited because of polymer degradation and capillary break of the threadline. The optimum spinning temperature increases with increasing molecular weight of the polymer. This means that higher molecular-weight polymers require higher temperatures for equivalent spinnability. The effect of molecular-weight distribution on spinnability has not been examined adequately. Kamide and Inamoto (ref.8) showed that the optimum spinning temperature shifts somewhat to a higher range as the polydispersity increases. The effect, however, is minor compared with that of average molecular weight of polymers (ref.9). The increase of optimum spinning temperature with increasing breadth of molecular-weight distribution indicates that higher-molecular-weight "tail" in distribution curves impedes the spinnability of the polymer. Industrial experience contributes to this conclusion as, in general, the narrow-molecular-weight polymers allow high spinning rates and yield a more uniform product than the broad-molecular-weight polymers of the same $[\eta]$ or melt-flow index.

There are indications also (ref.8) that a smaller molecular-weight fraction, particularly less than M_w of 3.2×10^4, is as detrimental to spinnability of the polymer as is the high molecular-weight "tail". Hence it is understood that information concerning the breadth of molecular-weight distribution M_w/M_n alone is not sufficient to judge the spinning quality of the polymer. It may transpire that the absolute value of the low and high-molecular-weight tail fractions is more important than the conventional M_w/M_n values.

III. MECHANICAL FACTORS

Two main mechanical factors influencing the stability of melt spinning process and affecting the uniformity of the threadline are the variations in screw back pressure of the extruder (surging), and the fluctuations in melt pressure in the spinnerette pack. Small variations in screw back pressure generally do not contribute to the instability of spinning. The metering pump normally used in a melt-spinning operation is positive displacement, rotating gear type apparatus. These pumps are built to withstand small variations in feed pressure without adversely affecting the uniformity of delivery. A large variation in screw pressure shows up in the form of extruder surging, which is easily detected and the corrective measures taken.

What escapes easy detection — and is probably not fully appreciated — is the pressure fluctuation at the delivery-end of the metering pump. This is caused primarily by worn-out or defective gear teeth of the pump and by the use of a pump of too large a capacity. Both of these factors contribute to a cyclic variation in flow rate Q, and thereby to a cyclic variation in melt pressure, ΔP. Figure 3 shows the pressure profile of the melt in the spinnerette pack. The upper curve is the melt pressure tracings for a pump of 20 cc capacity per revolution, while the lower curve depicts the pressure fluctuations using a 10 cc per revolution pump, delivering the same quantity of melt per minute. In both cases, the variations in pressure are cyclic and correspond to the spin pump drive rpm. However, the smaller capacity pump generated a much smaller ΔP than the larger capacity pump, which corresponds to a ΔQ reduction of from $\pm$ 1.5 percent to $\pm$ 0.15 percent and significant improvement in threadline uniformity.

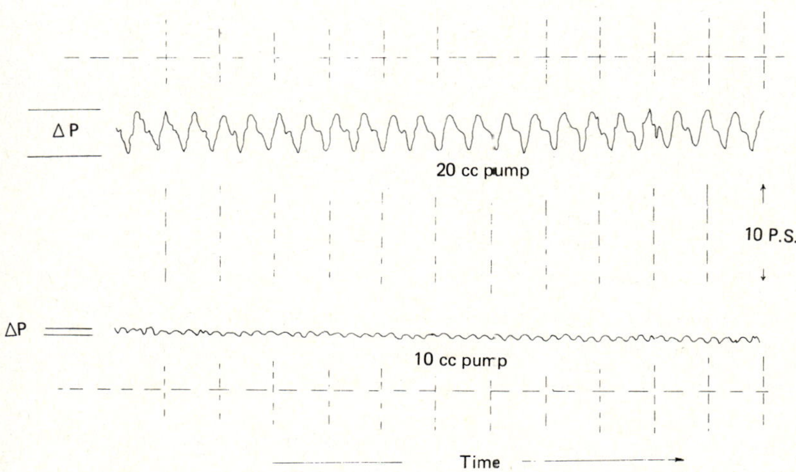

Fig. 3. Pressure Profile in the Spinnerette Pack Assembly

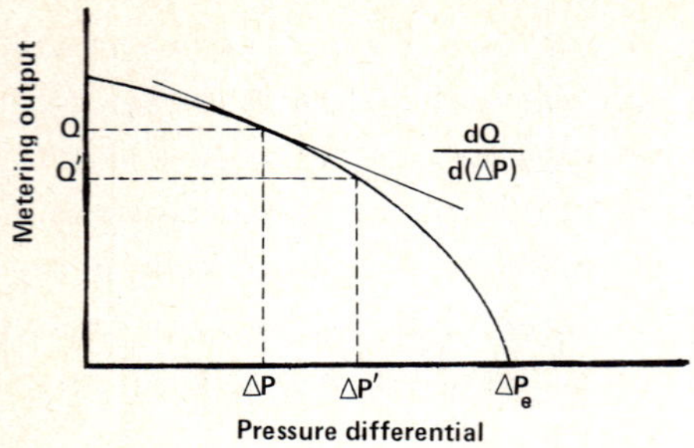

Fig. 4. Characteristic Curve of a Metering Pump

Another effect which is not often appreciated is the change in ΔP caused by gradual plugging of filter over the spinnerette. The increase in the spinnerette pack pressure changes the metering output Q of the pump, and decreases the denier of the threadline. This situation is schematically represented in Fig. 4 and 5. In the schematic of Fig. 5 the yarn experiences a sudden change in denier after the filter change. During the life of the filter, the denier change is rather gradual.

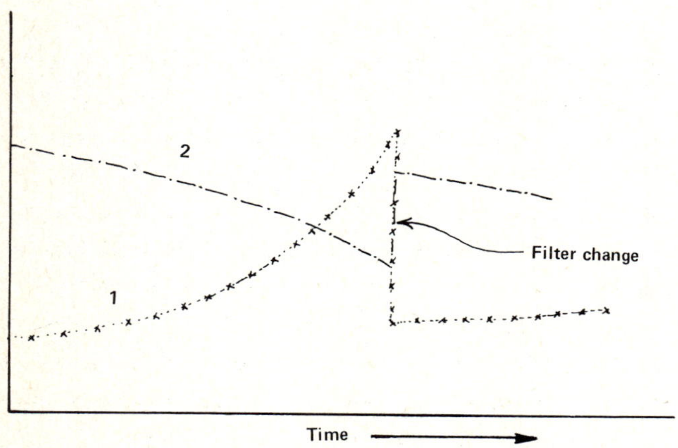

Fig. 5. Effect of Filter Plugging on Yarn Thickness (Schematic)
(1) Filter Pack Pressure
(2) Denier of the Yarn

The loose tolerances of spinnerette hole dimensions are often a source of non-uniformity in filament yarns and staple fibers. In acute cases, the stability of spinning also suffers. It has been shown (ref.37) on theoretical grounds that a 1 percent change in diameter d_o of the hole brings about 4 percent change of out-flow, while at 1 percent change of capillary length l_o corresponds

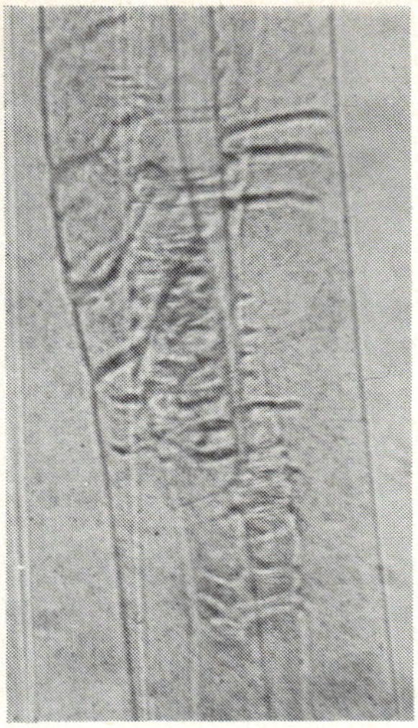

Fig. 6. "Married" Filaments Caused by Excessive Turbulence of Air in the Spinning Tube

to a 1 percent change in output only. This should be taken into account in spinnerette design, manufacture and handling. Some degree of variation in capillary dimensions is inherent in spinnerette manufacturing, but major changes arise from the use of the spinnerette. The cause of these changes is the presence of abrasive pigments in polypropylene melt, repeated cleaning operations and careless handling.

Mechanical disturbances of a multifilament spinline caused by excessive air blow or errant drafts in the spinning tube cause the individual filaments to melt-draw differently and thereby lead to uneven thickness of the fibers.

A uniformity or spin-continuity problem associated with multifilament yarn spinning is that of "married" filaments (Fig.6). These defects are caused by the fusion of two adjacent filaments. The turbulence produced by cross-blow of quenching air is the primary factor in inducing fusion between the filaments.

The action of a cross-blow on a threadline will cause it to bow. In the case of steady-state multifilament spinning, all filaments will be bowed in concert and will maintain their separation. If the individual filaments are nonuniform, then their relative displacement will change. The chances of them touching each other will increase with filament nonuniformity and with air turbulence. If the temperatures are right, "marriage" becomes inevitable. Subsequent to the "marriage", one of the filaments may break and produce a "slub". If the slub is too large to pass through the slub-catcher guide, a break in spin continuity would result. A slub is the several inches-long bristle which is produced by extrusion of the polymer in a free fall condition following the break of one of the married filaments.

It should be obvious that the smaller the displacement on bowing of the threadline, the less the chance of interfilament contact caused through either filament nonuniformity or random cross-blow fluctuations. The amount of displacement depends to a large extent on the spin tension; the higher the tension the less the threadline displacement, and the smaller is the chance of interfilament marriage. Unfortunately, there is no means available for increasing the spin tension without increasing the draw resonance (see Section VII). Thus, for a given combination of polymer, spinnerette hole geometry, and throughput rate, a decrease in temperature would result in higher threadline tension. But it will also result in larger die swell and consequently increased draw resonance. In view of these antagonistic tendencies, an optimum temperature must exist at which the uniformity is at the maximum.

IV. RHEOLOGICAL INHOMOGENEITIES OF THE MELT

If a body composed of elements of high and low viscosity is subjected to uniform shear rate, the high-viscosity elements are subjected to a high stress and the low-viscosity elements to a low stress. Assuming that the moduli of the two elements are the same, this stress variation results in a variation of shear and elastic strains. Following extrusion, this will result in nonuniform elastic recovery as the material leaves the die. These nonuniformities in the extrudate are further exaggerated in extensional flow as the filament thins and solidifies. Polymer materials with extra broad molecular-weight distributions are inhomogeneous on microscopic scale and may be able to respond as a mixture of low- and high-viscosity systems. The blending of two polymer grades of greatly differing molecular weights or intermixing of a polymeric additive with the bulk polymer (e.g., dye-promoting additives) further brings the material close to a mixture of high and low viscosities, referred to above.

Manabe (ref.24) discussed this problem, considering a simple model of deformation of a viscous fluid jet. His results can be summarized by the formulation:

$$(d^*/d_1)^2 = \lambda + D\,(1-\lambda)$$

where d^* and d_1 are diameters of thick sections and thin sections, respectively; $\lambda(= \eta/\eta')$ is the ratio of viscosity of regular parts of the polymer melt to that of the local inhomogeneities, the ratio λ being assumed constant throughout the solidification range; D (= V_1/V_o) is the deformation or "draw-down" ratio.

It should be noted that the irregularity ratio (d^*/d_1) increases with increasing rate of deformation.

Freeman and Coplan (ref.25) considered elastic oscillations in the die-swell region as one of the possible sources of nonuniformities in filament cross-section. These elastic oscillations may well be connected with fluctuations in viscosity on a microscopic scale, as discussed above.

TABLE I

Intrinsic Viscosities Along the Length of Two Monofilaments

Polymer Blend	I.V.	M.F.I.
Profax 6301/ Pigment A/ Stabilizer System A (Powder mixture)	1.90, 1.90, 1.45, 1.31, 1.52, 1.55, 1.60, 1.75, 2.17, 1.90	13.1
Profax 6301/ Pigment A/ N_2 blanket (no stabilizer) (Pelletized)	1.95, 1.95, 1.95, 2.00, 1.95, 1.95, 1.95, 1.95	10.6

Profax® is a registrade mark of Hercules Inc.

In addition to the inherent inhomogeneity of the polymer or the polymer mixture, mentioned above, melt inhomogeneity can arise from interactions with components of stabilizers and pigment systems, which can cause localized differences in degradation rates. Experimental verification of this effect is difficult to obtain, but industrial experience does point to the fact that some pigment or stabilization formulations contribute excessively to spinning problems that are not explainable rationally on any other basis. Random pressure fluctuations above the spinnerette hole, reflecting the fluctuations in melt viscosity, have not been noticed. Their frequency is probably too high to be detected by a pressure transducer or traced by the recorder. Table I, however, shows strong experimental evidence of melt inhomogeneity caused by the interaction of the polymer and the pigment or stabilizer system. Intrinsic viscosity measurements were made with 50 mg of fiber which corresponded to ~1.5 seconds worth of monofilament extrusion under the spinning conditions used. The values for MFl averaged ~6 minutes worth of extrusion. This type of melt inhomogeneity can also be caused by the interaction of isolated polymer particles with entrapped oxygen, or by localized hydrolysis of stabilizers components through the absorbed moisture content of the polymer.

In view of these considerations, it is advisable to start with a melt-compounded polymer, drying the polymer before loading, and utilizing a nitrogen blanket over the polymer chip.

Apart from the rheological inhomogeneities caused by inherent material inhomogeneities of the melt and those caused *in situ* by polymer-additive interactions, random fluctuations in rheological response may result by virtue of temperature and structure inhomogeneities of the melt. The former may be caused by the flow of molten polymer through channels of which some sections are too hot and some too cold. As the polymer melt is a poor conductor of heat, a thin outer layer of the polymer stream would acquire the temperature of the channel wall while the rest of the melt remains at the normal temperature. This thin layer of higher or lower temperature melt, when broken up and intermixed with the main polymer, would generate temperature-based inhomogeneities in the melt. The same results are obtained if the filter and spinnerette pack is at too high or too low a temperature compared with the normal temperature of the melt.

Another possible source of temperature inhomogeneities is the dead spots in the spinnerette pack assembly or in the counter-bore (the "approach") of the spinnerette hole. The polymer may stagnate in these dead spots and change in temperature. This stagnant polymer may occasionally streak out with the main polymer flow and create inhomogeneities in the melt. Within the extruder temperature gradients abound. Coupling this with the prevailing residence time distribution, one can imagine what can happen to a polymer system that is sensitive to both of them.

Structural inhomogeneities in the melt are also possible. Molten polypropylene still preserves some of the structure of the solid particles. This structure is, of course, destroyed at high temperature and by intensive mixing of the melt. In the case of polymers of broad molecular-weight distributions, however, some structure may prevail throughout the flow and may appear as rheological inhomogeneities in the die-swell region. Structural inhomogeneities are particularly of concern when the extruder is being used at the high end of its capacity, the temperature of the melt is low, and the polymer is of broad molecular-weight distribution.

It is quite obvious that all heterophase impurities in the spinning melt (solid particles, gas bubbles, etc.) should produce "knots", i.e. short irregular sections in undrawn fibers. Typical micro-photographs of heterophase knots in polypropylene fibers are shown in Fig.7.

V. MELT FRACTURE

It has been known for a long time that in the extrusion of polymer melts there is a critical value of shear rate (or shear stress) at which the extrudate's surface becomes rough, twisted, and distorted (Fig.8). This phenomenon is called "melt fracture" or "elastic turbulence". Polypropylene melts are very

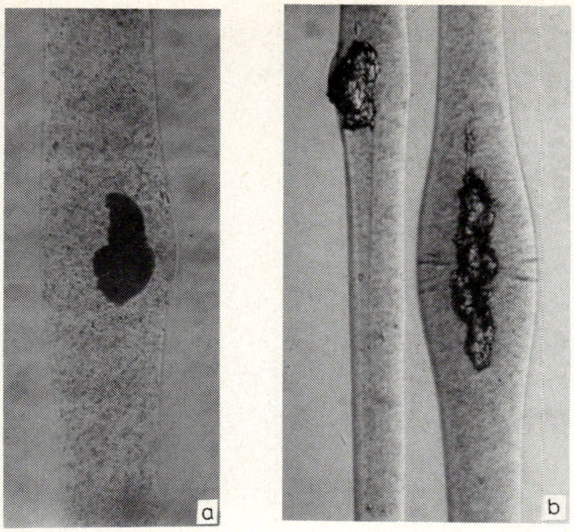

Fig. 7. Typical Microphotographs of Knots in Undrawn Polypropylene
(a) Knot With Agglomerated Pigment
(b) Knot With Cross-Linked Dye-Additive Polymer

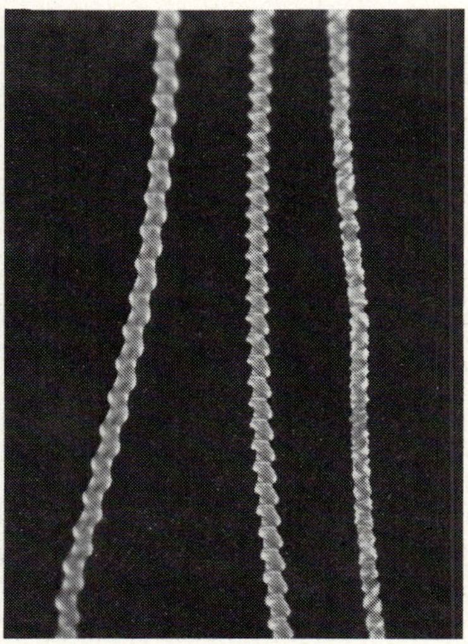

Fig. 8. Melt Fracture in Polypropylene

prone to this phenomenon.

The melt-spinning process can be thought of as consisting essentially of two consecutive processes: a controlled volumetric delivery of polymer melt to the bulge and a concurrent withdrawal of polymer from the bulge at a higher speed. It should be obvious that oscillatory variations in the mass output through the spinnerette hole and delivery to the bulge would have a corresponding effect on the output flow from the bulge. In the early stages of the occurrence of melt fracture, the fluctuations in mass output result in the uneven filament cross-section along its length: in the advanced stages of melt fracture the spinning of the polymer is not possible at all.

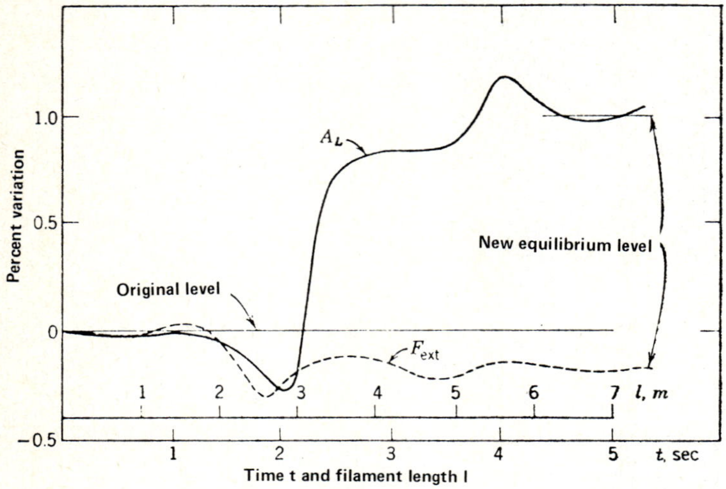

Fig. 9. Calculated Effect of a Sudden Change of Output Rate W on the Time Variation of Cross-Sectional Area A_L and Takeup Force F_{ext}. Polypropylene Monofilament Td = 122 den, V_L = 260 m/min, Relative Change of Output Rate W/W = 1% (ref.10)

Kase and Matsuo (ref.10) computed numerically the effect of instantaneous change of output rate W on the take-up force F and filament cross-section A. The results for 1 percent change in W as computed for 122 denier filament of polypropylene is shown in Fig.9. It is evident from Fig. 9 that a single change of W leads to some oscillation of fluctuation in F(t) and A(t) before a new equilibrium value is established. The periodic variation of output rate, as occurs under melt-fracture conditions, would therefore lead to gross periodic variations with accompanying secondary overtones in the filament cross-section.

Many researchers have made attempts to elucidate a possible mechanism for the onset of melt fracture. However, no unanimous opinion has yet emerged. The first problem is the site of the instability. The earliest idea, that of Spencer and Dillon (ref.4), was that an exit mechanism of extrudate buckling was involved. Now, with distortion usually damped out by increasing L/D ratio of the capillary, this idea no longer finds any support. Ballenger and White's (ref.15) observations of spiralling flow (see Chapter V) support the idea of a capillary inlet location for polypropylene. Similarly, Tordella (ref.16), on the basis of an L/D ratio effect plus his flow visualization studies (ref. 17) of low-density polyethylene, identified the die entry region as the site of failure. This has further been substantiated by Han and Lamonte (ref.18), who observed the decrease of pressure fluctuations along the length of the capillary.

With regard to more specific mechanisms, the opinions are rather confused. Some researchers have argued in favor of critical recoverable strain criteria for the inception of flow instability, while others contend that the mechanism must be closer to that of hydrodynamic stability. A popular visualization of the mechanism of onset of melt fracture is that, when the shear stress exceeds the shear strength of the melt, then the fracture occurs. Part of the melt slides past an adjacent part to give a discontinuous flow gradient. When the stress is relieved by the process, the fracture region "heals". This causes the stress to build up again and the process is repeated. As the stresses are highest at the entrance region of the capillary, it provides the locus of the melt fracture. In some cases the fracture-heal mechanism will propagate around the periphery of the capillary entrance to produce a corkscrew extrudate. In extreme cases, fracture occurs completely around the edge of the capillary and does not have time to heal, thus producing a plug flow.

Barnet (ref.19) has used melt fracture number N_{mf} to characterize a polymer melt for its tendency to melt fracture. It is $N_{mf} = \eta_0\dot{\gamma}/Q$, where η_0 is the viscosity at zero shear in poises, $\dot{\gamma}$ is shear rate $(4Q/\pi.r^3)$ in sec^{-1} and Q is M_w/M_n. For polypropylene, the average value of N_{mf} is 10.4×10^6. From the above equation it follows that for a given shear rate the tendency to melt fracture would increase with molecular weight and molecular weight distribution. The above equation, however, does not take into full account the effect of temperature and pressure differential across the spinnerette face. It appears, therefore, that a better way of predicting the onset of melt fracture would be shear stress rather than shear rate. In the $[\eta]$ range of 1-3 and at temperature varying from 200° to 300^oC, polypropylene exhibits melt-fracture behavior at practically constant shear stress, $0.8 - 1.2 \times 10^6$ dyns/cm^2 (ref.20).

To summarize the results of various reports on the topic of melt fracture and its dependence on various material and process parameters, it may be concluded that the melt fracture is intensified with:

(i) an increase in shear rate i.e., directly proportional to the throughput rate per spinnerette hole and inversely proportional to R_o^3,

(ii) an increase in molecular weight of the polymer,

(iii) an increase in polydispersity,

(iv) a decrease in melt temperature,

(v) a decrease in L/D ratio of the capillary,

(vi) a decrease in taper of the spinnerette hole entry.

VI. SHARKSKIN EFFECT

Sharkskin has the appearance of ridges running approximately perpendicular to the fiber axis (see Fig.10). Unlike melt fracture, the shape of the extrudate cross-section is essentially uniform and the surface defects may be so fine that the result appears to be merely a mattness or loss of gloss.

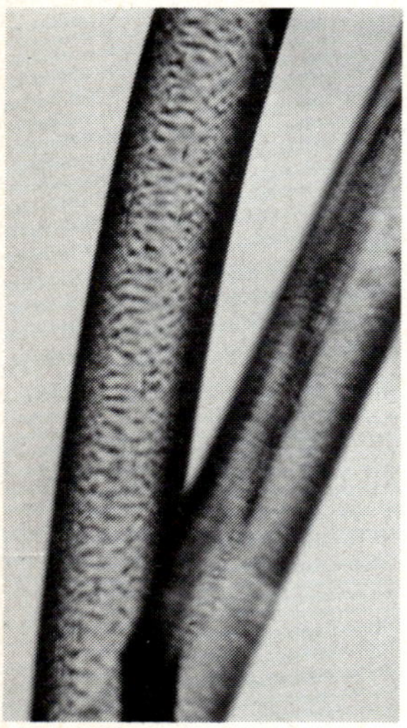

Fig. 10. An Example of "Sharkskin" in Polypropylene Fiber

Sharkskin has usually been distinguished from elastic turbulence or melt fracture for the following reasons (ref.21):

(a) Sharkskin has a perpendicular distortion whereas elastic turbulence in polypropylene often gives a helical pattern.

(b) Sharkskin can occur at lower extrusion rates than elastic turbulence and appears to be a function of the linear output rate ($Q/\pi R^2$) as opposed to the critical apparent wall shear rate ($4Q/\pi R^3$) for elastic turbulence (ref.22).

(c) Sharkskin is very dependent on temperature, being considerably delayed by increase in temperature (ref.23).

(d) Sharkskin appears to be unaffected by die entry angle, L/D ratio of the capillary, or the material from which the die is constructed.

Molecular weight appears to have very little influence on sharkskin, whereas molecular-weight distribution appears to be all-important. Both by comparing fractions with whole polymers (ref.22) and by artificially increasing the distribution by mixing polymers of widely different molecular weight, it has been found that the narrower the molecular-weight distribution, the greater the susceptibility to sharkskin.

Sharkskin seems to originate at the orifice exit and involves the surface layers, rather than the whole extrudate. The boundary conditions of laminar flow in a die require zero velocity at the die wall. After leaving the die, a uniform velocity profile is rapidly developed. This requires an extensional acceleration of the surface elements. The higher the viscosity of the material, the larger the stress so generated. In addition, because of the rate of this deformation, elasticity will be important and therefore high modulus materials will be most prone to give high stress. Thus, the greatest surface stress is encountered in materials with high viscosity and high modulus, or, high molecular-weight material of narrow molecular-weight distribution at low temperatures. If this stress exceeds some critical value, the surface ruptures, giving the well-documented rough surface appearance. The actual mechanism of rupture has not been established.

Sharkskin may be partly overcome by the use of low molecular-weight polymer, low spinning velocity, high temperature, and by incorporating some effective internal lubricant in the polymer mass.

VII. DRAW RESONANCE

"Draw resonance" is a term applied to cyclic gauge pulsations in a filament or a film when melt is drawn over a critical take-up speed. The cyclic pulsations of threadline thickness under draw-resonance conditions are different from the cyclic pulsations in the extrudate under mild melt-fracture conditions in that melt fracture starts to occur at some critical value of throughput rate even in the absence of threadline stretching, whereas draw resonance occurs only in the presence of thread stretching.

Draw resonance is probably the most important factor causing threadline instability and limiting the spinning speed in the production of polypropylene fibers. When observed longitudinally, the filaments spun under the conditions of draw resonance show a pronounced variation in their thickness, which is often periodic in nature, i.e., the thick and thin sections occur at regular intervals. The variation in filament thickness may be sinusoidal or pulse-formed. Polypropylene is particularly prone to this type of melt-spinning instability, although there is some evidence that high-viscosity polyester and nylon also show this rheological behavior to a lesser extent.

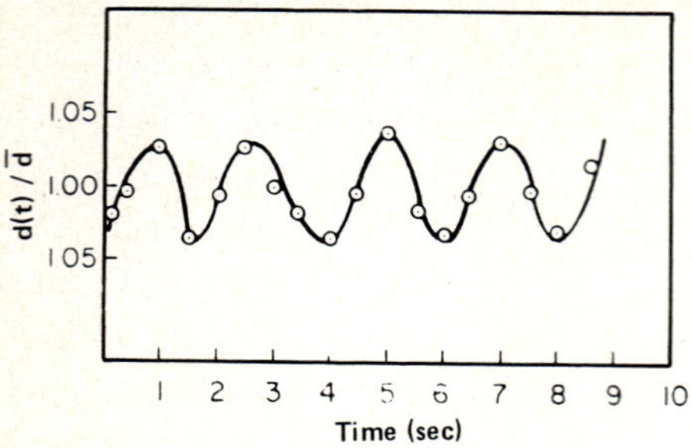

Fig. 11. Plots of Time vs. the Pulsing Fiber Diameter-to-Average Diameter Ratio. Die Dimensions $L/D = 0$; $D_R/D = 4$; $\alpha = 60^o$. Average Fiber Diameter $d = 1.988$ mm. (ref.38)

Figure 11 shows a typical plot of time versus the ratio of diameter-to-average diameter $d(t)/d$ for a pulsing threadline at an apparent stretch ratio V_L/V_0 of 58.5. Note that the figure was prepared from a series of motion pictures taken of the pulsing threadline by reading off the thread diameter from the projected image. It is seen that the thread diameter pulsates with a frequency and an amplitude whose values may depend on material characteristics and processing conditions. Figure 12 shows the Uster Eveness diagrams of two filaments of polypropylene, spun under two different sets of conditions, exemplifying draw resonance. A wave or pulse character of filament non-uniformity is again evident.

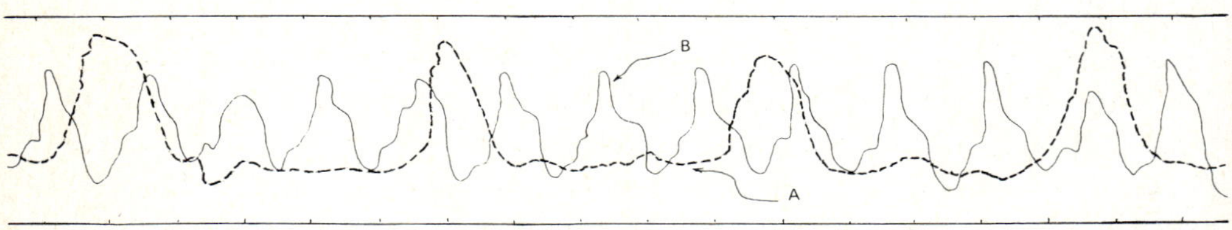

Fig. 12. Uster Diagram of Two Single Filaments, Spun Under Draw Resonance Conditions. (a) Normal Draw-Down Ratio; (b) High Draw-Down Ratio

As the individual filaments draw down under draw-resonance conditions, they do not thin down uniformly along their length. The thinning or melt-drawing appears to proceed by jumps or multiple necking. The extent of necking may differ from filament to filament at fixed time intervals and the points of necking may move up and down in space along the threadline. This kind of thinning behavior produces a fluctuation in filament cross-section with time which appears as "rain effect" in multifilament spinline. Similarly, it produces variations in filament cross-sections across the filament bundle at any fixed time. Thus, when the attenuating filament bundle freezes and solidifies, the individual filament may have sections which are thinner or thicker than the average value. The occurrence of these thick and thin filaments is, however, random across the filament bundle, and as a result the total denier of the multifilament bundle remains virtually constant.

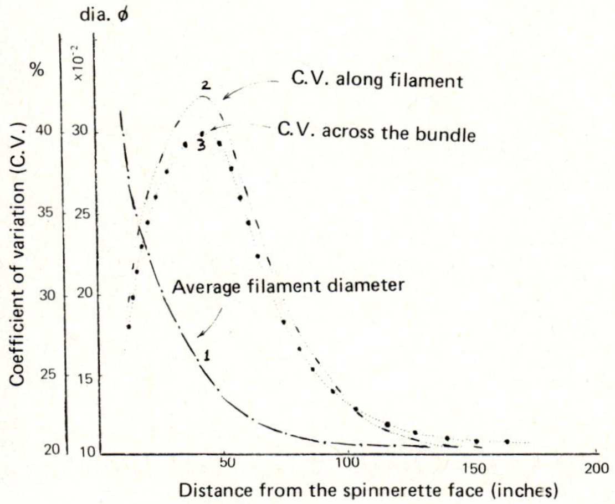

Fig. 13. Mode of Filament Attenuation Under "Rain Effect"

Figure 13 exemplifies the result of dynamically measured cross-sectional variation of polypropylene multifilament yarn spun under a set of semi-commercial processing conditions. Curve 1 represents the course of filament thinning plotted as the average of 50 filaments in the bundle. Curve 2 shows the cross-sectional variation on one filament with time at various distances from the spinnerette, and Curve 3 shows the interfilament cross-sectional variation along the spinline. These data show that variation in filament cross-section — whether intrafilament or interfilament — increases as the polymer jet draws down, reaches a maximum just before the solidification point, and then drops to a steady value after solidification. The point of maximum variations is where the rate of extensional deformation is maximum. This shows that the observed mass pulsations along the filament are closely connected with the rheological response of the melt under elongational deformation. The apparent improvement in filament cross-sectional uniformity after a region of maximum variation can be explained on the assumption that in the solidification region the thinner sections solidify first, leaving the thicker sections still in a plastic state. These are then drawn under applied tension.

Some experimental data concerning the draw resonance in polypropylene are presented in the following.

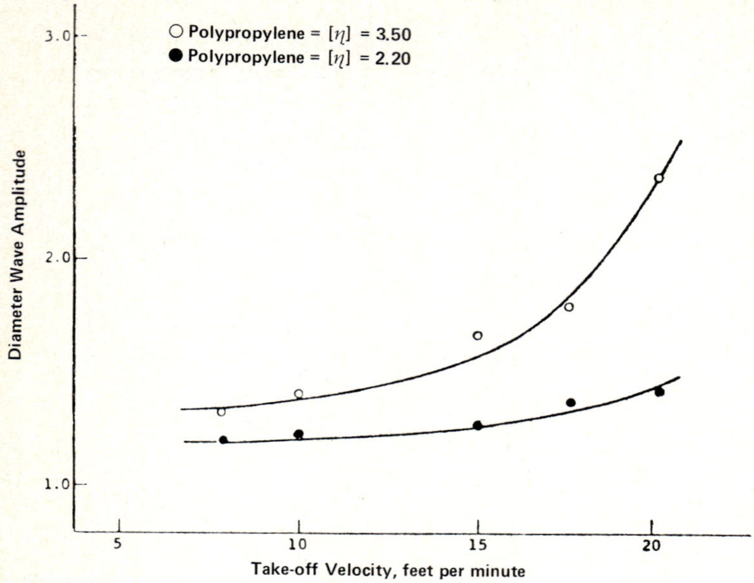

Fig. 14. Effect of Take-off Velocity on Diameter Wave Amplitude of a Monofilament. The Through-Put Increases With Take-Up Velocity to Keep the Denier of the Filament Constant (ref.29)

The draw resonance occurs at and above a certain throughput rate. At low throughput rates, increasing the draw-down ratio causes the threadline breakage before any filament pulsation is observed. However, with gradual increase in throughput rate, along with the corresponding increase in take-up speed, a point is reached when mass fluctuations along the threadline start abruptly. From this point onward, an increase in take-up speed with corresponding increase in throughput strongly increases the intensity of draw-resonance as measured by the wave amplitude. Figure 14 shows the effect of throughput (and take-up velocity) on the wave amplitude of thick-and-thin filament sections for polypropylene of two molecular weights. It is seen that the intensity of draw resonance increases as the throughput increases; the effect is more pronounced with higher-molecular-weight polymer than it is with lower-molecular-weight polymer.

Figure 15 shows plots of the pulsation frequency versus apparent stretch ratio V_L/V_0 at different melt temperatures. It is seen that thread pulsation increases with stretch ratio at a given melt temperature and that thread pulsation brings about an ultimate breakdown of the threadline.

Figure 16 gives typical plots of the ratio of minimum-to-maximum fiber diameter d_{min}/d_{max} versus apparent stretch ratio V_L/V_0. The diameters were determined from solidified fiber samples. Critical values of apparent stretch ratio $(V_L/V_0)_{crit}$ may be determined by extrapolating the curves in Fig. 16 to yield

d_{min}/d_{max} = 1.0. Of particular interest in Fig. 16 is the observation that, at 180°C, the severity of fiber nonuniformity increases as the stretch ratio is increased, whereas, at higher temperatures (200°C and 220°C), the severity of fiber nonuniformity first increases and then decreases as the stretch ratio is increased considerably above the critical value. In other words, it is possible to produce a more uniform fiber at stretch ratios very much above the onset point if the melt temperature (and possibly other processing variables as well) is properly chosen.

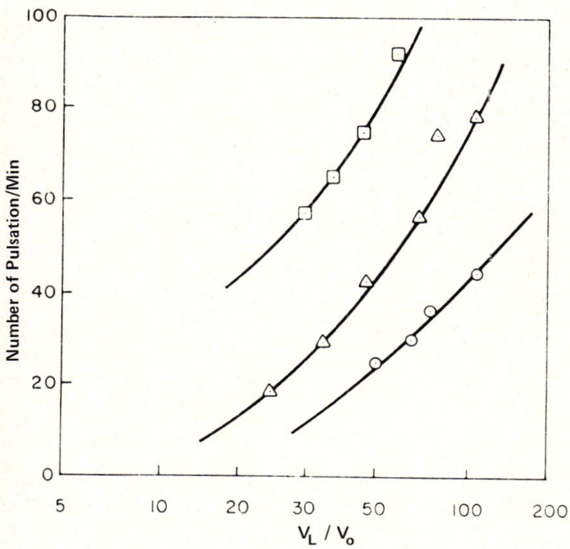

Fig. 15. Plots of the Number of Pulsations vs. Apparent Stretch Ratio of a Melt-Spun Threadline in the State of Draw Resonance at Different Melt Temperatures: (⊡) T = 180°C; (△) T = 200°C; (⊙) T = 220°C. Spinnerette Geometry: L/D = 1; D_R/D = 8; α = 180°. Spinning conditions: Melt-Spun into Ambient Air; Shear Rate in the Spinnerette is Approximately the Same for all Three Temperatures, $\dot{\gamma}_{app}$ = 380 sec^{-1} (ref.38)

Figure 17 shows plots of *true* critical stretch ratio $(V_L/V_j)_{crit}$ versus melt temperature for three values of L/D ratio. Here, V_j is the thread velocity at the position where the maximum die swell occurs. Using the mass balance equation, one can calculate V_j by:

$$V_j = V_o \ (D/d_j)^2$$

where V_o is the average velocity in the spinnerette hole, d_j is the relaxed thread diameter near the spinnerette, and D is the spinnerette hole diameter. Two important observations may be made from Figure 17: (1) at a fixed value of

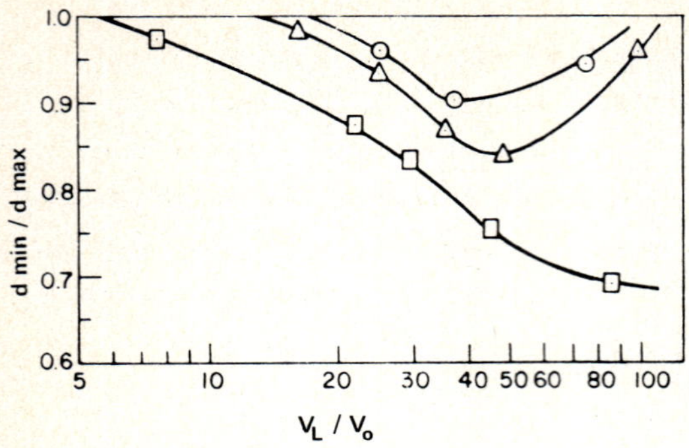

Fig. 16. *Plots of the Minimum-to-Maximum Diameter Ratio of the Melt-Spun Fibers vs. Apparent Stretch Ratio at Different Melt Temperatures: (□) T = 180°C; (△) T = 200°C; (○) T = 220°C. Spinnerette Geometry: L/D = 1.0; D_R/D = 4; α = 60° (ref.38)*

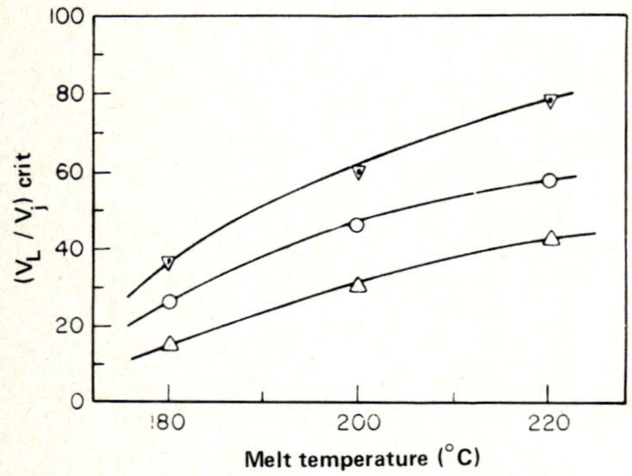

Fig. 17. *True Critical Stretch Ratio vs. Melt Temperature in Melt Spinning for Different Values of L/D Ratio: (▼) 4.0; (○) 1.0; (△) 0.0. Other Die Dimensions: D_R/D = 4.0; α = 60° (ref.38)*

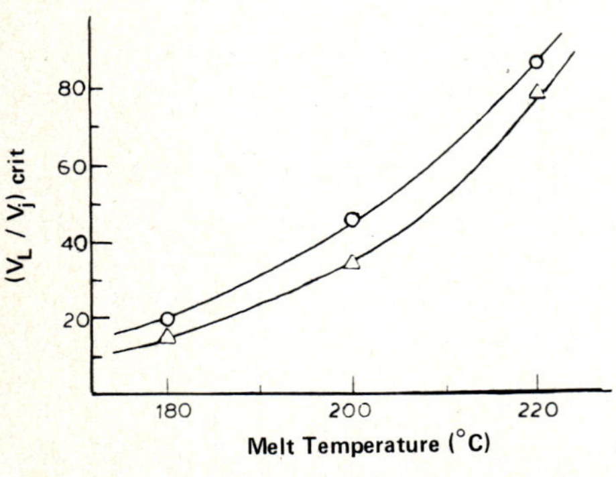

Fig. 18. *True Critical Stretch Ratio vs. Melt Temperature in Melt Spinning for Different Values of D_R/D Ratio: (○) 4.0; (△) 8.0. Other Die Dimensions: L/D = 1.0; α = 180° (ref.38)*

L/D ratio, the critical-stretch ratio is increased as the melt temperature is increased; (2) at a fixed melt temperature, the long die (i.e., the die having a large L/D ratio) gives rise to higher critical-stretch ratios than the short die.

Figure 18 gives plots of $(V_L/V_j)_{crit}$ versus melt temperature for two values of D_R/D ratio. It is seen that the critical-stretch ratio is increased as the D_R/D ratio is decreased from 8 to 4. It is also seen that, for a given die geometry, an increase in melt temperature brings about a higher critical-stretch ratio, implying less thread pulsation.

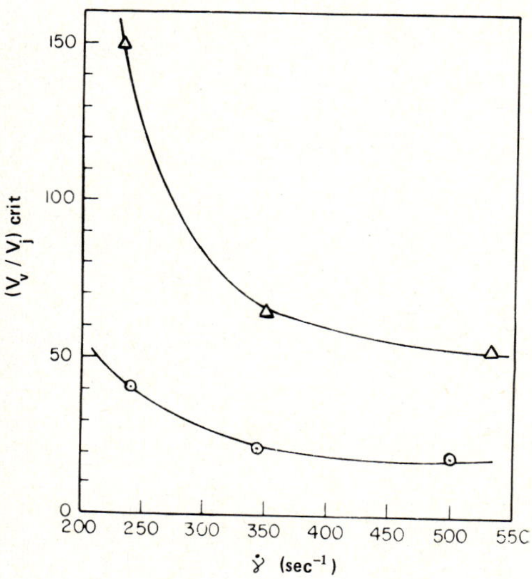

Fig. 19. True Critical Stretch Ratio vs. Apparent Shear Rate Within the Spinnerette Hole: (△) Spun into an Isothermal Chamber with Delayed Quenching; (⊙) Spun Directly into Ambient Air. Spinnerette Geometry: D_R/D = 8; L/D = 0.0; α = 60°. Melt Temperature = 180°C (ref.38)

Different methods of cooling the molten threadline are also seen to affect the critical-stretch ratio and, therefore, the uniformity of threadlines. Figure 19 illustrates the effect of two different methods of cooling. A delayed quenching (i.e., the threadline first travels through an isothermal chamber and then into ambient air,) gives rise to higher values of critical-stretch ratio (and hence more uniform fiber) than immediate quenching, in which the threadline passes directly into quiescent ambient air without passing through an isothermal chamber. This observation is consistent with the practical experience of commercial spinning operations.

Figure 19 also shows the effect of shear rate in the spinnerette hole on critical stretch ratio. The critical stretch ratio is decreased as the shear rate in the spinnerette hole is increased. This clearly indicates that the shear deformation (in terms of the residence time that the fluid experienced) influences the onset of draw resonance.

The intensity of draw resonance also depends on material parameters of the spinning melt. Figure 20 shows the effect of [η] of the polymer on the non-uniformity of multifilament yarns, measured as the coefficient of variation of filament cross-sections in the yarn bundle. It is seen that the non-uniformity of filament cross-section increases as the [η] of the polymer increases. Figure 20 also shows the effect of spinning temperature on the yarn uniformity. As the melt temperature increases, the non-uniformity of the yarn decreases.

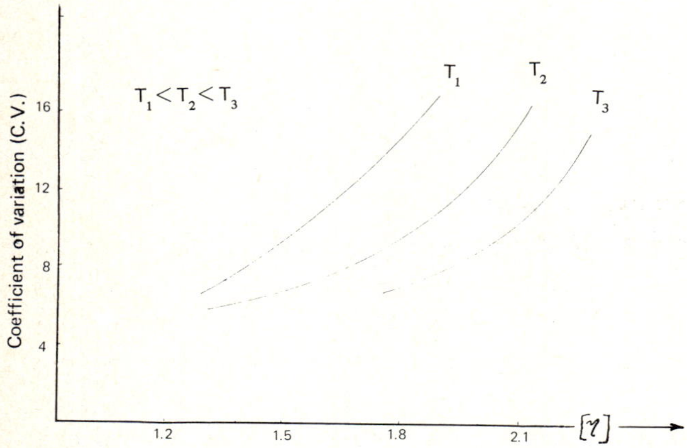

Fig. 20. Effect of [η] and Melt Temperature on Filament Non-Uniformity

Other material factors that affect the intensity of draw-resonance in polypropylene spinning are the molecular weight distribution of the polymer, isotacticity index, and the presence of additives in the polymer. Polymers of narrow molecular-weight distribution have a lesser tendency for draw-resonance when compared with the polymers of equivalent [η] but of broader molecular weight distribution. A decrease in isotacticity of the polymer decreases the pulse amplitude. Pigments and other additives, such as stabilizers, antistatic agents, processing aids, etc., when added in large quantities, increase the pulse amplitude.

Attempts have been made to correlate the degree of draw resonance with a set of measurable numbers. These efforts put major emphasis on die swell and draw-down ratio. The die-swell data would include the effect of molecular weight, MW distribution, throughput rate, temperature, and die geometry, while the draw-down ratio would take into account the rate of elongational flow as well as the extent of die swell.

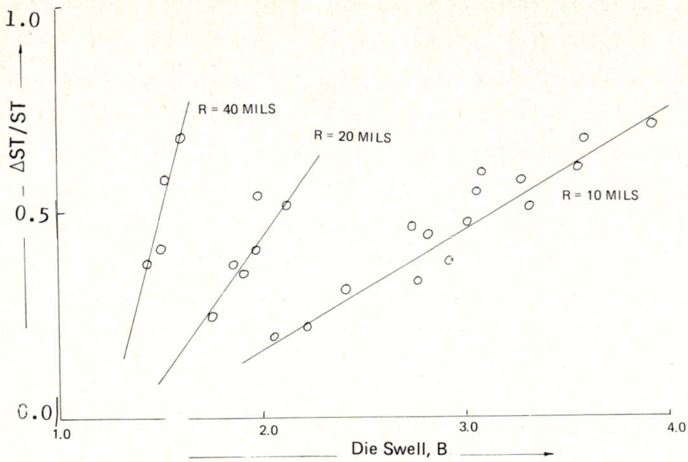

Fig. 21. Correlation of Die Swell Ratio B, with Variation in Filament Tension ΔST/ST (ref.40)

Figure 21 shows the dependence of draw resonance, as measured by variations in threadline tension ΔST/ST on die swell ratio B for three different capillaries, using single polymer lot but in each case varying L/D ratio of the spinnerette capillary, the throughput rate and the temperature of the melt. The final denier of the filament was kept constant, thus varying the draw-down ratio. A good correlation between ΔST/ST and die-swell ratio B is evident for each capillary radius.

Draw-down ratio E is commonly defined as:

$$E = \frac{\text{cross-sectional area of the spinnerette hole}}{\text{cross-sectional area of the fiber}} = (R_E/R_F)^2$$

The above draw-down ratio E is nominal as it does not take into account the die swell. Thus, a true draw-down ratio ε can be defined as:

$$\varepsilon = \frac{\text{cross-sectional area of the bulge ("swell")}}{\text{cross-sectional area of the fiber}} = (R_\varepsilon/R_F)^2$$

Here, R_ε is the radius of the bulge and R_F is the radius of the filament. As the die swell B is defined as:

$$B = \frac{\text{radius of the bulge}}{\text{radius of the spinnerette hole}} = R_\varepsilon/R_D$$

then $\varepsilon = E.B^2$

In Fig. 22 the data displayed in Fig. 21 are replotted in terms of ε yielding a family of parallel lines. Thus, although the true melt draw-down plays a major role in determining the level of nonuniformity, it is not the only factor. By inspection, one can see that there is a residual dependence on die radius, which approximates to $R^{-\frac{1}{2}}$. Replotting the ΔST/ST data versus $\sqrt{\varepsilon}/R_D$ (Fig.23) gives an excellent correlation with the majority of data points defining a common line. When one considers that the nonuniformity data used in Fig. 23 are derived from different polymers, throughputs, temperatures, and die geometries, the degree of correspondence is all the more remarkable. The additional dependence on $R_D^{-\frac{1}{2}}$ is no doubt related to the different cooling rates of threadlines. For a given final denier of the filament, $\sqrt{\varepsilon}/R_D$ is proportional to $B/\sqrt{R_D}$ and the latter than serves as useful parameter. The data plotted in Figure 24 testify to this conclusion.

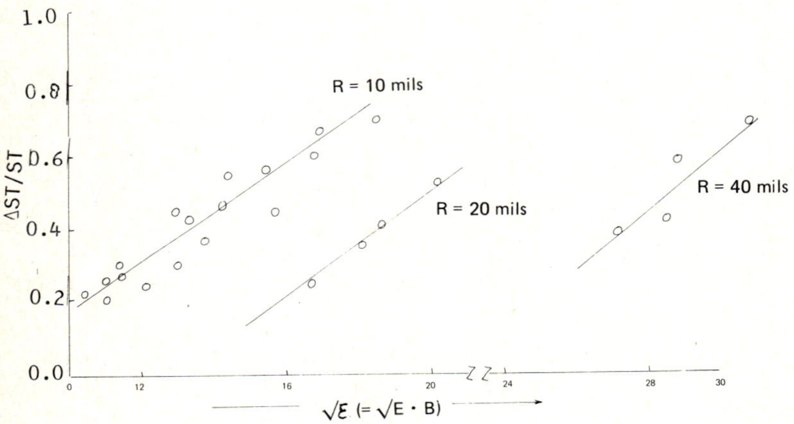

Fig. 22. The Correlation of Draw-down Ratio, E, and Die-Swell Ratio, B, with Filament Nonuniformity Measured in Terms of Tension Variation ΔST/ST (ref.45)

The above discussion can be summarized qualitatively as follows. As the instability in spinning due to draw resonance is proportional to $\sqrt{\varepsilon}/R_D$ or $(\sqrt{R_D} - B/R_F)$ it follows that the instability will increase with increasing capillary diameter, decreasing filament size, and increasing die-swell ratio. Die-Swell ratio in turn increases with increasing molecular weight, broadening MW distribution, increasing throughput, decreasing temperature, and decreasing capillary L/R ratio. These factors will influence the uniformity of spinning accordingly. The experimental data quoted previously, as well as extensive practical experience at the plant level, confirm these conclusions qualitatively.

Published information on the phenomenological aspects of draw resonance is very slight. Considering its industrial importance in melt spinning as well as in film-forming operations, it is rather surprising that such scant attention

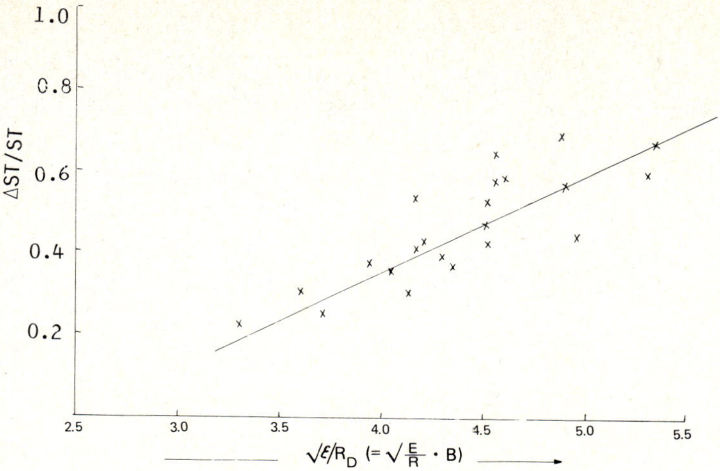

Fig. 23. Correlation of True Melt Draw-Down Ratio, ε, and Capillary Radius, R_D, with Filament Non-uniformity Measured in terms of Tension Variation, DST/ST (ref.40)

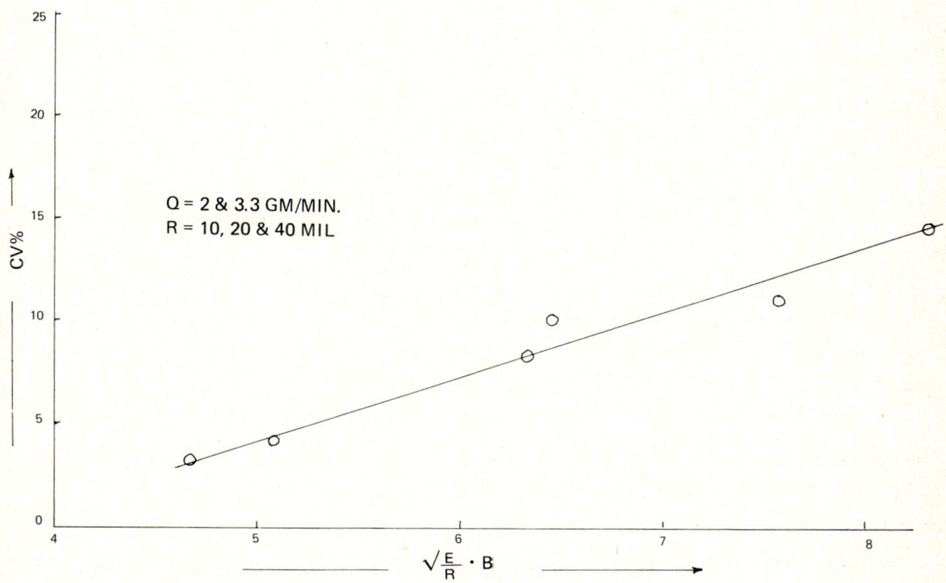

Fig. 24. Dependence of Filament Non-Uniformity (Measured as Coefficient of Variation, CV, in Filament Thickness), on Die Radius R and Die-Swell Ratio, B, at Constant Draw Down Ratio, E (ref.40)

has been paid to this phenomenon in the past. Although passing reference to draw resonance has been made in some publications (ref.24-27), the only systematic experimental study reported so far has been that of Bergonzoni and DiCresa, (ref.28,29), who investigated the role of draw resonance in the extrusion of polypropylene tapes. Thereafter, Shah and Pearson (ref.30-33) studied the theoretical stability of melt-spinning of Newtonian and non-Newtonian fluids under isothermal and nonisothermal conditions. However, experiments are lacking and the significance of cooling has not been taken into account. Kase (ref.34) and Han (ref.35) also undertook a rigorous mathematical simulation of draw resonance phenomenon, but the correlation with experimental data is again lacking. It is only recently that Kase and Ishihara (ref.36), tried to correlate the theoretical analysis with the experimental data, concentrating mostly on isothermal melt-drawing. These theoretical studies are useful, but they throw little light on the phenomenology of draw resonance and offer very little help to the process development engineer.

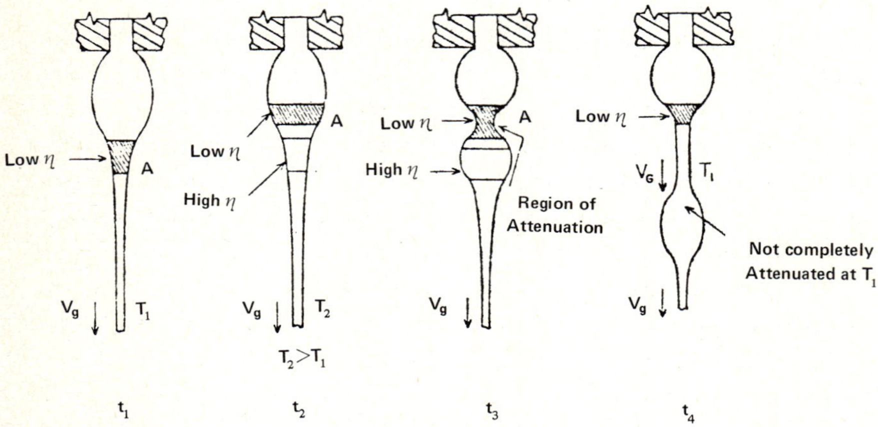

Fig. 25. Flow Model Showing "Rain Effect" (ref.25)

A theoretical model of elongational flow involving draw resonance can be constructed on the basis of mechanism of viscous flow coupled with the effect of pulsating cooling on the extrudate. During the extrusion of polymer through a spinnerette hole, a part of the work done on the polymer melt will be converted into heat and, if the process is assumed to be adiabatic, zone A of the extrudate in Fig. 25 will eventually become hot enough to show a decrease in its apparent viscosity. The filament will neck down at this point and cause the extrudate upstream to be pulled to a lesser extent, forming a thick section in the filament. Because of its smaller cross-sectional area, zone A will be cooling at a faster rate than the remaining material. As a result, further attenuation will be halted, deformation will be concentrated above the thicker section, and a new cycle started. The instability thus started can easily magnify many times, as shown by Pearson and others on the basis of their mathematical analyses.

The conclusions reached from experimental work and summarized earlier are consistent with this theoretical model. Thus, as the drawing speed is increased, a higher percentage of the work on the system is expended in sliding the molecules past each other, in disentangling them and generating more localized heated zones. Molecular mobility is increased by increasing melt temperature and this, in turn, allows the individual polymer chains or "melt elements" to slip past each other with diminished internal friction. An increase in MW obviously increases the internal friction; for a given rate of extension, more work will have to be supplied and therefore the draw response will increase. The effect of heptane solubles can be of a dual nature. Firstly, the smaller the heptane soluble content, the greater the isotacticity of polymer and the higher the tendency of the polypropylene to exist in "melt crystals". These "melt crystalline" regions are difficult to deform and offer more internal friction causing pulsation in melt drawing. Secondly, it is known that heptane-soluble content does not represent only the atactic content but also the low MW isotactic polypropylene. In this case, the smaller the amount of low MW polypropylene acting as internal lubricant, the greater will be the frictional force experienced in fiber extension. The narrow MW distribution polymer will tend to decrease the draw resonance because of the absence of very long-chain molecules generally present in nondegraded polypropylene. The presence of this fraction of extra long-chain molecules is expected to increase the internal friction of the melt because of the involved entanglement of the extra long chain with neighboring molecules. In addition, the presence of extra long-chain fraction of the polymer in the melt tend to give a flow pattern that is not that of molecular segments in motion but rather a melt elements flow. The latter pattern of polymer flow is more prone to instabilities than the normal homogeneous flow. Finally, the presence of low-molecular organic additives, such as stabilizers, processing aids, antioxidants, large amounts of pigments, etc., increases the tendency of the melt towards pronounced draw resonance because of its capacity to provide inhomogeneous spots for localized heating or viscosity fluctuations.

Han and Kim (ref.38) have recently attempted a rheological interpretation of draw resonance. In view of the fact that the critical stretch ratio is increased as the melt temperature is increased, together with the fact that the recoverable elastic strain (and also die swell ratio) decreases as the melt temperature is increased, it may be concluded that the increase of critical stretch ratio in the occurrence of draw resonance may be attributable to the corresponding decrease in fluid elasticity.

It has frequently been suggested in the literature that melt elasticity (in terms of die swell ratio) depends on die geometry. Furthermore, it is found that the recoverable elastic strain of a polymer melt decreases as the L/D ratio is increased and also as the D_R/D ratio is decreased. Therefore, it can be concluded that the observed increase in the critical stretch ratio as the L/D ratio is increased (and as the D_R/D ratio is decreased) may be attributable to the corresponding decrease in the amount of recoverable elastic strain of the melt upon exiting from the die. In other words, the onset of draw resonance may be related to the elastic properties of the exiting melt.

On the basis of their theoretical study, Fisher and Denn (ref.39) presented the view that the occurrence of draw resonance has nothing to do with the fluid elasticity, because even Newtonian fluids can exhibit the phenomenon of draw resonance at a critical stretch ratio of about 20, as was also predicted earlier by Pearson and Matovich (ref.30).

However, the onset of draw resonance in the spinning of viscoelastic polypropylene melts can indeed occur at stretch ratios much lower than the value of 20 predicted theoretically for Newtonian fluids, depending on the shear deformation history that the fluid experienced and on the melt temperature. Moreover, the onset of draw resonance can also be affected by the way cooling occurs.

The above results clearly indicate that fluid elasticity (or, more precisely, the amount of recoverable elastic strain in a fluid exiting from a die) tends to make the thread more unstable, an observation that is in conflict with some of the theoretical predictions (ref. 30,31,34). In other words, fluid elasticity plays an important role in the occurrence of draw resonance and it does indeed tend to destabilize the elongational flow behavior of viscoelastic fluids.

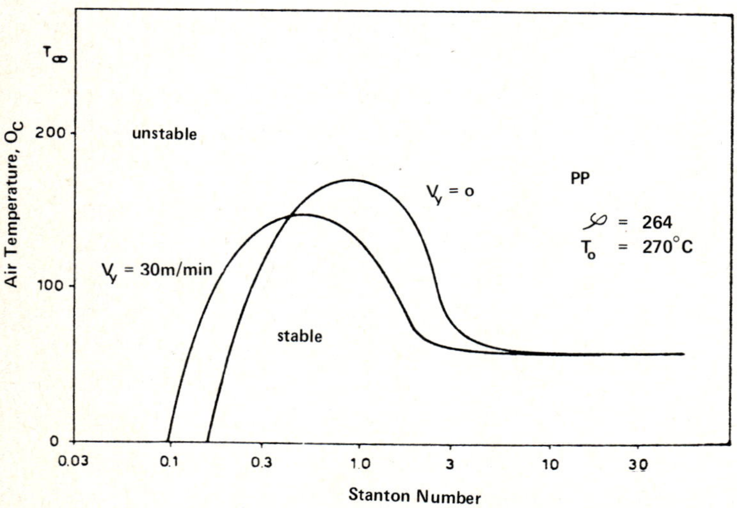

Fig. 26. Stability Regions for Melt-Spinning. Polypropylene. Spin-Draw Ratio $\mathcal{S} = V_L/V_o$, Extrusion Temperature, T_o, and Air Blow Velocity, V_y, Indicated. (ref.34)

The qualitative stability analysis of Pearson and coworkers (ref.31-33) has been extended by Kase (ref.34), who solved, numerically, linearized transient equations of melt-spinning and determined regions of unstable spinning, using all the available information about real materials and process variables. The results of this analysis for polypropylene are presented in Fig. 26. Three factors have been emphasized:

(1) Stanton number related to the heat transfer coefficient a* and the length L of the molten jet.

(2) Temperature of the ambient medium T_∞.

(3) Intensity of cross air blow.

It is evident from Fig. 26 that high T_∞ temperatures shifting the non-isothermal process in the direction of an isothermal one make spinning unstable. The effect of cross air velocity, V_y, is stabilization in the range of small S_t numbers and destabilization in the range of higher S_t values.

The ways and means of reducing "rain" in polypropylene spinning and obtaining a uniform and substantially loop-free yarn is a technological problem and largely a matter of compromises. Some material and process parameters leading to decreased rain effect have already been mentioned. For example, extensive mixing of the melt in the metering zone of the extruder, avoidance of dead spots in the flow path, avoidance of incompatible additives, use of smaller-diameter and longer-length spinnerette capillaries, low-to-medium-viscosity polymers of narrow-molecular-weight distribution, higher melt temperatures, low throughput per hole, and slow quenching rate, should be favorable factors, although some of these directions may not be desirable with regard to the tensile properties of the fiber or they may be unacceptable for economic reasons. It may be possible to decrease the "rain" substantially by slowing down the rate of extentional deformation at a given throughput with the help of an insulation collar installed below the spinnerette. In that case one would hope not only to increase the length of attenuation zone but also keep the melt viscosity at a very low level during the filament formation. Both of these factors should tend to decrease the draw resonance, with a consequent improvement in yarn quality.

VIII. INSTABILITIES CAUSED BY FLUCTUATIONS IN HEAT TRANSFER COEFFICIENT

Apart from the obvious mechanical vibration of spinning threadline induced by fluctuating air stream in the cooling chimney, one of the important causes of unsteady spinning conditions and poor quality of yarn is the fluctuating coefficient of heat transfer of the threadline. This can happen in two ways:

(i) change in velocity of cooling air,

(ii) change in temperature of cooling air.

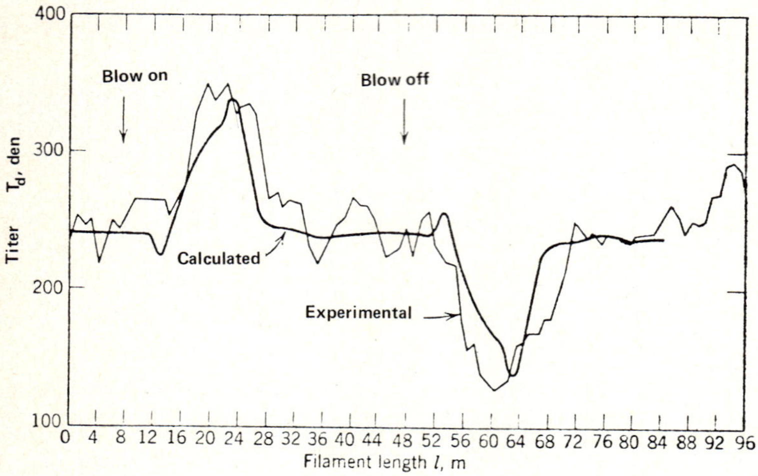

Fig. 27. Variation of Thickness of a Polypropylene Spinning Line Spun with Periodic Air Blow. Spinning Conditions: Td (total) = 240 den, n = 30, V_L = 500 m/min, T_o = 275°C, V_y = 40 cm/sec (blow), 0 cm/sec (Pause) (ref.10)

Kase and Matsuo (ref.10) computed the effect of instantaneous 1 percent change in heat transfer coefficient α on filament cross-section A and take-up force F. They showed that a sudden increase of α leads first to a slight decrease and then a strong increase in filament thickness followed by damped oscillations. Figure 27 shows the effect of periodic air blow on the thickness of polypropylene filament. Filament yarn becomes temporarily thick after a sudden increase in cooling air speed, and vice versa. The same behavior of spinning filament is expected when the temperature of the cooling air suddenly drops. In contrast to "rain" effect, where the cross-section of individual filaments changes but the total denier of the bundle essentially remains constant, the effect of fluctuating heat transfer coefficient is on the whole bundle. Thus, a marked change in Uster uniformity of the yarn is the result.

All the theoretical and empirical studies of melt-spinning show that stabilization of cooling conditions is of primary importance for the quality of the yarn produced. Temperature and pressure of cooling air and its natural and forced motion must be kept constant to guarantee that cooling conditions are kept as steady as possible. For this reason many special designs of spinning cell have been proposed (see Chapter VIII).

IX. REFERENCES

1. H. Nitschman and J. Schrade, Helv. Chem. Acta, 31(1948)297.
2. N. Hirai, J. Chem. Soc. Japan, Pure Chem. Sect., 75(1954)1019.
3. N. Hirai, Rheol. Acta, 1(1958)213.
4. A. Ziabicki and R. Taskerman-Krozer, Kolloid Z., 198(1964)60.
5. A. Ziabicki, in *Man-made Fibers,* Vol. I, Interscience, New York, 1968
6. J.C. Lee and H. Rubin, Rheol. Acta, 14(1975)427.
7. M.Reiner, in *Rheology,* Vol. I, Academic Press, New York, 1956
8. K. Kamide and Y. Inamoto, Sen-i Gakkaishi, 23(1967)79.
9. K. Kamide and Y. Inamoto, Sen-i Gakkaishi, 23(1967)578.
10. S. Kase and T. Matsuo, J. Appl. Polym. Sci., 11(1967)251.
11. M.T. Dennison, Trans. Plast. Inst., 35(1967)803.
12. G.A. Bialas and J.L. White, Rubber Chem. Tech., 42(1969)675.
13. J. White, J. Appl. Polym. Symp. No. 20(1973)155-174.
14. R.S. Spencer and R.E. Dillon, J. Colloid Sci., 4(1949)241.
15. T.F. Ballenger and J.L. White, J. Appl. Polym. Sci., 15(1971)1949.
16. J. Tordella, J. Appl. Phys., 27(1956)454.
17. J. Tordella, Trans. Soc. Rheol., 1(1957)203.
18. C.D. Han and R.R. Lamontte, Polym. Eng. Sci., 11(1971)385.
19. S.M. Barnett, Polym. Eng. Sci., 7(1967)168.
20. V. Capuccio et al, Chim. Ind., 44(1962)463.
21. J.A. Brydson, in *Flow Properties of Polymer Melts,* Van Nostrand New York, 1970.
22. E.R. Howells and J.J. Benbow, Trans. Plast. Inst., 30(1960)246.
23. P.L. Clegg, Brit. Plast., 39(1966)96.
24. T. Manabe, Sen-i Gakkaishi, 21(1965)107.
25. H.I. Freeman and M.J. Koplan, J. Appl. Polym. Sci., 8(1964)2389
26. J.C. Miller, SPE Trans., 3(1963)134.
27. J.R.A. Pearson and M.A. Matovich, Ind. Eng. Chem. Fund., 8(1969)605.
28. A. Bergonzoni and A.J. DiCresce, Polym. Eng. Sci., 7(1967)399.
29. A. Bergonzoni and A.J. DiCresce, Polym. Eng. Sci., 6(1966)45.
30. J.R.A. Pearson, M.A. Matovich, Ind. Eng. Chem. Fund., 8(1969)605.
31. Y.T. Shah and J.R.A. Pearson, Ind. Eng. Chem. Fund., 11(1972)145.
32. Y.T. Shah and J.R.A. Pearson, Polym. Eng. Sci., 12(1972)219.
33. Y.T. Shah and J.R.A. Pearson, Ind. Eng. Chem. Fund., 11(1972)150.
34. S. Kase, J. Appl. Polym. Sci., 18(1974)3279.
35. J. Appl. Polym. Sci., 16(1972)3307.
36. H. Ishihara and S. Kase, J. Appl. Polym. Sci., 19(1975)557.
37. A. Ziabicki, *Fundamentals of Fiber Formation,* Interscience, New York (1975)242-243.
38. C.D. Han, J. Appl. Polym. Sci., 20(1976)1555.
39. R.J. Fisher and M.M. Denn, ACS 159th National Meeting, Philadelphia, Pa., April, 1975.
40. J. Newport, Hercules Inc., personal communications.

X. BIBLIOGRAPHY

X.1. Books

1. H. Mark, Ed., Man-made Fibers - Science and Technology, Vol. I., Interscience, New York, 1968.
2. A. Ziabicki, Fundamentals of Fiber Formation, Interscience, New York, 1976.
3. Z.K.Walczak, Formation of Synthetic Fibers, Gordon & Breach, New York, 1977.

X.2. Journal Literature

1. A. Bergonzoni and A.J. DiCresce, Phenomenon of Draw Resonance in Polymeric Melts, I & II. Polym. Eng. Sci., 6(1966)45-59.
2. M.T. Dennison, Flow Instability in Polymer Melts: A Review, Plast. Inst. Trans. J., 35(1967)803-808.
3. S. Kase and T. Matsuo, Studies on Melt Spinning. II. Steady State and Transient Solutions of Fundamental Equations Compared with Experimental Results, J. Appl. Polym. Sci., 2(1967)251-287.
4. Den Otter, Melt Fracture, Rheol. Acta, 10(1971)200-207.
5. J.R.A. Pearson and Y.T. Shah, Stability Analysis of the Fiber Spinning Process, Trans. Soc. Rheol., 16(1972)519-533.
6. C.D. Han, Polymer Melt Flow Instabilities in Extrusion, Polym. Eng. Sci., 11(1971)385-394.
7. Y.T. Shah and J.R.A. Pearson, On the Stability of Non-Isothermal Fiber Extrusion, Ind. Eng. Chem. Fund., 11(1972)145-149.
8. J.L. White, Critique on Flow Patterns in Polymer Fluids at the Entrance of a Die and Instabilities Leading to Extrudate Distortion, Appl. Polym. Symp., 20(1972)155-174.
9. Y.T. Shah and J.R.A. Pearson, On the Stability of Non-Isothermal Fiber Extrusion - General Case, Ind. Eng. Chem. Fund., 11(1972)150-153.
10. C.D. Han and R.R. Lamontte, Melt-Spinning - III. Flow Instabilities in Melt Spinning, Melt Fracture and Draw Resonance, J. Appl. Polym. Sci., 16(1972)3307-3323.
11. J.R.A. Pearson, Stability of Isothermal and Non-Isothermal Fiber Spinning of Power Law Fluids, Ind. Eng. Chem. Fund., 13(1974)134-138.
12. S. Kase, Melt Spinning - IV. Stability of Melt Spinning, J. Appl. Polym. Sci., 18(1974)3279-3304.
13. S. Kase and H. Ishihara, Melt Spinning - V. Draw Resonance as a Limit Cycle, J. Appl. Polym. Sci., 19(1975)557-565.
14. J.C. Lee and H. Rubin, On the Breakup of Molten Polymeric Threads, Rheol. Acta, 14(1975)427-435.
15. C.D. Han, Studies on Melt Spinning - VI. The Effect of Deformation History on Elongational Viscosity, Spinnability, and Threadline Instability, J. Appl. Polym. Sci., 20(1976)1555-1571.
16. A.T. Serkov, Spinning Stability in the Preparation of Man-made Fibers, Faserforsch. u. Textiltech., 27(1976)397-406.
17. H. Ishihara, Studies on Melt Spinning - VII. Simulation of Draw Resonance Using Newtonian and Power Law Viscosities, J. Appl. Polym. Sci., 20(1976)169-191.
18. E. Boureaux, Polymer Flow Instability - A Review, J. Macromol. Sci. Reviews, C-16(1978)39-75.
19. J.L.White, Instabilities and Failure in Elongational Flow and Melt Spinning Process, J.Appl.Polym.Sci., 22(1978)3057-3074.

Chapter VII FUNDAMENTALS OF FIBER PROCESSING

I. INTRODUCTION

The mechanical characteristics of as-spun polypropylene fibers are far from those required by the textile industry. These fibers exhibit low tenacity, high irreversible elongation, and low moduli. The operation which improves these properties is *cold-drawing*. It consists of irreversible elongation of as-spun yarn in the solid state to 200 to 1000 percent of its original length. Cold-drawing can be undertaken at room temperature or at elevated temperature, up to the melting temperature of the polymer. Even at these high temperatures the process is called *cold-drawing* as opposed to *melt-drawing*, which occurs in the spinning of fibers.

The cold-drawing is accompanied by extension and parallelization of macromolecules and crystallites along the fiber axis. Orientation is always accompanied by a change in crystalline and morphological structure. In the fiber industry, drawing is carried out continuously and either directly follows the operation of spinning or else makes a separate operation starting with undrawn material previously collected on bobbins or in drums. Continuous filament yarns and tape or fibrillated film yarns are generally produced by combining melt-spinning and cold-drawing in one single-step process, while the production of staple fiber is generally accomplished through melt-spinning and a subsequent, separate step of cold-drawing.

The assumed draw ratio R, i.e., the ratio of drawing and feed velocities, $R = V_R/V_F$, can be achieved in one or more steps differing in draw conditions. Drawing conditions and the structure and the properties of the resulting fiber strongly depend on the characteristics of the original (undrawn) material. Therefore, in order to optimize any technical drawing process, one should know the relationships between undrawn material structure, drawing conditions and structure and properties of the drawn fiber. These relationships will be discussed in this chapter.

Solidification and mechanical treatment processes accompanying fiber-spinning and cold-drawing introduce some elements of instability into fiber structure and properties. Quenching of a molten polymer in melt-spinning and destruction of original crystal structure in cold-drawing lead to thermodynamically unstable systems which, given enough time, tend toward some equilibrium. The rate of these changes is controlled by the viscoelastic material characteristics. At room temperature the changes are slow. At elevated temperatures, equilibrium can be approached within a few minutes and fibers can be efficiently stabilized against subsequent changes during their application. This process step is often termed *heat-setting*.

Some degree of heat-setting is often inadvertantly accomplished in cold-drawing when using a heated drawing godet. Heat-setting is also an integral part of several texturizing processes, such as stuffer-box crimping, steam jet bulking and false-twist texturizing. When an added degree of heat-setting is desired, however, a distinct step of heat treatment of fiber and yarns is added to the process train. This applies particularly to the production of very low shrinkage yarns and fibers, to the enhancement of crimp permanancy and to the production of twist-set yarns. We will discuss some fundamental aspects of the heat-setting process in the later part of this chapter.

II. MECHANICS OF DRAWING

A general description of the cold-drawing process is most easily understood in terms of stress-strain behavior of the undrawn filament schematically depicted in Fig. 1. If an unoriented or partially oriented polypropylene fiber of cross-section A_o is stretched at a constant rate, a highly oriented fiber is obtained which finally breaks with an irregular crack. The normal stress-strain (σ_o - ε_o) or load-elongation (F_o - ℓ_o) or (F_o - λ_o) curve exhibits first a linear region, which after a brief nonlinear section forms a maximum - the "yield point" - characterized by a "nominal yield stress" $\sigma_{os} = F_y/A_o$ and strain ε_{os}. Concurrently, at some weak point a narrow constriction appears which soon develops into two necks propagating through the sample in opposite directions and transforming the original filament into much stronger material (strain hardening). During this propagation the nominal drawing stress $\sigma_{od} = F_d/A_o$ remains very nearly constant or increases slightly in approaching the final break characterized by the load F_b or nominal stress-to-break $\sigma_{ob} = F_b/A_o$. The ratio of cross sections of the original (A_o) to that of the transformed filament (A_∞) at the time of break yields the natural draw ratio $\lambda_\infty = A_o/A_\infty$, an important property of the starting material. It is usually derived from the ratio of the sample length at break to that before deformation. The natural draw ratio can also be defined dynamically as $\lambda_p = V/V_o$ where V is the velocity of drawing necessary to completely neck down the filament fed into the drawing zone at a speed of V_o.

The deformation involved in the drawing of polymer materials is not confined to small values. Consequently, the deformation ε and the resulting changes in cross section affect the actual tensile stress in the course of drawing. One should therefore distinguish between the *true stress*, σ, related to the actual cross section of the fiber A

$$\sigma = F/A$$

and the *conventional stress*, σ^o, related to the original cross sections of undeformed sample, A_o

$$\sigma = F/A_o$$

If the volume of the sample remains constant, both these characteristics are related by the following equation:

$$\sigma = R\sigma^o = \exp(\varepsilon)\sigma_o$$

When elongation proceeds at constant tensile force, F, the true stress σ increases with increasing elongation due to the reduction in cross section. On the other hand, σ^o remains constant independently of R or ε.

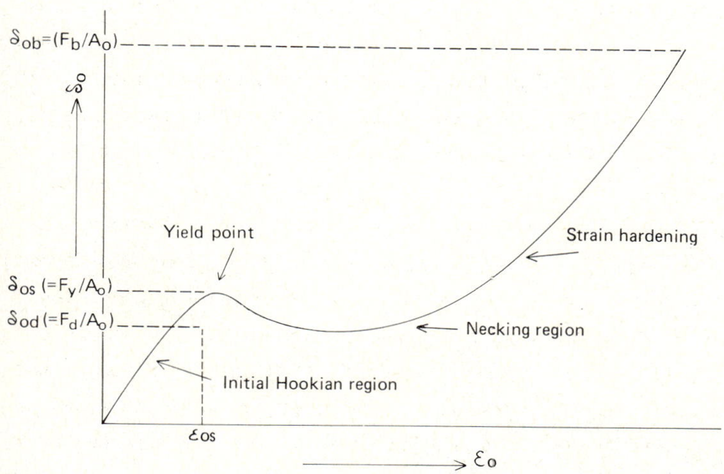

Fig. 1. Schematic of Drawing Behavior of Polypropylene Fiber

Figure 2 schematically illustrates the characteristics σ vs. ε for three basic types of solid materials.

The behavior shown in Fig. 2a is typical for plastic metals, while Figs. 2b and 2c show behavior typical of polymeric materials, depending on the drawing conditions and the material characteristics. This applies to polypropylene fiber as well.

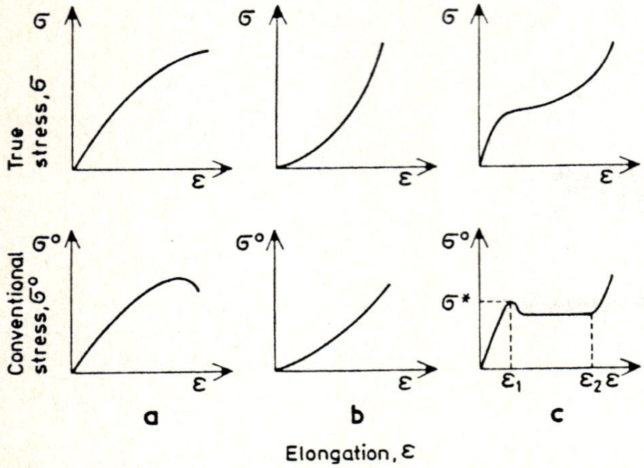

Fig. 2. Fundamental Types of Deformation Characteristics. (a) Material with Plastic Fracture, (b) Material with Deformation Strengthening, (c) Neck-Type Deformation (ref.20)

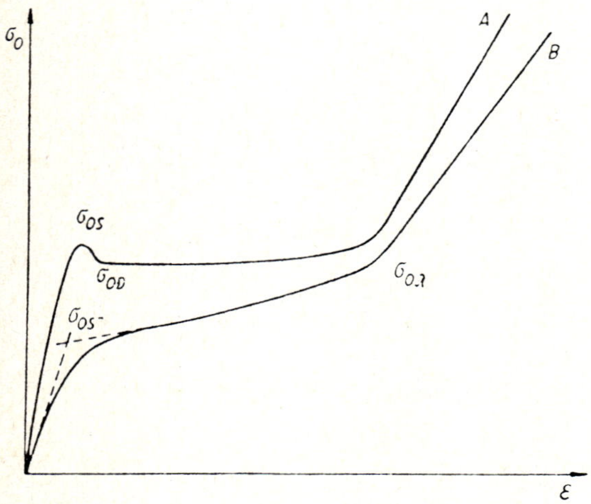

Fig. 3. Deformation curves of Inhomogeneous (A) and Homogeneous Drawing (B)

Figure 3 shows the two basic patterns of the dependence of σ_o on ε at constant deformation rate. The curve A represents the nonhomogeneous drawing; the curve B represents the homogeneous extension. Nonhomogeneous drawing is characterized by a pronounced necking effect while the homogeneous drawing occurs without any conspicuous neck formation. In physical form, the drawing filaments appear as

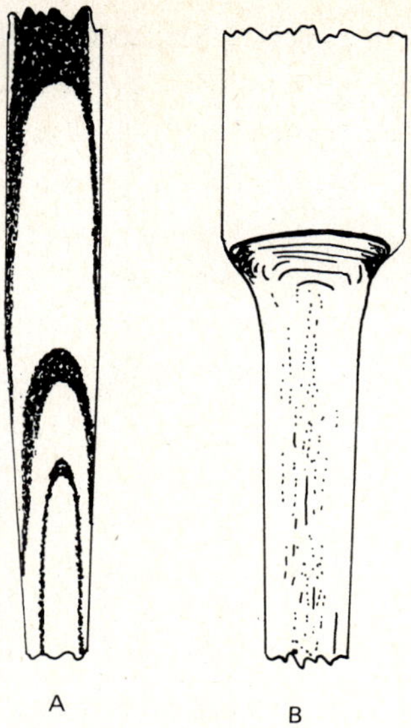

Fig. 4. Drawing of Polypropylene Filament
(A) Gradual Change in Orientation - Homogeneous Drawing;
(B) Abrupt Change in Orientation - Neck Formation

shown in Fig. 4.

The curves in Fig. 3 are characterized by the following points:

σ_{os} is the flow limit

σ_{od} is the stress of nonhomogeneous deformation

σ_{o} is the stress which corresponds to the end of deformation. The ratio of the lengths of the sample corresponding to the stresses $\sigma_{o\lambda}$ and σ_{os}, respectively, is the so-called natural draw ratio of the fiber $\lambda_{\alpha} = l_{o\lambda} / l_{os}$. It is also defined by $\lambda_{\alpha} = A_{o}/A_{\alpha}$.

Polypropylene fiber can deform homogeneously (without neck formation) or inhomogeneously (with necking), depending on the structure of the spun fiber and the deforming conditions. With an increase in temperature, the value of σ_{os} and σ_{od} are lowered and at the same time their ratio becomes smaller. In the interval ($\sigma_{o} = o, \sigma_{os}$), the $d\sigma_{o}/d\varepsilon$ value decreases and the section between σ_{os} and $\sigma_{o\lambda}$ becomes smaller. The heterogeneous deformation changes to homogeneous deformation and, consequently, the drawing process takes place without necking. With an increase in deformation rate, the values of σ_{os} and $d\sigma_{s}/d\varepsilon$ increase and an increase in these values often leads to a nonhomogeneous drawing. The relation between the rate of deformation V and the value of V_{os} can be expressed through an empirical equation:

$$\sigma_{s} = A + B \log V$$

where A and B are constants. A higher orientation of the spun fiber increases the values of σ_{os} and $d\sigma_{os}/d\varepsilon$ and leads the deformation to a homogeneous drawing. A larger cross section of the fiber favors necking. The value of σ_{od} does not depend on the drawing rate but it increases with the molecular weight of the polymer. On a structural basis, the change of paracrystalline form into thermodynamically more stable monoclinic structure leads the homogeneous drawing into an inhomogeneous type of deformation. An effect similar to that of increased temperature is also due to the presence of certain low molecular-weight substances in the material. They act as plasticizers and facilitate molecular motions.

III. STRUCTURAL CHANGES IN COLD-DRAWING

Cold-drawing is always connected with structural changes in the polymer. Although a large amount of structural data on the deformation of crystalline polymers has been collected in the last two decades, very little has been published about the molecular processes involved in stretching a crystalline filament far beyond its elastic limit to form highly oriented filament structure. This is because of the great difficulties encountered in trying to define the structure of the initial and final states, which must be understood before proposing a detailed mechanism for converting one into the other.

The study of the structural changes in cold-drawing of crystalline polymers is also handicapped by the absence of any consistent model of morphology. Earlier workers labored on the then-accepted model for crystalline polymers, the fringed miscelle. With the advent of electron microscopy and observations of the folded chain molecules, a number of investigators began to interpret the results of deformation in crystalline polymers in terms of more or less perfect crystals, utilizing the ideas and methods developed primarily in physical metallurgy. The crystal defect model appears to have been developed as a result of this basic reasoning. An equally useful model is the paracrystalline model by Hoseman and co-workers.

The morphological, and particularly the orientational aspects, of the initial microspherulitic structure, the intermediate structure in the neck, and the final fiber structure of polymers, have been investigated by electron microscopy, wide- and small-angle X-ray scattering, light scattering, birefringence, infra-red absorption, and dichroism, as a function of draw ratio, draw rate, temperature, and thermal history. The most commonly utilized methods for the study of deformation of crystalline polymers are, however, electron microscopy and X-ray diffraction. The information about the structural elements obtained through all these methods together permitted the development of a morphological model of plastic deformation of bulk polymers that gives a satisfactory explanation of the observed effects. However, specific experimental data are needed to clarify a great many details before they can be fully understood and applied to the process of cold-drawing of fibers.

A clear understanding of the state of crystalline polymers may be missing, but a qualitative and empirical description of the mechanisms of orientation - specially applicable to crystalline polymers - can be given by describing the polymer consisting of amorphous and crystalline regions as postulated in fringed miscelle or variations of chain folding models, described in Chapter V. In the unoriented state the crystallites are distributed in statistically random fashion in all directions. The chain molecules pass through a few amorphous regions and a few crystalline regions. When the polymer is subjected to an orientational force, the polymer segments in the amorphous regions begin to orient in the direction of the applied force and by so doing drag the crystallites along the fiber axis. As soon as the crystallites are oriented along the applied force, a further deformation necessitates a proportionally greater force. This pictorialization of the drawing process is obviously a simple abstraction of the true situation of structural changes during deformation, but it adequately explains the various process-properties relationships encountered in the cold-drawing process.

Cold-drawing of polypropylene fibers may include several kinds of structural and morphological changes in the fiber. The most important, although interdependent, are:

(1) Changes of paracrystalline or smectic order into more perfect crystalline forms during heating prior to deformation (if elevated temperatures are employed for drawing);

(2) Orientation of molecules in "amorphous" or less organized regions in the direction of the applied force;

(3) Orientation of crystalline and paracrystalline elements according to force vectors;

(4) Deformation and breakdown of super molecular structure into smaller elements;

(5) Alignment of remnants of destroyed morphological structure in the direction of applied force;

(6) Changes in crystallinity, crystalline size, degree of perfection, etc.

Of course none of the above changes takes place independently, as the whole process of structural changes is a result of several changes taking place simultaneously.

The most important factor determining the course of cold-drawing and the accompanying changes in morphology of polypropylene fibers is whether the

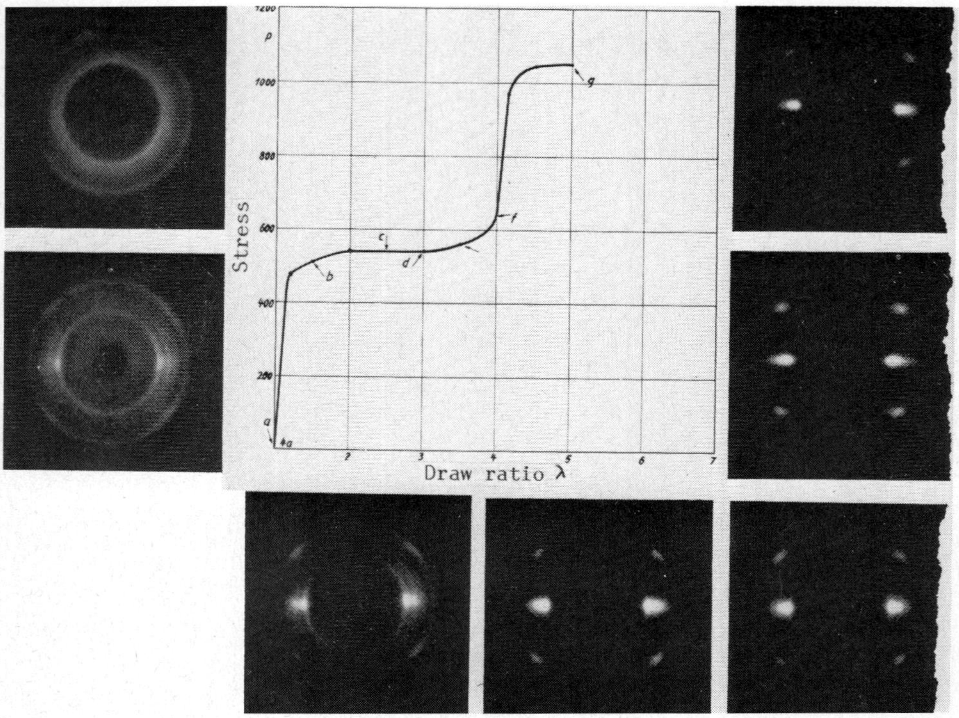

Fig. 5. Changes in X-Ray Diffraction on Drawing of Polypropylene Filament with Monoclinic Crystalline Structure at Room Temperature (ref.2)

structure of spun filament is predominantly monoclinic and spherulitic or predominantly smectic. Although these structures cannot be separately produced in purity, an understanding of the drawing process and its effect on the structural modifications can be obtained by separately discussing the drawing behavior of monoclinic/spherulitic structure and quenched/paracrystalline structure under unidirectional stress.

Jambrich and Diacik (ref.2) have investigated the deformation behavior of polypropylene fibers at ambient as well as at elevated temperature by measuring the drawing tensions at respective drawing conditions. Figures 5 and 6 depict

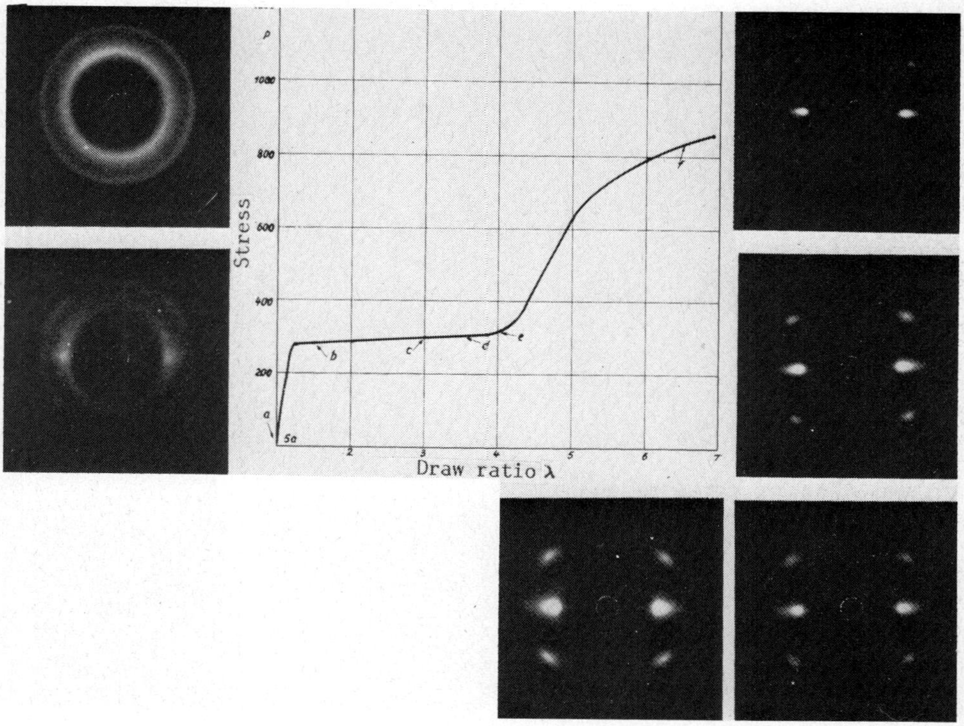

Fig. 6. Changes in X-Ray Diffraction on Drawing of Polypropylene Filament with Smectic Order at Room Temperature (ref.2)

the dynamic deformation curves of polypropylene fibers with the monoclinic and the smectic crystalline structure, respectively, drawn at room temperature. From these curves it can be seen that the fiber with thermodynamically stable crystalline structure (monoclinic crystalline order) in its undrawn state, requires considerably more force for deformation than that required to deform and extend a fiber of thermodynamically unstable structure (smectic or paracrystalline). These results should be expected, on the basis that monoclinic form has a three-

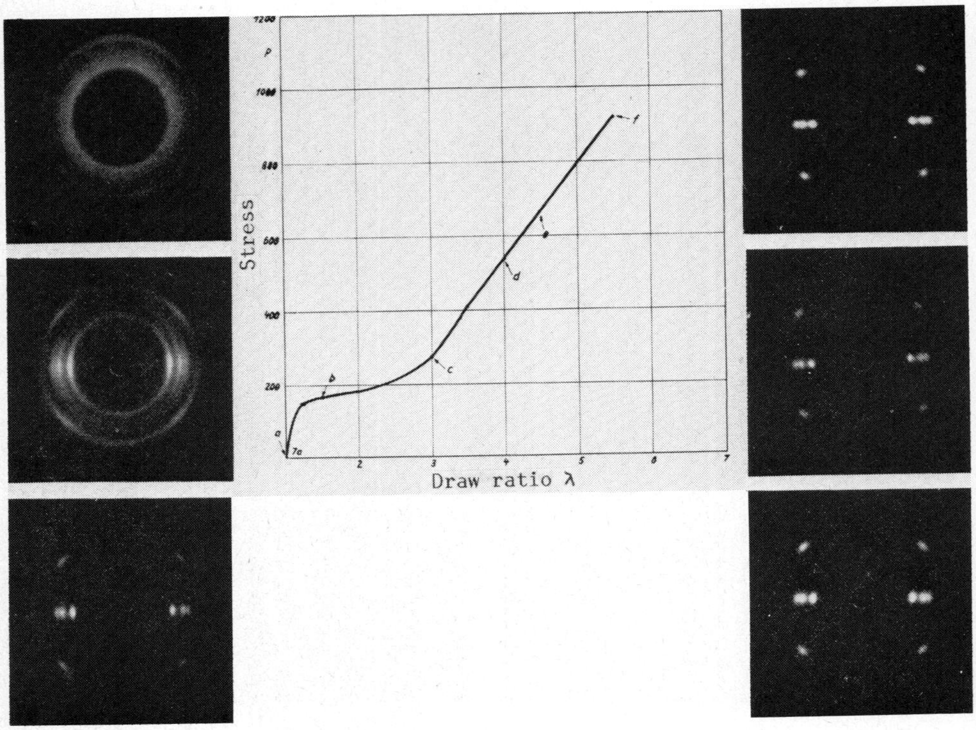

Fig. 7. Changes in X-Ray Diffraction on Drawing of Polypropylene Filament with Monoclinic Crystalline Order at 120°C Temperature (ref.2)

dimensional crystalline order with a large number of tie links by which it offers much more resistance to deformation than that offered by the less ordered smectic form. For the same reason, the degree of attainable deformation for smectic structure is much higher than that for more stable, monoclinic structure.

The structural changes taking place during the course of drawing are interesting. Figures 5 and 6 show the X-ray diffraction diagrams of the two fiber structures deformed to various extent at room temperature. In the case of thermodynamically stable crystalline order, the molecular orientation takes place up to a draw ratio of 3.5. This happens without any substantial change in the orig-

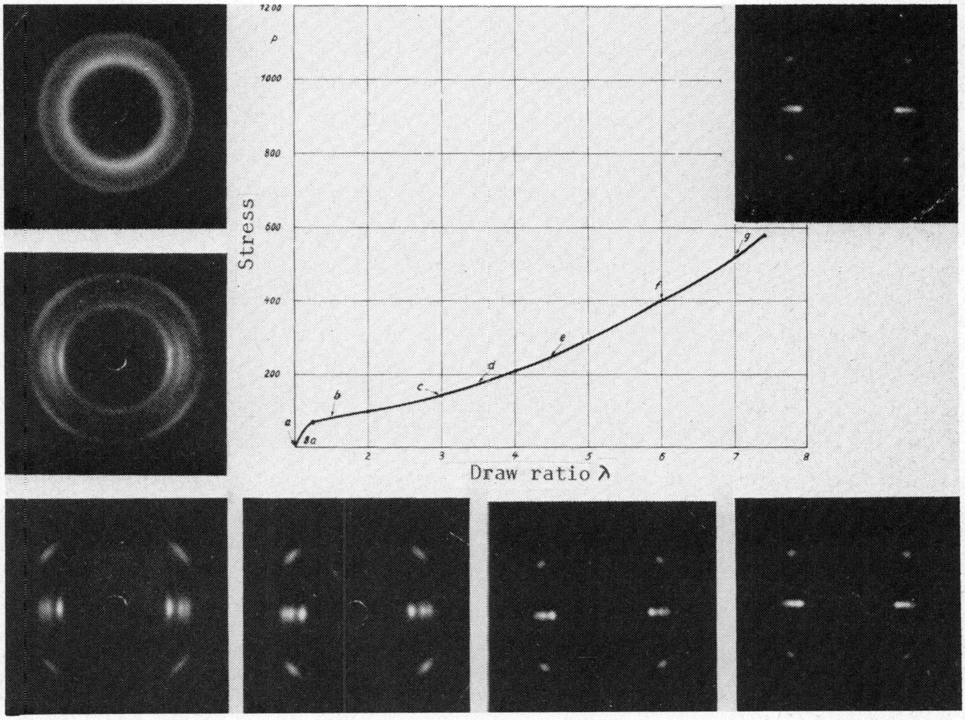

Fig. 8. Changes in X-Ray Diffraction on Drawing of Polypropylene Filament with Smectic Order at 120°C Temperature (ref.2)

inal crystalline order. On the deformation curve this range is seen as flow zone. After passing the flow zone, i.e., the zone of increased stress, a marked resistance to deformation is noticed. In the X-ray diagram this zone is characterized by a narrowing of equatorial reflexes (110), (040), and (130). In comparison, the less stable, smectic form manifests characteristic changes even below a draw ratio of 3.5. The diffraction ring completely disappears and (111) reflexes intensify. Beyond this draw ratio no significant changes in X-ray diffraction takes place.

Figures 7 and 8 depict the dynamic drawing curve of polypropylene fibers of the monoclinic and smectic fiber structures at 120°C, respectively. The curves show that an increase in temperature causes a significant decrease in drawing tension. This is evidently due to the increased mobility of the polymer chains which results in an increase in drawability. Figures 7 and 8 also show the X-ray diffraction diagrams in relation to the corresponding dynamic deformation curves. It is seen that in drawing at high temperatures, little orientation of crystallites takes place up to a draw ratio of 2.5. This is evidenced by no contraction in equitorial reflexes (110), (040), and (130). The input of thermal energy weakens the fiber structure so much that, up to a certain draw ratio, the internal friction of the polymer molecules is not enough to cause the crystalline orientation in the direction of the fiber axis.

The as-spun fibers of polypropylene with different morphological structures, when drawn at high temperatures, yield X-ray diffraction diagrams that are normally characteristic of a monoclinic structure. This is evidently due to the conversion of smectic structure, if present in the undrawn fiber, to thermodynamically more stable monoclinic structure above 70°C. Because of this possible conversion of smectic structure into monoclinic structure at high temperature of drawing, the drawing behavior of quenched structure at elevated temperatures is extremely complicated. The phenomena actually occuring are probably a combination of the stretching orientation of amorphous regions, orientation of crystallites, the conversion of the smectic portion to monoclinic crystalline structure, the orientation of converted monoclinic crystallites, the conversion of the more orderly crystalline region into less orderly fraction due to high draw ratios, etc. The X-ray diagrams in Figures 6 and 8 show that while no change from thermodynamically unstable structure to stable structure occurs on drawing at room temperatures at any draw ratio, the drawing at 120°C does produce changes in crystalline structure.

In the drawing of polypropylene fiber the degree of axial orientation, often characterized by optical birefringence, Δn, increases with increasing draw ratio R (Fig.9). In a first approximation, the observed birefringence Δn can be interpreted as the sum of contributions from crystalline and amorphous regions:

$$\Delta_n = \Delta_{n_c} + \Delta_{n_{am}} + \Delta_{n_f} = k_{cr}\ \theta_c\ \int_{cr} + K_{am}(1-\theta_c)\ \int_{am}$$

where $\int_{cr}$ and $\int_{am}$ are orientation factors of the crystalline and amorphous parts, respectively, θ_c is the degree of crystallinity, Δ_{nf} is the form birefringence, and K_{cr}, K_{am} are material constants.

It is evident that birefringence is a reliable measure of the orientation factors only when no phase transition (change of θ_c) takes place. Since the

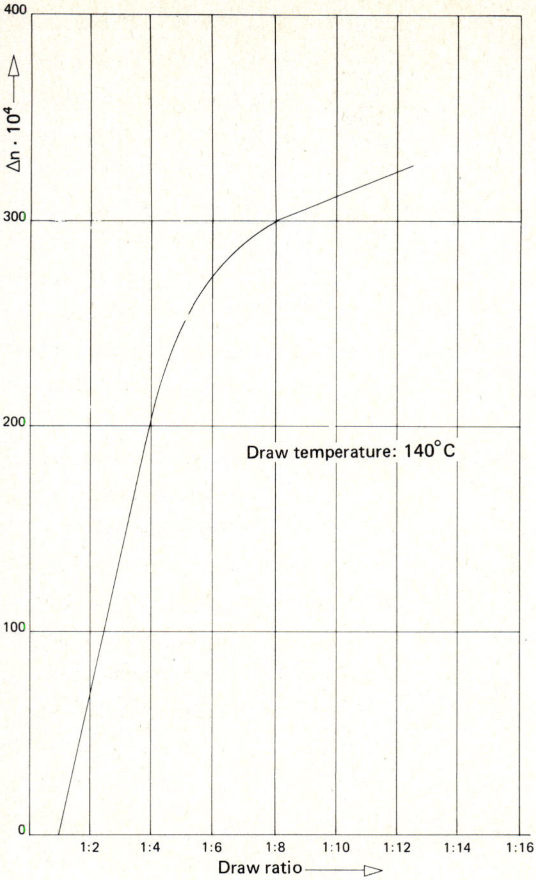

Fig. 9. Birefringences of a Polypropylene Tape Yarn as a Function of Draw Ratio (ref.1)

specific birefringence of crystalline regions can be higher than that of amorphous parts, considerable increase of the observed birefringence Δ_n can be due to crystallization without any change of orientation factors f_{cr}, f_{am}. This should always be taken into account in the interpretation of experimental results. Experimental data indicate that orientation characteristics are increasing functions of draw ratio but not a single value function. In many cases the observed birefringence is shifted toward lower Δ_n values when the drawing temperature is raised (Fig.10).

The orientation process in cold-drawing affects crystals much more than amorphous regions. It is important to note that the drawing of polypropylene fibers is always associated with a monotonic increase in the orientation of the c-axis and a monotonic decrease in the orientation of the a-axis and b-axis. This

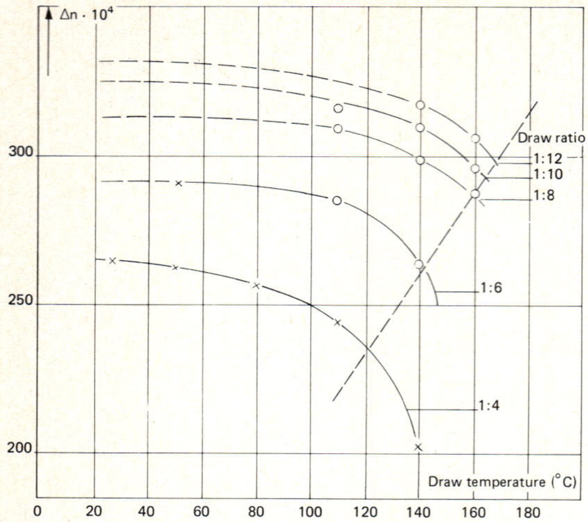

Fig. 10. Effect of Drawing Temperature on Δ_n *of Polypropylene (ref.1)*

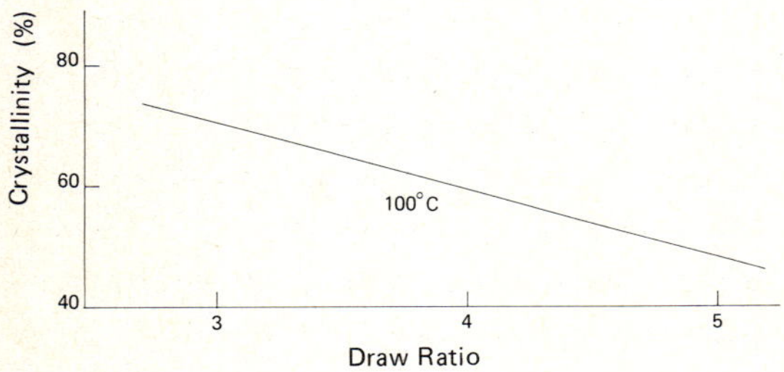

Fig. 11. Change of Crystallinity with Cold-Drawing of Polypropylene Filament (ref.6)

behavior, i.e., uniaxial orientation of axis c parallel to molecular chain axis and rotational disorder of perpendicular axes a and b, is different from the structural changes resulting in oriented crystallization discussed in the foregoing chapter of this book.

The question of the changes in the "degree of crystallinity" of polypropylene fibers on drawing has been often debated during the past few years. Conflicting results have been reported. In Fig.11, for example, is shown the dependence of crystallinity of a stretched fiber on draw ratio for a spun fiber of high crystallinity and for a moderate drawing temperature. As can be seen, crystallinity decreases with an increase in tension to which the fiber is subjected during cold-drawing. If drawing is carried out at higher temperatures, at which flow is facilitated, the reduction in crystallinity is less obvious. On the other hand, it has been shown that if the spun fiber is paracrystalline, cold-drawing is accompanied by an increase in crystallinity to a degree which depends on draw ratio, draw temperature, and rate of drawing. The influence of drawing temperature on crystallinity is shown in Fig. 12.

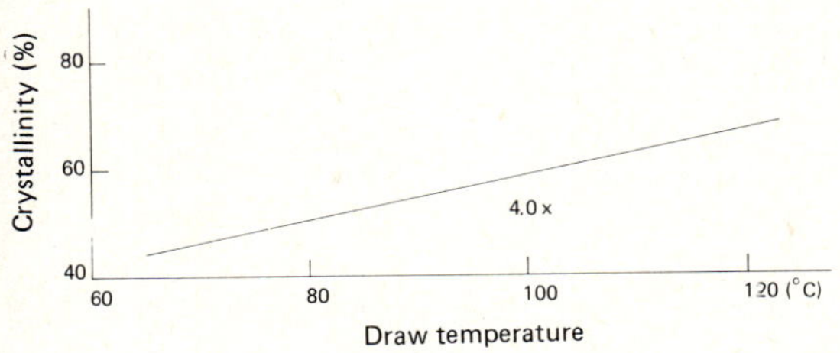

Fig. 12. Change of Crystallinity with Drawing Temperature at 4 D.R. (ref.6)

The confusion about the crystallinity of the fiber and its possible changes with drawing conditions stems from the lack of a proper definition ity and methods for its measurement. The two methods, i.e., densit and X-ray technique, yield two different measures of crystallinity and do not respond to drawing process in the same fashion (Fig. 13). Because of the microvoids produced in the fiber on cold-drawing, the apparent density of the fiber would change in a complex manner with the state of the spun fiber and drawing conditions. The changes occuring in density of the fiber are therefore not necessarily the result of packing density of the molecular chains, as is tacitly assumed to be the case. On the other hand, the X-ray method fails to take into consideration very small intermediate order (paracrystalline) which can certainly not be classified as amorphous.

The changes in filament density on drawing depends on the spun filament structure. A comparison of density measurement showed that in the case of rapidly quenched fiber, a density of 0.8803 g/cm^3 was obtained. On the other hand, a relatively slowly cooled fiber with developed crystalline structure reached a density of 0.8923 g/cm^3. On cold-drawing at room temperature the fiber of smectic structure changed in density from the original 0.8803 to 0.8976 g/cm^3 while that with monoclinic structure changed from 0.8923 to 0.9002 g/cm^3. On drawing at high temperature both filament structures ended up with almost the same maximum density, although at different draw ratios. The fiber density may not always change as described above. This is specially the case where the morphological structure of the fiber is composed of large spherulites, where the

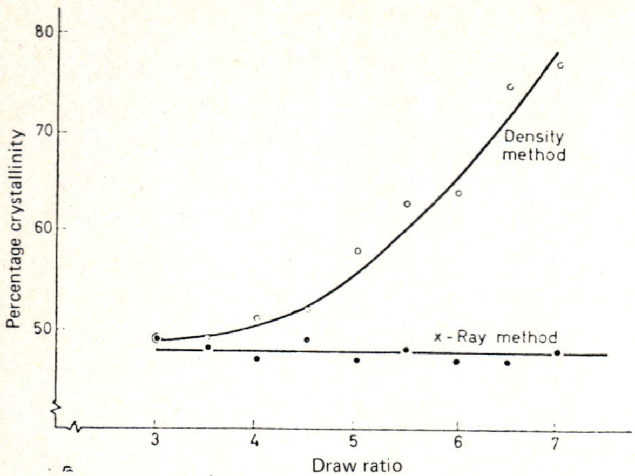

Fig. 13. Percentage Crystallinity versus Draw Ratio for Pin-and-Plate Drawn Polypropylene Fibers (ref.7)

spinning tension is high, where the filament shows a pronounced degree of cross orientation, where the cold-drawing takes place at relatively high speeds, where the molecular weight of the polymer is low and the molecular-weight distribution is broad. In such cases, the fiber density decreases in the initial stages of cold-drawing, reaches a minimum, then rises again, as shown in Fig. 14. The decrease in density with drawing is due to a multitude of microvoids formed through the fractures of microspherulites present in the spun fiber. To account for the increase in specific gravity with further increase in draw ratio, consider that at large draw ratios the spherulites completely disintegrate, the lamellae unravel and are transformed into microfibrillar structure which tend to heel or at least decrease the extent of microvoids in the fiber. These microvoids produced during the drawing of crystalline polypropylene fibers at low temperatures are generally of the order of magnitude of the wavelength of light. The polypropylene fibers of low apparent density thus become opaque and appear delustered. The "chalking" phenomenon is a common observation in drawing of polypropylene fibers in actual practice. The chalkiness produced by microvoids may be removed by heat setting or annealing the fiber.

On the molecular scale, the lamellar structure of unoriented polymer plays a significant part in the deformation process. It appears that the initial stages of draw consist primarily of orientation, through twisting and shear of lamellae, to some extent accompanied by an unfolding of the molecules. Groups of lamellae swing as units toward the draw direction, and slippage takes place between the lamellae. Some sort of fibrillation occurs, and it seems that fibrils so formed are composed of molecules from more than one lamellae.

Twisting, slip, shear, and phase transformation, which are analogous to those usually found in metals, have been observed in the study of deformation of single crystals. However, as the complexity of these units is increased by additional layers of material, the mechanism of deformation changes and becomes more complex, apparently requiring cooperative motion between units. As the sample size increases and we approach the bulk state, the problem is further compounded. Notwithstanding the complexity of the problem, the following structural model of the deformation process of a semicrystalline polymer, such as polypropylene, can be offered according to the visualization by Peterlin (ref.3).

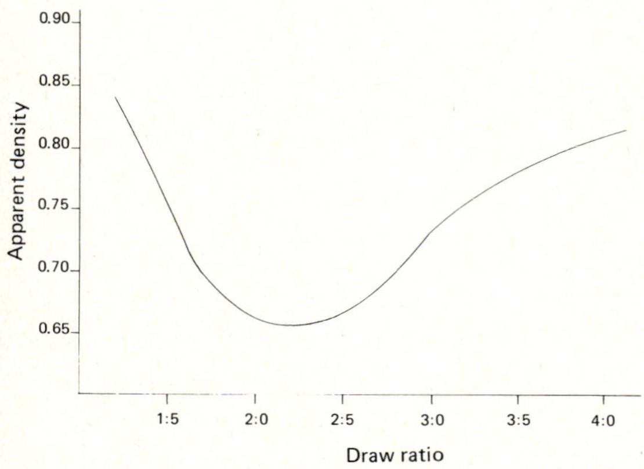

Fig. 14. Change of Apparent Density of Crystalline Polypropylene Filament with Drawing at Room Temperature

When the fiber is stretched the spherulites will pull out into ellipses. As spherulites are deformed under extensional stress, the internal structure within the spherulite area rearranges to relieve the imposed stress. It involves shear, slip, and rotation of stacked lamellae, phase transformation and twinning, chain tilt and slip inside each single lamellae. However, during this whole process, the individuality of spherulites, stacks of lamellae, and single lamellae, remains preserved. This spherulitic deformation process, or reorganization of crystalline and noncrystalline components, continues with increased elongation until the crystals cannot support the applied force. They then fragment into smaller components. When this happens the tightly interconnected, extended, elliptical spherulite structure is destroyed and a new microfibrillar structure is formed.

Further deformation continues primarily by conversion of the last remaining lamellar regions to fibrillar structure, by the sliding motion of microfibrils past each other, the stretching of interfibrillar tie molecules, and the unfolding of chain sections by which the tie molecules are anchored in the crystal blocks of adjacent fibrils. Finally, the structure as a whole cannot support the stress, flaws develop, and the sample breaks.

There are several possible mechanisms for such a process. Kubayashi (ref.4) considered complete unfolding of chains from the original crystals leading to bundle-like fibrils (Fig.15). On the other hand, Peterlin (ref.5) advanced the model (Fig.16) where the original chain-folded crystals are gradually melted, slipped, twisted, and partially unfolded in the transition region to yield defective fibrillar structure, involving chain fold, parallel to the fiber axis. Experimental data favor the latter model.

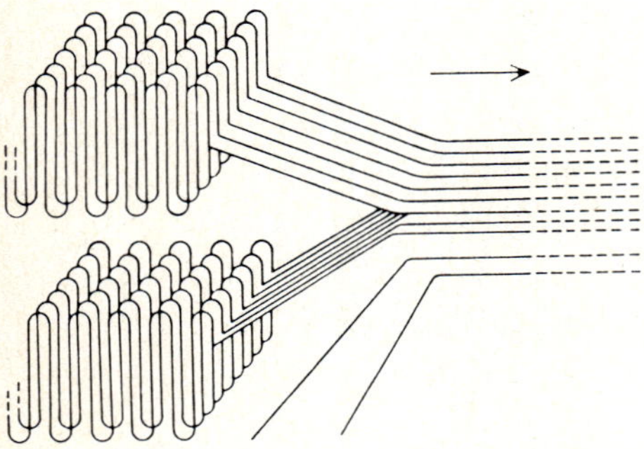

Fig. 15. Molecular Mechanism of Plastic Deformation of Polymer Crystals After Kobayashi (ref.4)

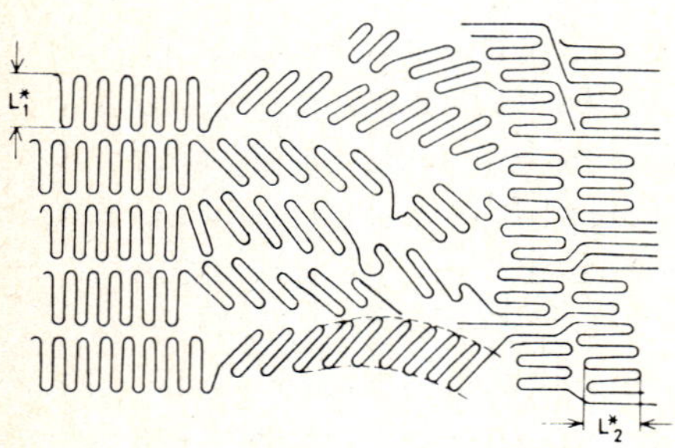

Fig. 16. Molecular Mechanism of Plastic Deformation of Polymer Crystals After Peterlin (ref.5)

Highly drawn fibers exhibit an almost complete orientation of the crystal lattice, a high orientation of the amorphous component, and a fairly good orientation of lamellae either perpendicular or at a finite angle to the fiber axis. The same anisotropy applies to mechanical properties with a high elastic modulus in the fiber axis and a smaller one perpendicular to it. Technically, the most important feature is the much higher values of the axial modulus, yield stress, and tensile strength, by one decade or even more, than observed in unoriented spherulitic material and a proportional reduction of draw range and elongation at break. As a rule, the effects increase linearly with the draw ratio. One is tempted to assume that this significant modification of mechanical properties is caused by chain orientation because both increase with increasing draw ratio. A closer analysis shows that such a parallelity obtains for the amorphous, but not for the crystalline, component. It is the consequence of the fact that in a polymer sample with increasing draw ratio the fraction of fibrous structure is steadily increasing and both the amorphous orientation and mechanical properties are proportional to this ratio.

The morphology of fibrous material reveals as the basic structural element the microfibril and not the lamella. Electron micrographs of surface replicas of slightly etched fibers and films with fibrous structure show exclusively highly aligned, very long microfibrils with lateral dimensions between 100 and 200 Å. The length is in the tens of micrometers. The microfibrils can be easily separated from each other. A higher organizational unit is the fibril, i.e., a bundle of parallel microfibrils formed during drawing from a stack of parallel lamella of the starting material and differing in draw ratio from the adjacent fibril originating from a stack with different lamella orientation. That makes the boundary between adjacent fibrils weaker and favors the formation of longitudinal voids.

During micronecking, folded chain blocks are broken off from the lamella and incorporated in the microfibril, which therefore consists of alternating crystalline and amorphous regions. A great many taut tie molecules, up to 30 percent of chains of the crystal lattice in highly drawn polyethylene, originating from partial chain unfolding during separation of the blocks, bridge the amorphous layer and connect the blocks, thus imparting to the microfibril great axial tensile strength and elastic modulus. Hence, the microfibril acts much more as a crystal with locally weakened areas than as an alternation of crystalline and amorphous layers.

The lamellar structure of the fibrous material is actually an optical artifact produced by the crystal blocks of adjacent microfibrils which tend to be rather well aligned laterally so that more or less wavy lamellae are formed (Fig.17). The waviness explains the two-point and the line diagram of small-angle X-ray scattering. A four-point diagram indicates a discrete orientation at a finite angle to the fiber axis. But there is practically no material connection between adjacent blocks of the same lamella in contrast with the great many tie molecules connecting blocks of the same microfibril. By prolonged etching one can remove the amorphous component in the layers between the blocks, and eventually the incompletely crystalline tie molecule bridges connecting the blocks in axial

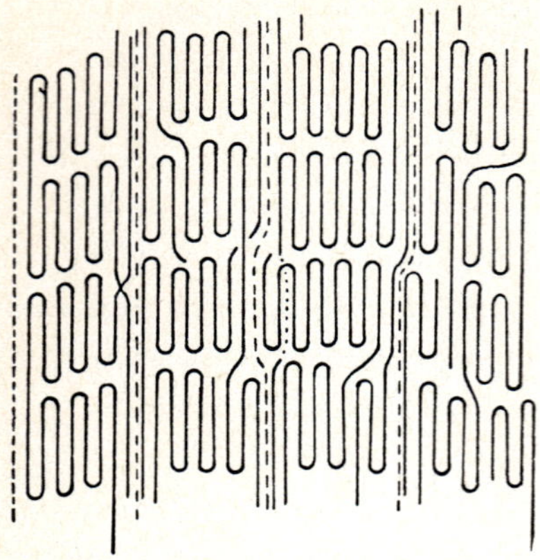

Fig. 17. Microfibrillar Model of Fibrous Structure (ref.12)

direction. That which remains is indeed the stacking of parallel lamellae without visible remains of microfibrillar structure. The persistence of lamellae also demonstrates that there is less amorphous material on the lateral than on the axial surfaces of the folded chain blocks.

The orientation of the crystal lattice increases very fast during drawing and soon reaches a value close to 1. But it is not a good parameter for the description of mechanical properties. It fails completely during annealing at higher temperatures, which only marginally reduces the crystal orientation, particularly if the sample is clamped so that it cannot shrink, but drastically reduces the mechanical properties down to the values of the unoriented material. Concurrently, IR dichroism shows a nearly complete loss of amorphous orientation, which seems to indicate that the taut tie molecules have almost completely relaxed and assumed a nearly equilibrium random conformation or have been reduced in number. Electron micrographs of surface replicas indeed show a rotation of lamellae around an axis perpendicular to the fiber axis and a lateral growth beyond the boundary of the fibril. One has the impression that the lamellae have attained a physical individuality not existing in the material as drawn and a substantial independence from each other which is only possible if the axial connection by tie molecules has been lost and the microfibrils have ceased to exist.

In a stress field the microfibril reacts almost as a rigid structure much stronger than any other element of the polymer solid. This is a consequence of the large fraction of taut tie molecules connecting consecutive crystal blocks like a quasicrystalline bridge and of the special dense packing of aligned microfibrils, but not a consequence of some higher crystal perfection of the blocks. Indeed, one knows that the crystal-defect concentration is increasing with draw ratio, yielding a lower than ideal crystal density and, consequently, a lower strength and elastic modulus of the crystals. The removal of crystal defects and restoration of crystal perfection during annealing make the crystals stronger but cannot prevent the drastic reduction of elastic modulus and strength of the fibrous material caused by the disappearance of tie molecules.

The dense packing of microfibrils with the enormous surface ($4l_{mf}d$) to cross-section (d^2) ratio ($4l_{mf}/d \sim 4000$) seems to overcompensate the role of crystal defects. The longitudinal displacement of a single microfibril is opposed by the sum of van der Waals forces along the whole microfibril, which adds so much that such a displacement is less easy than in the case of stacked lamellae. Moreover, it blocks the individual crystal deformation because such a deformation always demands some longitudinal displacement of the whole or a part of the microfibril. Consequently, the deformation of the fibrillar structure is much more a cooperative than a highly localized effect of a single crystal block. That not only makes the microfibril more resistant to deformation but also homogenizes the stress and strain field. The combination of all these effects results in the higher tensile strength and elastic modulus of fibrous material. As long as one can avoid longitudinal void formation, the elastic modulus in the transverse direction, too, is larger than in a spherulitic sample because, in spite of the fact that the crystal lattices of adjacent microfibrils are not in crystallographic register, the van der Waals forces between them are stronger than in the thicker amorphous layer between two stacked lamellae.

From the morphology of fibrous structure one can conclude that the plastic deformation will first involve the sliding motion of microfibrils which is opposed by the friction resistance in the boundary. In spite of the weakness of van der Waals forces, the friction resistance over crystal lamella that are 1,000 times longer than they are thick will be much larger than the forces opposing plastic deformation in spherulitic material. Hence the strain-hardening effect in the fibrous material is a straightforward consequence of the large surface-to-volume ratio of the strong microfibrils, which efficiently hampers shearing displacement of microfibrils and explains the increase of elastic modulus and stress to break by a factor nearly equal to the draw ratio.

The bundling of microfibrils into fibrils yields two interdependent modes of plastic deformation of fibrous structure under tensile load: sliding motion of fibrils upon each other (pure dilation as described by the affine transformation), and shear deformation of fibrils by sliding motion of microfibrils. The former mode is responsible for most of the macroscopically observed deformation and also for the final failure by microcrack formation along the outer boundaries of the fibrils. The coalescence of such microcracks leads to physical separation of individual fibrils which can be observed on the fracture surface of high-strength fibrous material. The sliding motion of the fibrils also produces a shear stress on more or less skewed fibrils. The shear stress is proportional to the skewedness, and the resistance to deformation proportional to the length of the fibril. Hence, one expects larger shear deformation in nylon with shorter and wider fibrils than in polypropylene with longer and thinner fibrils.

As a consequence of shear deformation of fibrils during plastic deformation of fibrous structure, one obtains a large fraction of interfibrillar tie molecules connecting adjacent microfibrils. The effect is larger, the smaller the draw ratio λ^* at which the transformation of spherulitic into fibrous structure is completed because the length of the fibrils is proportional to λ^* and the lateral dimensions to $\lambda^{*\,-1/2}$. The almost fully extended taut tie molecules are not in crystallographic register and hence form a gradually thickening noncrystalline boundary layer between the crystal blocks of adjacent microfibrils.

The ends of the microfibrils create about $10^{21}\ m^{-3}$ point vacancies in the microfibrillar superlattice. Under applied tensile load they may fail first, by microcrack formation, so that the adjacent microfibrils have to carry a heavier load than the rest of the sample. Hence they are the first candidates for rupture detectable by the radicals formed at the rupture of tie molecules in at least one amorphous layer of the microfibril affected. Depending on the ratio of axial strength to lateral adhesion of the microfibrils, the microcracks will grow parallel (high ratio) or perpendicular (low ratio) to the fiber axis yielding a large number of broken chains and radicals in the parallel and a small one in the perpendicular case. Nylon is an example of the former and polypropylene of the latter type.

In passing, a few more comments on tie molecules are worth considering. The existence of interlamellar tie molecules in the crystallized polymer has been referred to previously. Because the microfibrils at the single crystal crack are spatially separated from each other, in deformation process each microneck has to transform a section of lamella wider than the microfil. Hence, the folded chain blocks are not incorporated in the microfibril in exactly the same order as they were arranged in the lamella. Therefore, the tie molecules, as a rule, do not connect immediately adjacent blocks but rather blocks farther apart, thus crossing all intervening amorphous layers sandwiched between subsequent crystals. Their number increase with molecular weight of the polymer and the rate of crystallization of the spun fibers.

The interlamellar tie molecules add to the lateral cohesion of the fibrils, but also interfere with the longitudinal sliding motion of the microfibrils during the additional plastic elongation of the fiber structure. Interlamellar tie molecules of the original sample connect, after the transformation in the necking zone, blocks of different microfibrils. Their number increases with molecular weight and the rate of crystallization of the spun fibers. Such interfibrillar tie molecules may add to the lateral cohesion of the fibrils, but also interfere with the longitudinal sliding motion of the microfibrils during micronecking and the additional plastic elongation of the fiber structure. Their arrangement seems to be unidirectional so that they exert a finite tensile force and torque on the crystal blocks they connect. A manifestation of such a structure is the retractive force leading to fiber shrinkage by back sliding of microfibrils and the torque producing crystal orientation during annealing with free ends. The tendency of the tie molecules to contract reverses to some extent the deformational process.

IV. PRACTICAL ASPECTS OF DRAWING

In actual practice of polypropylene fiber production, the technologist is invariably confronted with the important question of choosing the proper drawing conditions for the desired fiber properties. These include (1) draw ratio, (2) draw speed and (3) draw temperature. Obviously, this choice is greatly influenced and often limited by various material variables and the structural state of the spun yarn. These include chemical structure variables (homopolymer or copolymer), polymerization variables (molecular weight, polydispersity, tacticity), additive variables (plasticizers, nucleating agents, dye additives), pretreatment variables (orientation, heat history), crystallization variables (percentage crystallinity, smectic or spherulitic morphology). None of these factors, however, independently defines the drawing behavior of the spun fiber and the ultimate fiber properties. It is rather the interaction of these parameters with each other and with the drawing conditions that defines the whole process and determines properties of the product. Defining a drawing process for a given spun fiber to achieve the desired physicomechanical properties is, therefore, a complex undertaking, and because of so many variables, largely an empirical exercise in checks and balances. The resulting process is mostly a question of various compromises in fiber properties and drawing conditions.

The effect of molecular weight and molecular-weight distribution on drawability of polypropylene fibers has been studied (ref.8,9). Drawability, as measured by maximum draw ratio, increases with molecular weight and becomes maximum at a M_v of 6.7 - 7.8 x 10^4, after which it deteriorates (Fig.18). The optimum temperature of drawing increases with increasing molecular weight of the polymer. Molecular weight of the spun polypropylene fiber also has a marked effect on the neck formation in drawing. At low molecular-weight level the neck formation occurs easily, but at high molecular weight the filament may draw without neck formation.

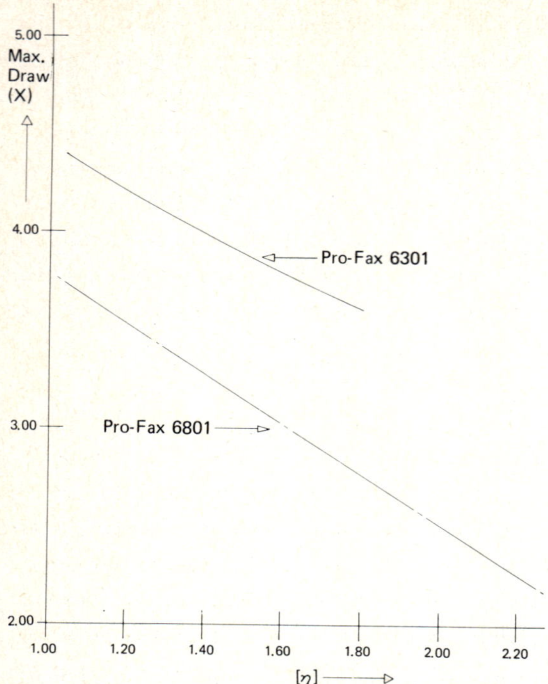

Fig. 18. The Influence of [η] of Yarn on its Drawability, i.e., the Maximum Draw Ratio at 90°C. The Respective Polymers Were Degraded by Organic Peroxides to the Given [η] Values (ref.21)

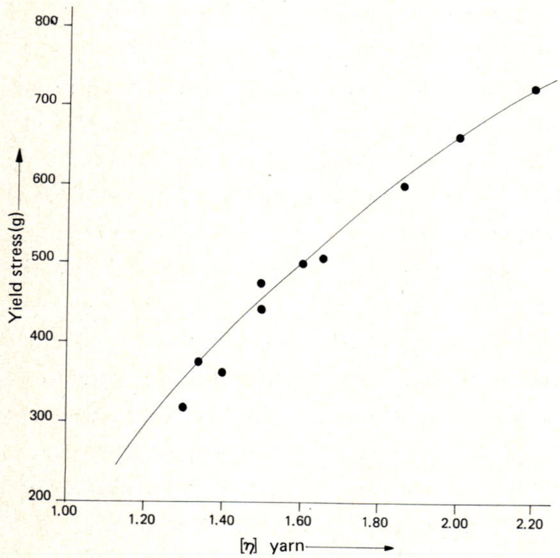

Fig. 19. Influence of [η] on Yield Stress (ref.21)

The stress required for neck formation increases with molecular weight (Fig.19).

In comparison with the effect of molecular weight, polydispersity seems hardly to affect the temperature dependence of drawability unless the width of the molecular-weight distribution becomes extremely excessive. The drawability of the filament with a broad molecular-weight distribution is higher than that with a narrow molecular-weight distribution at draw temperatures below 80°C, but at higher temperature the effect is minimal (Fig.20). Taking into consideration that the proportion of the low molecular-weight material increases with the widening of the distribution of molecular weight, it is believed that the low molecular-weight polymer acts as a kind of plasticizer for the bulk polymer at low temperature. The same effect is observed for yield stress. The polymers of broad molecular-weight distribution yield at lower stress than those of narrow molecular-weight distribution. Apart from the criterion of maximum draw ratio and yield stress as a measure of drawability, there is the practical consideration of the "ease of drawing" which determines the speed of drawing without excessive filament breaks. The polymers of narrow molecular-weight distribution generally yield cleaner yarns, albeit at low maximum draw ratios, than the polymers of broader distribution with equivalent $[\eta]$ values. This is probably due to the broad molecular relaxation spectrum of broad molecular-weight distribution. The large molecules thus wield certain breaking power to the process of deformation. The beneficial effect of some additives in the polymer has been mentioned in the patent literature for their effect on drawing. One class of additives is the processing aids mentioned in Chapter IV and elsewhere in this book. The inclusion of stearates of calcium, zinc, or sodium, for example, helps the drawing of the fiber considerably. This is expected on the basis that some of these processing aids act as mild prodegradants for commercial polypropylene and the ease of drawing is in effect the combined result of lower molecular weight and lower birefringence in the spun fibers. Another class of additives particularly effective for ease in drawing without affecting polymer degradation is paraffin waxes of various sorts. The inclusion of these waxes and paraffinic compounds in 2 to 5 percent concentration facilitates drawing at high speeds, increases the maximum draw ratio of the fiber, decreases the tendency of void formation and, in effect, results in the formation of high-tenacity fibers with improved luster. Silicon additives have also been variously mentioned in this respect. One effective way to improve the drawability of polypropylene fiber of high molecular-weight polymers is to increase the atactic content of the fiber. This can be done by intermixing with the bulk polymer the low molecular-weight, largely atactic fraction of the polymer. A tactic polymer is commercially separated from the bulk polymer and is available as a by-product. However, as the ease in drawing is not an end in itself, the concern over drawability should obviously be subsidiary to the desired tensile properties of the drawn fiber. Thus, the additive approach to facilitate the drawing is only feasible when high molecular-weight polymers are being dealt with and where such properties as resilience are not of particular importance.

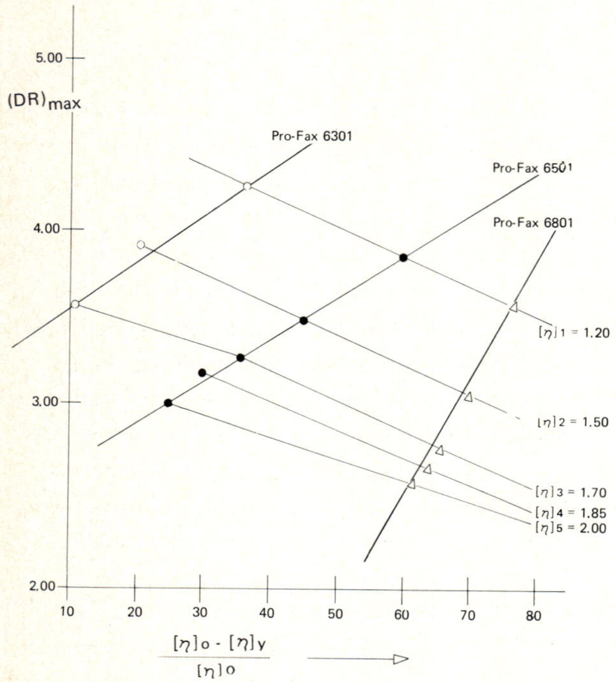

Fig. 20. The Effect of Polydispensity on Drawability of the Fiber at 60°C. $[\eta]_o$ and $[\eta]_y$ are the Intrinsic Viscosities of the Virgin Polymer and the Extruded Fiber, Respectively. According to the Theory of Polymer Degradation, the Higher the Value of $\frac{[\eta]_o - [\eta]_y}{[\eta]_o}$, the Narrower is the Molecular-Weight Distribution of the Extruded Polymer (ref.21).

The role of morphology and crystal types of spun fibers on its drawability has been extensively discussed in previous section. From these deliberations, a picture emerges according to which more perfect and uniform crystal structure, such as that obtained in heat-treated or aged polypropylene yarns, tend to decrease the drawability of the fiber. The nonuniform crystal structure, such as that obtained in most of the freshly spun yarns or rapidly quenched fibers, require less tension for drawing and normally draw to higher draw ratios.

Mechanical properties of undrawn fibers are very sensitive to spinning conditions. Figure 21 presents stress-strain curves for undrawn polypropylene

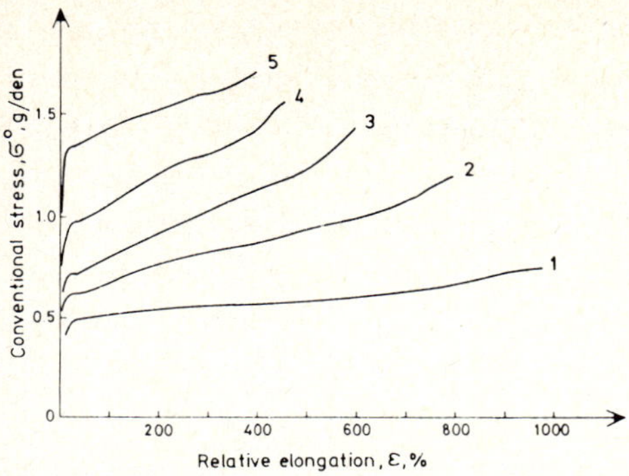

Fig. 21. Stress-Strain Characteristics for Undrawn Polypropylene Fibers Spun at Various Take-up Velocities. V_L: 1, 8.4 m/min; 2, 36 m/min; 3, 116 m/min; 4, 170 m/min; 5, 258 m/min. (ref.13)

fibers spun at velocities ranging from 8.4 to 258 m/min. The observed increase in stress and modulus and the reduction of elongation (maximum draw ratio) is due to increasing orientation. Figure 22 presents maximum draw ratios of undrawn polypropylene fiber plotted vs the orientation factor for crystalline regions. It is worth noting that R_{max} also depends on the crystal structure, and different correlations are obtained for fibers with prevailing hexagonal smectic and monoclinic modification.

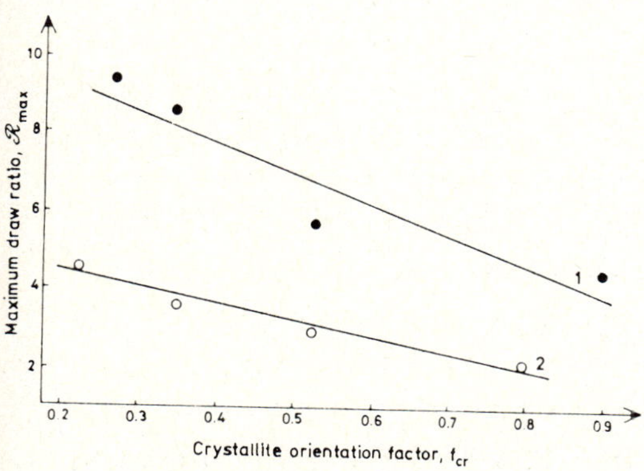

Fig. 22. Maximum Draw Ratio of Melt-Spun Polypropylene Fibers vs Orientation Factor of the Crystalline Phase. 1, Fibers with Prevailing Hexagonal Modification; 2, Fibers with Prevailing Monoclinic Modification (ref.11)

A great emphasis has been laid in the previous pages on the molecular orientation as a function of cold-drawing and the resulting improvement in the fiber strength in the drawn state. It may appear that the sole function of cold-drawing is to achieve the maximum orientation of the fiber and thereby the maximum tensile strength in the final product. This view of drawing is, however, erroneous in commercial practice. Although tenacity of the fiber is an important criterion for selecting the optimum drawing conditions for some uses, other mechanical properties and physical characteristics are as important for some other end uses.

In addition to the tensile properties of the final product, there may be several other considerations which affect the choice of drawing conditions for a given spun fiber. One such consideration, for example, is the ease of deformation of the drawn fiber which is an important property in all texturizing processes. Some other such considerations can be low extensibility (for warp yarns), high shrinkage (for bulky yarns), low shrinkage (for woven textiles), low abrasion resistance (for carpets and upholstery), high elasticity (for knitting yarns) and high resilience (for carpets). Coupled with various technical considerations is the ever-present consideration of economy and cost. The choice of drawing conditions is based on whether one or the other property is of primary importance and where a proper balance between quality and cost can be struck.

V. AGING AND HEAT SETTING

As indicated earlier, the fiber structure formed in melt-spinning is a result of uncompleted processes acting upon structure formation and is in a metastable condition. Given the opportunity, some of the structure-forming forces may continue to operate in post-spun condition and change the structure of the fiber in the solidified state. This is because the glass transition temperature for polypropylene is below room temperature and the possibility of limited molecular mobility at room temperature is theoretically not precluded. As the molecules or molecular segments change their positions in relation with each other, a partial recrystallization of the material or formation of new supermolecular structures is the result, the extent depending on the time and temperature given for such transformations. Similarly, it has been mentioned in previous sections that cold-drawing and texturizing processes cause the fiber to accumulate stresses by virtue of strained tie molecules. Given the opportunity, the strained molecular segments will try to rearrange themselves in a state of lowest potential energy. This process may result in fiber shrinkage and detectable changes in orientation and crystallinity.

These post-spinning and post-drawing or post-texturizing structural changes are very slow at atmospheric temperatures and are commonly known as "aging". At elevated temperatures, equilibrium can be approached within a few minutes and fibers can be efficiently stabilized against subsequent changes during their application. In this case, the process is called "heat stabilization" or "heat-setting". Because setting and aging produce essentially the same kind of results, differing in degree rather than in nature, both terms are sometimes fused into a single term, "thermal aging" for theoretical discussions. Heat setting is a technical equivalent of thermal aging.

V.1. Aging or Conditioning

When undrawn fibers are collected, spinning stress is gradually relaxed or spinning deformation is partially recovered. The most important changes, however, concern structural transformations during the post-spinning storage time, often called "conditioning". For polyamides, conditioning acquires an important significance because the atmospheric moisture accelerates structural changes in the fiber. For polypropylene, atmospheric moisture does not play a role and all the structural or mechanical changes are due to a temperature-time-load interaction.

The most noticeable change in an undrawn polypropylene fiber during storage is the change in its drawability. The freshly spun yarn is quite easy to draw, as shown by its maximum draw ratio and the yield stress. As the time passes, the maximum draw decreases and the yield stress increases (Fig.23). The effect is most rapid during the first day or two but the changes continue to be felt for weeks. A considerably aged yarn sometimes becomes so brittle that a technical drawing process becomes impossible to be carried out. The technical ramifications of this effect are three fold. Firstly, differently aged yarns would yield draw yarns of different mechanical quality (broken filaments). Secondly, the differently aged yarns would yield the drawn yarns of different relative shrinkage. Thirdly, the draw yarns from differently aged yarns will show quite different texturizing behavior, resulting in texture variations and streaks in woven or tufted goods. The same effects also cause variation in denier of textured yarns (Fig.24). The structural basis of these changes and mechanical behavior of undrawn yarns on aging is not clear. Probably a rearrangement of crystalline structure takes place under the effect of heat contained within the body of the filament.

Polypropylene fibers commonly obtained in technical spinning operation are partially in the so-called smectic or paracrystalline form. This smectic structure changes to more perfect monoclinic structure during the process of thermal aging if the temperature of treatment is above 70°C. Figure 25, for example, shows the effect of temperature on the smectic-monoclinic change. One can observe that this change takes place through a series of intermediate order states which depend on the temperature and duration of aging. As Figure 25 shows, the

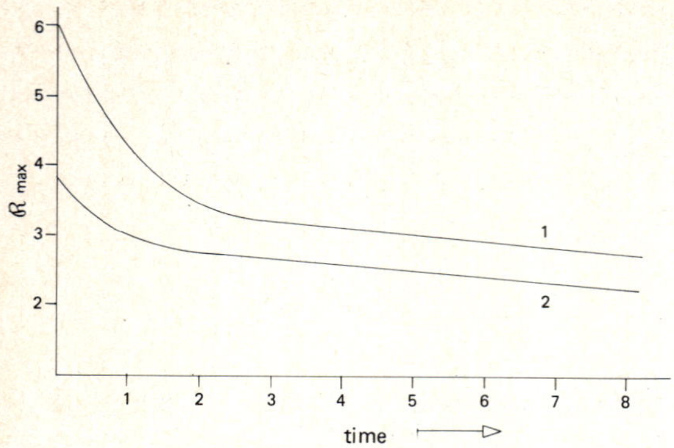

Fig. 23. Effect of Aging Undrawn Polypropylene Yarn at Room Temperature on the Drawability at 60°C. (1) Rapidly Cooled Sample, (2) Slowly Cooled Sample.

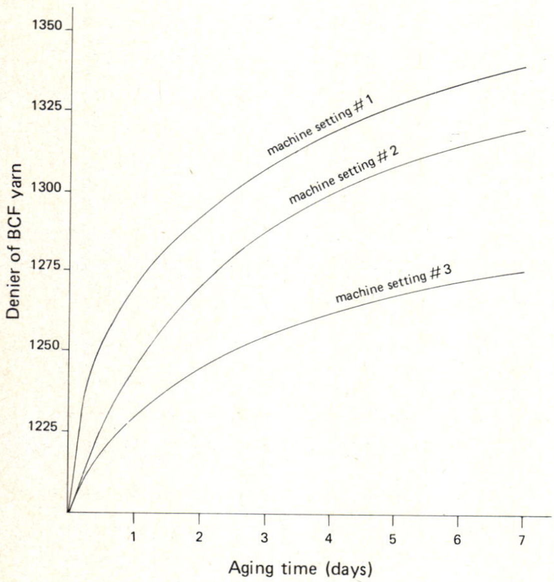

Fig. 24. Effect of Aging of Undrawn Polypropylene Yarn Before Texturizing on the Denier of Texture Yarn. Technisservice Stuffer-box Draw-Crimping Machine was used at Constant Machine Settings for each Series of Tests. The Change in Denier is Primarily Due to Change in Texture.

process does not begin below 60°C and for each temperature a saturation value, measured by the reciprocal of the half-width broadening of the (110) diffraction peak at about $2\theta = 14^{\circ}$, is reached which is practically proportional to temperature once the threshold temperature is reached. Although the process of smectic-monoclinic conversion process starts at 60°C, the rate of conversion is still very low. This conversion is only half complete when heat treatment takes place at 110°C but is largely completed if carried out at 145°C (ref.10). In comparison with the predominantly smectic form, however, the thermodynamically more stable monoclinic form does not experience any major changes in X-Ray diffraction pattern until a temperature of 170°C, at which the fiber begins to melt.

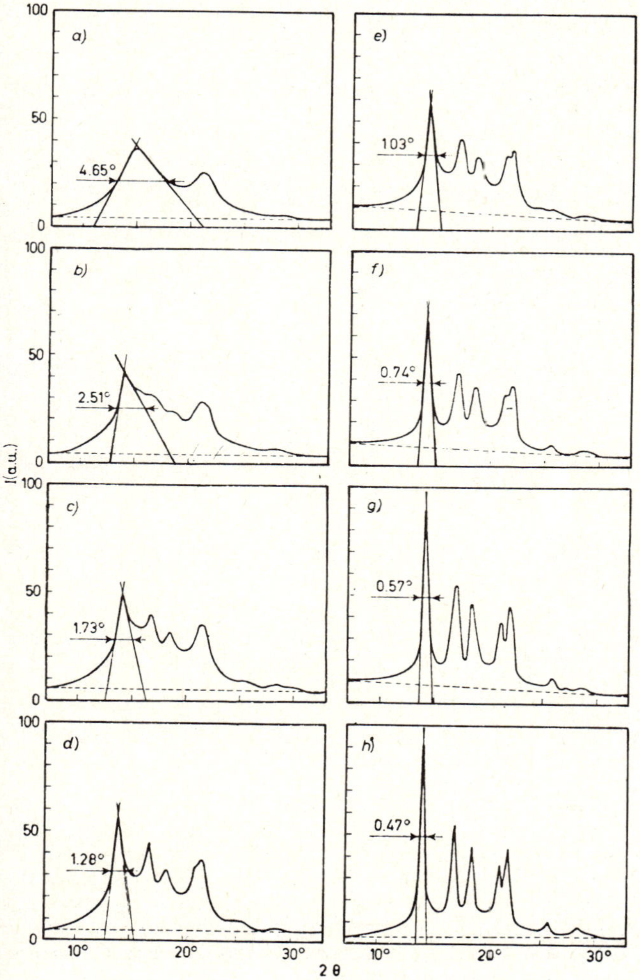

Fig. 25. Diffraction Scans (CuK$_\alpha$ Radiation, Proportional Counter) of the Paracrystalline form (a) and of the Same Annealed up to Saturation at 70°C (b), 80°C (c), 90°C (d), 100°C (e), 120°C (f), 140°C (g), and 160°C (h) (ref.14)

Aging also plays a role in drawn polypropylene fibers, especially when they are not heat stabilized or positively relaxed after drawing. As the drawn, unstabilized, or unrelaxed yarns are wound on a package, the heat contained in the interior of the filaments does not dissipate instantaneously. Under the influence of this body heat, the filaments begin to shrink and tremendous forces develop on the core of the package. The surface of the cylindrical package begins to bow inwardly as the sides of the package bulge outwardly. Sometimes, the shrinkage forces are so high that the whole package collapses. This phenomenon is common in high-speed drawing operations where the yarn does not have the opportunity to cool sufficiently before going on to the package. Heat-setting and controlled relaxation of the yarn before wind-up rectifies this problem, as is discussed in the next section.

Probably the most dramatic and the most damaging effect of aging is felt in texturizing of polypropylene yarns. All texturizing processes (stuffer-box crimping, hot-fluid jet texturizing, false-twist texturizing, gear crimping, etc.), are basically an attempt to impart to the fiber some desired configuration under

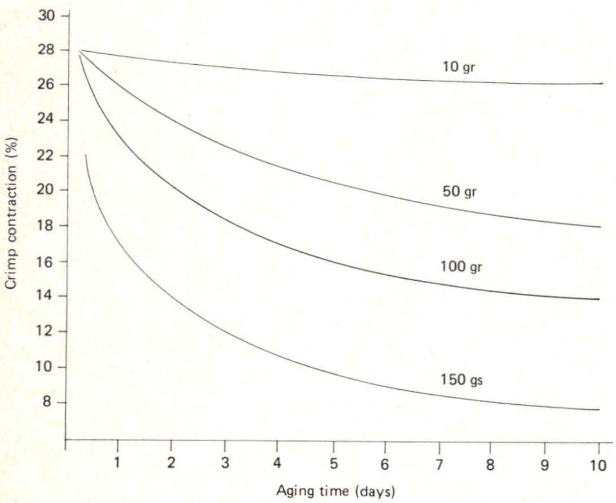

Fig. 26. Effect of Winding Tension and Aging Time on the Crimp Contraction of a 1800/70 Polypropylene BCF Yarn Produced Through Steam-Jet Texturizing

the influence of mechanical force and heat, stabilizing the newly imparted configuration through heat, cooling the heat-stabilized, crimped bundle of the yarn, unravelling the cooled, crimped bundle by the application of a pull, and winding the yarn under some desirable winding tension. In technological processes of texturizing, heat-setting and cooling of the crimped yarn bundle is never complete. As the yarn winds up on a package the same kind of changes begin to take place as those discussed in the case of drawn yarns. The crimped yarn tends to relax by contracting on the package but the winding tension impedes the process. The result of this interaction is some loss of crimp in the first place and then heat-setting of the yarn in this uncrimped state in the second place. The extent of crimp loss as a result of contraction depends on the degree of heat setting in crimping, winding tension, temperature of the yarn, and the time of aging before the yarn is put into the final product (Fig.26 and 27).

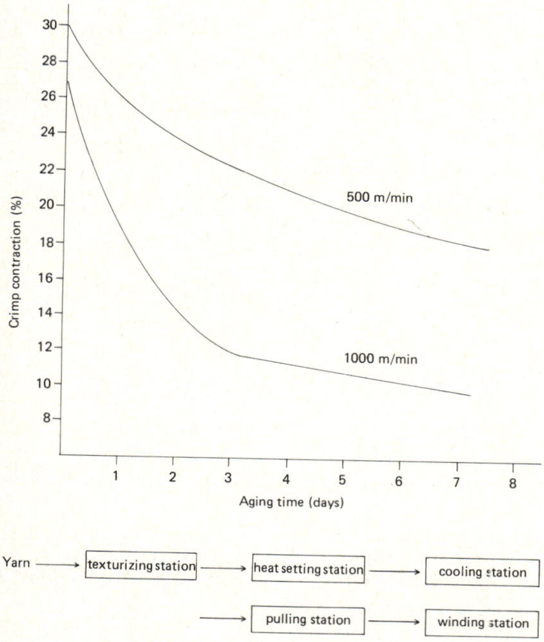

Fig. 27. Effect of Heat Setting and Yarn Temperature (Both of Which are Proportional to Texturizing Speed) on Aging of a 1800/70 Polypropylene BCF Yarn Produced Through Steam-Jet Bulking Process as above.

V.2. Heat Stabilization

Heat treatment of polypropylene fibers is aimed at stabilization of fiber dimensions (reduction of residual shrinkage or reduction of breaking elongation), stabilization of crimp and twist, and equilibration of crystalline structure and improvement of mechanical properties.

The changes of fiber structure in the stabilization processes are more explicit than those in drawing. Stimulation of molecular motions by heat treatment provides the conditions in which more perfect and more stable structures are developed with simultaneous relaxation of internal stresses. Figure 28 shows

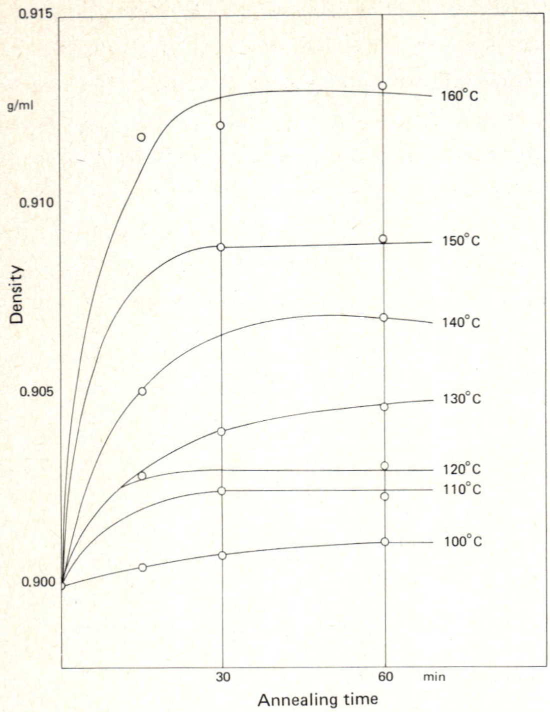

Fig. 28. Effect of Temperature and Duration of Annealing on the Density of Polypropylene (ref.15)

the effect of temperature and time of heat treatment on the density of polypropylene. If a straight relationship between density and degree of crystallinity is assumed, it may follow that heat treatment leads to an increase in degree of crystallinity of the fiber. Qualitatively the same results have been obtained through X-ray diffraction technique. Figure 29 represents the change in the degree of crystallinity as X-ray diffraction lines broadening of the (110) intensify with increasing annealing temperature. The degree of crystallinity appears to be a gradually increasing function of annealing temperature although not to the extent evidenced by density-measurement technique. The increase in crystallinity is probably due to the fact that annealing heals certain internal crystal defects and also permits some tie molecules and low molecular-weight fractions to be included in the crystal lattice. This process is obviously hindered by extensional tensions. The change in crystallinity is therefore more pronounced under freely shrinking conditions than under extensional tension (Fig.30).

The crystalline size increases with temperature of thermal treatment. The decrease in "line width" in Fig. 30 is interpreted as an increase in crystal perfection and/or in crystallite size. It is generally agreed that the changes in drawing characteristics of spun polypropylene fiber and changes in mechanical properties of drawn fibers upon aging can be traced back more directly to the change in crystallite size or crystallite perfection than to the change in the degree of crystallinity. Variations in birefringence are somewhat more difficult to interpret. The observed birefringence in oriented fibers depends on the orientation in crystalline and amphorphous regions as well as on the degree of crystallinity. Heating in the free state can hardly increase orientation, therefore an increase of Δn on heating (if observed) should be ascribed to crystallization. Reduction of birefringence with heating in a free state is a net result of partial

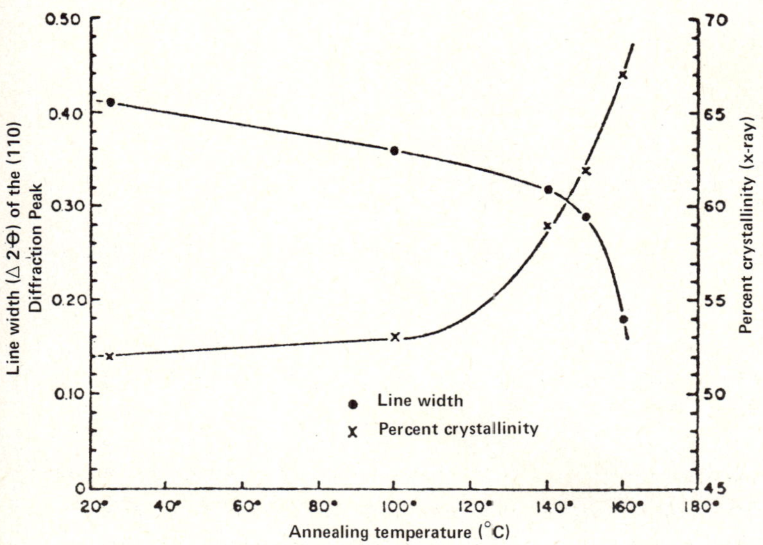

Fig. 29. Effect of Annealing Temperature on Line Width and X-Ray Crystallinity (ref.16)

disorientation of original crystalline and amorphous regions, possibly accompanied by recrystallization. When heat treatment is performed at constant length, or under tension, birefringence remains constant or increases as a result of crystallization. Typical relationships between birefringence and heat-treatment conditions for polypropylene fibers are shown in Fig. 31.

Disorientation of polymer crystals is a natural consequence of heat treatment in a free state, as associated with an increase in entropy of the system. Experimental data for polypropylene fibers have been published (ref.11). On the other hand, stabilization at constant length or under tension suppresses disorientation or even increases the crystallite orientation factor. Only at very high temperatures, close to the melting point, rapid disorientation takes place.

The heat-setting process can be divided into three steps:
(1) In the first stage, heat treatment decreases the intermolecular forces under the influence of elevated temperature. The bending stiffness of the molecules or molecular segments decreases and the motion of the structural elements

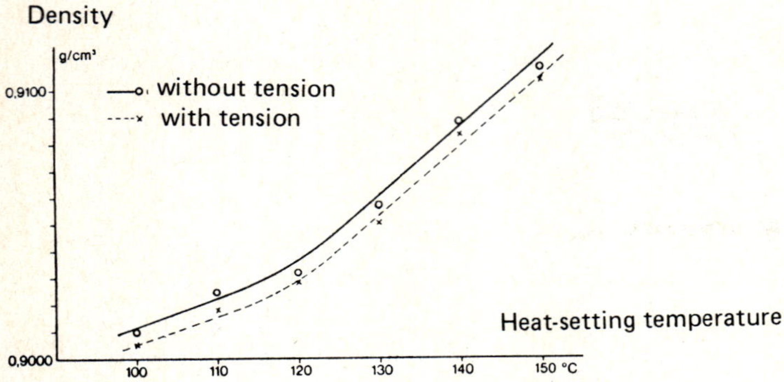

Fig. 30. Dependence of the Density on the Heat-Setting Temperature and Tension (ref.17)

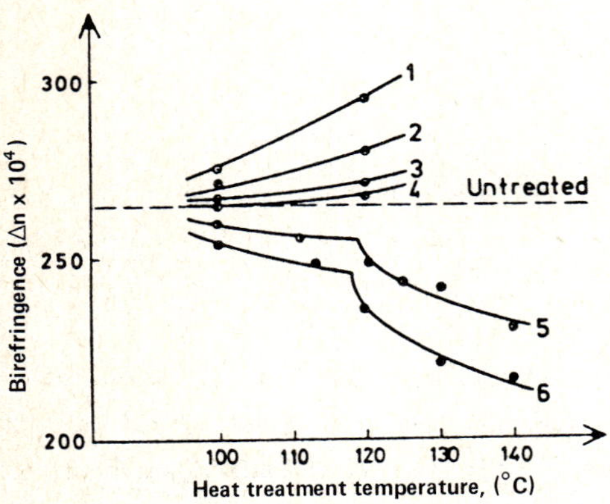

Fig. 31. Birefringence of Drawn Polypropylene Fibers Subjected to Heat-Treatment Under Various Conditions. (1) Fibers Elongated 10%, Heated in Ethylene Glycol; (2) Fibers Elongated 10%, Heated in Dry State; (3) Fibers Kept at Constant Length, Heated in Ethylene Glycol; (4) Fibers Kept at Constant Length, Heated in Dry State; (5) Fibers With Free Ends, Heated in Dry State; (6) Fibers With Free Ends, Heated in Ethylene Glycol (ref.18)

increases. This process is almost instantaneous and only limited in speed by heat-conduction rate.

(2) In the second stage, the increased vibrations of molecular segments make it possible for them to arrange themselves in a state of least potential energy which leads to increased crystallinity, enlargement of crystalline sizes, perfection of crystal aggregates, and increased uniformity in structural units. The changes taking place in this stage depend on temperature, duration of treatment, and the extent of external stresses.

(3) In the third stage, the more stable fiber structure, obtained in the second stage, freezes, and gross molecular motions stop. This stage, like the first, is very rapid and depends on heat-conduction rate.

The most important factor behind heat-setting processes on drawn fibers of various polymers is the way molecular flexibility depends on temperature. Each polymer has a hysteresis-loss pattern with a peak at a specific temperature. Although the quantitative interpretation of these loss peaks is far from clear, they are generally regarded in qualitative terms as being due to the onset of a new mode of molecular freedom as the temperature rises. In the case of poly-condensate polymers, such as nylon and polyester, this increase of flexibility above the upper transition is due not simply to a steady increase in flexibility, as for instance in heating a rubber; it is due rather to the rapid introduction over a narrow temperature range of a specific new set of possible molecular motions. Polypropylene, on the other hand, does not have any marked transition between room temperature and its melting point. At the temperature of use it is in an intermediate state of low stress-relaxation between two loss peaks. As a result, polypropylene does not approximate the behavior of an ideal type of doubled peak-relaxation response typical of nylon and polyester. Polypropylene fibers are not amenable to effective heat setting and are not noted for their good creasing and wrinkling properties. Isotactic polypropylene is, however, better than other polyolefins in this respect because of its very high crystallinity and because of its marked increase in molecular mobility above 70°C.

Mechanical properties as determined by structural characteristics are sensitive to the conditions of heat treatment. Tensile strength usually decreases when stabilization is carried out on free fibers and remains constant, or even increases, when tension is applied in the course of heating. These effects are evident from Fig. 32. Breaking elongation, which generally changes in the opposite direction to tenacity, increases during heat setting.

On the other hand, the residual shrinkage $\Delta\varepsilon_r$ is reduced more effectively by heat treatment in a free state. The residual shrinkage can be considered as the difference between the viscoelastic deformation frozen in the drawing process, ε_o, and the amount of shrinkage $\Delta\varepsilon$ in the stabilization process:

$$\Delta\varepsilon_r = \varepsilon_o - \Delta\varepsilon$$

Naturally, the higher $\Delta\varepsilon$ in the heat treatment, the lower the residual shrinkage $\Delta\varepsilon_r$. So the conditions of heat treatment should be chosen according to whether dimensional stability (fibers for textile usage with minimum residual shrinkage, $\Delta\varepsilon_r$) or minimum elongation (fibers for technical applications) are the most desirable deformation characteristics.

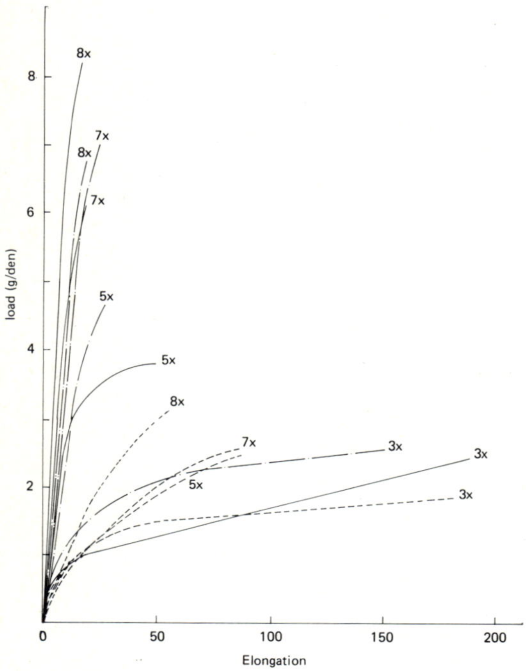

Fig. 32. Stress-Strain Curves for Drawn and Heat-Set Filaments (ref.19)

——— *Drawn*

—— · —— *Drawn and Stabilized at 130°C Under Tension*

- - - - *Drawn and Stabilized at 130°C Under No Tension*

The shrinkage temperature generally increases with stabilization temperature. Here again, the tension of heat stabilization process plays a significant role (Fig.33). The total work of deformation also increases with stabilization temperature, more so wher stabilization occurs without tension. The initial tensile modulus E, as related to molecular and crystal orientation, is reduced by heat treatment. Under tension the reduction is less pronounced, or the modulus increases. Figure 34 illustrates some mechanical characteristics of stabilized polypropylene fibers plotted against the tension applied in the heating process.

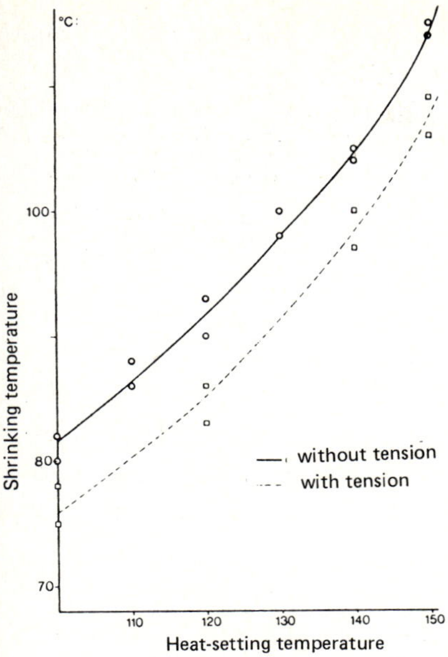

Fig. 33. Dependence of Shrinkage Temperature on the Heat-Setting Temperature of a Polypropylene Yarn (ref.17)

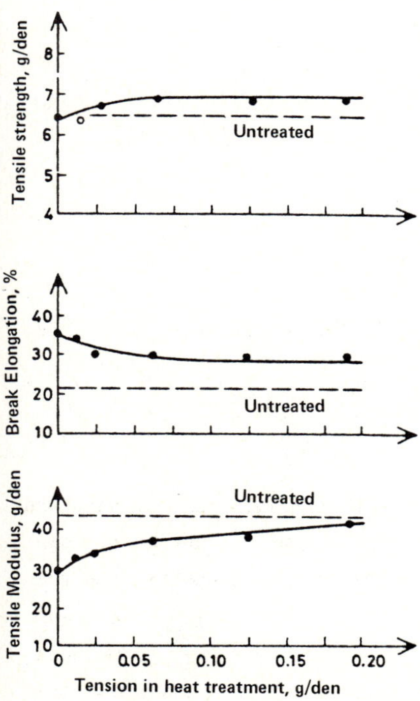

Fig. 34. Effect of Tension Applied in the Heat-Treatment on Some Mechanical Characteristics of Polypropylene Fibers (ref.11)

The resistance to multiple bending, multiple extension, and compression, as well as abrasion resistance, are expected to improve under heat treatment. It seems that all fatigue properties are related to nonequilibrium elements of fiber structure (frozen stresses,quenched amorphous regions) which, given appropriate kinetic conditions, undergo irreversible transformations. Equilibration of fiber structure in the process of heat treatment should, therefore, improve the resistance to dynamic stresses. Experimental results to these effects are, however, not available.

As both tensile strength and elastic recovery are improved by a simultaneous increase in crystallinity and orientation, practical ways and means have been devised to increase the degree of crystallinity and size of the crystallites of polypropylene fibers by heat setting with no decrease in orientation. One such technique is a two-stage drawing process discussed in Chapter IX. In this process the fiber is drawn in the first stage at elevated temperatures - between 110 and 120°C - to its maximum draw ratio. The drawn fiber is then further heated without relaxation and given a small additional draw in the second stage. The second stage draw in effect functions as a heat-setting step under high tension; the extent of heat-setting tension being a function of temperature and draw ratio. The fibers so produced are characterized by high tensile strength, improved modulus, moderate free shrinkage, and uniform dyeing characteristics.

Heat-setting plays an important part in texturizing of polypropylene filament yarns as well as in the crimping of staple fiber. In the common texturizing process, heat-setting under imposed stresses, as well as under free shrinkage, is involved. The aim of heat-setting in the texturizing process is to stabilize the crimped configuration of the filament in such a way that ordinary extensional stresses involved in the fabrication of the end product can not destroy the imparted crimp configuration. If the imposed extensional stresses should at all be successful in decrimping the filament in the course of fabrication, they do so by imparting large internal stresses to the fiber. These internal stresses are now capable of returning the filaments to their original crimped configuration when they relax in the finishing operation of the product. The technological implication of heat-setting to texturizing will be further indicated in Chapter IX.

VI. REFERENCES

1. W. Retting, Colloid and Polym. Sci., 253(1975)852.
2. M. Jambrick and L. Diacik, Faserforsch u. Textiltech., 15(1964)590.
3. A.Peterlin in *Structure and Properties of Oriented Polymers,* L.M. Ward Ed., John Wiley, New York, 1975.
4. K. Kobayaski, Proc. 4th Intern. Cong. on Rheology, Providence, R.I., 1963.
5. A. Peterlin, in *Man-made Fibers, Science & Technology,* H.Mark Ed., Interscience, New York, 1967.
6. V. Capuccio, Chim. Ind., 44(1962)463.
7. P.G. Stern, Polymer, 9(1968)471.
8. W.C. Sheehan and T.B. Cole, J. Appl. Polym. Sci., 8(1964)2359.
9. K. Kamide, Sen-i Gakkaishi, 23(1967)583.
10. H.W. Wyckoff, J. Polym. Sci., 62(1962)83.
11. M. Compostella, Angew, Chem., 74(1962)618.
12. A. Peterlin, J. Polym. Sci., 7(1969)115.
13. S. Oya, Sen-i Gakkaishi, 17(1961)1088.
14. R. Zannetti, Macromol. Chem., 128(1961)137.
15. J. Majer, Kunststoffe, 54(1967)578.
16. R.S. Schottland, Polym. Eng. Sci., 6(1966)244.
17. V. Hochman, Chemiefasern, (1973)337.
18. A. Nakai, Sen-i Gakkaishi, 17(1961)997.
19. G. Natta, Text. Ind., 69(1967)827.
20. A. Ziabicki, Fundamentals of Fiber Formation, Interscience, New York, 1975.
21. Personal Communication, G. Jurkiewich, Hercules, Inc.

VII. BIBLIOGRAPHY

VII.1. Books

1. P.H. Lindenmeyer, Ed., Supermolecular Structure of Fibers, Interscience, New York, 1967.
2. Encyclopedia of Polymer Science and Technology, Entry, Deformation of Polymers, Vol. 4, 620-647; Entry, Annealing, Vol. 2, 138-150; Entry, Orientation, Vol. 9, 624-648, Interscience, New York, 1968.
3. R.E. Wetten, Ed., Polymer Systems. Deformation and Flow, Macmillan, 1968.
4. R.R. Eirch, Rheology, Chapter on Deformation of Crystalline Polymers, by R. Stein, Academic Press, New York, 297(1969)351.
5. L.M. Ward, Structure & Properties of Oriented Polymers, Wiley, New York, 1975.
6. H. Mark, Ed., Man-Made Fibers, Interscience, New York, 1968.
7. J.W.S. Hearle and L.W.C. Miles, Heat Setting of Fiber & Fabrics, Merrow, Watford Herts Englad, 1971.
8. A. Ziabicki, Fundamentals of Fiber Formation, Interscience, New York, 1976.

VII.2. Journal Literature

1. G. Farrow, Measurement of Crystallinity in Polypropylene Fibers by X-Ray Diffraction, Polymer, 2(1961)409-441.
2. H.M. Morgan, Mechanical Model to Represent the Viscoelastic Properties of Polypropylene, Text. Res. J., 31(1961)993-994.
3. J. Power, Orientation Studies on Drawn Polyolefins, U.S. Dept. Comm. Off. Tech. Ser., 257(1961)562; 264(1961)894.

4. R.C. Laible, Viscoelastic Behavior of Isotactic Polypropylene Fibers, J. Appl. Polym. Sci., 6(1962)269-277.
5. S. Hushino, Orientation Studies on Drawn Polyolefins, J. Polym. Sci., 58(1962)185-204.
6. R. Buchadahl, Deformation Processes and the Structure of Crystalline Polymers, Angew. Chem., 74(1962)569-573.
7. M. Compostella, Fibers and Films of Isotactic Polypropylene. Preparation and Properties With Regard to its Crystalline Structure, Angew. Chem., 74(1962)618-624.
8. Z.D. Jastreebski, Annealing, Encyl. Polym. Sci., Tech., 2, 138-150, Interscience, New York. H.W. Wyckoff, X-Ray and Related Studies of Quenched, Drawn and Annealed Polypropylene, J. Polym. Sci., 62(1962)83-114.
9. G. Farrow, Crystallinity, Crystallite Size and Melting Point of Polypropylene, Polymer, 4(1963)191-197.
10. J. Jajen, Secondary Crystallization of Polypropylene, Kunststoffe Rundschau, 10(1963)9-11.
11. A. Weidinger, X-Ray Small-Angle Maxima in Oriented Polypropylene Fibers, Macromol. Chem., 62(1963)139.
12. M. Dole, The Glass Transition in Isotactic Polypropylene, ACS Polymer Preprint, 4(1963)
13. S.E. Ross, Relationship of Polypropylene Fiber Orientation as Determined by Optical Birefringence and Sonic Velocity Measurement, Text. Res. J., 34(1964)565-571.
14. N. Payne, Measuring the Density of Polyolefins with an Improved Gradient Column, Nat. Res. Stand., 4(1964)3-7.
15. E. Passaglia, Dependence of Mechanical Relaxation on Morphology in Isotactic Polypropylene, J. Res. Nat. Bur. Std., 68(1964)519-527.
16. I. Kirshenbaum, Heat and Entropy of Fusion of Polypropylene, J. Appl. Polym. Sci., 8(1964)2723-2730.
17. G. Budor, Determination of the Crystallinity of Polypropylene, Faserforsch. u. Textiltech, 15(1964)527-532.
18. S.G. Turley, Polypropylene Annealed at a Temperature Near Its Melting Point, J. Appl. Polym. Sci., 9(1965)2693-2706.
19. M. Jambrich, The Drawing Process of Polypropylene Fiber, Abhand. Deut. Akad. Wiss. Berlin, Bl. Chem. Geol. Biol., 3(1965)37-45.
20. D. Hansen, Mechanism of Cold Drawing in Crystalline High Polymers, J. Appl. Phys., 36(1965)332-334.
21. S.E. Ross, Some Observations Concerning the Behavior of Polypropylene Polymers and Fibers, J. Appl. Polym. Sci., 9(1965)2729-2748.
22. A. Peterlin, Polymer Deformation - Annealing of Drawn Fibers, J. Polym. Sci., 3(1965)151-155.
23. W.C. Sheehan, Relationship Between Structural Parameters and Tenacity of Polypropylene Monofilaments, Text. Res-J., 35(1965)626-637.
24. P.R. Pinnock, Mechanical and Optical Anisotropy in Polypropylene Fibers, Brit. J. Appl. Phys., 17(1966)575-586.
25. A.J. Rosenthal, T.E.½, An Index for Relating Fiber Tenacity and Elongation, Text. Res. J., 36(1966)593-602.
26. I.M. Ward, Mechanical and Optical Anisotropy in Polypropylene Fibers, Brit. J. Appl. Phys., 17(1966)575-586.
27. R.S. Schotland, Influence of Annealing on the Density and Tensile Properties of Polypropylene, Polym. Eng. Sci., 6(1966)244-272.
28. R.J. Samuels, Spherulite Structure, Deformation Morphology and Mechanical Properties of Polypropylene Fibers, J. Polym. Sci., 20(1967)253-284.
29. W.W. Cox, Morphological Transformations of Polypropylene Related to its Melting and Recrystallization Behavior, Polym. Eng. Sci., 7(1967)309-316.
30. L. Odon, Sorption and Diffusion Studies on Polypropylene Fibers, Macromol. Chem., 100(1967)11-21.
31. F. Rybnikor, he Determination of Crystallinity of Isotactic Polypropylene, Kunststoffe, 57(1967)199-203.

32. M. Reiner, Deformation of Polymers, Encyl. Polym. Sci., 4(1967)620-647, Interscience, New York.
32a. G. Natta, New Developments in Polypropylene Fibers, Lenzing. Rev. 24(1967)8-25.
33. M. Jambrich, Structural Changes of Polypropylene Fibers in the Two-Step Deformation Process, Faserforsch. u. Textiltech., 18(1967)331-335.
34. R.S. Stein, Physical Studies of Rates of Deformation of Crystalline Polymers, J. Polym. Sci., 20(1967)185-187.
35. R. Bunart, Molecular Deformation Mechanism in Partly Crystalline Materials, Kunststoffe, 57(1967)391-396.
36. R.J. Samuels, Spherulite Structure, Deformation Morphology and Mechanical Properties of Isotactic Polypropylene Fibers, J. Polym. Sci., 20(1967)253-284.
37. N.A. Brunt, Creep, Cold Drawing and Flow. A Theory of the Viscoelastic Behavior of High Polymers at Very Large Deformations, Kolloid Z. Z. Polym., 221(1967)12-27.
38. P.G. Stern, On the Structure of Polypropylene Fibers, Polymer, 9(1968)471-477.
39. D.M. Gezovich, Morphology of Quenched Polypropylene, Polym. Eng. Sci., 8(1968)209-219.
40. K.D. Pae, Effects of Thermal History on Isotactic Polypropylene, J. Appl. Polym. Sci., 12(1968)1901-1909.
41. R.J. Samuels, Quantitative Characterization in Drawn Polypropylene Fibers, J. Polym. Sci., 6(1968)1101-1139.
42. Z.W. Wilchinsky, Orientation in Polymers, Encyl. Polym. Sci. Tech., Interscience New York, 9(1968)624-648.
43. R.J. Samuels, Deformation Processes in Polypropylene Fibers and Film, Hercules Chemist, 56(1968)19-26.
44. D.M. Gezovich, Deformation and Aging of Quenched Polypropylene, Polym. Eng. Sci., 8(1968)210-215.
45. P.G. Stern, Structure of Polypropylene Fibers, Polymer, 9(1968)471-477.
46. R.J. Samuels, Quantitative Structural Characterization of the Mechanical Properties of Isotactic Polypropylene, ACS Polymer Preprints, 10(1969)1218-1223.
47. C. Balta, Plastic Deformation of Polypropylene - Influence of Temperature and Draw Ratio on the Axial Period, J. Mat. Sci., 4(1969)722-729.
48. D.D. Prevorsek, Nature of Thermal Plastic Deformation in Highly Oriented Polymers, ACS Preprints, 10(1969)1059-1066.
49. H. Baer, Structural Correlation in Chemical and Polymeric Fibers, Melliand Textiltech., 50(1969)1190-1194.
50. A. Peterlin, Plastic Deformation of Polypropylene - Small Angle X-Ray Scattering in the Neck Region, J. Appl. Phys., 40(1969)4238-4242.
51. V.G. Baranov, Relation of Initial Supermolecular Organization in Crystalline Polymers to the Organization Produced by Stretching, J. Polym. Sci., 8(1970)1015-1026.
52. G. Schoenfeld, Influence of Heat Treatment on the Structure and Strength of Polypropylene, Kunststoffe, 60(1970)177-184.
53. M. Kudra, Cold Drawing of Linear Polymers, Faserforsch. u. Textiltech., 21(1970)125-134.
54. P. Dryfuss, Simple Chain Folding Scheme for the Annealing Behavior of Polymer Crystals, J. Polym. Sci., 8(1970)253-258.
55. R. Chatterjie, Review on the Cold Drawing of Melt Extruded Synthetic Fibers, Silk Rayon Inds. India, 13(1970)233-240.
56. J. Juilfs, Meaning of Deformation Behavior of Drawn, Partially Crystalline Polymer Fibers, Kolloid 2, 241(1970)897-908.
57. A.A. Streltsov, Interrelated Influence of the Technological Process Parameters on the Abrasion Resistance of Polypropylene Monofilament, Fiber Chem., 2(1971)346-350.
58. R.S. Stein, Morphology and Deformation Mechanism of Crystalline Polymers, J. Polym. Sci., 32(1971)45-68.
59. K. Sakaoku, Electron Microscopy of Drawn Polypropylene, J. Polym. Sci., 9(1971)895-915.
60. J.L. Way, Deformation Processes in Fully Spherulitic Polypropylene, J. Mat. Sci., 6(1971)102-109.

61. C. Balta, Plastic Deformation of Polypropylene - Annealing of Drawn Polypropylene, Macromol, Chem., 141(1971)91-116.
62. A.B. Paksver, Influence of Preparation Conditions on the Thermal Shrinkage of Polypropylene Fibers, Faserforsch. u. Textiltech., 22(1971)329-334.
63. V.N. Sorokina, Mechanism of Abrasion of Polypropylene, Fiber Chem., 3(1971)627-630.
64. A.R. Duckett, Molecular Orientation in Polypropylene, J. Polym. Sci., Polym. Phys., 10(1972)2167-2172.
65. I.L. Way, Influence of Crystallization Temperature on the Deformation of Polypropylene Spheruletes, J. Mat. Sci., 7(1972)1345-1346.
66. S. Kapur, Aging of Quenched Polypropylene, J. Polym. Sci. Phys. Ed., 10(1972)2107-2124.
67. K.E. Prepelkin, Structure and Mechanical Properties of Highly Oriented Polymers, Angew Macromol Chem., 22(1972)181-204.
68. A. Peterlin, Molecular Model of Drawing Polyethylene and Polypropylene, J. Mat. Sci., 6(1971)490-508.
69. R.E. Robertson, Model for the Spherulite-to-Fibril Transportation, J. Polym. Sci., 10(1972)2437-2452.
70. N. Norosoff, Plastic Deformation of Polypropylene-Wide Angle X-Ray Diffraction in the Neck Region, J. Polym. Sci., 10(1972)1237-1254.
71. J. Juilfs, Stretching Behavior of Fibers and Yarns, Schweiz, Ach., 38(1972) 120-128.
72. R.A. Duckett, Molecular Orientation in Polypropylene, J. Polym. Sci., Polym. Phys. Ed., 10(1972)2167-2172.
73. N. Norosoff, Crystallite Size in Drawn and then Annealed Polypropylene, J. Polym. Sci., 10(1972)1221-1236.
74. Zh.G. Vasilenko, Structural Defects Produced in Polypropylene Monofils During Stretching, Fiber Chem., 4(1972)162-163.
75. S. Kapur, Aging of Quenched Polypropylene, J. Polym. Sci., Polym. Phys. Ed., 10(1972)2107-2124.
76. I. Diacik, Structural Changes in Synthetic Fibers in Drawing, Polypropylene Fibers, Faserforsch. u. Textiltech., 23(1972)70-74.
77. J.M.G. Cowie, Glass Transition Temperature of Stereoblocks, Isotactic and Atactic Polypropylenes of Various Chain Lengths, Eur. Polym. J., 9(1973)1041-1049.
78. A.J. Owen, Mechanical Anisotropy in Oriented Polypropylene, J. Macromol. Sci., Phys., 7(1973)417-430.
79. A. Peterlin, Structure and Drawing of Fibers as Spun, Appl. Polym. Symp., 20(1973)269-274.
80. W.A. Fraser, Thermally Induced Surface Deformations of Polypropylene Monofilaments, Textile Res. J., 43(1973)123-128.
81. A.J. Owens, Mechanical Anisotropy of Oriented Polypropylene, J. Macromol, Sci., Physics, 87(1973)417-430.
82. L. Kuezy, Influence of Certain Rheological Characteristics of Polypropylene Filaments, Ind. Textile, 1021(1973)155-160.
83. C.R. Desper, Structural Characterization of Isotactic PP Film of Ultrahigh Orientation, J. Macromol. Sci. Phys., 7(1973)105-119.
84. C. Lee, Formation of Translucent Cold-Drawn Polypropylene, J. Appl. Polym. Sci., 17(1973)3749-3761.
85. V. Hackman, Microscopic Determination of the Degree of Fixing Polypropylene Fibers, Chemiefasern, 23(1973)337-340.
86. G.N. Molkova, Properties of Undrawn Polypropylene Fibers as a Factor in its Post-Stretching Shrinkage, Fiber Chem., 5(1973)261-263.
87. E.S. Clark, Superdrawn Crystalline Polymers - a New Class of High Strength Fiber, Polym. Eng. Sci., 14(1974)682-686.
88. K. Dharod, Interaction of Melt-Spinning and Drawing Variables on the Crystalline Morphology and Mechanical Properties of High Density and Low Density Polyethylene Fibers, J. Appl. Polym. Sci., 18(1974)2539-2568.

89. G. Druzsbachzky, Shrinkage Behavior of Polypropylene Fibers of Different Molecular Weights, Faserforsch. u. Textiltech., 25(1974)392-396.
90. A. Peterlin, Plastic Deformation of Polymers with Fibrous Structure, Kolloid Z, 253(1975)809-823.
91. R.L. Rabkin, Contact Heating the Yarn in Orientation Stretching, Fiber Chem., 6(1975)692.
92. A. Peterlin, Molecular Aspects of Oriented Polymers, in Struct. Prop. Oriented Polym., L.M. Ward, Ed., Interscience, New York, 1975.
93. W. Retting, Orientation, Orientability and Mechanical Properties of Polymers, Kolloid & Polymer Sci., 253(1975)852-874.
94. V. Pechoe, Stress-Temperature-Time Relationship of Polypropylene Filaments, I. Faserforsch. u. Textiltech., 26(1975)619-623; II. 27(1976)27-31.
95. L.V. Zhuraviev, Review of the Mathematical Models of Technological Processes in Man-Made Fiber Production, Fiber Chem., 8(1976)384.
96. A. Akay, Processing Structure of Polypropylene Films, Plast. Rubber Inst., London, Conf. Polyp. Fibres & Textiles, York, Sept. 26-28, 1979.
97. A.F.Aspin, The Properties of Roll Embossed Yarns, Plast. Rubber Inst., London, Conf. Polyp. Fibres & Textiles, York, Sept. 26-28, 1979.
98. H.P.Nadella, Drawing and Annealing of Polypropylene Fibers: Structural Changes and Mechanical Properties, J. Appl. Polym. Sci., 22(1978)3121-3133.
99. R. Barthel, Dynamics of Fiber Drawing, Faserforsch. u. Textiltech. 29(1978)329.

Chapter VIII MECHANICAL ASPECTS OF FIBER PRODUCTION

I. INTRODUCTION

Production of polypropylene fibers can be schematically presented as outlined in Fig. 1. The whole process is roughly divided into six main steps:

(1) Compounding

(2) Melting the polymer

(3) Forcing the polymer melt through spinnerette orifices

(4) Elongation of the molten polymer jet, solidifying by cooling, and collecting the fibers

(5) Drawing, texturizing, staple cutting, baling, twisting, coning, etc.

(6) Waste reclamation

The steps 2, 3 and 4 are invariably combined into one operation and are collectively called spinning. Thus, the manufacturing plant activities are compartmentalized into compounding, spinning, post spinning, and waste recovery areas. With the advent of spin-drawing and draw-texturizing processes, however, this traditional breakdown of manufacturing activities is becoming less and less significant. In some processes, such as the production of tape yarns, compartmentalization is not possible at all. The manufacturing activities of a modern plant should, therefore, be viewed as a complete whole.

Just as the details of actual practices of fiber production are closely guarded secrets, the mechanical devices to undertake various manufacturing operations are proprietary information. The information discussed in this chapter is, therefore, of general nature. For detailed design information and procurement of available equipment, the machinery manufacturers and engineering consulting firms listed in Appendix I and II should be consulted. A wealth of design and process information is available in patent literature, some of which has been referred to in appropriate places in this book; the rest can easily be found by consulting an up-to-date file of the *Textile Technology Digest* , published each month by the Institute of Textile Technology, University of Virginia, Charlottesville, Virginia, and the *World Textile Abstract* , published by the Shirley Institute, Manchester, England. The Polymer section of the *Chemical Abstracts* is another excellent source of information in this connection. The monthly publication *Faserforschung und Textiltechnik,* East Berlin, Germany, carries an extensive section of patent abstracts. This can be used as an excellent supplementary source of information

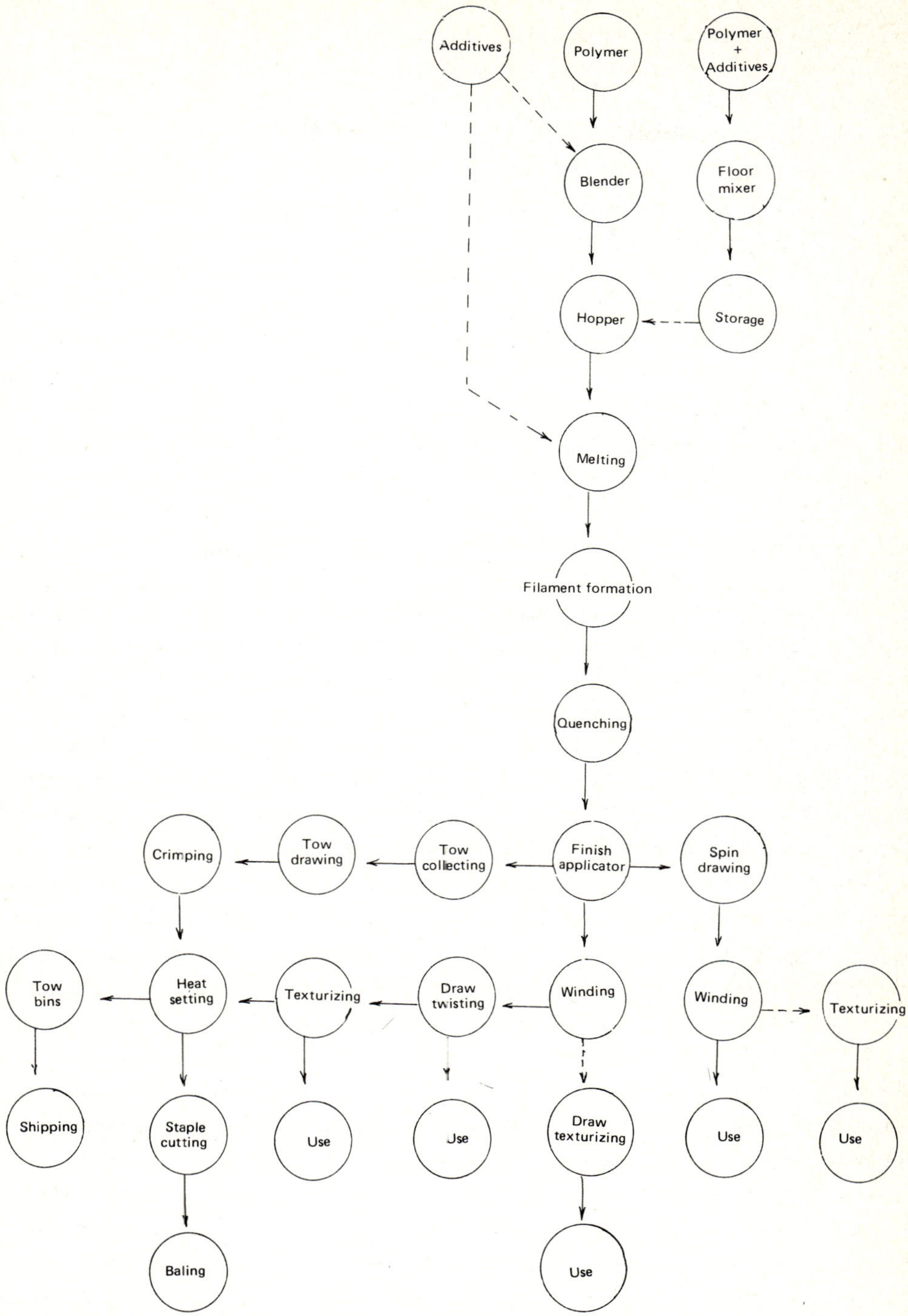

Fig. 1. Flow-Sheet Diagram of Fiber-Spinning Operations

for the developments in the socialist countries of Europe. The Bibliography at the end of this chapter lists a selection of United States patents from 1961 to 1979 concerning the melt spinning, yarn handling, fiber processing and support activities bearing upon the mechanical aspects of fiber manufacturing operations. A complete listing is obviously not possible. An attempt has, however, been made to include all pertinent patents applicable to the melt spinning and processing of polypropylene fibers and yarns.

II. MATERIAL HANDLING AND COMPOUNDING

The material-handling operations in a polypropylene fiber spinning plant include the procurement of raw materials, i.e., the polymer, pigments, stabilizers, auxiliary additives, etc.; storage and distribution; mixing; and hopper loading. A materials-handling system can be as simple as the procurement of pigmented polymer pellets in bags or drums and manual loading of the hopper, or it can be as complex as the procurement of polymer pellets or powder in carloads, storage in multiple silos, automated delivery to master blender, automated dosing of various ingredients into intensive mixers, and automated loading of hoppers.

In designing a system or choosing equipment for material handling, the starting point should be to determine the actual amount of polymer processed and the lot sizes of the extruded material. A small plant, obviously, cannot justify the purchase of resin in carloads (bulk purchasing) or the use of most automated equipment for moving, loading, and compounding polymer. On the other hand, bulk purchasing and automatic bulk handling resin systems command considerable economic advantage in medium to large fiber producing plants. Opinions may differ as to what constitutes a "medium" or "large" fiber producing plant and what the minimum capacity is at which an automated bulk handling system becomes economical. In the United States and Western Europe, where the resin can be easily shipped in bulk from the point of delivery, the labor is comparatively expensive and the capital is not scarce, the bulk purchasing and automatic bulk handling system may become feasible for a plant capacity of about 20 million pounds, whereas in other parts of the world a plant of this size may be better off relying on manual handling and batch compounding techniques.

The advantages of bulk resin handling systems are manifold. Savings are realized not only through bulk purchasing and low labor costs in materials distribution and hopper loading but also through the use of in-plant stabilization and coloration of the resin. Figure 2 schematically depicts one of the various possible layouts for bulk handling and compounding of the polymer. A brief description of various components follows:

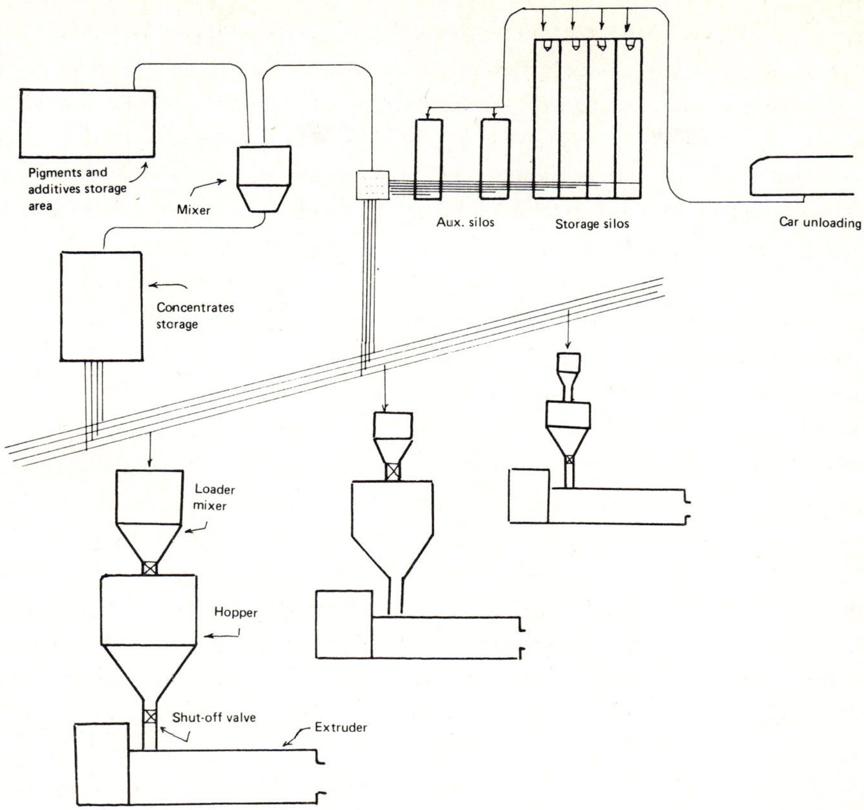

Fig. 2. A Typical Materials-Handling System for a Large-Scale Polypropylene Fiber Production Plant

The polymer pellets or powder received in bulk cars is air-veyed into a concrete structure containing four to six silos for each grade of polymer. Each silo should be large enough to hold the contents of a bulk car, with lots kept segregated in individual silos. Auxiliary silos are provided for special lots of the polymer. The material is taken simultaneously from multiple silos and air-veyed to the compounding station or to the upper hopper over the extruder in the production area. The flow of material out of several silos at a time helps to even out the lot-to-lot differences in the polymer.

Several types of loaders are available. Vacuum loaders with an integral motor are the most common for pellets while vacuum loaders with detached vacuum producing system is more suitable for powder handling. Positive-pressure loaders are used with some systems. Other types include the venturi loader and the auger loader. Efficiency of loaders depends on the principle of operation and the design. The efficiency of vacuum and pressure loaders is generally good but venturi-type loaders are less efficient. Auger material loaders have good efficiency but a long auger can present problems involving space and cleaning operations. The density of the polymer powder and the size and shapes of the pellets affect the performance of loaders and converging systems. When loading powders rather than pellets, all air-conveying loaders must be de-rated. In-plant, long distance conveying and loading is most commonly accomplished by carrying the pellets or powder in a fast-moving air stream. Both vacuum systems and positive-pressure systems are used, with vacuum the more popular.

The machinery for in-plant compounding can be divided into four classes:

(1) Discontinuous, batch-type operation using floor equipment for mixing

(2) Automatic loading-blending of various components over the hopper

(3) Continuously metering the additives concentrate in the molten form through a slave extruder into the main extruder

(4) Continuously metering the additives in the form of a slurry into the throat of the extruder.

The details of the batch-type mixing operation vary, depending on whether the polymer is in the powder or granular form and whether the additives are powders, pellets or liquids. In case the polymer is in powder form and the additives are powders or liquids, the mixing operation is rather simple and straightforward. The additives, whether liquids or powder, are mixed in heavy concentration with the polymer powder in a Henschel blender, Muller mixer, or sigma blade mixer in such a way that a free-flowing powder mixture is obtained. In case of thick, viscous additives, the mixer has to be heated while mixing. The free-flowing concentrate, so obtained, is then added to the bulk polymer powder and mixed in an air vortex mixer, auger-type mixer, or simple tumbler or drum mixer. The formulated resin can now be loaded with the help of appropriate loaders to the hopper of the extruder.

If the polymer is available in the form of pellets and the additives in the form of liquid or powder, the mixing becomes problematic. Tumbling of the polymer pellets in the presence of the powder or liquid additives has been tried in an attempt to coat the pellets with the additive powder or powder-liquid mixture. Surface active agents are sometimes employed to make the additive powder adhere to the pellets' surface. These procedures are, however, not efficient, and uniformly blended material is not obtained. In such cases, therefore, a poorly blended mixture is extruded into pellets and these blended pellets are used for fiber extrusion. Sometimes a poorly blended concentrate is melt-mixed and extruded into pellets of the size equivalent to that of the bulk polymer. These concentrate pellets are then used to form final polymer blends in tumble mixers or air vortex blenders. The batch operations of floor mixing of additives or pigments with the polymer tend to be expensive in operational cost but the capital outlay on equipment is generally low. Batch operations are more flexible than automated systems and are well suited to small and medium size production units.

Automatic dosing and blending of additives and the polymer directly on the hopper is an advanced compounding procedure. The chief advantage of additive mixing on the hopper is that the mixture drops directly into the extruder hopper. The mixture is not conveyed and, therefore, is not subject to separation. This can be especially important when the additive concentrate and the resin are powders or pellets of dissimilar sizes. The other advantage of mixing on the hopper is that when the color or additive is changed, it is not necessary to clean out the material lines and other loading equipment.

There are several kinds of equipment available for automatic, on-the-hopper compounding from a number of machinery manufacturers (Appendix II). Essentially they are self-contained, metal-fabricated units which can be conveniently located above the hopper of the extruder, and are equipped with hoppers and canisters holding resin, color dispersions, and free-flowing additive concentrates. They are provided with electronic control equipment to regulate resin and additive dosing as well as the mixing operation and final discharge into the processing equipment.

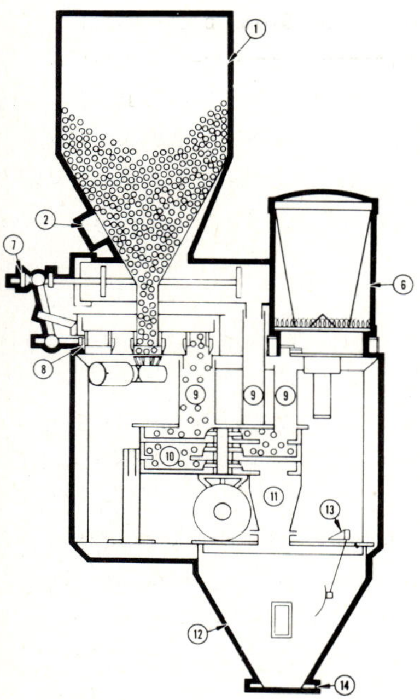

Fig. 3. Whitelock Over-the-Hopper Continuous Blender-Mixer

Both pellets and powders can be used for over-the-hopper compounding. These hopper mixers generally operate by gravity feed or positive displacement mechanism. The short-term accuracy of gravity feeders is limited because of varying bulk density of the materials. The accuracy of positive feeders is generally better. There are some feeders available that combine weighing with volumetric feeding. These feeders are believed to be more accurate than either gravity or positive feeders. The feed accuracy of minor additives such as pigments and stabilizers can further be improved by preparing a concentrate in the polymer, then employing the concentrate in automatic feed and blending devices. Figure 3 depicts a typical on-the-hopper blender.

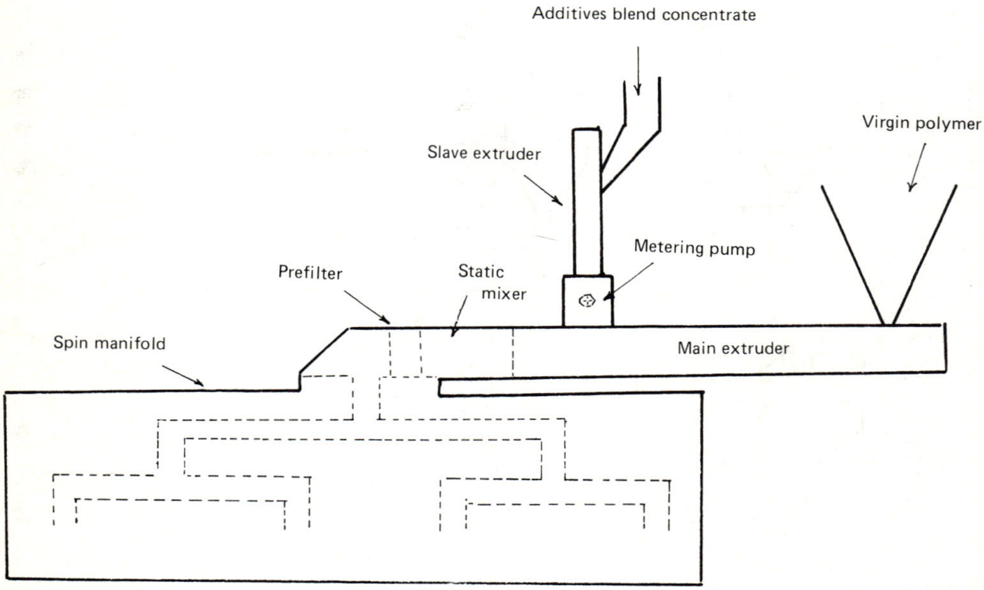

Fig. 4. Schematics of Melt-Proportioning System for Additive Blending in an Integrated Blending-Melting-Spinning Arrangement

Melt proportioning is another way of compounding the polymer. A concentrate of the additives and the colorants is prepared in a Henschel blender or Muller-type mixer and loaded in the hopper of a slave extruder. The molten concentrate is metered through a metering pump into the main extruder. The end of the compression zone of the extruder screw is generally chosen as the point of entry for the concentrate melt (Fig.4). The metering zone of the main extruder is often adequate to blend the concentrate melt with the polymer melt, but static mixing devices like those shown in Fig.5 and Fig.6 may be used at the end of the melter for improved mixing.

Another variation of the continuous mixing is the metering of pigment and stabilizer slurries in appropriate carriers into the feed zone of the extruder

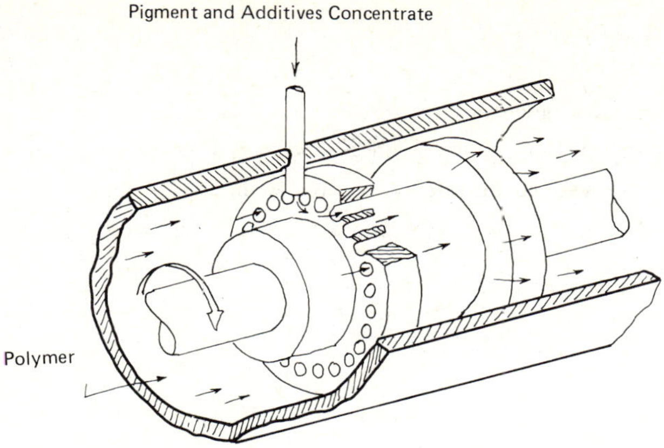

Fig. 5. Pigments and Additives Mixer for Melt-Proportioning Extrusion System

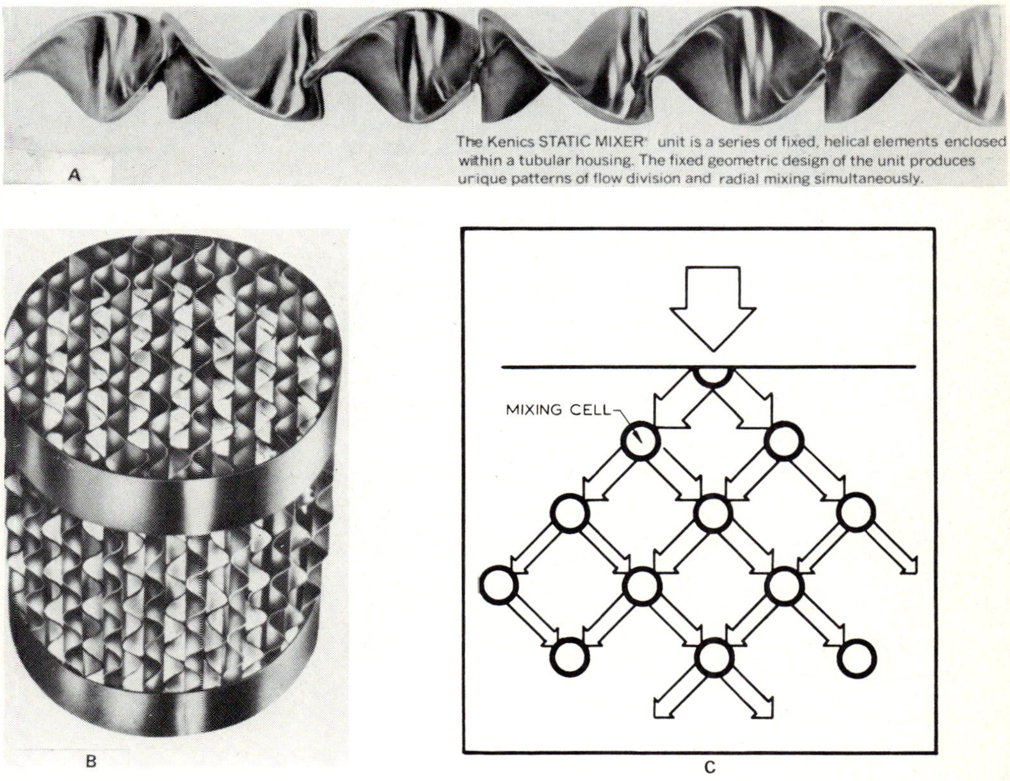

Fig. 6. Examples of a Static Mixer (A) *Kenics Static Mixer,* (B) *Koch Static Mixer Element. Several of these elements are arranged at right angle to each other to form a mixer;* (C) *shows schematically how each element divides the incoming fluid into two parts and how the next element combines the two different parts into one and then divides it into two.*

(Fig. 7). Liquid pigment dispersions of this sort are now commercially available from various suppliers. The same basic technology can be employed for the preparation of stabilizers, dispersions, or slurries.

The melt or liquid dispersion proportioning systems are the most accurate of all the automatic compounding systems discussed above although they tend to become rather high in capital costs and need more attention in maintenance.

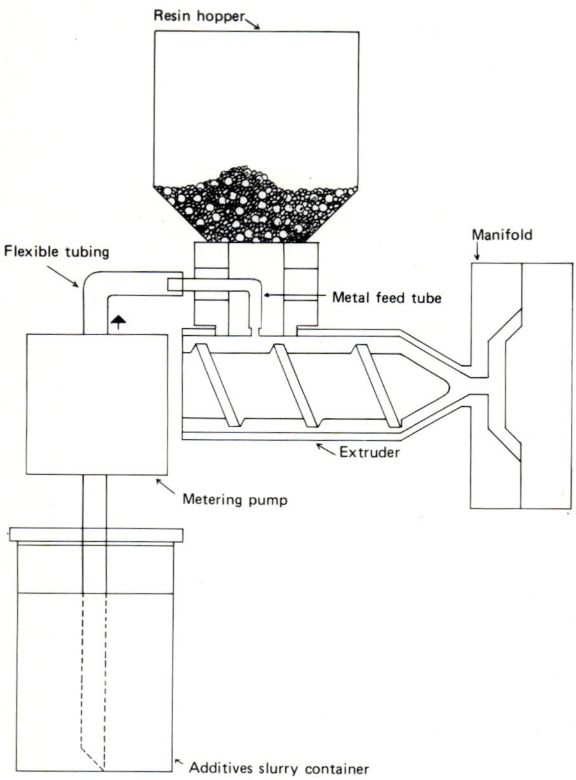

Fig. 7. Schematic Diagram of a Liquid-Colorant Supply System

III. EXTRUSION

"Grid" melting has been widely used for the production of thermoplastic fibers. In fact, a large portion of nylon and polyester fibers is still being produced through this process. A grid consists of a tubing heated electrically or by a suitable heat transfer fluid. The thermoplastic chip or powder falls on this heated structure, melts, then drips to a melt pool from where it is taken away by a metering pump for spinning. In the early days of development of polypropylene fiber, it was natural to adopt the grid melter for the spinning of fibers. The success was limited, however, because of the too high viscosity of polypropylene, which prevented the melt from dripping into the melt pool, and because of the sensitivity of the polymer to heat degradation. Forced grids were tried, wherein a positive pressure was applied through a screw to force the melt through the melting grid. These efforts were also largely unsuccessful, although Canadian Celanese Co. did produce commercial quantities of polypropylene fiber for a number of years through this technique. The continuous melting of polypropylene for fiber production has been successfully accomplished through extruders of one sort or the other and it is now the universal method for fiber production.

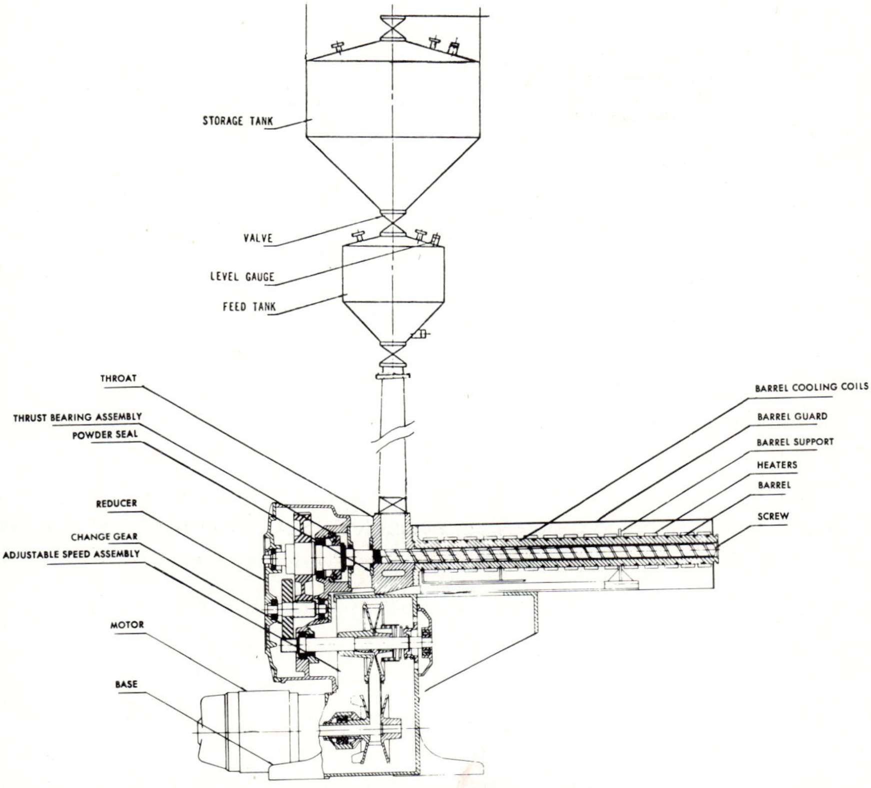

Fig. 8. Outline of a Common Melt-Spinning Extruder

Extrusion of thermoplastics consists of the melting and compressing of plastic granule or powder while it is being forced through a long passage by a screw conveyor turning in that passage. A feed opening is placed radially or tangentially at one end of this passage and an orifice at the other end is provided for the exit of the melt. The components of an extruder are shown in Fig. 8. These components and their functions are described briefly in the following pages.

For more detailed information, excellent reference works are available: some of them are listed in the Bibliography of this chapter.

Hopper: The most common form of feed hopper used on extrusion machines is the conical type with a cone angle of about 60-45 degrees. The material feeds into the screw merely by its own weight and, consequently, this type of hopper is frequently referred to as a gravity feeder. The hopper is connected with the extruder through a flexible tube. A shut-off gate is usually provided at the base of the hopper or just above the extruder feed, by means of which the supply of material to the screw may be stopped. This type of hopper feed has three disadvantages. First, the height of the material in the hopper is continuously varying and this produces slight changes in the pressure conditions in the feed end of the screw, affecting the overall quality of the melt. Second, there is always a danger of material bridging, which produces an erratic feed or even a complete stoppage of feed. Third, there is a danger of segregation of color-concentrate pellets or pigment particles from the polymer pellets or powder because of their different shapes and sizes. This danger is pronounced when dealing with pellet mixtures. One type of continuous feed is achieved by means of the vibrating feeder but this only enhances the segregation of the additive particles from those of the polymer.

The vacuum hopper is increasingly becoming popular for the extrusion of polypropylene powder blends. The hopper is of conventional design but is sealed and is placed under a vacuum. The material is fed into the hopper from a storage source in such a manner that the vacuum is not destroyed. A seal is also required in the extruder at the screw-bearing end. An air-operated, heavy-duty feed valve regulates the material feed to the hopper and the material level is maintained by means of a level-sensing device which activates the feed mechanism automatically. This type of feed arrangement is to a great extent free from the disadvantages mentioned above. Positive feed hoppers are also available. These consist of a conical hopper fitted with a rotating screw. These hoppers not only prevent the bridging in the feed throat but also keep the feed material in constant agitation, thus preventing it from segregating. Positive feed hoppers are, however, not very common in the fiber industry.

Drive: This is an adjustable-speed vari-drive motor varying in horsepower. Each horsepower can produce 5-10 lb/hr of extruded polypropylene depending on the back pressure and the extrusion temperature. The speed at which the extruder screw rotates can be continuously increased or decreased with the help of magnetic, electric, or mechanical devices actuated by screw pressure signals from a pressure transducer and control system, shown schematically in Fig. 9. The power source usually puts out speeds well above those needed for extrusion. Therefore, a transmission is needed to reduce the speed to a useful value while

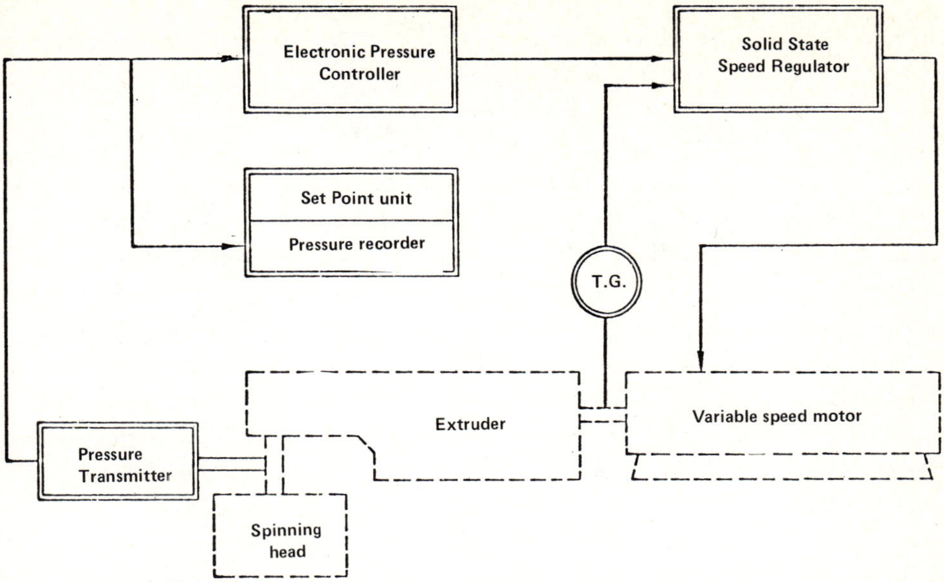

Fig. 9. Pressure Control Circuit in a Spinning Extruder

transmitting the power to the screw. Both belts and gear boxes are used. Common extrusion speeds are in the region of 10-50 rpm.

Extruder Mounting and Screw Support: Most extruders are mounted in a steel or cast-iron frame designed to be bolted to a concrete floor. In such mountings, the screw is in a horizontal plane. The extruders can, of course, be mounted vertically as shown in Fig. 10. This is, however, practical only in the case of small and medium sized extruders. The screw fits in a thrust bearing behind the hopper. Recently some extruders have been put on the market wherein the screw fits in the melt exit-end of the extruder, as shown in Fig. 11. This arrangement has certain advantages over the conventional extruders. The feed end of the conventional screw requires a deep feed-zone channel, resulting in a screw root of reduced diameter and low torque-bearing strength. With the discharge end of the extruder at the free rather than the driven end of the screw, as is the case in conventional feed-end driver screws, the extrusion pressure acts over the entire end of the screw to create an axial thrust that varies linearly with the extrusion pressure. Increasing extrusion pressure, therefore, requires superior thrust bearing. An extruder driven from the exit end of the screw, shown in Fig. 11, eliminates these two problems. The torque can now be applied at the end of the screw having the larger root diameter and higher torques can be accommodated. Since the discharge end of the metering section does not have the screw end exposed to the extrusion pressure, very little thrust is exerted upon the screw. Two additional features of this design are the reduced floor space when the screw is vertical, and the integral positive feed screw hopper.

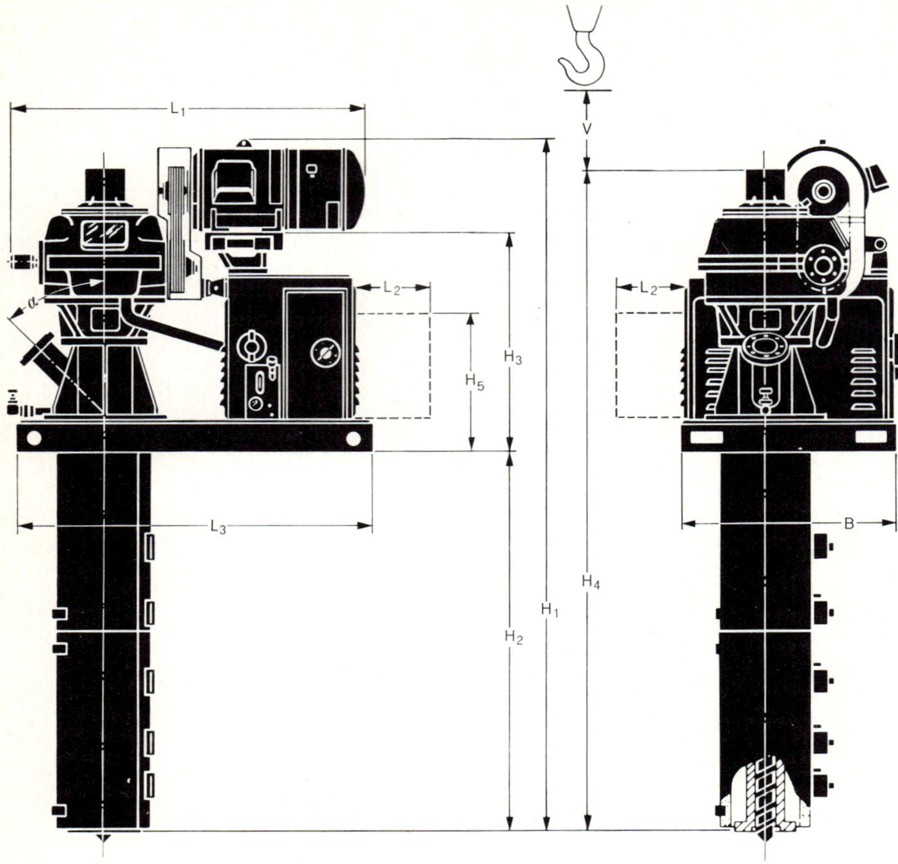

Fig. 10. A Vertical Single-Screw Extruder for Melt-Spinning of Polypropylene (Source: Barmag)

Barrel and Vents: The screw rotates in a barrel with a very hard and resistant metal inner surface. Actually, the barrel consists of heavy cylinders lined with a corrosion-resistant hard alloy; such as x-alloy, finished to a smooth surface. In the early days of polypropylene extrusion for fiber spinning, short length and large diameter extruders were commonly used. Figure 12 shows such an extruder popularized by Zimmer AG for the manufacture of polypropylene and polyester fibers. Barrel lengths have been growing longer and longer in the past decade. They are now fairly well set at 20 to 25 times their diameter. The exit-end driven extruders and vertically arranged conventional extruders may afford a higher l/d ratio. The end of the barrel lining is exposed and usually forms the material seal, along with the breaker plate.

Polymer blends may contain some surface moisture, and the melting of additives may give off some gases. Numerous systems have been proposed to let off these

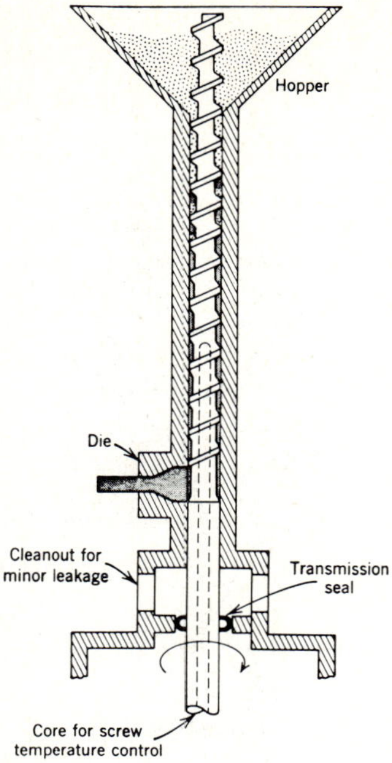

Fig. 11. Schematic Diagram of Egar Front-End-Drive Extruder

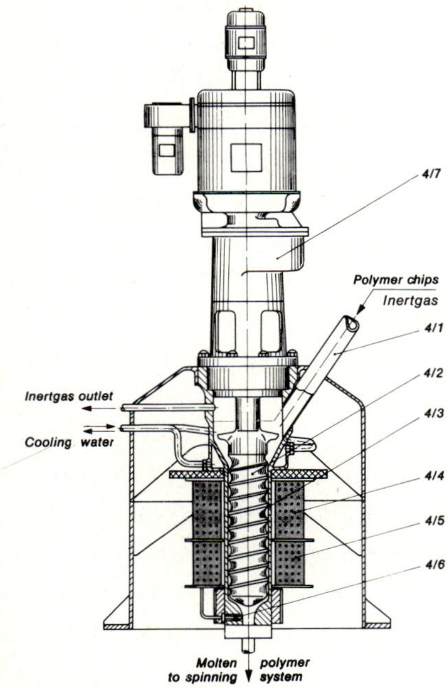

Fig. 12. A Short-Screw Melter for Polypropylene (Source: Zimmer)

gases. The vacuum hopper described previously is one such attempt. The "vented" barrels and screws are two other systems useful for this purpose. In the vented barrels, the gaseous material is vented off through the barrel at a high pressure area in the cylinder. In the vented screw, the volatiles are extracted through a hole drilled in the root of the cored screw. The venting systems require careful screw design, the operation is more critical, and variation of processing conditions is limited. These are not, therefore, commonly used in the spinning of polypropylene fibers.

Heating and Cooling Systems: Most machines are electrically heated through band-type resistance heaters around the barrel. Cast heaters (insulated resistance wire cased in metal and ceramic blocks) and tubular resistance heaters wrapped around the barrel are also used. A recent development that has grown in importance is induction heating. With this method the heat is actually produced in the barrel walls. It offers several advantages, such as ease of temperature control and rapid rate of cooling and heating. Even with the advances in the concept of adiabatic extrusion, which allows the heat to be produced within the polymer through compression and shearing, heating systems remain necessary. They are needed to permit startup and to insure temperature control in operation.

Cooling of the feed section of the barrel is as important in the extrusion of polypropylene as heating is in the subsequent sections of the barrel. This is to prevent premature melting of the polymer and bridging of the feed throat. The basic cooling system includes copper tubing carrying circulating cold water around the barrel.

Because of the current trend to faster speeds, larger machines, and more internal mixing, barrel temperatures may exceed the control points and initiate polymer degradation. Various cooling systems have been designed to overcome this kind of overheating. They include air blasts on the barrel, water cooling of the barrel and the water cooling of the cored screw. In extrusion operations for fiber spinning purposes, overheating does not generally occur. If this problem does arise, it is solved through the modification of screw geometry or lowering the throughput rate rather than cooling with water or air.

Screw: This is the heart of the extruder and the part that has the widest variation in design. The screws for single-screw machines are full-flighted, i.e., they have turns over their entire length. These turns are usually made with a constant pitch, equal to the screw diameter. An extruder screw can be visualized as performing four important functions.

(1) It is a conveyor. As a conveyor it can be imagined as screwing itself backwards out of the mass of material. Since the screw itself is held in place, the material does the moving.

(2) It is a compressor. Feed, whether pellets or powder, is higher in bulk

than the melt because of the included spaces between the feed particles. The screw must compress the feed together, pushing the air back out of the feed while conveying the mass forward along its axis.

(3) It is a melter. Probably the most important function of an extruder screw in melt spinning is that of a continuous melter, replacing the old-fashioned grid melter often used in the spinning of nylon and polyester fibers. The melting may occur through positive heat input across the barrel or through the frictional heat generated within the material.

(4) It is a metering device. Metering a precise amount of melt at a steady rate is an important and essential feature of an extruder used in the plastics industry. As the metering function is taken over by metering pumps in the spinning of fibers, the importance of metering by the screw may appear to be minimized. The metering action of the screw is needed, however, even in the presence of separate metering pumps because of the essential requirement of maintaining steady melt pressure upstream the metering pump.

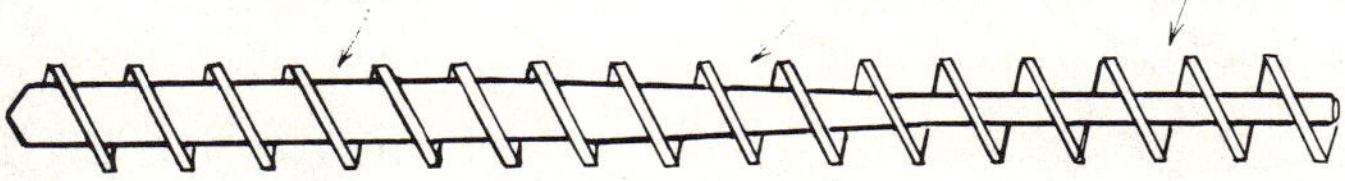

Fig. 13. A Full-Flight Extruder Screw

The entire length of extruder screw is conventionally divided into three main sections according to the functions which they predominantly perform (Fig. 13). They are:

(1) Feed zone

(2) Transition zone (sometimes also called compression zone)

(3) Metering zone

The polymer powder or pellets fall into the feed zone, which conveys them some distance down the barrel. Here the material is heated by conduction until it is ready for melting in the next zone. The feed zone consists of equally spaced channels of the same depth. Thus, no compression of the material takes place in this zone, which occupies generally one-third of the entire screw length. It is longer for pellets and shorter for powders. Feed materials differ widely in their physical form and may be supplied as free-flowing powders, regular cubes, random-cut chips with a percentage of fines, etc. The feed section has to deal with materials whose form can be any of these. It is important that the feed zone be capable of conveying sufficient material to keep the metering zone full. On the other hand, it is equally important to ensure that the supply of material from the feed zone is not too great to overrun the metering zone.

A pronounced departure from this balance either way will result in surging or pressure pulsation.

The transition zone, or compression zone, follows immediately after the feed zone. It can be formed by a gradual increase of the root diameter of the screw thread until the diameter of the metering section is reached. The compression zone is designed not only to compact the material but also to conform to the rate of melting and the volume change as the material passes from the solid to the viscous state. Heat is produced by mechanical working and aided by conduction through the barrel walls. This zone is commonly about one-third to one-half of the screw length for polypropylene, although much shorter transition zones have also been recommended. The selection of a proper compression ratio is mainly dictated by the bulk factor of the feed material. Compression for polypropylene may vary between 2:1 and 4:1, depending whether pellets and chips, or powder or small granules, are to be used as feed material.

The metering zone has constant cross section, like the feed zone, but it has a much smaller channel depth. In it, the melt is smeared and sheared to give a melt of uniform composition and temperature. The length and depth of metering zone varies widely. Longer metering zone and shallower channel depths help damp out any variation in the flow rate produced in earlier zones, cause good mixing, and deliver a melt of uniform composition and temperature. However, the heating that results from the high shear in such configurations tends to increase the melt temperature beyond the control limits. This produces excessive degradation of the polymer. On the other hand, shorter metering zone and deeper channels are inefficient mixers, melt pressure fluctuations are noticeable, and metering action is rather poor. However, the melt temperature can be controlled to a desirable value, output is generally higher for the given screw speed, and the degradation of the polymer is low. In view of these considerations it is sometimes advisable to utilize a screw with longer and shallower metering zone in the production of fine denier filament yarns, where good mixing and the uniformity of temperature is of utmost importance. A screw with shorter and deeper metering zone would probably be beneficial for the production of yarns for industrial applications and for the production of staple fibers.

As might be expected, it has been found that the screw configuration (helix angle, depth, pitch) most suitable for one form of material is not necessarily the best for another. The most efficient helix angle for the metering zone of an extrusion screw is 30°, but since the majority of screws are of the constant pitch variety, the usual compromise helix angle for the whole screw is based on the angle of the feed zone. A helix angle of 17.7° and pitch equal to the diameter is most commonly employed. The depth of the flight relative to that of the metering zone controls what is normally referred to as the compression ratio.

In addition to the screw design, the peripherial speed of the screw is an important variable in the performance of an extruder not only in the movement of material but also in establishing the amount of heat generated by shear and friction. The output of an extruder does not measurably increase in direct proportion to the increase in screw speed and power input. An economical speed which gives maximum output per unit of power input is, therefore, used as the operating speed. This will, of course, vary with material of different viscosities, screw-back pressure, melt temperature, and screw design.

Other Extruders: In addition to the single-stage, single-screw extruders described above, there are some other types of machines which are potentially of great interest for continuous melting, plasticating, and metering of polypropylene for the production of fibers. The most important among them are the Maillefer and the twin-screw extruders.

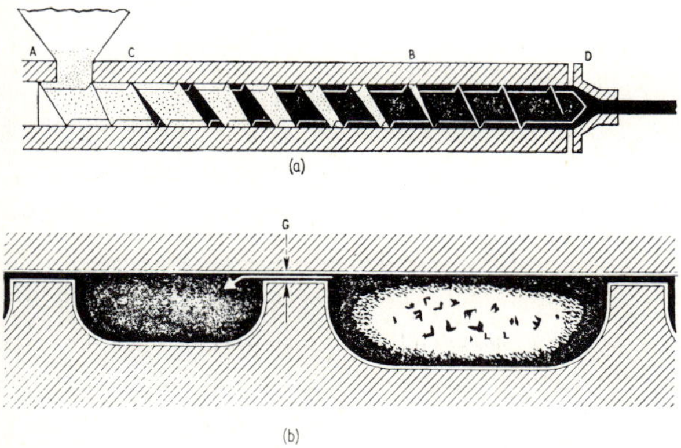

Fig. 14. Diagrammatic Arrangement of the Maillefer Screw showing (a), the flight arrangement, and (b), the transfer of the material between channels.

The Maillefer screw extruder (Fig. 14) resembles a common single-stage, single-screw extruder described thus far. However, the screw consists of two separate channels cut in the same shaft, one of which connects with the feed hopper while the other supplies the melt to the manifold. The feed channel width decreases progressively until it disappears entirely at a point about two-thirds along the screw length. This channel may be regarded as feed and part transition zones. The width of the output channel increases progressively from zero at a point just downstream of the feed hopper and attains full width at a point adjacent to the position at which the feed channel width is zero. The only connection between the two channels is over the flight lands which can perhaps be regarded as the transition zone. The viscous melt which forms in the feed channel in contact with the barrel is transferred to the output channel over the flight land so

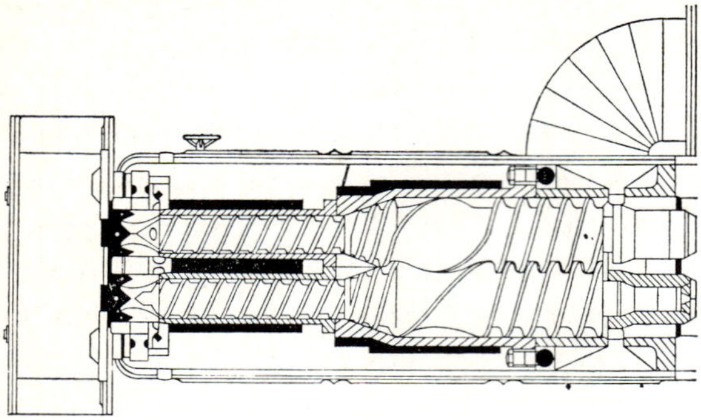

Fig. 15. A Twin Screw Extruder showing the arrangement of the Krauss Maffei DSM 11/150 machine

that in effect one channel is filled with fully melted material and the other with material in either granule or transition form. The heat transfer from the screw and barrel wall is, therefore, much increased and all material reaching the output channel will have been subjected to substantially similar heating and shear effects.

In twin screw extruders (Fig. 15), two intermeshing screws operate in opposite direction, forcing the melt forward at relatively low pressure while shearing it as it passes between them. The latter action is similar to a roll mill. The positive conveying allows very viscous melts to move through narrow channels at a relatively low pressure. This makes it possible to extrude polymers at low temperatures, despite their high viscosities at these temperatures. The melt can pass through the twin screw extruder at low shear rate, thus building up little heat and avoiding undue decomposition.

From a purely practical point of view the arguments for and against multiscrew extruders, as compared with single-screw machines, are difficult and controversial. Multiscrew machines will accept difficult feed materials more readily than will single-screw, heating can be better controlled, and compounding efficiency can be increased without polymer degradation. However, because of their higher cost, limitations of thrust bearings, drive complications, and the difficulty of avoiding dead spots in the extruder, their use has not been widespread except for compounding and pelletizing purposes.

IV. SPIN MANIFOLD

The molten polymer emerges from the exit end of the extruder and enters into a spin manifold. A spin manifold consists of distribution channels which supply the polymer to individual spin positions. Each spinning position comprises pump block, gear pump, filter pack, and spinnerette set. Theoretically, the melt can be divided into any number of spinning positions. Indeed, spin manifolds attached to one extruder and supplying the polymer melt to 32 or even 64 spin positions are common for large-scale production of polyester and nylon fibers. For the spinning of polypropylene fibers, however, such large spin manifolds are generally not employed. Manifolds of 4, 8 or at the most 16 positions are more common. This is because most of the polypropylene fibers currently being produced are pigmented products and the lot sizes of individual colors are rather small. At the excessively large extrusion rate realized by using large extruders and spin manifolds, the frequency of color change would be too high, excessive amounts of waste would be generated, and a significant portion of production time would be lost in down-times.

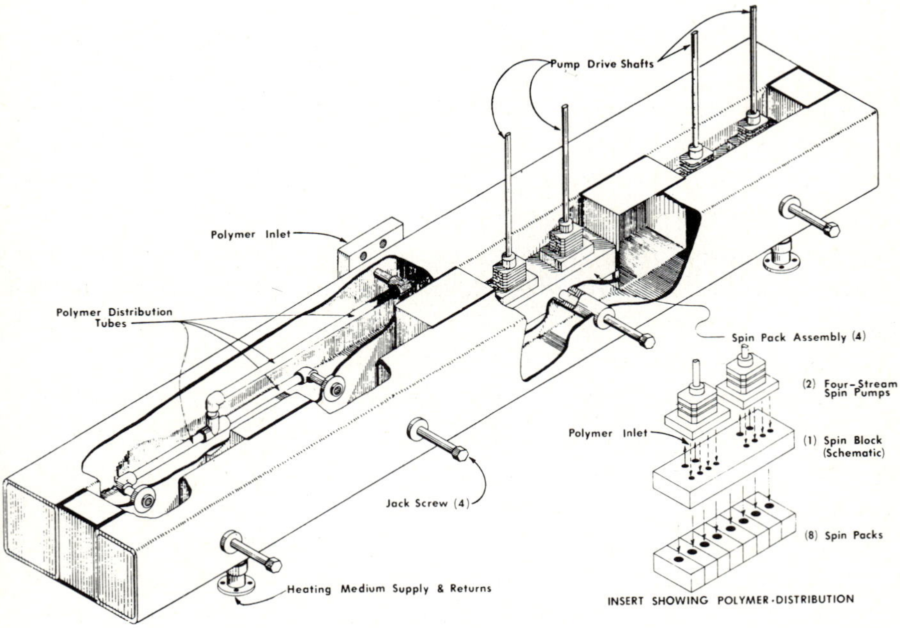

Fig. 16. Assembly of Melt Spinning Manifold. Four-Position Manifold With 8 Ends Per Position (cut-away showing spin pack arrangement). (Courtesy: Bouligny Co.)

Individual spinning positions may be equipped to take one, two, four, or eight spinnerettes and may be able to spin up to 16 threadlines (Fig. 16). For staple spinning, generally one or two threadlines per position are employed but for continuous filament yarns, a minimum of four threadlines per spin position is used. Spin manifolds capable of extruding eight threadlines per position are increasingly becoming common for the production of fine denier continuous filament yarns. The number of metering gear pumps per position for spinning of multiple threadlines depends on the design of spinnerette pack and type of pump used. Multiport metering pumps are now available which are capable of feeding one, two or four spinnerettes.

The whole spin manifold is uniformly heated by heat exchange vapors. The boiler can be an integral part of the spin manifold (generally for small units) or it can be a separate unit. In some designs, electric heating for individual spin positions is also provided. This is to allow the spinning at high temperatures without getting into polymer degradation problems.

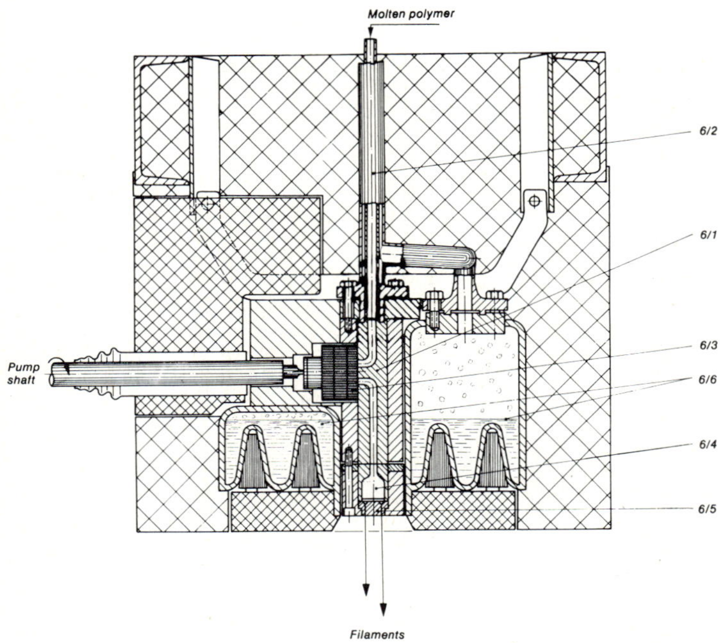

Fig. 17. A Typical Spinnerette Pack Assembly (Courtesy: Zimmer)
6/2 Polymer melt inlet, 6/1 Pump block, 6/3 Metering pump,
6/6 Dowtherm heating medium, 6/4 Spinnerette pack

Figure 17 shows the details of a typical spin beam (i.e., the spin manifold). A pump block (6/1) is flange-mounted on the spin position (6/2) of the manifold. One or more gear metering pumps (6/3) mounted on the pump block, meters the polymer to the spinnerette set (6/4). A filtering takes place before the melt reaches the spinnerette (6/5). Both the pump block and the spinnerette pack assembly are heated by the heat transfer medium (6/6) of the spinning system.

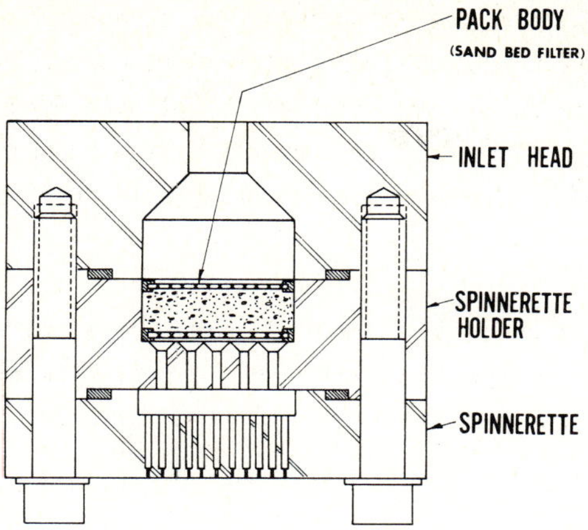

Fig. 18. A Typical Sand Pack Filtering System for Melt Spinning

The polymer melt generally contains appreciable amounts of impurities and pigment agglomerates which, if allowed to pass through the spinnerette holes, rapidly plug the spinnerette and make the spinning impractical. The melt has to be filtered, therefore, before it is extruded through the spinnerette. Many different types of filters are in use for the filtration process. Fine sand pack and stainless steel screens are particularly popular. A variety of configurations for the retention of sand have been proposed. One example is depicted in Fig. 18.

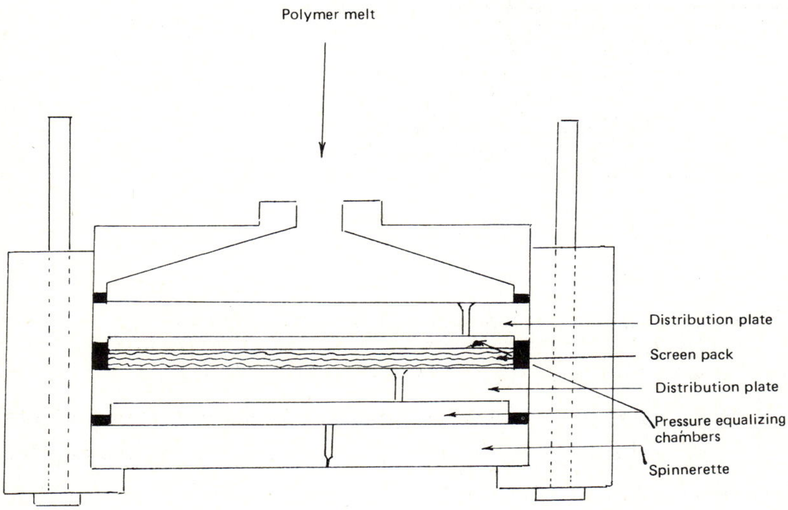

Fig. 19. A Typical Spin Pack Assembly With Screen Filter

Opinions differ as to the merit of sand pack over metal screen packs. It appears that graded screen packs have some functional and handling advantages over sand packs. The graded screen packs consist of metal screens which are stacked in such a way that the melt meets ever finer screens as it travels to the spinnerette. The finest screen size is generally of 200-400 mesh for fine denier filament, or 100 mesh screen for coarser filaments. The progressively fine screens are sandwiched between coarse screens for support. Figure 19 shows a typical spin pack with screen filter.

Another important filter medium is sintered metal discs or tubes of appropriate porosity (Fig. 20). Their use has not been widespread so far because of their high cost.

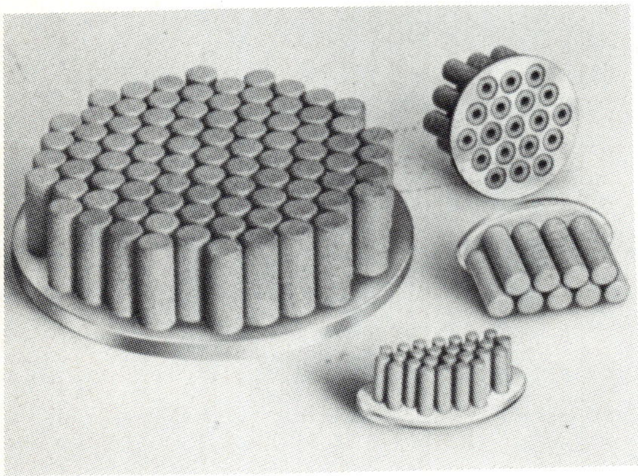

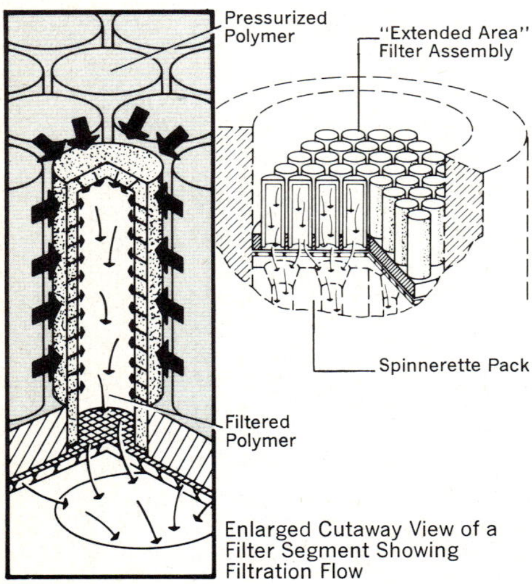

Fig. 20. Sintered Metal Tubular Filter Elements (Extended Area Filtration) (Courtesy: Mott Corporation)

As the filter area in the filter pack is limited and as polypropylene melt generally contains large amounts of particulate impurities and oversize pigment agglomerates, the rate of plugging is generally high. The excessive pressure buildup necessitates frequent change of filter packs which increases the downtime of the machine. To overcome this practical difficulty, a prefilter is commonly inserted at the exit of the extruder (Fig. 21). A large variety of these filtration systems are currently available in the form of automatic or semiautomatic screen changers.

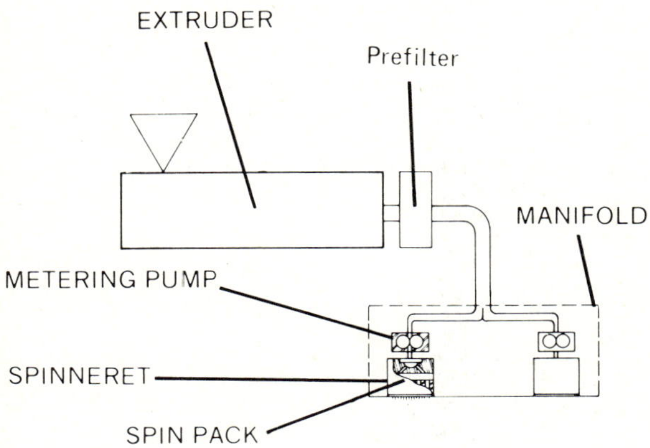

Fig. 21. A Prefilter Interposed Between the Extruder and the Manifold (Courtesy: Bouligny Co.)

These screen changers are automatic to the extent that as soon as a certain predetermined screw pressure is reached due to filter blinding, a sensor triggers the screen changing automat which pushes a new, clean filter into the path of the melt while the used filter is being pushed out of the way. When this happens, there is an immediate, dramatic change in operating pressure as the flow returns to its original pressure level. The entire line must now readjust, stabilizing momentarily at the lower pressure before the new filter too begins to blind and the entire pressure/temperature cycle begins anew (Fig. 22). The yarn produced during the stabilization period is often unacceptable due to the change in the denier of the threadline and to small variations in color. These automatic screen shifters, therefore, do not enable one to continue spinning while the filter change is occurring automatically. Nevertheless, these screen shifters, manual or automatic, are recommended on the basis that they considerably prolong the life of the spinnerette pack filters and drastically reduce the frequency of pack change.

Continuously advancing screen filters are also available. Figure 23 shows the principle of one type of such continuously advancing screen filters. A more advanced system is shown in Fig. 24. Although a continuously advancing

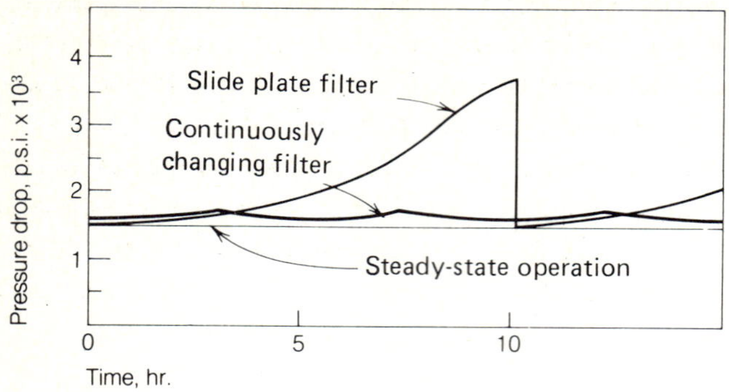

Fig. 22. Pressure profile of a typical automatic, slideplate screen changer. Note an abrupt change in ΔP at the end of a cycle. Compare this behavior with the pressure profile of a continuously changing filter. (Courtesy: Berlyn Corp.)

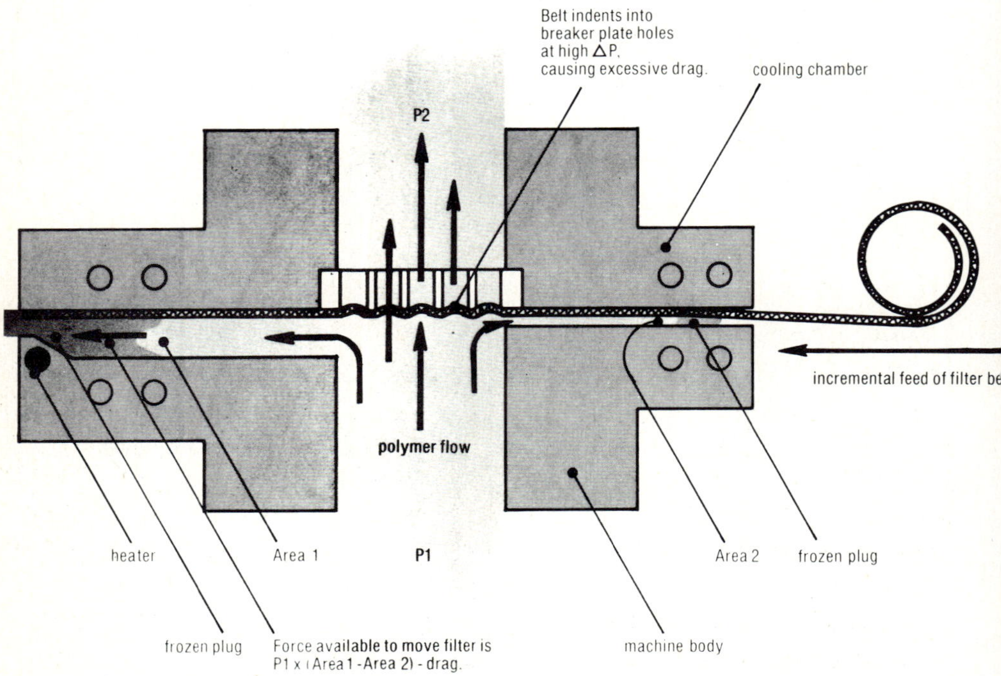

Fig. 23. The schematics of a screen belt type, continuously advancing, filter system (Courtesy: Berlyn Corp.)

filter system entails considerable capital expenditure, its use is strongly recommended on the basis of considerable reduction in waste and machine down-time. This applies especially to high capacity, multiple position machines.

The metering pumps used for the spinning of multifilament yarns are positive displacement, gear type pumps (Fig. 25). Their metering efficiency is relatively insensitive to minor changes in pressure differential, ΔP, between the screw pressure and the spinnerette pack pressure. However, if large variations occur either in the screw pressure or in the pack pressure, the metering displacement per revolution of the pump gear will change. This has its effect on the denier of the threadline, as discussed in Chapter VI. Displacement characteristics of gear metering pumps differ from manufacturer to manufacturer. A flatter curve of displacement vs ΔP is obviously more desirable than a steeper relationship.

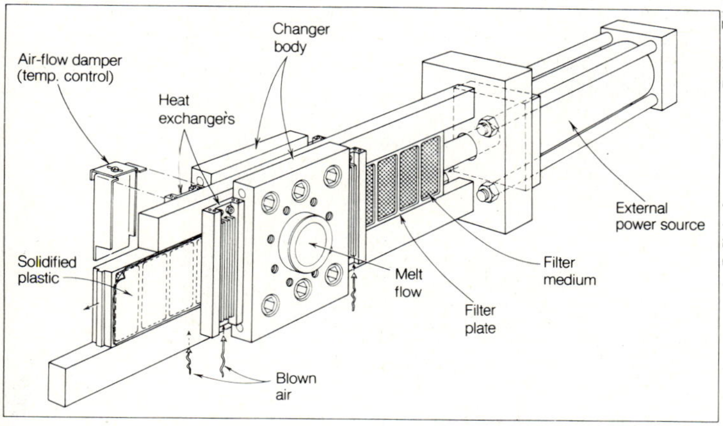

Fig. 24. Diagram of Berlyn Continuous Filter System (Courtesy: Berlyn Corp.)

Another factor bearing upon the metering uniformity of a gear pump is the choice of an appropriate drive. The metering pumps situated in the spinning head require appropriate driving power. The most important requirement of this drive is absolutely constant angular velocity, even when there are considerable fluctuations in the relatively high torque. It must be possible to switch individual pumps on or off while in operation and this must not involve even a temporary stoppage of the remaining spinning heads on the machine. Mechanical as well as electric drives are being used, the use of the latter being on the increase. In mechanical drive, one single motor drives a large number of spinning pumps by means of extensive gear boxes and shafts arrangement. Electric drives reduce the relatively high expenditure on mechanical drives and the costly array of pumps. The speed change is accomplished through the use of variable frequency inventors.

For melt spinning of polypropylene, the spinnerette is usually a flat plate, round or rectangular, flush or recessed in its mounting, although the use of thimble-shaped spinnerettes, with the cone end protruding into the melt, have also been described (Fig. 26). The flat spinnerettes are relatively thick, of

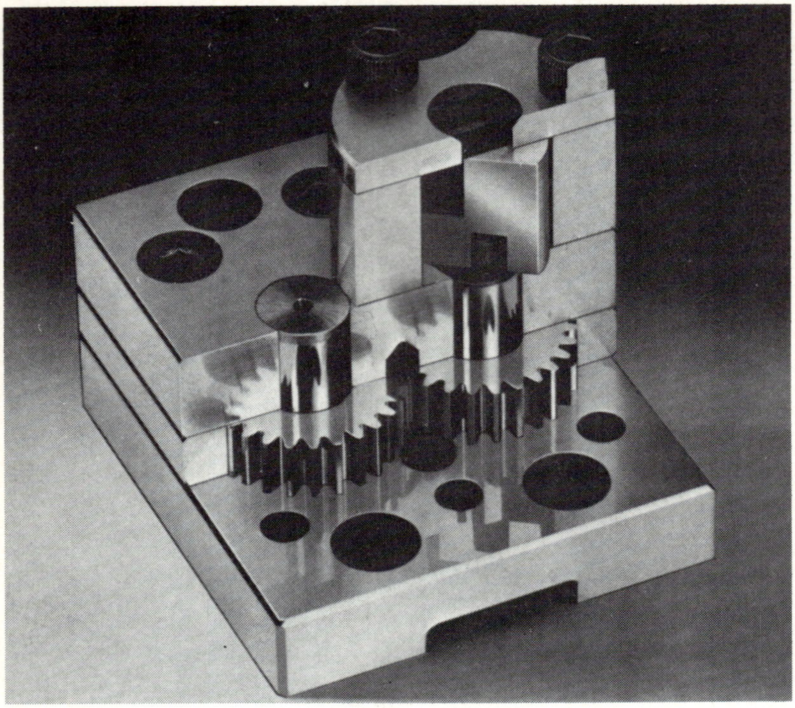

Fig. 25. Gear-Type Metering Pump for Melt Spinning

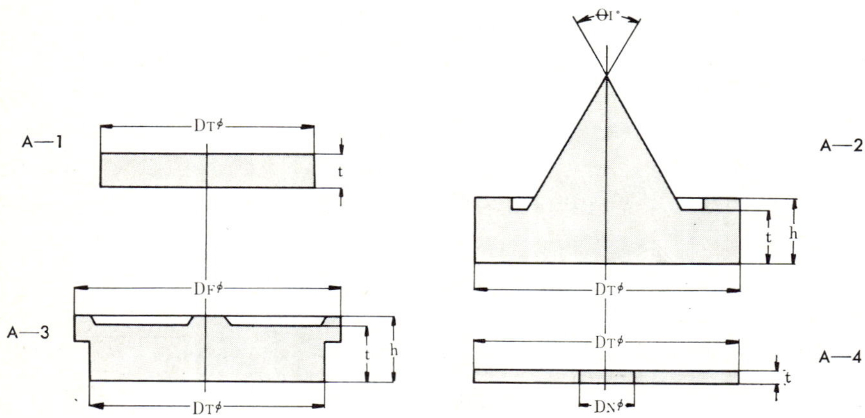

Fig. 26. Typical Forms of a Spinnerette (Courtesy: Nippon Nozzle)

the order of one-quarter to one inch thickness, but the cone-shaped spinnerettes may be thinner, between one-eighth and one-half inch. The individual holes are in the range of 0.010 - 0.030 inch equivalent diameter, depending on the throughput rate, melt viscosity, filament size to be produced, and the fiber cross section. The exact dimensions of spinnerette holes are normally a proprietary information: the discussion in Chapters V and VI can be used to reach a spinnerette design for any specific use. Higher throughput rates, higher viscosities, and lower spinning temperatures require larger-size holes. As discussed in Chapter V, the approach angle and length of the counterbore is as important as the capillary. Figure 27 shows some commonly used shapes. For holes of noncircular cross section, a long capillary is difficult to obtain. In such a case, a short capillary and relatively flat entry is used.

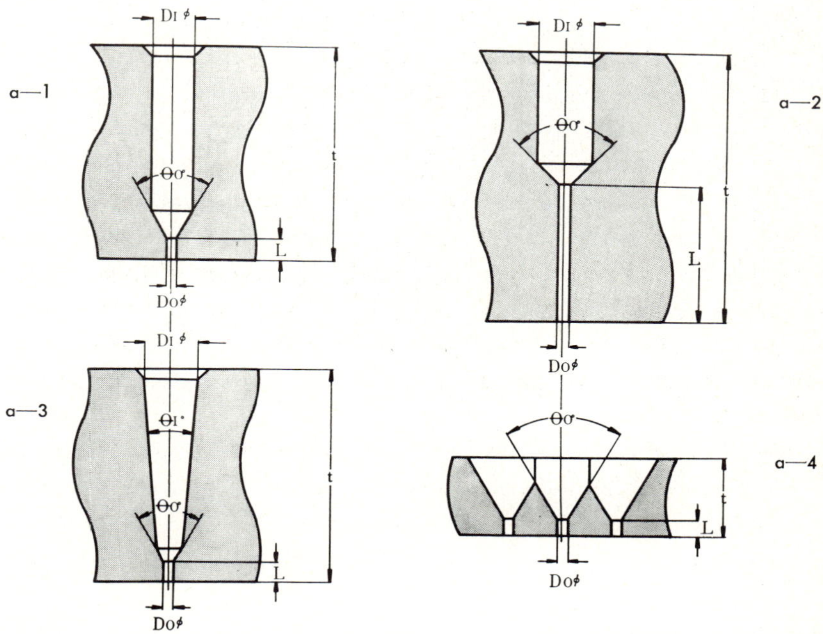

Fig. 27. Typical Shapes of "Counterbore" or "Capillary Approach" for Round or Simple Profiled Cross Section of Spinnerette Holes (Courtesy: Nippon Nozzle)

The number of holes in a spinnerette may range from a few for monofilaments to a few tens for filament yarns to a few hundreds for staple fibers. The maximum number of holes in a spinnerette of a given size is generally lower for polypropylene than that for polyester and nylon. This is because of two reasons: (a) the low melting temperature of polypropylene and the consequent difficulty in quenching, and (b) the relatively large die swell. However, by employing refrigerated air for cooling and using narrow molecular-weight polymer, a larger number of holes can be accommodated per unit area of the spinnerette. The pattern of holes in the spinnerette is important, particularly in the case of profiled filaments. Any particular pattern is dictated by the quenching arrangement employed and the direction of air flow. Thus, using a circular quenching arrangement, a concentric-circle geometry of holes layout may be used; for undirectional air quenching, rectilinear rows of holes are indicated. The angle between the rows of holes and the air flow direction would depend on the profile geometry of the filament cross section.

While we are still on the topic of spinnerettes and spin pack assemblies, it is worthwhile to make a few comments on the cleaning procedure for spinnerettes and other parts of the spin pack. The most commonly used methods incorporate a two-stage cleaning operation wherein the bulky polymeric material is removed in the first stage, and the final cleaning is done using "clean room" techniques.

Almost from the day of inception of polypropylene spinning, the large parts of the spin pack have been cleaned by burning off the parts in a furnace while salt bath and nitric acid have been used for the cleaning of spinnerettes. The salt bath is a molten mass of potassium nitrite, sodium nitrite and sodium nitrate mixture in 53:40:7 proportion. Nitric acid is used in seventy-five percent concentration. In both cases a 500°C temperature is used for a time period of about one to two hours. Although both of these methods accomplish satisfactory cleaning, the highly corrosive nature of the bath media result in a decreased life of the hardware. In addition, the spinnerette is exposed to high degrees of thermal shock which tends to distort the shape of the unsymmetrical cross section of the spinnerette holes.

In recent years three relatively new systems have been introduced as a result of the need for a better cleaning system. The T.E.G. or solvent system has proved to be successful for nylon and polyester but its effectiveness for polypropylene has been poor. The vacuum pyrolysis method was introduced in the early 1970s by Beringer Corp. Its effectiveness for nylon and polyester has been limited but for polypropylene it truly appears to be the solution. The vacuum pyrolysis jet cleaner, depicted in Fig. 28, utilizes heat and vacuum to remove the polymer from the spinnerette and the pack parts. The major portion of the polymer charge is trapped; the remaining polymer is vaporized and appropriately collected by means of water mist scrubbing. A fluid bed cleaning furnace was also made available (Procydyne Corp.) in the late 1960s. The system claims to offer the best features of both the molten salt bath and vacuum burn-off furnaces in that the heat transfer rates approximate those of a liquid salt bath and the cleaning is done in an oxygen starved atmosphere which will not support combustion or the resulting hot spots, particularly in the capillaries. The practical experience with polypropylene has, however, not substantiated these claims.

In the above-mentioned cleaning operations, the small holes in the spinnerette will contain materials inert to the burning-off and volatization process

Fig. 28. Vacuum Pyrolysis Jet Cleaner (Courtesy: Beringer Co.)

and must be removed by other methods. This is done with powerful air or steam blasts or, more effectively, with an ultrasonic cleaner. It is a device in which thorough and powerful mechanical agitation is provided to a liquid bath with the help of ultrasonic waves. Ultrasonic cleaners using magnetostrictive transducers are easier to operate and yield more consistent results as compared to transducerized tank - generator combinations. The ultrasonic waves, which are pressure waves, move through the bath and induce cavitation. It is this cavitation which produces a scrubbing action at the surface of the object being cleaned and releases energy, in the order of 15,000 psi, to loosen and lift dirt, pigment residues, or the contamination.

In summary, the pack cleaning operation consists of burning off the larger parts in a furnace and dusting off the ashes subsequently. The spinnerettes are treated in a salt bath or in a 75 percent nitric acid bath at 500°C for one to two hours, washed off with water and then treated in an ultrasonic cleaner for thirty to sixty minutes, depending on the power of the apparatus. Alternatively, the spinnerettes are burnt-off in a vacuum jet cleaner for one to two hours at 400°C and then finished in a ultrasonic cleaner.

V. SPINNING AND TAKE-UP SYSTEMS

The extruded polymer jets emerging from the spinnerette holes are drawn and simultaneously quenched with cold air. The filaments thin down to a desired denier and are solidified in about two to four feet from the spinnerette. The filament bundle passes over the finish applicator and is eventually collected by a suitable take-up arrangement. The attenuation force is supplied by a uniformly rotating take-up godet. All these elements are available in a variety of form and design, depending on the process used, patent coverage, machine manufacturers preferences, and the fiber manufacturers experience.

Control of the rate of cooling, and particularly uniformity of cooling between the filaments, is very important if consistent yarn quality is desired. Various arrangements have been described to achieve uniform quenching of the molten filaments with an air stream. The quench cabinet can be cylindrical, tapered cylindrical, half-moon, or rectangular. The air can be directed unidirectionally across the path of the filament (cross flow), radially directed inward (in-flow), radially directed outward (out-flow), along the threadline (draw-flow), or opposite to the threadline traveling direction (counter-flow). The most popular quench cabinets are rectangular with a unidirectional flow of air. One such arrangement is shown in Fig. 29. This quench cabinet can be divided longitudinally to accommodate multiple threadlines or shelved horizontally, with threadline slots, to ensure turbulence free air flow.

To protect the threadline from stray draft and promote uniform circulation, the entire filament bundle may be enclosed in a protective chimney. Unlike nylon, polypropylene fiber does not require steaming to stabilize the yarn and prevent it from "growing" on the package. The length of the chimney, if used at all, is therefore quite short for polypropylene fiber spinning.

The individual filaments are converged at a guide after solidifying, and come in contact with a finish applicator before reaching the forwarding roll. A finish applicator conventionally consists of a ceramic roll slowly rotating in a trough of finish (Fig. 30), or it may consist of a metal or ceramic slot into which the finish is pumped by a metering pump. The latter arrangement ensures a uniform and quantitative application of finish to the threadline although the system may be somewhat expensive to install. The schematics of such a system are given in Fig. 31.

The spin finishes are composed of suitable lubricating oils, antistatic agents, emulsifiers, and antibacterial agent. The lubricants may be of vegetable origin or completely synthetic chemicals. The mineral oils, commonly used in other synthetic fibers, are not suitable for polypropylene as these oils may swell the fiber and be absorbed into the body of the fiber.

The fiber finish has profound effect on the processability of the yarn as well as the tactile properties of the product. Spin finishes differ from producer to producer and often are carefully guarded secrets. In general, they are not simple treatments but rather complex mixtures. Some of the spin finish formulations developed for nylon and polyester fibers of comparative denier, fiber

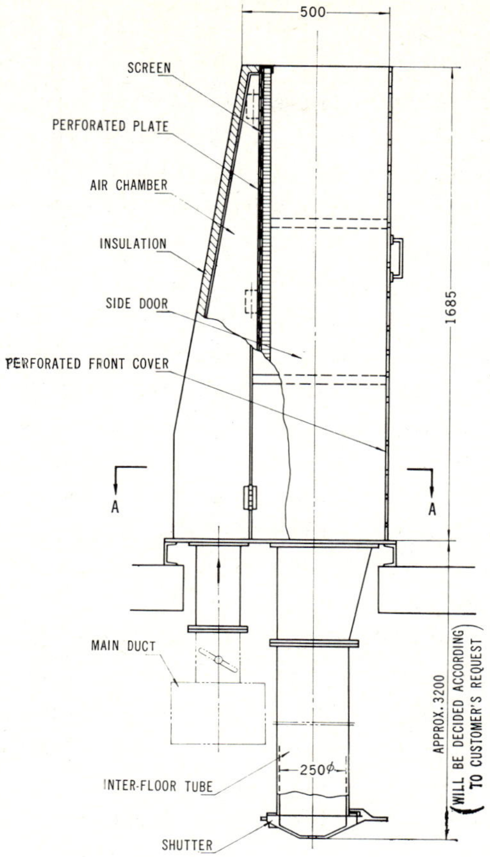

Fig. 29. A typical Cross-Flow Type Quench Cabinet (Source: Mitsubishi Heavy Industry)

cross section and use are also applicable to polypropylene, but in the majority of cases new formulations specific to polypropylene fibers have to be developed.

The spin finish must achieve a balance of properties which is not obvious at first glance. An effective finish must build a good package, make the fiber drawable at economical speeds, control the static accumulation, be stable to aging, be contributory to even dyeing and bulk developing, be nonyellowing, and be physically suited towards even application. It may be seen that simple materials may achieve one or perhaps two or three of these properties but be deficient in the other requirements. A balanced formulation of several different components is, therefore, needed. As no single formulation is suitable for all fibers, specific finishes are often developed for each end use and each denier of the fiber. The formulation of fiber finishes revolve around three fundamentals involved in the processing of staple fiber and yarns:

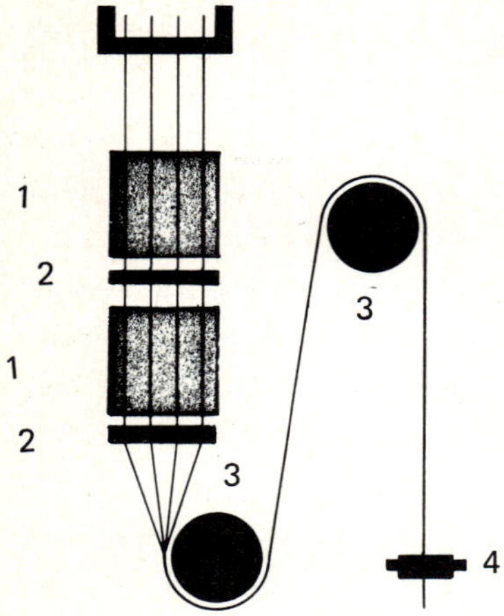

Fig. 30. A Conventional Finish Application and Take-up System
1. Lick roll finish applicator
2. Slub catcher and guide
3. Take-up rolls
4. Guide

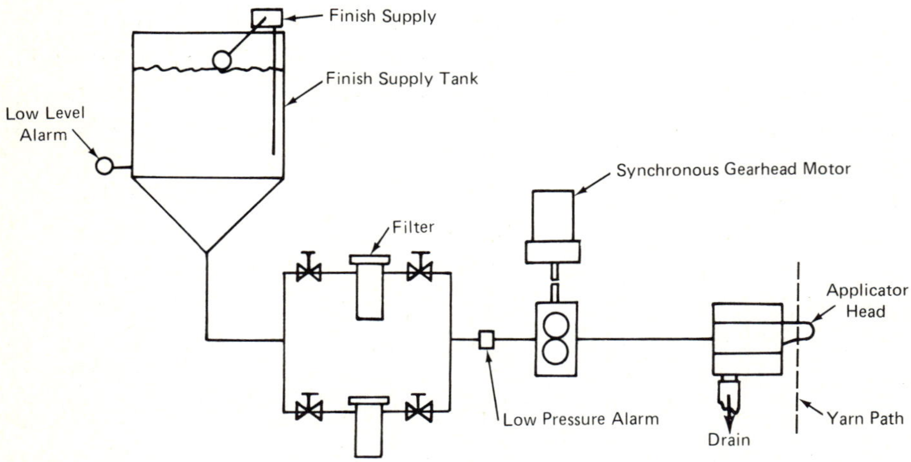

Fig. 31. Schematics of a Metered Finish Application System (Courtesy: Bouligny Co.)

(1) Fiber-to-fiber friction

(2) Fiber-to-metal friction

(3) Static

Fiber-to-fiber friction is important in such processing steps as picking, carding, drawing, drafting, spinning, and tow-to-top converting. It may be visualized that, in these operations, fibers are rubbing against their neighbors. The degree of friction, whether large or small, influences to a critical point the operating performance. Fortunately, a control is available in the form of fiber lubricants to establish a level of friction which embraces a wide latitude of acceptable variables and thus allows the production of an acceptable product.

A different area of fiber-to-fiber friction is concerned with continuous filament yarns. This may be illustrated by some examples within the fiber manufacturing plant: package building in spinning and filament drawing or tow drawing are the major steps where the fiber-to-fiber friction is of critical importance. In yarn processing, yarn delivery in coning, stitch formation in knitting, filament damage in braiding, strength and elongation in cordage, slippage of weave in fabric, yarn-to-fabric friction in sewing, are some of the areas where yarn-to yarn friction is important. It will appear obvious that a critical balance must be achieved to establish an acceptable performance.

Fiber-to-metal friction is often directly responsible for successful fiber and yarn performance. Tensions in coning and rewinding, tension of the ring travelers in spinning and twisting, friction of the warp in the loom, friction of the fiber on the card clothing, friction of the fibers in the pin drafter, and friction of the yarn on the knitting and tufting needles, are examples of yarn operations in which fiber-to-metal friction is important. In the fiber manufacturing plant, fiber-to-metal friction is important in yarn and tow drawing and several texturizing operations. A knowledge of the fiber-to-metal friction is essential in achieving advances in processability of the fiber. It is the task of the fiber engineer to blend all of the desired performing characteristics into a useful lubricating finish.

Polypropylene fiber, like other synthetic fibers, is prone to accumulate static charges that interfere not only in the spinning and drawing processes, staple cutting, and yarn texturizing, but also is the source of trouble in all subsequent processing of the fiber in textile mills. Both mechanical and chemical means are used as static control tools in the present-day plants. Radioactive eliminators and electric neutralizers are quite useful in handling continuous filament yarns in various stages of their processing. To a vastly greater extent, however, chemical means are used. The chemical antistatic agents are a part of the spin finish. These compounds may be anionic (e.g., alkylphosphates), cationic (e.g. quaternary ammonium salts) or nonionic (e.g., polyoxyethylene derivatives of sorbitol monolaurate and hexaglycol stearate) in nature. The choice of any particular group depends on several factors; the most important being the compatibility with other ingredients, odor, physiological reactions, and any deposition of solid material on the processing machinery. It should be noted that antistatic agents often bring about a change in frictional characteristics of the fiber.

The take-up system basically has two functions: First, it attenuates the molten filament to a desired denier and imparts orientation to the fiber. Second, it collects the solidified filaments on a suitable package or in a drum. Three types of take-up systems can be distinguished:

(1) Spin take-up which collects the individual spinning threadlines on high speed winders,

(2) Spin-draw take-up, which collects the individual threadlines on high speed winders after drawing the yarn to a desired draw ratio,

(3) Can take-up, which collects the individual spinning threadlines from all the positions of a spin manifold and lead the yarn in a can or a drum in the form of a tow.

Simple spin take-ups consist of two rolls rotating in opposite but equal speeds followed by a winder (Fig. 30). Spin-draw take-ups consist of a minimum of two, but preferably three, rotating godets. Each godet is electrically heated and temperature controlled and each godet is equipped with a freely rotating idler roll for separation of wraps around the godets. The second godet rotates about three to four times faster than the first godet. The third godet, if provided, may rotate at slightly lower speed (for controlled shrinkage of yarn), the same speed (for stabilization of yarn), or at slightly higher speed (for two-stage drawing). Instead of the combination of godet and idler, a pair of driven godets can also be provided. Figure 32 shows a typical spin-draw take-up for four-ends spinning.

The can take-ups consist of one rotating roll per spinning position delivering the yarn to a common forwarding nip roll, trio, or tooth gear roll system, which dumps the tow in a coiler drum (Fig. 33). This arrangement is obviously suitable for the production of staple fiber only.

A wide selection of winding machines, which are distinguished by the method of winding used and constructional details, is available for the various speed ranges for spin take-up and spin-draw take-up system. Basically, cross-wound spools can be formed either by friction drive or spindle drive. For frictional winding the spool lies on a friction roller by which it is driven at approximately circumferential speed. Generally, the friction rolls and traverse of a group of winding stations are driven by a common power source. The number of double strokes of the traverse remains constant and the spool rotation speed becomes less and less as the diameter is increased.

The package pattern resulting from frictional winding is known as random wind. A particular characteristic of random wind is manifested by a constant angle of lay with an increasing spool diameter, which results in an ever-differing traverse ratio. The advantage of frictional winding lies in the simple design

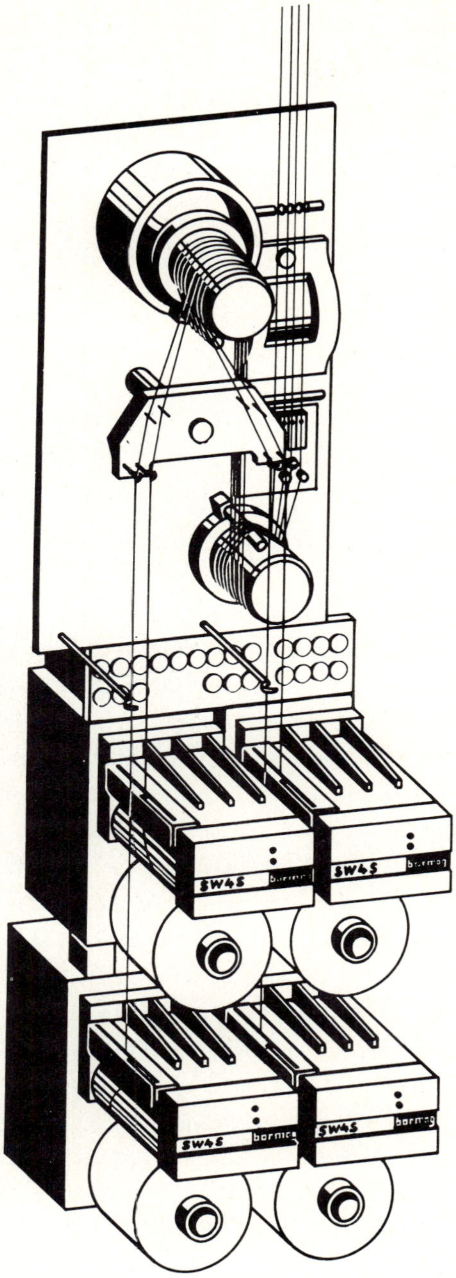

Fig. 32. A Conventional Spin-Draw Take-up Machine (Source: Barmag Machinenfabrik)

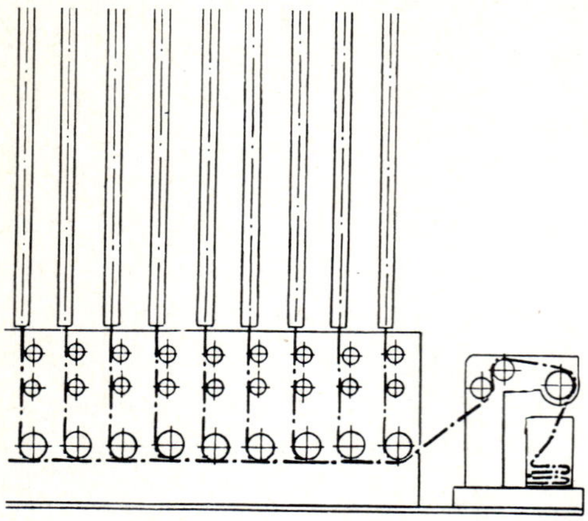

Fig. 33. A Tow Take-up System for Staple Fiber Production

of the winding unit and the low cost per station. The disadvantage of random winding is the uneven appearance of the package and somewhat less desirable unwinding properties.

With spindle-driven winding the spool is individually powered, and a constant relationship exists between the spool spindle speed and the traverse. With increasing spool diameter, the spindle speed reduces simultaneously with traverse speed. The result of the mechanism is a precision wound package.

A particular characteristic of the precision wound spool is a constant traverse ratio with a changing angle of lay as the spool diameter increases. The advantage of precision winding lies in the good-looking package and the excellent unwinding properties of the spool, even when unwound overhead at high speeds. Only the relatively high cost per winding station can be mentioned as a disadvantage. Filament tension regulation is included as a feature of precision winding. This tensioning is generally effected by a dancing arm which, according to its position, controls the setting of a sliding-armature motor or, with high winding speeds, a variation in current through the driving motor to regulate the motor torque and the winding tension of the yarn.

Equipment manufacturers are going to large spinning units and higher take-up speeds to improve the economics of fiber production. Recently, Leesona Corp. (USA) has introduced a precision winder capable of speeds up to 3,600 m/min and several machinery manufacturers in Europe have offered surface-driven, frictional winders capable of speeds of up to 4,000 m/min for spin-draw winding of textile yarns. Spinning take-up machines with as much as 16 winding positions per spinning cell have been made available. Other recent innovations in this field are the automatic end-transfer and doffing take-up machines. The capital cost for these sophisticated high speed winders is naturally very high.

It should be evident from the preceding discussion that a modern melt-spinning system is a complex array of sophisticated machine components. These components are arranged in such a way that a minimum of floor space is occupied and a minimum of length is traversed by the spinning threadline. A vertical arrangement is the most logical one in order to meet these requirements. Figure 34 shows a typical spinning position with four-ends per position take-up. The spinning department can be visualized as consisting of rows of 16, 32, or 64 of such take-up units, each take-up unit being fed by one spinning position and each extruder feeding to 4, 8, or 16 spinning positions. For large-scale production, the spinning blocks of all units can be heated and controlled through the same vapour heating system, and the pumps, finish applicator, take-up godets, and winders can be driven through common electric systems.

For the melt spinning of staple fiber and tow products, high productivity is achieved by using:

(1) As large a spin-filter pack and spinnerette as is compatible with spinning machine dimensions,

(2) As large a number of filaments per spinnerette plate as can be cooled effectively,

(3) An optimum number of spinning positions per machine matched to the melt supply capacity of the extruder,

(4) A high spinning speed, consistent with fiber orientation needs and handling capability.

Spinning units for staple fibers capable of polymer throughput rates up to 1.5 lb/min/pos are common. Spinnerettes of up to nine inches in diameter with up to 1,000 holes or rectangular spinnerettes with up to 3,000 holes have been developed. Tow collecting systems capable of linear fiber speed of up to 2,000 m/min have been reported.

Unlike polyester and nylon, which can be spun at as high a speed as the take-up system is capable of handling, the limit on spinning rate of polypropylene is reached much earlier than the capabilities of modern high-speed take-up units. If this spinning limit is exceeded, the filament cross section deteriorates, thick-and-thin filaments begin to appear, and spin continuity drastically suffers. Another limit on spinning speed of polypropylene is the relatively slow cooling and solidification rate of the filament. Thus, the high-speed take-up equipment for filament yarns and staple tow, important as it is in principle, in practice is not as important for polypropylene fibers as it is for nylon and polyester. Depending on the type of polymer used (melt index, narrow or broad molecular weight, additives, etc.), and denier per filament, 500 to 1,000 m/min spinning speed or 1,500 to 2,000 m/min spin-draw speed is probably an optimum range.

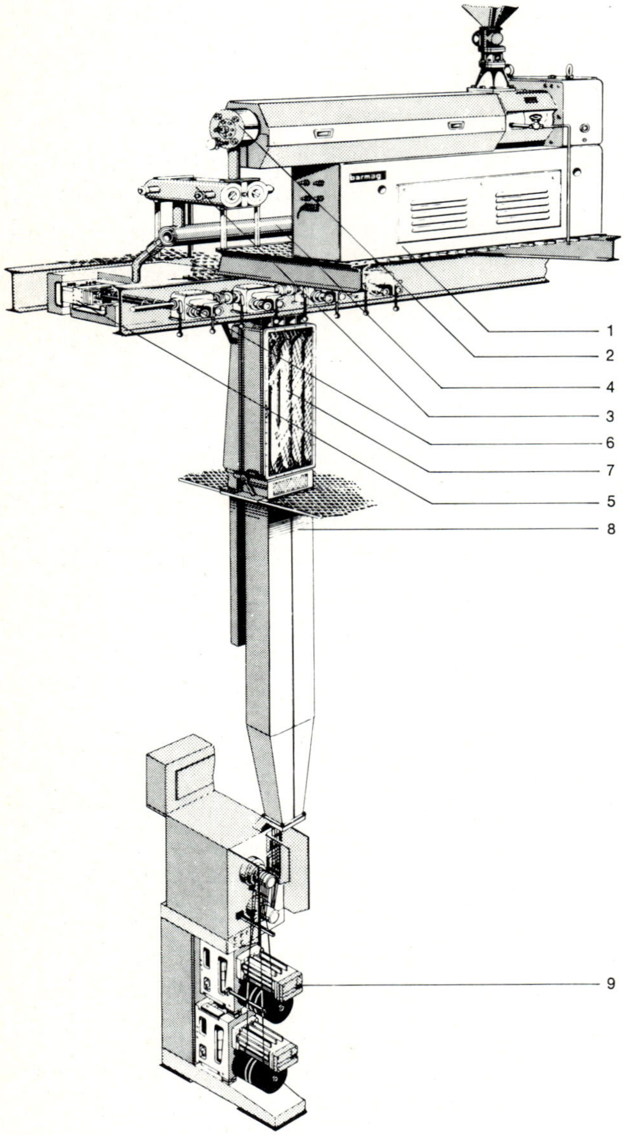

1. *Extruder*
2. *Temperature and Pressure Sensor*
3. *Prefilter*
4. *Melt-Distribution Manifold*
5. *Spin head*
6. *Pump drive*
7. *Cooling chamber*
8. *Spin chimney*
9. *Winders*

Fig. 34. A Conventional Melt-Spinning and Take-up System (Source: Barmag Maschinenfabrik)

VI. POST-SPINNING OPERATIONS

Filament drawing, heat setting, texturizing, twisting, and packaging are some of the important post-spinning operations for the production of filament yarns. Tow drawing, crimping, heat setting, staple cutting, and baling are the common steps involved in the production of staple fibers.

Yarns and tows are drawn by stretching between two sets of rolls. The draw ratio, which is expressed as the ratio of the surface speeds of the feed and draw sets of rolls, respectively, is generally obtained by mounting two sets of rolls of equal size on the driven shafts. Unlike nylon fibers, which can be drawn without the application of heat, polypropylene yarns and tows need the application of heat for cold drawing operation. This, of course, does not apply to specialty fibers, such as high-shrinkage fibers and conjugate yarns. Heat can be applied to the yarn and tow through heated godets, heated plate, hot oven, steam jet, and hot pins. The choice of the heating method is a matter of economics and depends on the nature of the product and the speed of the process. In common practice, heated godets are extensively used and have proved adequate. Hot ovens are used for the drawing of tows and steam jets are probably suitable for fine denier yarns. Although heated plates are common for the drawing of industrial yarns, the trend is away from them and toward heated godets. Heated pins are not recommended because of their changes in surface characteristics and the low speeds of drawing attainable with them.

A typical arrangement for drawing of filament of yarns is depicted in Fig. 35. For multistage drawing, one or two additional pairs of rolls are added in sequence. Of course, the yarn can also be taken up on winder packages (draw winding) instead of twister bobbins (draw twisting), as shown. The draw twisters or draw winders, equipped with more than two sets of godets, can be used for heat relaxation or annealing of the yarn after drawing. These operations are important in the production of low-shrinkage yarns and industrial yarns of high performance, described in Chapter IX.

Draw-twisting and draw-winding machines are commercially offered by several textile equipment manufacturers. Most recent models emphasize increased operating speeds and capacity, improved precision drives, and greater flexibility (staged draw zones). Draw twisters with spindle speeds of up to 12,500 rev/min and draw-winders with speeds of up to 1,200 m/min are on the market; machines with three separately controlled draw zones are also available. Draw twisters operable with industrial yarns of up to 3,000 denier at a draw speed of 700 m/min

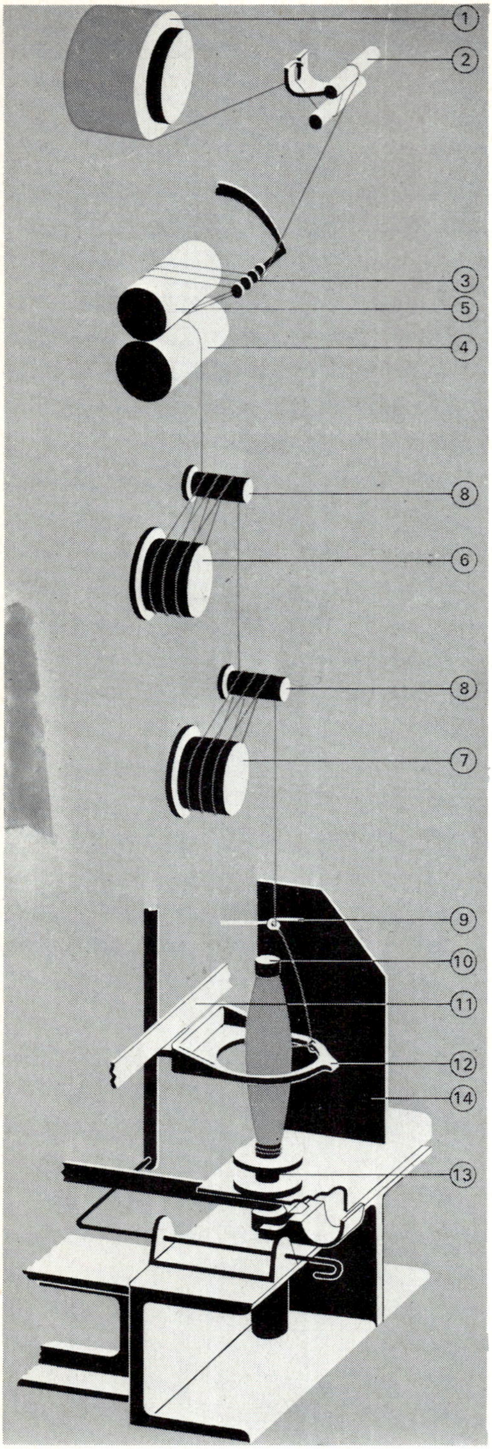

1. Undrawn, spun yarn
2. Pretension guide
3,4,5. Pretension rolls and guides
8. Idler roll for wrap separation
6. Heated feed roll
7. Heated draw roll
9-13. Twisting take-up

Fig. 35. A Draw-Twisting Arrangement (Courtesy: Barmag)

and yielding 11 lb packages are now being offered.

Draw-winding units and components have been developed for operation combined with the filament-spinning process (spin-drawing), where even higher speeds are required to reach an economically attractive melt-spinning capacity. A combination of a freely rotating idler roll with a positively driven heated roll is normally employed as a set of feed and draw rolls. For high-speed draw winders, a pair of positively driven heated rolls is sometimes used instead of roll-idler combination. This is more so in cases of spin-draw equipment. Draw twisters can handle only one end per position. The draw-winders, on the other hand, can process as much as four threadlines per position. Grooved idler rolls or grooved heated rolls are advisable in those cases where more than one end per position is being processed.

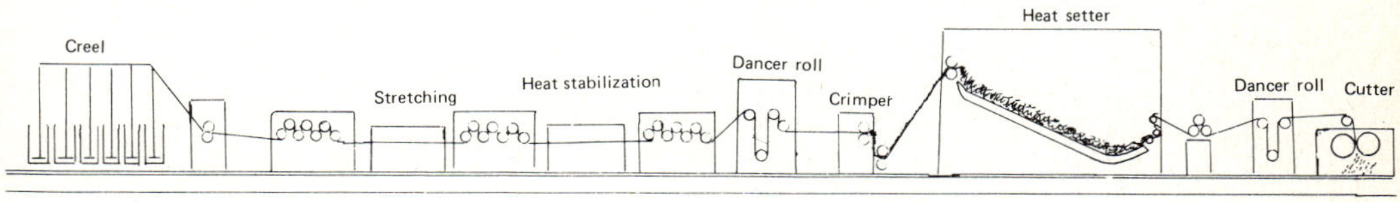

Fig. 36. A Typical Tow Processing Line

A common arrangement for tow drawing is shown in Fig. 36. Several ends of spun tow obtained in the melt-spinning process are combined to feed the drawing, crimping, heat-treating, or relaxing processes for the production of a drawn tow. The feed godet set is always heated, generally with steam, and where three sets are used, the second set is also heated. The heating of the third set is optional. The conduction of heat from the surface of the heated rolls is generally not enough at high speeds of tow drawing. A rectangular oven is therefore interposed between the feed sets of rolls and the draw set of rolls. The section between the second set of rolls and the third set of rolls is commonly used for tow relaxation or heat setting operation but, if desired, it can be used instead for second stage drawing. The drawn tow enters a stuffer-box crimper which delivers it to a can collector for batch heat setting, or to an oven for continuous heat setting. The tow is cooled and packaged for shipment to the textile mill to be converted into top on tow-to-top converting machinery, or cut into staple and bailed before shipment.

Polypropylene fibers are commonly given a setting or relaxation treatment at an elevated temperature in order to stabilize them against thermal shrinkage and in order to impart to them a permanency of crimp. Some degree of heat setting is achieved in the stuffer-box crimper where the heat generated in the stuffing process is adequate for the purpose. For better heat setting steam is injected into the crimper, which not only helps the heat setting process but also adds to the ease of crimping of the fibers. For a still higher degree of heat setting a heated J-box is normally utilized. The hot tow emerging from the crimper is collected in a J-shaped box which is heated by superheated steam. A temperature of 110° to $135^{o}C$ is maintained and a residence time of one to three minutes is strived for. The tow is continuously taken out of the J-box and fed into the cutter or plater, as the case may be. A cooling step is, however, interposed between the J-box and the cutter or plater. For a still higher degree of heat setting, a discontinuous process is used. The crimped tow is plated in large perforated boxes, which are then autoclaved at 120° to $130^{o}C$ under steam. A vacuum-steam-vacuum cycle is used for an efficient process. A steaming time of fifteen minutes is normally adequate. To meet high productivity rates, staple-fiber drawing machines have been developed to process tows of up to 1 million denier (drawn) at draw speeds of up to 400 m/min, with a 25 ton/day draw-line capacity.

Fig. 37. A "Gru-Gru" Staple Cutter
The two rotating wheels grip the tow and a flying knife cuts it. The cut staple is sucked away through a duct system to the baler. (Courtesy: Neumag Maschinenfabrik)

Cutting the tow into staple fiber may be done in a separate step, but it is usually integrated with the draw-crimp-heat treatment machines. Typical cutter speeds, in the range of 100-200 m/min, are usually the speed-limiting factor in integrated staple-fiber lines. New cutters, capable of processing tows of up to 2 million denier at speeds of up to 275 m/min, have recently been offered. Apart from speed consideration, the knife blunting and the ease of change of knife blades is of primary consideration in selecting a staple cutter. The so-called Gru-Gru cutter, depicted in Fig. 37, is very common in the industry. A more efficient and simple tow cutter is that offered by Lumus and depicted schematically in Fig. 38. Where integrated staple-fiber lines are being used, each line is equipped with two or three staple cutters so that the tow can be

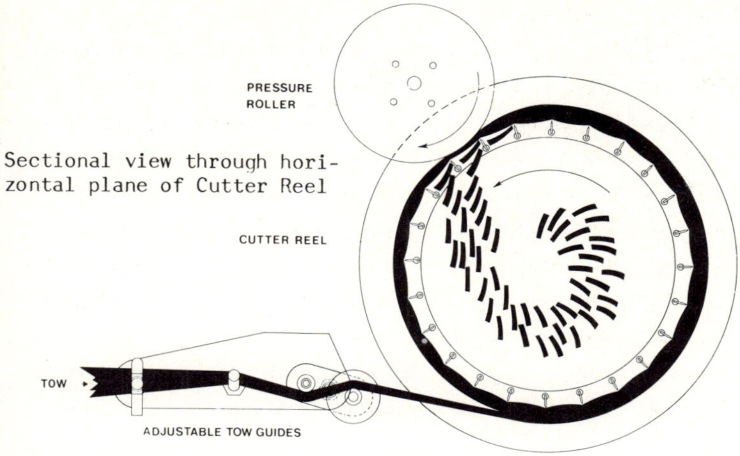

Fig. 38. Working Action of Lumus Cutter (Courtesy: Lumus Industries)

fed into one cutter while the blades of the other are being changed or sharpened.

Staple and tow may have superb physical properties and excellent potentialities, but if the fiber cannot be processed satisfactorily into yarn or nonwoven webs, it is worthless for textile use. Ease of processing is largely a matter of finish, crimp, and tow uniformity, not of inherent properties such as strength, elongation, modulus, dimensional stability, dyeability, etc. Spin finish is generally not adequate for trouble-free processing of the staple fiber in the textile mills. An over-finish is therefore applied in the crimping process.

The production of spun yarns from polypropylene staple fiber can be considered along the following three lines:

(1) The use of staple processing machinery, conventionally employed for cotton and wool.

(2) Tow processing procedures, whereby the starting material is a 200,000 to 1 million denier tow of uncut parallel filaments. The tow is cut (Pacific converter, Sydel converter, etc.) or broken (Turbo stapler), but the orientation of the fibers is not disturbed.

(3) Direct spinning of fiber bundle by utilizing tow processing principles on individual bundles of yarn.

All these processes are well known in the textile trade and need not be discussed here.

Filament and staple spun polypropylene fibers, as they come out of spinning, drawing, and cutting sections of the fiber manufacturing plant, are seldom used in this form for end-use products. A considerable variety of processing steps are generally required to convert such yarns into appropriate forms as to make them convertible into woven, knitted, and tufted goods. In general, the processing of filament yarn is much simpler and much easier than staple and tow. This statement does not apply across the board to the various texturizing and bulking processes for filament yarns nor does it apply to "direct spinning", which is closely related to stretch break or tow-to-top processing. Polypropylene filament yarns have been processed almost entirely on equipment that has been designed for the processing of acetate, nylon, and polyester yarns. These techniques of filament handling, twisting, winding, twist setting, warping, etc., are well established and the equipment used for these operations is common knowledge to one familiar with the textile trade. These topics need not, therefore, be discussed here. Polypropylene may differ significantly in its texturizing behavior from nylon and polyester yarns. A brief note in Chapter IX reviews the various techniques and equipment for texturizing of polypropylene filament yarns.

VII. SHORT-SPIN PROCESSES

The synthetic fiber industry is inherently a capital intensive industry. Because of heavy capital expenditure, the high-speed, closely controlled sophisticated equipment can be employed only for large production scale. This has been as true for polypropylene fiber production as it is for nylon and polyester fibers. The situation for polypropylene fiber, however, is changing, and it has become economically feasible to produce fibers on a small production scale and with considerably less capital cost. There are four reasons for this change: first, the polypropylene fiber producers do not have to undertake the capital intensive resin manufacturing operation which the nylon and polyester fiber producers have to; second, the technology of polypropylene fiber production is comparatively less complicated than the production of nylon and polyester fibers; third, the major market for polypropylene fibers and yarns is the nonapparel uses, which are not as critical or demanding as the apparel area; fourth, by the advent of short-spinning routes for fiber production, a production rate of as low as 50 to 100 lb/hr has become economically feasible.

It is anticipated that low-capacity operations will rapidly gain in popularity for the manufacturing of polypropylene fibers. This would give incentive to textile mills to install in-house production lines to meet internal requirements. This would be a major departure from the traditional synthetic fiber manufacturing practices. Some comments are, therefore, warranted on these so-called short-spinning routes for polypropylene fiber production.

"Short-spinning" is generally referred to as horizontal arrangement of spinning equipment which is characterized by low capacity and low processing speeds. These systems are offered by at least ten machinery manufacturing companies of Europe and the United States. A typical short-spinning system for the production of carpet or industrial yarns is shown in Fig. 39. A two-inch diameter single-screw extruder is employed to melt the polypropylene-pigments-stabilizers blend. The melt is transferred to an electrically heated block where it is picked up by two gear pumps which meter the polymer melt into two rectangular spinnerettes. Each spinnerette consists of four groups of holes, thus yielding eight ends per unit of production. All of these ends pass over a common finish applicator or two separate applicators, pass over to a take-up trio set of godets, and then to individual winders. A pair of heated godets is provided above the winder to cold draw the yarn before wind-up. Claims have been made by various machinery manufacturers for short-spinning equipment capable of handling as much as 32 ends per extruder. These claims, however, seem to be exaggerated: 16 ends per unit is probably a practical maximum.

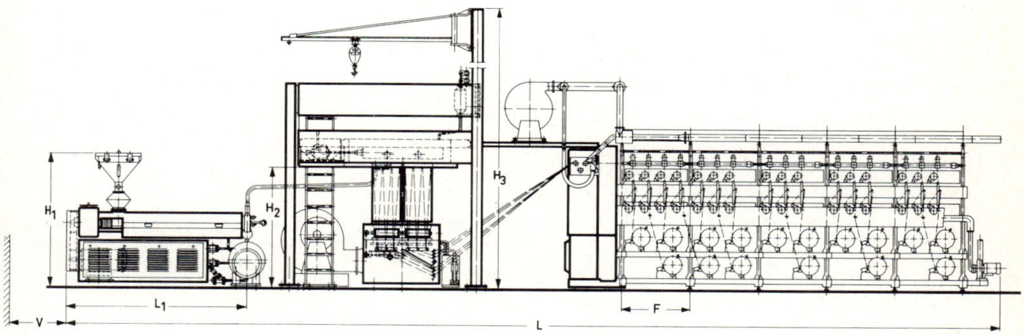

Fig. 39. A Typical Short-Spinning Unit for the Production of Continuous-Filament Yarns
(Courtesy: Barmag Maschinenfabrik)

In the short-spinning process, the extrusion rate is slow and the take-up rate generally remains in the range of 200 to 500 meters per minute. The slow spinning rate allows the filaments to solidify within a very short distance from the spinnerette: this renders the employment of long cooling chamber and the interfloor chimney unnecessary. Thus the extruder, the cooling chambers, and the take-up winders can be arranged horizontally on one floor. The slow spinning speed also makes possible the utilization of low-priced winders, thus further lowering the capital cost. As the number of ends per production unit is limited, a warp arrangement of drawing and winding is possible.

Short-spin processes are better suited to the production of staple fibers than continuous filament yarns. One such arrangement is shown in Fig. 40.

Three small melters feed the polymer melt to one rectangular spinnerette each. No metering pump is employed. The yarn bundles, consisting of as much as 2,000 filaments each, pass over a spin applicator roll, converge on a heated septet rolls arrangement, pass through a heated oven, and are drawn by another septet of rolls. The drawn tow then continuously feeds into a crimper and subsequently to a Gru-Gru cutter. The staple is blown to the baler where it is baled and stored.

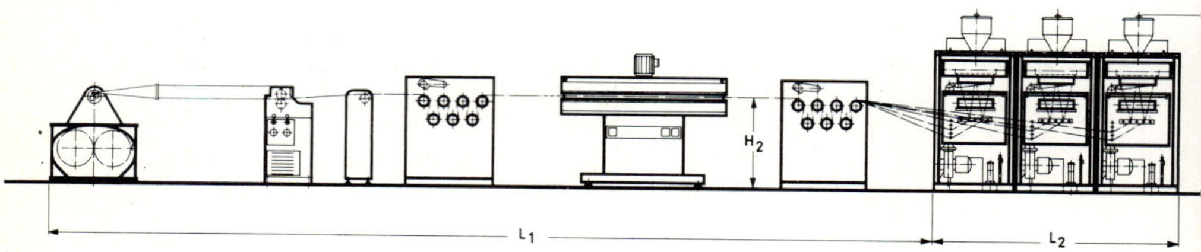

Fig. 40. A Typical Short-Spinning Staple Line (Courtesy: Barmag Maschinenfabrik)

It should be emphasized here that short-spin processes, although low in capital cost and comfortably low in capacity, are not as flexible as the conventional system of fiber production in terms of product range. Low-denier staple fiber is difficult to produce and the production of fine-denier filament yarns is out of consideration. The uniformity of the product and control of the process is also problematic. In spite of these limitations, the short-spin processes should be very attractive to carpet yarn manufacturers and to the producer or consumer of industrial yarns. These processes are ideal for developing countries where the capital is scarce, the labor is abundant, and the demand is low.

VIII. QUALITY CONTROL

Instituting a total quality control procedure is as important in the production of fibers as the manufacturing activity itself. The nature and the extent of quality control, however, differs from plant to plant and is mostly determined by such factors as the nature of the end products, total production capacity, and of course, the presence or absence of competition. A quality control organization can, therefore, be as simple as one person overseeing the general quality of the product and dictating the passing grades for each batch of the product, or it can be as extensive as available funds allow.

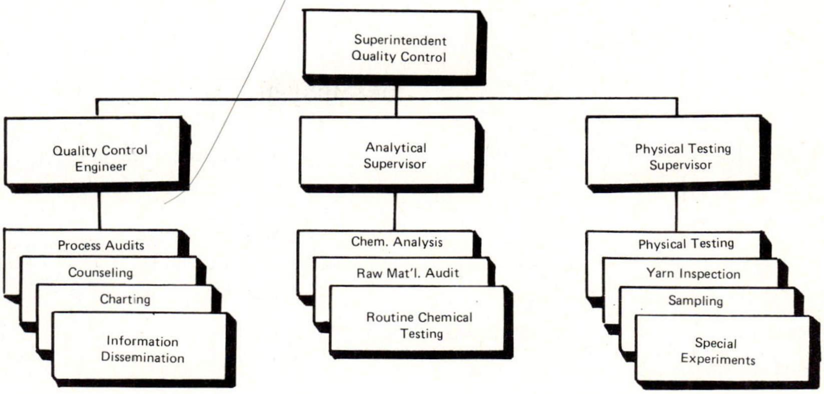

Fig. 41. A Typical Quality Control Organization for a Medium to Large Fiber-Production Plant

Figure 41 gives a functional organization chart of a typical quality control group serving a production facility of more than 20 million lb/year. Reporting to the Quality Control Superintendent is the Quality Control Engineer, Analytical Testing Supervisor, and Physical Testing Supervisor. The Physical Testing Supervisor is involved with actual testing of the yarn and staple, inspecting the product, sampling, and conducting special experiments. The Analytical Supervisor is involved with chemical analysis of raw materials and fibers. He is also involved in auditing raw material attributes and does certain other routine chemical testing. The Quality Control Engineer, sometimes also called Process Control Engineer, is involved with process audits, departmental counseling, charting, and dissemination of information.

Raw Material Control: Each incoming lot of the polymer is inspected for its color attributes, melt indices, die swell ratio, intrinsic viscosity, ash content, particle-size distribution, and processibility. The "processibility" is a nebulous term and is difficult to measure with precision. The evaluation of processibility generally boils down to an experimental spinning run under strenuous conditions. Each incoming lot of pigment is inspected for its color value, shade, quality of dispersion, and ease of dispersion in polypropylene. In the case of color concentrates, melt index is also measured. Moisture content of pigment dispersions and concentrates is also a vital control point. Finishes are difficult to characterize: viscosity, color, and dispersibility attributes may serve as control points. Stabilizers are tested for moisture content, particle-size distribution viscosity and chemical purity of the substance.

Physical Testing: Physical testing involves the sampling of the product on regular intervals and determining the denier, tenacity, elongation, shrinkage, and denier uniformity of filament yarns; fiber length distribution, crimp, neps, and tensile properties of staple products; denier, tensile properties, crimp, and bulk level of bulked continuous filament yarns. In addition, a visual observation is carried out for broken filaments, loops, thick filament sections, uncrimped sections, fused spots, etc. A statistical limit to these defects is applied and if any particular run exceeds this limit, the whole production run is down-graded. In some physical testing laboratories, certain prototype end-use tests are also performed. These include the determination of fiber-to-fiber friction, fiber-to-metal friction, finish level and small-scale carding operation. The determination of cross-sectional ratios of filament bundle (ratio of the largest to the smallest cross section in a filament bundle) and the aspect ratios of the profiled cross section is also routinely determined. In addition to the above tests, the physical laboratory is responsible for monitoring the heat and light stabilization of the product, color, and appearance ("dull" or "bright" or "fuzzy", etc.).

Process Control: The process control group makes daily checks to determine the degree of adherence to standard processing conditions. When a problem is found it is reported to the production group. The problem area is then reaudited to ascertain if the corrective action has been taken. A constant vigilence is necessary in spinning and drawing areas to make sure that all the temperatures are within the prescribed range and all the speeds are maintained as the process conditions specify. In order that the process control engineer performs his duties freely, he must not report to the production supervisor.

IX. WASTE RECOVERY

In the manufacture of polypropylene fibers some scrap material is practically always formed. Some of this scrap results from interruptions in the spinning process which are unavoidable when, for example, the bobbins are changed, the yarn breaks, extruder is being flushed for color change, etc., and some of it is produced just in the normal routine steps, such as string-up of draw twisters and tow-drawing machine. A considerable amount of waste is produced as edge trim in the production of tape and fibrillated yarns. Off-quality yarn which cannot be sold as second grade product becomes scrap. An average of 5 percent scrap is generally reached in a normal fiber producing plant. There is no sizeable market to absorb all the scrap fiber and tape produced. Since scrap is formed in large quantities, its cost is considerable and it is highly desirable to recover the scrap.

Most fiber producers manufacture either multifilament yarns and staple or film yarns and monofilament, but seldom both. The nature of the waste differs accordingly. In multifilament yarns and staple, if the drawn material could be collected separately from the drawn material it could be chopped, carded, and used as normal, mixed-length staple product. However, this kind of separation is not practical: the fibrous waste normally consists of drawn as well as undrawn fibers. Short of melting this waste and converting it into pellets, the mixed mass of textile material can be converted into a useful product by forwarding the mass at a low speed under restrained conditions, while withdrawing the mass at a high speed to separate, randomly stretch, and cool the filaments. The resulting mass of broken and drawn filaments can further be chopped or stretched broken to useful length of the fiber and used like normal staple product.

The above-mentioned simple treatment cannot be affected on the waste caused in monofil production, tape yarn production, and nonwoven fabrics manufacture. In these cases, the reclamation means changing the physical form of the waste to some new form which can be accepted by the extruder for processing into useful products.

The sprues, runners, short mouldings, etc., which are the waste produced by an injection moulding process, can be recovered merely by passing them through a granulator - a simple rotary mill in which the scrap material is sheared between stationary and rotating bars until the pieces are small enough to pass through a mesh of preselected size in the bottom of the cutting chamber. The pieces, though irregular in shape, can be mixed with virgin polymer and remoulded. Similar granulators can be used for film-yarn scrap if slight modifications are made to the feed and withdrawal sections of the machines. However, the cut material is flake-like and of such low bulk density that it cannot be fed to an extruder. The cut pieces have somehow to be compacted to raise the bulk density from about 0.07 g/ml to a value much closer to the 0.5 g/ml of normal polymer. Fortunately this is not a problem unique to polypropylene yarn; the same problem has been faced by polyethylene and polypropylene film manufacturers and successfully solved by fusing the flake produced by the granulating mill. A broad division may be made between (a) those units in which complete fusion and homogenization is achieved by the use of an extruder and (b) others in which only partial fusion of small clumps of the cut flake is attempted, so that little if anything must be done in the way of particle-size reduction.

Extrusion Technique: The sequence of the operation is as follows (Fig. 42):

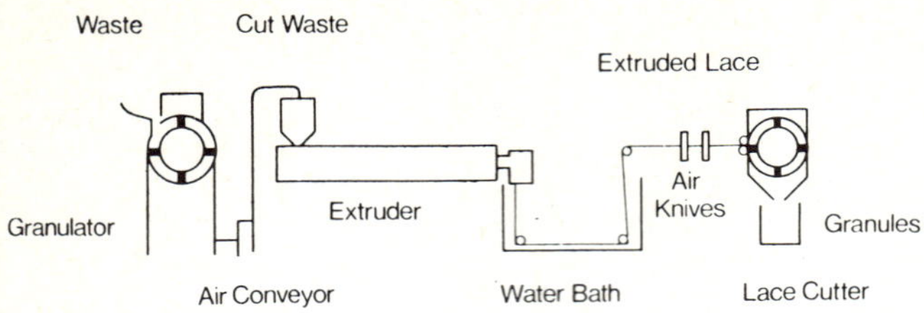

Fig. 42. Line Diagram of Extrusion Recovery Process (Courtesy: The Plastics Institute, London)

(1) Cutting the scrap in a granulator,

(2) Conveying the scrap to the hopper of a specially designed extruder. To achieve a reasonable output, particular attention must be paid to the feed section of the extruder. One version of the equipment available has the extruder mounted vertically to allow greater access of the flake to the screw, while in another the barrel is enlarged at the feed end and has particularly deep flights in the screw. The great advantage of recovery by means of an extruder system is that it becomes possible to remove contaminants from the waste by using fine filter packs in the extruder. To do this, automatic screen chargers, actuated by pressure transducers situated behind the screen, are recommended.

(3) Extrusion of one or more filaments of about 3/16 inch in diameter,

(4) Cooling in a water-quench bath,

(5) Drying by passing through circular air knives,

(6) Cutting, using a combined haul-off and cutting unit.

Incomplete Fusion Technique: In one form of the apparatus offered, the process can be described as follows (Fig. 43):

(1) Cutting the scrap in a granulator,

(2) Conveyance to fusion chamber and fusion by passing the material between rotating and stationary disks. The amount of heat generated is controlled by adjusting the gap between the disks or by regulating the flow of water used to cool them, or by both methods,

(3) Conveyance of the fused lumps produced to a second granulator.

Fines produced in the second granulator are recycled to the fusion chamber. By suitable selection of the screen in this granulator, particles of chosen size, reasonably free from fines, can be produced. Unlike the extrusion method, the bulk density of the material produced is still not as high as that of the normal polymer: a typical value obtained is 0.35 g/ml.

In another form of incomplete fusion apparatus the waste is put into a large bowl fitted with high-speed rotating blades which shear the scrap, and in doing so generate enough heat to melt it and enough air turbulence to keep the individual fused particles in suspension. The process, a batch one, is terminated

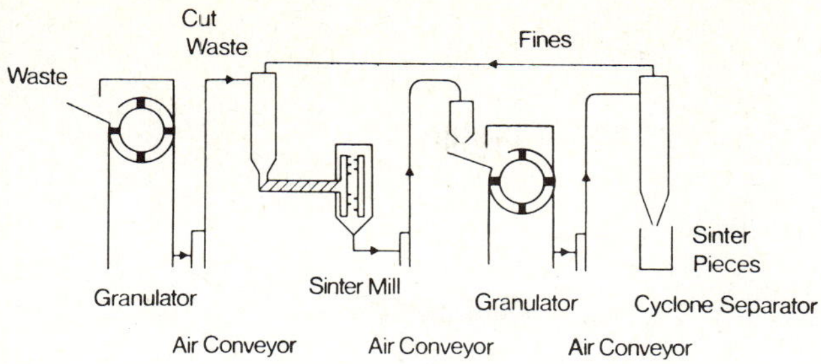

Fig. 43. Line Diagram of One Form of Sinter Recovery Process (Courtesy: The Plastics Institute, London)

by injecting just sufficient water into the bowl to solidify the particles. By careful metering of the water the particles can be cooled sufficiently to prevent their aggregation when they are discharged from the machine, yet to keep them perfectly dry all the moisture is flashed off as steam. The final part of the equipment is a granulator to break up the few agglomerates which are formed to a size suitable for subsequent extrusion (Fig. 44).

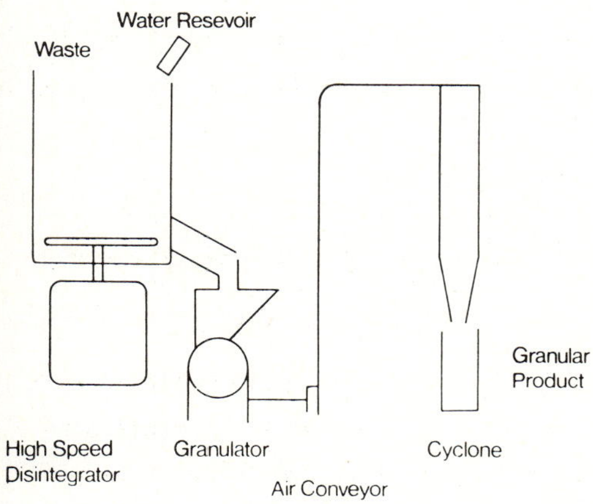

Fig. 44. Line Diagram of Batch-Method Sinter Recovery Process (Courtesy: The Plastics Institute, London)

Another variation of the sintering technique consists of chopping up the waste, mixing it with about 25 percent of polyethylene powder, sintering the mixture in a screw extruder, and feeding the sintered polypropylene-polyethylene mixture to a pelletizing extruder.

Because of the critical nature of fiber spinning, it is generally not possible to reuse the reclaimed material for the production of fiber. The reclaimed material can, however, be used in moulding processes wherein the off-color pelletized material is mixed in small proportions with the virgin polymer.

X. BIBLIOGRAPHY

X.1. Journal Literature

1. E. Rhodes, Role of Fiber Lubricants, Text. Merc., 144(1961)469-471.
2. J.E. McIntyre, Synthetic Polymers Fibers - Production, Rev. Textil Prog., 15(1963)109-131.
3. Small Scale Fiber Plants Popular Abroad, Chem. Eng. News, 42(1964)26-27.
4. F. Fourne, Melt Spinning Processes for Synthetic Fibers and Filaments, Textil-Praxis, 19(1964)1169-1178.
5. F. Fourne, Laboratory Machines and Pilot Plant for the Extrusion and Stretching of Synthetic Filaments, Chemiefasenu, 14(1964)654-656.
6. N.R. Cohen, Lubricants for Synthetic Fibers, Mod. Textil Mag., 46(1965)60-65.
7. N.R. Cohen, Lubricants for Polypropylene, Text. Inds., 129(1965)87-91.
8. Anon., Melt Spinning Acetate and Polypropylene, Chem. Eng., 72(1965)106, 108.
9. W. Albrecht, Quality Control in Fiber Production and Processing, Z. Ges. Textilind., 67(1965)773-778, 67(1965)863-868.
10. H. Bohringer, Definitive Quality Controlling Properties of Fibers and Methods for Testing, Abh. Dt. Akad., Wiss. Berlin Ks. Chem. Geol. Biol., 3(1965)223-231
11. F. Fourne, Melt Spinning Processes for Synthetic Fibers and Yarn, Textil-Praxis 1(1965)1-11.
12. F. Fourne, A New Melt-Spinning Process for Synthetic Yarns with Coarse Single Titers, Textil-Praxis, 20(1965)633-634.
13. W. Gresch, Buss-ko-kneader for the Thermomechanical Degradation of Polyolefins, Kunststoffe Plastics, 13(1966)166.
14. R.A. Lolley, Measurement of Melt Pressure in Plastics Extrusion, Plast. Inst. J., 34(1966)187-188.
15. M.R. Harris, A New Concept in Melt Pressure Measurement, SPE J., 22(1966)69-70.
16. N.M. Bikales, Extrusion, Encycl. Polym. Sci. Tech., 6(1967)466-467.
17. F. Martelli, Single-and-Twin-Screw Extruders. A Technical Comparison, SPE J., 23(1967)53-58.
18. A.M. Willer, Instrumentation for Temperature Control (in the Plastics Industry) Plast. Des. Process, 7(1967)10-15.
19. H. Domininghaus, Extruders of Special Design, I. Kunststoffe Rundschau, 13(1966 645-649, II. 14(1967)14-17.
20. H. Ennecking, Developments in Take-up Machinery, Chemieforsern, 17(1967)504-508
21. K. Rangarajan, Recent Trends in Melt Spinning, Petrol. Hydrocarbons, 3(1968) 8-12.
22. J.E. McIntyre, Manufacture of Man-Made Fibers, Encycl. Polym. Sci. Tech., 8(1968)374-404.
23. H. Ennecking, Spinning Take-Up Today, Textil. Ind., 132(1968)81-84, 86.
24. K. Rangaranjan, Recent Trends in Melt Extrusion, Petrol Hydrocarbons, 3(1968)8-12.
25. D.W.H. Roth, Impact of Chemical Engineers on Man-Made Fiber Technology, Chem. Eng., 75(1968)86-94.
26. G.A. Bialas, Screw Extrusion of Polymer Melts, An. Tech. Conf. SPE, 27th, 15(1969)126-131.
27. P.E. Hime, Fibers Plant Engineering, Chem. Eng. Progr., 65(1969)45-51.
28. R. Rub, Air Conditioning Units for Man-Made Fiber Production and Processing Plants, Chemieforsern, 19(1969)919.
29. K. Pestel, Technical Problems in Spin Stretch Winding, Deut. Textiltech., 19(1969)78-81.
30. H.J. Collmick, Inaudible Sounds Solves Difficult Cleaning Problems, Chemiefaser, 19(1969)275-280.

31. W. Wegener, The Distribution of Pressure in Draw Twister Packages, Textil-Praxis 24(1969)(4), 448-453; (5), 301-307; (6), 369-376; (7), 448-453; (8), 521-525; (9), 584-588.
32. F. Rub, Air Conditioning Units for Man-Made Fiber Production and Processing Plants, Chemiefersern, 19(1969)919-922.
33. G. Illing, Reworking of Man-Made Fibers and Sheets to Granulated Compounds, Kunststofftechnik, 8(1969)196-198.
34. K.J. Baum, Profit from Screw Design, Plast. Polym., 38(1970)411-417.
35. R.B. Gregory, Extrusion and Mixing: Flow and Heat Transfer, Plast. Polym., 38(1970)120-124.
36. G. Menger, Processing of Polyolefin Powders on Single Screw Extruders, Plastverarbeiter, 21(1970)5.
37. H.T. Kim, Measuring True Melt Temperature in Extruders, SPE J., 26(1970)31-36.
38. I.M. Khomenko, Reliability of the Spinnerette Assemblies in the Synthetic Fiber Sector, Fiber Chem., 2(1970)488-490.
39. Tow Cutter Has Only Two Moving Parts, Textile Month, 60(1970).
40. S. Kleinheins, Reclamation of Synthetic Fibers, Chemiefasern, 20(1970)1058-1063.
41. K.E. Ergasher, Analysis of the Staple Fiber Crimping Process, Fiber Chem., 3(1971)482-484.
42. W. Weisser, Electrical Equipment for Extruders, Chemiefasern, 21(1971)634-636.
43. U.M. Kosel, Novel Concept of Single Screw Extrusion, Plast Polym., 39(1971)319-327.
44. A.G. Yaskov, Analysis of the Causes of Spinnerette Failures, Fiber Chem., 3(1971)631-633.
45. V.I. Levsch, Analysis of the Staple Fiber Crimping Process by Factorial Experiments, Fiber Chem., 3(1971)482-484.
46. H.O. Schiedrum, Advanced Screw Design for Extruding High Viscosity Polyolefins, Plasterarbeiter, 22(1971)639-642.
47. W. Funk, Spinnerette Components for Man-Made Fiber Industry, Chemiefasern, 22/74(1972)518-522.
48. D. Semier, Threadguides for Stretching and Crimping Synthetic Fibers, Chemiefasern, 22(1972)895.
49. J.P. Redston, Emulsifier Choice in Design of Spin Finishes for Man-Made Fibers, Colourage, 19(1972)25-30.
50. V.A. Sukharev, Regulation of the Take-up Speed in Synthetic Fiber Extrusion, Fiber Chem., 4(1972)85-88.
51. R.H. Woodward, Quality Control in Man-Made Fiber Manufacture, Text. Inst. Ind., 10(1972)294-296.
52. W. Dohrn, Finishes in the Manufacture of Man-Made Fibers, Part 2, Textilverinigung, 5(1972)158-159.
53. New Staple Cutting Machine, Textilbull. Spinning, 4(1972)393.
54. I.M.Kohmenko, Reliability Index of Machines and Equipment for the Manufacture of Synthetic Fibers, Fiber Chem., 2(1972)482-484.
55. I. Klein, Extrusion. Update on Melt Technology, Plast. Des. Process, 3(1973)(8), 10-15; (9), 22-24; (10), 10-15.
56. H.O. Schiedrum, Problems in Extruding Polyolefin Powders, Plastverarbeiter, 24(1973)341-349.
57. Development of New Extrusion Techniques - Freeze Spinning, Textile Machinery, 62(1973)17.
58. K. Roggert, New Equipment from the Melt-Spinning Process, Chemiefasern, 23/75(1973)1183-1186.
59. J. Piso, Denier of Man-Made Yarns Monitored Within 1%, Mod Textiles, 54(1973)46-47.
60. E.N. Shubina, Statistical Quality Control in Fiber Production, Fiber Chem., 5(1973)612-618.
61. G.M. Kovaleva, et al, Product Quality Attestation in the Man-Made Fiber Industry, Fiber Chem., 5(1973)600-602.
62. Directory of Leading Japanese Manufacturers of Textile Machinery and Related Equipment, J. Textile Mach. Soc., 19(1973)1-4.

63. Fourne Expands Fiber Equipment Potential, Modern Textiles, 54(1973)70.
64. Designing Electrical Systems for a Fiber Manufacturing Plant, Textile Ind., Supplement Aug. 1973.
65. R. Barthel, "Possibilities of Reduction of Fiber Break Frequency in Man-Made Fiber Production, Faserforsch. u. Textiltech., 25(1974)325-332.
66. W.F. Bernholz, Spin-Finish Usage and Compounding for Man-Made Fibers, Polym. Plast. Tech. Eng., 3(1974)1-27.
67. M.P. Eparkhin, Maintenance of Man-Made Fiber Production Equipment, Fiber Chem., 6(1974)100-102.
68. Reliable Denier Control Instrument for Use at Extrusion, Textilbetrieb, 92(1974)49.
69. G.P. Rossinskaya, Spare Parts Consumption Norms at Man-Made Fiber Establishment Fiber Chem., 6(1974)103-104.
70. H. Samaus, Continuous Regranulation of Film, Fiber and Wastes, Kunststoffe-Plastics, 12/74(1974)33-35.
71. F.G. Kollman, Continuous Reclamation of Film Wastes, Kunststoffe-Plastics, 12/74(1974)21-23.
72. J.S. Singleton, Continuous Polymer Filtration, Fiber Producer, 2(1974)128-130.
73. N.H. Gordon, Feedback Pressure Control, Fiber Producer, 2(1974)124-128.
74. J.E. Obetz, Spinning: The Finish is Critical, Fiber Producer, 2(1974)123-127.
75. Portable Sensor Measures Denier of Man-Made Fibers, Mod. Textiles, 55(1974)54.
76. A.T. Serkov, Fiber Extrusion Through and Air Gap, Fiber Chem., 7(1975)40-47.
77. N.R. Schrott, Motionless Mixers in Plastic Processing, Chem. Eng. Progress, 71(1975)54-58.
78. J.T. Marcus, Gel Removal Through Porous Media, SPE ANTEC, Atlanta, 1975.
79. R.L. Powell, New Developments in Metering, Mixing and Dispensing of Multi-Component Reactive Resins, SPE ANTEC, Atlanta, 1975.
80. M. Mayer, Modern Machine for Spinning and High Speed Winding of Synthetic Filaments, SPEC ANTEC, Atlanta, 1975.
81. M. Jambrich, Conventional and Continuous Preparation of PP, PA and PET Fibers with Respect to Structural and Morphological Changes, Faserforsch u. Textiltech., 26(1975)361-368.
82. P.N. Samosyuk, Efficiency of the Heating System for Spinning Heads for Synthetic Fibers, Fiber Chem., 7(1975)197-200.
83. E. Welfers, Methods for the Production of Man-Made Staple Fibers, Chemiefasern, 26/78(1976)1079.
84. Anon., Reprocessing Plant for Man-Made Fiber Waste, Int. Text. Bull. Spinn., 3(1976)262.
85. J.K. Nigam, Recent Advances in Melt Spinning Techniques, Synth. Fibers (India), 5(1976)6.
86. F. Fourne, High Speed Spinning and Cutting of Synthetic Fibers, Chemiefasern, 26/78(1976)1098-E188.
87. E.W. Davis, Application of Inverter Drives, Fiber Producer, 5(1977)9-19.
88. W.H.J. Badrian, Reclamation of Waste Polypropylene Fibers, Carpet and Rug Ind., 5(1977)18-22.
89. R.T. Fenner, Developments in the Analysis of Steady Screw Extrusion of Polymers, Polymer, 18(1977)617.
90. Anon., Staple Equipment Update, Fiber Producer, 5(1977)9.
91. H.C. Lummus, Developments in Man-Made Fiber Baling, Fiber Producer, 5(1977)58.
92. H.R. Billica, Role of Finishes in Processing Man-Made Staple Fibers, Chemiefasern, 27/29(1977)E56-59.
93. R. Kleber, Chemistry and Tribology of Spin Finishes, Chemiefasern, 27/29(1977)E53-56.
94. J.K.L. Bajaj, Waste Reclamation, Fiber Producer, 5(1977)30.
95. C. Hobing, Finish Application Systems, Fiber Producer, 5(1977)44.
96. R.H. Wagner, Reprocessing of Synthetic Fiber Production Wastes, Chemiefasern, 27/29(1977)E-90.
97. R.E. Parker, Inverter Vs. Rotating Drive for Man-Made Fiber Spinning, Fiber Producer, 5(1977)35.

98. Anon., Spinnerettes: Selecting Steels and Cleaning Methods, Fiber Producer, 5(1977)22.
99. Anon., New Spinnerette Inspection Equipment, Chemiefasern, 27/29(1977)E-131.
100.M.Schwab, Spinnerettes for Man-made Fiber Spinning - Shapes, Materials, Methods of Manufacture, Chemiefasern, 27/29(1977)267-275.
101.A.M.Kirichenko, Drawing off Heat from the Spinning Cabinet, Fibre Chem., 9(1977) 98-101.
102.P.G.Moore, Survey of Polypropylene Lubrication, Plast.Rubber Inst. London, Conf. Polyp.Fibers & Textiles, York, Sept. 26-28, 1979.
103.J.S.Mason, Pneumatic Conveying of Powders and Granules, Plast. Rubber Inst. London, Conf. Polyp. Fibers & Textiles, York, Sept. 26-28, 1979.
104.G.Schubert, Development Trends in Machines for Synthetic Fibers, Melliand Textilber. Intern, 59(1978)702-704.
105.H.L.Davis, Continuous Monitoting of Yarn Uniformity During Melt Spinning, Text. Res.J., 48(1978)329-334.
106.F.Fourne, Yarn Cooling in Melt Spinning, Chemiefasern, 28/80(1978)E77-80,315-323.
107.W.H.Hill, Spin Pack Problems, Fiber Producer, 6(1978)20,23,75.
108.I.L.Langley, Concepts of Spinnerette Desighn, Fiber Producer, 6(1978)14-18,74,75.
109.T.B.Baldridge, Spinnerette Cleaning, Fiber Producer, 6(1978)48.
110.E.Shanholtzer, Spinnerette Technology, Fiber Producer, 6(1978)52.
111.D.E.Hansen, Polymer and Spinnerette Design for Multifilament Polypropylene Fiber Production, Plast.Rubber Inst.London, Conf. Polyp. Fibers & Textiles, York, Sept. 26-28, 1979.
112.H.C.Podrigues, Optimization of Process Variables in the Production of Polypropylene Fibers, Plast. Rubber Inst. London, Conf. Polyp. Fibers & Textiles, York, Sept. 26-28, 1979.
113.R.A.Gill, The Influence of Process Variables on Morphology and Properties of Melt Spun Polypropylene Fibres, Plast. Rubber Inst., London, Conf. Polyp. Fibers & Textiles, York, Sept. 26-28, 1979.
114.J.P.Praussen, The Relationship between the Properties, Processability and Molecular Structure of Uniaxially Oriented Polypropylene, Plast. Rubber Inst., London, Conf. Polyp. Fibers & Textiles, York, Sept. 26-28, 1979.
115.R.McIlhagger, Extrusion of Polypropylene Continuous Filament Yarns, Plast. Rubber Inst., London, Conf. Polyp. Fibers & Textiles, York, Sept. 26-28, 1979.
116.D.P.Bash, Additives and their Blending for Polypropylene, Plast. Rubber Inst., London, Conf. Polyp. Fibers & Textiles, York, Sept. 26-28, 1979.

X.2. U.S. Patents

U.S. 3,092,891, Monofilaments From Viscous Polyolefins, Montecantini, June 11, 1963
U.S. 3,119,212, Tow Packaging Apparatus, Am. Cyanamid, Jan. 28, 1964
U.S. 3,123,889, Tow Packaging Apparatus, Monsanto, March 10, 1964
U.S. 3,137,033, Draw Stretching Apparatus, Akzu, June 16, 1964
U.S. 3,143,764, Heat Stretching Device for Synthetic Filaments, Hoechst, Aug. 11, 1964
U.S. 3,144,187, Pneumatic Conveyor For Use in Lacing-Up of Freshly Spun Filaments, Am. Cyanamid, March 11, 1964
U.S. 3,149,515, Cutting and Suction Device for Tow Processing, DuPont, Dept. 22, 1964
U.S. 3,150,552, Cutting Apparatus for Producing Staple Fibers From Tow, Am. Enka, Sept. 29, 1964
U.S. 3,161,100, Tow Cutting Apparatus, DuPont, Dec. 15, 1964
U.S. 3,162,080, Tow Feed Means For Staple Fiber Cutter, Dec. 22, 1964
U.S. 3,172,186, Draw Stretching Apparatus, Bayer, March 9, 1965
U.S. 3,176,374, Draw Stretching Process for Polyester or Polypropylene Tow, ICI, April 6, 1965
U.S. 3,188,239, Spinnerette Cleaning Process, Am. Enka, June 8, 1965
U.S. 3,191,479, Helical Cutter For Tow, Fuji Bo, June 29, 1965
U.S. 3,217,578, Beria Cutter With Tow Straightened by Air Currents, Monsanto, Nov. 16, 1965
U.S. 3,264,481, Method and Apparatus for Inspecting Clogged Holes, FMC, Aug. 2, 1966
U.S. 3,266,084, Filament Guide and Feed for Heat Stretching Apparatus, Thery, Aug. 16, 1966
U.S. 3,271,493, Melt Spinning Process, Am. Enka, Sept. 6, 1966
U.S. 3,271,818, Quenching Apparatus, DuPont, Sept. 13, 1966
U.S. 3,273,200, Distributor Plate for Melt Spinning of Filaments, Allied Chemicals, Sept. 20, 1966
U.S. 3,274,644, Adjustable-Flow Quenching Chamber for Polymer Filaments, DuPont, April 27, 1964
U.S. 3,276,095, Thermoplastic Yarn Stretching and Heating
U.S. 3,277,228, Heat Stretching Method and Apparatus for Olefin Filaments, Chemcell, Oct. 4, 1966
U.S. 3,277,537, Guide For Shaping Tow, DuPont, Oct. 11, 1966
U.S. 3,280,424, Apparatus for the Manufacture of Synthetic Filaments, Am. Enka, Oct. 25, 1966
U.S. 3,281,900, Extrusion Apparatus, Phillips Petrol., Nov. 1, 1966
U.S. 3,286,569, Helical Tow Cutter and Honing Device, DuPont, Nov. 22, 1966
U.S. 3,286,992, Fluid Mixing Device, Arthur D. Little, Nov. 22, 1966
U.S. 3,299,469, Melt Spinning Apparatus, DuPont, Jan. 24, 1967
U.S. 3,307,216, Annular Spinning Apparatus, Fiber Ind., March 7, 1967
U.S. 3,308,507, Extrusion Apparatus, Rexall, March 14, 1967
U.S. 3,311,687, Synthetic Monofilament Yarns, F. Scragg, March 28, 1967
U.S. 3,317,980, Machine for Heat Stretching Synthetic Filaments, Ricten, May 9, 1967
U.S. 3,320,343, Melt Spinning Synthetic Linear Polymers, VEB Chem. fabrik Schwarza, May 16, 1967
U.S. 3,328,850, Patterned Roll Opening Apparatus for Crimped Tow, Celanese, July 4, 1967
U.S. 3,440,685, Methods and Apparatus for Quenching Melt Spun Filaments, Monsanto, April 29, 1969.
U.S. 3,388,442, Lacing Device for Filament Treatment Device, FMC, July 18, 1968.

U.S. 3,334,532, Apparatus for Cutting Crimped Tow Into Staple Fiber, Monsanto, April 8, 1967
U.S. 3,334,533, Mechanism for Cutting Staple Fiber From Tow, Monsanto, Aug. 8, 1967
U.S. 3,335,210, Melt Spinning Apparatus for Uniform Filament Quenching, Monsanto, Aug. 8, 1967
U.S. 3,339,362, Method of Joining Multifilament Strands, DuPont, Sept. 5, 1967
U.S. 3,339,819, Forwarding Roll For Transporting Freshly Spun Filaments to Take Up Spool, Glanzstoff, Feb. 15, 1967
U.S. 3,341,914, Air Jet Process For Drawing Filaments, Brit. Nylon Spinners, Sept. 19, 1967
U.S. 3,351,992, Tow Packaging Method, Eastman, Nov. 14, 1967
U.S. 3,363,387, Modular Draw Stretching Machine, Turbo Mach., Jan. 16, 1968
U.S. 3,363,577, Shaft Sealing For Man-Made Fiber Spinning Gear Pump, Am. Enka, Jan. 16, 1968
U.S. 3,381,336, Melt Extrusion Spinning Head System, S. C. Wells, May 7, 1968
U.S. 3,381,869, Filament Threading Device, DuPont, May 7, 1968
U.S. 3,385,491, Crimped Tow Feed For Staple Cutting Apparatus, DuPont, May 28, 1968
U.S. 3,386,145, Quality Monitoring System For Melt Spinning Apparatus, Eastman Kodak, June 4, 1968
U.S. 3,387,521, Tow Cutter Apparatus, Teijin, June 11, 1968
U.S. 3,388,442, Lacing Device For Filament Treatment Apparatus, FMC, July 18, 1968
U.S. 3,389,429, Melt Spinning Apparatus, Allied Chem., June 25, 1968
U.S. 3,407,437, Spinning Head For High Pressure Melt Spinning, Barmag, Oct. 29, 1968
U.S. 3,409,958, Apparatus For Continuous Heat Stretching and Heat Shrinking of Filaments, Zinser, Nov. 12, 1968
U.S. 3,413,397, Draw Stretching Process For Polypropylene Filaments, Eastman, Nov. 26, 1968
U.S. 3,414,939, Apparatus For Quenching Melt-Spun Fibers, Am. Cyanamid, Dec. 10, 1968
U.S. 3,422,713, Granulating Apparatus For Cutting Freshly Extruded Plastic Strips, Zimmer, Jan. 21, 1969
U.S. 3,426,632, Vacuum Waste Removal System For Flying Knife Type Tow Cutter, DuPont, Feb. 11, 1969
U.S. 3,433,055, Method of Measuring the Metering A-curacy of a Spinnerette, DuPont, March 18, 1969
U.S. 3,436,448, Method and Apparatus For Spinning Uniform Fibers, Am. Cyanamid, Apr. 1, 1969
U.S. 3,436,789, Spinnerette Cleaning Apparatus, DuPont, April 8, 1969
U.S. 3,436,792, Apparatus For Producing Polymer Strands or Granules From Liquid Material, Hench, April 8, 1969
U.S. 3,436,823, Spinnerette Wipe Tool, Monsanto, April 8, 1969
U.S. 3,437,725, Filter Pack For Melt Spinning Apparatus, DuPont, April 8, 1969
U.S. 3,439,382, Melt Spinning Spinnerette Assembly, Am. Enka, April 22, 1969
U.S. 3,440,685, Method and Apparatus For Quenching Melt Spun Filaments, Monsanto, April 29, 1969
U.S. 3,444,587, Shaft For The Air Flow Quenching of Melt Spun Fibers, Polovets (USSR), April 8, 1969
U.S. 3,456,857, Filament Wrap Detection and Feed Interrupter For Continuous Filament Extrusion, DuPont, July 22, 1969
U.S. 3,458,890, Cross-Flow Jet For Removing Liquids From Running Filaments, Aug. 5, 1969
U.S. 3,459,910, Yarn Transfer Aspirator For Continuous Filament Winding, Leesona, July 1, 1969
U.S. 3,460,201, Cabinet For Air-Stream Cooling of Filament Spun From a Polymeric Melt in a Spinning Machine, Zaitsev (USSR), Aug. 12, 1969
U.S. 3,464,300, Coordinated Filament Cutting Device and Aspirator, DuPont, Sept. 2, 1969
U.S. 3,466,703, Spinnerette Assembly, DuPont, Sept. 16, 1969
U.S. 3,471,608, Process For The Continuous Heat Stretching and Heat Setting of Filaments, Barmag, Oct. 7, 1969

U.S. 3,472,107, Tow Cutting Apparatus With Knife Resharpening Means, Am. Enka, Oct. 14, 1969
U.S. 3,475,527, Process For Destroying Melt Crystalline Order In Fiber-Forming Polymers, Monsanto, Oct. 28, 1969
U.S. 3,487,144, Apparatus and Method For Cooling Extruded Molten Filaments, FMC, Dec. 30, 1969
U.S. 3,488,806, Melt Spinning Pack Assembly, DuPont, Jan. 13, 1970
U.S. 3,491,405, Apparatus For Producing Filaments and Yarns By Melt Extrusion, Chemcell; Jan. 27, 1970
U.S. 3,492,691, Melt Extrusion, ICI, Feb. 3, 1970
U.S. 3,496,261, Supplying Molten Polymers to a Number of Spinning Heads, W. G. Parr, Feb. 17, 1970
U.S. 3,497,586, Process and Apparatus For Uniform Melt Spinning of Thermoplastic Yarn, FMC, Feb. 24, 1970
U.S. 3,503,100, Method of Processing Large Denier Tow, Eastman, March 31, 1970
U.S. 3,504,078, Melt Extrusion Process, DuPont, 1970
U.S. 3,508,296, Melt Extrusion Apparatus, Teijin, April 28, 1970
U.S. 3,508,432, Spinnerette Capillary Inspection Apparatus, Monsanto, Apr. 28, 1970
U.S. 3,509,598, Spinnerette Pack Assembly, Am. Enka, May 5, 1970
U.S. 3,509,709, Pneumatically Bundled Yarn Without True Twist, Nippon Rayon, 1970
U.S. 3,516,199, Nozzle Block For Extrusion Device, Zimmer, June 23, 1970
U.S. 3,523,827, Spinnerette Reconditioning Method, Monsanto, April 11, 1970
U.S. 3,540,281, Method For Measuring Surface Temperature of Filament Heating Rolls, Teijin, 1970
U.S. 3,549,743, Multistage Drawing Process For Polypropylene, Chemcell, 1970
U.S. 3,551,971, Heat Stretching Apparatus For Man-Made Filaments, Zimmer, 1971
U.S. 3,555,537, Monitoring Method For Filament Extrusion Process, DuPont, 1971
U.S. 3,559,860, Pneumatic Filament Handling Device, ICI, Feb. 2, 1971
U.S. 3,564,958, Yarn Transfer Aspirator, Leesona, Feb. 23, 1971
U.S. 3,570,339, Yarn Cutting and Transfer Aspirator For Continuous Filament Winding, Leesona, 1971
U.S. 3,581,487, Yarn Drawing Mechanism, Courtauld, June 1, 1971
U.S. 3,594,342, Introduction of Additives Into Plastics and Wax Melts, WEB Leuna July 20, 1971
U.S. 3,600,492, Process and Device For The Manufacture of Monofilaments of Circular Cross Section Made of Synthetic Linear Polymers, Hoechst, 1971
U.S. 3,609,805, Cooling Apparatus For Extruded Synthetic Material, Scheer & Cie, Oct. 5, 1971
U.S. 3,611,485, Extrusion Chimney, Monsanto, Oct. 12, 1971
U.S. 3,618,667, Apparatus For The Melt Extrusion of Linear Fiber-Forming Thermoplastic Polymers, Glanzstoff, 1971
U.S. 3,621,088, High Production of Water Quenched Filaments, Phillips Petrol, 1971
U.S. 3,626,041, Apparatus and Process For Making Continuous Filament, Monsanto, 197
U.S. 3,626,156, Temperature Control of a Jacketed Chamber of a Melt Extrusion Machine, Teijin, Dec. 7, 1971
U.S. 3,630,384, Filter Bed Element For Use in Melt Extrusion, Teijin, 1971
U.S. 3,633,256, Orientation Drawing Chamber For Fibers, Monsanto, Jan. 11, 1972
U.S. 3,634,573, Method For Producing Fibrous Structures, Celanese, Jan. 11, 1972
U.S. 3,639,557, Process and Device For Filtering Melts of Fiber Forming High Polymers, Hoechst, 1972
U.S. 3,691,232, Process For Making Multifilament Yarns, Am. Enka, 1972
U.S. 3,647,330, Apparatus For Forming Filaments, Phillips Petrol., 1972
U.S. 3,650,645, Spinning Chimney, Teijin, 1972
U.S. 3,655,314, Extrusion Apparatus Composed of Modular Extrusion Units on Common Heating Beam, Barmag, 1972
U.S. 3,656,383, Apparatus For Automatically Cutting Connecting Tow Between Tote Boxes, Eastman, April 18, 1972
U.S. 3,659,980, Apparatus For Melt Extruding of Synthetic Filaments, Phillips Fibers, May 2, 1972

U.S. 3,672,013, Apparatus For Production of Polymer Fibers, Shell Oil, 1972
U.S. 3,672,801, Spinning Quench Chamber Having a Conical Flow Direction, DuPont June 27, 1972
U.S. 3,679,786, Method and Apparatus For Melt Extruding of Synthetic Filaments, 1972
U.S. 3,684,416, Adjustable Spinning Shaft For Melt Spun Yarns, Barmag, 1972
U.S. 3,684,648, Multicomponent Filamentary Material Formed From Waste Material, Kunkel, Aug. 15, 1972
U.S. 3,686,386, Multifilament Extrusion Method and Apparatus, Beloit, Aug. 22, 1972
U.S. 3,696,184, Melt Spinning, Concord, Oct. 3, 1972
U.S. 3,704,971, Spinnerette Assembly, DuPont, Dec. 5, 1972
U.S. 3,705,227, Process and Apparatus For Quenching Melt Spun Filaments, DuPont, Dec. 5, 1972
U.S. 3,706,407, Piddler Mechanism For Strand Material, Bouligny, Dec. 19, 1972
U.S. 3,720,382, Process and Apparatus For The Continuous Production of Synthetic Fibers, Vickers - Zimmer, March 13, 1973
U.S. 3,724,492, Distributor For Viscous Fluid Spinning Melts or Solutions, Barmag, April 3, 1973
U.S. 3,725,523, Extracting and Drawing a Tow, DuPont, 1973
U.S. 3,729,279, Spinnerette With Head Filter, Mott, 1973
U.S. 3,730,662, Spinnerette Assembly, Monsanto, 1973
U.S. 3,732,678, Apparatus For Joining Filaments to a Tow, DuPont, 1973
U.S. 3,759,010, Screw Jet Pack For Textile Fibers, Allied Chem., 1973
U.S. 3,761,559, Opposed Flow Spinnerette Blanketer, DuPont, 1973
U.S. 3,762,850, Spinning Head, Akzona, 1973
U.S. 3,763,520, Method and Apparatus For Transferring Tows, Teijin, 1973
U.S. 3,765,852, Apparatus For Producing Fibers From Melts, Shule, 1973
U.S. 3,768,355, Apparatus For Cutting Tow Into Staple, Faremer, 1973
U.S. 3,780,516, Process For Forming Yarn of Improved Properties, Phillips Petrol., Dec. 25, 1973
U.S. 3,785,018, Hot Stepwise Drafting With Independent Free Wheeling Idler Rolls, Celanese, Jan. 15, 1974
U.S. 3,787,265, Process and Apparatus For Producing Fibrous Structure, Celanese, Jan. 22, 1974
U.S. 3,796,119, Staple Fiber Cutter, FMC, March 12, 1974
U.S. 3,799,509, Mixer For a Melt Spinning Apparatus, DuPont, March 26, 1974
U.S. 3,801,242, Apparatus For Treating and Winding Up Spun Filaments, Neumag, 1974
U.S. 3,802,821, Spinnerette Head, 1974
U.S. 3,803,282, Method of Preparing Synthetic Continuous Filament Yarns By The Coupled Spinning-Drawing Process, Teijin, 1974
U.S. 3,804,569, Spinnerette Wiper, DuPont, 1974
U.S. 3,806,017, Apparatus For Forwarding Filamentary Material, DuPont, 1974
U.S. 3,817,672, Spinning Apparatus With Vaporous Heating Jacket, Barmag, 1974
U.S. 3,824,050, Apparatus For Extruding Synthetic Resin Filaments, Reifenhauser, July 16, 1974
U.S. 3,829,543, Melt Spinning of Polymers, ICI, 1974
U.S. 3,830,617, Melt Spinning Apparatus, ICI, 1974
U.S. 3,834,600, Apparatus For The Transport of Yarns, Rohne-Poulene, 1974
U.S. 3,847,522, Spinnerette Blanketing Apparatus, DuPont, 1974
U.S. 3,847,524, Spinnerette Head Assembly With Porous Metal Filter and Shear Elements, Mott, Nov. 12, 1974
U.S. 3,855,453, Apparatus For Controlled Quenching of Melt Extruded Filaments, Allied Chem., Dec. 17, 1974
U.S. 3,881,230, Tow Processing Apparatus, DuPont, May 6, 1975
U.S. 3,902,833, Spinning Machine, Automatik, Sept. 2, 1975
U.S. 3,902,834, Blow Shaft, Hagenburg, Sept. 2, 1975
U.S. 3,902,835, Man-Made Fiber Extrusion Apparatus With Two Part Openable Head, Theysohn, 1975
U.S. 3,908,048, Method of Improving Interfiber Cohesion of Filament Yarn, Kao Soap, Sept. 23, 1975

U.S. 3,912,439, Strand Extrusion Apparatus, Dow Chem., 1975
U.S. 3,917,141, Process and Apparatus For Transporting the Yarn At High Speed Rhone-Poulene, Nov. 4, 1975
U.S. 3,925,525, Spinning Method, Celanese, 1975
U.S. 3,930,292, Devices For Threading Filaments on Rollers, Barmag, 1976
U.S. 3,932,576, Apparatus and Method of Melt-Spinning, Concorde, 1976
U.S. 3,936,253, Apparatus For Melt Spinning Man-Made Fibers, Allied Chem., 1976
U.S. 3,938,924, Thermoplastic Melt Apparatus, Celanese, 1976
U.S. 3,938,925, Spin Pack Assembly, Allied Chem., 1976
U.S. 3,939,639, Spin Stretching and Spin Stretch Texturizing of Multifilament Yarns, Bayer, Feb. 24, 1976
U.S. 3,947,202, Apparatus For Filtering Molten Plastic Strands, Goeller, 1976
U.S. 3,948,592, Spinning Unit For Melt Spinning High Polymers, Zimmer, 1976
U.S. 3,949,043, Manufacture of Monofilaments, BASF, 1976
U.S. 3,955,906, System For Transporting a Filament Bundle From Extrusion to Draw Process, Teijin, May 11, 1976
U.S. 3,975,475, Wiping Spinnerette Face With Cooled Wiper, DuPont, Aug. 17, 1976
U.S. 3,978,751, Apparatus For Cutting Fiberous Tow, Keith, 1976
U.S. 3,988,086, Melt Spinning Apparatus With Convergence Guide, Allied Chem, 1976
U.S. 3,988,882, Method and Apparatus For Simultaneously Texturing and Cutting Continuous Yarns, Rhone-Poulene, Nov. 2, 1976
U.S. 3,992,864, Method and Apparatus For Automatic Reattachment of Yarns in Spinning Machines, Zinser, Nov. 23, 1976
U.S. 3,999,910, Filament Quenching Apparatus, Allied Chem., 1976
U.S. B,506,624, Filament Breakage Detection and Correction, Celanese, 1976
U.S. 4,003,195, Filamentary Tow Packaging Method, DuPont, 1977
U.S. 4,043,718, Spinning Apparatus With Retractable Suction Gun, Teijin, 1977
U.S. 3,965,511, Process For Continuous Heat Setting and Shrinking Synthetic Fibers, Fleissner, 1976
U.S. 4,037,288, Filament Liquid Quenching Apparatus, Phillips Petrol., 1977
U.S. 4,035,127, Melt Spinning Apparatus, Toray Ind., 1977
U.S. 4,035,790, Monitoring Device For Fiber Cutter, U.S. Gypsum, 1977
U.S. 4,017,249, Melt Spinning Apparatus, Barmag, 1977
U.S. 4,000,238, Method For Producing Synthetic Yarns, F11, 1976
U.S. 4,001,367, Method For Permanently and Uniformly Incorporating an Additive Into An Undrawn Fiber, M&T Chem., 1977
U.S. 4,001,359, Testing and Correcting Metering Accuracy of Multihole Spinnerettes, Allied Chem., 1977
U.S. 4,050,866, Apparatus For Melt Spinning, Akso, 1977
U.S. 4,011,640, Yarn Entanglement Nozzle, Milliken, 1977
U.S. 4,005,566, Process and Apparatus For Imparting Coherence to Tow, DuPont, 1977
U.S. 4,000,960, Apparatus For Manufacturing Synthetic Tow For Stretch Cut Spinning Process. Mitsubishi Rayon, 1977
US 4,059,668, Method of Tow Stretching, Monsanto, 1978
US 4,069,566, Take-up Method for Continuous Filament Bundles of Synthetic Fibers and Apparatus Thereof, Toyobo, 1978
US 4,070,730, Production of Chopped Fibers, Owen-Corning, 1978
US 4,088,431, Melt-Spinning Filters, ICI, 1978
US 4,098,564, Apparatus for protecting a Spinnerette in a spinnerette Pack, Phillips Fibers, 1978
US 4,102,625, Spinnerette Holer, A.Mauer SA., 1978
US 4,125,922, Jet Tangler for Multifilament Yarns, Textured Yarns, 1978
US 4,133,620, Polymer Filament Manufacturing Device having Reduced Vertical size, Diedier, 1979

APPENDIX I

Suppliers Directory - Turn-Key Plants, Know-How and Consulting Services

Asahi Engineering Co. Ltd., 3-25-23 Takanawa Minato-Ku, Tokyo, Japan

Alex James and Associates, P.O. Box 5882, Greenville, South Carolina 29606, USA

Bouligny Co., P.O. Box 10158, Charlotte, North Carolina 28237, USA

Chemtex Inc., 850 Third Ave., New York, New York 10062, USA

Diedier Engineering GmbH, P.O. Box 360, 4300 Essen, Germany

Emil Blaschke and Co. GmbH, P.O. Box 2207, 7057 Endersbach, Germany

Fourne KG, D-5305 Impekoven/Bonn, Germany

Hills R & D, 631 Nasa Blvd., Melbourne, Fla. 32901, USA

Inventa AG, P.O. Box 2701, CH-8023 Zurich, Switzerland

International Fibers Inc., 201 E Morehead Street, Reidsville, North Carolina , 27320, U.S.A.

James Mackie and Sons, P.O. Box 149, Belfast, Ireland

Karl Fischer Industrieanlagen, Holzhauser Str. 161, 1 Berlin 27, Germany

Lurgi Minraloltechnik, Fiber Div., P.O. Box 119181, 6 Frankfurt, Germany

Mitsubishi Heavy Industries, 6-3 Marunuchi 2 Chome Chuoda-ku, Tokyo, Japan

Neumag GmbH, P.O. Box 54, D-2350, Neumunster, Germany

Reifenhauser KG, P.O. Box 1345, 5210 Triodorf, Germany

Syntex Chemie GmbH, Gwinnerstrasse 30 (a), 6000 Frankfurt, Germany

Tessiltechnica, P.O. Box 129, 50100 Firenze, Italy

Teijin Heavy Industries, 7-3 Kitahama 3-Chome Higashiku, Osaka, Japan

Zimmer AG, Borsigallee 1, 6 Frankfort/M., Germany

APPENDIX II

Suppliers Directory - Machinery and Equipment

Automatik Apparate - Machinenbau, Postfach 25, 8754 Grossestheim 2, Germany	(17,18)
Alucolor, Berrenrather St. 511, 5032 Efferen/Koln, Germany	(9)
Amacoil Machinery Inc., 111 Plain Ave., New Rochelle, New York 10552, USA	(2,3)
Automated Energy Systems Inc., P.O. Box 667, Matthews, North Carolina 28105, USA	(1)
Atlantic Research Co., P.O. Box 38, Gainesville, Virginia 22065, USA	(3)
American Pulverizers Co., 1249 Macklind Ave., St. Louis, Mo. 63110, USA	(18)
ARCT, Rue Cuvier, F-42300 Roawne, France	(6,8,9,10,11,12,13,15)
Bolton Emerson Co., 9 Osgood St., Lawrence, Mass. 01842 USA	(18)
Buss-Condux Inc., 2450 Delta Lane, Elk Grove, Ill. 60007, USA	(18)
Bouligny Co., P.O. Box 10158, Charlotte, North Carolina 28237, USA	(1,4,8,9,10,13,17,19,20,21)
Burlington Engineering Co., P.O. Box 31 Graham, North Carolina 27253, USA	(23)
Barmag-Barmer Machineufabrik, D0563 Remscheid-Lennep, Germany	(1,4,5,6,7,8,9,10,12,13,14, 15,16,17,20,21,22)
Batson Machinery, Box 3978 Greenville, South Carolina 29608 USA	(4)
Beloit Italia, SpA., Pinerolo, Italy	(13,16)
Blaschke and Co., GmbH, Postfach 2207, 2056 Endersbach, Germany	(1,4,8,9,12,13,16,17,19)
Berlyn Corp., 93 W. Main St., Milbury, Mass. 01527 USA	(3,4,18)
Beringer Co., Inc., P.O. Box 485 Marblehead, Mass. 01945, USA	(4,11)

Black Clawson Co., Rte. 3W, Fulton, New York 13069 USA	(13,16)
Cumberland Engineering, P.O. Box 6065, Providence, Rhode Island 02940 USA	(18)
Chattanooga Boiler Co., 1037 E. Main St., Chattanooga, Tenn. 37401, USA	(23)
Covema SRL, Fontana 1, Milan, Italy	(9,12,13)
Conair Inc., 2006 Raymond St., Bay City, Mich. 48706, USA	(2,3)
Chemiefaser Lenzig AG., 4860 Lenzing, Lenzing, Austria	(2,9,10,13,14,15,16,20,21)
Cognatex SA, Via Celice 94, 40026 Imola (Bo) Italy	(9)
Crosol, Inc., P.O. Box 1328 Greenville, South Carolina 29601, USA	(5,8)
Cambridge Wire Cloth Co., P.O. Box 399, Cambridge, Md. 21613, USA	(6)
Cleveland Wire Cloth Co., 3573 E. 78th St., Cleveland, Ohio 44105, USA	(6)
Cerrini, S.A.S. Bust Arsizio, Italy	(16)
Cosmoplastics, S.V.l. Milan, Italy	(16)
Dolci Spa, Via Zucchi 18, 1-20095 Cusano Milanino, Italy	(1,16)
Dienes-Honeywell GmbH, Postfach 96, Muellerweg II, Mullheim/M, Germany	(1,9,13)
Dobson and Barlow Mach., Bradley Fold, Nr. Bolton, Lancashire, England	(12)
Day Mixing, 4932 Beech St., Cincinnatti, Ohio 45212, USA	(3,18)
Dietze and Schell Mach., Postfach 279,D-8630 Coburg, Germany	(12)
Davis Standard, U.S. Hwy. 1, Pawcatuck, Conn. 02891, USA	(16,20)
Egan Machinery Co., S. Adamsville Rd., Sommerville, New Jersey 08876, USA	(4,16,20
Engelhard Industries, CH-8039, Zurich, Switzerland	(7)
Farrel Co., 25 Main, Ansonia, Conn. 06401, USA	(16)

Fleissner GmbH, Rt. 5, Box 1005, Charlotte, North Carolina 28208, USA	(13,14,15,19)
Fluid Dynamics, P.O. Box 10, Cedar Knolls, New Jersey 07927, USA	(4,6)
Feinpruef GmbH, P.O. Box 119, 3400 Guettington, Germany	(5)
Filterite Corp., Greenspring Dr., Timanium, Md. 21093, USA	(6)
Feldmuehle, Fabrik Str. 23/28, 731 Plochingen, Germany	(8)
Fourne KG, Industriegebiet, Impelcoven/Bonn, Germany	(1,8,9,10,11,12,13,14,15, 17,22)
Fycon, 3609 Triclan Dr., Charlotte, North Carolina 28210, USA	(1)
Gaston County, P.O. Box 308 Stanley, North Carolina 28164, USA	(23)
Gavazzi Spa, Via G. Ciardi q, 120148 Milano, Italy	(9)
Gruenberg Oven Co., Inc., P. O. Box 1183, Williamsport, Pa. 17701, USA	(11)
Hartig Plastics Mach., P.O. Box 791, New Brunswick, New Jersey 08903, USA	(4,16,20)
Hills R & D, 631 Nassa Blvd., Melbourne, Fla. 32901, USA	(17)
Hobbs, 666 Lincoln St., Worchester, Mass. 01605, USA	(16)
Husky Manufacturing Co., 201 E. Morehead St., Reidsville, North Carolina 27320	(10,13,21)
IWKA, P.O. Box 3409, Karlsnuhe, D-7500, Germany	(9,10)
Johnson Div. Leesona Corp., 178 Production Rd., Chippewa Falls, Wisc. 54729, USA	(13,16,20)
Krupp GmbH, P.O. Box 710360, D-282, Bremen, Germany	(15)
Kleinwefers Industries Co., Postfach 1560, 415 Krefeld, Germany	(1,16)
Kenics Corp., Kenics Park, N. Andover, Mass. 01845, USA	(3)
Koch Engineering, 4111 E. 37th St. N., Wichita, Kans. 67220, USA	(3)

Kaufman SA, BP 8 27600 Gaillou,
Gaillou, France (4)

Kawasaki Heavy Industries, 16-1 Nakamochi-Dori
2 Chome Ikuta-Ku, Kobe, Japan (5)

Kasen Nozzel Manufacturing Co., Kabata Cho Koto-Ku,
Osaka, Japan (7)

Killion Extruders Inc., 56 Depot St.,
Verona, New Jersey 07044, USA (13,16,20)

Krauss-Maffei Ag., 9 Munchen 50, Germany (16)

Liasson Co., P.O. Box 1231,
White Plains, New York 10602, USA (7)

Laumatex Corp., Rock Meadow Rd.,
Unbridge, Mass. 01569, USA (9)

Lummus Ind., P.O. Box 1260
Columbus, Ga. 31902, USA (15)

Leesona Corp., 333 Strawberry Rd.,
Warwick, Rhode Island 02887, USA (9)

Modern Plastics Machinery Corp., 64 Lakeview Ave.,
Clifton, New Jersey 07015, USA (4,13,16,20)

Mekitterick Co., 13 Acton Rd.,
Chelmsford, Mass. 01824, USA (23)

Mackie and Sons Ltd., P.O. Box 149,
Belfast BT 12 Ted N. Ireland (10)

Mitsubishi Heavy Industries, 6-3 Marunouchi 2
Chome Chiada-Ku, Tokyo, Japan (1,6,9,10,12,13,14,16,17,20,22)

Marzoli Fratelli SpA, Corso Matteotti 11,
Milano, Italy (12)

Mott Corp., F. Ind. Park,
Farmington, Conn. 06032, USA (6)

Murata Machinery Ltd., 3 Minami-Vehiaicho Kisshoin,
Minami-Ku Kyoto, Japan 601 (9,12,13,14,17)

Neumag GmbH., Postfach 54,
D-2350 Neumunster, Germany (1,8,9,10,13,14,15,17,22)

Mautamix bv, P.O. Box 773,
Haarlem, The Netherlands (3)

NRM Corp., 47 W. Exchange St.,
Akron, Ohio 44308, USA (4,16,20)

Neudai Inc., Box 6476,
Greenville, South Carolina 29606, USA (5,6,7)

Nissho-Iwai Co., 1211 Ave. of the Americas, New York, New York, USA	(7,9)
Omanra International, Box 685, Pineville, North Carolina 28134, USA	(5,6,7,12,15)
Plasticizers Ltd., Dringhlington, Bradford, Yorkshire, England	(12,13,14,16,20,21)
Patterson Industries, 1250 St. George St., East Liverpool, Ohio 43920, USA	(3)
Prodex, Ford, New Jersey, USA	(16)
Reifeuhausen, Postfach 150, 5210 Troisdorf, Germany	(2,3,4,6,13,16,17,20)
Rapid Machinery Co., P.O. Box 286, Marian, Iowa 52302, USA	(3)
Rieter, P.O. Box 4383, Spartanburg, South Carolina 28303, USA	(9,12)
Rosemount-Nachville Inc., Iou Heil Quaker Blvd., Lavergne, Tenn. 37086, USA	(1)
Sprout Waldron and Co., Inc., 98 Logan St., Munchy, Pa. 17756, USA	(2,3)
Seydel Co., P.O. Box 7740, 4800 Bielefield, Germany	(13,15)
Sterling Extruder Corp., 901 Durham Ave., S. Plainfield, New Jersey 07080, USA	(4,16)
Slack and Paar Ltd., Kegworth near Derby, England	(5,8)
Schimadzu Seisakusho Ltd., 14-5 Uchikandarl-Chrome Chiyada-Ku, Tokyo 101, Japan	(7,10,12)
Santa Andrea Nivara SpA, P.O. Box 175, 28100 Novora, Italy	(9,10,12)
Toray Engineer Co., 5-2 3- Chome Nakanoshima Kita-Ku, Osaka, Japan	(2,7,9,13,15,17,23)
Thoreson McCosch Inc., 1885 Thunderbird St., Troy, Mich. 48084, USA	(2,3)
Teijin Co., 14-4 Ginzo 6-Chome Chnu Ku, Tokyo, Japan	(9,10,13,14)
Thermo Kinetics Industries, P.O. Box 6747 Sta. B, Greenville, South Carolina 29606, USA	(16)
Torii Winding Machine, Chiekoin-Nishi, Koyoto, Japan	(9)

Toshiben Machine Co., 2-11 4-Chome-Ginza Chuo-Ku, Toyko, Japan	(10,12,14,15)
Tanaka Kikinzoku Kogyo KK, 14 Nihonbashi Kayabacho 2 Chome, Chuoku, Tokyo, Japan	(7)
Trent Inc., 201 Leverington Ave., Philadelphia, Pa., 19127, USA	(11,13,14)
Turbo Machine Co., P.O. Box 12527, Gastonia, North Carolina 28052, USA	(10,12,14,23)
Utita SPA, Via Moscova 36, Milano, Italy	(4,5,6,7,9,10,13,14,16, 17,18,20)
Vibra Screw, 755 Union Blvd., Totowa, New Jersey 07512, USA	(2,3)
VEB Germania, Postfach 162, 901 Karl-Marx-Stadt, Germany	(7)
Whitlock Inc., 23231 Industrial Park Dr., Farmington, Mich. 48024, USA	(2,3)
Whitin Machine Work, P.O. Box 1889, Charlotte, North Carolina 28233, USA	(9,12)
Wright Ind., 707 Spencer Lane, Nashville, Tenn. 37210, USA	(13,14)
Warner and Swasey Co., Drawer 815, Bessemer City, North Carolina 28016, USA	(15)
Zenith Products Co., 432 Cherry St., West Newton, Mass. 02165, USA	(5,8)
Zinser, Postfach 1480, 7333 Ebersbach, Germany	(9,10,12)

(1) Heated rolls

(2) Polymer handling systems

(3) Colors and additives blenders

(4) Screen changers

(5) Metering pumps

(6) Melt filters including screen, sand and sintered metal

(7) Spinnerettes

(8) Finish applicator (wheel type and metered type)

(9) Take-up (filament) including spin-draw assemblies

(10) Tow take-ups

(11) Burn out ovens and salt baths

(12) Draw twisters and draw winders

(13) Draw stand (Tow)

(14) Tow crimpers

(15) Tow cutters

(16) Extruders

(17) Spin packs and spin manifolds

(18) Dicers

(19) Annealing and heat setting systems

(20) Film casting equipment

(21) Delustering equipment

(22) Spin-draw-texturizing machines

(23) Autoclaves

APPENDIX III

Suppliers Directory - Spin Finishes

AZS Chemicals Co., 762 Marietta Blvd., N.W., Atlanta, Ga. 30318, USA

Ashland Chemicals Co., P.O. Box 2219, Columbus, Ohio 43216, USA

Diamond Shamrock Corp., Napco Div., 350 Mt. Kamble Ave., Morristown, N.J. 07960, USA

Dow Corning Corp., S. Saginaw Rd., Midland, Mich. 48640, USA

Emery Industries Inc., 1300 Caren Tower, Cincinnatti, Ohio 45202, USA

GAF Corp., 140 W. 51st St., New York, New York 10020, USA

George A. Goulston Co., Monroe, North Carolina 28110, USA

Hoechst AG, P.O. Box 800320, 6230 Frankfort/M, Germany

PVO Internatinel Inc., 416 Division Ct., Boonton, N.J. 07005, USA

Syntex Chemie GmbH, Gwinnerst 30A, 6000 Frankfort/M, Germany

Stockhausen & Cie, P.O. Box 570, 4150 Krefeld, Germany

Chapter IX FIBER ENGINEERING

I. INTRODUCTION

This chapter deals with the engineering or designing of polypropylene fibers and yarns to meet specific end-use requirements. Polypropylene fiber has already come a long way from the early stages of development common to all synthetic fibers. The results of process research and development endeavors during the past decade enabled polypropylene fiber to compete favorably with other natural and synthetic fibers. Recently, more and more attention is being given to product development, or *fiber engineering* , for specific end uses, including some previously known for polypropylene fiber as well as those which polypropylene fiber is striving to penetrate. As a result of these extensive research efforts, we may expect in forseeable future a large number of *third generation* polypropylene fibers, tailored for specific end uses.

Fiber engineering may be based on chemical modifications of the spinning mass or on modifications of spinning and fiber processing techniques. Or it may consist of both chemical and physical manipulations. The obvious examples of chemical modification are those connected with stabilization of polypropylene fiber against heat and light degradation, mass coloration, and the development of dyeable polypropylene fibers. These subjects are the topics of separate chapters of this book. Other examples of chemical modification, but in no way less important, can be seen in the improvement concerning static charge, flamability characteristics, bacteriostatic properties, optical properties, abrasion resistance, stain resistance, frictional characteristics, and other such properties of the fiber.

Among the examples of engineered fibers obtained through manipulation of spinning and post-spinning operations can be cited the development of conjugate fibers and split-film yarns. These products have been discussed elsewhere in this book and form the topics of separate chapters. Other examples of specially tailored polypropylene fibers are the various types of textured and novelty yarns, high-shrinkage fibers, high-tenacity yarns, elastic yarns, fibers of special cross sections, and pilling-resistant product.

In addition to these possibilities, fiber engineering efforts have resulted in the development of processes which are considerably different from conventional spinning, drawing, and fabricating techniques commonly known to the fiber producer. Examples are the development of microfibers, spun-bonded fabrics, and synthetic pulp and paperlike structures. We cannot review and evaluate here all the possibilities which may exist in the realm of fiber engineering. A brief discussion of some of these developments, as they specifically apply to polypropylene, will be given in the following pages; for specific details the indicated references should be consulted. A selected number of patents has been listed in the Bibliography for additional ideas in process and product development.

II. CHEMICAL MODIFICATIONS

II.1. Biconstituent Fibers

The concept of manufacturing melt-spun fibers from mixtures of two or more chemically distinct polymers is relatively new. Two such biconstituent fibers, based on mixtures of nylon and polyesters were commercialized by Allied under the trade name "Source" and by AKZO under the trade name "Enkatron". These fibers consist of a multitude of very fine fibrils of polyester embedded in a continuous matrix of nylon. Fibers of this nature have been named, variously, biconstituent fibers, bicomponent fibers, fibril-reinforced matrix-type fibers, matrix fibers, or simply, composite fibers. None of these nomenclatures, however, offers a full description of the structural nature of such fibers.

This class of fibers has been found to have several desirable properties, such as unique luster, anisotropic optical reflection, improved plyability, softness, antistatic properties, and improved dyeability. Biconstituent fibers based on polypropylene as a major component have been produced. In fact, a large part of research and development efforts in connection with dyeable polypropylene fibers directly relates to the production of biconstituent fibers. Some of the minor polymeric components in polypropylene matrix have been shown to impart to the fiber a unique sheen which is normally not obtainable in 100 percent polypropylene fibers. Low molecular-weight polyethylenes, various ethylene copolymers, and other emulsifiable large molecular-weight hydrocarbon-based polymers, fall in this category. Most of these fibril-matrix type fibers have pronounced anisotropic optical properties which make the transmission, absorption, and scattering of light from the filament strongly dependent on the angle of incidence. This directional effect can be put to good use in cut-pile fabrics and sheared carpets. The light incident on the side of the filament is mainly transmitted or reflected but the light incident on the cut end of the fiber is mainly diffused and scattered. This produces an extraordinary contrast between the appearance of the sheared and unsheared portions of the fabric. The useful minor components for this purpose are those which do not greatly differ from polypropylene in refractive index, are essentially noncrystalline, and show a good compatability with polypropylene.

If the refractive index of the minor component differs significantly from that of polypropylene, the included fibrils would impart to the fiber a subtle delustering effect which is also anisotropic in optical properties. A majority of polymeric materials fall into this category. Some polymeric materials are known to impart an antistatic properties when dispersed in polypropylene matrix as microfibrils. Polyethylene oxide has been mentioned as an example. Still other polymers, notably chlorinated vinyl polymers, are being tried to improve the flammability characteristics of polypropylene fibers.

The essential feature of the biconstituent fibers discussed above is that the added component rarely exceeds 15 percent of the total fiber composition, the continuum of polypropylene matrix remains intact, fibrillation of the filament is avoided, and the mechanical properties are essentially equivalent to those of pure polypropylene fiber. In another group of composite fibers, no such limitations are imposed. The goal of these endeavors has been to produce a staple look plyable yarn through composite monofilament spinning, drawing, calendaring, and microfibrillating. These structures have been discussed in Sections III.6. and IV.1. of this chapter.

Still another variation on the theme of biconstituent fibers is to incorporate into the polypropylene mass a comparatively low molecular-weight polymeric material which can be partially extracted in an acidic or alkaline water bath. The elongated crevices and channels, left behind by the extracted additive polymer on the surface of the fiber, may impart to the fiber a slight luster or subdued dullness, depending on the type and the quantity of the additive component incorporated and extracted respectively. Examples of such additives are low molecular-weight polyamides and other basic polymers as well as amides and esters of long-chain fatty acids. The presence of microcrevices on or near the surface of the round cross-sectional filament effectively reduces the magnifying lense effect of filament curvature and thereby helps to reduce the soiling of the product.

In the development of biconstituent fibers, difficulties arise from the fact that polymers of different chemical types are often incompatible with each other and many of them have low resistance to thermal degradation or thermoxidation. This renders the spinning and further processing of biconstituent fibers problematic.

II.2. Fibers with Improved Frictional, Abrasion, and Soil-release Properties

The admixing of some oligomers or high molecular-weight fatty compounds with polypropylene prior to spinning imparts to the fibers certain desirable properties such as improved frictional characteristics, abrasion resistance, antisoiling, soil release and static control. The number of such additives prescribed in the patent literature goes into hundreds, but precise information as to the extent of their effect and the mode of their action is lacking. In general terms it can be stated that microcrystalline waxes, soaps, silicon compounds, fatty esters of organic acids, high molecular-weight fatty alcohols, and wax-like materials containing fluoroalkyl groups, are useful as internal lubricants. They do so by exuding to the surface of the fiber and forming a thin layer of low frictional characteristics.

The incorporation of atactic polypropylene, low molecular-weight polyethylene, fatty oils and waxes in 5 to 10 percent concentration, are reported to improve the abrasion resistance of polypropylene fibers and are useful in the production of yarns to be used in ropes and cordage. The improvement in abrasion probably comes from increased flexibility of the fiber and the decreased fiber-to-fiber and fiber-to-metal friction. Amides and acidic esters of long-chain fatty acids, such as those commercially used as antiblocking agents in the manufacture of polypropylene and polyethylene film, serve as good antisoil and antistatic agents. In such compositions, the fatty "tail" is embedded in the body of the fiber and the functional "head" protrudes to the surface, forming a permanent layer of hydrophilic groups. Various oxidized and emulsifiable microcrystalline waxes are also useful for this purpose. The fiber remains lubricated and maintains soil-release and antistatic character through repeated washing. As the surface layer is washed out, a new layer quickly forms through exudation from the interior of the body of the fiber. The antisoiling action comes about because of the antistatic nature of these agents and the soil-release action is based on the ease of emulsification of the soiled surface layer of the additive.

II.3. Antistatic Fibers

Because of their hydrophobic nature, synthetic fibers are prone to electrostatic charge accumulation. The static charges cause processing problems such as "ballooning" of yarns due to "like charge" repulsion between individual filaments, excessive filament breaks, yarn distortion, clinging of fibers to metal, and tactile shock upon contact with processing equipment. Charged fabric is susceptible to increased contamination and dust attraction; in apparel, the charges are associated with annoying shock, uncomfortable cling, and potentially dangerous sparks and hazardous explosions.

There are currently three basic approaches to overcome the problems of static in synthetic fibers, namely, surface treatment, internal antistatics, and admixing of conductive fibers.

In surface treatment, thermoplastic polymers are formed into filaments by the melt-spinning process. An antistatic finish, applied to the freshly extruded filaments, suppresses friction and static problems in subsequent processing. However, such treatment lacks durability and is scoured off easily during the finishing operation of the final product. Thus, the surface treatment is effective for the processing of the fiber but does not provide any protection from static charge in the course of use.

Another approach that has been under extensive investigation for some time is the incorporation of antistatic agents in the polymer mix prior to extrusion. The antistats embodied in the polymer remain active subsequent to extrusion without impairing the bulk properties of the polymer.

It is appropriate here to examine briefly the functional mechanism and requirements of internal antistats. An internal antistat will have polar groups oriented toward the external surface of a polymer, thus forming a conducting layer on the polymer exterior which will reduce the electrical resistance and accelerate the dissipation of electrostatic charges. To achieve maximum effectiveness, the distance between the conductive globules of the antistat (dispersed as a discontinuous phase in the polymeric matrix) should be kept as short as possible. This is done by maximizing the number of closely, uniformly spaced conductive particles in the fiber cross section. Current flow is expedited by reducing the high resistance offered by the nonconductive fibers in the current path.

The requirements of a successful antistat may be summarized as follows:

Efficacy: An antistat incorporated in the fiber should be effective at low enough concentration levels so that polymer processing efficiency remains unimpaired. High concentrations of internal antistats may promote polymer reactivity and lead to difficulties in controlling both the polymerization process and polymer characteristics. Therefore, an internal antistat must be highly effective, and must show pronounced surface active response. Most of the internal antistats currently in use either increase fiber conductivity by hydrogen bonding with moisture, or contribute significant polarity to the fiber surface.

Thermal Stability: Internal antistats are exposed to elevated temperatures at the various processing stages of extrusion, drawing, texturizing, heat setting, dyeing, and curing. An antistat which is vulnerable to rapid volatilization or degradation upon heat exposures will be lost from the system, resulting in deficient fiber performance. Thermally unstable antistats may also cause fiber discoloration. A highly volatile antistat may generate unpleasant working conditions and potential environmental pollution through liberation of excessive smoke. Accumulation of antistat from vapor condensation on chamber ceilings or air ducts poses a fire hazard. Thermal stability is an important consideration in selection of internal antistats.

Compatibility: Some antistat compatibility with polymer is imperative in order to minimize potential polymer coagulation and subsequent processing problems, but it should be emphasized that antistats, for maximum effectiveness, should be no more than moderately compatible with the polymers. Preferably they should remain dispersed and not dissolve in the polymer. This permits slow exudation of the antistat to the polymer surface, thus forming a conductive monomolecular layer which adheres tenaciously enough to provide durable static protection.

An internal antistat with high polymer incompatibility will show rapid exudation to the polymer surface. The resulting heavy antistat concentration makes the polymer sticky, attracting dust. A film of the antistat thus formed is susceptible to easy abrasion, and the desired effect may be lost in a relatively short period.

Distribution: Uniform distribution of an antistat in a polymer is essential to minimize variations in physical and optical properties and also the streaking that follows fiber dyeing. Interfacial tension, or the difference between the surface tensions of antistat and polymer, is a key parameter in controlling distribution of the antistat in the polymer; a low interfacial tension assures uniform additive dispersion.

When the polymers are melt blended, proper mixing must be assured to achieve uniform antistat distribution. Excessive mixing - or high shear rate tending to reduce the antistat particle size below effective range - should be avoided. On the other hand, very large particles, exceeding 20 microns, may impede fiber spinning and drawing operations. Particle size should be preferably optimized to 2 to 5 microns. Particles become attentuated during spinning and drawing; resulting filaments thus contain elongated particles which insure effective performance.

Durability: Progress in internal antistat technology has been slow because of the scarcity of compounds possessing adequate durability. An internal antistat must have marked surface activity and show resistance to removal by multiple washings and dry cleanings.

In order to retain efficacy subsequent to heat setting, dyeing, and curing, an internal antistat must resist removal by steam distillation or vaporization, and deactivation by chemicals used in processing.

Effects Upon Polymer Properties: An ideal internal antistat should not impair the physical properties of a polymer through embrittlement, discoloration, or odor development. The antistat should not interfere with monomer or comonomer propagation which would lead to reduced polymerization and loss of fiber tenacity. The antistat contact time with the polymer melt prior to extrusion should be carefully controlled to prevent possible interaction or polymer interchange. Chemical interaction between the polymer and antistat can reduce the antistat effectiveness and, more importantly, cause reduction of desirable properties of the fiber. Generally, the contact time should be brief and consistent with efficient uniform mixing; not exceeding 30 minutes.

The antistats should not render fibers sensitive to photolytic or thermal degradation or other contaminant effects. The antistat choice often represents compromise between potential effectiveness and fiber modification. However, constantly advancing internal additive technology should allow development of both improved fiber properties and enhanced antistatic effect.

Toxicity and Pollution: In view of the increasingly stringent controls mandated by the regulatory agencies, toxicity and pollution are most important considerations in any product development. Consequently, during processing, antistats should not liberate excessive smoke or volatiles which could endanger the health of operating personnel, and they should not cause irritation or sensitization of skin on contact with the fiber. The antistat or its components remaining in air or water effluents should pose minimal health hazards and be innocuous to aquatic life.

Many types of chemicals have been reported as effective internal antistats for synthetic fibers. Ethylene oxide adducts predominate, however, because they are readily available in great variety, may be easily modified for desired characteristics such as solubility, and are versatile performers. These materials and various other antistatic agents will be compared with regard to advantages and limitations.

Polypropylene can be rendered antistatic by use of polyglycols, polyglycol monoethers, polyglycol monoesters, alkylalkanolamines, and alkoxylated allylene diamines. These products provide good static protection, but exude much too rapidly to the polymer surface. On the other hand, salts of fatty acids, due to their high molecular weights, diffuse too slowly to the surface and are inefficient antistats. Fatty acids of molecular weight lower than caproic potentially improve the diffusion rate, but evolve objectionable odors. The hydroxyethyl-dodecyloxypropylamine salt of hydroxypropionic acid reputedly gives permanent antistatic effect, shows an optimum diffusion velocity, and keeps the bulk polymer properties intact.

Incorporation of quaternary ammonium salts such as stearyl polyether acetal ammonium chloride, used with an inorganic "modifier" such as aluminum oxide, produced polypropylene filaments with antistatic properties durable to more than 30 wash cycles. Other fiber properties such as tenacity, elongation, and modulus remained.

Metal salts, such as the calcium salts of various amphoteric surfactants obtained by the reaction of dodecyl trimethylene, nonylamine or stearyl imidazoline with maleic anhydride, impart permanent static protection for the life of the fiber.

Schiff's bases (azomethane derivatives such as the reaction product of propionic aldehyde and lauryl amine) are reported as suitable internal antistats for polypropylene.

In view of the numerous problems encountered with active internal antistats, emphasis is recently being placed on making fibers more conductive with chemical inert substances. To achieve freedom from static one does not need to have all of the fibers conductive but only to blend the static-prone fiber with a conductive fiber. Metallic fibers such as stainless steel or copper in amounts as small as 0.5 percent, are sufficient to provide static protection. This protection, however, deteriorates as fibers mat under repetitive compression, and handling of metallic fibers on conventional textile equipment requires special precautions. An alternate procedure is to modify the fiber to impart conductivity and then use this fiber as the blending fiber or sometimes as the principal fiber. This approach has been quite successful in the case of nylon and polyester fibers, and there is no reason why it should not prove successful for polypropylene fibers.

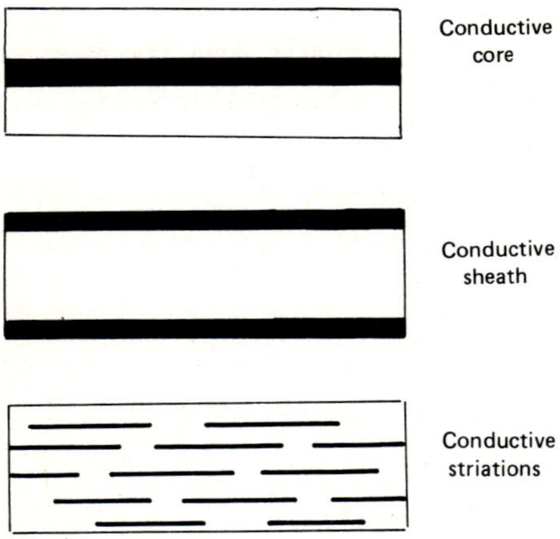

Fig. 1. Fiber Modification for Static Resistance

This has been done in three ways (Fig. 1). One approach is spinning a fiber with a conductive core. This has been accomplished in Antron III carpet fiber where the fiber has a sheath core structure with a core composed of polyethylene containing high loadings of conductive carbon black. The nylon sheath, which comprises 96 percent of the fiber, imparts all the good mechanical and durability features of nylon.

The second approach is to spin a fiber where the conductive coating is attached to the surface of the fiber. This technique has been used by ICI in the preparation of "epitropic fibers". Bicomponent fibers are used where one of the components has a lower melting point than the other. The lower melting point component can be softened preferentially and particles imbedded in it while the second component remains unaffected by the heat treatment. The particles become an integral part of the fiber surface. Approximately 2.5 wt-percent of imbedded carbon decreases the electrical resistance of polyester fiber by a factor of one million. These two approaches are finding application in blends such as staple yarns for carpets.

The third approach involves introduction of overlapping conductive segments. Many attempts have been made to incorporate conductive particles such as carbon black or metals, but this is not attractive because loadings exceeding 30 percent are needed to achieve useful conductivity. If the conductive segments are elongated and overlapping, much smaller loadings are sufficient. One of the most intensively explored approaches of this technology is incorporation of poly-(ethylene oxides) into the fiber. As mentioned before, the additive forms a dispersion in the bulk polymer and thereby a structure is obtained upon spinning and drawing which contains elongated striations of poly(ethylene oxide). This markedly enhances the conductivity of the fiber. While some of the poly(ethylene oxide) is removed when the fiber is dyed and washed, enough remains to provide a conductive layer at the striations. This technique has been extended to other additives which form incompatible conductive polymer blends.

II.4. Fibers of Improved Fire Retardancy

The flammability hazard of textile materials has received recognition for a considerable period of time. Much of the impetus for conducting research work on flame-retarding textile materials has been generated by enactment of federal, state, and local legislations, as well as concern by various textile organizations. Of particular importance in the United States and Europe are the recent combustibility tests for home furnishing materials and children s sleepwear. It is anticipated that the need for better flame resistance for home furnishing applications will be a definite requirement in the not too-distance future not only in the United States and Europe but also in other parts of the world.

Although polypropylene is a comparatively slow-burning material, carpets and fabrics made of polypropylene fiber do not meet the current ASTM D-2859 Methamine Pill test and the imminent federal flammability standards in some constructions. Considerable interest is therefore being expressed to upgrade the flame retardancy of these products. The flammability standards for home furnishing products are important for polypropylene, as most of the multifilament and staple fiber is currently being consumed for carpeting and upholstery fabrics.

Principally, the following three approaches can be taken to improve the flame retardancy of a fabric:

(1) Flame-retardant finishing of surface fibers

(2) Application of flame-retardant latexes for carpet and upholstery backing

(3) Modification of polymer composition for better flame retardancy.

The first approach is not applicable to polypropylene fibers because of the inert nature of the polymer which makes it resistant to any chemical treatment generally used for other natural and synthetic fibers. The second approach is applicable only where a back-coating is required. This is the case with tufted carpets and upholstery fabrics. A number of fire-retardant latex formulations has emerged. These are based on the use of halogenated latexes, such as polyethylene-vinylchloride latexes, or the use of ordinary latexes mixed with fire-retardant additives. These generally fall into two categories: metal hydrates (such as alumina) and compounds of halogen, phosphorous, antimony, or boron. The former group functions by releasing water of hydration at high temperatures, while the latter group functions by char formation and flame poisoning, as discussed elsewhere in this section.

The back-coating approach is evidently limited in scope and depends on the expertise available in the finishing mill. A more basic approach would be to modif the polymer composition in order to improve the burning characteristics of the fiber itself. Certain chemicals are notable for their ability to impart flame resistance to polymeric materials. Some of these have been tried for polypropylene and through these efforts a knowledge of the mechanism of polymer combustion and fire retardancy is slowly emerging.

The combustion of flammable high polymers, including polypropylene, takes place in the following manner (ref.3):

(i) Heating: As a result of heating, the temperature rises, the rate determined by geometric factors of the product and the environments.

(ii) Deterioration: Thermal oxidation takes place, according to the mechanism discussed in Chapter III. Polypropylene is oxidized to yield low olefins, aldehyde hydroxyl radicals, etc. Hydroxyl radical oxidizes carbon monoxide to carbon dioxide with the evolution of heat.

(iii) Pyrolysis: When most of the bonds in polypropylene are broken, the polymer is degraded yielding gaseous products and polymer fragments. The char formation is minimal. Among gaseous products, pentane, pentene, hexene, butane, ethylene, propylene, formaldehyde, and carbon monoxide are flammable. Polymer fragments of low molecular weight form smoke and escape while higher molecular-weight fragments form a liquid pool around the burning spot or drip down the product.

(iv) Combustion: Combustible gases formed by pyrolysis are ignited in the presence of oxygen, and combustion takes place. When a portion of the material burns, the heat of combustion gives rise to thermal decomposition of other portions. Formed gases are heated and burn.

The fire retardant chemicals of interest for incorporation into the polypropylene spinning mass are halogen compounds, phosphorous compounds, antimony oxides, energy quenchers, and miscellaneous synergists. Flame-retardant compositions have a close analogy with stabilization formulations, i.e., often a complex recipe is involved wherein one component acts as primary flame retardant and the others as synergists. The art of flame-retardant formulation is still in its embryonic state and no commercial successes have yet been reported. The following comments about the flame retardants of interest are meant as guidelines to further research and development.

It has not so far been possible to prove experimentally exactly how halogen flame retardants work when incorporated in polypropylene. Several theories are current. According to one of these theories, the fire retardant causes the burning substrate to follow an abnormal course of decomposition, yielding less flammable gaseous products. According to another mechanism, the fire retardant itself decomposes to gaseous products which mix with the normal gaseous products from the decomposing polymer to yield a less flammable mixture. In the latter case the gaseous products of the flame retardant decompose under heat in such a way that radicals and atoms are produced that act as chain terminators.

The second mechanism is more popular and experimental results show that it is the case at least with bromine compounds. The process of combustion is assumed to be a radical reaction. When the chain interruptions outweigh the chain branching, the reaction dies down (ref.2).

In details, one starts out from the reactions in a hydrocarbon flame, illustrated by the following scheme:

Exothermic reaction	$HO^{*} + CO \longrightarrow CO_2 + H^{*}$
Chain branching	$H^{*} + CO_2 \longrightarrow HO^{*} + CO$
Inhibition	$RX \xrightarrow{Q'} R^{*} + X^{*}$
	$X^{*} + H^{*} \xrightarrow{Q'} HX$
	$HX + OH^{*} \longrightarrow H_2O + X^{*}$
	$X^{*} + HR \longrightarrow HX + R^{*}$

The polymer RH absorbs heat Q' and yields R* and H*. Oxygen attacks R* to start oxidation. The fire retardant RX absorbs energy Q and decomposes to yield X*, which reacts with R*. X* reacts with H* to form HX; it also reacts with OH* formed by oxidation reactions. Accordingly, the exothermic reactions are suppressed and the combustion system is cooled.

According to this mechanism, the bond energy of carbon-halogen bonds in organic compounds must be lower than the bond energy of carbon-hydrogen bonds in the polymer. The flame retardation effect is greater when the difference between the bond energies is greater. In other words, when a heated polymer RH dissociates into R* and H*, organic halogen compound RX has already yielded X*. Before oxygen attacks R*, X* attacks it to terminate decomposition. As Br-C bond energy is less than Cl-C bond energy, it is readily understood that bromine compounds are more effecti than chlorine compounds. Experience bears out this expectation. Some investigator (ref.3) are not convinced of this mechanism of flame retardancy of halogen compounds. Their results show that the *structure* of the halogen compounds, rather than the *content* of halogens, are important factors. Experience shows that aliphatic halogen compounds are more effective than aromatic halogen compounds.

It is known that polyolefins may be made self-extinguishing by admixing chloroparaffins with them. These compounds, however, tend to eliminate hydrogen chloride at temperatures far below those normally considered suitable for melt-spinning of polypropylene. The stabilizers are rapidly depleted and the polymer experience a severe degradation. In addition, the process machinery is attacked by HCl and corroded. Aliphatic bromine compounds are even worse in this respect. Halogenate aromatics are better as far as heat stability is concerned but they are in general poorer in their fire-retardancy effect. Cyclic halogenated compounds seem to be the best compromise. Several fire-retardant compounds of this sort have been commercialized and many more halogenated compounds based on cyclic ring structures and aromatic-aliphatic molecules have recently been described in patent literature

The utility of most of the commercially available halogenated fire retardants is limited for polypropylene fibers because of the excessive degradation of the additives and their adverse effect on the heat stability of polymer at the high processing temperatures needed for melt-spinning. The processing temperature limitation for these flame-retardant polymeric compositions is said to be raised i incorporating small amounts of an aromatic ester, such as di-n-octyl phthalate (ref.4). The inclusion of HCl fixing agents, particularly epoxide-containing structures, may also be helpful.

Certain phosphorous compounds, specially organic phosphates, have been found effective as fire retardants for polypropylene. They are effective in decomposition rather than in the subsequent combustion of the polymer. They cause the carbonization of polymers and prompt the formation of char. Accordingly, phosphorous compounds hinder oxidation, which is a major heat source for the combustion of the polymer. Polypropylene burns without forming char. It therefore requires a large amount of phosphorous compound for effective char formation.

The phosphorous compounds are always used with brominated compounds with which they synergize. Hydrogen bromide is the most effective compound for flame poisoning. However, in the absence of phosphorous, it does not remain for a long time in the region of combustion because its boiling point (-67°C) is low. In the presence of phosphorous, hydrogen bromide promotes the formation of char and forms high boiling PBr_3 (b p, 173°C), PBr_5 (b p, 106°C) and $POBr_3$ (b p, 189.5°C). These compounds repeat the process of bromination and dehydrobromination. Accordingly, the effect increases with the residence time of these compounds in the combustion zone. A similar effect appears in the case of phosphorous and chlorine compounds as well. Most of the known phosphorous fire retardants are unstable to hydrolytic degradation but active research continues to synthesize improved structures. Several phosphorous compounds having bromine atoms as integral part of the molecular structures have recently been synthesized. They show an *autosynergistic* effect.

Antimony trioxide *per se* shows no flame-retardation effect. When it is used simultaneously with halogens, a marked fire-retardance effect appears. Presumably, $SbBr_3$, having a high boiling point (b p, 280°C), is formed and stays in the combustion zone for a long time. The use of antimony trioxide is somewhat problematic in spinning of polypropylene because of its dulling effect on the extruded fiber, high plugging rate, and high prices. These factors have spurred the search for other materials which could be used for Sb_2O_3. Antimony glycoxide, zinc borate, sodium tetraborate and several phosphorous compounds have been mentioned.

Another group of synergistic compounds with the following general formula has been described by BASF (ref.5).

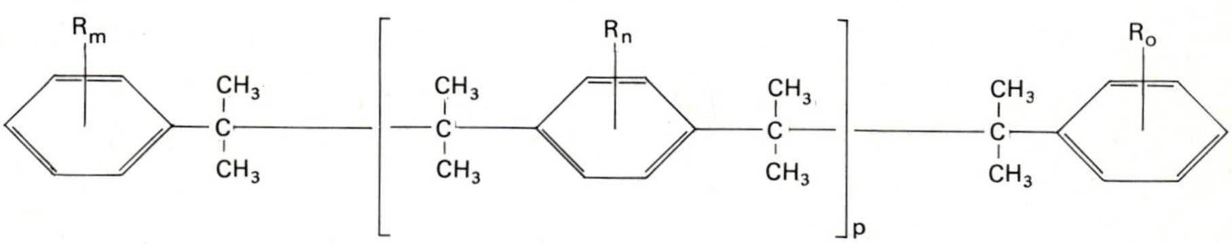

m ≠ o ≠ to 5, n = 0 to 4,
p = 0 to 200 , R = Alkyl, Cl, Br

|I|

The addition of halogen compounds and antimony trioxide can be significantly decreased by the use of the compounds of the above formula. This synergistic effect can be explained as follows, (ref.6):

$$Sy - Sy \dashrightarrow 2\ Sy^*$$

$$Sy^* + HR \dashrightarrow HSy + R^*$$

$$HSy + Br \dashrightarrow HBr + Sy^*$$

The polymeric synergists form radicals in the flame. These radicals, together with the hydrocarbon HR, form stable intermediate product SyH which continues to react, combining X^* and H^* at high speed. The two-part reactions depicted above are characterized by the fact that they take place together and at a greater speed than direct reaction

$$X^* + HR \dashrightarrow HX + R^*$$

The synergist, therefore, essentially acts as a transfer catalyst. This is in agreement with the fact that only small amounts of polymeric synergist |I| are necessary to achieve a synergistic effect (ref.6).

In addition to the conventional and well known classes of fire-retardant compounds discussed above, there is a possibility of utilizing common antioxidants for imparting some degree of fire retardancy to polypropylene fibers. The phenolic free radical inhibitors would not be efficient in this respect because, in combustion reactions, the rate of initiation of free radical formation is very high and chain length short. However, the energy quenchers, such as UV 1084 or Ferro AM 110, may be quite effective in quenching the excited states of polymer molecules in the condensed phase and preventing a rapid oxidation of the polymer. The research in this area of fire retardants is still in its infancy.

The problem of imparting fire retardance to polypropylene fiber is similar in nature to that of stabilization against light and heat. Thus, not only the efficiency of the fire retardant is important but also other factors such as compatability, color, chemical reactions with additives, effect on polymer stabilization, toxicity, odor, dispersibility, economy, etc., have to be taken into account. There are few additives known which meet all these requirements. The fiber containing fire retardants is generally opaque and dull in appearance and its thermal and UV stability is generally low. In view of the large amount of research and development work currently being undertaken in this area, however, significant improvements can be expected in the near future.

II.5. Bacteriostatic Fibers

Efforts have lately been made to introduce the concept of wall-to-wall carpeting in hospitals, schools, convalescent homes, nursing homes, and sanitoria. In spite of the evident advantages in the cost of floor installation, maintenance and comfort, success toward this objective so far has been only modest. One major objection has been the danger of bacterial growth on the carpeted floor. Several antibacterial finishes have been suggested but these treatments generally show only temporary bacteriostatic properties. On shampooing, the finish is removed and the bacteriostatic action of the fiber surface severely diminished. To overcome this difficulty, the incorporation of a bacteriostatic agent in the polymer prior to spinning has been suggested. Advantage has been taken of the fact that the bacteriostatic agent present within the body of the fiber would slowly but continually exude to the surface and maintain a fraction of its concentration on the fiber surface. The quantity of bacteriostatic agent removed from the fiber surface in service would be, therefore, continually compensated by the reservoir present in the body of the fiber.

Hexachlorophene has proved to be an efficient bacteriostatic agent and is amenable to blending and extrusion with polypropylene (ref.7). Its tendency to exude to the fiber surface can be adjusted by choosing a proper concentration in the formulation. The concept of formulating thermoplastic materials with bacteriostatic agents prior to fabrication into products has been explained by Hopfenberg and Tulis in detail (ref.7). Derivatives of 5-nitrofuran and their mixtures with trichlorocarbonilide has also been mentioned as bacteriostatic agents suitable for incorporation within the body of the fiber (ref.8). New basic developments include the formulations with synergists which enhance the effectiveness of the primary bacteriostatic agent in the composition and provide additional properties such as antistatic behavior. The choice of bacteriostat or bacteriostat/synergist combination is obviously dictated not only by long-term bacteriostatic effect but also by human safety criteria in compliance with the current health and safety requirements.

III. PHYSICAL MODIFICATIONS

III.1. Textured Filament Yarns

The highest concentration of development efforts in filament yarn of polypropylene during the past few years has been in textured yarns. The growth of this market in the last decade has been truly phenomenal.

The effect is to impart stretch, luxurious bulk, greater absorbency, and improved hand, combined with the rugged performance, on continuous-filament yarns. Texturizing is the permanent introduction of crimp, loops, coils, or crinkle into otherwise smooth, continuous filaments.

Two distinct categories of textured yarns are:

(1) Stretch Yarns

(2) Bulk Yarns

Stretch yarns are characterized by high elongation, rapid recovery, and moderate bulk per unit weight. Bulk yarns combine great mass and bulk with limited stretch and impart to the end-use product a soft, full hand and adequate recovery from extension or compression. Nylon stretch yarns are extensively used in growth-type and stretch-to-fit garments. Polypropylene has just begun to penetrate the apparel field and no significant quantities of polypropylene stretch yarns have yet been produced. Due to its slow recovery and somewhat higher creep than nylon, polypropylene stretch yarns might not be able to capture any sizeable market in stretch garments. Bulk yarns, on the other hand, are important for upholstery fabrics and carpeting - a traditional area of polypropylene consumption. Thus, in the textured-yarns area, polypropylene bulk continuous-filament yarns are much more important than stretch yarns.

Polypropylene does not differ very much from nylon and polyester in the ways and means of texturizing. It can therefore be textured by any one of the ingenious techniques which have been successfully employed for texturizing of nylon and polyester yarns. Edge crimping (Agilon), stuffer-box crimping, gear crimping, false twist texturizing, air-jet bulking (Taslan), and steam-jet bulking, are the most important methods suitable for texturizing polypropylene yarns. The success of all these texturizing processes, with the exception of air-jet bulking, depends on the thermoplasticity of polypropylene and its ability to be deformed, heat set, and developed. These methods have been adequately discussed in connection with nylon and polyester yarns by several workers and considerable amount of technical literature is available on this topic. Some selected references are listed in the Bibliography of this chapter. For patent literature, an up-to-date file of the "Textile Abstracts" published by Shirley Institute, Manchester, UK, and the "Textile Technology Digest", published by the Institute of Textile Technology, University of Virginia, Charlottesville, Va., USA, should be consulted.

From the point of view of the fiber producer, the production of polypropylene bulked-continuous-filament (BCF) yarns for carpets, upholstery, drapery, and crafts is very important commercially. Traditionally, stuffer-box and gear-crimping techniques have been employed for the production of these yarns. Practical experience has shown, however, that the texture uniformity is very difficult to attain with these techniques and the overall quality of the product is not up to the mark. The cover, bounce, crush resistance, and general appearance of the carpets made of the stuffer-box or gear-crimped yarns is inferior to the commercial nylon carpets with which they are meant to compete. Because of the texture nonuniformity of these yarns, the tufted or woven products are often streaky. The hand is generally harsh and the processibility of the yarns into finished products is problematic due to broken filaments and other defects in the yarn. Because of these reasons, more and more BCF yarns are now produced through fluid-jet texturizing techniques.

Very little is known in the public regarding this promising technique. A brief review is therefore in order here.

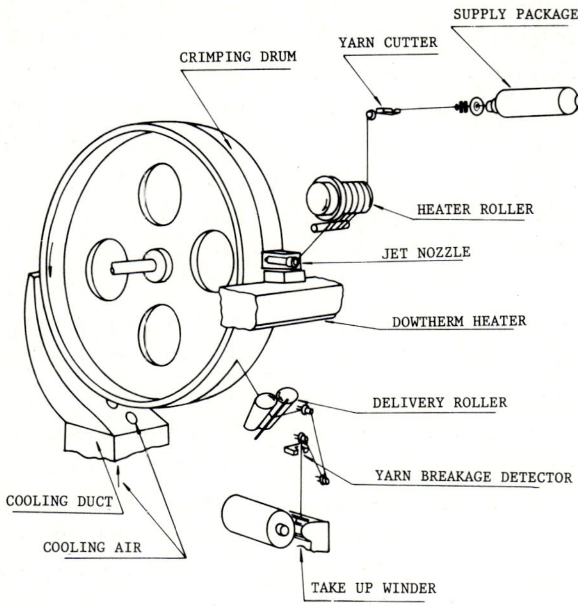

Fig. 2. A Fluid-Jet Bulking Arrangement for the Production of BCF Yarns (Courtesy Mitsubishi Heavy Industries)

The fluid-jet bulking process for the production of BCF yarns basically consists of feeding the filament yarn into a hot fluid (steam or hot air) jet at a speed of X_1 and taking up the bulked yarn at the other end at a speed of X_2. X_1 is higher than X_2; the difference constitutes the degree of bulk achieved. In almost all successful fluid-jet bulking techniques the filament yarn emerging from the exit end of the fluid jet is given an opportunity to flare open and then impinge onto a restriction, which may be a perforated plate, a screen, a card-cloth, or even a wad of already collected filaments. As the filaments impinge on a perforated element and onto a wad of loosely collected filaments, the hot fluid is allowed to escape from the system and, concurrently, the crimped filaments are withdrawn at a controlled rate. Figure 2 shows one such arrangement. The bulking process can, of course, be integrated with a cold-drawing or even with a spin-drawing process as depicted in Fig. 3.

The fluid jet has three distinct functions to perform:

(1) It aspirates the yarn into the hot fluid-treatment environment within the jet.

(2) It heats up the individual filaments of the yarn to make them easily bend and kink while they strike an obstacle in their path at the exit of the jet. In the

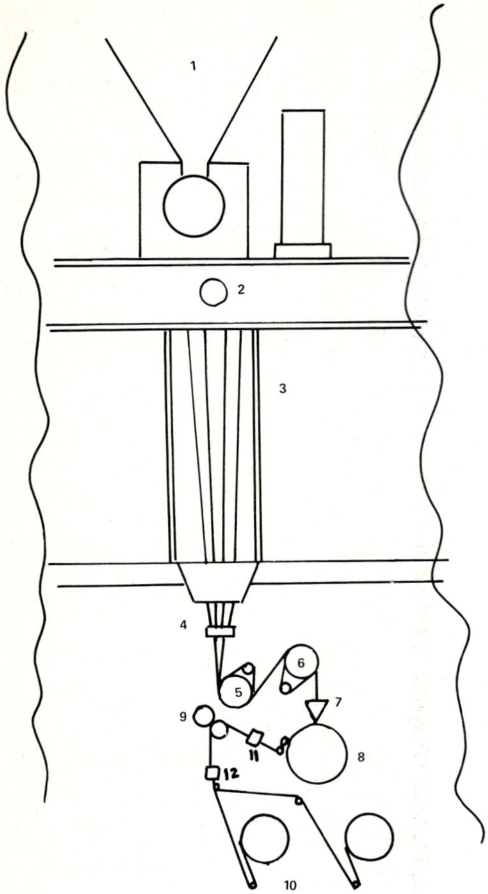

Fig. 3. Production of Polypropylene BCF Yarns Through One-Step Spin-Draw-Bulk Arrangement. (1) Extruder, (2) Metering Pump, (3) Quenching Cabinet, (4) Finish Applicator, (5) Feed Roll, (6) Draw Roll, (7) Steam Jet, (8) Bulking Roll, (9) Take-Up Nip-Roll, (10) Winder, (11) Interlacer Jet, (12) Finish Applicator. (Courtesy International Fibers)

patent parlance, this function is often referred to as "plasticization". An additional heating element, such as a heated feed roll, may sometimes be necessary to achieve effective plasticization.

(3) It opens up the yarn bundle into individual filaments and propels them to an obstruction with a force sufficient for each filament to get randomly kinked along its length. Unlike gear or stuffer-box crimping, the crimp so produced is random and irregular in shape and size and is three-dimensional in character. These factors impart to the fluid-jet textured yarn a high degree of bulk and cover.

In addition to imparting the crimp to the individual filaments of the yarn, two additional factors are important in making any fluid-jet bulking process a success.

(1) The crimped yarn must be "set" by the application of heat in order to impart on the filaments a strong crimp memory. Some degree of crimp setting is inadvertently achieved, especially at slow processing speeds, by the heat retained in the interior of the filaments as they emerge from the fluid jet. Some incidental heat setting is also realized by the constant impinging of hot fluid on the wad of already crimped filaments. For more effective heat setting or for high processing speeds, however, a positive heat-setting step is often required. This is especially necessary for polypropylene yarns for two main reasons. Firstly, polypropylene fibers are significantly more crystalline than nylon and polyester fibers. The molecular processes involved in heat setting and annealing are considerably slower in crystalline regions than those in the amorphous regions. Secondly, the glass transition temperatures of nylon and polyester are above room temperature while that of polypropylene is below room temperature. This means that once the filaments of nylon and polyester are given a shape at high temperatures, they maintain that shape at room temperature. On the contrary, polypropylene still has some molecular mobility below the bulking temperatures and is capable of altering its shape under stress at room temperature. This tendency of polypropylene to undo the crimping process at room temperature cannot be eliminated altogether but it can be minimized to a considerable degree through effective heat setting. (See Chapter VII for more on this topic.)

(2) The crimped and heat-set yarns cannot be tensioned for wind-up if the yarn is still hot, otherwise the crimping process would be largely undone. The crimped-yarn tow must be cooled as near as possible to room temperature before it is tensioned for opening and winding. Some degree of cooling is incidentally achieved at slow processing speeds as the yarn travels through the air to the winding station. The interlacing and finish application steps further help to cool the yarn. However, the cooling of polymers is a very slow process. At high speeds of processing a positive step of cooling must be introduced to achieve a high degree of bulk in the finished goods.

The properties of feed yarns have an overbearing effect on the success or failure of any fluid-jet bulking process. They also determine to a large extent the quality of the finished product. The first and foremost criterion of a feed yarn to be converted into a BCF yarn product is its ability to deform permanently into an imposed configuration. This criterion requires that the constituent material of the filaments be able to experience a plastic flow under heat and stress. The higher the elongation of the filament, the easier it is to deform permanently under crimping forces. Low elongation yarns (high draw ratios), on the other hand, experience plastic flow to a limited degree: they are therefore difficult to kink and deform. Similarly, highly crystalline yarns are difficult to bend or compress, nor is it easy to impose on them a plastic flow: they are therefore difficult to crimp.

In heat-setting operation, the amorphous or less ordered regions of the filaments tend to rearrange themselves into more ordered regions. The less perfect crystallites tend to perfect themselves and the small crystallites endeavor to grow in size. If the feed yarn does not have enough of such less ordered regions, the opportunity for the polymer material to reorganize itself in the imposed configuration of the crimp is greatly reduced and the heat setting becomes difficult. Thus, high-shrinkage yarns (indicating a large proportion of amorphous or less ordered regions) are generally more amenable to heat setting and thereby to registering the crimp memory in the filaments.

The yarns of too-high elongation and too-high linear shrinkage cannot, however, be used as appropriate feed material. This is because of their tendency to fuse and flatten in the crimping device. Moreover, the finished product obtained from such feed yarns may lack in certain desirable properties. For example, the resistance to creep of the bulked yarns under continued tension, stability to heat shrinkage of the fabrics, resistance to roll-crushing of the carpet pile, and dyeing uniformity, are adversely affected by high shrinkage and high elongation of feed yarns.

In addition to the stress-strain properties and morphological structure of the filaments of feed yarn, the efficiency of a bulking process also depends on the geometry of the fiber cross section. It is common knowledge in the industry, for example, that a yarn of trilobal filament cross section bulks up much more easily than the cross section of a solid filament, such as round. The geometry of filament cross section is important from several aspects. Firstly, the solid cross section, such as round and delta, has an effectively thicker core than that of a pronged or lobed cross section, which has to be bent inelastically to form a crimp. Secondly, a thicker core takes more time to heat thoroughly for effective plasticization. Thirdly, a cross section of low surface area, for example round or delta, takes more time to cool and lock in the crimp configuration. Within the lobed or pronged cross sections, the symmetry of the cross section is also important. A symmetrical cross section, such as equilateral Y, would have more propencity to bend in all directions than asymmetrical Y. The higher probability of bending of the former cross-section would result in higher frequency crimp under comparative processing conditions.

The last, but not the least, factor relating to the feed yarn and having a bearing on process parameters and bulk-yarn quality is the spin finish on the yarn. The volatile components of the spin finish are a nuisance in operation, the depositing finishes foul the yarn path and impair the continuity of operation, too low a level of spin finish leads to spot fusion and light spots in bulked yarns and too high a level interferes in heat transfer to and from the filaments.

III.2. High-Tenacity Industrial Yarns

Synthetic fibers are finding increasing use in industrial applications. Polypropylene continuous-filament yarns have a considerable share of this market. The volume of polypropylene yarns for industrial uses is expected to grow rapidly as the synthetic fibers in general replace natural fibers in such uses as fishnets, cordage, marine ropes, agricultural twine, conveyor belting, tenting material, support fabrics, and several other novel uses where natural fibers could not be used because they lack strength. Polypropylene could command a major share of this market.

Polypropylene textile fibers generally have a tenacity of 3 to 5 g/den with 15 to 50 percent elongation. Products of higher strength are possible, however, and tenacities as high as 14 g/den have been obtained. In general, the higher the tenacity, the more acceptable a yarn is for industrial end uses. Considerable research and development work has been carried out by several research laboratories and fiber manufacturing companies to investigate the material and process parameters which influence the tensile properties of polypropylene yarns, and to optimize these factors at commercially acceptable production rates. Some of the results will be discussed in this section.

Fiber tenacity is considered the most important property of industrial yarns, but other yarn properties cannot be neglected altogether. This applies especially to toughness, impact resistance, the work of rupture, extension at break, modulus, and strength conversion efficiency. Added to this list are also such properties as fatigue resistance and abrasion resistance of the product.

TABLE I

Interchangeability of Tenacity and Elongation in Polypropylene Fibers (ref.10)

	D.R.	Ten. (g/d)	Elong. %	$T.E^{\frac{1}{2}}$
Set A	4.8	4.5	68	37
	5.5	5.5	40	35
	6.6	6.5	30	36
	6.9	7.5	26	38
Set B	2.0	3.3	260	52
	2.7	4.5	170	59
	3.0	5.3	150	65
	3.3	5.7	125	63
	3.6	6.0	95	59

It is well known that drawing of a melt-spun fiber can increase its ultimate tenacity, usually at the expense of its elongation. This exchange function has been analyzed for various fibers (ref.10), including polypropylene. It has been shown that the tensile data of polypropylene as well as some other fibers can be fitted into the following empirical equation for a given parent spun yarn:

$$(\text{Tenacity}).\ (\text{Elongation})^{\frac{1}{2}} = \text{Constant} \quad \text{or} \quad T.E.^{\frac{1}{2}} = K$$

The constant of the above equation, often called the *tensile factor* , can be considered as an approximate measure of toughness and may be used instead. Table I shows the constancy of this factor for two sets of polypropylene yarns. Strictly speaking, toughness is proportional to the area under the load-extension curve of a fiber. Toughness of a fiber is of prime importance for most industrial uses. It is closely related to the impact resistance of the yarn. Toughness also determines the useful life of the fibrous product. In the development of high-tenacity industrial yarns, it is important that toughness increases in at least the same proportion as the tenacity. As will be shown in the forthcoming discussion, polypropylene fibers of high tenacity and adequate toughness can be prepared by a proper choice of polymer and processing conditions. These yarns compare very well with commercially available polyester and nylon high-tenacity yarns (Table II)

TABLE II

Tensile Properties of Some Representative Industrial Yarns (ref.10)

Yarn	Tenacity (g/den)	Elong. (%)	$T.E.^{\frac{1}{2}}$
Polyester I	7.4	15	28.8
Polyester II	8.5	9	25.5
Nylon I	7.9	14	29.2
Nylon II	9.0	15	35.1
Polypropylene I	7.5	25	37.5
Polypropylene II	9.3	21	42.8

The modulus, usually measured on continuous filament yarn, is the 2 percent secant modulus. This is the stress (in g/den) to produce a 2 percent extension multiplied by 50 to give the stress at a theoretical 100 percent extension. The 2 percent modulus is regarded as being most meaningful, since most products are designed with a safety factor of from 5:1 to 10:1 and the actual extension of the yarn during use will be of the order of 2 percent. High modulus is a desirable property in most applications, but it may be necessary to sacrifice toughness or extension at break to achieve high modulus in a fiber. Polypropylene high-tenacity yarns produced by normal spinning and drawing techniques generally do not have adequate modulus for critical low-extension end uses, but through the employment of suitable polymer and processing conditions, yarns with high modulus values can be prepared. If the tenacity is increased without a corresponding increase in modulus, any product designed to a tensile specification would be a disadvantage as far as extension at working load is concerned.

As well as having good properties in single-yarn form, it is important for a high-tenacity industrial yarn that the tensile properties are utilized advantageously in the final product. In general, it is necessary to process all yarns to final product before the conversion efficiency can be calculated. It would obviously be an advantage if some measure of the conversion efficiency could be obtained from the properties measured on the continuous-filament yarn in single form. The two most important factors in determining the conversion efficiency are considered to be the extension at break of the yarn and the shape of the curve before break (ref.11).

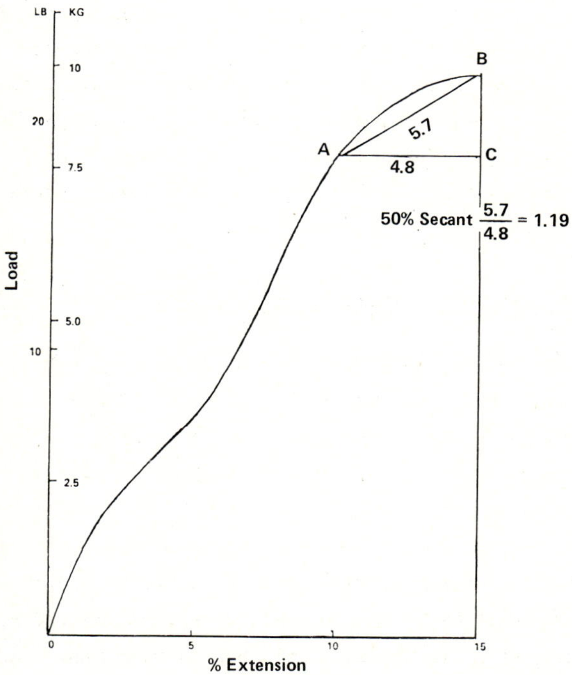

Fig. 4. Method of Calculation of 80 Percent Secant From Load-Extension Curve (ref.12)

Glen (ref.12) has derived a relation between the conversion efficiency and the secant at 80 percent of the breaking load for polyester two-ply cabled yarns. The higher the 80 percent secant value, the lower is the conversion factor. Although no such relationship has been reported for polypropylene high-tenacity yarns, it is to be expected that the relation would hold for polypropylene fibers as well in qualitative terms. The method of making this measurement is shown in Fig. 4. It involves drawing a line from the break point to a point representing 80 percent of the breaking load. If the load-extension curve has a relatively flat top, then the conversion efficiency would be good and lines AB and AC would be of a similar length. The closer the ratio AB/AC to unity, the better the conversion efficiency should be.

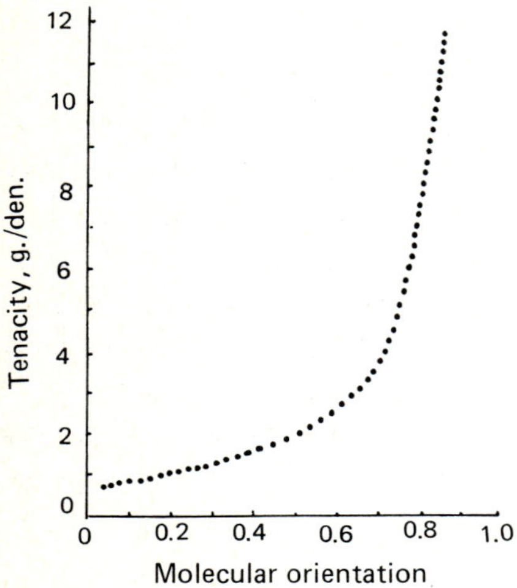

Fig. 5. Relationship Between Molecular Orientation and Tenacity of Polypropylene Monofilaments of Different Molecular Weights. (ref.14)

Excellent studies have been conducted in the past (ref.13-18) on the relationship of structural parameters and tensile properties of polypropylene fibers. It has been well recognized that the tenacity of the fiber increases as its molecular orientation increases (Fig.5). The overall fiber orientation is composed of crystalline orientation factor and amorphens orientation factor: both of these factors are influenced by melt drawing as well as cold drawing. Molecular orientation increases with increasing melt draw-down ratio and cold-draw ratio. Cold drawing is much more effective for orientation than melt drawing. Polypropylene fibers spun with low spin orientation are capable of attaining a higher degree of total orientation on cold drawing than those spun with higher spin orientation. Similarly, rapidly quenched polypropylene fibers allow much higher total orientation on cold drawing than the slowly cooled fibers (Fig.6). These peculiarities can be easily explained on the basic structural considerations discussed in Chapters V and VII.

It has been mentioned previously, in Chapter VII, that the undrawn crystallized polymer consists of folded chain lamellae joined by a small but important fraction

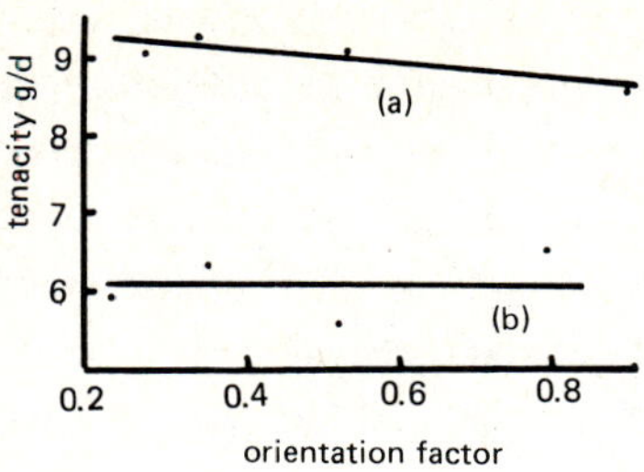

(a) Smectic Fibers

(b) Crystalline Fibers
(Crystallinity 50-60 Percent)

Fig. 6. Maximum Tenacity of Drawn Fiber as it Depends on Orientation and Crystalline Nature of Undrawn Fiber (ref.16)

of tie molecules. On drawing, the lamellae break-up into folded chain blocks, of the order of 100-300 Å in diameter, which subsequently aggregate to form the fundamental units of drawn structure, the fibril. The fibril is comprised of alternating crystal blocks and amorphous layers, the latter consisting of chain folds, free chain ends, and tie molecules. Thus, the folded chain blocks which were neighbors laterally in the underformed structure could become neighbors longitudinally in the drawn fibril. The tie molecules which connect the blocks in the longitudinal draw direction provide the principal source for strength and modulus of the fiber. According to this model, illustrated in Fig. 7, the tensile strength and modulus will increase with the fraction of the tie molecules and their alignment along to draw direction.

High melt draw-downs in the course of spinning invariably lead to high spinline tensions which give rise to row-nucleated crystalline morphology. This type of morphological structure resists orientation in cold drawing because the mode of stress concentration is such that little breakdown of lamellae into folded chain blocks takes place and fibrillar structure is not fully developed. The slowly cooled filaments are characterized by a higher degree of structural perfection, a higher degree of crystallinity, larger crystal sizes, and above all, a smaller number of tie molecules than obtained by rapid cooling. As a result, a slowly cooled filament would not only resist cold drawing but also yield a drawn structure with a low level of load-bearing tie molecules as compared to a rapidly quenched fiber.

Several investigators (ref.13-20) have considered the dependence of fiber tenacity on molecular weight and molecular-weight distribution of the polymer. The result of these investigations can be expressed in the form of the following empirical equation (ref.21):

$$\text{Ten.} = K_1 + K_2 (M_v)_f + K_3 \frac{(M_v)_p}{(M_v)_f} - K_4 \text{ (C.V.)}$$

Where K_1, K_2, K_3, and K_4 are constants, depending on process conditions; C.V., the coefficient of variation, reflects the cross-sectional uniformity of the filament bundle, $(M_v)_f$ is the weight-average molecular weight of the extruded fiber, and $(M_v)_p$ is the weight-average molecular weight of the starting polymer. In the above equation, the second term, $(M_v)_f$, reflects the contribution of the molecular weight of the polymer to the tenacity of the fiber while the third term, $(M_v)_p/(M_v)_f$, reflects the contribution of molecular-weight distribution. According to the theory of random scission, the higher the value of $(M_v)_p/(M_v)_f$, the narrower is the molecular-weight distribution of the polymer in the fiber.

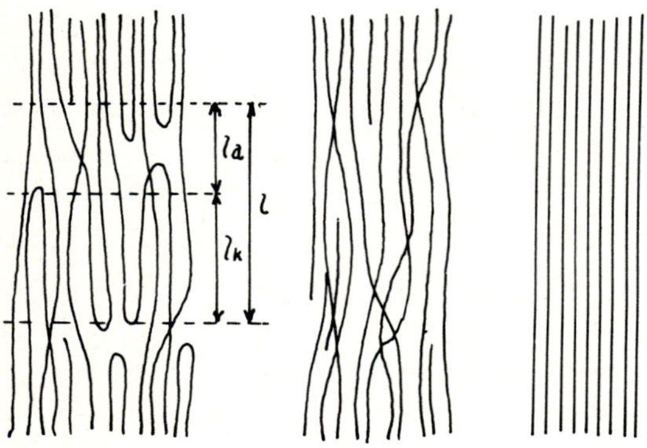

A B C

Fig. 7. Structure of Highly Oriented Fibers (ref.65)
A. Amorphous-crystalline fibrillar structure
l_a length of the amorphous region
l_k length of the crystalline region
l long period
B Partially ordered unfolded fibrillar structure
C Ideal structure of highly oriented fiber

The above equation can be represented schematically in the form of Fig. 8 which shows that polymers of narrow molecular-weight distribution (degraded) would yield fibers of higher tenacities than the polymers of broader molecular-weight distribution (virgin polymer). The experimental results presented in Table III confirm this conclusion.

It has been amply demonstrated (Fig. 9), that the tenacity of the drawn polypropylene fiber is proportional to the number-average molecular weight, $\overline{M}_n$, of the polymer in the fiber without making reference to the original polymer. The reason for the apparent dependence of attainable fiber tenacity on the molecular-weight distribution of the polymer can be seen in the relation of the commonly utilized viscosity-average molecular weight, $\overline{M}_v$, with number-average

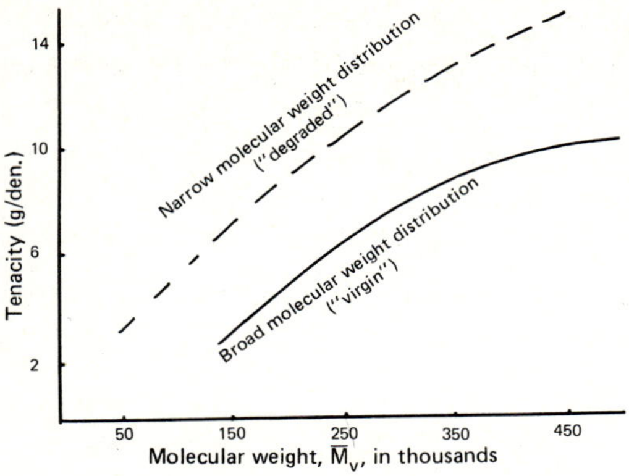

Fig. 8. Schematics of the Effect of Molecular Weight and Molecular-Weight Distribution on the Tenacity of Polypropylene Filaments

TABLE III

Effect of Fiber Molecular-Weight Distribution on Tenacity (ref.15)

Polymer[a]	Original molecular weight $\overline{M}_{wo}$	Spinning polymer molecular weight	Fiber molecular weight $\overline{M}_w$	$\overline{M}_w/\overline{M}_{wo}$	Tenacity at various draw ratios, g./den. 3.70	4.20	4.70
T	330,000	190,000	170,000	0.51	5.8	7.1	7.7
U	230,000	185,000	172,000	0.75	5.1	6.1	6.6
V	195,000	195,000	180,000	0.92	4.7	5.4	6.4

[a] Polymers T and U were pre-degraded to the spinning polymer molecular weights shown. All polymers were then extruded at 245°C. to form fibers with the monoclinic crystal structure.

molecular weight, $\overline{M}_n$, and weight-average molecular weight, $\overline{M}_w$. It has been shown in Chapter I that the values of $\overline{M}_v$ of a polymer is closer to $\overline{M}_w$ than to $\overline{M}_n$. It has also been mentioned there that in thermal degradation of polypropylene, $\overline{M}_n$ does not change appreciably while $\overline{M}_w$ experiences a drastic change. Thus, for a given $\overline{M}_v$, the degraded polymer (narrow molecular-weight distribution) is in fact of higher $\overline{M}_n$ than the undegraded (broad molecular-weight distribution) polymer.

It is seen from Figures 8 and 9 that in the lower range of molecular weight, the tenacity of the fiber increases linerally with molecular weight. It levels off to a constant value at higher values of molecular weight. A further increase in molecular weight generally decreases the tenacity of the fiber instead of increasing it. The leveling off of the fiber tenacity with increasing molecular weight beyond a certain limit is due to the inability of these fibers to attain a high degree of orientation in the cold-drawing process. A decrease in fiber tenacity at very high molecular weights reflects the poor cross-sectional uniformity of the yarn spun from these viscous polymers under industrial processing conditions.

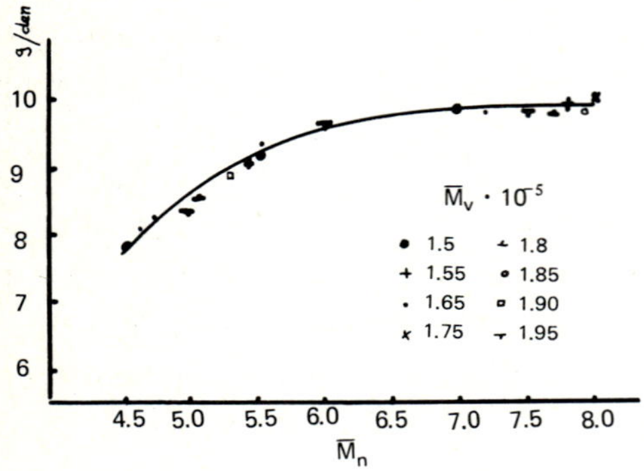

Fig. 9. Variation of Tenacity with Number-Average Molecular Weight (ref.18)

There are three aspects of filament uniformity reflecting on the strength of a multifilament yarn:

(1) Structural uniformity across the filament cross section. It has been discussed in Chapter V that conventional polypropylene filaments prepared under commercial spinning conditions show a pronounced degree of skin-core structure, which originates primarily through the preferential orientation of the outer surface and the peripherial cooling of the filament. On drawing such a structure, most of the stress is concentrated on the skin portion, thereby rupturing or cracking the filament surface at high draw ratios and leaving only the core portion to sustain the load.

(2) Filament-to-filament uniformity. If the filaments of a multifilament yarn differ from each other in their thickness and cross-sectional shape, they would also differ in their crystallinity and orientation and, consequently, in drawability. At high draw ratios the filaments with a higher degree of preorientation would break and be lost for load-bearing purposes. At low draw ratios, the filaments of lower preorientation would not draw to their maximum draw ratio and would exhibit tenacities lower than the other fibers in the filament bundle. This situation is generally encountered when a faulty spinnerette is in use or where the melt-pressure distribution over the spinnerette face is not uniform. Some filaments, especially those on the periphery of the spinning multifilament threadline, may receive more cooling air than those in the interior of the bundle and thereby attain different drawing characteristics.

(3) Configurational and structural uniformity along the filament. The configurational and structural uniformity along the filaments, caused by "rain effect", is as detrimental to high tenacity as the filament-to-filament nonuniformity in the bundle. Both of these configurational nonuniformities can be measured by the coefficient of variation of filament cross section across the filament bundle. It has been observed that the larger the coefficient of variation, the less potential there is for achieving high-tenacity values in drawing. This is exemplified in Fig. 10. It is likely that a decrease in tenacity with an increase in molecular weight of the polymer, observed by various industrial workers in the early days of polypropylene fiber development, has been due to intrafilament nonuniformity. Higher viscosity polymers are more prone to "rain effect" (Chapter VI) than the lower viscosity polymers. Thus, the beneficial effect of high molecular weight of the polymer on tenacity is nullified and sometimes overshadowed by the adverse effect of structural and configurational nonuniformities along the filaments.

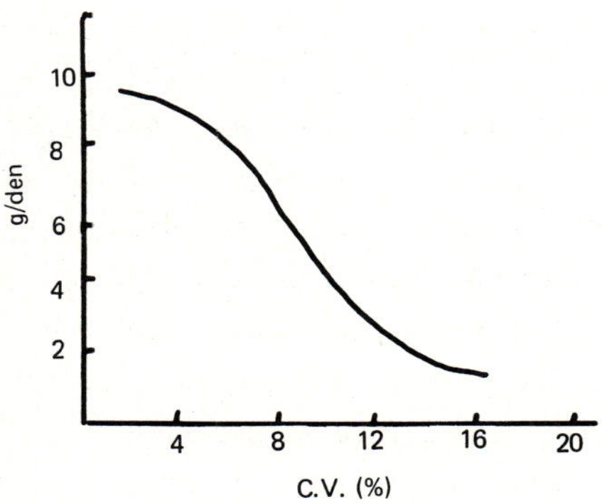

Fig. 10. Dependence of Tenacity on Coefficient of Variation of Filament Cross Section

It should be apparent from the foregoing discussion that, in the production of high-tenacity yarns, it is preferable to use high molecular-weight polymers of narrow molecular-weight distribution, apply extrusion conditions so that the fibers with low spin orientation and predominantly smectic structure are obtained, and draw to a draw ratio which is close to their maximum. In addition to these basic structural factors, the production of high-tenacity yarns requires the spinning of filaments with high interfilament and intrafilament uniformity without sacrificing production economies. The details of processing conditions that affect the filament-structure formation and cold-drawing behavior have been discussed in Chapters V, VI, and VII.

As is apparent from the discussions in Chapters V and VII, several of the processing conditions favorable to the formation of a desirable fiber structure for high-tenacity fibers are contradictory to each other. In such cases, either a compromise has to be worked out or some other means have to be adopted to offset the undesirable side effects of the given process conditions on the structure of the fiber. The following notes may be helpful in designing a successful process for the production of high-tenacity yarns.

(1) Use highly stabilized polymers of high molecular weight (I.V. range of 2.0 to 3.0), narrow molecular-weight distribution, and high tacticity (low heptane solubles). A good starting point for the choice of polymer is to begin with the highest molecular weight material available. Thermally degrade it to a workable viscosity and stabilize it with a combination of phenolic, phosphite, and amine-type antioxidants.

(2) Melt the polymer in a low-shear extruder at the lowest workable temperature to prevent unnecessary degradation.

(3) Keep the spinning block temperature about 10 to 20 degrees higher than the temperature of the melt emerging from the extruder in order to increase the fluidity of the melt without unduly degrading the polymer.

(4) Undertake high-pressure spinning. The principle of high-pressure spinning involves the conversion of mechanical energy into heat during a pressure drop, thus causing a sudden temperature increase in the melt. In this way, the fluidity of the melt is increased without causing molecular degradation. Besides the advantage of minimizing thermal degradation, the high-pressure spinning technique offers considerably improved temperature uniformity of the melt, which is likely to result in a uniformly spun product. Pressure drop can occur across the filter pack distribution plate on the spinnerette. But it is usually advisable to cause a major pressure drop across the spinnerette. This can be done by utilizing spinnerette holes of large L/D ratios and rather narrow counterbores.

(5) The combination of high viscosity and high spinning speed invariably leads to a spun yarn with high birefringence and pronounced row-nucleated crystallinity which is not desirable for the attainment of high tenacities in the final product. The delayed quenching technique, in which the filaments are exposed to quenching atmosphere after they are fully attenuated to the desired filament denier in fluid state, yield a yarn of low birefringence and minimal row-nucleated crystallinity. For this purpose, a heated shroud is mounted below the spinnerette so that it forms a cylindrical protective cover around the spinning threadline and prevents it from cooling and solidifying within a given distance from the spinnerette (Fig. 11). The recommended length and temperature of the heated shroud depends on the melt temperature, throughput rate, melt viscosity, spinning speed, the desired spin orientation in the filaments, and tolerable limit of cross-sectional nonuniformity, as measured by the coefficient of variation of the filament cross section in the bundle. Generally speaking, the higher the length of

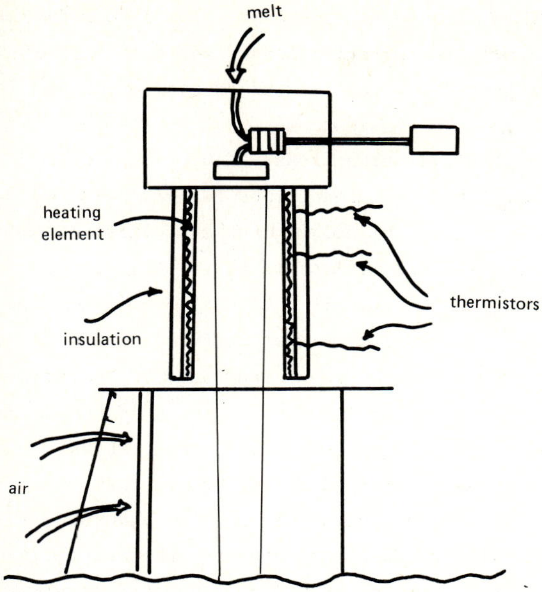

Fig. 11. Delay Quenching Collar

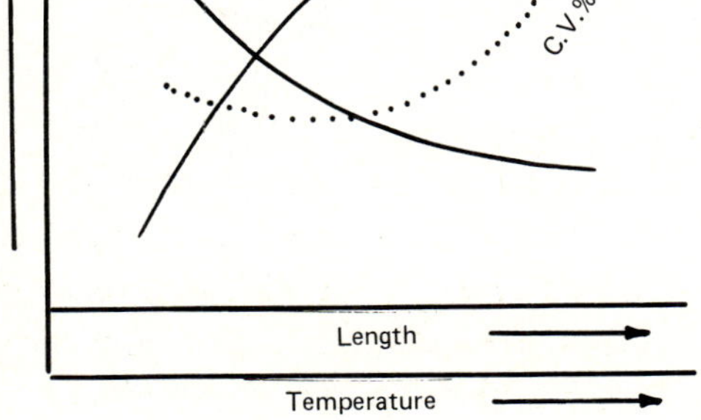

Fig. 12. Effect of Length and Temperature of Frlsy Wurnvh Vollst on Birefringence, Cross-Sectional Uniformity, and Ultimate Tenacity of Polypropylene Yarn (Schematic)

the shroud and the higher the shroud's temperature, the lower the spin orientation of the filament attained at a given throughput rate (Fig. 12). The shroud's length and temperature also influence the cross-sectional uniformity of the bundle. Uniformity generally decreases with increasing length and increasing temperature of the heated shroud (Fig. 12). The filaments may distort severely and spinning may become difficult under extreme conditions. Table IV shows typical interaction of polymer viscosity and quenching conditions on yarn properties. Further details of the construction and use of the delay-quenching collar or heated shroud can be found in various patent disclosures listed in the Bibliography.

(6) Quench the filament bundle rapidly with chilled air or preferably with chilled water as the filament emerge from the delay-quenching collar. A large variety of arrangements for rapid quenching is possible and these techniques are the subject of numerous patent specifications. One such arrangement, using chilled water, is shown in Fig. 13. As explained in the previous section, the rate of filament quenching determines to a large extent the crystal structure of polypropylene fiber and consequently its ease of cold drawing. However, care is to be taken that rapid quenching is not achieved at the expense of filament uniformity, otherwise the higher draw ratios will not translate into a higher tenacity of yarn.

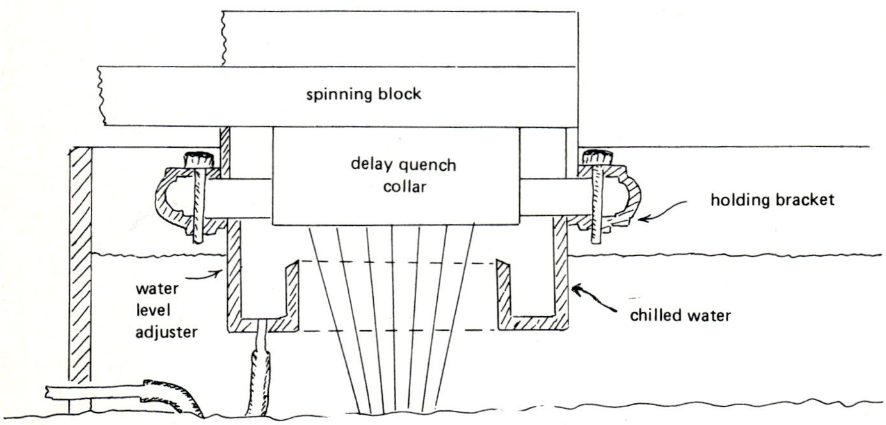

Fig. 13. Controlled Rapid Quenching of Filament Yarns with Chilled Water (ref. 66)

(7) Draw the yarn immediately without any significant aging. Aging at room temperature tends to improve the crystalline order of the undrawn filament which renders the subsequent drawing difficult and the desirable high-draw ratios unattainable. Similarly, as has been mentioned in Chapter V, the smectic structure obtained through rapid cooling of the filament gets readily converted into

TABLE IV

Interaction of Spinning and Quenching Conditions on the Tenacity of Polypropylene Yarns (ref.67)

Example	1	2	3	4	5	6
Composition of the mix:						
a) polypropylene intrinsic viscosity (determined in tetrahydronaphthalene at 130°C.)	3.00	2.50	3.00	3.00	2.00	3.00
ash content	0.01%	0.02%	0.01%	0.04%	0.02%	0.01%
residue after heptane extraction	94.00%	94.5%	94.00%	95.6%	96.1%	94.00%
b) calcium stearate	0.3	0.3	0.3	0.3	0.3	0.3
c) stabilizer %						
laurylthiodipropionate	0.6	0.6	0.5	0.6	0.5	0.6
4,4[1]-thiobis-3-tert.butyl-methacresol	0.2	0.2	0.2	0.2	0.2	0.2
Spinning conditions: screw temperature	230°	230°	230°	240°	230°	230°
head temperature	235°	230°	230°	240°	240°	235°
spinneret temperature	285°	270°	250°	295°	285°	285°
spinnaret type*	40/0.8X16	40/0.8X16	50/1.3X26	20/0.4X8	20/0.4X0.8	40/0.8X16
temperature of the heated tubular device placed beneath the spinneret	160°	140°	130°	280°	270°	110°
max pressure (kg per sq. cm.)	30	25	30	45	35	30
winding speed (m per min.)	600	600	80	240	320	600
Stretching conditions: temperature	130°	130°	150°	130°	135°	130°
medium	steam	steam	steam	warm air	warm air	steam
stretching ratio	1:8.3	1:8.5	1:10	1:6.28	1:10	1:8.3
Properties of the stretched fiber: tenacity (g/den.)	9	8.8	9.5	14.5	11.5	5.8
elongation (%)	18	20	12	23	20	14

* A 40/0.8X16 spinneret has 40 holes each 0.8 mm. in diameter and 16 mm. long.

monoclinic crystalline structure at temperatures above 70°C. It is, therefore, advisable to draw the spun yarn with smectic structure without giving it an opportunity to convert into monoclinic structure. However, as drawing at temperature below 70°C does not allow maximum possible orientation, a second drawing at higher temperatures becomes inevitable.

(8) Draw the moderately cold-drawn fibers again at high temperatures to achieve the maximum orientation. This second-stage draw ratio is generally in the range of 1.2 to 1.8 and the temperatures used normally range from 110 to 140°C. Several different arrangements between the first-stage draw and the second-stage draw ratios and the respective draw temperatures have been proposed as the optimum processing conditions. In addition to the high tensile strength, the yarns drawn in two subsequent stages are characterized by high modulus, low elongation, and high toughness values.

(9) Anneal the fully drawn fibers under tension. The highly drawn filaments would generally tend to shrink when relaxed or when subjected to high temperatures in use. To prevent this shortcoming, it is desirable to anneal the yarn at temperatures in the range of 135-140°C. The annealing should be done under high tension (no contraction).

(10) Cool the drawn and annealed yarn without relaxation. This is to prevent the disorientation which may possibly occur on winding.

(11) Wind up under controlled tension.

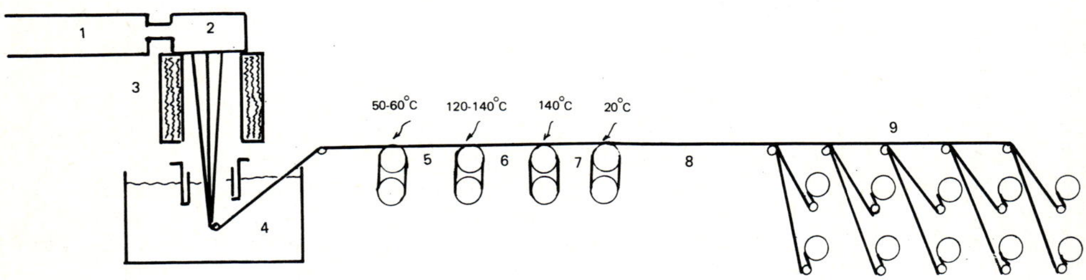

Fig. 14. Schematics of a Warp-Type Spinning Arrangement for the Production of High-Tenacity Industrial Yarns
(1) Extruder
(2) Spinning Head
(3) Heated Shroud
(4) Quench Tank (Chilled Water)
(5) Cold Draw Zone (2.0 - 4.0 x)
(6) Hot Draw Zone (1.2 - 2.0 x)
(7) Stabilization Zone (1.01 to 1.10 x)
(8) Cooling Zone
(9) Winding Station

The process described above can be undertaken with several mechanical arrangements. Two such arrangements are shown in Figures 14 and 15.

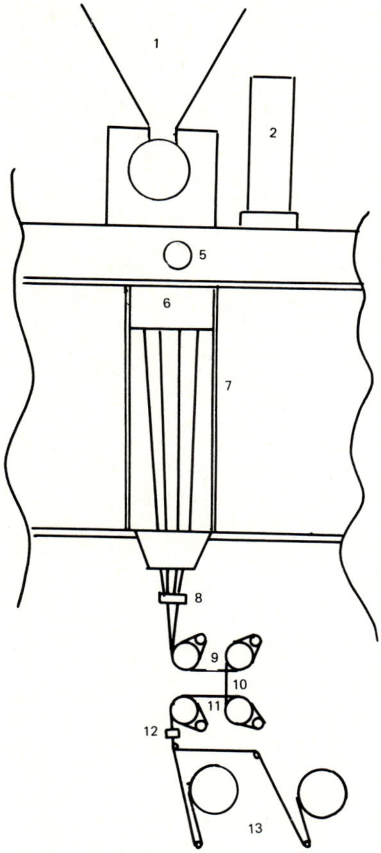

Fig. 15. Production of High Tenacity Industrial Yarns Through One-Step Spin-Draw-Stabilization
(1) Extruder
(5) Metering Pump
(6) Insulating Collar
(7) Quench Box
(8) Finish Applicator
(9) Cold Draw Zone
(10) Hot Draw Zone
(11) Heat Setting (Stabilization) Zone
(12) Air Interlacer
(13) Winder

III.3. Fibers With Profiled Cross Section

The spinnerette holes through which the extrusion takes place are normally circular in cross section. During the subsequent cooling of the fluid threadline the surface-tension forces in the melt tend to maintain this cross section. This is in marked contrast to fibers spun from solution, where coagulation and evaporation processes produce much more complicated shapes. However, these more irregular cross sections confer various useful properties on the fibers, mainly in luster, covering power, and hand. Extensive efforts have therefore been made in the past decade to obtain noncircular cross sections in melt-spun fibers as well. In recent years, the fibers having various cross sections other than circular have indeed been manufactured and made available to the trade. They are called *shaped fibers,* or fibers with *profiled* or *modified* cross-section.

Starting in 1954, when probably the first patent application was filed on profile spinnerette designs and cross-sectional shapes of profiled filament, the number of patents granted in this field has become so large that the issue of proprietary rights has become very confused. Besides the profile cross sections with genuine advantages in some properties, a proliferation of patent disclosures has emerged, the aim of which seems only to circumvent the previously disclosed patent claims. Some of these patents are listed in the Bibliography of this chapter.

The techniques which have been used for the production of profiled cross sections from nylon and polyester have also proved successful for the production of profiled polypropylene filaments. A wide variety of such modified cross sections is available in *second generation* polypropylene yarns for carpets and upholstery fabrics. The important properties that have been achieved are greater bulk, improved optical properties, improved soiling resistance, dryer hand, along with "silkiness", "crunchiness", "softness", etc.

For polypropylene fibers, the noncircular cross sections offer special opportunities. This is because polypropylene fibers of round cross section have a rather "waxy" hand and, by using suitable noncircular cross sections, this waxy hand can be minimized or altogether eliminated. The reasons for the relatively crisp and crunchy hand are found in the bending geometry of the fiber which is restricted in noncircular fibers only to certain preferable directions. A feeling of a dry hand is probably obtained because of the sharp corners and a limited area of contact in the case of profiled cross sections.

One of the most interesting facts in the design of spinnerette holes for fiber of profiled cross section is that the shape of the finished filaments can be made

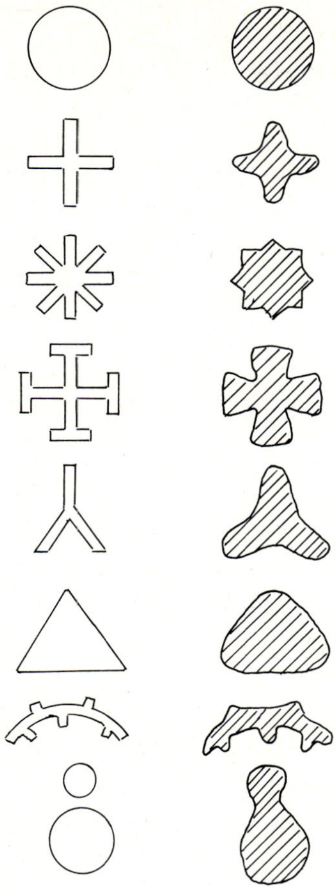

Fig. 16. Some Spinnerette Hole Profiles and Corresponding Filament Cross Sections

quite different from that of the spinnerette hole, depending on rheological properties of the material being extruded and depending on the spinning conditions, such as the temperature of the melt, jet stretch, and cooling rate. It is known, for instance, that fibers of an elliptical cross section having various aspect ratios can be produced from a rectangular spinnerette hole. Figure 16 shows the profile of a few spinnerette holes along with the cross section of the corresponding extruded fibers.

Three primary considerations in designing a spinnerette hole for the spinning of profiled filaments are the die-swell phenomenon, melt viscosity, and the surface tension of the molten filament. As polypropylene differs in these material parameters from nylon and polyester, it is obvious that the guidelines for the spinnerette hole design for polypropylene profiled fibers would differ from those utilized for nylon and polyester. Very little appears to have been mentioned in technical literature to account for a change of extrudate shape in spinning, let alone the differences between various polymers.

The changes in extrudate cross section occur while the extrudate is still in the liquid state. This implies that the significant region lies within a relatively short distance from the spinnerette face. It is also this region where the extrudate swell occurs in the emerging stream. The distance at which maximum swell occurs depends on the velocity of the fluid in the spinnerette and also on a characteristic *recovery time*. Most of the published literature on die swell referred to in Chapter V is concerned with circular capillary holes. Han (ref.9) has recently published a brief account of the die-swell phenomenon from a rectangular slot and showed, that unlike the extrudate swell from a circular capillary, the extrudate from a rectangular die swells nonuniformly at two adjacent sides: there is more profound swell on the long side than on the short side. At the corners of the rectangular slots the shear rates are zero and so are the normal stresses, giving rise to no extrudate swell.

The magnitude of die swell of polypropylene is known to be far greater than that of nylon and polyester. The surface-tension force can be a contributing factor to the changes in extrudate cross section, converting it into a more rounded configuration. However, the magnitude of surface-tension force for polypropylene is considerably smaller than that for nylon and polyester. This renders the formation of profiled cross section from polypropylene melt a lot easier than that from other thermoplastic materials.

Melt spinning of noncircular filaments requires not only the use of special spinnerette but also a judicial choice of polymer and a careful adjustment of spinning conditions. Given a profiled spinnerette hole, fibers of various degrees of sharpness in cross-sectional contour can be obtained by adjusting material and processing parameters in spinning. Efforts are often made to obtain fiber of well-defined and sharply profiled cross section. Higher viscosity, lower spinning temperatures, lower throughput, rapid cooling, and low draw-down ratios are beneficial to this end. Fibers of low viscosity or those spun at high temperatures tend to deform in cross section before they solidify. High draw-down ratios, by virtue of the increased degree of "draw resonance", may contribute not only to the differences in the cross-sectional shape of the filaments, but also to their relative size. The spinning of modified cross sections is as a rule more difficult than that of round cross section.

An increased rate of cooling, associated with the increased surface of profiled filaments, leads to the shortening of the deformation zone, increased orientation, and lower drawability of spun filaments. Therefore, application of a spinnerette with noncircular orifices makes it necessary to make adjustment in drawing conditions. They would be different from those found optimal for cylindrical fibers.

Similarly, changes in bending characteristics of the profiled filaments necessitate the development of new processing conditions in texturizing and converting into yarns and fabric.

Polypropylene profiled filament yarns are relatively new on the market, mainly dominating in BCF carpet yarns. Because of the relative ease of spinning of polypropylene into profiled filament yarns as compared to nylon and polyester and because of the desirability of minimizing the somewhat waxy hand of polypropylene, a great surge of profiled cross sections is currently underway in the industry. Textured yarns for carpets, knitted goods, and upholstery fabrics are of special interest in this respect.

Apart from hand, bulk, luster, and brightness, the soiling propensity of the product is an important consideration for an increasing interest in modified cross sections. It is generally recognized that, all else being equal, fiber with round, smooth cross sections will retain a minimum quantity of soil. However, a round, transparent cross section fiber can serve as a lens to magnify the soil deposited upon its surface (ref.23-25). Multilobal, profiled cross sections strongly disturb this "lens effect" and restrict the passage of light to certain angles about the axis, the light entering from other angles being internally reflected. This has the effect of soil-hiding in the product.

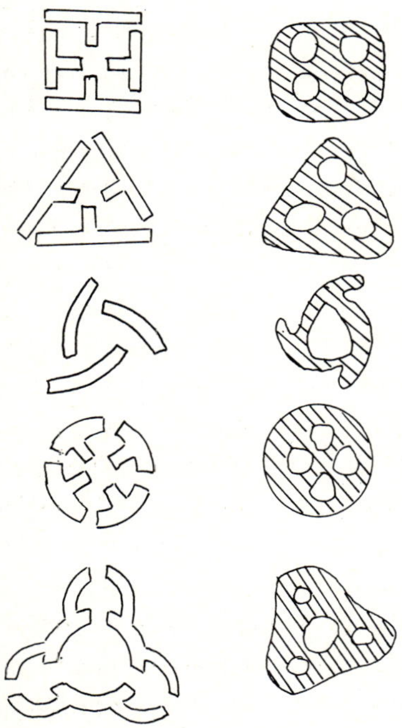

Fig. 17. Profile of a Few Spinnerette Holes and the Extruded Hollow Fibers

Use of multilobal, modified cross sections in decreasing the apparent soiling is quite effective in low to normal soiling environments but becomes less effective as the saddles between the lobes become filled with particulate soil. These uses therefore require those cross sections which provide less opportunity for soil to be trapped on the surface of the fiber. One way to obtain a smooth fiber surface, along with light scattering without magnification, is to design filament cross sections with relatively smooth surfaces, such as circles, deltas, and rectangles, but with internal holes (Fig.17). The fibers of such cross sections are generally characterized by a subdued luster or opacity along with the property of low soiling and easy cleanability.

Apart from low apparent soilability, hollow fibers described above command certain other advantages over solid round or solid profiled cross sections. Some of these advantages are improved bulk, high buoyancy, improved thermal and acoustical properties, reduced pilling, special optical effects, and greater dielectric strength (ref.26). Hollow filaments have also been reported to have less tendency to fibrillate under flexing conditions than the corresponding solid filaments.

A large number of patents have been issued for the spinnerette design and production techniques for hollow filaments from melt-spinnable polymers. Most of these techniques can be applied to the production of hollow fibers from polypropylene. In fact, polypropylene is better suited to these techniques because of its high viscosity, high die-swell ratio, and low surface tension. Tubular, hollow fibers are not of much importance for use in textile and carpeting, as they have the tendency to flatten under pressure. Filament cross sections, round or profiled, with two, three, or more holes, have better properties. The design principle of such cross sections is to extrude the polymer through a set of curved or straight slots arranged in such a way that their ends are separated from each other with a gap not more than about 5 times the average width of the slot. The polymer melt emerging from each slot swells and coalesces with the near end of the neighboring slot. Because of the high melt-swell ratio of polypropylene, a larger gap is possible between the slots than that possible for nylon and polyester. This renders the construction of spinnerette easier and makes it possible to bore sturdier holes. Polypropylene fibers of such cross sections are not yet available but extensive efforts are being made toward this end.

Along with various advantages, the profiled cross sections also bring some disadvantages with them. Experience has shown that dye uniformity of profiled filaments is not as good as that with round cross sections. A slight modification of average fiber cross-sectional shape along the length of a filament bundle may result in visible streaks or *barre* effects in the product. The use of delustering agents, higher twist level in the yarn, and a larger number of filaments in the bundle, tends to decrease this kind of optical nonuniformity in the product (ref.27).

III.4. High-Shrinkage Fibers

Shrinkage of synthetic fibers in dyeing, finishing, and washing has been a point of concern for both the fiber and the textile industry. Efforts have always been made to produce fibers and yarns with a maximum of 2 percent shrinkage in boiling water. This low shrinkage range is necessary for ordinary textile fibers used in most apparel as well as in industrial fabrics. On the other hand, there are some uses where a comparatively high-shrinkage fiber is employed with particular advantages. Among such products can be mentioned felts, base fabrics for synthetic leather, effect yarns, and high-bulk yarns. Polyvinylchloride fibers of very high shrinkage characteristics were probably the first to be offered commercially to the textile and carpet trade. Later on, moderately high shrinkage acrylic staple fibers were offered for the production of high-bulk spun-yarns, and more recently some polyester high-shrinkage fibers are being offered for various end uses. Polypropylene fibers of high-shrinkage characteristics are beginning to appear in limited quantities and market potential seems quite high. Polypropylene is specially ammenable to the production of high-shrinkage fibers; fibers with 30 percent or more shrinkage in boiling water can easily be obtained.

As has been mentioned in Chapter VII, the origin of shrinkage in polypropylene fibers can presumably be traced to the existence of tie molecules which connect different lamellae or blocks of neighboring microfibrils in the supermolecular structure of the fiber. Such interfibrillar tie molecules may add to the lateral cohesion of the fibrils, but they also interfere with the longitudinal sliding motion of the microfibrils during micronecking and the additional plastic elongation of the fiber structure. Their arrangement seems to be unidirectional so that they exert a finite tensile force, or torque, on the crystal blocks they connect (ref.28), affecting the retractive forces and leading to fiber shrinkage by backsliding of microfibrils on relaxation.

The number of interlamellae and interfibril tie molecules increases with molecular weight, molecular-weight distribution, and the rate of crystallization of the spinning material. The dependence of tie molecules on the molecular weight of the polymer is a necessary consequence of the greater probability of large molecules to crystallize in more than one block of lamellae than that of shorter molecules, which can easily be accommodated within one lamellae by chain-folding mechanism. The increase in number of tie molecules with increasing polydispersity of molecular weight is presumably due to the presence of a long tail of high mol molecular-weight chains which may statistically provide an increased number of tie molecules to the supermolecular structure. An increase in the number of tie molecules with increasing rate of crystallization is a natural outcome of the theory of lamellae formation by chain-folding mechanism, discussed in Chapter V. A definite crystallization time is required for a molecule to fold into a regular array and form a lamellae. If this required time is not allowed, the crystallization will be imperfect and the chain molecules may fold in sections and be a part of two or more lamellae. The rate of crystallization is increased in practical spinning operation by utilizing low melt temperatures, lower quench-air temperature higher quench-air speed, higher rate of spinning, and low fiber deniers. Molkova, et al, (ref.29) have examined the properties of undrawn polypropylene fibers and its effect on post-stretching shrinkage.

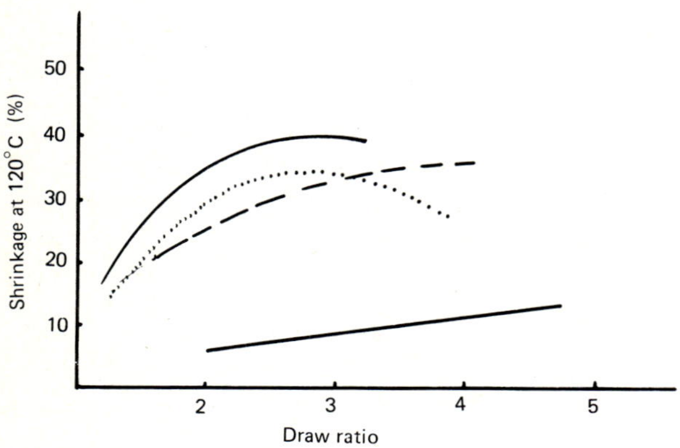

Fig. 18. Effect of Draw Ratio (50°C) on Linear Shrinkage of Differently Spun Polypropylene Filaments

Apart from the material characteristics and spinning conditions, the shrinkage of polypropylene fiber is also dependent on drawing and relaxation conditions. Several studies have been reported examining the changes in structural elements of polypropylene fibers on drawing and its shrinkage characteristics (ref.29-30). Fiber shrinkage generally increases with increasing draw ratio. For some spun fibers it may reach a plateau beyond certain draw ratio, while for some other fibers it reaches a maximum in the range of 2.5 to 3.0, as shown in Fig. 18. The temperature of drawing is as important as draw ratio in affecting the shrinkage of the drawn fibers. Table V shows that the lower the drawing temperature, the higher the heat shrinkage of drawn fiber.

Among the process and finishing conditions which affect the development of fiber shrinkage in the drawn fiber are the relaxation temperature, relaxation

TABLE V

Effect of Drawing Temperature on Heat Shrinkage of Polypropylene Fiber

Drawing temperature	Fiber shrinkage at 120°C	
	Fiber A	Fiber B
°C	%	%
25	52	45
50	50	45
70	48	40
90	45	38
120	40	30
130	35	20

time, and the applied load during relaxation. The dependence of fiber shrinkage on temperature can be presented by the equation:

$$S_t = a.T.\exp(-\Delta E/RT)$$

where S_t is the total shrinkage after a time t, a is a constant, E is the activation energy of shrinkage, and R and T are the gas constant and absolute temperature, respectively (ref.31). The reported values of E for polypropylene range from 8 to 37 kcal/mole for differently drawn yarns (ref.30). These values are considerably lower than those reported for polyamide (ΔE=40 Kcal/mol) or those for polyester (ΔE=56 Kcal/mol). The effect of time on shrinkage is not very significant after a minute of relaxation at all temperatures. The fiber reaches its maximum shrinkage almost instantaneously at about 80°C (ref.30). The effect of load during relaxation is very significant and often a key to various product developments in practice. It can be represented as

$$S_t = a_1 \exp.(-b\theta)$$

where a_1 and b are constants and θ is the load (ref.30).

One of the most promising uses of high-shrinkage polypropylene fibers is in the production of high-bulk staple spun yarns. If high-shrinkage staple fiber is blended with low-shrinkage staple, whether natural fiber or normal polypropylene staple, and spun, the yarn would look just like any other spun yarn. However, when these yarns are subjected to temperatures which would cause the shrinkage to take place in the high-shrinkage component, the low-shrinkage fibers would be forced to arc up around the shrinking fibers and the yarn would be puffed up. Because of this inherent capability of these yarns for bulk development on heat relaxation, these blend yarns are sometimes called *pop corn* yarns. Such yarns have proved a brilliant success in acrylics and are finding increasing use in polyesters.

The blending of high-shrinkage and low-shrinkage fibers can be done utilizing either staple fibers or tows. The former method is generally preferred as it also allows the utilization of natural fibers or acrylics for nonshrinking component. The proportion of high-shrinkage polypropylene fiber in such blends rarely exceeds 50 percent and is generally in the range of 20 to 30 percent (ref.32).

In addition to their use in the production of high-bulk yarns, the high-shrinkag polypropylene fibers can also be used to produce novelty yarns and pucker effects in woven fabrics. Their use in the production of bulky nonwoven fabrics is also very promising.

III.5. Elastic Polypropylene Fibers

Upon drawing a crystalline polymer into a highly oriented state, it is expected that the elastic modulus and tensile strength increase while elongation to fracture decreases. In addition, the traditional expectation is that no appreciable elastic recovery will occur when the material is subjected to large deformations. Thus, it is extremely interesting that several crystalline polymers, especially isotactic polypropylene, are capable of being processed into highly oriented fibers which shows increased modulus and tenacity and which also show an unexpectedly large extensibility and high elastic recovery from large extensions.

At the outset, it should be recognized that a semantic problem exists in describing these materials. The combination of high elastic modulus, high extensibility and high elastic recovery from large extensions is not a normal mode of crystalline polymeric behavior. The term "high modulus - high recovery", "hard elastic" and "entropy elastic" have been used. Cannon and coworkers (ref.71) in an excellent review of these materials have used the simple term "springy" to describe these speciality materials in an analogy of a metallic spring which possesses the same combination of properties. Here we shall follow the textile terminology of "elastic" fibers with the conditions and pre-qualifications set fort before.

The first elastic polypropylene fiber was described by Herrman in 1966 in a US patent to E.I. DuPont de Nemours and Co. (ref.72). The patent states that the fiber produced under specified conditions exhibits an elongation to break of 100 to 700 percent and shows a tensile recovery of 82 percent after an elongation of 25 percent.

In addition, Herrman, in the above patent, identifies a highly oriented morphology as responsible for the unique elastic properties of the polypropylene. The patent states that the gamma condition, as determined by x-ray diffraction, must be between 10° and 30°. The processing conditions required to obtain the elastic form of polypropylene are extensively described in the patent. Briefly, the desired morphology (as measured by gamma orientation) is achieved by temperature regulation and windup speed once spinnerette parameters are known. The desired orientation is achieved by specifying a suitable spin stretching (ratio of windup speed to the calculated speed of polymer leaving the orifice) at the proper quenching conditions and/or suitable drawing of the solidified fiber. Herrman found that the drawing may be done under any suitable conditions, but the draw ratio must not exceed 2.5 and preferably no drawing is performed. Quenching conditions are controlled by a cool air flow over the fibers as they leave the spinnerette.

The final processing requirement for the development of the springy form of polypropylene is annealing. The annealing can be carried out at temperatures of $105^{o}C$ to $160^{o}C$ for times ranging from 30 to 60 minutes.

Since the appearance of the above-mentioned Herrman's patents, a number of other patents have appeared dealing with several aspects of fiber spinning and processing technology for the production of elastic polypropylene fibers. In addition, several research reports have been published dealing with the morphology, mechanical properties and physical mechanisms of behavior of these fibers. This list, included in the Bibliography of this chapter, shows the intense industrial and academic interest in the basic search required to develop hard elastic polymeric fibers over the past decade. In view of this intense interest, we shall deal here with polypropylene elastic fibers in some detail.

The essential morphological feature of the elastic materials as revealed by X-ray diffraction and electron microscopy is the presence of stacked crystalline lamellae. Their lamellar surfaces are aligned normal to the fiber extrusion direction (row nucleated morphology). The mechanism of elasticity is based on a splaying-apart of these lamellae, involving their reversible bending and torsional deformation during macroscopic deformation of the material (Fig. 19 and Fig. 20). Thus, in contrast to rubber elasticity, where the kinetic units are flexible chain segments, the kinetic units for the elasticity of a "hard" elastic material are the lamellar crystals. Due to the orientation of the lamellae normal to the fiber extrusion direction, the elasticity is exhibited almost exclusively in that direction.

Figure 21 represents a typical stress-strain diagram for springy polypropylene. Upon loading, the material exhibits Hookean behavior (line AB) to point B which is usually referred to as a yield point. Path BCD represents the continuation of sample deformation to rupture (point D). Figure 22 recalls the comparison between the stress-strain behavior of spring polypropylene and other materials. A characteristic feature of the tensile behavior of springy polypropylene fibers

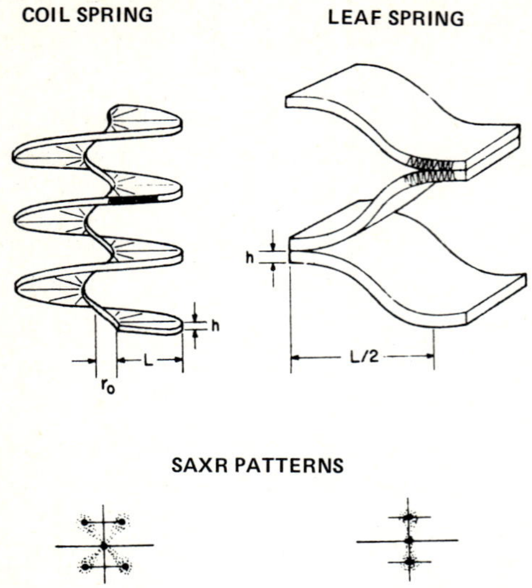

Fig. 19. Schematic Structural Models for Energy Driven Elasticity of Springy Polymers (ref.74)

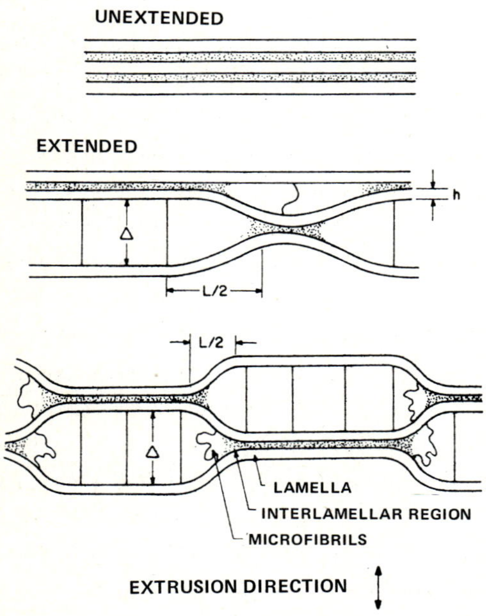

Fig. 20. Morphological "Leaf Spring" model for Springy Polymer Behavior

is that they undergo a large extension without occurrence of "necking" and plastic flow. A relatively high, immediate elastic recovery as measured by the intercept of the retraction curve with the elongation scale, indicates the dominance of instantaneous elasticity in the elastic deformation of an elastic "hard" material, although a small additional recovery in length is observed after a 2 minute delay time.

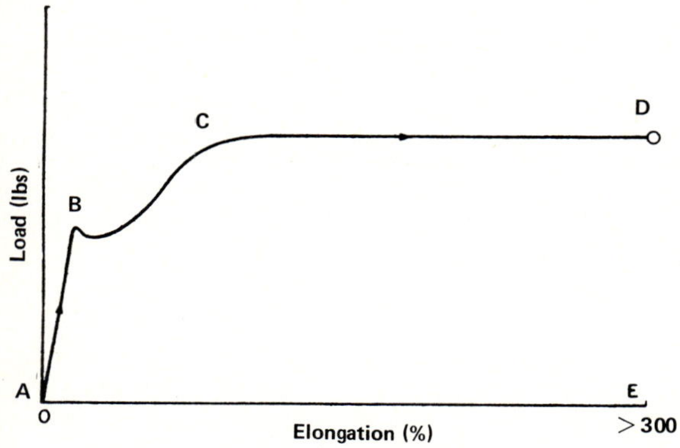

Fig. 21. Typical Load-Elongation Diagram for Springy Polypropylene (ref. 71)

The value of the initial elastic modulus (slope of line AB, Fig. 21) for springy polypropylene (SPP) is approximately 400,000 psi versus 130,000 to 200,000 for bulk polypropylene, and 850,000 for a high tenacity fiber. The yield stress at point B of Fig. 21 is in the neighborhood of 7500 psi versus 5000 psi for bulk polypropylene, and 65,000 psi for the high-tenacity fiber. Thus, we observe slight increases in elastic modulus and yield stress over the spherulitic form of polypropylene, but values lower than those observed in the more highly oriented, high tenacity fiber.

In terms of ultimate strength, however, a more important increase is obtained for the maximum stress supported by the fiber. As seen in Fig. 23, this exceeds 10,000 psi versus 5,000 for the bulk form, and it increases with increasing strain rate.

Figure 24 represents a loading, unloading and reloading cycle for springy polypropylene. Path ABC represents the loading history of the sample to 50 percent elongation, while CA' indicates the path taken upon unloading. The permanent deformation (AA') experienced by the sample is much less than would be expected from the loading path ABC. This amazing ability of elastic polypropylene to recover immediately (elastically) from extensions of up to 30 times greater than the yield strain has been examined in detail by several researchers. Some of these results will be discussed in the following.

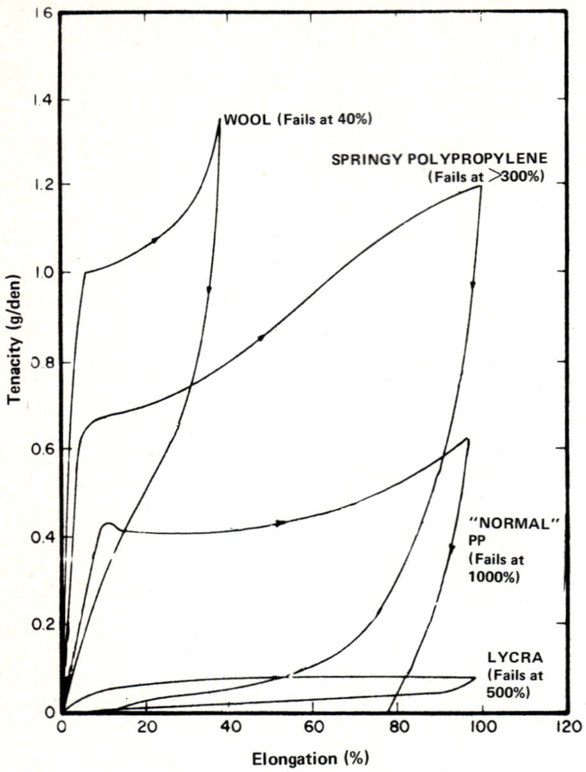

Fig. 22. Comparison of Stress-Strain Behavior of Some Springy, "Normal", Polymeric and Natural Fibers

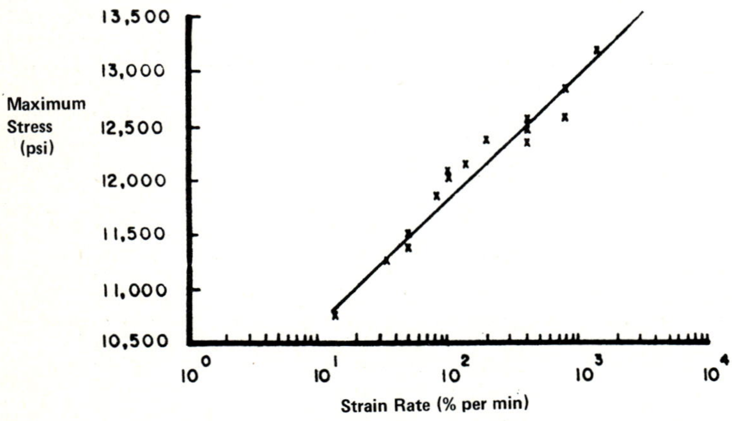

Fig. 23. Maximum Stress Supported by Springy Polypropylene as a Function of Strain Rate (ref.75)

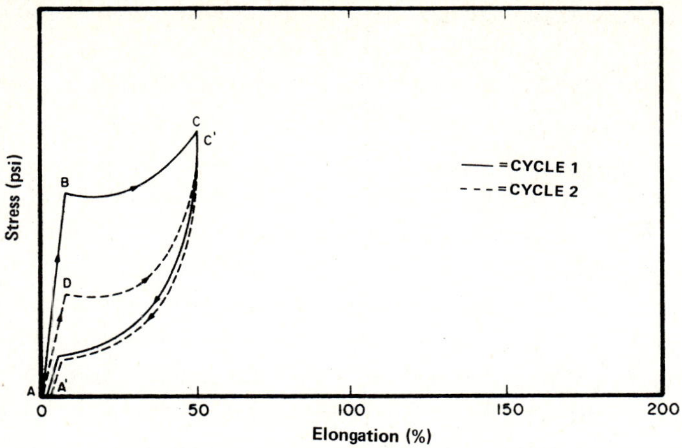

Fig. 24. Typical Stress-Strain Curves for Two-Cycle Loading of Springy Polypropylene to 50 Percent Elongation at a Strain Rate of 2 In./Min. (ref.75)

TABLE VI

Elastic Recovery of Elastic Polypropylene as a Function of Conditioning Treatment and Percent Elongation (ref.71)

Treatment	Extension (%)		
	25	50	100
Dry	92.2	94.3	90.2
0.15N NaCl[a]	87.2	92.0	85.5
0.90N NaCl[a]	88.5	95.0	84.2
Autoclaved	93.8	93.4	87.5
Isopropanol Extracted	94.8	95.6	93.0
Ethanol Extracted	94.9	96.2	92.8
Stress Relaxed 600 sec	83.6	87.0	81.7

[a]Samples placed in wet cell filled with saline solutions and tested immediately thereafter.

TABLE VII

Elastic Recovery of Springy Polypropylene as a Function of Cyclic Extension and Strain Rate (ref.75)

Strain Rate (%/Min)	Cyclic Extension		
	25	75	150
	Elastic Recovery (%)		
33	95.3	94.7	76.8
50	94.0	94.7	81.2
100	95.2	94.9	80.7
200	95.6	94.3	77.5
500	100.0	96.0	77.8

The recovery values of an elastic polypropylene sample are listed in Table VI, as a percentage of the designated extension at three levels of extension (25%, 50%, and 100%). That this fiber does not perform as a classical highly oriented crystalline polymeric fiber is clear: 85 to 90 percent recovery from 100 percent extension is possible.

Further details of the nature of the elastic recovery appear in Table VII. These data indicate that the elastic recovery is independent of strain rate at extensions of 25%, 75%, and 150%.

The time-dependent recovery or rehealing of elastic polypropylene is an important aspect of its behavior. Figure 25 shows rehealing as measured by the regain of second-cycle yield point as a function of time between stress cycles. Figure 26 shows rehealing as examined by time-dependent recovery. Elastic PP was extended by 10, 25, 50, and 100 percent and released. As can be seen, the recovery of the initial set, i.e., distance AA' of Fig. 24, is linear in the logarithm of time.

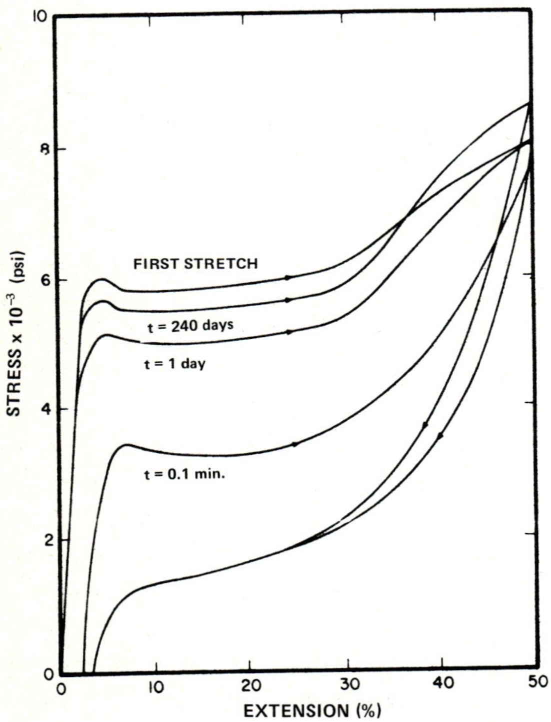

Fig. 25. Rehealing Effect (Recovery of Yield Point) in Springy Polypropylene as a Function of Time (Elongation: 50%; Temperature: 23°C) (ref.75)

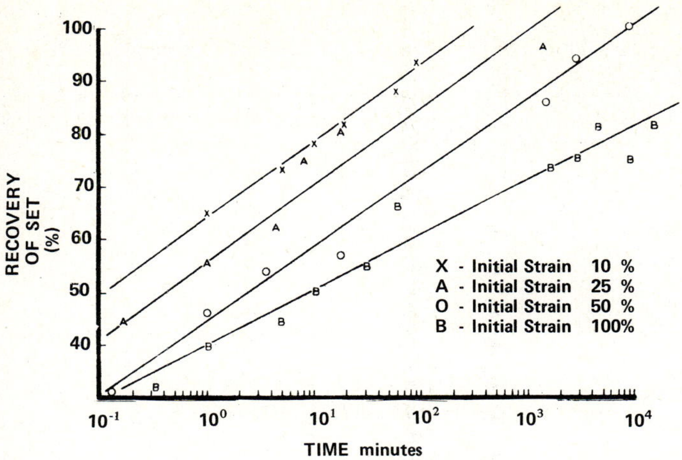

Fig. 26. Recovery of Initial Set for Springy Polypropylene as a Function of Time for Various Elongations (ref.75)

The effects of cyclic loading on springy polymers are complex. Figure 24 shows a typical loading cycle to 50 percent extension. As can be seen, subsequent loading results in a decrease in yield point B to point D. Figure 27 shows similar results upon extension to 150 percent. From these figures it can be seen, schematically, that as the extension on the first cycle increases, the second cycle yield point decreases. Figure 28 shows the magnitude of the decrease in the yield point with extension and Fig. 30 shows the general trend of the decrease in yield point with repeated extension.

Figure 28 also shows that the second cycle modulus, as measured by the initial slope of the stress-strain curve, decreases as the amount of first-cycle extension increases. It is interesting to note that the decreases in modulus and yield point vary in essentially the same manner.

Another interesting aspect of extension cycling of the springy polymers is that the maximum load supported, i.e., point D in Figure 21 decreases only slightly with cyclic extension, reaching a maximum of about 22 percent at 100 percent extension (see Fig.29). Thus, the large decrease in the yield point upon cycling does not necessarily mean a weaker fiber. In addition, the elongation to failure of springy polymers after cycling does not change significantly. Table VIII shows a change from 335 percent to 382 percent over a decade of cycles and from 25 to 300 percent cyclic extensions.

Park and Noether (in press) report data for modulus decrease for springy polypropylene with cyclic loading. Figure 30 shows that the modulus loss reaches a constant value after about 4-5 cycles. Thus, when a dynamic cyclic loading condition is applied, the material properties rapidly come to an equilibrium.

TABLE VIII

Residual Extension to Rupture after Cyclic Loading for Springy Polypropylene as a Function of Cyclic Extension (ref.75)

Cyclic Extension (%)	Number of Cycles	Elongation at Rupture (%)
0	0	335
25	9	343
50	11	366
100	9	327
150	10	340
200	10	336
300	10	382

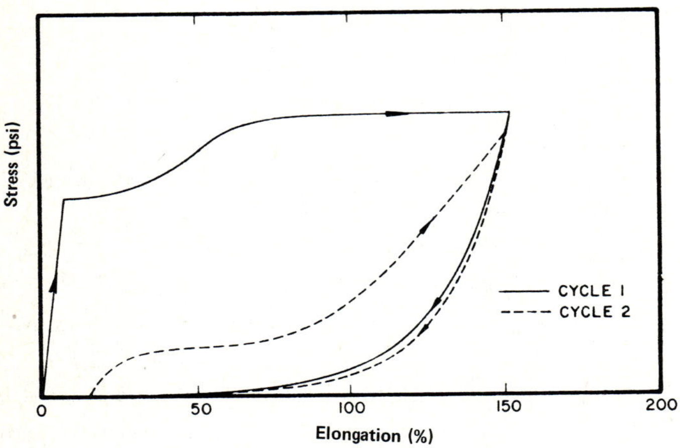

Fig. 27. Typical Curves for Cyclic Loading of Springy Polypropylene to 150 Percent Elongation (ref.75)

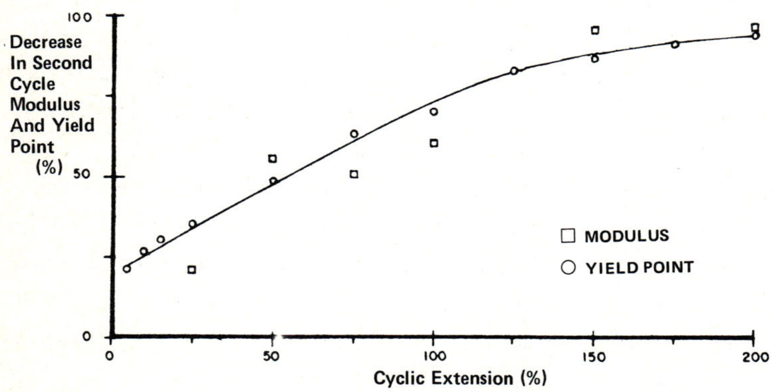

Fig. 28. Decrease in Second Cycle Modulus and Yield Point as a Function of Cyclic Extension for Springy Polypropylene (ref.75)

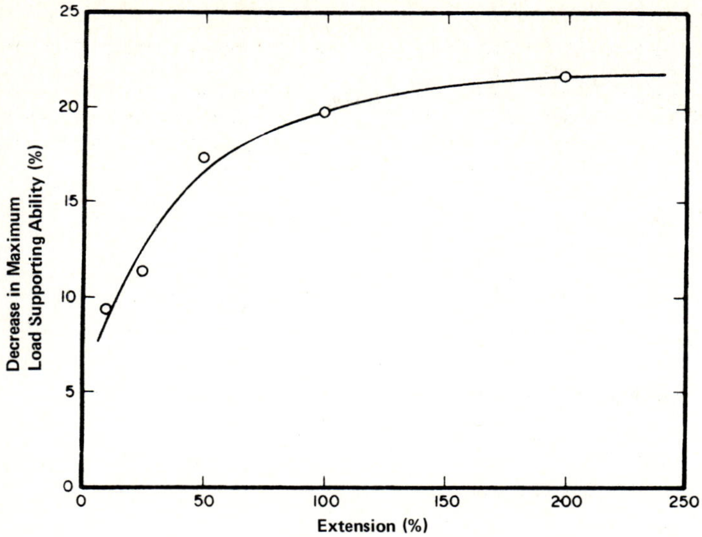

Fig. 29. Decrease in Second Cycle Maximum Load Supporting Ability of Springy Polypropylene as a Function of Extension (ref.75)

The effect of repeated loading cycles on elastic recoveries of an *as-spun* elastic polypropylene fiber (ER_{100} of 80%) and its annealed counterpart (ER_{100} of 96%) is shown in Fig. 31. After an initial decrease, recovery of the annealed fiber tends to level off at large numbers of loading cycles, whereas elastic recovery of the unannealed sample seems to continue to decrease over the observed range of loading cycles. This indicates that the structure of the annealed elastic material is more stable under repeated loading and unloading, and its fatigue resistance is less strain-sensitive.

While the elastic recovery (ER_{50} and ER_{100}) is of the order of 95 percent on the first cycle and gradually reaches a constant level of about 85 to 95 percent, depending on extension (50 to 100%), the work recovery in the initial cycle is relatively low, 30 to 40 percent (Fig. 32). This is due to the large amounts of energy required to overcome the van der Waals forces in the intercrystalline regions during the first 5 to 10 percent extension.

The work recovery increases on cyclic loading to about 60 percent within a few cycles (Fig. 32). Eventually, both elastic and work recoveries reach practically constant values, the paths of extension and retraction becoming constant. The major change in stress-strain curves occurs in the initial stage of extension (up to 25%) between the first and second cycle. The stress level at the extension limit (50%, 100%), however, is not greatly affected. Equally, the change in the shape and position of the retraction curve is relatively minor. As seen from Fig. 30, initial moduli of *as is* and annealed fibers show a complimentary behavior, also reaching a constant level after a few cycles.

Considering the highly elastic behavior of the "hard" elastic fibers in the whole range of extension and the vast difference in the time of recovery on both sides of the "yield" point, it becomes apparent that we are dealing here with two types of elastic mechanisms; (a) a relatively slow process restoring the original stress level of the "yield" point after relaxation to a low or zero load level (relaxation time of hours), and (b) a very fast, practically instantaneous recovery mechanism beyond this point.

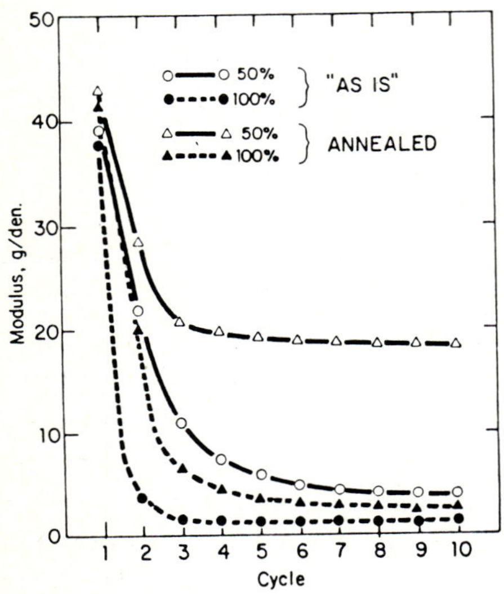

Fig. 30. Modulus of Elastic Polypropylene Fibers as Function of Cycle for 50% and 100% Extension. "As Is" Fibers: ○ 50% Extension, ● 100% Extension; Annealed Fibers: △ 50% Extension, ▲ 100% Extension (ref.68)

Stress relaxation and creep behavior are important factors in prediction of polymer performance under sustained tensile loading. Figure 33 shows the relaxation behavior of elastic polymer.

Figure 34 depicts the tan δ versus temperature at 3.5 Hz profiles for springy polypropylene and a normal polypropylene (NPP) fiber of higher orientation as shown in x-ray diffraction. It is interesting to note that there is a difference of 10 to 15° in the peak location of the SPP and that the SPP peak is higher. Thus, from the rheovibron data, it can be seen that at small deformations the work loss of SPP is greater than for NPP, occurs at a lower temperature, and is narrower than the corresponding NPP curves. This is an indication of the change of thermomechanical behavior of polypropylene with orientation.

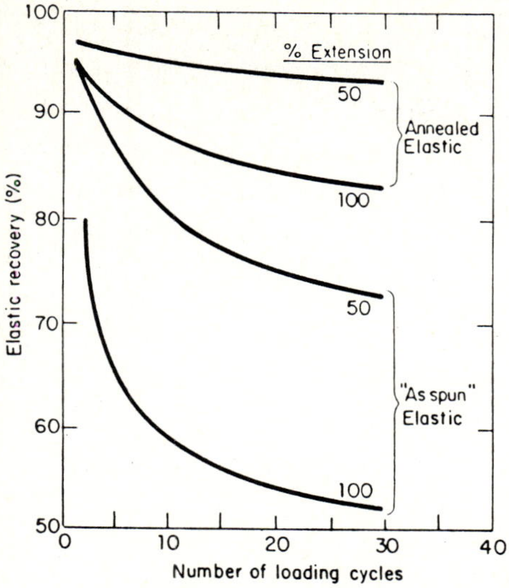

Fig. 31. Effect of Repeated Loading Cycles on Elastic Properties of Elastic Polypropylene Fibers. Elongation Rate: 100%/min. 50 and 100% Extension for "As Spun" and Annealed Fibers (ref.68)

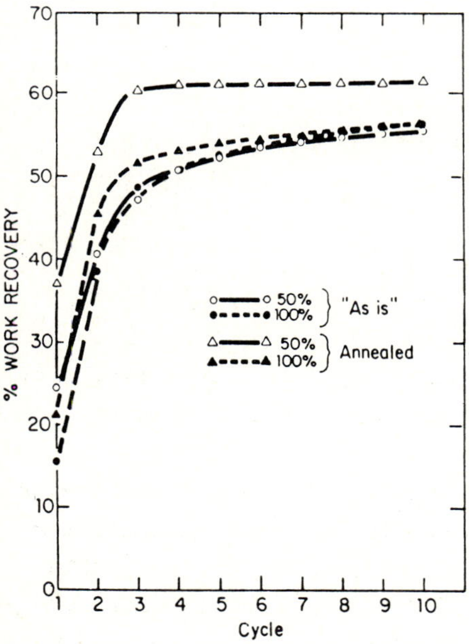

Fig. 32. Work Recovery of Elastic Polypropylene for Cycling to 50% and 100% Extension. "As Spun" Fibers: o 50% Extension; o 100% Extension; Annealed Fibers: Δ 50% Extension, Δ 100% Extension (ref.68)

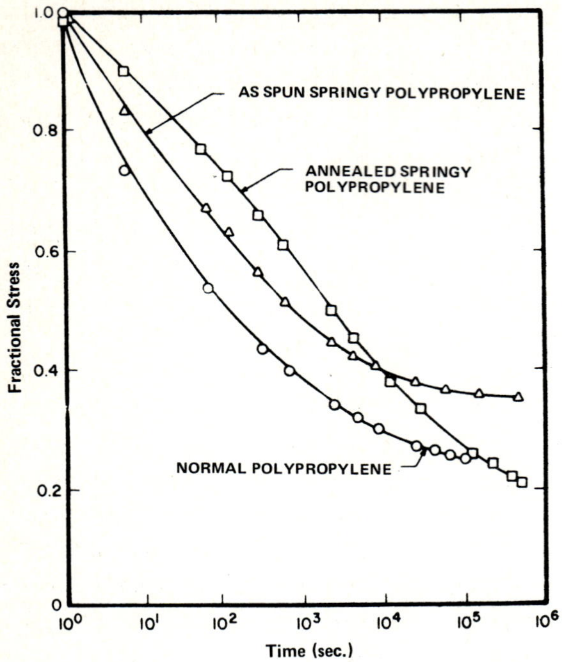

Fig. 33. Stress Relaxation Behavior of Springy and Normal Forms of Polypropylene Fibers at 23°C

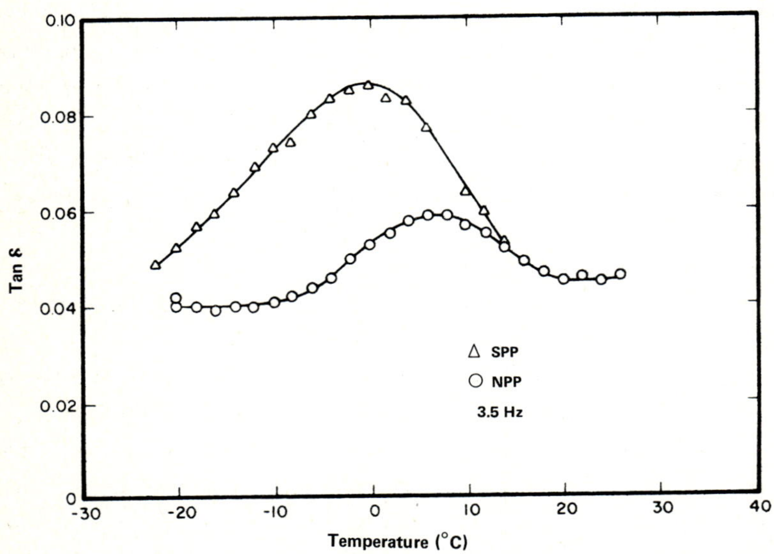

Fig. 34. Work Loss Measurements for Springy Polypropylene and Normal Polypropylene Fibers (ref. 75)

The temperature dependence of the tensile characteristics, break elongation, tenacity, initial modulus, and yield stress, for the same three fibers are shown in Fig. 35. Modulus, tenacity, and yield stress behave as normally expected. The maximum in elongation at 90°C is an effect of measurement conditions; at and above this temperature some crystallization occurs while the sample reaches temperature equilibrium for the tensile test and therefore the break elongation approaches the values of annealed fibers. The "yield" stress seems to be independent of annealing conditions; as will be shown below, it correlates with the onset of void formation and the transition to a different super-molecular deformation mechanism.

What sharply distinguishes the elastic polypropylene fiber from spandex and rubber is a large, mainly reversible reduction in apparent density on stretching. Figure 36 shows the relative apparent density measured as a function of stretching. For comparison, see the similar data for a spandex fiber, known to be true elastomer. The results, showing about 50 percent reduction in unstretched density at 100 percent extension for elastic polypropylene fiber, demonstrate that the elastic fibers differ fundamentally from most other true elastomeric fibers, which either increase slightly or do not change their density on stretching. The change of apparent density of elastic polypropylene fibers is reversible to the extent the fibers recover from the linear extension. These results lead to the speculation that submicroscopic cracks or cavities are formed on extension of the fibers, i.e., that some sort of submicroscopic accordian-like behavior is involved. The measurements by high-pressure mercury porometry indicate that this is indeed the case. Electron microscopy of the abraded surfaces of elastic hard fibers indicates that the lamellae appear to tilt and splay apart, creating gaps or crevices between them. These interlamellae spaces are believed to represent the internal volume created on stretching. The elasticity of the fiber, therefore, derives primarily from a cumulative bending of the thin, crystal lamellae, as described before.

The basic process for developing elasticity in polypropylene fiber is dependent on forming the row-nucleated morphological structure described previously, and annealing of such a structure under no-load conditions.

The effects of processing on the morphology of springy polymers is significant. Obviously, without a stress crystallization, the row-nucleated structure essential to springy polymers does not occur. However, even within the regime of row-nucleated structure, other variables can affect the ultimate springiness of the polymer. Table IX shows an example of the effects of various processing parameters on some physical properties of springy polypropylene. Notice that spinnerette temperature, quench rate and windup speed all affect the elongation to break, tensile recovery, tenacity, etc. Other parameters which affect the

TABLE IX

The Effects of Varying Process Conditions on the Physical Properties of Springy Polypropylene (ref.72)

Spinning Conditions				Physical Properties						
Item	Spinnerette Temperature (°C)	Quench, (cfm)	Windup, (ypm)	Orientation Angle	Orientation Type	TR 25 (%)	Tenacity (GPD)	Elongation at break (%)	Mi. (GPD)	Dpf
V-A	290	3.6	100	16°	gamma	93	1.6	480	10.0	12.0
V-B	290	6.7	100	10-21°	gamma	95	1.7	396	7.5	13.6
V-C	270	6.7	180	14°	gamma	87	1.3	147	10.0	4.4
V-D	290	3.6	90	15°	gamma	93	1.7	342	12.3	12.5
V-E	290	2.6	90	11°	gamma	85	1.9	190	11.3	8.2
VI-F	295	1.6	100	43°	normal	74	1.2	497	10.0	16.6
VI-G	295	3.6	100	10-21°	gamma	89	1.6	520	9.7	18.5
VI-H	295	6.0	100	18°	gamma	93	1.6	414	10.3	18.0
VI-I	290	2.0	100	...	normal	77	1.2	477	9.3	13.3
VI-J	290	6.0	100	10-30°	gamma	84	1.6	469	9.3	13.0
VI-K	260	6.0	100	10-21°	gamma	92	1.6	417	11.3	10.7
VI-L	250	3.6	137	10-21°	gamma	90	1.8	312	12.7	4.1
VI-M	250	3.6	300	10-21°	gamma	85	2.0	243	13.0	3.8
VI-N	280	4.4	245	10-21°	gamma	90	...	...	...	.....
VI-O	300	3.2	123	...	normal	55	1.0	509	4.7	7.4

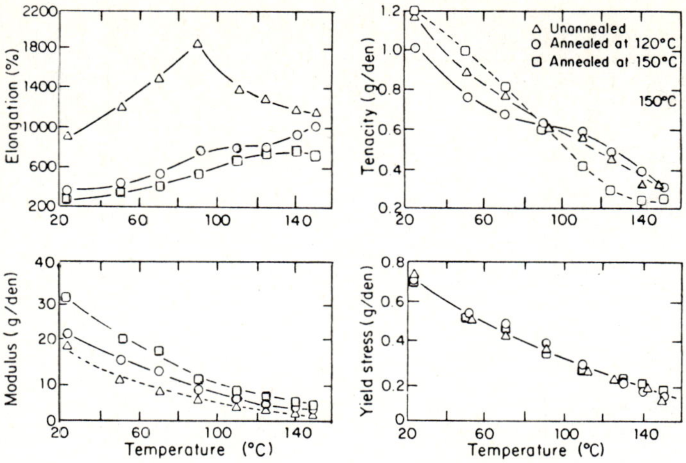

Fig. 35. Temperature dependence of Tensile Properties of Elastic Polypropylene Fibers. Δ *Unannealed;* o *Annealed at 120°C; Annealed at 150°C. Elongation Rate: 100%/min. (ref.68)*

properties of the polymer include draw ratio. Thus, for SPP it is desirable that the draw ratio be less than 2.5 and preferably 1.0.

In summary, the favorable processing conditions are the employment of low spinning temperatures, high melt draw-down ratios, high spinning speed, and high quenching rate. High molecular-weight polymers of broad molecular-weight distribution are specially suitable for the development of stress-nucleated morphology. Cold drawing is a negative factor.

The annealing of as-spun, stress-nucleated, polypropylene fibers is an essential part of the process for the production of elastic polypropylene fibers. As shown in Table X, a good *as-extruded* elastic polypropylene fiber has a relatively high degree of orientation of the lamellar crystals as revealed by both X-ray diffraction and birefringence measurements. The overall effect of annealing, as observed by X-ray small-angle scattering, indicates increased superstructural perfection and order.

The variations in elastic recovery of an elastic polypropylene fiber (from 100 percent extension at room temperature) with annealing temperature is shown in Fig. 37. The elastic recovery increases rapidly with annealing temperature between 80° and 120°C, leveling off at a value of approximately 95 percent. Annealing at 120°C for 60 minutes introduces a sufficient degree of structural perfection for the maximum attainable value of elastic recovery (Curve 1).

The degree of perfection of the structural elements, which are responsible for the elasticity of the fiber, depend on the duration of annealing and, more importantly, on the temperature of annealing. Figure 38 shows the permanence of elastic character by measuring this property at varying temperatures. The

TABLE X

Physical Properties of Polypropylene Fibers of Varying Elastic Recovery (ref.68)

Sample		ER_{50} (%)	Density[a] (g/cc)	X-Ray small-angle Spacing ( )	Bire-fringence[b]	Orientation by X-Ray[c]
A	As-extruded	24.5	0.8864	110	0.0026	None
	Annealed	42.0	0.9086	200	0.0055	
B	As-extruded	34.5	0.9004	130	0.0072	Fair
	Annealed	84.7	0.9074	180	0.0103	
C	As-extruded	40.6	0.8996	140	0.0185	Highly
	Annealed	93.2	0.9086	180	0.0176	Oriented

All samples annealed at 140°C for 45 min. in air
(a) Determined with a density-gradient tube containing isopropanol-water mixture
(b) Measured in silicone oil (Dow Corning-550)
(c) By wide-angle X-ray diffraction

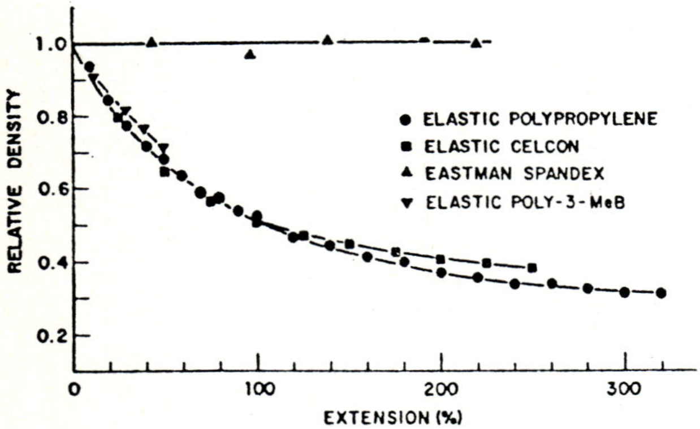

Fig. 36. Relative Density Measurements on Slow Stretching Elastic Poly-3-MeB, Elastic Celcon, Elastic Polypropylene, and Experimental Eastman Spandex Fibers (ref.69)

elastic recovery of the unannealed sample starts to decrease immediately with an increase in temperature of measurement reaching a constant value of about 30 percent at 70°C. With the 120°C annealed sample, due to its relatively more perfect structure, the elastic recovery decreases only slightly over the temperature region up to 100°C, and drops rapidly in the neighborhood of the annealing temperature. The elastic recovery of the sample annealed at 150°C is essentially constant at a high value of 95 percent at temperatures up to 140°C.

The stress-strain and elastic properties of elastic polypropylene yarns are somewhat between those of true elastic fiber and an oriented hard fiber (polypropylene textile fiber). Normally, it is necessary to draft the true elastic fiber several hundred percent during core spinning and fabric contraction to reduce

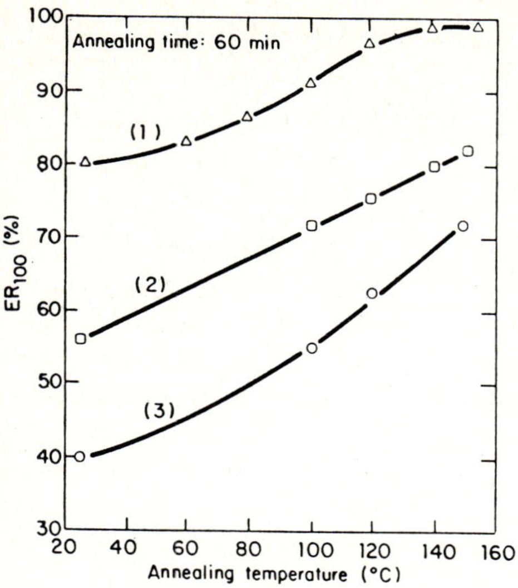

Fig. 37. Dependence of Elastic Recovery on Annealing Temperature of Polypropylene Fibers Spun Under Various Conditions. Nominal Spin Temperature (1) 230°C; (2) 260°C; (3) 280°C. Annealing time: 60 min. (ref.68)

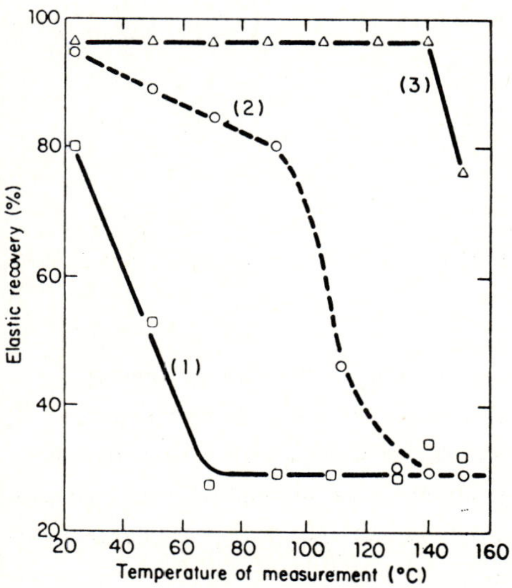

Fig. 38. Temperature Dependence of Elastic Recovery of Elastic Polypropylene Fibers (1) "As spun", (2) Annealed at 120°C, (3) Annealed at 150°C. Testing Conditions: Temp.: 23°C, Extension 100%, Elongation Rate: 100%/min. (ref.68)

its elastic extension to a desirable range of 40 to 50 percent. This operation is, of course, not required in the case of elastic polypropylene yarn, as the working elasticity of the fiber is already in that region. Although the extensibility of elastic polypropylene fiber is as high as that of most stretch fabric, the problem is to construct the fabric in such a way that the extensibility of the fiber is still available. Because of this, elastic polypropylene yarns cannot be used simply to replace rubber or spandex in stretch garments. New fabrications procedures that would take into consideration the high modulus and low extensibility of polypropylene elastic yarns have to be worked out. As the modulus of elastic polypropylene is about ten times that of Lycra, a 15-denier monofil might be used where a 140-denier spandex is indicated. In addition to its use in stretch fabrics, elastic polypropylene may have utility in furniture tapes, automotive seat covers, support webbing, etc.

III.6. Monofilaments and Related Products

By common trade terminology, the term *monofilament* is applied to filaments having cross-sectional diameters of about 0.006 inch (150 denier) and larger. With few exceptions, the markets for monofilament and multifilament yarns do not overlap. Until recently, most manufacturers produced either monofilament or multifilaments, seldom both.

Like multifilaments, monofilament products have to be specially engineered for specific uses. They can be the "splitting" type or "nonsplitting" type: the former is employed in products where flexibility of structure is a primary requirement, the latter is used where stiffness of the product is of no consequence. One important quality requirement for both types of products is for the monofilaments not to curl spontaneously.

Theoretically, monofilaments can be produced by the same spinning and drawing techniques as those used for the production of multifilament yarns. However, the monofilaments so produced have a marked tendency to curl. They are, therefore, unsuitable in many applications and are difficult to handle in many other applications. Because of this difficulty, the production of monofilaments is generally carried out at a low speed, utilizing the machinery arrangement similar to that used for the production of split-film yarns, discussed in Chapter XII. The monofilaments are extruded at a relatively low speed through a spinnerette (often called a "die"), quenched in cold water, heated in an oven or on heated multiple driven rolls under nominal tension, drawn, quenched again in cold water under tension, and then taken up (ref.33).

The monofilaments produced through the above general process have some tendency to split in repeated flexing. In some uses this property is intolerable. For such uses, a high molecular-weight polymer of broad molecular-weight distribution is generally utilized with advantage. The quenching of the monofilament is delayed to the point where all the hot drawing has already occurred. This melt-drawn filament is then quenched rapidly in cold water. The crystallization under tension is thereby avoided. The quenched monofilament yarn may not be cold drawn for some uses, especially for the production of bristles, but instead annealed under tension. The fibrillating tendency of these bristles is further reduced by utilizing a sharp, heated knife for cutting the monofilaments into bristles so that a sealed, rounded end is obtained.

If the production of fibrillating monofilaments is desired, polypropylene is mixed with an incompatible fiber-forming polymer in more than 10 percent concentration, extruded, melt drawn at high speed, quenched rapidly while the filament is being attenuated, and drawn in an oven or over heated rolls. The fibrillation can be accomplished through calendering the cold-drawn monofilament and flexing with turbulent fluids utilizing suitable fluid jets (ref.34). The incompatible fiber-forming polymers suitable for this purpose are nylon, polyester, polystyrene, and copolymers of ethylene.

The resulting monofilament is a composite structure of a multitude of very small, discontinuous, low-denier, interconnected fibrils. The fibrils of the assembly either do not adhere to each other at all or do so only for a short part of their length, but remain as part of the same assembly by virtue of being interlocked with each other in a network structure. The fibrillated monofilament resembles with a staple spun yarn. In some applications the fibrillation is allowed to develop in use. The applications include cleaning brushes for industrial and household purposes. The tips of these bristles fibrillate into very fine fibrils of limited length while the stems remain intact.

Monofilaments obtained from random polymer blends have a structure wherein the fibrils are not continuous along the fiber axis. For this reason, the properties such as tenacity and elongation of each polymer are not fully utilized. It has also been experienced that the small fibrils of the added polymer considerably reduce fatigue resistance of the composite monofilament. This may be due to the ends of the fibrils cutting into the matrix polymer. To overcome these difficulties, the two components are fed separately into a mixing device capable of arranging the polymers into laminated streams, which can further be subdivided and broken into a very large number of fine continuous-melt streamlets. The composite melt flow thus produced is fed to the spinnerette and spun into monofilaments. A number of such mixing devices have been described (ref.35). The resulting monofilament consists of a multitude of continuous fibrils of component

A embedded in matrix B or, if the composition is 50/50 by volume, the monofilament may consist of a loosely held bundle of essentially parallel fibrils of component A and component B, as depicted in Fig. 39. These monofilaments are especially suitable where strength is as important as the flexibility of the yarn.

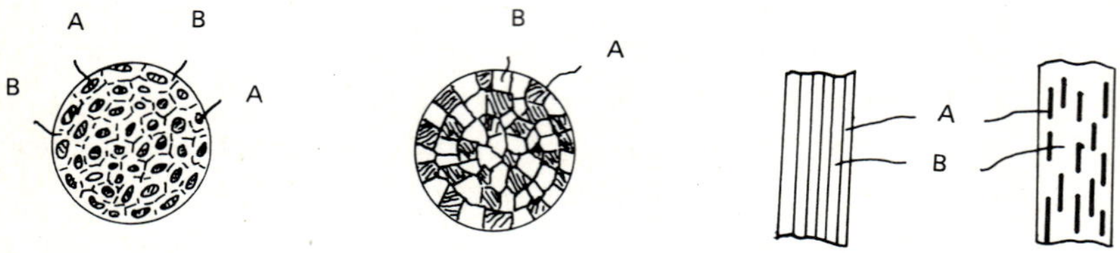

Fig. 39. Schematic of Blended Monofilament Structure

III.7. Pilling-Resistant Fibers

The pilling process involves entanglements of surface fibers held to the base fabric by a few anchor fibers. If these fibers are not broken by normal surface abrasion, pills accumulate and become unsightly. Hence, for pill-prone textiles, it is important that the component fibers have relatively low toughness. One route that has been widely investigated to prepare fibers of improved pilling resistance is the use of low molecular-weight polymers. Abrasion resistance and toughness decrease with decreasing molecular weight. The reduction in molecular weight must be carefully balanced with orientation to yield the necessary performance characteristics.

Experience with other synthetic fibers has indicated that fibers with high elongation show a marked tendency for pilling. In polypropylene fibers, the elongation to break can be significantly decreased by heat setting the staple in relaxed state at 130 to 140°C. The annealing of fibers after stuffer-box or gear crimping is an especially effective way of decreasing the pilling tendency of polypropylene fibers. This improvement probably comes about because of the preferential disorientation of the deformed section of the fiber and thereby creating weak spots along the fiber. In addition to these two measures, there is a third approach to the problem. Some profiled cross sections, such as "T" or "Y", are less pill-prone than the normal cross sections of delta or round. The combination of low viscosity with profiled cross section has proved to be a successful way of achieving fibers of low pilling tendency.

III.8. Resilient Products

One of the main deficiencies of polypropylene fiber in the early stages of its development was its relatively low resilience, i.e., the lack of ability to recover from compression, which made it rather a low-quality fiber for such uses as fiberfil and floor coverings. This deficiency has been removed to a great extent by utilizing relatively high molecular-weight polymers and heat setting of the fiber under low tension after crimping. The fiber may be cold drawn before crimping and heat setting (ref.36), or for some products an undrawn fiber can be used (ref.37). Table VII illustrates the effect of heat setting on the resilience of cold-drawn crimped polypropylene staple useful as fiberfil.

Crimping and heat setting can be done in one operation like that accomplished to some degree in steam-jet texturizing. This is probably one of the reasons that early carpets made from nonheat-set, stuffer-box, or gear-crimped polypropylene yarns fared poorly in walk-out test compared with carpets made from steam-jet textured yarns. The failure of stuffer-box and gear-crimped staple as fiberfil can also be traced to the lack of "life" which is a characteristic of heat-set products.

In a batch process, the heat setting can be done by steaming the tow or staple in high-pressure autoclaves, commonly used in textile industry. It has been discussed in Chapter VII that polypropylene fibers are difficult to heat set and require relatively long periods of treatment. A temperature as close as possible to 140°C is recommended for such processes (ref.38).

Table VII. Effect of Heat Setting on Resilience of Polypropylene Staple Fiber after Cyclic Loading (arbitrary units)

Draw Ratio	Heat Setting Temperature (°C), 3 min.			
	110	120	130	140
Undrawn	50	50	60	65
2.5	75	75	85	90
3.0	80	85	95	95
3.5	75	75	90	95
4.0	60	60	80	90
4.5	60	65	80	85
5.0	60	65	75	85

IV. MISCELLANEOUS PRODUCTS

IV.1. Superfine Fibers

Superfine or microfibers are fibrous materials less than about 5 micron in diameter. Such fibers are known in nature in the form of spider silk (0.5dpf) and pineapple leaf fiber (0.3dpf). Manmade microfibers have been researched in many countries and at least one United States company is known to be marketing a submicron glass fiber for filters. The production of superfine fibers from nylon and other polymers has been accomplished by United States Naval Research Laboratory, Exxon Research and Engineering, Arthur D. Little Inc., and 3M Company (ref.39,40). One of the most important commercial organic microfibers, known as "Textryls", was produced by DuPont in the early sixties but it was later withdrawn from the market (ref.41). Microfibers are known to possess aerosol filtration efficiencies far in excess of ordinarily used materials. Furthermore, superfine fibers represent a new area of research where completely altered physical properties might be anticipated as a result of the nearly colloidal state of the substance. Superfine polypropylene fibers combine the extraordinary properties common to all microfibers with an excellent resistance to common chemical and oleophilic properties of the polymer. Several companies and research laboratories are presently known to be active in this area of product development.

Several techniques have been mentioned in patent literature for the production of microfibers. The following three methods merit a brief discussion:

(1) Centrifuge spinning

(2) Spray and flash spinning

(3) Incompatible blends

Centrifugal spinning consists of propelling polymer melt from the perimeter of a rotating disc onto a moving screen, and collecting the fiber in the form of batts, roving, or nonwoven fabrics. Several patents have been issued relating to this technique, mainly to Monsanto and Corning Glass (ref.42). The basic technique is an extension of well-established techniques for the production of "glass wool".

Spray spinning means ejecting a mixture of air and polymer from a spray nozzle onto a moving screen. This technique has been extensively researched by United States Naval Laboratories, Esso Research, and 3M Company. Several improvements to this technique have lately appeared (ref.43). A good description of the basic technique is given in reference 39 and more recently in reference 44. Because of its large potential for polypropylene fibrous products, a summary of the process, as recommended by Esso Research, is given here (ref.44). Further details can be obtained from the references quoted and the literature mentioned in the Bibliography of this chapter. Figure 40 shows a schematic of the basic process. This is essentially the same technique as that utilized by United States Naval Laboratories in the early 1960s.

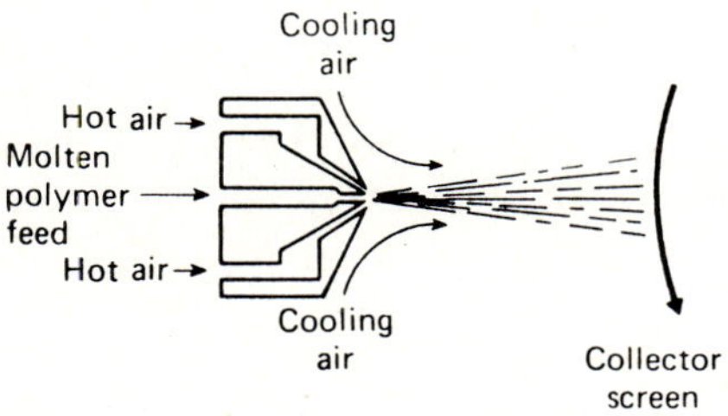

Fig. 40. Schematic Diagram of Melt-Blowing Process (ref.44)

High-velocity hot air is used to attenuate molten polymer into very fine fibers. Resin is fed into an extruder where it is melted and heated to the appropriate temperature required for fiber formation. The extruder feeds a special melt-flowing die which is a key to the successful operation of the process. As the resin emerges from the die orifices, it is contacted by the high-velocity air which attenuates the polymer into a blast of fine fibers. The fibers are quenched by a large volume of cooling air which is aspirated into the primary jet. This normally solidifies the fibers an inch or so from the die tip. The fibers are collected on a moving screen placed in front of the blast. As the fibers land on the screen they entangle to form a cohesive web. Some of the air passes through the screen, but most is carried away along the screen surface.

In the normal operation of the process, fiber breakage occurs continuously. As the melt emerges from the die tip the high-velocity air attenuates each stream so rapidly that separate discontinuous fibers are blown away from the die face. This is significant. Fiber lengths are on the order of a few inches. Each fiber has associated with it what we refer to as "shot", a particle of polymer, considerably larger in diameter than the fibers, which is formed by the elastic "snap back" of the fiber ends upon breaking. With proper operation, shot size is quite small and has little effect on final product properties.

The freedom from the need to maintain continuous fibers is in great contrast to typical melt spinning-based processes, where the entire operation depends on avoiding breaks. Fiber breakage in these processes usually means line shutdown, so that breaks are avoided at all costs. Thus, other processes are limited to fiber sizes of about 1 denier, whereas in melt blowing, ultrafine fibers are possible, typically 1/20 denier. Plugged or partially plugged polymer orifices are a minor problem in melt blowing but are intolerable in continuous-fiber processes. Melt blowing can tolerate nonuniform polymer melts and mixtures of polymers which cannot be handled by other processes. Thus, the discontinuous fiber operation makes for versatility, stability, and overall ease of operation.

The fibers are deposited on the collector screen as a continuous web of entangled fibers. The good integrity of the web is due primarily to fiber entanglement and surface attraction between the very small fibers. Melt bonding is believed not to be important except in the special cases of very close web collection or when very large fibers are produced. In any case, the "as-formed" web has good integrity and can be readily slit, wound, or further processed without additional bonding.

Basic weights as high as 1000 g/m^2 or as low as 5 g/m^2 can be produced. The normal method of changing basic weight is by changing collector speed. Because of the high-fiber velocities relative to the collector speed, the nature of the web does not change over a wide range of collector speeds. This means that rapid changes in basis weight are possible with little or no adjustment to the die. Of course, if takeoff equipment is limited in speed, it may be necessary to reduce die throughput to achieve very low basis weights.

The fibers are normally collected at relatively short distances from the die, on the order of 6 to 12 in. This short path gives a positive control of the fiber blast and also good edge control. These factors lead to stable operation.

The extrusion process is normally carried out at high temperatures. The softness of the web is highly dependent upon the speed of the air flow. This is illustrated by an example from a patent specification.

Polymer melt-flow rate (polypropylene):	33.6 g/min
Extruder Temperature:	310°C
Die Temperature:	276-285°C
Air Temperature:	266-282°C
Polymer Rate:	7.1 g/min
Collector Distance:	8 in.

	Air Rate (kg/min)	Web
(1)	0.131	Coarse, brittle mat composed of large, ropy fibers
(2)	0.215	Softer mat; finer fibers
(3)	0.299	Soft, pliable mat of fine continuous fibers

(4) 0.320 Mat contains large shot particles, which are rough and of poor appearance.

It is suggested that the higher the die temperature employed, the lower the optimum air rate. In addition, at very high air rates the shot is so small as to be undetectable. For example, the following conditions are quoted in a patent.

Extruder Temperature:	310°C
Die Temperature:	341°C
Air Temperature:	304°C
Polymer Rate:	18.7 g/min
(a) Air Rate:	0.789 kg/min good web
(b) Air Rate:	0.948 kg/min excellent web

Figure 41 shows a schematic of a typical melt-blowing unit. The unit has three 40-in. wide dies fed by a single extruder. The use of multiple dies has several advantages. First, production rate is increased beyond that possible with a single die. (Normal die throughput is 1.3 lb/hr/in of die width.) At higher rates, fibers become larger and shot more pronounced, although webs produced at these higher rates are still acceptable for many applications. By the use of multiple dies, high web-production rates can be achieved without increasing die throughput above the optimum for the particular product.

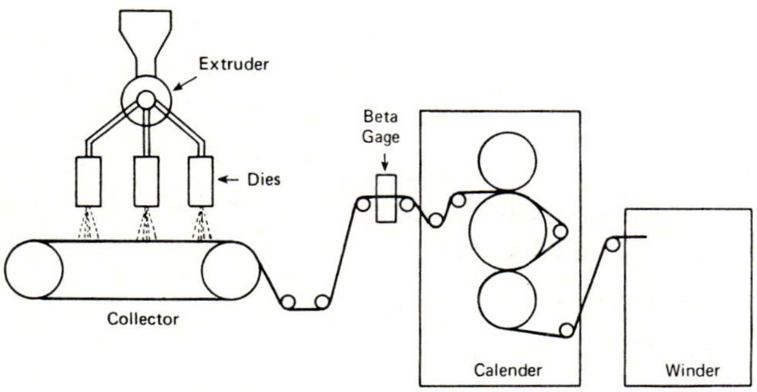

Fig. 41. Schematic of Three-Die Melt-Blowing Line (Air Lines and Ducts not Shown (ref.44)

Second, multiple dies produce more uniform webs with less problems with gage bands. Pilot operation with a single die has produced cross-sheet basis weight uniformity of $\pm$ 10 percent using the present die design. New designs currently being developed should reduce this to about $\pm$ 5 percent. The use of multiple dies in series gives a further randomization of gage nonuniformities.

Third, multiple dies give the flexibility to produce layer structures with different web characteristics in each layer. If multiple extruders are also used, different polymers can be extruded to produce a wide variety of laminated structures. This will be discussed later.

A disadvantage of multiple-die operation is that, for heavier basis weights, the layers produced are not well bonded to each other. Although there is enough interlayer bonding so that the web stays together and can be handled as a single layer in subsequent processing operations, in some end uses delamination can be a problem. However, the web is often further compacted and bonded, sometimes by calendering. In this case the layered product is quite satisfactory. On the other hand, if individual layer weight is less than about 20 g/m^2, delamination is not a problem, even without additional bonding.

In a multiple-die line, provision must be made to supply cooling air to the individual dies. A single die aspirates the air it needs from its surroundings. Multiple dies are equipped with cool air ducts (not shown in the figure) to supply the needed cooling air to the vicinity of the high-velocity jet. In this way, the proper temperature is maintained at the web surface.

Exhaust ducts are also designed to remove most of the air from the web formation area. Only a small amount of air is taken through the collector screen. Proper air control is another important key to successful operation.

Also shown in Fig. 41 are the beta gage, calender, and winder, which make up the rest of the present pilot line. The beta gage is used to measure web basis weight and to check web uniformity. The calender is used to make dense, compacted sheets for specific end uses. Trimmed rolls of calendered or uncalendered material are wound on the winder to diameters of 24 in.

TABLE VIII

Typical Web Properties of Melt-Blown Polypropylene Webs (ref.44)

Basis weight	5-1000 g/m^2
Tensile strength	1 lb/in. oz/yd^2
Tear strength	70 $g/oz/yd^2$
Air permeability	High, controllable
Porosity	75-95%
Covering power	Excellent
Hand	Very soft, drapable

Table VIII shows typical web properties for the melt-blown polypropylene material. As mentioned previously, basis weight can be varied over a very wide range. Of special interest are the very lightweight materials which have good integrity and fiber coverage. These materials are possible only because of the small fiber diameter.

The values shown for tensile and tear strength are typical for medium-weight base webs. These values can be varied somewhat by changes in blowing conditions and collector distances. These values are low compared to highly bonded webs composed of textile fibers. For this reason, the melt-blown material is not being considered for applications where strength is the most important requirement, such as, for example, in carpet backing or industrial bagging. The melt-blown webs are aimed at applications where small fiber diameter and the unique properties that come from this are paramount.

Small pore size is an example of such a property. Depending on the degree of compaction, pore sizes range from a few microns upward. Yet, because of the fibrous structure, air permeability or breathability is high. Thus, with hydrophobic polypropylene, webs can be made so that water drops will roll off and not penetrate, but water vapor and air will readily pass through. Porosity expressed as percent void fraction is normally around 85 percent in the base web.

TABLE IX

Important Process Variations (ref.46)

Variation	Type of product
Low air-rate operation	Continuous fibers, large diameters
Multiple-die operation for multiple-component webs	Laminated products with different polymers--fiber sizes in layers
Coating on substrates	Reinforced webs, laminates, coated films, fabrics
Collection on patterned surfaces	Novel, bulky structures
Secondary air additive addition	Filled webs, chemically-treated webs
Roving process	Cylindrical products

Because of the small fiber's diameter, the covering power of even very lightweight sheets is excellent. The webs are normally very soft and drapable. These properties can be adjusted over a reasonable range by adjusting blowing conditions which affect fiber size and/or by changing collecting conditions. Thus, the basic product is a fine-fibered, soft, drapable web suited for applications where small fiber diameter gives unique advantages and only moderate strength is needed.

Several variations of the basic melt-blowing process further extend the range of products that can be made. Table IX lists the most important of these. First on the list is low air-rate operation. As stated previously, the major advantage for melt-blowing is that it produces very fine fibers in a simple, stable process. The die, however, can also be operated with a low air rate to make continuous filaments with sizes in the textile fiber size range and above. These fibers show very little orientation and therefore have very low tensile strength. However, they can be made to have very high elongation, resulting in webs with high tear strength. Other web properties which can be achieved with the coarse fibers are abrasion resistance, rigidity, and controlled, open-pore structure. This type of web may be most useful in combination with the fine fiber product.

In the multiple-die operation described previously, the dies in series are operated at different conditions to produce laminated structures with different properties in each layer. For example, the coarse fiber webs described above can be combined in a single operation with fine fiber webs to produce a graded pore-size filter structure. The use of several dies with individual extruders for each makes possible structures with different polymers in the layers. For example, a lightweight adhesive polymer layer can be combined with a layer of higher-melting, fine fibers for a nonwoven web with one side suitable for bonding to a variety of surfaces.

Coating on substrates also provides many possibilities. Fine fibers have been blown directly onto both sides of open scrims to produce reinforced webs. The scrim gives additional strength and the fibers provide the desirable surface and coverage properties. Many kinds of substrates can be coated, e.g., fabric, film, and other nonwovens. In many cases, especially with very light layers, there is sufficient adhesion so that no additional bonding is required. The use of a substrate allows extremely lightweight webs to be formed. Polypropylene has been deposited in fiber layers as light as 0.1 g/m^2.

Collection of fibers on a patterned surface has been demonstrated, and a variety of dimpled or waffle-patterned webs are possible. Such webs are made at fairly close die-to-collector distances where the force of the air-fiber blast causes the web to conform to the collector surface.

Finally, the roving process is a major departure in die design, but uses the same techniques of fiber formation. In this variation, instead of a flat die with the orifices in a straight row, a circular die with the orifices in a circle is used. The die is operated so that the conical fiber blast is focused roughly on a point where the fibers are formed into a continuous cylindrical roving. This technique has been studied as a way to make fine-fibered cigarette filters, but may also have applicability for other rod-shaped or rope-like products.

The flash-spinning method is a variation of the spray spinning technique. It consists of spray spinning a polymer solution in highly volatile solvent into a vaporizing chamber wherein the microfibers are collected on a moving screen or any other collecting device and the solvent is removed. Several patents have recently appeared relating to this technique (ref.45). Chlorofloroaliphatic compounds have been mentioned as suitable solvents for the production of polypropylene microfibers (ref.46). This technique is evidently more expensive than the centrifugal spinning and spray spinning of polymer melt, but with this method fibers of submicron diameters can be produced.

The spinning of incompatible polymer blends into a tow of microscopic fibers has been already discussed under Section III.6. The same technique can be used to produce a sheet structure of interconnected microfibers. Polypropylene may, for instance, be blended with polystyrene or any other incompatible polymer in 50/50 volume proportion, formed into a sheet, and the added polymer extracted by a suitable solvent. A porous, suede-like polypropylene fabric of good tear strength is obtained. The tear strength is increased if the sheet is calendered before extraction. For the production of fibrous tow, extraction may not be necessary. Numerous composition-of-matter, process and products patents have lately appeared on this subject (ref.47) (see Bibliography).

Mats of polypropylene microfibers serve as excellent filter medium for fine aerosols. They may find extensive use in anti-air pollution devices, cigarette filters, masks for mining, medical and industrial operations, strainers, microfiltration of beverages, etc. The oleofilic properties of polypropylene makes these webs especially suitable for filtering out oily contaminants from air and water. The superfine fiber structures can be used as carriers for ion exchange resins. A very promising use of polypropylene microfiber sheets is as battery separators in Ni-Cd batteries. Because of their excellent porosity and flexibility, the sheets of microfibers can find excellent use in the preparation of suede-like materials and other nonwoven fabrics. Electric insulation is another promising use of paper-like structures prepared from microfibers from polypropylene.

IV.2. Synthetic Pulp and Papers

The increasing scarcity and cost of wood pulp, coupled with the rapidly decreasing cost of synthetic polymers, motivated research for the development of manmade pulp and paper. The interest is especially high in Japan where no local supply of wood pulp is available; most of the research impetus has therefore come from there. Synthetic paper is of interest to other industrialized nations for such uses as electric wire wrapping, filtration, and speciality printing papers. Because of its high strength, cheap raw material base, and low cost of conversion, polypropylene fiber has great potential for the production of synthetic paper and paper-like products. In this use, the only other candidates are polyethylene, polystyrene, and polyvinyl alcohol fibers.

Three basic approaches are being pursued:

(1) Synthetic paper produced via polymer film structure
(2) Synthetic paper produced via synthetic pulp, using common paper-making machinery

(3) Synthetic paper via spun bonded techniques.

Polyolefin films are widely used in place of cellulose-based paper for many packaging applications. However, normal polymer films lack some of the required qualities for printing stocks. Synthetic papers used for the printing arts must possess a number of properties to compete with cellulosic paper on the basis of performance, aside from the matter of competitive cost. Synthetic papers must be opaque, bright, ink receptive, lightweight, stiff, and have the same handling characteristics as natural paper. The approaches that have been employed to impart these properties to plastic films can be grouped into two broad categories: internal methods and surface methods. Internal methods are compounding, foaming, stretching, leaching; surface methods are coating, solvent-swelling/coagulation, particle embedding, surface foaming, and mechanical restructuring. Many paper-like properties are developed simultaneously as film is biaxially oriented by a processing step that relates several internal methods.

A number of synthetic paper products are produced by stretching thermoplastic films containing fine particle-size fillers. Some products are predominantly high polypropylene modified with other polymers such as polystyrene and ethylene-vinylacetate, and are surface treated in addition to being biaxially oriented (e.g. Yupo FP produced by Oji-Yuka Synthetic Paper Co., Japan).

In a mixture of two incompatible polymers, one polymer will generally form a continuous phase with the other polymer dispersed on either a micro- or macro-scale depending on the degree of incompatibility and the intensity of mixing during preparation of the compound. When the dispersed polymer is present as microscopic particles, it can behave in the same way as a filler and cause cavitation when a film of the compound is drawn beyond the yield point of the continuous phase. Polystyrene (ref.47-49), polyacetates (ref.48), copolymers of ethylene (ref.48,49), and methymethacrylate polymers (ref.50) have been particularly mentioned as second polymer in the blend. No such product based on polypropylene has yet emerged commercially but it is expected that, because of the unidirectional effect in strength and fibrillating tendency of such structures, a superposition and binding of two sheets in perpendicular directions would be required for a satisfactory product. In contrast with the etched or filled sheet material, the synthetic paper made from incompatible blends of polymers are fibrous in nature. They can be folded easily and some can be rejected in paper recycling machines.

Some plastic papers based on mixtures of incompatible polymers are prepared by a process that involves a combination of orientation and solvent treatment. Photomicrographs of such products show raised nodules of a dispersed resin on the surface and a high level of porosity in the cross section.

Synthetic pulps, or *synpulps* as they are called, are wood-pulp substitutes made from plastics. They are projected to be much less expensive than the modified textile fibers which have been available to specialty papermakers for many years. Synpulps are made for existing paper machines, and as such could prove attractive to customers. Synpulp production basically consists of extruding a sheet or monofilaments of polypropylene mixed with some incompatible polymer and fillers, drawing it unidirectionally, cutting it into small lengths, and fibrillating (ref.5). The fibrillating and pulping of mixed polymer structure can usually be accomplished in conventional beaters and shredders used in the paper industry. The proportion of added incompatible polymer in polypropylene need not be higher than a few percent because there is the option of mechanical fibrillation of the sheet or filament structure. This is probably the most promising approach at the present time as it is compatible with the conventional art of paper making. It renders the blending of different types of pulp possible.

One route to the production of synpulp is by way of polymerization reaction. Propylene has been polymerized under such conditions that the polymer is obtained in the form of swollen elongated microfibers. Fibrous structure can also be formed in post-polymerization inline treatment. The Mitsui-Zellerbach project in Japan and the Solvay project in Kaly are concerned with reaction synpulps. Projected costs are expected to be comparable to that of polypropylene made in the same volume.

The production of spun-bonded sheets of microfibers loaded with caoline or other fillers, then calendering and finishing, seems very promising for large-scale industrial production although it departs from the conventional art of paper making and does not allow mixing with pulp from natural sources. The research efforts in this direction have been rather limited so far.

In addition to the efforts to produce paper from predominantly synthetic fibers, attempts have been made to use synthetic fibers as a reinforcing component. The DuPont "Textryls" was one such product. Polypropylene is reportedly in this capacity in some high-grade papers, mainly used as wrapping for electric wires.

IV.3. Spun-bonded Fabrics

Manufacturing a textile fabric without going through the painstaking work of fiber preparation, carding, spinning and knitting, or weaving, has remained a dream for decades in the textile industry. The recent proliferation of wet-laid or dry-laid nonwovens and felt-like materials is the result of research efforts directed toward this goal. These activities were undertaken primarily by the textile manufacturers. The advent of spun-bonded technology brings this dream closer to reality and makes it possible for the fiber manufacturer to produce textile fabrics directly under the spinnerette. Polypropylene, along with polyethylene and polyester, remained the choice starting material for such structures.

Spun-bonded fabric is an array of continuous filaments (as opposed to staple fibers in conventional wet-laid or dry-laid nonwovens) randomly arranged into a flat sheet of uniform thickness. The filaments are "melded" to each other at crossover points, which imparts structural integrity, dimensional stability and strength to the fabric. These fabrics were developed in the early 1960s and the introduction of commercial products was led by DuPont soon after. A large number of patents was filed by this company. More recently, a number of other companies have been involved in this field. A list of the most important producers of polypropylene spun-bonded fabrics is given in Table X, along with the trade names of their spun-bonded products.

TABLE X

Spun-bonded Continuous-filament Polypropylene Nonwoven Fabrics

Producer	Country	Trade-name	Process
J.H. Benecke	West Germany	Corovin	Docan (Lurgi)
Celanese/FH	U.S.A.	Mirafi 140	Heterofil
Chemie Linz	Austria	Polyfelt	Docan
Crown Zellerbach	U.S.A.	Fibertex	Docan
		Crowntex	
DuPont	U.S.A.	Typar	Own
		Tyvek (PE)	
ICI	U.K.	Cambrelle	Heterofil (own)
		Terram	Heterofil
Kimberly-Clark	U.S.A.	Kimkloth	Docan
Lutravil Spinnvlies	West Germany	Lutrasil	Own
MPC-Spunbond	Japan	Tafnel	-
Sodoca	France	Sodospun	Docan
'Wilhelm Pieck'	East Germany	Sualen PT (PE)	-

The production of polypropylene spun-bonded fabrics can be divided into five steps:

(1) Polymer extrusion through spinnerettes

(2) Formation of filaments

(3) Laying the filaments in random fashion in the form of a fabric

(4) Bonding the filaments

(5) Finishing

The process of melting polymer, forcing the melt through the spin manifold, and metering it onto the spinnerette under high pressure, is similar to that described in Chapter VIII for the manufacture of staple and filament yarns. The details of filament formation differ from process to process. The so-called DuPont process calls for the utilization of mechanical devices, such as nip rolls

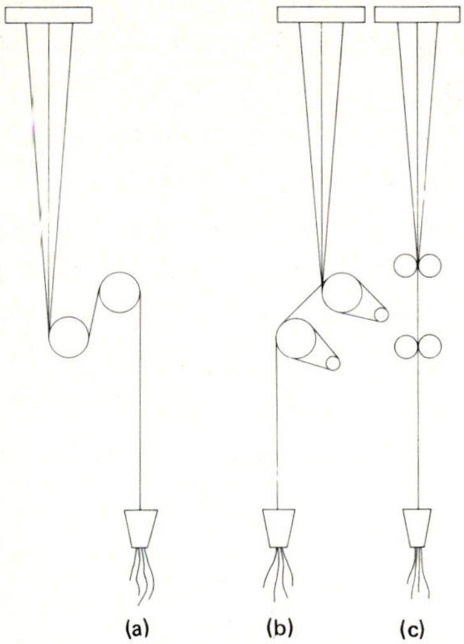

Fig. 42. Mechanical Means of Filament Formation for the Production of Spun-Bonded Fabrics

(Fig.42a), denier-setting rolls (Fig.42b), and pairs of spin-drawn rolls (Fig.42c). Other processes utilize aerodynamic means to draw the melt into suitable filaments. Figure 43 shows three alternative systems.

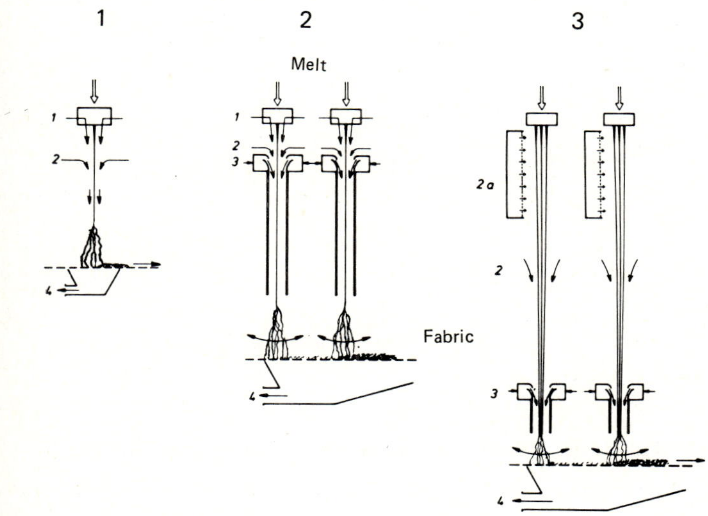

Fig. 43. Aerodynamical Methods of Filament Formation and Web Laying for the Production of Spun-Bonded Fabrics (ref.70)

The prime requirement in the filaments-laying step is the necessity of separating the filaments from each other and laying them uniformly in the form of a sheet, whether randomly or in a prearranged fashion. According to DuPont's process (Fig.44) (ref.52), the filament bundle is aspirated through an air jet which is electrostatically charged to high voltage. The electrostatically charged filaments repel each other; they are simultaneously impelled towards the fabric-forming belt for deposition. The depositing frame is charged with opposite charge or a high vacuum is applied to facilitate the filament deposition and fabric formation. Another variation of this theme is the "coanda effect" (ref.53), which is a hydrodynamic principle for filament separation. According to Carl Freudenberg's process (ref.54), the filaments are not allowed to form a bundle at any stage of spin-drawing: they are kept separated in the form of a warp and are impelled in this form to the web-forming screen (Fig.45). More recent patents suggest the use of supersonic velocity of air streams (ref.55), deflection of the filaments from side to side by the use of air jets (ref.56,57), a mechanical device for splaying the filaments out to width (ref.58), and the control of the height of the sprays above the collection to hide joints in the web (ref.59).

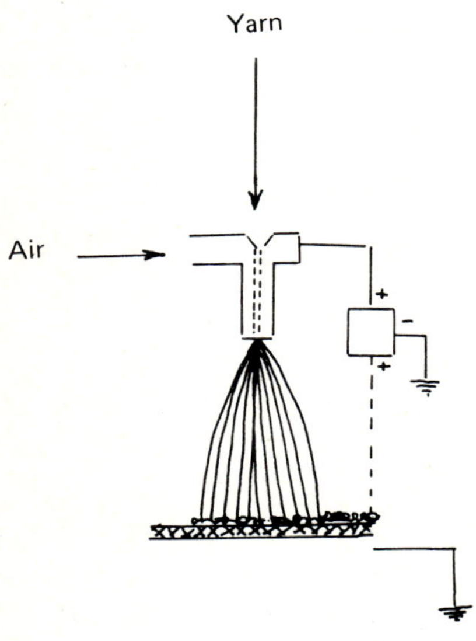

Fig. 44. Schematics of DuPont Web-Laying Process for the Production of Spun-Bonded Fabrics

Collection of the filaments can be effected by means of a moving flat surface; this is usually perforated to prevent the air stream from being deflected in an uncontrolled fashion. A suggestion that is designed to result in a more drapable

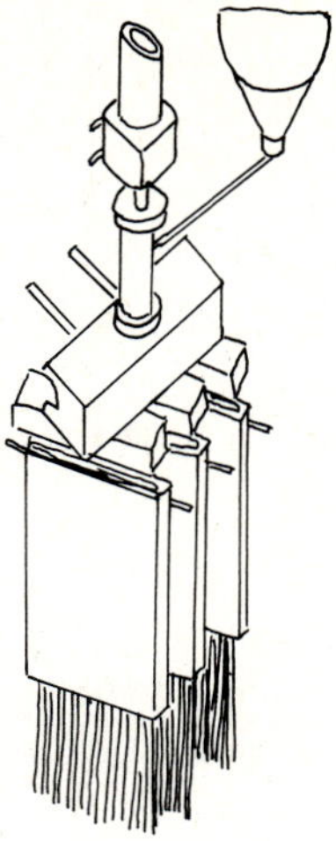

Fig. 45. Schematic of Carl Freudenberg's Process for the Production of Spun-Bonded Fabrics

fabric is to employ a collecting device having a surface moving in a direction counter to the direction of movement of the filaments (ref.60).

The filaments collected in the form of a uniform cross-laid or randomly laid sheet on the web-forming screen lie loosely upon each other: they must be bonded to impart strength and structural integrity to the web. A variety of means can be applied. For example, the use of heat and pressure between heated rolls or endless screens has been mentioned in various DuPont patents. The welding or melt bonding of filaments takes place, preferably at crossover points. The use of embossing rolls or hot needles is another variation of this method (ref.61). The strength of the fabric depends on the extent of the bonding but this also increases the stiffness of the fabric. The balance of strength and plyability is therefore achieved through a careful control of temperature and pressure or the density of the bosses. The admixing of low melting filaments, such as those of polyethylene or ethylene copolymers, into the mass of polypropylene filaments has been proposed to facilitate bonding and easy control of processing conditions (ref.62). The same objective can also be attained by intermixing drawn polypropylene filaments with some undrawn filaments. In this case, the less oriented filaments will melt at somewhat lower temperature than fully drawn filaments and thereby function as bonding centers. The binder filaments can be incorporated into the web as separate low-oriented fibers or as random sections of undrawn filaments present in the same bundle. Separate binder and matrix filaments can be produced by using separate spinning and drawing machines and combining the filaments prior to or during web laydown. They can also be produced by splitting the threadline from the spinnerette so that a part bypasses the drawing operation. Another method involves the use of a spinnerette with varying capillary geometry which produces filaments with varying response to the drawing operation. The filaments with segments having different levels of orientation can be produced by pulsing throughput or by pulsing the draw ratio, such as by using a fluted feed roll in drawing operation. The use of external binding agents, such as rubber latices, resin solutions, low-melting powdered polymers, etc., has also been suggested and pioneered by Carl Freudenberg, but the use of this technique for polypropylene is rather limited.

Finishing of the bonded web comprises a heat stabilization process and the application of lubricant. Heat setting is necessary to prevent the fabric from shrinking in down-stream processes. It is done by heating the fabric with steam under restrain. Lubrication is necessary for easy penetration of tufting needles into the fabric and preventing the pierced filaments from breaking.

Excellent reviews are available on the properties and uses of spun-bonded fabrics. A good review of the variety of products from the spun-bonding process has been given by Hartmann (ref.63). This paper is concerned with the products of the firm Lutravil Spinnviles, a jointly owned venture of BASF and Carl Freudenberg. The latter of these companies claimed to be the first in Europe to have developed a spun-bonded product. The review includes a discussion of artificial leather substrates, interlinings, filters, needle-punched carpet backings, and tufted-carpet backings. This list illustrates the main products in this field.

Polypropylene spun-bonded products have become very popular substrates for carpet primary backing. Large quantities of these fabrics are also being used in soil stabilization. There are, of course, several other uses of spun-bonded polypropylene fabrics.

V. REFERENCES

1. S. Yamazaki, Internat. Chem. Eng., 11(1971)367.
2. J. Eichhorn, J. Appl. Polym. Sci., 3(1964)2497.
3. C.P. Fenimore and F.J. Martin, Combustion and Flammability, 10(1960)135.
4. US Patent 3,624,028, Phillips, 1972.
5. Belgian Patent 737,979, BASF, 1970.
6. A. Hoffman, Kunststoffe, 61(1971)811.
7. H.B. Hopfensberg and J.J. Tulis, Modern Plastics, (1970)110.
8. USSR Patent 256,934, 1969.
9. C.D. Han, J. Appl. Polym. Sci., 15(1971)1091.
10. A.J. Rosenthal, Symp. on Polypropylene Fibers, Southern Research Inst., Birmingham, Ala., 1964.
11. M.M. Platt, et al, Text. Res. J., 22(1952)641.
12. W. Glen, J. Text. Inst., 62(1971)600.
13. W.C. Sheehan, et al, J. Appl. Polym. Sci., 8(1964)2359.
14. W.C. Sheehan, et al, Text. Res. J., 35(1965)626.
15. S.E. Ross., J. Appl. Polym. Sci., 9(1965)2729.
16. V. Capuccio, et al, Chim. Ind. (Milan), 44(1962)463.
17. M. Compostella, et al, Angew. Chem., 74(1962)618.
18. G. Natta and M. Compostella, Z. Ges. Textiliand., 69(1967)827.
19. PWO Wijga, Physical Properties of Polymers, Soc. Chem. Ind. Monograph No.5(1959) London.
20. J. Van Schooten, et al, Polymer, 2(1961)161.
21. G.Jurkiewtch, Hercules Inc., Private Communications.
22. US Patent 3,437,725, DuPont, 1969.
23. J.M. Swanson, Sixth Ann. Synth. Fiber Symp., Williamsburg, Va.
24. A.I. Chem. Eng., April, 1969.
25. A. Breuninger, Melliand Textilber, 12(1971)1444.
26. O.E. Bruck, Text. Inst., and Ind., 3(1965)118.
27. H. Biesser and R. Hesse, Chemiefasern, 17(1967)262.
28. A. Peterlin, J. Mat. Sci., 6(1971)490.
29. G.N. Molkova, et al, Fiber Chem., 5(1973)261-263.
30. A.B. Parksver, Faserforsch. u. Textiltech., 22(1971)329.
31. A. Ribnick, Text. Res. J., 39(1969)428, 742.
32. US Patents 3,330,897; 3,256,258; 3,323,190; British Patents 962,231; 935,809; 1,052,550; Belgian Patents 650,890;684,614; French Patent 1,327,327.
33. Canada Patent 704,412, Whitin 1965.
34. US Patent 3,506,535, Allied 1970; US Patent 3,604,196, Allied 1971.
35. British Patent 1,248,293, Kanegafuchi, 1972; US 3,613,173, Kanegafuchi, 1971; US 3,531,368, Toyo Rayon 1970; US 3,641,232, Am. Enka, 1972.
36. US Patent 3,152,380 DuPont, 1964.
37. British Patent 966,498, ICI 1964.
38. US Patent 3,686,848, Uniroyal, 1972.
39. D.E. Till, Modern Text. Mag., 40(1959).
40. Van A. Wente, Ind. Eng. Chem., 48(1956)1342.
41. Fibers and Plastics, June, July, Aug., 1961.
42. US Patents 3,409,938; 3,031,717; 3,358,322; 3,031,707; German Patent 1,199,431.
43. US Patents 2,357,392; 2,483,404; 2,483,405; 2,483,406; 2,810,426; 3,595,245; 3,689,342.
44. R.R. Buntin, Tappi, 56(1973)74.
45. US Patents 3,227,784; 3,227,794; 3,169,899; 3,504,076; 3,081,519; British Patent 1,225,574.
46. US Patents 3,467,744; 3,564,088.
47. German Patent 2,019,335.

48. Japan Patent 71/43,468.
49. Japan Patent 71/43,468.
50. Japan Patent 71/43,470.
51. German Patent 2,064,932.
52. US Patents 3,117,053; 3,490,115; 3,163,753; 3,459,627; 3,477,103; 3,341,394; 3,481,005; British Patents 1,087,410; 1,067,039; 1,072,445; 1,164,921; 1,170,639; to DuPont.
53. US Patent 3,314,122 to DuPont.
54. US Patent 3,502,763.
55. US Patent 3,565,729, Carl Fruedenburg.
56. British Patent 1,218,009 to Bayer.
57. British Patent 1,244,753 to DuPont.
58. British Patent 1,219,921 to Monsanto.
59. US Patent 3,734,803 to Allied Chemical.
60. US Patent 3,708,365 to ICI.
61. US Patent 3,689,342 to Celanese Corp.
62. US Patents 3,459,627; 3,360,421 to DuPont.
63. US Patents 3,502,538; 3,322,607; British Patent 1.072.445 to DuPont.
64. L.Hartmann, Melliand Textilber.Int., 54(1973)460,605.
65. K. West in Carpet Substrales, Textile Trade Press, Manchester, 1973.
66. K.E.Peripelkin, Faserforsch. u. Textiltech., 25(1974)251.
67. US Patent 3,664,782 to Phillips Petrol., 1969.
68. British Patent 941,199 to Montecatini 1969.
69. I.K. Park and H.D. Noether, Colloid and Polymer Sci., 253(1975)824.
70. R.G. Quynn and H. Brody, J. Macromol, Sci. Phys., 5(1971)721.
71. L. Gerkin, Melliand Textilber, 4/1974(1974)326.
72. S.L. Cannon, et al, J. Polym. Sci., 11(1976)209,276.
73. US Patent 3,256,258, to DuPont, 1966.
74. C.A. Garber and E.S. Clark, J. Macromol. Sci. Phys., 4(1970)499; Int. J. Polym. Mat., 1(1971)31.
75. B.S. Sprague, J. Macromol. Sci. Phys., 8(1973)157.
76. S.L. Cannon, et al, Polym. Eng. Sci., 15(1975)633.

VI. BIBLIOGRAPHY

VI.1. Books

1. E.A. Hutton, A Survey of the Literature and Patents Relating to Bulked Continuous Filament Yarns, Shirley Inst., Manchester, England (1963).
2. L. Roth, Synthetic Fibers for Paper Making, Bibliographic Series No. 223, Inot Paper Chem., Appleton, Wisc., (1965).
3. G.D. Wilkinson, Textured Yarn Technology, Vol. I to III, Monsanto Textile Div., Decatur, Ala., (1967-1974).
4. Textured and Novelty Yarn Processes, Noyes Data Corp., Park Ridge, N.J.,(1967)
5. M.G.Halpern, Synthetic Paper from Fibers and Film, Noyes Data Corp.,N.J. (1975)
6. K.Johnson, Antistatic Copositions for Textiles and Plastics, Noyes Data, N.J.(197
7. D.R.Paul, Ed., Polymer Blends, Vol. I & II, Academic Press, New York,(1978)

VI.2. Journal Literature

1. Anon, Fiber Crimping Survey, Mod. Text Mag., 42(1961)62-63.
2. S. Englhardt, Determination of Crimping and Crimp Durability, Duet. Textiltech, 12(1962)321-324.
3. Anon, Developments in Bulked Yarns and Yarn Bulking Techniques, Textil Recorder, 80(1962)71-73.

4. F. Bolland, The Influence of the Cross-Section of Synthetic Filaments on the Appearance and Performance of the Finished Articles, Royonne et Fibres Synthetic, 19(1963)855-864; 935-940.
5. W. Morawek, Bulked Yarns and Their Relationship to Wool Yarns, Melliand Textilber 44(1963)566-572.
6. G.C. Tesoro, Antistatic Agents, in Kirk-Othmer Encycl. Chem. Technol., 2(1963)649-72.
7. A.J. Hall, A New Method of Reducing Pilling on Synthetic Fiber Fabrics, Textile Recorder 81(1963)65-66.
8. W. Friedmann, Measurement of Bulkiness in Man-Made Fiber Assembly, Faserforsch. u. Textiltech. 15(1964)340-352.
9. F. Burkle, Bulked and Crimped Yarns, Melliand Textilber. 45(1964)878-881; 971-975; 1091-1096.
10. W.C. Sheehan, Production of Super Tenacity Polypropylene Filaments, J. Appl. Polym. Sci., 8(1964)2359-2388.
11. G.M. Jeffrey, Current Trends in the Development of Man-Made Fibers, Textile Recorder, 81(1964)44-64.
12. W. Meissner, On the Problem of the Production of Man-Made Fibers From Polymer Alloys, Faserforsch. u. Textiltech., 16(1965)400-408.
13. S.M. Atlas and H.F. Mark, Factors Modifying the Essential Properties of Fibers, Chem. Ind., 93(1965)413-418.
14. W. Morawek, The Latest Position in Texturizing Technology, Textil-Praxis, 20(1965)793-806.
15. J.A. Jackson, Tennessee Eastman Technique for Measuring the Comparative Bulk of Textile Yarns, Mod. Text. Mag., 46(1965)47-49.
16. Dutch Research into Hollow Fibers, Skinner's Record, 39(1965)846.
17. G. Behle, Preparation and Characteristics of Textured Yarns, Z. ges. Textilind., 67(1965)447-453.
18. L. Roth, Synthetic Fibers for Paper Making, (Bibliography) Int. Pap. Chem., Appleton, Wisc., (1965)172.
19. J. Hall, Pilling Problems With Synthetic Fiber Materials, Textile Recorder, 84(1966)73-76.
20. E.A. Tippets, Fiber Engineering in Order to Achieve Certain Final Properties, Z. Ges. Textilind., 69(1966)639-647.
21. H. Mark, On the Modification of Certain Fiber Type by Chemical and Physical Methods, Chemiefasern, 16(1966)258-264.
22. P.V. Papero, Fundamental Properties Considerations in Tailoring a New Fiber, Textile Res. J., 37(1967)823-833.
23. R.B. Mumford, Multipolymer Fibers, Mod. Text. Mag., 48(1967)51-53.
24. G. Natta, et al, New Developments in Polypropylene Fibers, Z. ges. Textilind., 69(1967)827-835.
25. H. Bieser, Cross-section of Man-Made Fibers. Part I, Chemiefasern, 17(1967)262-268.
26. M. Baumgartner, Texturizing Processes for Synthetic Yarns With Special Reference to the Carpet Section, Textil Veredlung, 2(1967)831-840.
27. J.W.S. Hearle, Yarn Bulking,the Fundamental Mechanics. Part I, Skinner's Record, 41(1967)93-94, 106.
28. E.A. Tippets, Fiber Engineering to Meet End Use Requirements, Text. Res. J., 37(1967)524-533.
29. H. Boisen, Cross-Section of Man-Made Fibers With Reference to Processing and Applications, Chemiefasern, 17(1967)262-268.
30. O.L. Shealy, Spunbonded Fabrics, Osterr. Chem. Ztg., 68(1967)147-152.
31. J. Lunenschloss, Texturizing Machine, Textil Praxis Int.,2(1968)25-31
32. M.D. Rosenzweig, Silklike Synthetic Fibers Break Out of Their Cocoons, Chem. Eng., 75(1968)94,96.
33. Silklike Fibers Throughout the World, Chemiefasern, 19(1969)520-523.
34. H.P. Weider, Special Problems in the Testing of Textured Yarns, Melliand Textilber. 49(1969)1264-1267.
35. P. Lennox-Kerr, Prospect of More Fibers Via Phase Separation Spinning, Textil. Ind., 133(1969)104-107.

36. O.L. Shealy, Spunbonded Polypropylene Carpet Backing, Text. Res. J., 39(1969)254-260.
37. W. Stein, Measurement of Crimp Contraction and Shrinkage Forces on the Running Filament, Chemiefasern, 20(1970)748-754.
38. R. Hametz, Flame Retardant Synthetic Textile Fibers. Ind. Eng. Chem., 62(1970)41-53
39. W. Berger, Synthetic Fibers of Polymer Blend, Faserforsch, u. Textiltech., 21(1970)288-294.
40. K.H. Wallhaeusser, Antimicrobial Finishing of Textiles, Textilveredlung, 5(1970)3-14.
41. R. Chatterjie, Hollow Filament, Silk & Rayon Ind. India, 13(1970)323-325.
42. R.C. Nametz, Flame Retarding Synthetic Textile Fibers, Ind. Eng. Chem., 62(1970)41-53.
43. A. Brewninger, New Methods of Producing Carpets That Retain their appearance by Means of Fiber Modification, Melliand Textilber. Int., 52(1971)1444-1446.
44. R.G. Quynn and H. Brody, Elastic Hard Fibers, J. Macromol Sci., Phys. B, 5(1971)721-738.
45. A.H. von Arnim, Economic Aspects of Spun Webs, Melliand Textilber. Int., 52(1971)52-55.
46. A. Breuningen, New Method for Low Soiling Carpets Through Modifications of Fiber Cross-Section, Melliand Textilber, 12(1971)1444-1446.
47. A.B. Parksver, Influence of Production Conditions on the Thermal Shrinkage of polypropylene fiber, Faserforsch. u. Textiltech., 22(1971)329.
48. A.H. von Arnim, Economic Aspects of Spun Webs, Melliand Textilber Int., 52(1971)52-54.
49. R. Krome, Problems and Perspectives of Spun Webs Produced Directly From Polymer, Deut. Textiltech., 22(1972)414-417.
50. C. Michels, Consideration of Some Special Problems in the Development of Spun-Bonded and Split Fiber Non-Wovens, Deut. Textiltech., 22(1972)41-414.
51. J. Starita, Microstructure of Melt Blended Polymer Systems, Trans. Soc. Rheol., 16(1972)339-367.
52. Synthetic Polymers for the Production of Paper, Chemiefasern/Textil. Ind., ??(1972)1150-1153.
53. A.Schippin, New ways to produce Textiles, Textile Inst., Manchester, 1972.
54. E.S. Clark, Row Structure Elasticity; An hypothesis for Elastic Behavior in Crystalline Films, ACS Div. Org. Coat. Plast. Chem., 32(1972)19-26.
55. R.W. Wright, Applications of Synthetic Fibers to Paper Making Process, Paper Trade J., 156(1972)38-40.
56. O.I. Nachinkin, et al, Cross-Sectional Shape of Man-Made Fibers, Fibre Chem., 5(1973)158-160.
57. A.C. Tanquary, Hollow Fibers, Bull. South Res. Inst., 26(1973)10-13.
58. E.S. Clark, Mechanism of Energy Driven Elasticity in Crystalline Polymers, Polym. Sci. Tech., 1(1973)267-282.
59. G.N. Molkova, Properties of Undrawn Polypropylene Fibers as a Factor in its Post-Stretching shrinkage, Fibre Chem., 5(1973)261-263.
60. L. Hartman, Spun Bonded Technology and Industrial Uses of Spun Bonded Non-Wovens, Melliand Textilber.Int., 54(1973)460-462, 650-657.
61. K. Kirchenberger, Applying Spun Bonded and Wet-Laid Technologies to the Production of Multi-Layered Fabrics, Melliand Textilber. Int., 2(1973)257-461.
62. J.P. Sibilia, Surface Composition of Biconstituent Fibers, J. Appl. Polym. Sci. 17(1973)2911-2914.
63. B. Morris, ICI Develops Fiber to Conduct Electricity, Daily News Record, 3(1973)9.
64. L. Hartman, Spunbondeds' Production and Industrial Applications, Melliand Textilber. Int., 2(1973)355-357.
65. R.R. Buntin and D.T. Lohkamp, Melt Blowing - A One-Step Web Process for New Non-Woven Products, Tappi, 56(1973)73-77.
66. Newton and Ford, Non-Woven Fabrics, Textile Progress, 5(1973)48-86.
67. W.H. Hoge, Making it Big With Plastic Papers, Plastic Engineering, (1973)51-53.

68. B.S. Sprague, Relationship of Structure and Morphology to Properties of Hard Plastics, Fibers and Films, J. Macromol Sci. Polym., 8(1973)157-187.
69. Surface Composition of Biconstituent Fibers, J. Appl. Polym. Sci., 17(1973)2911-2914.
70. Riethmayer, Antistatic Agents - A review, Gummi, Alst., Kunst., 26(1973)76-78. 80,88,182-184,1298,300,302,304,306,308,419,20,25.
71. Enter the Epitropics, Textile Mfr., (1973)100, 28.
72. Chemicals Used as Spin Finishes for Man-Made Fibers, Text. Res. J., 43(1973)325-335.
73. M.J. Schiek, Friction and Lubrication of Synthetic Fibers, Text Res. J., 43(1973)198-204, 254-259, 342-347.
74. L. Gerking, Spun-Bonded Fabrics, Melliand Textilber 4(1974)326-330.
75. Von V. Frauzer, Synthetic Papers, Angew Macromol Chem., 40(1974)305-323.
76. A.S. Chegolya, Problems in the Modification of Synthetic Fibers, Lenzinger Ber., 36(1974)194-196.
77. D. Goritz and F.H. Muller, Recovery Mechanism in Elastic Hard Fibers, Colloid & Polymer Sci., 252(1974)862-870.
78. H. Berg, Chemical or Physical Modification of Synthetic Fibers, Lenzinger Ber., 36(1974)36-47.
79. T.M. Holliday, Spunbonded Fabrics, Modern Textile, 55(1974)40-46.
80. R. Penissen, Modification of Chemical Fibers, Lenzinger Ber., 36(1974)24-36.
81. L. Gerking, Methods of Producing Spun Bonded Fabrics and Their Special Features, Melliand Textilber Int., 3(1974)232-236.
82. V.S. Ellis, Epitropics - Third Generation Conductive Fibers, Textile Mfr., 101(1974)19-21, 23.
83. D. Machalaba, Spun Bonded Fabrics Burgeoning, Daily News Rec., 4(1974)16.
84. G. Druzsbaczky, Shrinking Behavior of Polypropylene Fibers of Different Molecular Weight, Faserforsch u. Textiltech., 25(1974)392-396.
85. R.J. Bercan, Recent Development in Man-Made Flame Resistant Fibers, Textile Flammability, (1974)212-238.
86. H.W. Fink, Antistatic Agent for Plastics, Kunststoffe, 64(1974)64-67.
87. R.A. Hentschel, Spunbonded Sheet Products, Chem. Technol., 4(1974)32-41.
88. H. Bery, Chemical or Physical Modification of Synthetic Fibers, Lenzinger Ber., 36(1974)36-47.
89. E.R. Blair, Development and Commerciallization of Melt Blowing Process for Manufacture of Fine Fibered Webs, Formed Fabrics, 5(1974)39-42, 50.
90. G.M. Venkatesh, Soiling of Textiles and Development of Antisoiling and Soil Release Finishes; Review, Text. Res. J., 44(1974)352-362.
91. Von Falkai, Are Polymer Mixtures of Significance to the Man-Made Fiber Market?, Part I and II, Chemiefasern/Textilber. Eng. Ed., 27(1974)37-41.
92. K. West, Carpet Substrates, Textile Trade Prem, Manchester, 1975.
93. R. Natha, Internal Antistats, Am. Dyestuff Rep., 64(1975)41, 44-49.
94. The Mechanical Behavior of Springy Polypropylene, Polym.Eng.Sci., 15(1975)633-645.
95. S.L. Cannon, Mechanical Behavior of Springy Polypropylene, Polym. Eng. Sci., 15(1975)633-645.
96. I.K. Park, Crystalline Hard Elastic Materials. Part II, Mechanical Properties, Kolloid Z., 253(1975)824-839.
97. D. Goeritz, Morphology and Mechanical Behavior of Elastic Hard Fibers, Colloid Polym. Sci., 253(1975)844-851.
98. H.D. Noether, Low-Modulus "Hard" Elastic Materials, Text. Res. J., 46(1976)467-478.
99. F. Mares, Modification of Fiber Surfaces, Fiber Producer, (1976)57,59,61.
100. T.N. Seregina, Antimicrobic Activity of Polypropylene Yarn Modified With Medicinal Preparations, Fiber Chem., 8(1976)227-228.
101. H. Joerder, Spunlaced Non-Wovens - Production, Properties and Potential Application, Melliand Textilber. Eng., 5(1976)717-718.
102. S.K. Braunman, Sb_2O_3 - Halogen Fire Retardance in Polymers, I, II, III, J. Fire Retard. Chem., 3(1976)117-163.

103. K. Dietrich, Structure of Matrix-Fibril Filaments, Faserforsch u. Textiltech., 27(1976)491-495.
104. P.C. Limburg, Man-Made Fiber Alloying. The Researcher's Horn of Plenty, Chemiefasern, 26(1976)1075.
105. B. Von Falkai, Are Polymer Mixtures of Significance to Mass Produced Fiber Market, Chemiefasern, 26(1976)1071.
106. D.M. Bigg, Review of Techniques for Processing Ultra-High Modulus Polymer, Polymer Eng. Sci., 16(1976)725-734.
107. D. Cansfield, Preparation of Ultra-High Modulus Polypropylene Fibers and Films, Polym. Eng. Sci., 16(1976)721-724.
108. S.L. Cannon, Hard-Elastic Fibers (A Review of Novel State for Crystalline Polymers), J. Polym. Sci., Macromol. Reviews, 11(1976)209-275.
109. W.N. Taylor, Superdrawn Filaments of Polypropylene, ACS Polymer Preprints, 18(1977)332-337.
110. A.S. Brown, Antistatic Yarns From Synthetic Fibers, Chemiefasern, 27(1977)219
111. D.Swanson, Antistatic Agents Within the Fiber, Knitting Ind. 97(1977)20.
112. R.G.Mansfield, Spunbonded Nonwovens Eye Road Building, Textile Ind.,127(1977)81
113. Ann., Spunbondeds Growing in Versatility, Fiber Producer, 5(1977)42.
114. G.Balbach, Antistatics for Polyolefins, Kunststoffe, 67(1977)154.
115. J.P.Bex, Polyolefin Synthetic Pulp, SPE RECTEC "Polyolefins II" Houston, Texas, Feb. 14,15, 1978.
116. I.M.Okamoto, Ultrafine Fiber and its use, Chemiefasern/Textilind. 29/81(1979)30.
117. I.M.Okamots, Ultrafine fiber and its use, Chemiefasern Textilind. 29/81 (1979)94.
118. N.V.Tamazina, Fibers from PP/PET Mixtures, Fibre Chem., 9(1977)485-486.
119. A.V.Genis, Aerodynamic Melt Spinning, Fibre Chem., 10(1978)6-10.
120. L.E.Seidel, Birth of a Man-made Fiber Variant, Fiber Producer, 6(1978)44-46.
121. A.Judge, High Shrinkage Polypropylene Fiber, Plast. Rubber Inst., London, Conf. Polyp. Fibres & Textiles, York, Sept. 26-28, 1979.

VI.3. Patents

High Shrinkage Fibers and Products

US 2,921,360, US 3,317,335, US 3,377,794, US 3,388,547, US 3,439,393, US 3,555,136, US 3,570,136, US 3,950,587,

Antistatic Compositions

US 2,525,691, US 2,891,028, US 2,891,029, US 2,891,031, US 2,891,032, US 2,922,773, US 3,190,763, US 3,183,202, US 3,206,429, US 3,220,276, US 3,177,174, US 3,223,695, US 3,299,006, US 3,316,232, US 3,329,557,

US 3,669,736, Teijin, 1972
US 3,475,898, DuPont, 1969
US 3,454,519, Nat. Dist. Chem. Corp., 1969
US 3,803,453, DuPont, 1974
US 3,515,698, Hercules, 1970
US 3,652,473
US 3,738,864, DuPont, 1973
US 3,804,875
US 3,922,249, Phillips Petrol, 1975
US 3,969,559, Monsanto, 1976
British 839,869
British 972,244, ICI, 1964
British 731,728
British 867,176
British 907,016

British 993,530, Lio Fat & Oil, 1965
British 972,244
British 993,530
British 982,268, Cassella, 1965
British 1,019,115
British 1,019,670
British 1,013,960, General Mills, 1965
British 1,023,524
British 1,020,298
British 1,018,262, Chem. Werke Huels, 1966
British 1,020,302
British 1,012,234
British 1,019,670, Chem. Werke Huels, 1966
British 1,203,869, Bayer, 1970
British 1,115,069, Hoechst, 1968
British 1,126,016, Kao Soap, 1968
British 1,230,453, Kanegafuchi, 1971
British 1,237,589, DuPont, 1971
British 1,250,917, Asahi Kasei, 1971
British 1,240,011, Toray, 1971
British 1,305,144, Beaunit, 1973
British 1,355,232, Chem. Werke Huels, 1974
British 1,362,533, Sandoz, 1974
British 1,391,262, ICI, 1975
British 1,399,141, Monsanto, 1975
British 1,399,142, Monsanto, 1975
British 1,369,911, Chem. Werke Huels, 1974
British 1,425,420, Bayer, 1976

Elastic Polypropylene Fibers

US 3,256,258, DuPont, 1966
US 3,323,190, Hercules, 1967
US 3,330,897, Chem Cell, 1967
US 3,432,590, Nat. Plast. Prod., 1969
US 3,485,706, Hercules, 1969
US 4,006,208, Daicel, 1977
British 887,673, Montecatini, 1962
British 962,231, Celanese, 1964
British 1,040,366, Hercules, 1966
British 1,087,045, Hercules, 1967

High Tenacity Industrial Yarns

US 3,090,997; 3,091,015; 3,092,891; 3,161,709
US 3,361,859, DuPont, 1968
US 3,366,782, Phillips Petrol., 1969
US 3,437,725, DuPont, 1969
British 811,349, 1962
British 931,417, ICI, 1963
British 941,199, Montecatini, 1963
British 982,630, 1964
British 1,057,538, BASF, 1967
British 1,075,022, Uniroyal, 1967
British 1,079,986, F11, 1967
British 1,117,946, Allied Chem., 1968
British 1,220,424, Toray, 1971

Filament Cross-Section

US 2,999,296, DuPont, 1961
US 3,095,258, DuPont, 1963
US 3,097,416, DuPont, 1963
US 3,109,278, DuPont, 1963
US 3,109,768, DuPont, 1963
US 3,135,813, ICI, 1964
US 3,156,607, DuPont, 1964
US 3,169,089, Celanese, 1965
US 3,164,949, DuPont, 1965
US 3,194,002, Eastman, 1965
US 3,217,074, Gould, 1965
US 3,219,074, Gould, 1965
US 3,266,087, Allied Chem., 1966
US 3,259,676, DuPont, 1966
US 3,239,865, Whiting, 1966
US 3,251,729, Shaw, 1966
US 3,259,674, DuPont, 1966
US 3,308,221, Allied Chem., 1967
US 3,313,000, DuPont, 1967
US 3,330,721, Gould, 1967
US 3,350,488, DuPont, 1967
US 3,413,706, DuPont, 1967
US 3,419,936, Monsanto, 1967
US 3,357,048, DuPont, 1967
US 3,438,087, ICI, 1969
US 3,439,381, Akzu, 1969
US 3,457,341, DuPont, 1969
US 3,465,618, Monsanto, 1969
US 3,478,389, Monsanto, 1969
US 3,508,432, Monsanto, 1970
US 3,493,459, Monsanto, 1970
US 3,585,684, Monsanto, 1971
US 3,508,390, Allied Chem., 1970
US 3,616,633, Soc. Rhodiaceta, 1971
US 3,555,600, Moore, 1971
US 3,626,045, Courtaulds, 1971
US 3,630,824, Barmag, 1971
US 3,635,641, Allied Chem., 1972
US 3,537,135, Celanese, 1970
US 3,640,670, F.I.I., 1972
US 3,650,659, Akso, 1972
US 3,652,753, F.I.I., 1972
US 3,659,983, Dow, 1972
US 3,686,377, DuPont, 1972
US 3,724,672, Baslaw, 1973
US 3,728,428, Allied, 1973
US 3,734,993, Biere, 1973
US 3,738,789, F.I.I., 1973
US 3,724,672, Leonard, 1973
US 3,728,842, Allied Chem., 1973
US 3,738,789, Fiber Ind., 1973
US 3,745,061, DuPont, 1973
US 3,734,993, Paliyanko, 1973
US 3,811,984, Joonas, 1974
US 3,814,561, Kanegafuchi, 1974
US 3,839,526, Celanese, 1974
US 3,857,012, 1974

US 3,851,036, Dow, 1974
US 3,256,334, DuPont, 1975
US 3,884,754, Kanebo, 1975
US 3,887,747, Kanebo, 1975
US 3,924,988, DuPont, 1975
US 3,948,004, O'Casey, 1976
US 3,940,469, Standard Oil, 1976
US 3,994,122, DuPont, 1976
US 4,001,369, DuPont, 1977
US 4,020,229, Hercules, 1977
US 4,056,418, Shell Oil, 1977
US 4,055,696, Mitsubishi Rayon, 1977
US 4,044,089, DuPont, 1977
British 859,325, Inst. Text Tech. Chem., 1961
British 859,814, Triko, 1961
British 886,313, Inst. Text. Tech. Chem., 1962
British 904,805, Reifenhauser, 1962
British 950,213, ICI, 1964
British 945,183, Holebet, 1964
British 964,459, Monsanto, 1964
British 979,199, DuPont, 1965
British 986,718; 986,719, DuPont, 1965
British 1,008,381, Monsanto, 1965
British 986,990, N. V. Onder, Res., 1965
British 1,009,625, VEB Chem. Schwarza, 1965
British 1,033,543, Brit., Nylon Spin, 1966
British 1,017,137, Monsanto, 1966
British 1,054,904, Celanese, 1967
British 1,057,181, Snia Viscosa, 1967
British 1,071,909, Dow, 1967
British 1,086,873, ICI, 1967
British 1,088,020, DuPont, 1967
British 1,096,093, DuPont, 1967
British 1,096,316, ICI, 1967
British 1,096,317, ICI, 1967
British 1,099,258, Inventa, 1968
British 1,123,139, VEB Chem. Schwarza, 1968
British 1,136,689, VEB Chem. Schwarza, 1969
British 1,144,117, DuPont, 1969
British 1,153,543, Snam Progetti, 1969
British 1,167,622, Teijin, 1969
British 1,168,759, Glanzstuff, 1969
British 1,170,552, Snam Progetti, 1969
British 1,171,027, Snam Progetti, 1969
British 1,171,028, Snam Progetti, 1969
British 1,187,710, Snam Progetti, 1970
British 1,207,408, Allied Chem., 1970
British 1,211,510, Marmag, 1970
British 1,226,726, Allied Chem., 1971
British 1,266,182, F.I.I., 1972
British 1,267,330, F.I.I., 1972
British 1,267,517, Scragg, 1972
British 1,277,783, Rohm & Haas, 1972
British 1,292,388, DuPont, 1972
British 1,340,846, DuPont, 1973

Spun Bonded and Microfibers

US 3,013,936, DuPont, 1961

US 3,099,067, Union Carbide, 1963
US 3,286,896, DuPont, 1964
US 3,116,589, DuPont, 1965
US 3,156,752, DuPont, 1965
US 3,174,271, DuPont, 1965
US 3,283,366, Johnson & Johnson, 1966
US 3,358,322, Monsanto, 1967
US 3,549,734, Yasuda, 1967
US 3,382,305, DuPont, 1968
US 3,384,944, DuPont, 1968
US 3,444,863, Celanese, 1969
US 3,467,744, DuPont, 1969
US 3,484,899, DuPont, 1969
US 3,485,708, DuPont, 1969
US 3,497,918, DuPont, 1970
US 3,510,389, Kendall, 1970
US 3,542,615, Monsanto, 1970
US 3,543,332, Celanese, 1970
US 3,546,063, DuPont, 1970
US 3,549,734, Yasuda, 1970
US 3,881,003, DuPont, 1970
US 3,627,863, FMC, 1971
US 3,554,854, Carl Freudenberg, 1971
US 3,560,318, Union Carbide, 1971
US 3,615,995, Esso Res., Dev., 1971
US 3,627,863, FMC, 1971
US 3,723,238, Enka, 1973
US 3,738,884, Celanese, 1973
US 3,740,302, Celanese, 1973
US 3,743,272, Owens-Corning, 1973
US 3,762,896, Owens-Corning, 1973
US 3,767,523, Kimberly Clark, 1973
US 3,796,619, Celanese, 1974
US 3,797,074, DuPont, 1974
US 3,798,100, Rhone-Poulenc, 1974
US 3,825,380, Exxon Res. Eng., 1974
US 3,848,027, Crown Zellerbach, 1977
US 3,849,241, Exxon Res. Dev., 1974
US 3,853,651, Rhone-Poulenc, 1974
US 3,882,095, Crown Zellerbach, 1975
US 3,878,014, Beloit, 1975
US 3,891,610, Crown Zellerbach, 1975
US 3,914,501, Union Carbide, 1975
US 3,928,416, Toray Ind., 1975
US 3,932,081, Tanag, 1976
US 3,920,507, Crown Zellerbach, 1975
US 3,920,508, Crown Zellerbach, 1975
US 3,928,496, Toray Ind., 1975
US 3,920,509, Crown Zellerbach, 1975
US 3,214,925, Gulf Research, 1976
US 3,994,258, Bayer, 1976
US 3,954,361, Beloit, 1976
US 3,996,324, Metalgesellschaft, 1976
US 4,ool,157, BASF, 1977; US 4,002,796, Montedison, 1977
US 4,007,247, ICI, 1977; US 4,010,229, Solvay, 1977
US 4,025,593, Solvay, 1977; US 4,081,226, Solvay, 1978
US 4,111,620, Celanese, 1978; US 4,125,584, Gulf Oil, 1978
US 4,134,931, Gulf Oil, 1978
British 1,061,337, DuPont, 1967

British 1,126,026, ICI, 1968
British 1,126,990, DuPont, 1968
British 1,142,253, Stami Carbon, 1969
British 1,157,299, Ugine Kuhlmann, 1969
British 1,156,321, Brit. Cellophane, 1969
British 1,157,437, ICI, 1969
British 1,168,016, FMC, 1969
British 1,178,438, Carl Freudenberg, 1970
British 1,214,786, FMC, 1970
British 1,215,537, Carl Freudenberg, 1970
British 1,215,679, Bayer, 1970
British 1,215,847, FMC, 1970
British 1,218,009, Bayer, 1971
British 1,245,078, Toray Ind., 1971
British 1,247,373, 3M Co., 1971
British 1,260,427, Metallgesell Schaft, 1972
British 1,262,531, Mitsubishi Rayon, 1972
British 1,282,176, Metallgesell Schaft, 1972
British 1,263,221, Toray Ind., 1972
British 1,268,861, Akzu, 1972
British 1,282,176, Metalgeselschaft, 1972
British 1,285,216, VEB Chemiefaser Comb. Schwarza, 1972
British 1,285,381, Toray Ind., 1972
British 1,287,917, Crown Zellerbach, 1972
British 1,295,267, Esso Res. Dev., 1972
British 1,268,823, Mitsubishi, 1972
British 1,300,813, Bayer, 1972
British 1,302,908, Hitachi, 1973
British 1,323,174, Toray Ind., 1973
British 1,334,117, Wiss Tech., Zent. Tech. Text., 1973
British 1,335,607, Esso Res. Dev., 1973
British 1,413,256, BASF, 1973
British 1,350,487, Crown Zellerbach, 1974
British 1,359,317, OJI Yuka, 1974
British 1,372,115, Gulf Res. Dev., 1974
British 1,372,116, Gulf Res. Dev., 1974
British 1,373,708, Gulf Res. Dev., 1974
British 1,374,640, Gulf Res. Dev., 1974
British 1,367,577, Mitsubishi, 1974
British 1,379,672, OJI Yuka, 1975
British 1,381,852, Stami Carbon, 1975
British 1,385,134, Gulf Res. Dev., 1975
British 1,387,233, Gulf Res. Dev., 1975
British 1,391,579, Gulf Res. Dev., 1975
British 1,392,667, Monticatini, 1975
British 1,400,982, Pall Mall Canada, 1975
British 1,401,599, Gulf Res. Dev., 1975; British 1,416,416, Solvay, 1975
British 1,446,034, Crown Zellerbach, 1976
British 1,443,581, Hoechst, 1976
British 1,441,883, Solvay & Sie, 1976
British 1,439,514, Mitsubishi Rayon, 1976
British 1,440,737, Toray Ind., 1976
British 1,435,488, BP Chem., 1976
British 1,450,892, ICI, 1976
British 1,451,421, Stami Carbon, 1976
British 1,461,794, BASF, 1977; British 1,452,041, CFRGC, 1976
British 1,469,157, Snam Progetti, 1977; British 1,471,097, Montedison, 1977
British 1,477,378, BASF, 1977; British 1,480,259, Anic SpA, 1977
British 1,478,839, A.Keller, 1977; British 1,480,329, BP Chemicals, 1977
British 1,484,584, Battelle, 1977; British 1,491,050, BASF, 1977

British 1,502,717, Montedison, 1978; British, 1,505,490, ICI, 1978
British 1,509,878, CFR, 1978; British 1,511,034, Sumitomo, 1978
British 1,520,382, Toray, 1978; British 1,523,501, DFR, 1978

Flame Resistant Compositions

US 3,640,946, Hercules, 1972
US 3,640,949, Michigan Chem., 1972
US 3,650,300, Johnson & Johnson, 1972
US 3,700,625, Phillips Petrol., 1972
US 3,749,600, Dow Chem., 1973
US 3,775,365, Phillips Petrol., 1973
US 3,927,145, Union Carbide, 1975

Miscellaneous

US 3,092,891, Production of Olefin Filaments for Brush Bristle, Montecatini, 1963
US 3,122,508, Polypropylene Brush Bristles, Whiting, 1963
US 3,118,161, Foamed Polypropylene Filaments for Brush Bristles, Whiting, 1964
US 3,145,525, Polypropylene Polyethylene Ropes, Wall Rope, 1964
US 3,132,194, Preparation of Microporous Olefin Filaments, Phillips Petrol., 1964
US 3,143,584, Spinning of High Tenacity Polypropylene Filaments, ICI, 1964
US 3,165,947, Plastic Composition for Rope Manufacture, Wall Rope, 1965
US 3,173,163, Polypropylene Flagged Brush Bristles, Whiting, 1965
US 3,327,339, Composite Filaments for Brush Bristles, Lemelson, 1965
US 3,214,234, Oriented Foamed Polyolefin Fibers, Phillips Petrol., 1965
US 3,164,947, Plastics Compositions for Rope Manufacture, Wall Rope, 1965
US 3,173,163, Polypropylene Flagged Brush Bristles, Whiting, 1965
US 3,349,052, Thermoplastic Polyethylene Wax Fibers, Sun Oil, 1967
US 3,383,276, Extruded Synthetic Brush Bristle Filaments, Gould, 1968
US 3,454,519, Self-lubricating Polypropylene Fiber, Nat. Dist., 1969
US 3,551,279, Synthetic Fibers Having a Silklike Surface Luster and Light Transparency, Snia Viscosa, 1970
US 4,054,025, Process for the Production of Filament Yarns with Statistically Distributed Broken Individual Filaments, Bayer, 1977
US 3,677,881, Open Celled Polypropylene Filament of Improved Uniformity, Chem. Cell, 1972
US 3,646,153, Synthetic Fibers Having Improved Soil and Stain Repellency, Allied Chem., 1972
US 3,837,022, Fibers for Pillow Stuffing, Phillips, 1974
US 3,867,187, Polypropylene Filaments Having Improved Soiling Characteristics, Phillips, 1975
US 3,896,812; 3,896,813, Sutures Having Long Lasting Biocidal Properties, Suture Inc., 1975
US 3,991,549, Staple Look Filament Yarns, Hoechst, 1976
US 3,995,044, Process for the Manufacture of Filament Yarn Having Protruding Filaments, Hoechst, 1976
US 3,867,187, Polypropylene Filaments Having Improved Soiling and Croching Characteristics, Phillips Petrol., 1975
US 3,993,834, Light Transmitting Filament, Mitsubishi Rayon, 1976
British 927,317, Treating Polypropylene Filaments for Use in Brushes, ICI, 1963
British 966,498, Resilient Undrawn Polypropylene Fibers, ICI, 1964
British 988,068, Polypropylene Brush Bristles, Whiting, 1964
British 1,008,670, Polypropylene Flagged Filaments for Brushes, Whiting, 1965
British 1,039,488, Decorative Polypropylene Yarns, Hercules, 1966
British 1,093,370, Propylene/Ethylene Copolymer Ropes, ICI, 1967
British 1,088,506, Composite Hollow Polyolefin Fibers Filled with Foam for Bristles and Mop Fibers, J. C. Lewis, 1967
British 1,089,525, Polypropylene Fibers with Improved Abrasion Resistance and Dyeability, Mitsubishi Rayon, 1967

British 1,105,609, Modified Fibers with Hydrophilic Surfaces, Courtaulds, 1968
British 1,135,542, Polyolefin Fibers Having a High Degree of Thermal Shrinkage, Asahi Chem., 1968
British 1,167,038, Fine Denier Microporous Foamed Filaments, ICI, 1969
British 1,272,076, High Strength Polypropylene Foam Fibers, Saleem, 1972
British 1,325,776, Composite Fibers, Toray Ind., 1973
British 1,442,499, Method for Producing Hairy Multifilament Yarns, Hoechst, 1976
British 1,442,281, Method for Producing Hairy Multifilament Yarns, Hoechst, 1976

Chapter X DYEING OF POLYPROPYLENE FIBERS

I. INTRODUCTION

Most of the polypropylene yarn and staple currently available in the market is colored through pigmentation. However, the large number of shades required by the trade has created problems in two areas: production economy and inventory. Especially affected by the proliferation of shades is the apparel industry, which is still the most promising market for volume consumption. It is, therefore, imperative that familiar dyeing techniques used for stock dyeing, hank dyeing, piece dyeing, and printing, be made available for polypropylene fiber to make it an acceptable textile material.

Because of its nonpolar structure, polypropylene is said to be undyeable. But this had been said also about the older synthetic fibers. Polypropylene presented dyestuff manufacturers and dyeing technologists with a problem reminiscent of that caused by the introduction of cellulose acetate in the 1920s and acrylic fibers in 1950s. In both cases, the new fiber was found to have little substantivity for existing dyes. With cellulose acetate, the problem was solved by development of a new class of dyes - the disperse dyes; with acrylics, the fiber itself was modified to make use of one class of existing dyes.

Dyeing methods useful for polypropylene fibers have been investigated by many laboratories throughout the world. Several laboratories have expended considerable effort in synthesizing dye structures specially meant for hydrocarbon fibers while others have tried to modify the fiber itself. The problem of dyeability of polypropylene fiber is still far from solved. The extensive research efforts during the last decade have, however, provided some definite leads around which more than one dyeable fiber are developing. To appreciate these successes or near-successes, an appreciation of the past efforts is in order. The following will provide a brief review of these efforts.

It is not surprising that Montecatini and its associates, ACNA, were among the first to investigate the methods of dyeing of polypropylene fiber with common dyestuffs. Later, several other fiber and dyestuffs manufacturing companies of the United States, Europe and Japan joined the race. The most noteworthy among them are ICI, Ciba, Bayer, Sumitomo, Hodogaya, Eastman, Allied, Hercules, and Hoechst. It is estimated that over 200 patents have been issued so far in the United States alone using more or less original ideas for the dyeing of polypropylene fiber. A cursory look at patent literature shows that the problem has been attacked from three different directions.

1) Attempts to apply conventional methods for dyeing of unmodified polypropylene fiber with appropriate dyes through the use of various dye accelerators, dye carriers, fiber swelling agents, etc. This approach has proved successful in the past for cellulose acetate and polyester fibers.

2) Modification of the polymer by attaching dye-receptive monomers or functional groups to the polymer molecules (copolymers or graft polymers). This approach has been successfully used in the past for polyacrylonitrile fibers.

3) Mixing dye-receptive additives with the polymer prior to fiber spinning. This approach has not been tried in the past for any of the commercially successful fibers. For polypropylene, however, it promised quite a good chance of success from the very start.

There are three major requirements which have to be met to make a fiber dyeable.

1) The dye molecule must be able to penetrate the body of the fiber.

2) There must be present some receptor sites in the fiber so that the dye molecules are retained in the body of the fiber and resist washing out.

3) There must be preferential mass transfer of the dye from the dye bath or print paste to the fiber.

The third requisite is generally met if the first two conditions are satisfied. In addition to the above-mentioned basic requirements for dyeability, there are several other demands which are important from the commercial point of view: most important, that the fiber be dyeable and printable with equipment and procedures commonly known to the textile trade, the dyestuffs and auxiliary chemicals be common and not too expensive, the dye bath exhaustion be close to completion, the coloration be level, the palette of shades be wide, and the fastness properties of the dyed goods be adequate.

The penetration of dyestuffs into the fiber depends mainly on its diffusion coefficient which is not only affected by the structure of the dye, but also by the structure of the fiber. Because of its hydrocarbon nature, unmodified polypropylene fiber is likely to be penetrated only by oleophilic dye molecules. In order for smaller amounts of oleophilic dyestuffs to diffuse into the fiber, either the pure hydrocarbon nature of the polymer should be changed or some submicroscopic discontinuities with high interfacial energy should be created within the body of the fiber. As described subsequently in this chapter, most of the research laboratories around the world have mainly concentrated their attention on the importance of dye sites in the fiber and quite often have neglected the importance of diffusion of the dye into the fiber. It will be seen that, even if adequate dyesites are present in the fiber, the fiber may still remain undyeable. This happens if the dye molecules are unable to diffuse into the fiber.

A dye site is a functional group on a polymer molecule which exerts a binding force on a dye molecule. There are three general types of binding forces in a fiber:

1) Chemical (ionic, convalent, coordinate)
2) Polar
3) Nonpolar, or van der Waal, forces

An example of the forces of chemical combination are: the ionic bond between an acid dye and a basic group in wool; a covalent bond between a reactive dye of chlorocyanurate type with hydroxy groups of a cellulosic fiber; a coordinate bond between the metal atom and the dye molecule in mordant dyeing. An example of polar forces is the hydrogen bond formed between a direct dyestuff and the hydroxyl groups in cellulosic fibers. The weakest and least understood of the three types are nonpolar, or van der Waals forces which, for instance, hold a disperse dye in a fiber of cellulose acetate or polyester. The relative role of each of these forces in any fiber cannot be fully defined. A combination of all three forces is probably involved in the dyeing of most fibers.

In trade, dyes are usually classified according to the mechanism by which they dye. Although the definitions are not very clear, for the purpose of this discussion we may follow the same classification and loosely divide the available dyestuffs into the following groups:

1) Anionic or acid dyes
2) Cationic or basic dyes
3) Disperse dyes
4) Reactive dyes
5) Ingrain dyes
6) Mordant dyes

The largest class is the acid dyes, which primarily dye by the ionic mechanism, forming a salt with basic groups in the fiber, such as in wool, silk, and nylon. They are further divided into acid, acid milling, 1:1 and 1:2 premetallized dyes. All of these dyes are soluble in hot water to varying degrees. This class is distinguished by a wide palette of shades, by a generally high light fastness, high drycleaning fastness, moderate wash fastness, and is comparatively low in price.

The second largest group is the disperse dyes, commonly known for the dyeing of manmade fibers in general and nylon, polyester, and cellulose acetate in particular. These dyes are only sparingly soluble in water and have a strong affinity for hydrophobic fibers. This group is distinguished by their large number, fairly good palette of shades, high wash fastness, good leveling, moderate light fastness, and adequate solvent fastness.

Cationic dyes are primarily of interest only to modacrylic fibers, although they are also being used to dye basic dyeable polyester and nylon fibers. They are bright in shade, have good wash fastness, good drycleaning fastness, but only moderate in light fastness.

Ingrain dyes and reactive dyes are primarily of interest to cellulosic fibers although some of the reactive dyes are also used for nylon, wool, and silk. They are distinguished by their very high ratings of wash fastness.

Mordant dyes have been known in the trade for the dyeing of wool and acrylic fibers although their use is now quite limited. The dye is diffused into the fiber which is then treated with a metal salt solution, such as potassium dichromate. The metal ions penetrate into the fiber and form a complex, large dye molecule in the body of the fiber. Mordant dyes attained some revival in the development of dyeable polypropylene fiber. However, the metal ion, instead of diffusing into the fiber from the dye bath, was contained within the fiber body and ready for complexing with the infusing dye molecule.

II. DYEING UNMODIFIED POLYPROPYLENE FIBER

Before polypropylene was declared "undyeable", every attempt was made to dye this fiber with conventional dyes. It can be seen from the patent literature of the late fifties and early sixties that everything known in the book and all possible modifications of the art had been tried. Some specific dyes, mainly those with linear long-chain hydrocarbon groups pendant to the dye molecule, may have produced a colored fiber, but in no case did anything approaching a full color range result from these endeavors. The fastness properties were also poor. Among the dyestuffs tried for the dyeing of unmodified polypropylene were a) Vats: usually applied by a vat acid procedure; b) Azoics: usually applied from an acidified naphthol-base dispersion with subsequent diazotization in situ; c) Disperse and solvent dyes. Only disperse dyes and chemically related compounds showed any promise.

While the dyestuff manufacturers concentrated on the selection of suitable dyes from the available dyestuff structures or on the synthesis of new dyes, and while the polymer chemists tried to modify the polymer characteristics, the dyeing technologists explored new ways to apply the available dyestuffs to unmodified polypropylene fiber. A wealth of patent literature has been accumulated describing the use of various dyeing methods. A number of dyeing auxiliaries has been mentioned to accomplish satisfactory penetration of the fiber by the dyestuff. Of special importance are the use of swelling agents and carriers in the dyebath. High temperature dyeing is another approach which has been extensively considered for the dyeing of unmodified polypropylene. The use of various organic and inorganic acids in the dyebath has also been mentioned. Table I lists a few of these attempts. Almost all of these approaches are expensive and not suitable for low cost fiber such as polypropylene. Moreover, fastness properties still remain poor.

TABLE I

Attempts to Dye Unmodified PP Fiber With Modified Dyeing Methods

Chemicals Added to Dyebath	Patent References
Hydrocarbons and Chlorinated hydrocarbons	U.S. 3,046,076, Burlington Ind; U.S. 3,069,220, Allied; Brit. 921,063, ICI; Brit. 896,318, ACNA; Brit. 809,496, ICI; Brit. 809,495, Glazstoff; Brit. 921,430, Courtaulds; Brit. 814,582, Glanzstoff;
Polyalkylene glycol	Brit. 900,934, Brit. Nylon;
Long chain alkyl phenols	U.S. 3,157,457, Onderzoekings;
Organic acids	U.S. 3,128,146, Am. Aniline; Brit. 921,063, ICI; U.S. 3,120,423, Crown Z.; Belg. 600,176, Ciba;
Mordant agents	U.S. 3,098,044, Geigy
Reducing agents (vat dyes)	U.S. 3,056,643, Hercules; U.S. 3,104,150, Hercules;
Soluble thiocyanates	U.S. 3,201,189, GAF;

Treating the fiber surface to make it more receptive to dyestuffs was another approach very eagerly pursued for the dyeing of polypropylene fibers. Table II is a partial list of the numerous pretreatments of the fiber surface prior to dyeing. These can be placed in two general categories: the reactive type, in which the polymer backbone is chemically modified on the fiber surface; and the so-called infusing type, which consists merely of infusing a material into the fiber with the hope that it will stay there through the dyeing process and act as a dye-receptive site. Many of these endeavors were the offshoots of surface treatment processes for polyolefin films to make them susceptible to printing and coating or increasing their adhesion to other substrates.

Fiber treatments in general tend to be costly. They involve treatment of

TABLE II

Chemical Treatments of Fiber Surface for Dyeing

Treatment	Patent References
Swelling or cracking agents	U.S. 3,179,484, ACNA; U.S. 3,102,323, DuPont; U.S. 3,159,449, Hercules; U.S. 3,135,662, Montecatini; Brit. 905,833, DuPont; Brit. 1,046,525, Hoechst; Brit. 1,046,527, Hoechst; France 1,388,253, Hoechst;
Strong oxidizing agents	U.S. 3,101,989, Rhodiaceta; U.S. 3,305, 297, A.G. Ras; U.S. 3,150,917, Gagliardi; Brit. 985,294, Courtaulds: Brit. 886,701, IEI; Belg. 626,518, BASF;
Sulphonation and chlorosulphonation	U.S. 2,937,066, Dow; U.S. 3,148,017, Gagliardi; U.S. 3,150,917, Gagliardi; Brit. 893,603, Montecatini; Brit. 921,059,059, Courtaulds;
Halogenation	Brit. 876,045, ICI;
Phosphorylation	U.S. 3,039,840, Hercules; U.S. 3,116,966, Montecatini; U.S. 3,145,073, Gagliardi; Brit. 907,763, ICI; Brit. 881,981, Montecatini;
Organic N-compounds	U.S. 3,112,983, ACNA; U.S. 3,098,692, Gagliardi; U.S. 3,140,194, Gagliardi, U.S. 3,254,934, Bayer;
Reducing agents	Brit. 982,627, ICI
Werner chrome complexes, chromyl compounds, metal salts	USP 3,169,423, Gagliardi; U.S. 3,186,976, In.Fr.Pet. U.S. 3,169,043, In.Fr.Pet;

fiber with chemical reagents, as well as drying, handling, and the recovery of excess materials. The cost of these operations must be added to the usual cost of dyeing which would make the overall cost of dyeing too high to be commercially feasible. An additional problem is the fact that the treated fibers only yield to ring dyeing (dyeing on the surface), which is not desirable from the viewpoint of color fastness.

In summary, untreated polypropylene fiber does not take enough dyestuff to yield deep colors, the range of shades is limited, and wash and drycleaning fastness is very poor. The use of auxiliary chemicals or modifications in dyeing techniques does not improve the situation significantly either. The surface treatments are expensive and tend to affect the physical properties of the fiber unfavorably.

III. COPOLYMERIZATION AND GRAFT POLYMERIZATION FOR DYEING

The attempts to dye natural polypropylene having been unsuccessful, attention was turned to modification of the polymer to impart dyeability. These attempts constituted the putting of dye-receptive groups in the chain of polypropylene molecule (exemplified in Fig. 1) or attaching a segment of a reactive polymer to the backbone of polypropylene (exemplified in Fig. 2). The first approach is called copolymerization, the second is called graft polymerization.

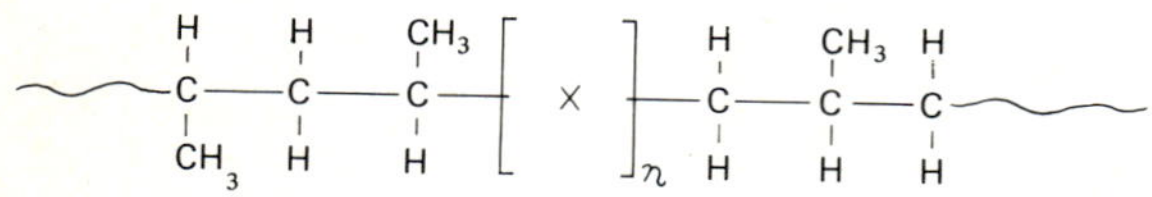

Fig. 1. Improving Dyeability by Copolymerization

Copolymerization in the main chain has been successfully utilized for improving the dyeability of acrylic and other fibers. For acrylic fibers methylvinyl pyridine is used as a comonomer to provide a basic pyridine nitrogen which serves as dye-receptive site. Some polyethylene terephthalate polymers, in which some of the benzene rings are substituted with sulphonic acid groups to provide basic dyeability, have been described in patent literature and recently commercialized by a few fiber producing companies.

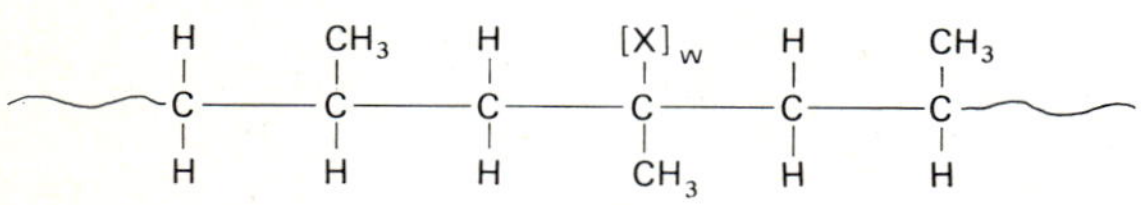

Fig. 2. Improving Dyeability by Graft Polymerization

Copolymerization is difficult with polyolefins. This is because the Ziegler-Natta type catalysts, used in the production of polyolefins (Chapter I), are inactivated by polar materials. There are reports in patent literature of copolymerization of such materials as vinyl-pyridine, styrene, acrylic acids, acrylates, sulphonic acid derivatives, carbon monoxide, naphthols, etc., with propylene. None of these processes, however, is expected to result in a commercial product because of the two main reasons:

1) The efficiency of copolymerization of propylene with polar compounds is

quite low, if at all possible.

2) Copolymerization strongly impairs the crystallization behavior of polypropylene and lowers its melting point. The adverse effects of copolymerization on physical and mechanical properties alone would make it an undesirable approach for fiber production.

Graft polymerization would seem to be another theoretically attractive approach to achieve enhanced dyeability of polypropylene. Although the dye-receptive segments are not part of the main polymer chain, they are attached to the polymer backbone with primary bonds. In grafting of polypropylene, the polymer is subjected to the action of an initiator, which severs some of the vulnerable tertiary hydrogens from the polypropylene backbone, leaving free radical sites. These free radicals are capable of initiating polymerization of the added monomer.

TABLE III

Graft Polymerization of Polypropylene

Grafted Monomers	Patent References
Vinyl monomers (styrene, substituted styrenes, vinyl pyridines, vinyl esters, etc.)	U.S. 3,166,607 Montecatini; U.S. 3,529,038 Rhone Poulance; U.S. 3,218,117, Am. Cyanamide US 3,049,508, Dow; US 3,058,950, Dow; US 3,020,174, Montecatini; US 3,073,667, Montecatini; US 3,093,615, Montecatini; Brit. 1,166,089, Mitsubishi Brit. 1,032,505, Dow; Brit. 1,080,254, Chev. Res. Brit. 875,758, Montecatini; Brit. 997,093, Courtaulds Brit. 946,824, ICI; Brit. 883,793, Polyplastics; Brit. 1,025,474, Geleji;
Acrylic monomers	US 3,162,697, Phillips; US 3,205,156, ToyoRayon; US 3,254,541, Dow; US 3,632,678, Dart; US 3,177,269, Dow; US 3,022,191, Montecatini US 3,093,615, Montecatini; US 3,131,990, Montecatini Brit. 1,166,089, Mitsubishi; Brit. 891,944, DuPont Brit. 928,156, Montecatini; Brit. 1,025,474, Geleji
Unsaturated N-bases, sulphonic acids, epoxies	US 3,049,507, Dow; US 3,314,743 Gagliardi; Brit. 938,041, Shin Nippow;
Polyadehydes, polyamides (mechanograft)	Brit. 993,434, Rommler;

Several methods have been used for producing free radical sites on polypropylene molecules: chemical, such as conventional peroxide or azo polymerization initiators; peroxidation or ozonization; electron and gamma radiation, etc. Many types of monomers have been used for grafting. A partial list of these attempts is given in Table III. As is apparent from this list, the entire range of monomers is covered: acidic types, such as acrylic acids; basic types, such as vinyl pyridines and vinyl-pyrrolidones; and neutral types, such as styrene, vinyl acetate, etc. Sometimes graft polymerization has been used as fiber treatment. One may, for example, graft-polymerize methacrylate monomers or vinyl pyridines in vapor phase on polypropylene yarns or on fabricated goods. Some attempts have been made to combine grafting with other fiber treatments. For instance, the polypropylene fiber can be grafted with chloromethyl-styrene and then after-treated with an amine to replace the chlorine of the chloromethylstyrene with an amine group, which would be the actual dye receptor.

Grafting is technically more appealing in the case of polypropylene than copolymerization. In spite of the numerous inherent difficulties in graft polymerization, this approach has been extensively pursued by various research laboratories. The Polymer Institute of Hungary, for example, has done extensive work on graft polymerization of styrene and other monomers on polypropylene. The fibers so obtained were very dyeable with disperse dyes (ref.2). Canada Research Council had an active program on radiation-initiated grafting. As much as 35 percent methyacrylic acid, for example, could be easily grafted on polypropylene fibers. In a few cases grafted polypropylene fibers were even brought to the stage of pilot plant production. Montecatini (ref.1), for example, was once reported as having a dyeable fiber of such a type in experimental production. The failures were primarily on the basis of economics rather than technical setbacks.

IV. DYE-RECEPTIVE ADDITIVES

The admixing of dyeable additives with the polymer prior to extrusion was early recognized as a practical approach to dyeing polypropylene fiber. In effect, the scheme calls for coloring the additive particles, dispersed uniformly in the body of the fiber, with the result that the whole fiber appears to be dyed macroscopically. It was soon found that incorporation of a dye-fixing additive into the polymer melt prior to fiber spinning was technically much easier and economically more feasible than the feat of creating dye-fixing groups on the polymer chain by copolymerization, graft polymerization, or fiber treatment with chemical means. Moreover, it was observed that the effect of the incorporation of dye additives in the form of simple mixture was far less harmful to the fiber properties than was copolymerization or graft polymerization.

The extensive research efforts during the late fifties and throughout the sixties have produced a large number of patent disclosures pertaining to specific classes of dye additives. In principle, these additives fall into three classes:

a) Metallic compounds

b) Polymeric additives

c) Low molecular-weight organic and inorganic compounds.

The first two categories will be reviewed here in detail. The last one is a catchall, in which a variety of low molecular-weight organic and inorganic materials has been used to impart dyeability, but no practical use has so far emerged.

From the point of view of end product, the additives approach can be seen to concentrate into four major areas of activities:

a) Development of mordant dyeable polypropylene fiber using chelating-type disperse dyes (mordant-disperse dyes)

b) Development of disperse dyeable fiber using common disperse, vat, and azoic dyes

c) Development of acid-dyeable fiber using anionic dyestuffs, including wool mordant dyes.

d) Development of basic dyeable fiber using cationic dyestuffs commonly used for modacrylic fibers and basic dyeable polyesters.

Although the concept of admixing reactive additives with the polymer sounds relatively simple, there are several requirements which must be simultaneously met to make a prospective additive a desirable candidate for serious consideration. Some of these requirements are summarized below:

1) Additive must be inexpensive

2) Additive concentration should not exceed 10 percent in the fiber for adequate dyeability

3) Additive should be at least partially miscible with polypropylene melt for ease of dispersion and processing

4) Additive should be stable at spinning temperatures

5) Additive should soften above 140°C (maximum heat setting temperature for the fiber) but should have its glass transition temperature below 90°C (temperature of the dye bath)

6) Additive should be insoluble in dilute acids to avoid extraction from the fiber under scouring and dyeing conditions. It should also be insoluble in organic solvents to avoid extraction in drycleaning solvents

7) Additive and its degradation products must be nontoxic and odorless

8) Additive should have low tendency of migration to the surface of the fiber in processing and in use

9) Additive must not adversely affect the thermal or light stability of polypropylene

10) Additive must not react with any stabilizers present in the polymer to

deactivate them or produce colored products.

The satisfaction of these requirements simultaneously is, of course, a difficult task and herein lies the specificity of successful dye additives.

IV.1. Mordant-Dyeable Fibers

One of the most successful methods for modifying polypropylene to make it dyeable is by the addition of small amounts of metal compounds to the polymer, the metal being selected on the basis of its ability to form chelates or coordination complexes with polar organic compounds. Aluminum, chromium, nickel, and zinc are some of the metals which are effective in this respect. Polypropylene fibers modified to provide dyeability in this manner are being commercially produced by Hercules, Vectra, and Phillips in the United States and by several companies in Japan.

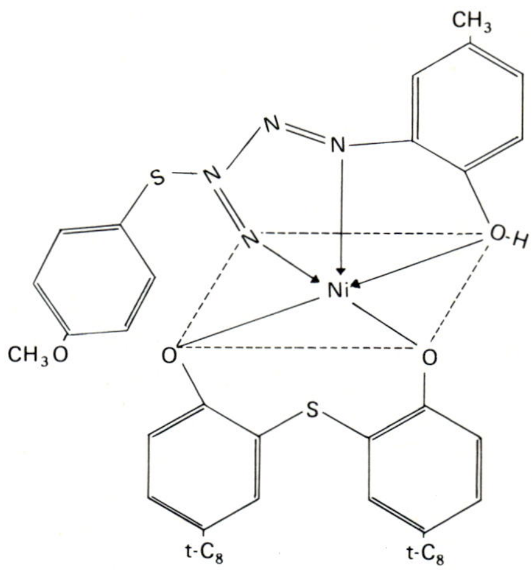

Fig. 3. A Suggested Structure for the Chelated Dye-Dyesite Product

The dyeing mechanism of these fibers is based on the formation of colored Werner complexes of metals with mordant dyes, as shown in Fig. 3. Heavy metal ions have empty orbitals which can accept an unshared electron pair from electron-rich atoms in the dye. This type of bond is usually known as coordinate bond.

Since these complexes have very small dissociation constants, the bond formed is quite strong. Chrome mordant dyes are held in wool by this mechanism, and it is from these that the idea of incorporating the metal in the fiber evolved.

The common mordant dyes are of acid type, very soluble in water and small in molecular weight. These factors work to the disadvantage of diffusion of the dye into polypropylene. To overcome this difficulty, new mordant dyes were synthesized. They retain the essential character of metal coordination but diffuse into the fiber-like disperse dyes. The palette of this group of disperse mordant dyes, commonly known as metal complexing dyes or PP dyes, was broadened by inclusion of some of the common metal-complex disperse dyes, already known and widely used for dyeing of polyester and cellulose acetate fibers. By so doing, a fairly sizable group of disperse-type metal complexing dyes was made available to the trade and eventually attained a great significance for the dyeing and printing of polypropylene fiber.

The structures of metal complexing dyes meant for polypropylene are adequately described in trade journals (ref.9,10,11,12) and patent literature. Basically, these compounds are mordant dyes devoid of any strongly ionizable water solubilizing groups. A sample structure is that of PP green BM (Fig.4).

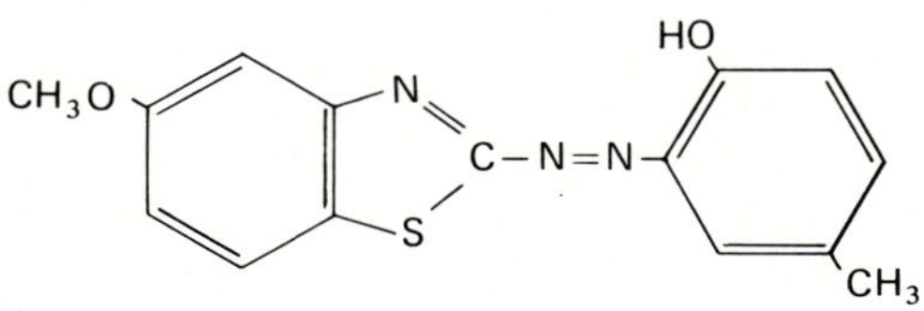

Fig. 4. Structure of PP Green BM

All of these dyes must contain functional groups capable of supplying electrons to the hybridized orbitals of the metal atom present in the fiber. A majority of these dyes satisfies this requirement by virtue of the following groupings:

1. 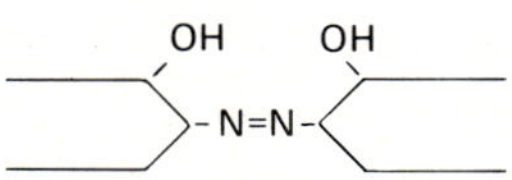

4. 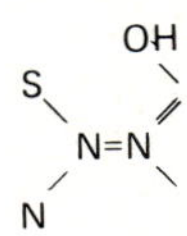

2. 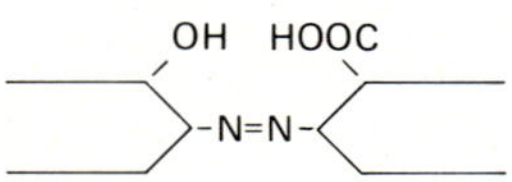

5. 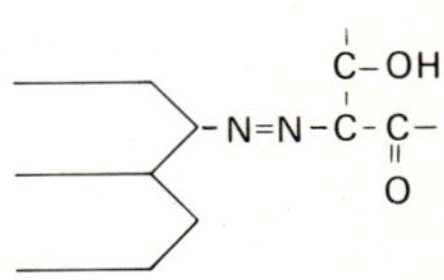

3. 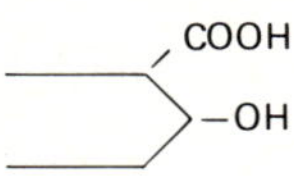

6. 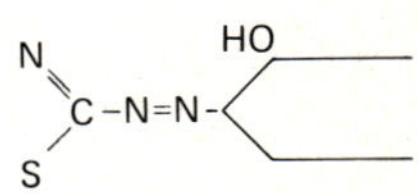

In the dyeing of metal-modified polypropylene with chelating dyes, an actual chemical reaction takes place. The dyeing behavior of metal chelating dyes, therefore, resembles to a great extent the behavior of reactive dyes. Once the dye molecule has reacted with the dyesite, it cannot be dislodged. The chelating dyes differ from reactive dyes, however, in that the former change color on reacting with the dyesite while the latter do not.

The rate of diffusion of the dye depends mainly on its structure, while its rate of reaction with the metal and the fastness properties of the metallized dye structure depend both on the dye structure and on the nature of the metal ion used. Of course, the reaction conditions, i.e., the temperature and the pH value of the bath and the concentration of the dye in the bath, also play a role in determining the reaction rate and fastness properties of the dyed goods. Significant changes in shade and fastness properties are often noticed at different pH levels within a narrow range.

The idea of using metallic compounds for dyeing of polypropylene fibers dates back to the observation made at the research laboratory of one of the fiber companies that some mordant dyes were showing unexpectedly good color fastness properties on polypropylene fibers. This was traced to the presence of metal catalyst residues from the polymerization step (ref.3). Active development work on this approach started quite simultaneously in the laboratories of practically all important fiber producing companies. This interest may have been sparked to a degree by certain nickel compounds which were found to be excellent light stabilizers for polypropylene and served dual purpose (see Chapter II). These compounds were commercially introduced in the United States by Ferro Corp., and popularized by Hercules, Inc.

Three types of metallic compounds have been mentioned as dyesites for mordant dyeable (metal complexing) polypropylene fibers (ref.4,5,6,7):

1) Inorganics, such as aluminum, chromium, nickel, and zinc salts
2) Salts of organic acids, such as aluminum and nickel stearate
3) Organometallic materials, such as nickel organic compounds primarily used for the stabilization of polypropylene against light (refer to Chapter II for structures).

Although a variety of metal-containing compounds has been disclosed in the patent literature, (see Table IV), only aluminum and nickel compounds have been researched to a developmental stage. In the United States, the nickel system is presently the favored one, while in Japan the aluminum system seems to be more common. Suitable dyes are available for both systems.

Combinations of metal compounds with other polymeric materials have been investigated by several research laboratories. The primary objective of these efforts has been to make the fiber dyeability more versatile, but in effect

TABLE IV

PP Fibers Dyeable Through Admixture With Metal Compounds

Metal Compounds	Patent References
Polyvalent Chelatable metals (Al, Zn, Ni, Cr, Co, Zr) as organic salts of fatty acids	US 3,310,548, Esso; US 3,248,378, Hoechst; US 3,410,799, Esso; US 2,984,634, Eastman US 3,424,716, Esso; Brit. 1,069,890, ToyoRayon; US 992,379, Hercules; Brit. 1,085,327, Mitsubishi; Brit. 1,071,004, Mitsubishi; Brit. 932,897, DuPont; Brit. 934,667, Montecatini; Brit. 970,919, Hercules France, 1,418,743, Montecatini; France 1,418,744, Mont. France 1,426,176, ToyoRayon; France 1,416,413, Mitsub. France 1,427,260, Montecatini; France 1,427,261 Mont. Belg. 630,351, Ethyl; Belg. 619,493, ICI; Belg. 617,280, Reeves; Japan 16,301/66 ToyoRayon; Japan 16,302/66 ToyoRayon; Japan 16,303/66 ToyoRayon; Japan 16,304/66 ToyoRayon; Japan 7,450/66 ToyoRayon; Japan 20,103/66 Mitsubishi; Japan 20,102/66, Mitsub.
Metal complex of a triazole	France 1,393,883, Bayer; Germany 1,469,600 Asahi
Metal (Ni or Zn) phenolates	US 3,000,886, Esso; US 3,332,732, Esso; US 3,072,601, Hercules; US 3,072,909, Hercules; US 3,072,910, Hercules; Brit. 1,049,020, Montecatini; Brit. 1,137,381, Mitsubishi; Brit. 1,102,215, Montecatini; Brit. 996,109, Hercules; Belg. 674,123, Mitsubishi; Belg. 657,488, Mitsubishi; Belg. 579,636, Ferro
Metal thiocarbamates	Brit. 1,070,944, Bayer
Metal alcoholates and related compounds	Japan 19,815/66, Mitsubishi; Japan 4354/67, Mitsubishi; Japan 18,580/67 Mitsubishi; Japan 25791/64, Daiwa
Metal Salts of Piperazines and vicinal diamines	France 1,427,257, Montecatini US 3,284,428, Esso
Complexes of Ni, Al, Co, etc.	US 3,268,476, Hercules; US 3,284,428, Esso; US 3,296,239, Esso; France 1,416,413, Mitsubishi; France 1,427,257, Montecatini; Brit. 1,052,967, Mitsubishi
Metal salts of substituted phosphinic acids	Brit. 1,162,462, Geigy
Metal compounds plus polymeric additives	US 3,240,552, Eastman; US 3,163,492, Hercules; Brit. 1,007,753, Eastman; Brit. 1,064,404, Montecatini; Brit. 1,079,225, Montecatini; Brit. 1,013,042, ICI; Brit. 1,070,944, Bayer; Brit. 1,007,337, Eastman

these efforts have resulted in enhancing the diffusion rate of the metal complexing dyes into the fiber thus making the process of dyeing and printing faster. Most of these improvements have, however, been nullified with a decrease in wash and drycleaning fastness.

In the United States, nickel-modified fibers have been made available by Enjay, Vectra, Phillips, and Hercules. A variety of nickel compounds has been disclosed as dyesites for polypropylene. In practice, however, the nickel complexes intended as UV stabilizers are primarily used because of their dual functions. Several of these compounds are available commercially and are described in Chapter II. Sometimes a mixture of one of these nickel complexes and nickel stearate is used because of economy reasons.

It has been noticed (ref.8) that different organo-nickel dyesites yield different color value and brightness when dyed with metal-complex dyes. This indicates that the dyesite retains some or all of its original identity after chelating This is further confirmed by the fact that the heat and light stability of dyed fibers containing such complexes is not significantly affected after dyeing.

Just as there are differences in the color value and brightness of dyed or printed polypropylene with different dyesites, there are differences in the fastness properties obtained. Aliphatic nickel compounds generally yield dyed fiber of lower light fastness than those containing aromacity in the original dyesite molecule. The structure of the dyesite influences the ability of the chelation product to dissipate absorbed light energy with minimum bond breakage. Resonance energy in aromatic ligands is probably the primary factor determining the light fastness of the dye-metal complex.

Different dyesites also cause differences in drycleaning, gas fading and crock fastness. It has been found, for instance, that the dyed fibers containing nickel stearate as dye additive have significantly lower drycleaning fastness than those containing nickel chelate UV stabilizers. Similarly, the fibers containing UV 1084 (Ferro) show somewhat more gas fading than those containing Irgastab 2002 (Geigy). Wash fastness of nickel dyeable fibers is generally not a problem because of the large size of the dye-nickel complex molecule.

From the viewpoint of piece dyeing, metal-complex dyeing presents problems in that:

1) Exhaustion is extremely dependent on temperature of the dye bath.
2) The rate of dyeing is different for each dyestuff.
3) The chelated dyes have very poor migration properties; leveling is therefore a problem.
4) Dyeing is rapid at elevated temperatures.
5) The complex formation is strongly dependent on temperature and pH of the dye bath. Significant changes in shade are often noticed at different pH levels

within a narrow range.

6) Shade correction is very difficult.

7) Stripping of color is problematic and never complete.

Brightness of many shades attainable is similar to that of disperse dyes on polyester fibers, although there has been a general complaint in the past with regard to the dullness of shades obtained with metallizable dyes. This concern proved to be somewhat unwarranted because most of the nylon carpets, with which polypropylene carpets have to compete, are being dyed with metallized dyes. These are similar in shades as those obtained by mordant dyes on nickel-modified polypropylene.

Light fastness of these dyes, as is the case with other classes of dyestuffs, depends on individual dyes. It increases with depth of shade and is affected by the pH of the printing paste or the dye bath. Washing and drycleaning fastness of chelating dyes as a class are very good and do not depend, unlike other classes of dyes, on depth of shade. Resistance to gas fading is generally good. It depends significantly on the structure of the nickel compound used as dye additive. Simple compounds, like nickel stearate, are not as good as complex molecules such as UV 1084 or Irgastab 2002. Crock fastness is generally very good provided care is taken in after-scouring. Crocking problems arise if the fabric is heated above 120°C, such as in drying and latexing of carpets and upholstery fabrics. It appears that the metallized dye complex migrates to the surface of the fiber at high temperatures. The use of large-size, high-melting Ni-compounds as dye additive may be useful to suppress this tendency of exudation. Tables V and VI list some fastness properties of metal-complexing dyes on polypropylene fibers of various origins. For further listings, references 6,14,20 and trade publications of various dyestuff manufacturing companies may be consulted.

The metal-complexing dyes are applied from an acid dye bath containing a dispersing agent as well as a sequestering agent. The use of 0.1 to 0.2 percent (owf) ethylene diamine tetra-acitic acid or other suitable sequestering agent is required where metal ions are present from hard water or steam lines. Metals such as Cu, Fe, Mn, etc., if not chelated by a suitable sequesterant, will complex with a portion of the dye, causing weak or off-shade dyeing with poor crock fastness. Pastel and light shades are particularly sensitive to this condition. The conventional dyehouse equipment is used for stock, skein, and piece dyeing. The usual dyehouse precautions for control of time-temperature relation should be maintained and strictly observed for repeatability. It is recommended that the temperature be increased slowly to obtain controlled exhaustion and maximum levelling. Dyestuff exhaustion and penetration into the fiber are enhanced by dyeing at the boil and using a suitable carrier. Salicylate-type carriers are particularly useful. In order to obtain optimum fastness properties, an

(Continued on p. 482)

TABLE V

Fastness of Nickel Chelating Dyes on Natural Polypropylene (ref.8)

Name of Dye	%	Crocking AATCC-116-1966T Wet	Crocking AATCC-116-1966T Dry	Xenon-Arc AATCC 16E-1964 AATCC SFH to Appreciable Fade	Gas Fading AATCC-23-1962 3 Cycles	Ozone AATCC-109-1963T
Polypropylene Brill. Yellow GM	0.5	5	5	>240	3	4-5
Polypropylene Brill. Yellow GM	2.5	4-5	4-5	>240	3	4-5
Petramin Yellow GL	0.5	4-5	5	>240	4	5
Petramin Yellow GL	2.5	4	4-5	>240	4	5
Petramin Yellow FRL	0.5	5	5	>240	4-5	5
Petramin Yellow FRL	2.5	4-5	5	>240	4-5	5
Petramin Yellow F3RL	0.5	4-5	5	>240	3-4	5
Petramin Yellow F3RL	2.5	4-5	4-5	>240	4-5	5
Polypropylene Brill. Orange RM	0.5	4-5	5	>240	4	4-5
Polypropylene Brill. Orange RM	2.5	4-5	5	>240	4-5	4-5
Petramin Orange GR	0.5	4-5	4-5	160	4	4
Petramin Orange GR	2.5	4	4-5	120	3-4	4-5
Petramin Orange FRR	0.5	4-5	5	>240	4	4-5
Petramin Orange FRR	2.5	4-5	4-5	>240	4	4-5
Petramin Orange 5R	0.5	4-5	5	240	4	4-5
Petramin Orange 5R	2.5	4	4-5	>240	4	4-5
Polypropylene Scarlet RBM	0.5	5	5	>240	4	4-5
Polypropylene Scarlet RBM	2.5	4-5	4-5	>240	4	4-5
Petramin Red F2B	0.5	4-5	5	160	4	4
Petramin Red F2B	2.5	3	4	160	3-4	4
Petramin Red BL Paste	0.5	4-5	5	>240	3-4	4-5
Petramin Red BL Paste	2.5	4	4-5	>240	4	4-5
Polypropylene Brill. Violet 3RM 200%	0.25	4-5	4	160	3-4	4-5
Polypropylene Brill. Violet 3RM 200%	1.25	4-5	5	>240	4	4-5
Petramin Violet R	0.5	4-5	5	80	3-4	4-5
Petramin Violet R	2.5	4	4-5	240	4-5	4-5
Polypropylene Violet 3BRM	0.5	5	5	160	3	4-5
Polypropylene Violet 3BRM	2.5	4-5	5	240	3-4	4-5
Petramin Dark Blue GL	0.5	4	4-5	240	4	4-5
Petramin Dark Blue GL	2.5	3	3-4	>240	4	5
Petramin Blue GGL	0.5	4-5	5	>240	3-4	4
Petramin Blue GGL	2.5	4-5	5	>240	4	4-5

TABLE V (continued)

Name of Dye	%	Crocking AATCC-116-1966T Wet	Crocking AATCC-116-1966T Dry	Xenon-Arc AATCC 16E-1964 AATCC SFH to Appreciable Fade	Gas Fading AATCC-23-1962 3 Cycles	Ozone AATCC-109-1963T
Polypropylene Dark Blue 2 BM 200%	0.25	4-5	5	240	3-4	4
Polypropylene Dark Blue 2 BM 200%	1.25	4-5	5	240	4	4-5
Petramin Blue B	0.5	4-5	5	80	2	4
Petramin Blue B	2.5	4	4-5	120	2-3	4-5
Polypropylene Brill. Blue BM 200%	0.25	4-5	5	60	2-3	3-4
Polypropylene Brill. Blue BM 200%	1.25	4	4-5	60	2-3	4
Polypropylene Green BM	0.5	4-5	5	240	4	4
Polypropylene Green BM	2.5	4	4-5	240	4-5	4
Petramin Green 5B Paste	0.5	5	4-5	240	3-4	-
Petramin Green 5B Paste	2.5	5	4-5	240	4	-
Petramin Brown 3R	0.5	4-5	5	240	4-5	4-5
Petramin Brown 3R	2.5	4-5	5	240	4-5	4-5
Altco-lene Yellow 3RP	1.0	4-5	5	240	4-5	-
Altco-lene Orange RP	1.0	5	5	240	4-5	-
Altco-lene Red 2BP	1.0	4-5	5	240	4	-
Altco-lene Violet BP	1.0	5	5	80	4	-
Altco-lene Blue NP	1.0	5	5	240	4-5	-
Altco-lene Green TP	1.0	5	5	160	4-5	-
Altco-lene Brown BRP	1.0	4-5	4-5	160	4-5	-

TABLE VI

Fastness Properties of Dyeable PP Fiber Dyeings (ref.20)

Fastness properties				Washing		Dry Cleaning		Perspiration		Sublimation		Rubbing	
Dyestuff	Dyeing depth	light	Gas fade	Effect on the dyeing	Staining of ad-jacent	Effect on the dyeing	Staining of ad-jacent	Effect on the dyeing	Staining of ad-jacent	Effect on the dyeing	Staining of ad-jacent	Effect on the dyeing	Staining of ad-jacent
Dainal	0.5	4	5	5	5	5	5	5	5	5	5	5	5
Yellow P-3G	5	5	5	5	5	5	5	5	5	5	5	5	4-5
Dainal	0.5	5	5	5	5	5	5	5	5	5	5	5	5
Yellow PGR	5	5	5	5	5	5	5	5	5	5	5	5	5
Dainal	0.5	4	5	5	5	5	5	5	5	5	5	5	5
Golden													
Yellow P-5R	5	5	5	5	5	5	5	5	5	5	5	5	5
Dainal	0.5	4	5	5	5	5	5	5	5	5	5	5	5
Orange PB	5	5	5	5	5	5	5	5	5	5	5	5	5
Dainal	0.5	5	3	5	5	5	5	5	5	5	5	5	5
Orange P-3R	5	5	5	5	5	5	5	5	5	5	5	4-5	4-5
Dainal	0.5	3	4	5	5	5	5	5	5	5	5	5	5
Scarlet PG	5	4	5	5	5	5	5	5	4	5	5	4	4
Dainal	0.5	4	2	5	5	5	5	5	5	5	5	5	5
Beige PA	5	5	3	5	5	5	5	5	5	5	5	4	4
Daika	0.5	4	5	5	5	5	5	5	5	5	5	5	5
Red AB	5	5	5	5	5	5	5	5	5	5	5	4	4
Liopal	0.5	5	5	5	5	5	5	5	5	5	5	5	5
Red PBR	5	5	5	5	4-5	5	5	5	5	5	5	4-5	4-5
PP	0.5	2	5	5	5	4-5	5	5	5	5	5	5	5
Blue No. 2	5	4	5	5	5	4-5	5	4	4	5	5	4	4-5
PP	0.5	4	3	5	5	5	5	5	5	5	5	5	5
Blue No. 3	5	5	5	5	5	5	5	5	5	5	5	4	4-5
PP	0.5	4	2	5	5	5	5	5	5	5	5	5	5
Blue No. 4	5	5	4	5	5	5	5	5	5	5	5	5	5
PP	0.5	2	5	5	5	4-5	5	5	5	5	5	5	5
Blue No. 7	5	4	5	5	5	4-5	5	4	4	5	5	4	4-5
Daika	0.5	3	3	5	5	5	5	5	5	5	5	5	5
Blue A-3G	5	5	-	5	5	5	5	5	5	5	5	5	5
Daika	0.5		3	3	3	5	5	5	5	5	5	5	5

TABLE VI (continued)

Fastness properties				Washing		Dry Cleaning		Perspiration		Sublimation		Rubbing	
Dyestuff	Dyeing depth	light	Gas fade	Effect on the dyeing	Staining of adjacent	Effect on the dyeing	Staining of adjacent	Effect on the dyeing	Staining of adjacent	Effect on the dyeing	Staining of adjacent	Effect on the dyeing	Staining of adjacent
Brill Blue													
A-FG	5	4	-	5	5	5	5	5	5	5	5	4-5	4-5
Daika	0.5	4-5	5	5	5	5	5	5	4-5	5	5	4	4
Navy Blue													
PR	5	5	5	5	5	5	5	5	4	5	5	4	4
Lipal	0.5	3-4	4	5	5	5	5	5	5	5	5	5	5
Violet PR	5	5	-	5	5	5	5	5	5	5	5	5	5
Daika	0.5	3	5	5	5	5	5	5	5	5	5	5	5
Violet													
P-2B	5	5	-	5	5	5	5	5	5	5	5	5	5
PP	0.5	5	3	5	5	5	5	5	5	5	5	5	4-5
Brown No.1	5	5	5	5	4-5	5	5	5	4	5	5	5	4
Matsulan	0.5	5	5	5	5	5	5	5	5	5	5	5	5
Brown 2RY	5	5	5	5	5	5	5	5	5	5	5	4-5	4-5
PP	0.5	3	5	5	5	5	5	5	5	5	5	5	5
Brown No.6	5	4	5	4	5	5	5	5	5	5	4	4	4
PP	0.5	3	5	5	5	5	5	5	5	5	5	5	5
Black No.1	5	4	5	4-5	4-5	5	5	5	5	5	5	4	4

Method for the determination of the color fastness of textiles.

1. Light JIS L-1044
2. Burnt gas fade JIS L-1055
3. Washing JIS L-1045 (MC-3)
4. Dry cleaning JIS L-1003 (perchlorethylene)
5. Perspiration AACTT (natural perspiration)
6. Sublimation Scorch tester (at 130°C)
7. Rubbing JIS L-1048 (crock meter)

Technical data of Daiwa Spinning Company Limited.

effective after-scour is necessary. A detergent-alkali sequestering agent scour, at a temperature ranging from 180° to 205°F, is common, but an acidic afterscour has been suggested for enhancing the light fastness of the dyed goods. A typical dyeing procedure is outlined in Table VII.

TABLE VII

Typical Dyeing Procedure for Ni-Modified Polypropylene Fibers (ref.13)

Prescour:		
	Alkali (soda ash or tetrasodium pyrophosphate)	pH 8.0-9.0
	Nonionic or anionic detergent	1.0-2.0%
	Sequestering agent	0.3-1.0%
	Temperature	160-180°F
	Time	10-20 min
	Cooling rinse until clear	
	Dry (if necessary)	220°F (maximum)
Dyeing:		
	Set the dyebath at 80°-100°F with nonionic detergent and EDTA or NTA. Adjust pH 3.5-4.5 with acetic acid, tartaric acid, or citric acid. Add dye, bring to 200°-210°F. Dye 30 minutes. Rinse and dry.	

While the rapid strike rate of the chelate dyes and their poor migration properties are the main problems in piece dyeing of metal-modified polypropylene goods, these properties are a natural advantage in printing. It is, therefore, quite apparent that a chelate dyeing system is destined more for printing than for beck dyeing.

In the early stage of development of dyeable polypropylene fibers, attention was paid mainly to piece dyeing aspect. It is only recently that attention is being diverted to printing and those processes of coloration which use printing principles (e.g., space dyeing). The interest in piece dyeing of metal-modified polypropylene so far may have not been solely based on the obsession of polypropylene fiber manufacturers to imitate nylon in all its aspects. In part it was due to the lack of availability of printing capacity, specially for heavy gauge fabrics such as carpets and upholstery fabrics. The same applies to other processes of dyeing which use printing principle for coloration. As more and more printing and space dyeing capacity is made available, nickel-modified polypropylene receives more and more attention.

The print paste used for the printing of metal-modified fiber is composed of chelating dyes, a thickener, carrier, acid for pH control, sequestrant, and water. A homogenous paste is obtained by preparing the thickener in a conventional manner and adding the carrier, acid, and sequestrant. An antifoam agent may be used if needed. To this paste is added the dye which has previously been dispersed with a small amount of anionic surfactant. The printing operation is conventional and requires only the standard equipment. After printing, the goods are steam aged at 215 - 225°F (ref.14), scoured to remove thickener and other unused chemicals, extracted, and dried. Typical print paste formulations are shown in Table VIII.

TABLE VIII

Typical Printing Formulations for Ni-Modified Polypropylene Fiber

Screen Printing	*Roller Printing*
x% Dyestuff	x% Dyestuff
2% Citric acid	1% Citric acid
0.75% Cibaphasol AS or Progalan X-13	0.75% Cibaphasol AS
30.0% Kelzan (2% Sol.)	20.0% Seatex x-HV (2% Sol.)
15% Polygum 260	--
Balance water	Balance water

As is the case with beck dyeing, fastness properties differ from dye to dye. The same is true with the rate of exhaustion, color buildup, and the brightness of shade. A compromise must be reached, therefore, in the selection of dyes for the preparation of print paste and the achievement of compound shades. A typical combination of dyes with good fastness properties to achieve most of the desired shades is listed in Table IX.

TABLE IX

Suitable Dyes for Printing of Compound Shades on Ni-Modified Polypropylene Fiber

Petramin Yellow FRL	Verona Dyestuffs
Petramin Yellow F3RL	Verona Dyestuffs
Petramine Dark Blue GL	Verona Dyestuffs
Petramine Red BL	Verona Dyestuffs
Polypropylene Green BM	Allied
Polypropylene Yellow GM	Allied
Polypropylene Blue 2BM	Allied

There are four approaches to styling that can be used in achieving different effects in printing with chelating dyes:

1) Pad-dyeing a ground shade on natural fiber and then overprinting,
2) Overprinting of pigmented fiber,
3) 100 percent coverage print on natural fiber,

4) Space-dyeing of natural or lightly pigmented yarn.

IV.2. Disperse-Dyeable Fibers

In view of the comparatively greater affinity of disperse dyes for polypropylene, observed in the early efforts of dyeing polypropylene fiber, it was natural that extensive development work would be conducted to improve this potential. In addition to the early promise for improved dyeability, a disperse-dyeable polypropylene fiber was considered to have certain advantages over mordant-dyeable fiber, described above. For example:

1) A large variety of disperse dyes was readily available, thus there was no need to develop new dyes,

2) Levelling characteristics of disperse dyes were very good, resulting in uniform dyeing,

3) In addition to the common disperse dyes, some other classes of dyestuffs, notably vats and azoics, could be used on disperse-dyeable fibers by slightly modifying the dyeing or printing procedure,

4) Because of their insolubility in water, the wash fastness of disperse dyes was excellent,

5) The low resistance of disperse dyes to organic solvents was not a serious shortcoming for dyed or printed polypropylene goods because most of the fiber was destined to carpeting, which need not be drycleaned.

During the last decade a considerable amount of research and development work has been conducted to produce disperse-dyeable fibers for the specific use of carpeting. The approach, almost universally followed, has been to intermix some appropriate polymeric compound with polypropylene prior to spinning. A variety of polymeric compounds, ranging from cellulosic esters to wholly synthesized polymers, has been investigated (see Bibliography). In fact, it appears that every conceivable polar compound, when mixed with polypropylene and spun, will improve the dyeability of the fiber with disperse dyes. In practice, however, the choice has been considerably limited. This is because of the following factors:

1) Most of the polar polymeric compounds are not compatible with polypropylene, thus rendering the spinning and drawing operations impractical. It has been found that the more polar a compound is, the less compatible it will be with polypropylene. Thus, simple polyamides and polyesters with strong polar groups in their molecular chains are less compatible with polypropylene than polystyrene and copolymers of ethylene.

2) Many of the polar compounds are higher melting, necessitating a high

temperature of spinning which might result in the degradation of polypropylene.

3) Most of the polar compounds are thermally unstable at temperatures needed for satisfactory spinning of polypropylene fibers. Thus, the use of cellulose ethers and several other compounds would be marginal in their utility as dye additives. Several of the polymeric additives, otherwise suitable as dye additives, break down into oligomers, which later tend to exude to the fiber surface and impair the crock fastness of the dyed or printed goods.

4) Many of the polymeric additives, suitable from the processing point of view, are not polar enough to yield wet and drycleaning fast dyeing. This applies, for example, to polyesters, polycarbonates, polyethylenevinylacetate, and ethylene-alkylacrylate copolymers.

5) Most of the additives require incorporation in polypropylene in amounts over 10 percent for satisfactory dyeing. Such high quantities of incompatible or only partially compatible additives severely impair the physical and mechanical properties of the resulting fiber.

In view of these problems, the choice of a suitable additive with all-around adequate properties becomes a practical difficulty. An ideal additive for improving the disperse dyeability of polypropylene fiber is a polymeric compound with polar groups pendent to its chain, amorphous or only slightly crystalline in nature, partially compatible with polypropylene, melting above 140°C (the maximum temperature encountered by polypropylene in drawing, heat setting, and texturizing and having its glass transition temperature below 90°C (temperature of dyeing and printing). Several polymers, such as saturated linear copolyesters, copolyamides, copolymers of vinyl pyridines with styrene, copolymers of ethylene with vinyl acetate, and alkyl acrylate, may approach this kind of ideal additive. A few systems based on some of these additives have reached semicommercial and commercial stages of development in the United States at one time or the other. The current interest is low, however, because of the greater emphasis on the development of acid-dyeable fibers.

Although disperse-dyeable polypropylene fibers are relatively simple to produce and easy to dye in level shades with or without the help of carriers, the fastness properties of dyed goods do not meet all the requirements of the textile trade. This is specially true for their generally poor drycleaning fastness and only fair resistance to light fading. In addition, the shades obtainable with disperse dyes are not particularly bright in comparison with those obtainable with acid and basic dyes. Because of these reasons the development of a general purpose disperse-dyeable polypropylene fiber has been slow, and research efforts have been virtually abandoned. However, in certain uses - for instance carpets, where drycleaning fastness is not an important requirement - disperse-dyeable polypropylene fiber can satisfactorily be used.

Two United States companies introduced disperse-dyeable polypropylene fibers for the carpet trade in the mid-sixties: U.S. Rubber Company presented its Polycrest SDR dyeable BCF yarn, designed to be dyeable with premetallized acid dyes but actually more easily dyeable with common disperse dyes; and Hercules, Inc., introduced its Herculon-2, specifically meant to be a staple fiber dyeable with disperse dyes. Both of these systems are reportedly based on homopolymers of vinyl pyridines or their copolymers with styrene. In Japan, Toyobo Co. developed its disperse-dyeable fiber in the late sixties to a semicommercial stage, mostly in staple form. These fibers are reportedly based on copolymers of alkylacrylates with ethylene.

All these fibers were very dyeable with disperse dyes, displayed the typical fastness properties of disperse dyes, and compared well with nylon. Their acceptance was, however, not instantaneous, and as more and more efforts were diverted to the development of acid-dyeable fibers, the disperse-dyeable fibers were withdrawn from the market. Apart from the customer's traditional resistance to a new fiber, the main factor contributing to the failure of the disperse-dyeable fibers has been the competition with nylon BCF yarn - polypropylene had to compete with no advantage except low price. Once this difference became only marginal, polypropylene lost its main selling point.

The other important factor contributing to the diversion of development attention away from disperse-dyeable polypropylene and focusing it more on acid-dyeable fiber was the commercial necessity of having one dyeing system for both carpet, and for apparel and upholstery end uses. It was evident that the trade would not support two kinds of dyeable fibers - one for carpets and the other for upholstery and apparel.

IV.3. Acid-Dyeable Fibers

In view of the limitations noted for disperse-dyeable fibers, efforts have recently been diverted by major fiber producing companies to develop acid-dyeable polypropylene fibers. Acid-dyeable fibers have the following advantages:

1) The cheapest dyes can be used,
2) A wide range of colors is available,
3) Bright shades are possible,
4) A broad palette of lightfast dyes is readily available,
5) The possibility of union dyeing with wool and nylon exists,
6) Acid-dyeable polypropylene fibers would also be dyeable with disperse dyes in some of the cases.

In order for a fiber to be dyeable with acid dyes, it must contain basic sites which can combine with the acid group in the dye molecules. The task, therefore, is to introduce these groups into polypropylene. Various polymeric compounds containing basic nitrogen groups have been examined and a large number of patents has been issued on this subject (see Bibliography). All conceivable polymeric compounds with basic groups at the chain end or along the molecular chain have been investigated. Examples of such compounds are basic polyamides, polyureas, polyurethane, polyamines, polyimines, condensation products of epichlorohydrin with amines, polyacrylamides, and polyamineacrylates. Polymers with hetrocyclic N-groups along the chain, such as pyridine groups, are another important class of compounds extensively investigated for acid dyeability of polypropylene. Because of the limited compatibility of the above-mentioned additives with polypropylene, various copolymers of basic monomers with unsaturated hydrocarbons have been tried. The most important of this class of additives is the copolymers of ethylene and aminoalkyl-acrylate or the copolymers of vinylpyridine with long chain alkylacrylates.

The anionic dyes are limited in their ability to diffuse into polypropylene. To overcome this difficulty a number of after-treatments of basic nitrogen-containing fibers has been suggested or the employment of auxiliary additives has been recommended. The latter approach has been named in patent literature as "synergistic mixtures" of dye additives. Such multiple additive systems are exemplified by the use of polyvinyl acetate/ethylene copolymers in combination with basic polyamides. The auxiliary nonbasic additive probably serves to open up the fiber structures or to act as "emulsifier" for the additive dispersion in the polypropylene matrix.

Because of their good heat stability, polyamides were probably the first group of polymeric compounds to be recommended for the production of acid-dyeable polypropylene fibers. Polyamides contain amine groups at the end of the chain which can act as dyesite. However, the number of side amino groups thus introduced in the fiber is limited, and the filaments containing these additives are dyeable only to pastel shades. The lowering of molecular weight would increase the number of amino end groups in the polymer but this would also increase its solubility in acidic dyebath. To enhance the basicity of the fiber without decreasing the molecular weight of the additive, the use of basic polyamides has been recommended. These additives are the polycondensation products of dicarboxylic acids and polyamines having two primary and one or more secondary amino groups (such as diethylene triamine and triethylenetetramine). Thus, the amino groups not participating in the amidation reaction, generally the secondary amino groups, provide the strongly basic dyesites for acid dyes.

Among other possibilities are the copolyamides containing tertiary amine or piperazine structures along the chain. These polymers can be prepared by polycondensation of suitable diamines (such as hexamethylene diamine) and dibasic acids (such as adipic acid or sebacic acid) with the addition of suitable quantities of a triamine or a piperazine (such as bis-amino-propylpiperazine). They can also be prepared by reacting diamines with a dibasic ester containing piperazine ring as part of its structure. A number of basic copolyamides have been synthesized and used as effective dye additives for polypropylene by ICI, Montecatini, Toyo Rayon, Nippon, and Hercules.

Closely related to acid dyeable systems based on basic polyamides or copolyamides is the approach based on the utilization of condensation products of epichlorohydrin with various amines. Montecatini has been particularly active in the development of such a system and its acid-dyeable fiber (Mareklon D) may well be based on such additives. The epichlorohydrin basic condensation products face the same processing problems and hold the same kinds of promises as those discussed for basic polyamides.

Polymers and copolymers which contain recurrent pyridine rings in the chain is another group of compounds very extensively researched to impart acid dyeability to polypropylene. The impetus for development work probably came from the success of acid-dyeable polyacrylonitrile fiber produced by copolymerizing a minor quantity of vinyl pyridine with acrylic monomer. The same success has not been realized in the case of polypropylene. The copolymerization of vinyl pyridine monomers with propylene monomers, as mentioned earlier, was not possible and the additive approach did not produce any deep-dyeing fiber. Permeability of the fiber to acid dyes and the incompatibility of these additives with polypropylene appears to be at the heart of the matter. As indicated before, it is not enough to put a dye acceptor in the fiber for successful dyeability. There must, in addition, be a way of getting the dye molecules into the fiber. Vinyl pyridine polymers apparently do not render the modified fiber diffusible to acid dyestuffs. The difference in cohesive energy density of polyvinylpyridine and polypropylene is so large that the former is very incompatible with polypropylene. Thus, the additive polyvinyl pyridine polymers end up in the fiber as discrete foreign particles encapsulated by a matrix of pure polypropylene, instead of being present as interconnected microfibrills which help to open up the fiber structure and provide a path for dye diffusion. Higher amounts of the additive will largely act as filler and impair the physical properties of the fiber. U.S. Rubber (Uniroyal) overcame this problem by after-treating the fiber with prescribed Lewis acids in water solutions. Permeability of the dyes presumably increased by a change in the surface structure of the fiber which made the additive particles available to the dye molecules. Polycrest D, dyeable polypropylene fiber of U.S. Rubber (Uniroyal), was reportedly based on this technology.

Apart from the question of compatibility, heat stability, and cost, the appreciable degree of solubility of vinylpyridine polymers in acid dye bath causes extraction of the additives and poor fastness properties of the dyed goods. To overcome these difficulties, some laboratories - specially Hercules, Uniroyal, and Mitsubishi - have researched the copolymers of alkyl vinylpyridines with other monomers, such as N-substituted acrylamides, styrene, alkyl acrylates. These copolymers are noncrystalline in nature and somewhat more miscible with polypropylene. However, the substantivity of these fibers for acid dyes was rather limited and they proved to be more suitable as disperse-dyeable fibers.

The permeability of acid dyes into polypropylene fibers is much more limited than that of disperse dyes. Two broad approaches are emerging to enhance the diffusibility of acid dyes into fibers along with the provision of dyesites. One approach is seen in a number of patent disclosures utilizing the copolymers of ethylene or long chain alkyl acrylates with strongly polar monomers. The second approach is the utilization of synergistic mixtures of dyesite additives, whereby one component, by virtue of its compatibility with polypropylene, helps to open up the fiber structure for better permeability of the dye, while the second additive, by virtue of its strongly basic character, supplies the actual dyesites for dyeing.

Two promising examples of the first approach to the problem is the recently developed additive system by Mitsubishi based on the copolymers of vinyl pyridine and alkylacrylic esters and the system developed by Toyobo and DuPont based on the copolymers of ethylene with alkylaminoalkylacrylates. Because of their good compatibility, the dye additives of this group can be blended with polypropylene to a greater concentration than those based on basic polyamides, epoxy condensation products, or polyvinylpyridine, mentioned above. The resulting fiber is consequently more permeable to acid dyes. In view of their relatively low number of functional points, such as pyridine or amino nitrogen groups, an addition of at least 10 percent of the additive is generally necessary to achieve adequate depth of shade. Aside from their dyeability with acid dyes, these fibers are also dyeable with common disperse dyes. Although no commercial fiber is yet available based on these systems, Toyobo and Phillips are known to have developed an acid-dyeable polypropylene fiber based on this approach to semi-commercial stage. In the United States, DuPont is known to be actively working on such an additive, presumably in collaboration with Toyobo of Japan.

The use of synergistic mixtures of additives is a comparatively recent approach to solve the problem of introducing dyesites and simultaneously render the fiber structure permeable to water soluble dyes. Several laboratories are known to be working on this approach. One typical example is Canadian Celanese approach, employing a mixture of polyethylene vinylacetate or polyethylene vinylalcohol and basic polyamides (ref.15), or Sumitomo's approach utilizing a mixture of ethylene vinylesters and amino compounds (ref.16). Polyethylene vinylesters of proper ethylene/vinylester ratio are good additives for disperse dyeability and the basic polyamides or amino compounds are good dye-fixing additives for acid dyes. The incorporation of ethylene vinylester copolymers alone would not impart any acid dyeability to the fiber because of the absence of basic dyesites. Similarly, the incorporation of basic nitrogen compounds in small quantities alone would not impart dyeability to the fiber because of the poor diffusion of dyes into the fiber. The larger quantities of the basic additive would be expensive, render the spinning difficult, and impair physical properties of the fiber. However, if small quantities of the basic additive are combined with an appropriate proportion of ethylene vinylester copolymers, and the mixture is used in about 10 percent proportion, the modified fiber is easily dyeable with acid dyestuffs.

In spite of the feverish research activities of several fiber producing companies all over the world, no truly commercial acid-dyeable polypropylene fiber is yet in widespread use in the textile trade. Some acid-dyeable polypropylene fibers have, however, been produced in development quantities and field tested for carpet, upholstery and apparel end uses. A few of them, after showing considerable promise, have, for various reasons, faded away in the development history of dyeable polypropylene, but some have survived to a point that they can be expected to have strong chances for commercialization. Acid-dyeable fibers from Montecatini (Meraklon D), Hercules (Herculon acid dyeable), Beaunit, Phillips, and Toyobo, are the examples of successful dyeable fibers nearing full scale commercialization.

The first truly acid-dyeable polypropylene fiber, Montecatini's Meraklon D, was announced in early 1962 (ref.18) but commercialization could not be realized until recently. While Polycrest® of Uniroyal and other semicommercial fibers introduced during the same period were satisfactorily dyeable only with metallized dyes and only to medium shades, Meraklon D was dyeable with all anionic dyes and to very deep shades. Merkalon D is presently available only in staple form for carpet and knitting yarns. Its dyeing behavior resembles that of wool, one reason the use of Meraklon is increasing in combination with wool. Because of its strong affinity for acid dyes and fast rate of dyeing, the levelling is reported to be a problem in piece dyeing. Typical dyeing procedures for various groups of acid dyes are shown in Table X. Wash and drycleaning fastness of dyed goods is very good and light fastness depends on individual dyes. Although the fiber dyes like wool, the light fastness of various dyes has little in common with that of wool. Table XI lists light fastness data for some anionic dyes on Meraklon D. The gas fading is also reported to be problematic with some dyes. In spite of a few limitations, Meraklon D seems destined to achieve a respectable place among general purpose fibers in the West European carpet trade and

TABLE X

Typical Dyeing Procedures for Meraklon D

1:2 Premetallized dyes:	The bath is set at 40:1 liquor ratio with 2% Albegal CL (ciba) 4% Invadine AR cons. PH adjusted to 4 with acetic acid 80%. Goods enter at 60°C, run for 15 minutes. Add dyestuffs, bring to boil in 30 minutes. Dye 90 minutes, rinse, and aftertreat in 1g/1 ultravow JM. Tow 15 minutes at 60-70°C. Rinse and dry.
1:1 Premetallized dyes:	The bath is set at pH 2 with 2% sulphuric acid 80% and dyed as above.
Acid dyes:	The bath is set at 40:1 ratio with 1.5% acetic acid 80%, 5% Glaubers salt, 8% Invadine AR cons. Goods enter at 40°C, run 10 minutes. Add dyestuffs, bring to boil in 30 minutes. Dye 60 minutes. Add 2% formic acid 85%. Dye 30 minutes. Rinse and dry.
Chrome dyes:	3% dyestuff, o.5% Levegal FTS, 3% formic acid 80%, 90 minutes at 90°C. x% Potassium bichromate (according to recommendation of the dyestuff manufacturer), 2 x% formic acid 80%, 90 minutes at 90°C. Wash with 0.5 g/1 Levapon 150, rinse and dry.
Meraklon D/wool blends:	Pretreat in 1 g/1 Invadine JFC, 1cc/1 formic acid 80%, 30 minutes at 60°C. Rinse. Set bath 4% Invadine AR cons, 5% Glanber's salt, 1.5% acetic acid 80%. Goods enter 40°C, 15 minutes. Add dyestuffs, bring to boil in 30 minutes. Dye 60 minutes. Add 2% formic acid 85%. Dye 30 minutes. Rinse and dry.

knitting industry.

In the United States (ref.19), Hercules has recently announced the commercialization of an acid-dyeable fiber. It is available in staple as well as apparel denier textured yarn. A similar acid-dyeable fiber was announced by Beaunit Corporation. In dyeing properties, these new acid-dyeable fibers very much resemble nylon except that the pH of their dyebath must be maintained at considerably lower level. Table XII details the dyeing procedure. Good light fastness, measured by the Xenon Fade-Ometer is obtained. Fastness in Florida exposures is significantly better. The wash fastness of the scoured goods is acceptable for apparel and home furnishings. Some further improvement in wash fastness with deep shades is obtained by using classical after-treatment method of tannic acid and tartar emetic. Drycleaning fastness is excellent and there is no problem with crock fastness, either wet or dry. Some evidence of gas fading is observed with the blues, and care is warranted in dye selection, especially in light shades involving a blue. No problems are encountered in obtaining

(Continued on p. 494)

TABLE XI

Light Fastness of Some Anionic Dyes on Meraklon D

1:2 Premetallized Dyes	
Dyestuff (3%)	Xenotest
Isolan Yellow GGL extra conc.	6-7
Isolan Yellow GL	6
Isolan Yellow RLS	6
Isolan Orange GGL	5
Isolan Orange GL	5
Isolan Orange RLS	5-6
Isolan Scarlet GLS	5-6
Isolan Red 3GLS	5-6
Isolan Red GLSN	5
Isolan Rubine FB	5
Isolan Bordeaux RLS	5
Isolan Bordeaux BBLS	5-6
Isolan Bordeaux FBS	4-5
Isolan Violet RLS	4-5
Isolan Dark Brown RL	4-5
Isolan Brown GL	4-5
Isolan Brown RLS	4
Isolan Brown BLS	4-5
Isolan Brown TLS	5
Isolan Green FG	4
Isolan Violet BL	5
Isolan Grey BRLS	5
Isolan Blue RLS	4-5
Isolan Blue FBN	4
Isolan Navy Blue GL	4-5
Isolan Navy Blue RRL	5
Isolan Black GL (8%)	6-7
Isolan Black RL (8%)	6-7
Levalan Olive GL	6
Levalan Dark Brown TL	6
Levalan Brown BRL	5-6
Levalan Bordeaux GTL	6
Levalan Navy Blue RL	6
1:1 Premetallized Dyes	
Neolan Yellow BE	7
Neolan Orange GRE	7
Neolan Bordeaux BE	6
Neolan Red GRE	6
Neolan Red REG	5-6
Neolan Violet 5RF	5-6
Neolan Violet 3R	5
Neolan Navy Blue RL	5
Neolan Green BL	5-6
Acid Dyes	
Benzyl Fast Yellow GW	4-5
Benzyl Fast Orange G	5
Tectilon Yellow 2R	5-6
Tectilon Red 3B	4-5

TABLE XI (continued)

Acid Dyes (continued)

Dyestuff (3%)	Xenotest
Tectilon Blue R	4-5
Tectilon Blue 4G	5
Kiton Fast Red BL	5
Kiton Fast Red 4BLN	5
Kiton Fast Blue 2BRL	5
Kiton Fast Blue 4GL	5
Alizarine Fast Violet FRL	5-6
Alizarine Fast Violet 6R	5-6
Alizarine Fast Blue BE	5-6
Alizarine Fast Blue R	5
Alizarine Fast Blue RFE	4-5
Alizarine Fast Green 2GW	6
Alizarine Fast Green G	5-6
Chrome Dyes	
Telon Chrome Yellow T	6
Telon Chrome Yellow R	4-5
Diamond Chrome Yellow BN	6
Telon Chrome Orange G	4-5
Diamond Fast Red 3BG	4-5
Telon Chrome Red C	4
Monochrome Red 5G	4-5
Diamond Chrome Fast Bordeaux BT	5
Telon Chrome Green B	4-5
Telon Chrome Brown GG	4-5
Telon Chrome Brown R	6
Telon Chrome Brown C	6
Diamond Chrome Fast Brown TBL	6
Telon Chrome Grey FR	5
Diamond Blue FBG	4
Diamond Blue BHG	5
Diamond Indigo Blue R	4-5
Diamond Fast Blue BL	6
Diamond Navy Blue RRN	5
Diamond Fast Dark Blue RRL	5-6
Diamond Chrome Brilliant Blue BL	5
Telon Chrome Navy Blue R	5
Diamond Black PVT conc. (4.5%)	5-6

deep dyeings, including blacks. Fastness properties of compound shades are reported to be quite satisfactory as well. Table XIII, for example, depicts the color fastness properties of some single compound shades. The recently commercialized acid-dyeable staple fiber by Phillips and Toyobo's acid-dyeable fiber resemble each other in their dyeing and color fastness properties.

TABLE XII

Dyeing Procedure of Acid-Dyeable Herculon®

Piece Dyeing

1. Scour at 150°F for 15 minutes with 0.5% nonionic surfactant.
2. Rinse well at 100°F.
3. Add 1-2% Alkanol ND* (or comparable product) and 2% Tanapol JN, run at 100°F for 5 minutes.
4. Add dyestuff and run 5 minutes.
5. Add approximately 4% formic acid (90% and run 5 minutes (pH should be 3.0-3.5).
6. Raise temperature to 205°-210°F at 3°F/min. and hold 45-60 minutes.
7. Rinse at 100°F until clear.
8. Scour with water or mild alkali at 150°F for 15 minutes.
9. Rinse at 100°F until clear.

If an aftertreatment of the fabric appears necessary to improve wet and/or perspiration fastness the following procedure may be carried out immediately after the dyebath. No afterscour should be necessary.

1. Set bath at 100°F with 1% tannic acid.
2. Raise to 205°-210°F at 4-5°F/min. and hold for 15 minutes.
3. Add 2% tartar emetic and hold 15 minutes.
4. Cool slowly to 100°F and rinse thoroughly at 100°F.

Treatment with Mesitol NBS can also be used in place of the tannic acid/tartar emetic treatment. It has the advantage of being a one-step process and is cheaper in cost.

Package Dyeing

1. Set machine cycle 4 minutes in-out; 2 minutes out-in.
2. Rinse with water at 100°F for 5 minutes.
3. Scour at 150°F for 15 minutes with 0.5% nonionic surfactant.
4. Cool to 100°F.
5. Add 1-2% owf Alkonol ND (or comparable product) and 2% Tanapol J.N. over one cycle and run one cycle.
6. Add dissolved and strained dyestuffs over one cycle and run one cycle.
7. Add approximately 4% owf formic acid (90%) over one cycle and run one cycle (pH should be 3.0-3.5).
8. Raise temperature to 180°F at 3°F/min. and hold 10 minutes. Then raise temperature to 205-210°F at 1.5°F/min. and hold for 45 minutes.
9. Cool to 175°F slowly.
10. Cool to 100°F and rinse at 100°F until clear.
11. Scour with water or mild alkali at 150°F for 15 minutes.

Printing of commercially available acid-dyeable polypropylene fibers is similar to printing nylon or wool. A typical print formulation for Herculon fiber is given in Table XIV. The same formulation, but at somewhat higher pH range, can be used for Meraklon D. The printing of polypropylene requires special care in the selection of a proper thickener. Most of the thickeners used in textile printing of manmade fibers create a problem on polypropylene, i.e., the thickener does not come off completely in washing.

TABLE XIII

Typical Fastness Properties of Some Selected Dyes on Herculon Acid Dyeable (ref.19)

Dyes	Crock Fastness		Gas (2 cycles)	Xenon	Wash#2	Drycleaning
Dyes (0.5%)	Wet	Dry				
Single Dyes						
Cl Acid Blue 80	5	5	4	5	5	5
Cl Acid Yellow 151	5	5	5	>6	5	5
Cl Acid Red 191	5	5	5	5-6	5	5
Cl Acid Red 337	5	5	5-4	5-6	5	5
Cl Acid Yellow 174	5	5	5	6	5	5
Cl Acid Blue 40	5	5	4-3	5	5	5
Compound Shades						
Gold	5	5	5	>6	5	5
Avocado	5-4	5	4	6	4-3	5
Green	5	5	5-4	5-6	4	5
Red	5-4	5	5	5-6	4	5
Beige	5	5	3	5	5	5
Blue	5	5	4	5	4	5

TABLE XIV

Typical Print Formulation For Herculon Acid Dyeable

1.5% Keltex S (buffered sodium alginate)
5.0% Urea
5.0% $(NH_4)_2SO_4$
5.0 Stearic acid
0.5 Sodium stearate
x% Acid or premetallized dye
3.0% Methyl cellosolve
2.0% Kromfax (thioglycol)

IV.4. Basic Dyeable Fibers

Following the lead of basic dyeable polyester, nylon, and acrylic fibers, attempts were made to modify polypropylene by incorporating into the spinning polymer some anionic additives such as long chain sulphonic esters, polybasic acids, acid anhydrides, ion exchange resins, etc., so that the cationic dyes could be used for dyeing of the fiber. None of these efforts, however, have met any degree of success. This is because most of the acidic additives are sensitive to heat and decompose severely in the process of spinning; most of them are soluble in water and, therefore, easily leach out in the dyebath; and almost all of them react with the stabilizers present in the formulation, leading to severe degradation of polypropylene. In cases where these shortcomings can be controlled to some degree, the wash fastness of the dyed goods is extremely poor. Moreover, it has been observed that the basic dyes, which otherwise show very good light fastness on acrylic, and polyester fibers, fade readily on polypropylene.

Because of the difficulty of finding a suitable additive for basic dyeability and because of the inherently poor color fastness of basic dyes on polypropylene, the development of a successful basic dyeable polypropylene fiber appears for the time being a remote possibility.

IV.5. Mechanism of Dyeing Polypropylene Fibers

The mechanism of dyeing hydrophobic fibers, such as nylon and polyester fibers, is not yet well understood. This applies even more to the dyeing of modified polypropylene. There are basic differences between the dyeing mechanisms for chelating dyes, disperse dyes, and anionic dyes. Moreover, there are additive and structural differences within one class of dyeable fibers. None of the commercial dyeable polypropylene fibers has been available for long enough to allow the accumulation of data about its dyeing behavior. The few scattered studies reported in this connection have been empirical in nature. The following outline summarizes these thoughts.

Shilova, et al (ref.21,22), examined the diffusion of metal-chelating disperse dyes into polypropylene fibers and found the diffusion rate of Disperse Blue N I H dye approximately the same for all fibers containing different amounts of metal compounds, with a value of 2.10^{-9} cm^2/sec. They found that the equilibrium dye concentration increased with increasing metal additive in the fiber, and that the absolute dyeing rate increased more slowly than the concentration of active sites in the fibers, so that the relative rate of dyeing fell. The decrease in the relative rate of dyeing with increase in dye sites was thought to be due to the presence of a saturation front on the fiber surface which effectively acted to slow the relative rate of complex formation with the metal ion.

Although no analytical data are available, practical experience in printing and dyeing shows that the establishment of the saturation front is a function of diffusion rate of the individual dye. This is evidenced by the fact that the saturation front can be pushed deeper into the fiber core as the temperature of dyeing increases and if certain carriers are used. Thus, contrary to the results of Porter (ref.23), the overall rate of dyeing seems to be controlled by the rate of diffusion under practical dyehouse conditions rather than by the rate of complex formation.

It is widely accepted that the dyeing of hydrophobic fibers with disperse dyes occurs through solution mechanism, i.e., the dye forms a "solution" with the polymer in its amorphous regions. The transfer of the dye from the dye bath to the polymer phase is governed by the partition coefficient of the dye. All the fragmentary information available indicates that the same general mechanism of dyeing applies to polypropylene fibers also. The diffusion-controlled rate of dyeing of polypropylene fibers with disperse dyes has been indicated by Katayama, et al (ref.24). Bird and Patel (ref.25) showed that beyond the saturation point of the polymer phase, addition of further dye to the dye bath did not increase the amount of dye on the fiber. An increase in temperature did not change the saturation value.

Polypropylene fiber is more highly crystalline than any other fiber; as little as 30 percent may be accessible to dyes. Saturation values for the fiber are, therefore, very low. The admixing of a suitable dye additive into the polypropylene matrix causes the amorphous portion of the fiber to increase, so that proportionally more "solvent" is available to the dye to form a solid solution. Higher saturation values are thus obtainable.

The additive-polypropylene system can be pictured as a large number of fine, elongated fibrils of the additive embedded in the matrix of polypropylene. These fibrils may or may not be interconnected with each other, depending on the additive concentration, its compatibility with polypropylene, and the spinning conditions. The number of fibrils (and the interphase area) would increase as the compatibility of the additive with polypropylene, the amount of shearing in the fabrication of the fiber, and the relative amount of the additive, increases. The interphase area thus created is important in that it represents high energy surfaces. The dye molecule, in addition to forming the usual solid solution in the additive phase, would be prone to preferentially concentrate on these interphase high energy surfaces and would yield saturation values which will be higher than those justifiable on purely solution mechanism.

The diffusion of dye from the fiber surface to the interphase area or into the body of the additive fibrils may take place through one or both of the following mechanisms:

1) The molecule at the surface of the fiber diffuses through the amorphous

region of polypropylene or through submicroscopic cracks and voids on the fiber surface and gradually gets concentrated on the interphase area or forms a solid solution in the body of the fibrils.

2) The molecules diffuse through a wicking mechanism, the fibril ends on the fiber surface acting as wicks. They penetrate into the body of the fiber along the additive fibrils, polypropylene matrix playing a negligible role.

It appears that both of these mechanisms are at play in the dyeing of modified polypropylene fibers with disperse dyes. The fact that the former mechanism plays a role is evident from the fact that disperse dyes would dye a polypropylene fiber which contains only a small amount of an additive. Some disperse dyes, as mentioned earlier, would dye polypropylene even without any modification of the filament composition. If the dye penetration would occur only through wicking mechanism, then such fibers would not be dyeable at all, or the dyeing would occur only on the surface. The importance of the latter mechanism is indicated by the fact that the rate of dye diffusion increases with the amount of dye additive present in the fiber and the amount of shear to which the polymer mixture is subjected. The extent of the contribution by each of these mechanisms probably depends on the nature of the dye and the additive.

The dyeing of basic modified polypropylene fibers with anionic dyes is an ion exchange reaction. Before such a reaction can take place, the dye must reach the functional group of the additive. However, the partition of the ionized dye between the dyebath and the polymer phase is exclusively in the direction of water. The diffusion of dye into the fiber to reach the dyesite is, therefore, difficult. There are three possibilities for dye molecules to diffuse into the body of the fiber and reach the functional groups of the additive.

1) Pores and cracks occur in polypropylene phase during drawing, crimping, heat setting,scouring, and dyeing; these submicroscopic pores and cracks may allow diffusion of dyes into the fiber.

2) Diffusion may occur along interconnected fibrils through a wicking process.

3) Dye diffuses through polypropylene in the form of a unionized molecule and thus behaves like a disperse dye.

The fact that low temperature drawing of most of the basically modified fibers increases its dyeability and that the dye absorption is maximum where the apparent density of the fiber is minimum, lends credence to the mechanism of diffusion through voids and microcracks. The hypothesis that dye penetration into the fiber occurs mainly along the interconnected additive fibrils also seems plausible. Electron microscopic studies have shown that some of the additive microfibrils are at the surface of the fiber. These microfibrils are highly polar and are readily swollen by water. Diffusion of the dye then occurs through the microfilaments. The crossover points between these microfilaments serve as the path for a complete penetration of the fiber. Several experimental facts point to this mechanism of dyeing.

1) The dye absorption is higher at the ends of the fiber than across the fiber axis.

2) The more compatible a dye additive is with polypropylene, the better dyeing properties it imparts to the fiber for equivalent numbers of dyesites.

3) The dye absorption does not increase linearly with the amount of dye additive present in the polymer blend. At first the dye absorption increases slowly with increasing amounts of the additive. At a certain concentration of the additive, depending on its compatibility with polypropylene and its fiber forming capability, the dye absorption increases abruptly. It appears that at this point the microfibrils begin to interconnect and form a continuous network in the polypropylene matrix. Such a structure is obviously very favorable to wicking of the dye into the body of the fiber.

The penetration of acid dyes in the form of a unionized molecule offers another possibility to explain the dyeing mechanism of acid-dyeable fibers. According to this postulate, the acid dye may behave as disperse dye in the undissociated form and may penetrate the fiber by a polysolution mechanism described previously in connection with disperse dyeability. This mechanism does not need the support of the additive fibrils or interphase boundary area for the penetration of the dye into the interior of the fiber. This aspect of dye diffusion gets support from the fact that organic acids, such as acetic acid, are better absorbed by acid-dyeable polypropylene fibers than are strongly ionized inorganic acids, such as HCl (ref.26). Moreover, it is true with most of the acid-dyeable polypropylene systems that dyeability with acid dyes increases with decreasing pH of the dye bath. It is evident that low pH decreases the ionization of the dye molecule and may force it to dye the fiber through disperse dyeing mechanism. This is specially true with low ionizing acid dyes, such as 1:2 premetallized dyes.

In the absence of specific experimental data, it is difficult to speculate as to which one of the three modes of diffusion of ionic dye into the modified polypropylene fiber is predominant. Probably all of these mechanisms play a role, depending on the dye, the nature of fiber modification, and the pH of the dye bath.

V. CONCLUDING REMARKS

The problem of dyeing polypropylene fiber is still far from solved. The solution, however, seems closer at hand than ever before. After the initial euphoria about the Ni-and Al-dyeable polypropylene fibers in the mid-sixties, the enthusiasm decreased to virtually no interest in the late sixties. As more and more printing, TAK dyeing, and space dyeing capacity is made available in the textile finishing industry, Ni-dyeable fibers are capturing the fancy of major fiber producing companies again. Increasing quantities of Ni-printed and space-dyed polypropylene textile goods are currently being made available to the consumer. This activity is expected to continue and a substantial portion of polypropylene fiber will be consumed as Ni-printed carpets and upholstery goods in the foreseeable future.

The real battle is in the field of acid-dyeable fibers. Meraklon D of Montecatini has already established itself in European markets in the form of staple goods. Substantial quantities of this fiber are presently being used in carpet and knitted goods, both as 100 percent polypropylene and as PP/wool blends. A BCF yarn has not yet made its debut. This development, if successful, will be a very significant event in the history of polypropylene fibers. Phillips and Hercules's acid-dyeable staple fibers and filament yarns are still in the development stage, but judging from their advanced level of mill testing during recent years, it is reasonable to expect that these fibers will play an important role in broadening the use of polypropylene in apparel and upholstery fabrics. The development of an acid-dyeable carpet backing fabric, commercially offered by Thiokol, is an off-shoot of this development. There is no acid-dyeable polypropylene fiber yet available in the Japanese market although a few companies have been testing their products on mill scale in recent years. A winner will eventually emerge but it may take some time before a successful acid-dyeable is commercially available in this market.

There is no doubt that the developmental efforts outlined in this chapter for commercialization of conventionally dyeable polypropylene fibers are bound to be successful. However, the trade has become so familiar with the excellent color fastness properties of pigmented polypropylene fiber that it is not likely to replace pigment-colored fiber by a dyeable fiber in its entirety. There will always be a certain portion of polypropylene fiber that will be pigment colored rather than dyed. In fact, the pigmented fibers offer so many advantages over the dyeable products that a major portion of polypropylene fiber may continue to be colored through pigmentation. The merits of dyeable fiber lies primarily in its potential as apparel fiber and as printed fabrics in upholstery application. Low pile printed carpets is also a large scale potential user of dyeable polypropylene fiber.

APPENDIX I

Identification of Dyeable PP Fibers

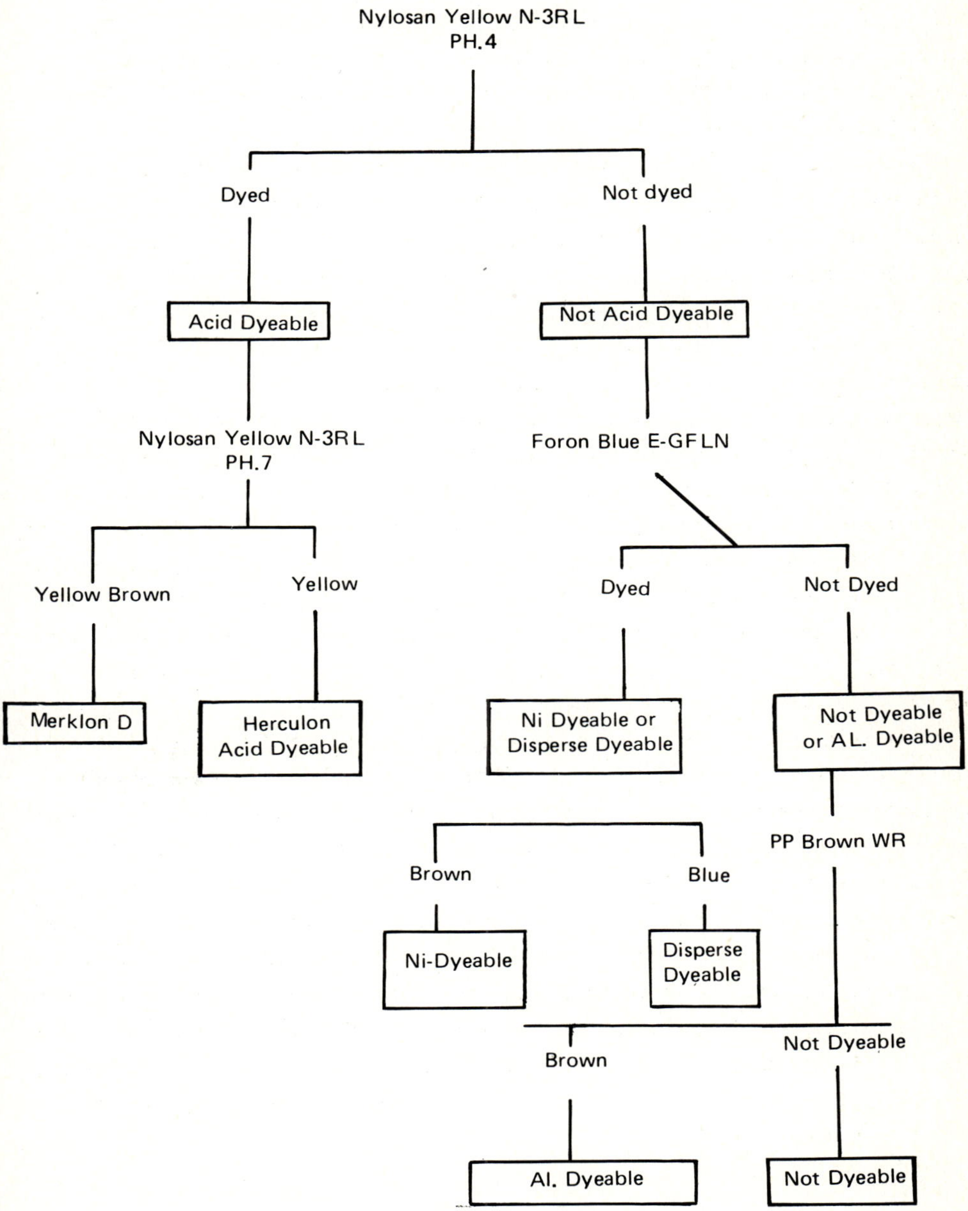

VI. REFERENCES

1. V.L. Erlich, Symp. on Polypropylene Fibers, Southern Res. Inst., Birmingham, Ala., 1964.
2. G.L. Odor, Abhand. Deut. Akad. Wiss. Berlin, Chem. Geol. Biol. 1(1963)427.
3. M. Farber, Symp. on Polypropylene Fibers, Southern Res. Inst., Birmingham, Ala., 1964.
4. H.P. Bauman, Am. Dyes Rep., 52(1963)527.
5. E. Fordemwalt, et al, Am. Dye. Rep., 54(1965)107.
6. A.J. Hall, Text. World, 109(1965).
7. V.L. Erlich, Modern Text., 36(1965).
8. R.R. Haynes, Text. Chem., 28(1970)282.
9. P.L. Scright, Am. Dye. Rep., 53(1964)300.
10. H.P. Bauman, Am. Dye. Rep., 52(1963)527.
11. A.F. Turbak, Text. Res. J., 37(1967)350.
12. N.L. Anderson, et al, Am. Dye. Rep., 52(1963)48.
13. F. Fordemwalt, et al, Am. Dye. Rep., 54(1965)107.
14. R.R. Haynes, et al, Text. Chem., 3(1969)70.
15. Canadian Patent 890471, Can. Celanese, 1972.
16. British Patent 1,071,367, Summitomo, 1971.
17. R.G. Curtis, et al, Am. Dye. Rep., 53(1964)380.
18. Chem. Eng. News, 29(1962)24.
19. M. Levine and R.P. Weimer, Text. Chem. Col., 1970.
20. ECAFE Seminar on Development Prospects of Man-Made Fibers in Asia and Far East, Tokyo, 1966.
21. G.I. Shilova, et al, Zh. Prikl. Khim, 44(1971)592.
22. G.I. Shilova, et al, Zh. Prikl. Khim., 44(4)(1971)592.
23. R.A. Porter, Master's Thesis at Georgia Tech., Atlanta, Ga., 1967.
24. A. Katayama, et al, J. Soc. Text. Cell. Ind., Japan, 19(1963)828.
25. C.L. Bird and A.M. Patel, J.S.D.C., 84(1968)560.
26. J. Cegarra and P. Peunte, Text. Res. J., 39(1969)611.

VII. BIBLIOGRAPHY

VII.1. Journal Literature

1. E.M. Eddington, A Status Report on Dyeing of Polypropylene, Am. Dyestuff Reptr., 52(1963)31.
2. Piece-Dyeable Polypropylene Fiber Goes Commercial, Chem. Eng., 71(1964)54-56.
3. M. Farber, Theoretical Aspects of Polypropylene Dyeing, Proc. Symp. PP Fibers, Birmingham, Ala., (1964)100-120.
4. V.L. Erlich, Development of Commercial Polypropylene Fiber. II. The Dyeing Problem and Efforts to Solve it, Mod. Text. Mag., (1965)46-78.
5. D.D. Gagliardi, Chemical Modification of Polypropylene, Am. Dyestuff Reptr., 54(1965)41-45.
6. L. Alexandru, et al, Synthetic Fibers From Mixtures of Polypropylene With Other Polymers, Faserforsch. u. Textiltech., 16(1965)433-438.
7. B. Beghelli, Dyeable Polypropylene Fiber, Rayonne et Fibres Synth., 21(1965)287-292.
8. B.C.M. Dorset, Dyeing Polypropylene: Some Recent Developments, Text. Mfr., 91(1965)334-338.
9. J.J. Press, New Dyeable Polypropylene Fibers, Mod. Text. Mag., 46(1965)68-76.
10. M. Farber, Latest Developments in Polypropylene, Am. Dyestuff Rept., 55(1966)536-541.
11. P. Arnaud, Dyeing of Polypropylene Fibers, Rev. Belge Mat. Plast., 9(1968)535-547.
12. M. Farber, Polymeric Dye Receptors for Polypropylene Fiber, SPEJ., 24(1968)82-87.
13. J. Cegarra, Reactivity of Meraklon DR With Acids, Textile Research J., 39(1969)611-617.
14. M. Levine, R. Weimer, Acid Dyeable Polypropylene, Text. Chem. Colorist, 2(1970)269-272.
15. V. Filbert, et al, Improving the Dyeability of Polypropylene Fiber by Modification With Metal Compounds., Fibre Chem., 3(1971)541-542.
16. R.R. Haynes, Printing Nickel-Modified Polypropylene for Indoor Use, Text. Chem., 2(1970)279-282.
17. Acid-Dyeable Polypropylene, Text. Inds., 139(1975)75.

VII.2. U.S. Patents

3,037,835	Montecatini	1962
3,073,667	Montecatini	1963
3,090,769	Eastman	1963
3,093,615	Montecatini	1963
3,098,697	Montecatini	1963
3,099,515	ICI	1963
3,102,869	Eastman	1963
3,112,159	Montecatini	1963
3,112,981	ACNA	1963
3,115,478	Montecatini	1963
3,116,966	Montecatini	1964
3,131,990	Montecatini	1964
3,148,936	Esso Res.	1964

U.S. Patents (continued)

3,151,185	Montecatini	1964
3,151,928	Montecatini	1964
3,156,943	Eastman	1964
3,156,745	Hoechst	1964
3,163,492	Hercules	1964
3,170,004	DuPont	1965
3,172,723	Celanese	1965
3,192,007	Esso Res.	1965
3,205,156	Toyo Rayon	1965
3,215,487	Montecatini	1965
3,228,744	Am. Cyanamide	1966
3,231,530	Monsanto	1966
3,236,918	Asahi Kasei	1966
3,238,163	ICI	1966
3,240,552	Eastman	1966
3,274,312	Montecatini	1966
3,248,378	Hoechst	1966
3,284,428	Esso Res.	1966
3,287,082	Eastman	1966
3,296,239	Esso Res.	1967
3,299,030	Esso Res.	1967
3,299,185	Ube-Nitto	1967
3,315,014	Uniroyal	1967
3,316,328	Iowa	1967
3,317,633	J.P. Stevens	1967
3,320,226	Montecatini	1967
3,322,704	Shell	1967
3,326,826	Esso Res.	1967
3,326,847	Mobil Oil	1967
3,328,484	Soc. Rhodiaceta	1967
3,332,732	Esso Res.	1967
3,332,906	Ethylene-Plastique	1967
3,332,927	Cabot Corp.	1967
3,337,651	J. J. Press	1967
3,337,652	J. J. Press	1967
3,355,416	Mobile Corp.	1967
3,366,710	J. J. Press	1968
3,371,982	Hercules	1968
3,375,213	J. J. Press	1968
3,388,190	Union Carbide	1968
3,397,032	ICI	1968
3,402,222	Union Carbide	1968
3,410,799	Esso Res.	1968
3,413,076	Nippon Rayon	1968
3,419,638	Beaunit	1968
3,423,481	Tukuyama Soda	1969
3,431,322	Eastman	1969
3,433,853	Hercules	1969
3,433,573	Monsanto	1969
3,439,066	Uniroyal	1969
3,453,248	Dow Corning	1969
3,468,864	Continental Can	1969
3,487,453	Phillips	1969
3,501,371	Montecatini	1970
3,507,605	Phillips Petroleum	1970
3,510,437	Montecatini	1970
3,523,315	Asahi Kasei	1970

U.S. Patents (continued)

3,530,201	Uniroyal	1970
3,551,521	Hercules	1970
3,558,579	Celanese	1971
3,585,256	Uniroyal	1971
3,607,815	Esso Res.	1971
3,621,075	Phillips	1971
3,632,678	Dart Ind.	1972
3,637,601	Phillips	1972
3,639,505	Phillips	1972
3,639,528	Esso Res.	1972
3,642,955	Ciba-Geigy	1972
3,652,734	Uniroyal	1972
3,658,937	Montecatini	1972
3,668,280	Montecatini	1972
3,703,392	Celanese	1972
3,707,351	Am. Cyanamide	1972
3,742,104	Celanese	1973
3,752,868	Chisso	1973
3,765,829	Ethylene Plastique	1973
3,819,758	Phillips	1974
3,822,144	Phillips	1974
3,842,147	Mitsui Petroleum	1974
3,873,646	Lubrizol	1975
3,907,935	Montecatini	1975
3,928,269	Phillips Petroleum	1975
3,928,504	Shell Oil	1975

VII.3. British Patents

866,920	Montecatini	1961
873,830	Montecatini	1961
876,045	ICI	1961
879,198	Montecatini	1961
881,981	Montecatini	1961
884,665	Montecatini	1961
886,701	ICI	1962
889,534	Continental Can	1962
892,379	ICI	1962
893,603	Montecatini	1962
893,604	Montecatini	1962
983,605	Montecatini	1962
894,178	Montecatini	1962
904,998	Montecatini	1962
976,320	Hercules	1962
919,863	ICI	1963
921,059	Courtauld	1963
921,663	ICI	1963
922,471	ICI	1963
923,407	Montecatini	1963
931,646	Can. Celanese	1963
931,889	Montecatini	1963
932,897	DuPont	1963
934,667	Montecatini	1963
935,125	ICI	1963
938,041	Chisso	1963

British Patents (continued)

940,952	Montecatini	1963
941,646	Montecatini	1963
942,131	Montecatini	1963
950,852	Montecatini	1964
960,789	Montecatini	1964
968,387	Hoechst	1964
973,821	Celanese	1964
975,918	Toyo Rayon	1964
979,980	Union Carbide	1965
980,971	Kurashiki Rayon	1965
981,584	PCMPC	1965
983,434	DuPont	1965
982,953	Montecatini	1965
985,143	Hoechst	1965
985,937	Nippon Rayon	1965
987,421	J. J. Press	1965
990,406	Asahi Chem. Ind.	1965
992,011	Montecatini	1965
992,012	Montecatini	1965
995,802	Union Carbide	1965
999,721	Eastman	1965
1,003,005	ICI	1965
1,004,478	Montecatini	1965
1,005,725	BASF	1965
1,006,850	Montecatini	1965
1,006,581	Montecatini	1965
1,007,337	Eastman	1965
1,007,753	Eastman	1965
1,009,907	Montecatini	1965
1,009,616	Montecatini	1965
1,017,454	Montecatini	1965
1,025,474	F. Geleji	1966
1,030,554	Nippon Rayon	1966
1,031,852	Montecatini	1966
1,053,806	Dow Chem.	1967
1,057,812	F. Geleji	1967
1,070,944	Bayer	1967
1,078,167	Montecatini	1967
1,079,225	Montecatini	1967
1,085,327	Mitsubishi Rayon	1967
1,095,116	Hoechst	1967
1,101,609	Dynamite Nobel	1968
1,102,215	Montecatini	1968
1,104,059	Montecatini	1968
1,111,210	Chemcell	1968
1,118,685	Uniroyal	1968
1,132,645	Tokuyama Soda	1968
1,134,121	ICI	1968
1,143,944	ICI	1969
1,145,689	Asahi Kasei	1969
1,148,649	Asahi Kasei	1969
1,155,474	Asahi Kasei	1969
1,157,500	Montecatini	1969
1,162,465	Geigy	1969
1,163,266	ICI	1969
1,178,996	Montecatini	1970
1,181,727	ICI	1970

British Patents (continued)

1,183,369	Mitsubishi Rayon	1970
1,197,277	Vicker-Zimmer	1970
1,219,277	Vicker-Zimmer	1970
1,219,639	Montecatini	1971
1,226,049	Chisso	1971
1,252,672	Montecatini	1971
1,320,172	Montecatini	1973
1,355,837	Montecatini	1974
1,362,703	Montecatini	1974
1,374,265	Shell Internat.	1974
1,387,597	Snia Viscosa	1975
1,393,918	Ciba-Geigy	1975
1,394,459	Phillips Petroleum	1975
1,478,047	Shell Internat.	1977
1,451,607	Shell International	1978
1,523,206	Text. Inst. France	1978

Chapter XI CONJUGATE FIBERS

I. INTRODUCTION

The crimping of polypropylene fibers - as well as other melt-spun fibers - is mainly done by thermomechanical means (Chapter IX). These methods have in common the mechanical deformation of straight filament into crimped form, followed by a heat setting of the deformed configuration. The essential feature of thermomechanical texturizing is that the new, crimped configuration produced mechanically is made permanent by thermal relaxation of the stresses produced in forced configuration. For effective thermomechanical texturizing, this stress relaxation must occur within the time scale of commercial production rate. For polypropylene, the stress-relaxation spectrum is quite large; the crimp-setting process is, therefore, quite slow. Moreover, the glass-transition point of polypropylene is below ambient temperature; the setting of crimped configuration is, therefore, not as permanent as, for example, for nylon or polyester, whose glass-transition temperatures are considerably above the temperature of normal use.

As an alternative to these traditional thermomechanical techniques, a method is available for producing fibers which possess crimp as an integral part of their structure, somewhat analogous to that of wool. Wool fiber contains two components in a side-by-side relationship. These components differ chemically and are produced at different growth rates so that, in the relaxed state, the two sides are of different length. The resultant fiber looks microscopically like an imperfect coiled spring and responds mechanically like such a structure.

In synthetic fibers it is not possible to grow the two components at different rates but the differential in length can be accomplished by various means, including different shrinkage characteristics of the two components. The fiber, on acquiring the differential lengths in its two components, will crimp. The crimp produced by this technique is somewhat analogous to that of wool. Polypropylene is specially amenable to this treatment among the melt-spun synthetic fibers. The production of "self-crimping" polypropylene fiber is therefore of great commercial importance.

The fibers of two components with different shrinkage potentials in side-by-side or asymmetrical sheath-core relationship are known as *bicomponent* or *conjugate* fibers. The latter term is favored because of the possible confusion of *bicomponent* with *biconstituent* fibers, which are the fibers of two different

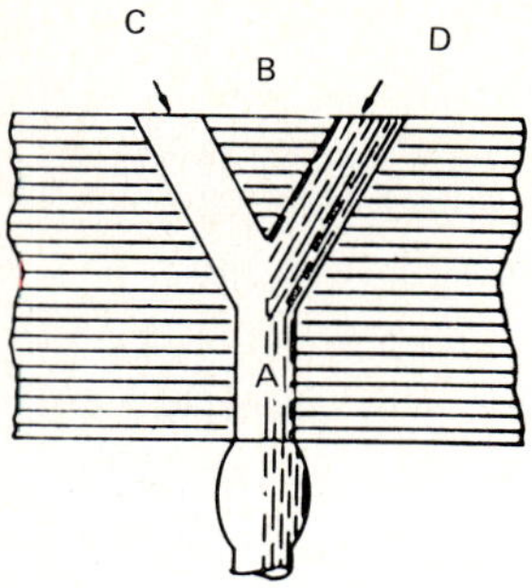

Fig. 1. Schematics of Classical Arrangement for Conjugate Spinning

polymers intimately mixed with each other. The use of the term "bicomponent" fiber should further be discouraged because it tacitly implies that the fiber is made of two distinct components, which might not always be the case with self-crimping polypropylene fibers. The use of the term "conjugate fiber", on the other hand, focusses attention on structural aspects rather than on chemical composition.

The first distinctly conjugate synthetic fibers were prepared by Sisson and co-workers (ref.1) from viscose solution. Their discovery of the new type of fiber was an outgrowth from studies of amorphous "skin" and crystalline "core" structure in viscose. These suggested spinning a fiber with two different parts arranged side by side in one continuous coherent fiber. A system is described for such fiber, as illustrated in Fig. 1. The different solutions in channels C and D are kept separated by a septum B, until they are allowed to merge immediately above the spinnerette hole A. The two components thus emerge together in side-by-side relationship. The Sisson discoveries led to a generic patent (ref.2). Fibers with quite different chemical compositions in the two components were included as well as those differing mainly in physical conditions. These discoveries were quickly followed by Nicoll of DuPont (ref.3) and others and extended to melt-spinnable polymers. Polypropylene proved specially amenable to this technology. In fact, a self-crimping yarn could be produced without going through the pains of conventional conjugate spinning as described by Sisson and afterwards by others.

During the early days of the development of polypropylene fibers, it was noticed that polypropylene fiber spun at lower temperatures than usual developed a helical crimp on being cold-drawn to comparatively low draw ratios and then heat-relaxed under low tension. The same effect was encountered when high-viscosity polymers were spun under high threadline tension. It seemed probable that the cause of the crimp development was some asymmetry introduced in the spinning stage of the process, and this was confirmed. Patents were issued as early as 1960 (ref.4) describing the product and outlining the process to manufacture such self-crimping fibers of polypropylene. Hercules, Canadian Celanese, and ICI did extensive work in this field and Canadian Celanese was in commercial production of self-crimping carpet yarns of polypropylene in 1962.

Most of the early efforts for the production of self-crimping polypropylene fibers were based on a combination of material characteristics and spinning and quenching conditions that involuntarily produced a weak conjugate structure in polypropylene fiber in the same way as certain coagulating conditions brought about a differential thickness in the skin of the viscose fiber at the hand of Merion and Sisson (ref.5). As the technology of conjugate spinning for nylon developed in DuPont's laboratories and elsewhere, it opened up a way of enhancing the conjugate structure of polypropylene fibers by employing two different states of polymer melt and by voluntarily adjusting their specific proportions in the fiber. Numerous patents were issued which described the production of polypropylene fibers, one side of which was different in some way from the other side, using more or less the original Sisson's approach. This unleashed a steady stream of patent disclosures encompassing all the possible methods of producing structural and material differences between the two sides of the conjugate fiber and describing several mechanical ways to put the two desired structures in side-by-side or asymmetric sheath=core configuration. Some of these methods will be discussed here. For further details, a scrutiny of the selected patents listed in the Bibliography of this chapter should be useful.

II. MECHANICS OF CONJUGATE FIBERS

The demonstration of the two-component nature of wool and the advent of manmade conjugate fibers led to several investigations into the mechanics of crimp-related properties, e.g., the geometry of fiber cross-section and the distribution of components, and created substantial interest in the theoretical aspects of crimp formation. All these analyses are the adaptation of the analogy with bimetallic theory of Timoshenko (ref.6), which considers the factors controlling bending of a bimetallic strip of uniform width and thickness along its length as it is subjected to changes in temperature. The differential in coefficient of expansion of the two metals provides the force of curvature of the bimetallic strip.

In analogy with bimetallic element, the shrinkage differential of the two components provides the necessary force of curvature in a conjugate fiber. Thus, two primary requirements must be fulfilled to achieve self-crimp in conjugate fibers. Firstly, the components must be arranged in an acentric or asymmetric distribution in the fiber cross section, i.e., either the components should be arranged side-by-side or in sheath-core configuration where the core is displaced from the center of the fiber cross section. Secondly, the components should be capable of shrinking unequally during heating or other aftertreatments.

The crimp frequency C_f (defined as the number of helical turns per length of fiber) is given (ref.7) by

$$C_f = k.C_p.\Delta a/h$$

where k is a constant (slightly dependent on the ratio of the moduli of the two components), Δa is the differential shrinkage of the two components, h is the average fiber thickness, and C_p is the crimp potential function reflecting the relative amounts of the two components and their distribution in the fiber cross-section.

Curvature of the crimp can, therefore, be predicted from considerations of the shrinkage and bending properties and the relative amounts and geometry of the components in the fiber. For a given polymer pair the amount of crimp can be varied by varying the amount -nd geometrical distribution of the two components.

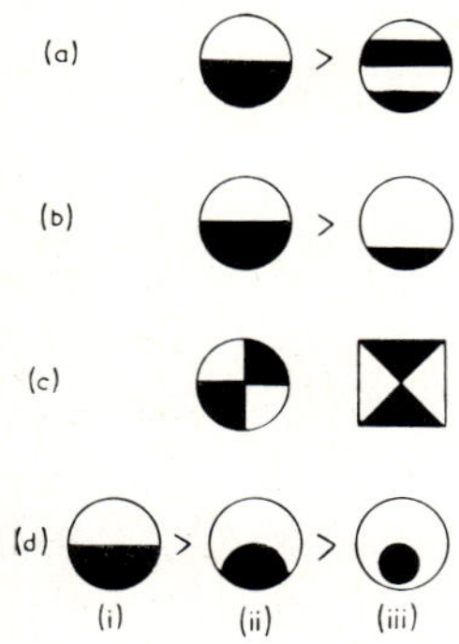

Fig. 2. A Qualitative Indication of the Dependence of Crimp Potential on Component Distribution and Interphase Geometry (ref.9)

Crimp potential function, C_p, has been vigorously analyzed for the case of round cross sectional fiber (see Bibliography). This analysis leads to certain general conclusions applicable to all cross sections. These conclusions, some of which are intuitively obvious, can be summarized as follows (ref.8):

The maximum crimp potential is found in fibers with only one interface (Fig. 2a).

The crimp potential is at maximum when the interface passes through the center of area of the cross-section (Fig. 2b). The crimp potential is zero for any distribution with a center of symmetry (Fig. 2c).

There are, obviously, other considerations. For example, the interface should be straight or nearly so for maximum crimp. Also, a fiber with cross-section of Fig. 2d(i) will have a greater crimping potential than the one shown in Fig. 2d(ii) which in turn will have a greater crimping potential than that shown in Fig. 2d(iii).

In the important special case when the cross-section is equally divided between the two components and the interface is straight and it passes through the center of cross-sectional area, the crimp potential term, C_p, becomes unity and the curvature R is given by:

$$\frac{1}{R} = \frac{K}{h} \cdot \frac{\Delta L}{L}$$

where K is a constant, h is the fiber thickness, L is the extended length of the fiber, and ΔL is the contraction on free relaxation of the coil.

It can be shown that for an elliptical cross-section the constant K takes the value 1.7 and for a rectangular cross-section K = 1.5 (h and R being measured in the same units). The cross-section of many fibers can be approximated to one or the other of these shapes and an estimate of fiber curvature close enough for most practical purposes can be calculated from the expression:

$$\frac{1}{R} \simeq 1.6 \frac{\Delta L}{Lh}$$

When the line of junction of the components is straight but does not pass through the area center, the curvature is reduced in proportion to a crimp contraction term according to Fitzgerald & Knudsen (ref.7).

It has already been shown that the curvature developed in a fiber is proportional to the differential shrinkage between the components during relaxation from a drawn-oriented state. The more highly shrunk component takes the inside of the helix. The shrinkage of the components will depend upon the physical nature of the components (relative degree of crystallinity and type of morphology) or chemical differences in the component (homopolymer/copolymer, polypropylene/other polymer, polypropylene/polypropylene + additive, polypropylene/polypropylene alloys, etc.), upon the degree of stretch imposed on the fiber in spinning and drawing, and on the conditions of fiber relaxation.

Polymers differing either in chemical composition or structural state would be expected to orient and shrink to different extents when subjected to a particular stretching and relaxation cycle, and therefore demonstrate self-crimping behavior when combined into a side-by-side or assymetric sheath-core fiber.

Calculation indicates (ref.9) that for a good crimp the two components should differ in length by 10 to 20 percent after shrinkage. A limit in crimp level is rapidly reached, however, beyond which a further increase in differential shrinkage does not contribute to the increase in crimp level. Theoretically this limiting point is reached when the crimp curvature becomes about twice the filament size. In practical terms, however, an increase in crimp level becomes insignificant as the crimp curvature attains a value of about 4 to 5.

For the development of latent crimp, the polymer chains in the fiber have to be freed to some extent and allowed to rearrange in a more relaxed configuration. Polypropylene chain molecules are capable of relaxation, albeit at a very slow rate, even at room temperature. Thus, a drawn conjugate polypropylene fiber develops a significant degree of crimp when allowed to relax at ordinary atmospheric conditions. The process of relaxation is considerably accelerated by the application of heat or by the treatment with hydrocarbon liquids. Often the rate of crimp development is quite high: the crimp development is, therefore, spontaneous.

Bulkiness and stretch are inherent properties of a crimped fiber. The bulkiness may be defined as its space-filling ability. As this property depends to a very large degree on factors outside the fiber itself, no unique expression to describe this property can be given. In a yarn of helically crimped fibers the bulkiness will clearly tend to be reduced by any interpenetration and interlocking of adjacent helical coils. Interpenetration depends to a large extent on the crimp frequency of the fibers in a yarn and the number of helix reversals occurring in the fibers. Obviously, it is easy for two helical fibers of the same pitch and same rotational direction to interpenetrate. Reversal points, however, tend to reduce interpenetration.

Where maximum bulkiness is required it is advisable to intermingle fibers with a wide range of crimp frequencies, as helical coils widely differing in pitch do not interpenetrate readily. Conjugate fibers, as they are produced industrially, do have different frequencies of crimp and different pitch because of unequal shrinkage differential along the length of the individual fiber and because of the structural differences from filament to filament. Several patents have been obtained which seek to increase this inhomogeneity in crimp frequency and pitch in order to enhance the bulk of self-crimping fiber assemblies.

The stretch or extension-relaxation properties of self-crimped conjugate fibers is usually analyzed in terms of a helical-coil model. By definition, the stretch, S, of such a model is given by:

$$S = \frac{l_f - l_i}{l_i}$$

where l_i is the initial length of the coil as measured along the axis of the cylinder enclosing the coil and l_f the maximum extended length equal to n/C_f. Here n is the total number of helical turns, and l_i is equal to nP. With these substitutions, the stretch equation is:

$$S = (1/C_f - P)/P$$

Here C_f is crimp frequency and P is the helix period. This analysis is, however, not based on reality, as is shown below.

If a uniform bicomponent fiber were allowed to develop its crimp in a completely free condition, it would curl into a completely flat helix. The radius of the helix would be the radius of crimp curvature of the fiber (R) and the number of turns in the helix would be equal to:

length of fiber/$2\pi R$

In practice, conjugate fibers do not develop their crimp in a completely unrestrained condition but in a condition where the ends of the fibers are usually prevented from rotating and where the fibers are, to a greater or lesser extent, held in a state of longitudinal restraint. A fiber developing crimp under these conditions will ideally take up a helical form but it will be extended into a helix of greater or lesser pitch depending on the particular restraints. As only latent bending forces are present in the fiber before crimp development, no real twist can exist in the crimp-developed fiber. The fiber will thus be divided into sections of alternate clockwise and counter-clockwise helix rotation. The frequency of reversals in the fiber will depend on the frequency of rotational restraints along its legth with, in general, at least one reversal between each pair of restraints.

Fig. 3. The Helical Crimp Configuration in a Conjugate Fiber with Several Reversal Points

Thus, a conjugate fiber developing crimp under practical conditions (i.e., in the presence of some longitudinal restraints and with the fiber ends usually prevented from rotating) will be divided into sections of alternate clockwise and anticlockwise helix rotation (Fig. 3). The points of reversal rotate about the filament axis as the filament is extended or relaxed, thus effectively increasing or decreasing the number of helical turns. The crimp frequency of a fiber in a yarn, therefore, depends upon three factors:

(a) the potential crimpability of the individual fiber

(b) the conditions of restraint during development of the crimp, and
(c) the extension subsequently imposed on the yarn.

In view of these complexities, no simple theory of springlike behavior can be employed to explain the extension-relaxation properties of bicomponent fibers.

Self-bulking conjugate fibers differ in tactile properties as well as in extension-relaxation behavior from other types of crimped fibers. The zig-zag crimp of the stuffer box (Chapter IX) process gives a product of relatively low contractive power and therefore a yarn of soft bulky character without a great deal of stretch, a product quite different from crimped conjugate yarn. The same is true with gear-crimped and knit-deknit yarns.

The crimp in these yarns is rather passive, just like stuffer-box crimp, in comparison with the lively crimp of conjugate yarns. Edge-crimp yarns, on the other hand, are closely comparable with the crimped conjugate yarns because their crimp is due to a structural asymmetry and is helical in character. False-twist yarns and conjugate fibers both have a basically helical structure. However, whereas the equilibrium state of conjugates is a flat helix of fixed radius and individual filaments are torque-free, false-twist filaments have an equilibrium state which is an untwisted helix varying in radius, and the filaments are torque-lively. As a result of this, the reactive power of false-twist yarn would in general be greater than bicomponent yarn of the same filament denier. The frequent reversals and variations of helix radius in false-twist yarn tend to discourage interpenetration of the helical coils, and little yarn bulkiness is lost on this account.

III. STRUCTURAL BASIS OF POLYPROPYLENE SELF-CRIMPING FIBERS

As emphasized earlier, the phenomenon of self-crimping in conjugate fibers of polypropylene is the result of structural asymmetry across the filament cross section induced in spinning and enhanced in subsequent processing. This asymmetric structure of the fiber makes the component parts respond differently to extensional deformation and thermal shrinkage, thereby imparting a coiled configuration to the entire fiber. The structural asymmetry of polypropylene fiber can have its origin in orientation and crystallinity differential, which can be augmented through material differences of the two components.

When viewed in the interference microscope, polypropylene spun fibers with conjugate structure show a distorted fringe pattern which suggests a region of lower birefringence on one side than the other. The different drawing behavior of polypropylene fibers of different molecular orientation has been discussed in Chapter VII. On cold drawing, higher stresses are accumulated on the higher birefringence side than on the lower birefringence side. On relaxing, the highly stressed side would recover closer to its original length than the less stressed material. To relieve this length differential, the fiber twists into a helix with reversal points: the crimp frequency and crimp amplitude depending on the geometric factors and the contractional force differential, as discussed earlier.

Although birefringence asymmetry is a necessary condition for self-crimping polypropylene fibers, it is not a sufficient condition for potential crimpability. This is obviously due to the other structural and material factors which may contradict or reinforce the birefringence influence on the drawing and relaxing behavior of the two conjugate components. One of these basic factors is the crystallinity, including not only the degree of crystallinity but also the form of crystallites, the size of crystallites, crystalline orientation, and the degree of cross orientation (row nucleated crystallinity) in the two sides.

The overall crystalline orientation of *as-spun* self-crimping polypropylene fibers is higher than that of the normally spun fibers and differs appreciably from that of precursors of "high tenacity" yarns (Chapter IX). The spun fibers show a pronounced degree of "cross-orientation", described in Chapter IX. The presence of this "cross orientation" in at least one component is a necessary, although not a sufficient, condition for self-crimping. As this crystalline structure is favored by high stresses in attenuating polymer jet during spinning, a stress differential in the two components will produce the asymmetric crystalline distribution in the fiber cross section. The required stress differential can be produced through a differential in molecular weight, molecular-weight distribution, temperature, rate of cooling, etc.

In addition to the differential in degree of crystallinity and cross orientation, structural differences between the conjugate components may arise due to the size of crystallites and morphological details. Sometimes these structural differences are so pronounced that they can be observed through an ordinary phase-contrast microscope (Fig. 4).

The major objective of all exercises in the spinning of polypropylene conjugate fibers is the production of a fiber structure that is conjugate in crystal structure or conjugate in molecular orientation, or both. In practice it is impossible to separate these two parameters and most often a structure which is conjugate in crystal structure is also conjugate in molecular orientation.

Fig. 4. Conjugate Structure of Undrawn Polypropylene Filaments as Seen Through a Phase-Contrast Microscope

IV. FORMATION OF CONJUGATE STRUCTURE IN POLYPROPYLENE FIBERS

Numerous methods are feasible for creating structural and property differential across the filament cross section. Patent literature is replete with examples wherein practically every conceivable material and process parameter affecting the structure of filament has been tried to achieve a differential in deformability, stress concentration, and shrinkage. Some of these possibilities with relevant examples will be briefly discussed in the following pages. For detailed information, the related patents listed in the Bibliography should be consulted.

IV.1. Differential Cooling of Individual Filaments

Conjugate structure is known to form in regenerated cellulosic fibers when coagulated in strong solutions in such a way that one side of the fiber coagulates earlier than the other. The same phenomenon is observed in polypropylene when the molten spun fiber is quenched in such a way that one side solidifies more rapidly than the other (ref.10). In fact, polypropylene structure is probably the most sensitive to the cooling conditions among the common fiber-forming materials. This is due to the fact that the rate of cooling of molten polypropylene determines whether the crystalline structure of the solid mass would be predominantly smectic or predominantly monoclinic (Chapter I). As discussed in Chapter V, the rate of cooling also determines the molecular orientation, the size of the crystallites, and the degree of cross orientation.

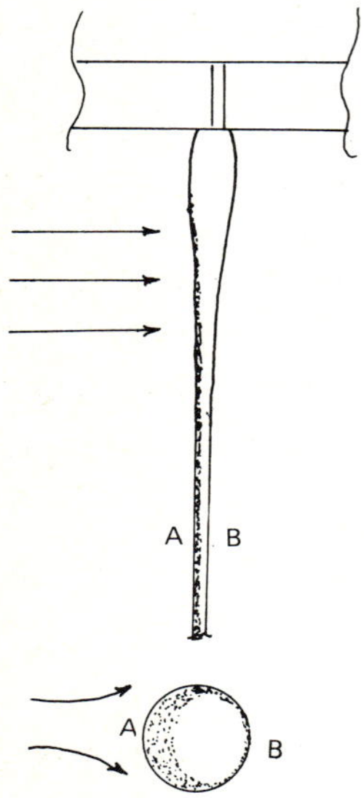

Fig. 5. Unidirectional Cooling of Polypropylene Filament for Conjugate Effect

Imagine a polymer jet being attenuated and simultaneously quenched by a unidirectional blast of cooling air, as depicted in Fig. 5. A skin-core structure of the filament is being formed, as is normal in filament spinning. However, in this particular case, the skin structure is being frozen earlier on side A than on side B. At some stage of filament solidification, side A becomes difficult to attenuate while side B can still yield under spinning tension. As a result, the spinning stress is concentrated more on side A than on side B. The side A acquires higher orientation, smaller crystallite size, and more cross orientation. Thus, a filament is formed which has a differential skin structure with respect to its orientation and crystallinity. This conjugate structure is further enhanced in cold drawing.

IV.2. Differential in Flow Birefringence

It has been mentioned in Chapter V that when a polymer melt flows through an analus, it develops a preferential molecular orientation along the direction of the flow, referred to as flow orientation or flow birefringence. This property can be used advantageously in the production of conjugate structure in the fiber. The flow orientation depends on the velocity of flow and the constriction of the flow channel, molecular weight, molecular-weight distribution, and the temperature of the melt. All of these factors have been used in the production of conjugate polymer streams of different flow birefringence (ref.11). To preserve the differential in flow birefringence of the filament, it has to be quenched and solidified as rapidly as it emerges from the spinnerette hole. The birefringence differential produced in shear flow is, however, far less pronounced than that produced in extensional flow. This is because of the presence of the die swell area just below the spinnerette hole and the practical difficulties in freezing the originally obtained flow birefringence in the fiber. The practical utility of flow birefringence differential in producing conjugate fibers of polypropylene is, therefore, limited.

IV.3. Differential in Melt Temperature

An important route to creation of a structural differential in conjugate fibers is to spin two polypropylene melt streams in side-by-side relationship, each stream being different in temperature. The difference in temperature causes the two sides to cool at different rates which are reflected in differences of crystallinity and molecular orientation (ref.12). The differential in melt temperature can be produced by utilizing two separate extruders for melting the polymer, or by subdividing one single stream into two, one stream being passed through a path at higher temperature than the other. The temperature differential alone is often not enough to yield a commercially feasible process except in those cases where a polymer of high viscosity is being used or where the filaments of low denier are being produced. Most often the differential in temperature is viewed and utilized as reinforcing or secondary effect in conjunction with other conjugate effects produced through differential cooling, utilization of asymmetric fiber cross section, differential heat degradation, etc. Sometimes a differential in temperature is a product of necessity. For example, in the conjugate spinning of two polymers widely different in melt viscosity, the utilization of higher temperature will be necessitated for the polymer stream of higher viscosity.

IV.4. Asymmetric Filament Cross Section

A conjugate structure in polypropylene fiber can also be produced by employing a bilobal cross section of the fiber in such a way that the fiber cross section is symmetrical only in one plane. Several different geometrical configurations have been proposed but probably the most promising cross section seems to be that resembling a figure eight or modifications thereof (ref.13), as depicted in Fig. 6. It is evident that under normal conditions of quenching, the smaller lobe of the filament will cool much faster than the larger lobe and in effect the same kind of structure differential is developed as that created by unidirectional quenching. On relaxation of asymmetrical cross section fibers, the smaller lobe side is expected to be inside the coil.

Fig. 6. Some Examples of Asymmetric Cross Section With Self-Crimping Characteristics

The asymmetric cross section can be profitably used in the production of self-crimping fibers utilizing high-viscosity polymers. However, with low-viscosity polymers, the structural effect produced by asymmetric cross section is generally not enough to produce a commercially viable process. In such cases, other reinforcing effects, such as differential cooling, differential melt temperature, differential in melt viscosities, or chemical effects, are required (ref.14).

IV.5. Differential Degradation and Heat Stabilization

As has been discussed in Chapter II, the changes in molecular weight and molecular-weight distribution resulting from thermal degradation affect the flow properties of the melt, the drawing behavior of spun yarns, and the response of drawn yarns to heat treatment. This factor has been very ingeniously employed in the production of polypropylene conjugate fiber, one side of which consists of more degraded polymer than the other side. The differences in mechanical properties of the two components and the differential response to stress concentration in drawing would impart to the fiber a self-crimping characteristic. Starting with a broad molecular-weight distribution of intrinsic viscosity $[\eta]$, and degrading it to a polymer of intrinsic viscosity $[\eta']$, the degree of depolymerization or degradation can be defined as $[\eta]/[\eta']$. A value of $[\eta]/[\eta']$ between 1.5 to 2.0 is generally adequate for the degraded component to attain a good self-crimping polypropylene fiber (ref.15).

The differential in heat degradation can be achieved by utilizing two separate melters, one operated at higher temperatures, higher back pressure, or even utilizing two different kinds of screws, or degrading with the help of prodegradants such as benzoyl peroxide or dicumyl peroxide (ref.16). The same function can also be achieved by stabilizing the two components to different degrees. These methods of creating two melt streams of differential degradation are in effect equivalent to utilizing two polymers of different molecular weights and molecular-weight distribution, as discussed in the next section.

Utilizing one single extruder, streams of different degrees of degradation can be created by dividing a high molecular-weight polypropylene melt stream into two sections and subjecting them to different heat histories before merging them into a conjugate filament (ref.17). The differential heat histories can be achieved by making one stream flow to shorter distance than the other or subjecting one stream to higher temperature path than the other, or both. The melt stream which stays longer in the system or flows through a path of higher temperature will degrade more than the other.

Another way to cause differential degradation without the necessity of even splitting the main polymer stream is based on the fact that polypropylene melt is a poor conductor of heat. If the polymer is made to flow in lamellar fashion through a heated pipe, the outer portion of the melt stream would be subjected to higher temperature and longer residence times than the inner core. This would tend to degrade the outer portion of the polymer stream more than the inner portion. The differentially degraded sheath-core polymer stream, thus obtained, can be directly spun into conjugate filaments through the lamellar spinning technique discussed in Section V.1., or the inner core may be separated from the sheath and spun through the conventional two-component systems.

In all these methods, the following relevant points are pertinent and worth noticing:

(1) A high molecular-weight polymer should be used as the starting raw material. This is because high molecular-weight polymer is more sensitive to thermal degradation than the low molecular-weight polymers.

(2) A polymer of broad molecular-weight distribution is preferable as starting material. As degradation generally tends to narrow the molecular-weight distribution of polypropylene, the differentially degraded components of the melt would not only differ in molecular weights but also in their polydispersities.
(3) The stabilization system should be specifically developed for the process. Some of the antioxidants are more sensitive to temperature than the other. A guideline principle in the choice of the processing stabilizer system should be that little degradation occur at normal processing conditions but a drastic effect be realized at higher temperatures or at longer residence times.

IV.6. Differential in MW or MW Distribution

Creating a differential of molecular weight and molecular-weight distribution between the two conjugate components through differential thermal degradation or differential stabilization is rather an uncontrolled process. The differential is heavily dependent on the quality of the raw material, needs very closely controlled operating conditions, and is based on a delicately balanced stabilization system. A theoretically simpler and technically sound principle for the attainment of a conjugate structure based on molecular parameters of the two components is the employment of two separate sources of polymer melts, differing in molecular weight or molecular-weight distribution, or both. In using two separate polymers, three possibilities are at hand:
(1) The two components differ in molecular weight but essentially have a similar molecular-weight distribution (for example, Profax 6301 and Profax 6501).
(2) The two polymers are essentially of the same weight-average molecular weight, M_w, but they differ in polydispersity (Profax 6301 versus polymer obtained from Profax 6801 or 6501 degraded to an equivalent melt-flow level).
(3) The two components differ both in molecular weight and in polydispersity (a pair of randomly chosen polymers taken from two different producers).

The use of two polymers of different average molecular weight in side-by-side configuration was probably the first to attract attention for the manufacture of conjugate fibers. Patent applications were filed as early as 1958 in Britain, the United States, and several other countries (ref.18).

The difference between the melt viscosities of the two components is manifested in the fiber structure in several different ways. The high molecular-weight portion develops a higher degree of spin birefringence, it accumulates more stresses in deformation, and subsequently it shrinks more on heat relaxation than the lower viscosity component.

In working with polymers of two divergent melt viscosities, a bending phenomenon of unsolidified filament immediately below the spinnerette hole is noticed. When the bending tendency of the fluid jet is very pronounced, the fluid jet may "knee" to one side to such an extent that it touches the hot spinnerette surface, ceases to flow as a filament, and forms instead a "drip". In such cases, successful spinning operation cannot be effected. This difficulty can be alleviated by utilizing special spinnerette construction where the area around the spinnerette hole is recessed (ref.19), keeping the spacing between the spinnerette hole large and spinning speed low. A convenient method to avoid kneeing of the filaments is to utilize different melt temperatures on the two polymer systems in such a way that the melt viscosities of the two components more or less become equivalent (ref.20). The method has, however, its own limitation as a drastic adjustment of melt temperatures to balance the melt viscosity of the two components may create difficulties in bulk development.

The conjugate fibers obtained by side-by-side spinning of two polymers widely differing in melt viscosities are characterized by a curved boundary between the two components. This is because the low I.V. polymer is more fluid than the higher I.V. polymer and tends to encircle the latter in a crescent-shaped cross section. This tendency of encircling the higher molecular-weight component by the low molecular component can be suppressed to some extent by spinning the former at lower temperature than the latter so that the melt viscosity of the two components is in the same range.

The structural differences similar to those mentioned above can also be produced by using two polymers of approximately the same molecular weight but differing in molecular-weight distribution (ref.21). The basis of structural differences between the two conjugate components can be traced back to the differences in melt-flow behavior of the two polymers, the mode of crystallization, and cold deformation. The polymers of different polydispersity are available commercially.

If the polymers of required specifications are not readily available they can be produced from high molecular-weight polymers of broader polydispersity through thermal degradation. To aid such a process a prodegradant such as benzoyle peroxide or dicummyl peroxide can be profitably utilized. Of course, the differential fiber structure can be obtained by using two polymers which differ not only in their average molecular weights but also in their molecular-weight distributions. This situation is probably nearest to practice and the easiest to manage. In such cases, the degraded or narrow molecular component is generally of lower molecular weight than the undegraded or broad molecular-weight distribution polymer.

The following factors may be noted about the differential in molecular weight and molecular-weight distribution and its effect on crimp development.

(1) The larger the difference of $|\eta|$ between the components, the higher is the

crimp potential, as shown below:

$[\eta]_1/[\eta]_2$	M_w/M_n	Crimp Contraction %
2.05/1.20	7-10	180%
2.05/1.70	7-10	105%
2.05/1.50	3-5	60%
1.80/1.50	3-5	20%

(2) The higher the combined molecular weight of the fiber, the higher the crimp potential:

$[\eta]_1/[\eta]_2$	M_w/M_n	Crimp Contraction %
2.05/1.70	7-10	105%
1.50/1.20	7-10	70%

(3) The higher the polydispersity of the components, the higher the crimp potential:

$[\eta]_1/[\eta]_2$	M_w/M_n	Crimp Contraction %
2.05/1.20	7-10	180%
2.05/1.20	3-5	60%

(4) In contrast to the poor or virtually no-crimp potential of conjugate fiber composed of components of equivalent $|\eta|$ and equivalent polydispersity, good crimping yarn results from components of equivalent $|\eta|$ but different polydispersity:

$[\eta]_1/[\eta]_2$	M_w/M_n	Crimp Contraction %
1.85/1.85	3-5/3-5	no crimp
1.80/1.85	7-10/3-5	150%

(5) If a difference in molecular weight as well as in polydispersity is involved in the two components of the conjugate structure, a combination of higher molecular weight plus broader polydispersity and lower molecular weight plus narrow polydispersity would yield a better crimping yarn than vice versa:

$[\eta]_1/[\eta]_2$	M_wM_n	Crimp Contraction %
1.50/1.20	7-10/3-5	210%
1.20/1.50	7-10/3-5	80%

IV.7. Conjugate Structure Developed by Chemical Means

Differential shrinkage behavior of conjugate components of the fiber can also be achieved by inducing differential crystallization behavior in the filament by chemical means, increasing the melt viscosity of one stream by partial cross-linking in the process of extrusion or by admixing a plasticiser or viscosity depressant in one component, or even admixing a second polymeric material with polypropylene in one stream.

A differential in crystallinity can be brought about by the addition of a nucleating agent in one stream of the polymer (ref.22). The conjugate side, which contains a nucleating agent dispersed in the polymer, would crystallize prior to the other side which does not contain any added nucleating agent. The higher rate of nucleation is reflected in the smaller average size of the crystallites and, in turn, in the higher shrinkage of drawn yarns. Sodium benzoate or benzoic acid is a useful nucleating agent for polypropylene, and many others have been mentioned in the patent literature (ref.23). Normally the differential in crystallization is not enough as to form a base for a satisfactory commercial process of conjugate fiber production; other differential treatments, such as differential cooling, differential in viscosity, etc., are combined to enhance the effect.

Another example of achieving conjugate structure through chemical means is the use of bridging agents in one of the polymer streams (ref.24). By bridging is meant a slight cross-linking of the polymer. Bridging agents for polypropylene are polyfunctional azido compounds exemplified by sulphonazides and azido formates of dimer acids. A slight cross-linking of polypropylene molecules induced by bridging agents increases the effective melt viscosity of the polymer and leads to higher flow birefringence of the corresponding component in the fiber. On relaxation the bridged side would have considerably higher shrinkage potential than the other side.

A third possibility of chemically modifying the shrinkage behavior of one component of the conjugate fiber is to admix a plasticizer or an internal lubricant in one of the conjugate streams (ref.25). The presence of any slipping, lubricating, or plasticizing agent in one side of the fiber alters the flow behavior of the corresponding component and changes its deformation behavior in cold drawing. The plasticized component will normally draw without chalking and without accumulation of high internal stresses. A number of suitable compounds are commercially available ranging from paraffin waxes, oligomers, soaps, long chain fatty amides, esters of long chain fatty alcohols, to silicone greases and plain heavy oils.

The deformation characteristics and shrinkage behavior of polypropylene can also be changed by incorporating in it another thermoplastic in 10 to 30 percent proportion. The use of polyisobutylene (ref.26) and atactic polypropylene and propylene/ethylene copolymers (ref.27) have especially been mentioned. Conjugate fibers can be prepared wherein one side consists of polypropylene alone and the other side consists of a mixture of polypropylene with additive polymer in a way as described by ICI for nylon/(nylon-PET) conjugate fibers (ref.28). Some of these additive polymers, such as copolymers of ethylene-vinylacetate or amino-alkylated ethylene and acrylates, may function not only as modifying agents for mechanical properties of the polymer but also as dye-site additives. Thus, a conjugate fiber can be prepared, one side of which is dyeable and considerably different in shrinkage potential from the other side.

Still another ingenious way of modifying the deformation and heat shrinkage properties of polypropylene through chemical means is by "stereoplasticization" (ref.29). Highly isotactic polypropylene can be easily converted into a more flexible and substantially elastic *randiotactic* polymer by mixing in it small amounts of a suitable organic peroxide and some bromine-phospherous compounds and extruding under normal conditions of temperature and throughput. For example, polymers of varied physical properties, particularly with respect to deformation behavior and thermal shrinkage characteristics, can be prepared by utilizing 0.01 to 0.05 percent di-cup, 1 to 5 percent tris (2.3 - di-bromo-propyl) phosphate and 0.2 to 1.0 percent stearic acid and extruding at 230° to 280° C. The resulting polymers consist of blocks of isotactic and atactic chain segments arranged in random order. They are characterized by somewhat lower isotacticity, lower melt temperatures, low crystallinity, high thermal shrinkage, and higher elasticity, depending on the amount and relative proportions of the modifying chemicals. A conjugate fiber consisting of isotactic polypropylene on one side and randiotactic polypropylene on the other side would deform irreversibly on cold drawing in the isotactic component but would stretch predominantly reversibly on the *randiotactic* side. On relaxation, the fibers would curl spontaneously.

IV.8. Differential in Polymer Structure

A convenient and obvious way to achieve a differential in mechanical properties and shrinkage characteristics of a conjugate structure is to use two chemically different polymers in conjugate spinning. One such possibility with polypropylene is the use of polyethylene as the other conjugate component. Such conjugate fibers have a tendency to split at the interphase boundary and thereby loose bulk in use. The use of copolymers of polypropylene with minor amounts of ethylene is more suitable, both from the point of view of bulk retention and mechanical properties of the fiber. Copolymers (ref.30) as well as block polymers (ref.31) are suitable.

The copolymers of propylene which are most suitable for conjugate spinning are those which have 2 to 25 percent ethylene copolymerized, either randomly or in blocks, with them. Block polymers are more suitable because of their superior mechanical properties and their high shrinkage potential on heat relaxing. Several commercially available "high impact" polypropylene resins are, in fact, copolymers of propylene and ethylene and can be used for conjugate spinning. The guiding principle in the choice of a suitable copolymer should be that the melt viscosity is approximately the same as that of polypropylene, used as the other component of the conjugate fiber.

V. SPINNING OF CONJUGATE FIBERS

It should be evident from the foregoing discussion that the mechanics of spinning conjugate fibers from polypropylene will vary from product to product and will depend on the principle employed for the formation of conjugate structure. A large number of patents has been issued in the United States, Western Europe, and Japan describing the spinning of conjugate fibers. Some of them are especially meant for polypropylene while the others are general. Most of these methods can be discussed under two headings:

(1) Production of self-crimping yarns through conventional spinning equipment;

(2) Production of self-crimping fibers through bicomponent spinning.

A brief survey of each technique follows.

V.1. Self-Crimping Through Conventional Spinning

Early successes in the production of self-crimping polypropylene yarns were largely the results of capitalization upon the incidental discovery that high molecular-weight polypropylene, when spun conventionally under certain conditions of melt temperature, spinning speed, draw-down and quenching conditions, yields a yarn with substantial self-crimping potential. This capability of self-crimping could be amplified by aging and cold drawing. The unique combinations of material and process parameters were latter analyzed in terms of their respective contributions to the formation of asymmetric or conjugate fiber structure. A cursory look at patent literature, along with an understanding of the role of molecular and process parameters in fiber structure formation and the mode of deformation of different structures on cold drawing, would show that successful processes for the production of self-crimping polypropylene fibers can be based on one or more of the following effects:

(1) Asymmetric quenching to create and arrest conjugate structure in the fiber structure;

(2) Radial differential degradation of the polymer stream and its spinning through a circular spinnerette so that a portion of the more degraded material is included in each filament, thus creating a base for asymmetric structure formation.

(3) Use of asymmetric spinnerette hole cross section, thus creating differential streaming orientation and differential cooling.

Polypropylene fibers with helical crimp are indeed available which have been produced by a combination of more than one of the above effects. The asymmetry in fiber structure produced through asymmetric cooling, differential degradation, temperature differential, or asymmetric cross section, can be easily realized by using polymers of high viscosity and broader molecular-weight distribution and spinning at relatively low temperature at maximum speed. All these conditions produce polypropylene fibers of high birefringence and pronounced cross orientation. Thus, self-crimping polypropylene fibers have been produced in such a way that average birefringence in the range of 15 to 20 x 10^{-3} develops (ref.32). The effects of polymer viscosity, spinning temperature, spinning speed, and quenching rate on yarn birefringence is interrelated and, consequently, a number of suitable combinations is possible which would yield a fiber of high and asymmetric birefringence and potential self-crimping ability. This is probably the reason for the proliferation of patent disclosures describing the production of self-

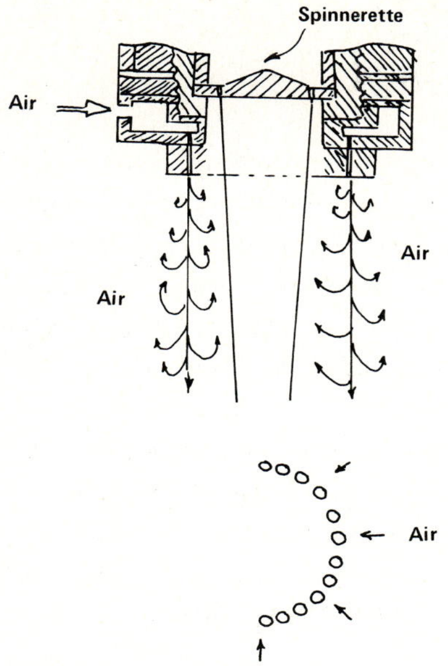

Fig. 7. Asymmetric Quenching With Circular Pattern of Holes (ref.68)

crimping yarns with the help of conventional spinning arrangements (ref.33).

One obvious method of asymmetric quenching is with the help of a cold blast of air impinged perpendicular to the direction of filament attenuation (ref.34). This method presupposes the use of a rectangular type of spinnerette with one straight line of holes. Few such spinnerettes are used in practice because of the very small number of holes which can be accommodated in one line. A better way of asymmetric cooling is through a circular cooling ring which either impinges the cold air perpendicular to the threadline or forms a circular envelope of cold air on the outside of the filament bundle emerging from a round spinnerette, as shown in Fig. 7.

Other means of asymmetric cooling have also been mentioned in the patent literature. For example, a parallel filament bundle is brought in contact with a solid surface, continuously chilled by circulation of cold fluid within its body, as near to the spinnerette as possible (ref.35). Filaments can also be asymmetrically quenched by a film of liquid so thin that only one side of each filament contacts the film of liquid (ref.36). The film is formed on the surface of a perforated hollow body filled with a cold liquid, which penetrates through the pores. Solid and liquid contact surfaces are more efficient in asymmetric cooling of the filament which results in a sharper boundary between the rapidly quenched portion and that of the delayed cooled portion than that produced by air quenching. But the methods suffer from the disadvantage that the cooling cannot be directed to the filament side of one's choice. Thus, it is not possible to combine the asymmetric cooling with asymmetric cross-section effect.

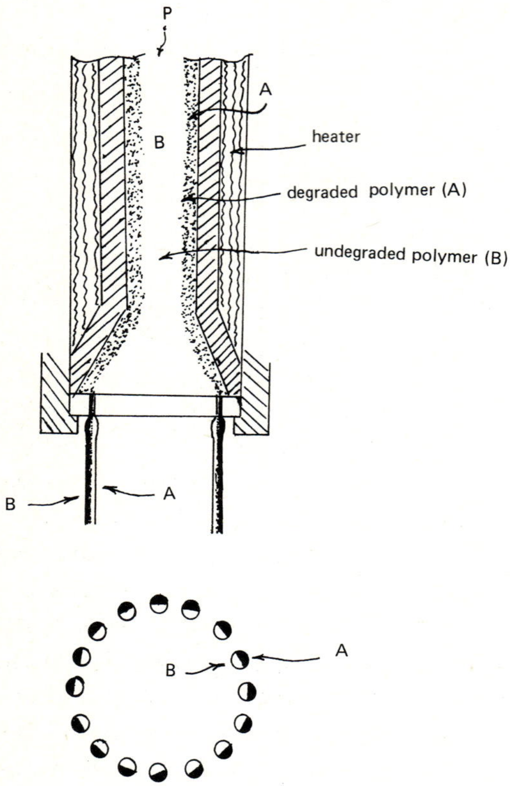

Fig. 8. Differential Degradation of Polymer Stream P into More Degraded Component A and Less Degraded Component B

The conjugate fiber structure produced by asymmetric quenching can be enhanced by an asymmetry in filament temperature or degree of polymer degradation. This can be conveniently done in conventional spinning arrangement by passing the polymer stream emerging from the metering pump through an annular conduit maintained at relatively high temperature and spinning it through a circular spinnerette with one circle of holes, as depicted in Fig. 8. Because of the poor heat conductivity of the molten polymer and because of the laminar flow of the melt, the outer layer of the melt stream would acquire higher temperature than the inner core and, if sufficient time is allowed, the outer layer would degrade a little more than the inner core. On spinning through a circular spinnerette of one single row of holes, each filament will emerge as composed of one part of the high temperature, more degraded sheath and one part of the low temperature, less degraded core material of the main polymer stream. This principle of conjugate spinning is further illustrated in Section V.2.2. The effects of asymmetric quenching or asymmetric degradation can further be enhanced by the use of an asymmetric fiber cross section as described in Section IV.4. Care is taken, however, that the small lobe is always on the outer periphery of the spinning threadline or pointing towards the oncoming air stream.

V.2. Bicomponent Spinning

The process of bicomponent spinning for the production of conjugate fibers is basically the question of forming the component streams and joining them in such a way that each fiber is composed of two components joined in an asymmetric arrangement. The multitude of patent disclosures and process details develop, however, because of the different ways of producing the component streams and different ways of joining them in the conjugate structure. For example, component streams differing in every conceivable way in material parameters can be formed using two separate melting devices, or two component streams can be realized from one melting device by subdividing the main stream into two partial streams and treating them differently along their way to the common spinnerette before merging. Similarly, component streams can be merged into conjugate fibers in side-by-side arrangement or in asymmetric sheath-core relationship. Moreover, the component streams may be allowed to join with each other right before the spinnerette hole or they may be merged well ahead of the spinnerette. Further differences in merger techniques emerge from the fact that the conjugate fiber can constitute different proportions of the components. In this section, these aspects of process variations will be mentioned in brief: for details the respective patent disclosures referred to in the text and listed at the end of the chapter should be consulted. The choice of an appropriate technique for the formation of component streams, merging methods for creating conjugate structure, the precision and uniformity of the components within the filament and between the filaments, the choice of the material parameters, processing conditions, and equipment design is not only a factor of the product quality and intended use, but also of the production economics and capital outlay on the machinery.

V.2.1. Formation of Component Streams

The formation of component streams using two separate melters (extruders) is simple and needs no detailed discussion. The polymers, differing in molecular weight, molecular-weight distribution, tacticity, additive content, copolymerization, etc., are melted in separate extruders and pumped through separate metering pumps into the bicomponent spinning manifold. The method is quite universal in that it affords a wide variety of polymers to be handled and gives complete control in tailoring component streams used for any desired effect in the conjugate fiber. However, it necessitates the installation, operation, and maintenance of two extruders instead of one needed for conventional fiber spinning.

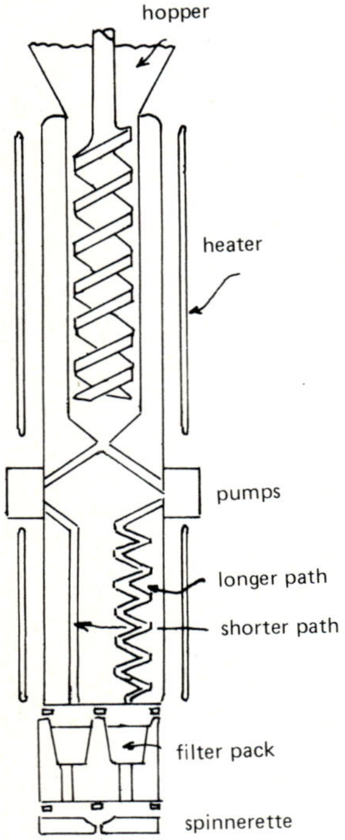

Fig. 9. Formation of Two Component Streams from a Single Melter (ref.50)

Polypropylene is unique among the common fiber-forming polymers in that by treating the polymer in molten state to different thermal conditions, widely divergent properties of the products can be achieved. This sensitivity of polypropylene melt to processing conditions makes it feasible to create two or more component streams of different physical and mechanical properties from one single melt stream emerging from the extruder. Thus, the main melt stream can be divided into two substreams, each flowing through two paths differing in temperature, length, and cross-sectional area, as shown in Fig. 9, creating a differential in molecular weight and molecular-weight distribution between the two substreams; one consisting of high molecular weight and broader molecular distribution and the other consisting of lower molecular weight and narrow distribution. The two streams can, in addition, differ from each other in melt temperature. The extent of material and state differential so created between the two substreams would depend on four main factors:

(1) The molecular weight and polydispersity of the original polymer (Fig.10).

(2) Extent of process stabilization (Fig. 11).

(3) Residence time differential (Fig. 12).

(4) Temperature differential (Fig. 13).

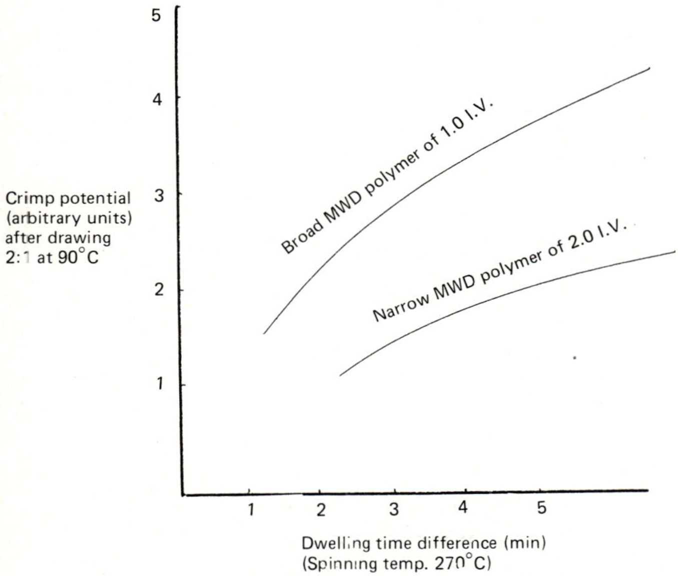

Fig. 10. Effect of Polydispersity of the Original Polymer and the Differential in Residence Time on Crimp Potential of Drawn Yarn

The hydrogen-stopped virgin polymers (unpellitized) are especially suitable for this treatment. The viscosity of the polymer is so chosen that enough differential in melt viscosity between the two substreams is created and the combined polymer viscosity is such that the mass is easily spinnable. The degree of thermal stabilization of polymer is important. If the polymer is too well stabilized, no substantial degradation of the polymer would take place. On the other hand, if the polymer is too poorly stabilized, both streams would be susceptible to easy degradation. In both cases, therefore, differential in molecular weight and molecular-weight distribution would not be obtained. Given the proper starting material with adequate stabilization, the differential degradation of the two

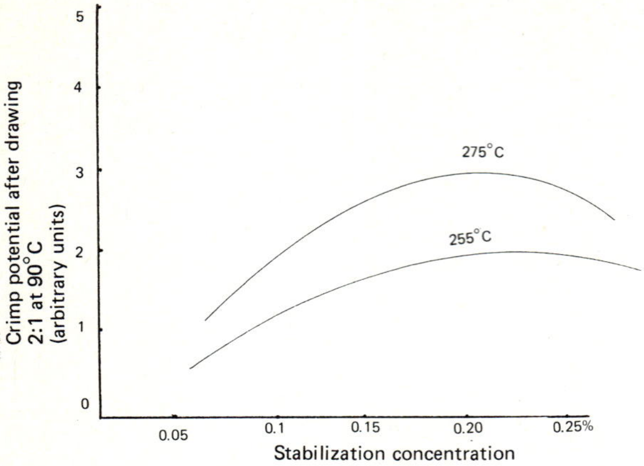

Fig. 11. Effect of Thermal Stabilization on Crimp Potential

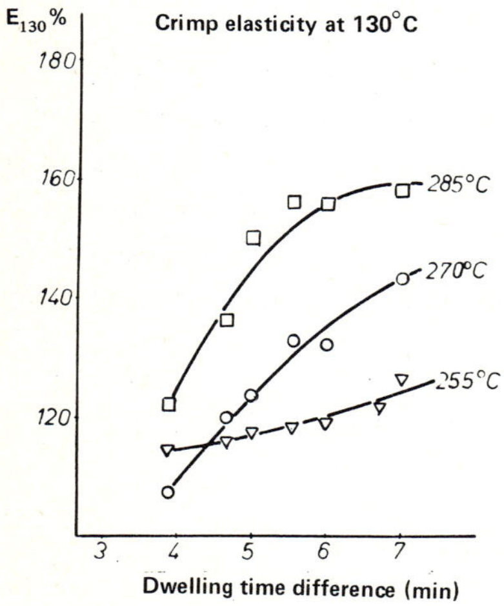

Fig. 12. Effect of Spinning Temperature and Dwelling Time Difference on the Crimp Development of Polypropylene Fibers Produced by bifurcating a single Polymer Stream into Two Flows of Unequal Passage Length and Then Combining Them in Side-by-Side Configuration in the Filament.

substreams would be dependent on the temperature and residence times in the respective flow paths.

Once the two component streams, obtained either through separate melters or through subdividing one main stream into two, are on hand, they can be branched to multiple channels and fed to the spin manifold. Spin manifolds used to handle conjugate spinning accommodate two separate sets of melt streams, making them somewhat more complicated than those designed for monofilament spinning. The basic principles of branching are, however, the same.

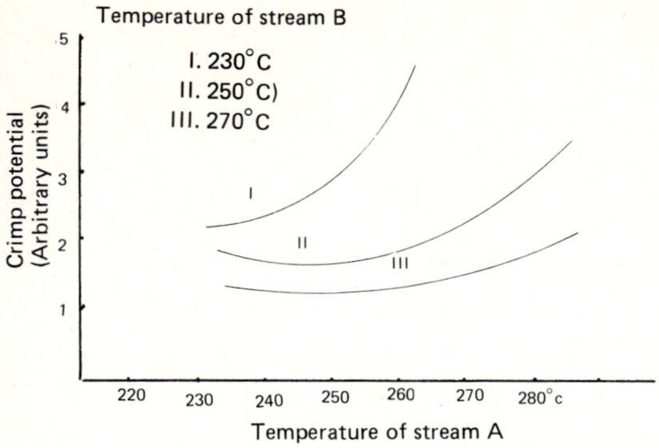

Fig. 13. Effect of Temperature Differential on Crimp Potential

V.2.2. Merging of Component Streams Into a Conjugate Fiber System

The subject of merging two component streams into conjugate fibers is extensively covered by hundreds of patents. Most of them apply to the production of polypropylene fibers as well as fibers from other polymers. The cardinal principle of most of these techniques is the division and subdivision of each component stream into a multitude of lamellar flows and then combining the two lamellar flows just before or on the counterbore of the spinnerette hole, as depicted in Fig. 1.

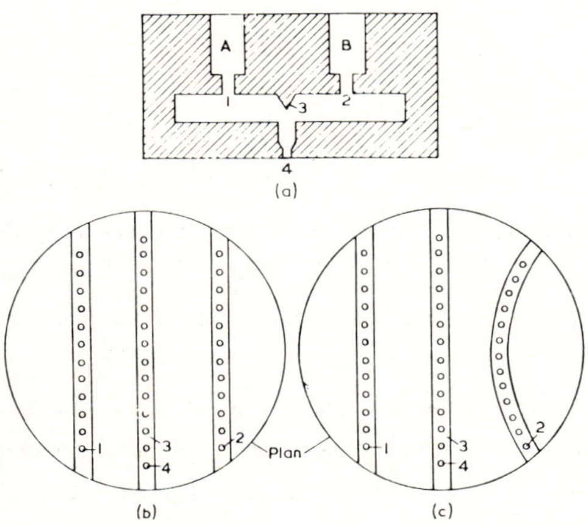

Fig. 14. An Example of Commercial Spinning Arrangement for the Production of Conjugate Fibers, Using a Septum Concept (ref.52)

As should be evident from the schematic of Fig. 1, the spinning of bicomponent fibers requires the use of novel and precise equipment to bring together two component streams in a predetermined conjugate arrangement. The shape of the line between the two components can be controlled by adjusting the shape and position of the separating element with relation to the spinnerette hole. The ratio of the components in the fiber can be adjusted by adjusting the pumping speed of the respective pump. Thus, by adjusting the lateral and vertical position of the separating element in relation to the spinnerette hole, and by adjusting the ratios of the components in the fiber, a wide variety of cross-sectional arrangements of the components can be achieved, ranging from sheath-core to perfect half-moon arrangement. Figures 14 and 15 show the polymer distribution and merging devices commonly recommended for the production of conjugate fibers. In Fig. 14, polymers A and B are supplied via rows of channels 1 and 2 to either side of the septum 3 situated just above the row of orifices 4. By arranging the orifices in the spinnerette in straight rows, conjugate fibers are produced in which each filament is equivalent to the other in fiber composition and cross-sectional geometry of the components. If, however, a random distribution is desired, it can be obtained by arranging the orifices in curved rows. In Fig. 15, channels 1 and 2 supply polymers A and B to orifices 3; there is, of course, a multiplicity of channels and orifices arranged, for example, in the form of a circle.

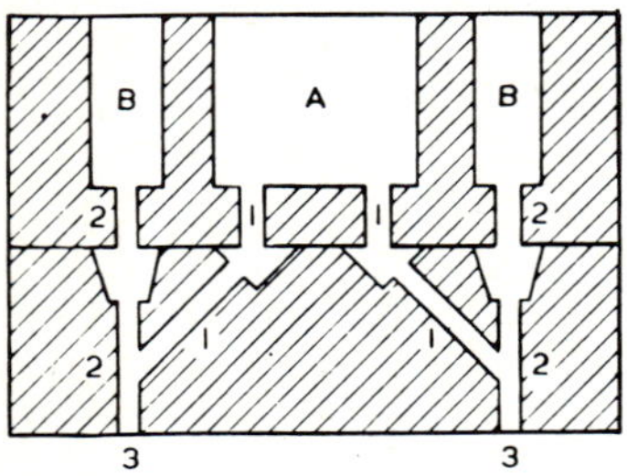

Fig. 15. Example of Conjugate Spinning Using a Septum Concept (ref.53)

The utilization of a relatively sharp and precisely positioned separating edge has been necessitated in the conjugate spinning of low-viscosity polymer solutions because of the need to prevent them from mixing with each other until the spinnerette hole entry, to ensure definite lateral arrangement of the components. This principle was adopted also in the melt spinning of conjugate fibers. A variation of this theme has been mentioned in the early patents pertaining to melt spinning of conjugate fibers. It was, however, later recognized that polymer melts generally tend to flow in lamellar fashion and tenaciously refuse to diffuse or mix with each other without added work performed on them. This property of melt-flow behavior has been taken advantage of in the form of simplified designs of component merging devices, such as the one depicted in Fig. 16. In such designs, no solid separating edge is needed; the polymer flow develops in such a pattern as though a separator were present between them. The relative proportions of the two components in the conjugate fibers are controlled by the rates of feed of the two components to the orifices. The production of uniform fibers by this method depends on maintaining the pumping pressures at constant level, maintaining constant viscosities for the two components, and having a very precisely constructed system of channels leading to the spinnerette orifices.

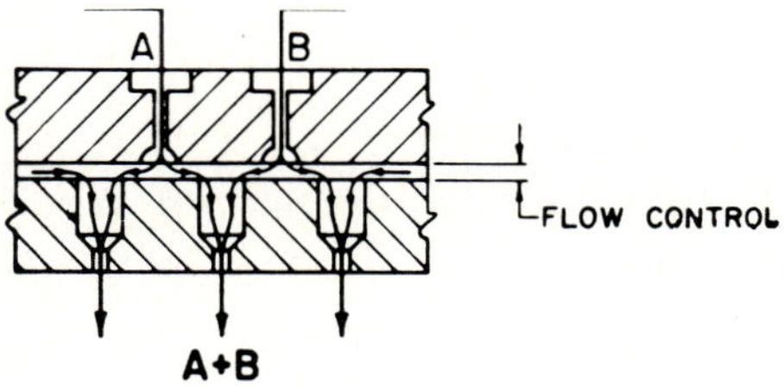

Fig. 16. Component Merging arrangement for Conjugate Spinning Without Using a Septum

The production of a few fibers by this method is probably reasonably straightforward for anyone skilled in the art of fiber extrusion. But for the production of large numbers of fibers, as required for economic viability, the situation is much more difficult. A large number of distribution systems have been described in the patent literature. Two representative systems are outlined in Figures 17 and 18.

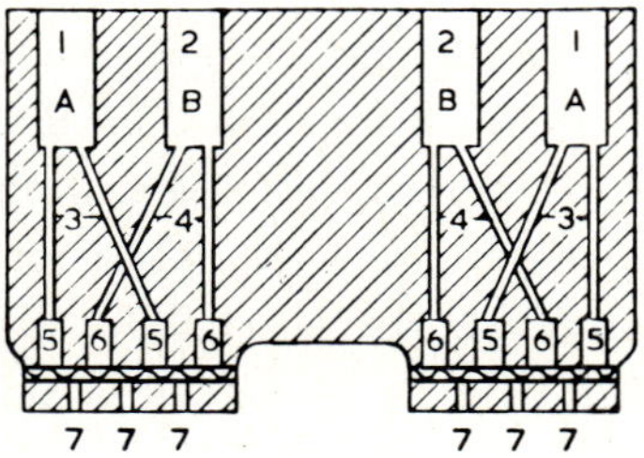

Fig. 17. An Example of Conjugate Spinning Using no Septum for the Separation of the Components. Three Circular Rows of Conjugate Filaments Emerge from the Spinnerette. (ref.54)

Using this principle, the orifices in the spinnerette can be arranged in concentric circles in conjunction with circular melt-distribution plates or they can be arranged in straight lines in conjunction with rectangular distribution plates. The basic design feature of these arrangements is the provision that the pattern of spinnerette holes is precisely aligned with the interphase of the two components.

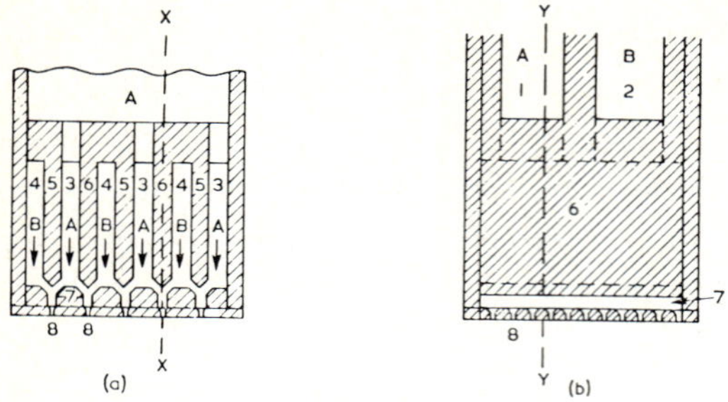

. (a) Longitudinal view taken on cross-section Y-Y
(b) Lateral view taken on cross-section X-X

Fig. 18. An Arrangement of Supplying Parallel Rows of Orifices with Two Components of Polymer Melt. The Arrangement Does Not Use A Septum. (ref. 55)

Fig. 17 is an example of a distribution arrangement supplying three concentric rows of orifices. Main channels 1 and 2, supplying polymer A and polymer B, are annular cavities. Four sets of secondary supply channels 3 and 4 supply pairs of annular spaces 5 and 6, which are symmetrically arranged on either of each of the three rings of orifices 7. The number of rings of orifices in the spinnerette can be increased by dividing and subdividing the two component streams separately and then combining them over the spinnerettes as mentioned. An example of such systems is depicted in Fig. 19. An arrangement supplying parallel rows of orifices is shown in Fig. 18. Polymers A and B supplied from the main channels 1 and 2 pass down channels 3 and 4 respectively between alternate vertical plates 5 and 6. The two components merge side-by-side in slits 7 and proceed to be extruded through rows of orifice 8.

The concept that component streams do not mingle, when merged in laminar flow, has been extended to "mixed stream spinning" of conjugate fibers. In this arrangement the component streams are merged well ahead of the spinnerette face, in the form of concentric rings or "pipe-in-pipe" mixing elements. The resulting composite melt stream is then spun into fibers in conventional manner without any efforts being made to keep the component layers separate prior to extrusion through the spinnerette holes. Fitzgerald and Knudsen (ref.37) have described this approach in detail and several United States and other patents are based on this technique (ref.38).

Figure 20 illustrates the concept of a "pipe-in-pipe" mixing element. One of the component streams envelopes the other component stream at the exit end of the tube of the inner stream. The resultant stream contains in cross section the individual streams in concentric rings. The component streams can be necked down or blown up in cross section without any change in the concentricity or relative proportions of the two component streams. Additional streams can be added by successive mixing elements to provide merged streams of alternating concentric

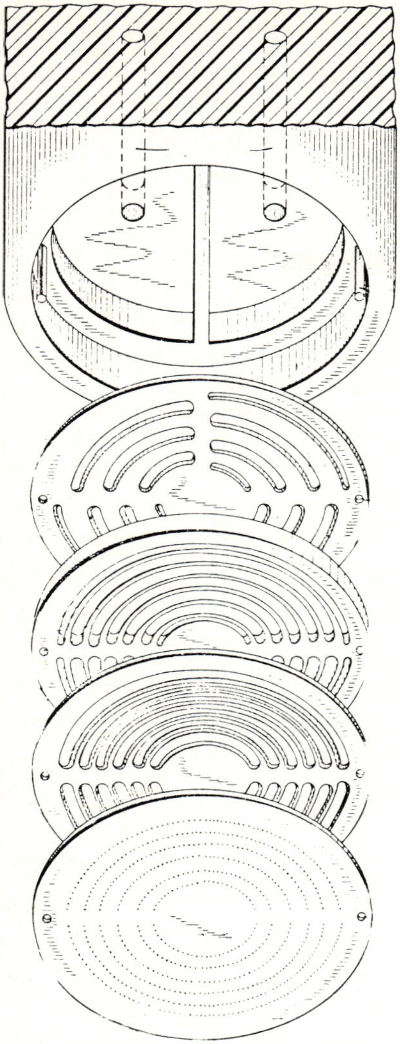

Fig. 19. Melt-Distribution Plates for Conjugate Fiber Spinning (ref.56)

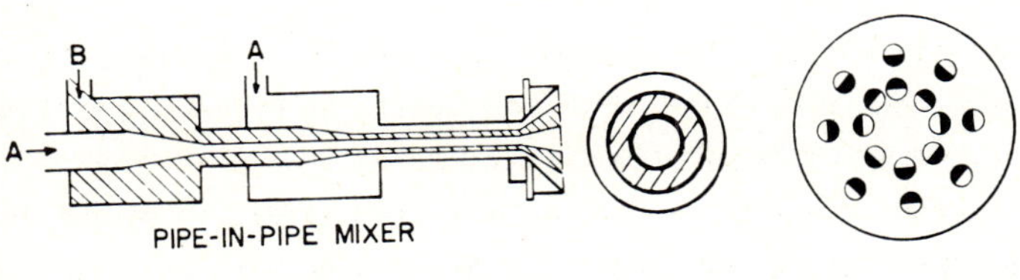

Fig. 20. Schematics of Lamellar Spinning for Conjugate Fibers (ref. 57)

layers of the two components. The relative amount of material in each stream is determined by the volumetric rate at which each successive stream is pumped.

Component streams combined well back of the spinnerette do not intermingle or mix as they flow to the spinnerette holes. The interface between the component streams is well defined. When the composite stream of two concentric component layers is supplied to a spinnerette, with a single ring of holes concentric with the interface between the component streams, a conjugate fiber is produced. This fiber has the same distribution of components as that produced with conventional distributors. Each filament contains a single interface between the components, and the relative amount of the two components is in proportion to the rates at which component streams are supplied to the spinnerettes. When a component stream made up of several concentric layers of melt is fed to a spinnerette with multiple concentric rings of holes, conjugate fibers can be produced at high throughput rate. A necessary requirement for this arrangement is that the number of rings of holes corresponds to the number of interfaces between the component streams. To obtain uniform conjugation, the volume flow of each layer must be in proportion to the number of holes it supplies. The spacing between rings of holes is not critical, since the flow pattern of the streams will adapt to exit through the spinnerette holes (ref.37).

This technique is not restricted to a particular geometry of layers such as the concentric rings. A laminar device for producing composite streams containing parallel layers of the individual streams has also been constructed (ref.39). In this device, alternate plates provide new streams or layers to the composite stream. Spacer plates are used to maintain the integrity of the supply streams. Each stream of a given component is supplied from the same source. The relative amount of each supply stream going into each layer is determined by the relative dimensions of the channels connecting the supply streams with the composite stream. Providing different geometries and component distributions in the merged streams is only a matter of ingenuity in design of the metering device. Conjugate fibers are formed by aligning rows of holes with the interfaces of parallel layers of component streams. Fiber produced by careful matching of melt merging devices and spinnerettes have the same uniform distribution of components as those produced with the conventional knife edge mixer devices or variations of it.

A special case of mixed-stream spinning of conjugate fiber is described in Fig. 8 and mentioned in Section V.1. Polypropylene melt of large molecular weight and broad molecular-weight distribution is metered by a pump through a suitable length of a metal pipe P which is electrically heated to a temperature considerably higher than the original melt temperature. Because of the poor heat conductivity of the polymer melt and because of the susceptibility of the large molecules to heat degradation, the peripherial layer of the polymer stream would preferentially degrade to lower average molecular weight polymer of narrow molecular weight distribution. The differential in molecular weight and molecular-weight distribution is not gradual but rather a definite layer of degraded material is formed enveloping the inner core of the essentially undegraded material. This sheath-core type mixed stream can be blown without turbulence and distortion in concentricity for filtration to a larger cross section and then necked down to be spread on a spinnerette with one ring of holes, as described earlier.

V.2.3. Bicomponent Spinning for Conjugate Staple Fibers

One of the disadvantages of controlled dozing of polymer components to individual spinnerette holes is that because of the complexity of construction, only a limited number of holes can be accommodated in the spinnerettes. This, of course, limits production. The aforementioned processes for the spinning of conjugate fibers may be appropriate, therefore, for the production of bulked or stretch filament yarns with comparatively small numbers of filaments in the bundle, but few of them can be economical enough to justify their adaptation for large scale staple production. In addition to the limitation in production rate, conjugate fibers of exactly the same composition arranged in identical geometrical configuration would tend to crimp as a bundle and resist opening. This would lead to poor cohesion in yarn spinning and contribute to the appearance of "shiners" or bundles of "married filaments" in the finished fabric.

These difficulties can be minimized by spinning a large number of filaments in such a way that a random mixture of filament composition is obtained. Thus, while in the tow there may be present some filaments which are composed purely of component A and some filaments which are composed purely of component B, a majority of filaments would be composed of components A and B arranged in conjugate fashion in random proportions. A complete spectrum is, therefore, obtained ranging from 100 percent A to 100 percent B. After cutting and drawing, the stock fiber would be constituted of a random mixture of fibers, each having a different shrinkage and crimp characteristics. As precise dozing of the components in each hole is not necessary, a large number of holes can now be accommodated in the spinnerette.

The basic principle of random bicomponent fiber spinning for staple fiber production is the formation of a multitude of longitudinal layers of two different polymer components along the direction of the flow. This multilayer melt stream is then fed to a conventional spinnerette with a large number of randomly spaced holes. Because of the high viscosities and low liner velocities of flow, the lines of demarcation between the adjacent layers remain fairly well defined.

Where one of these lines of demarcation intersects on an orifice of the spinnerette a conjugate filament is formed. Since the layers can be fairly thin, there will be a large number of these lines of demarcation and, consequently, a large number of composite filaments will be produced. There will be some orifices in the spinnerette which will receive only component A or component B; they would, of course, not exhibit a conjugate structure. The presence of these filaments is not harmful to the quality of the spun yarns. In fact, they contribute to the bulkiness of the yarn structure. However, these low-crimp homofilament fibers should not exceed 35 percent of the composition of the staple; otherwise, spinning difficulties might occur and the resulting yarns may lack adequate loft and bulk.

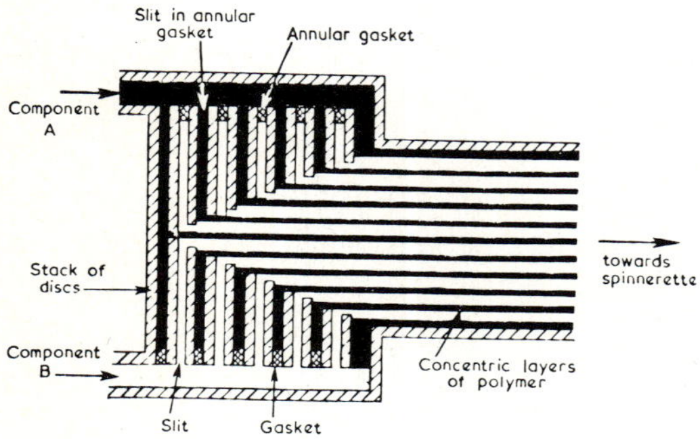

Fig. 21. A Circular Lamellae Multiplication Device for the Production of Randomly Conjugated Filaments (ref.58)

A number of possible devices can be envisaged for producing multiple layers of polymer flow and spinning them through a multiple hole spinnerette in random composition. One arrangement is the aforementioned mix stream spinning system. A mixing device such as those depicted in Figures 21 and 22, can be used so that a large number of thin layers of components A and B are produced as circular rings or as straight leaflets. In the first case a spinnerette of noncircular arrangement of holes is employed and in the latter case a spinnerette of circular arrangement of holes is used for receiving a random composition of the components. It should be evident that no efforts for holes alignment with the component streamlets are made.

Another method of producing a random composition conjugate staple fiber is through an arrangement depicted in Fig. 23. The component A is extruded through a large hole spinnerette or a distribution plate into component B, so that component A streams are surrounded by component B and the former is uniformly distributed into the latter without losing its individual identity or continuity. This mix stream is then extruded through a normal spinnerette as described above.

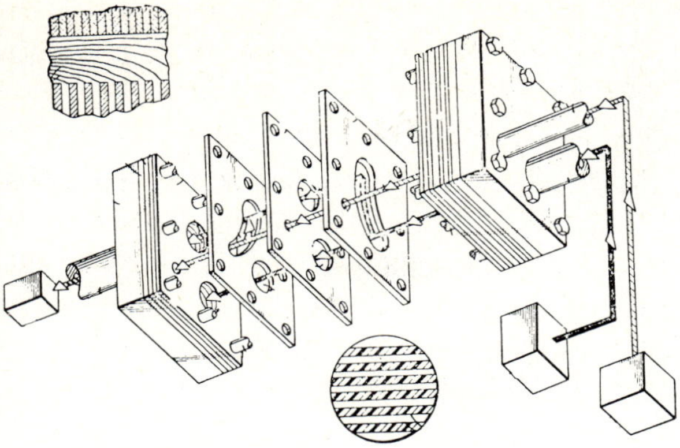

Fig. 22. A Leaflet-Type Mixing Device for the Production of Randomly Conjugated Filaments (ref.59)

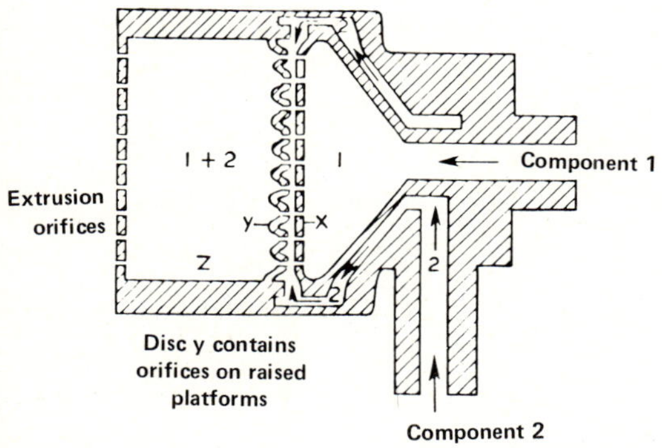

Fig. 23. A Double-Spinnerette Type Merging Device for the Production of Homofil/ Conjugate Mixed Filaments (ref.60)

Another simple arrangement for the production of random mix flow is through the use of a static mixer, such as the Kenics static mixer described in Chapter VIII. The number and relative cross-sectional dimensions of each component streamlets would depend on the number of mixing elements in the static mixer. The bundle of filaments having varying degrees of extensibility and breaking strength may introduce some difficulties in tow drawing. Generally speaking, this kind of mixed tow is drawn to draw ratios that do not exceed the breaking elongation of the least extensible filaments.

V.2.4. Nonsplitting Conjugate Fibers

The difference in shrinkage specificity of the components in a single composite fiber, which is a necessary condition for the development of three-dimensional crimps, necessarily results in the generation of internal stresses at the interphase of the composite fibers, thereby inducing a tendency of the fiber to split along the interphase. This is especially true when two polymers of poor adhesion properties are spun in side-by-side relationship.

Even such chemically similar polymers as polyethylene or propylene-ethylene copolymer are not sufficiently coadhesive with polypropylene to yield satisfactory composite filaments. The problem of splitting sometimes arises even in those cases where both the components are polypropylenes of greatly differing molecular weight or where one component contains a large proportion of other additives, such as those used for dyeing and flame retardance. Three approaches can be taken to circumvent this problem:

(1) Asymmetric sheath-core structure,
(2) Nonlinear contoured interconstituent boundary,
(3) Placement of a third component, having good adhesion to both components.

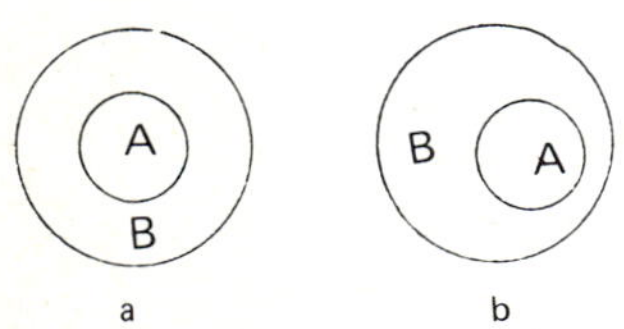

Fig. 24. Filament Cross Section With Sheath-Core Structure. (a) Symmetrical, Noncrimping; (b) Asymmetrical, Self-crimping

Sheath-core conjugate fibers are those in which one of the components is completely surrounded by the second component in an accentric arrangement, as shown in Fig. 24. The most common ways of manufacturing these fibers described in the patent literature employ one or the other principle of component merging depicted in Fig. 25. Here again, the systems which employ a septum or "inner orifice" are more precise but at the same time more complex than those relying on "mixed-stream spinning. Figure 26 is an example of the former system, Figure 27 diagrams the latter system.

In Fig. 26, the core component is supplied from reservoir 1 via a row of channe 2 to a row of inner orifices 3. The sheath component in reservoir 4 is supplied

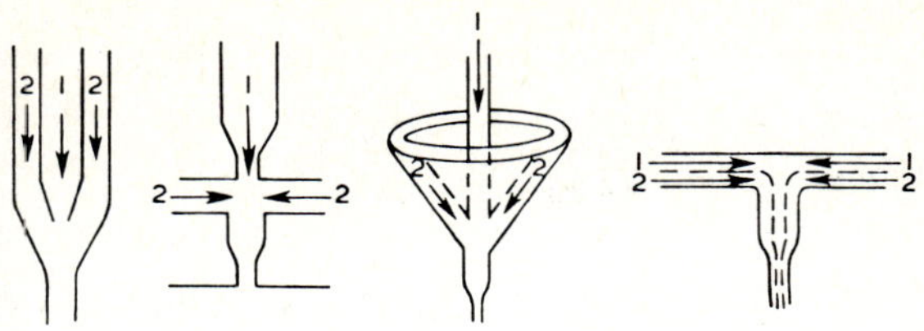

Fig. 25. The Principles of Spinning of Sheath-Core Filaments (ref.9)

through channels 5 to surround the core component as extruded at outer orifice 6. The inner orifices 3 can be offset relative to the outer orifices, giving an accentric sheath-core conjugate fiber. In Fig. 27, the core component is supplied via three concentric rings of channels 1 to the annular space 2, whence it flows to the outer annular space 3. Here it combines with the sheath component, fed to space 3 through the outer ring of channels 4. The two components proceed together as two layers into the flat region 5 towards the four rings of orifices 6 and the central orifice 7. The two components then extrude through the orifices in the form of sheath-core filaments of varying accentricity.

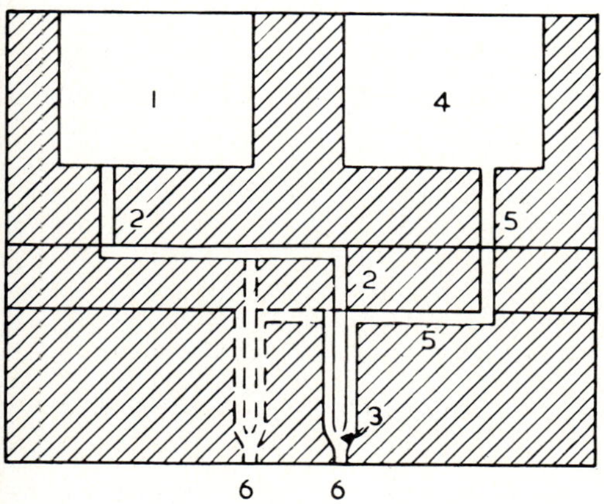

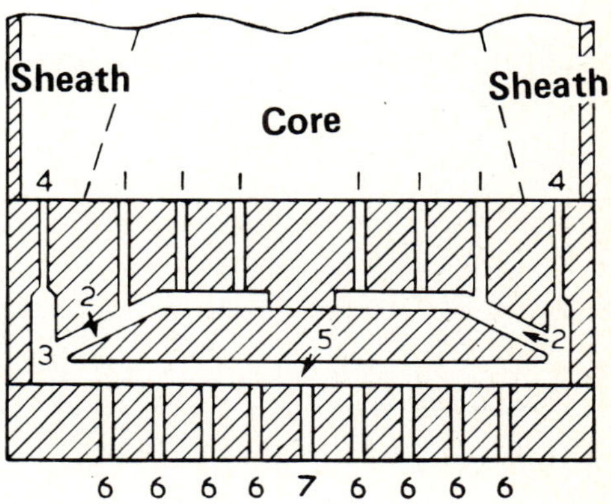

Fig. 26. An Arrangement for Spinning of Sheath-Core Conjugate Fibers Utilizing the "Inner Tube/Outer Tube" Concept (ref.61)

Fig. 27. An Arrangement for the Spinning of Sheath-Core Filaments According to the Concept of Fig. 25 d (ref. 62)

Sheath-core conjugate fibers can also be produced through side-by-side spinning arrangements, described in the previous sections, where one component is considerably lower in melt viscosity than the other component. The lower viscosity component encircles the high viscosity component in a crescent-like cross section.

Although the acentric sheath-core structure does prevent the conjugate fiber from splitting, the discussion in Section II shows that it also reduces the bulking potential of the fiber. An alternative approach to prevent the splitting of conjugate fibers without drastically reducing its bulking potential is to distort the interphase line in such a way that one component protrudes into the other and forms an interlocking system, as shown in Fig. 28. Such specific conjugate filaments can be manufactured effectively by using an improved spinnerette, described by Kanegafuchi Boseiki and depicted in Fig. 29.

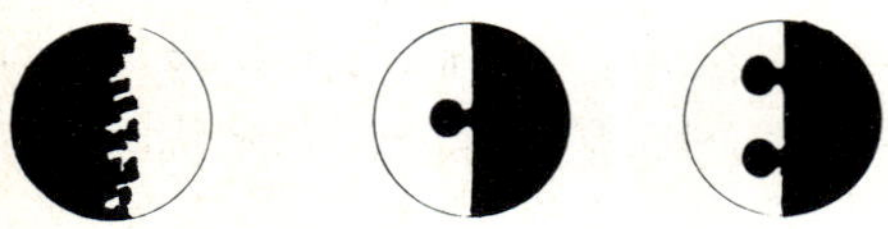

Fig. 28. Some Examples of Nondelaminating Heterofilaments

Interlocking of the two components can also be achieved by intermingling the two components slightly at their interface. This is done either by a very fine mesh inserted close to the spinnerette orifice (ref.40), or by arranging a controlled amount of turbulence as the two components are brought together just prior to extrusion (ref.41), or by allowing the two polymers to remain in contact for longer than usual prior to extrusion (ref.42). Side-by-side conjugate fibers with wedged, wavy, or zigzag interface have also been described (ref.43).

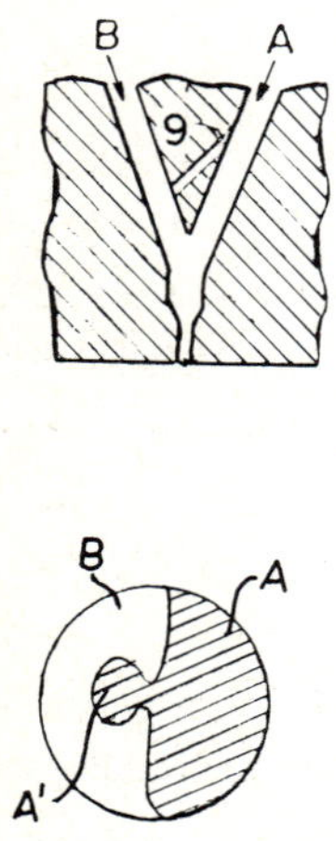

Fig. 29. Spinnerette Orifice for Producing a Conjugate Filament with Interlocking Interphase Boundary (ref.63)

The third possibility of overcoming splitting is by having two components spun in such a conjugate configuration that a thin line of a third component is introduced between the two main components. The third component is characterized by good adhesion to both of the main components. The third phase can be an entirely distinct polymer or it can be a 50/50 mixture of the two main components (ref.44).

VI. CRIMP DEVELOPMENT

The "as-spun" conjugate fibers rarely have enough crimp potential to make them useful for intended products. Also, undrawn fibers have poor tensile strength and excessive elongation which make them unsuitable for either textiles or floor covering products. The spun conjugate fibers, therefore, need further processing in order to develop adequate crimp and tensile properties. Four basic steps are necessary to achieve this objective. They are:

(1) Aging of spun yarn to enhance and stabilize the conjugate structure produced in spinning,

(2) Cold drawing of aged yarn to enhance the stress differential in the filaments and to improve the tensile properties of the fiber,

(3) Heat relaxing the cold-drawn yarn or staple to develop crimp and stabilize the crimped fiber structure.

(4) Preserving crimp in continuous-filament yarns and tow to facilitate the working of yarns and staple in further processing operations.

These processing steps can be undertaken separately or they can be combined in a continuous process. The following comments are meant as guidelines for process development.

Aging: The fundamentals of aging and heat setting have been discussed in Chapter VII. It has been mentioned there that the fiber structure produced in the course of spinning does not freeze completely on solidification. As the glass transition temperature of polypropylene lies below room temperature, the possibility of subtle changes in crystalline order of the fiber still exists at room temperature. Indeed, changes do take place on aging whereby the spun-fiber structure (in the case of conjugate fibers, the respective structures of the conjugate parts) strives to perfection.

In order to develop the optimum crimp potential in the filament, it is necessary that the undrawn yarn be aged prior to cold drawing. At room temperature aging, the development of optimum crimp potential takes place at an exponentially decreasing rate over about two to ten days. The rate of aging can be significantly increased by reheating the yarn to about 70 to 120°C for a short time.

Several attempts have been reported to devise practical methods of applying this heat treatment uniformly. The use of heated godets is one such successful approach; radiant heating of yarn, as it is being wound up on a package, is another method (ref.45). Probably the use of heated godet is preferable for a continuous process and the use of radiant heat on winding package is more suitable for a discontinuous process. Heating the spun yarn package in an oven is not advisable because the poor heat transfer characteristics of yarn results in nonuniform aging within the package. A tension gradient across the package also makes this method unsuitable.

Cold Drawing: The primary objective of cold drawing of filament is to improve its tensile properties. In the case of conjugate fiber, however, the cold-drawing process is also necessary to enhance the crimp potential of the fiber. The change in crimp potential on cold drawing depends on the differential in birefringence and molecular parameters of the conjugate components, draw ratio, draw temperature, and speed of drawing.

It has been noticed in Chapter VII that extension of polypropylene can happen by void formation or by molecular orientation. It has also been noticed that the fibers which have been extended predominantly through void formation have much higher shrinkage potential on heat relaxation than those extended through molecular orientation. Therefore, efforts are made to spin the conjugate fiber under those conditions which favor the structure formation of one component readily susceptible to void formation and the other component readily susceptible to molecular orientation. This differential is further accentuated in the process of aging. Starting with a given conjugate fiber structure, the drawing conditions are chosen in such a way that one component is extended through molecular orientation while the other predominantly extends with the formation of voids. The differential so achieved depends on draw temperature, draw ratio, and to some extent on drawing speed.

A too low drawing temperature may be detrimental to crimp potential of the fiber because void formation would occur in both the components. Similarly, high temperatures of drawing are detrimental to crimp potential because the molecular orientation would occur in both components. In both cases a decrease in differential shrinkage will result. The optimum drawing temperature is a function of the starting differential in fiber structure and the overall molecular weight of the fiber. Figure 30 shows the effect of draw temperature on the crimp potential of three different conjugate yarns drawn to 3.0 draw ratio.

Apart from the question of crimp development, too low a drawing temperature results in poor drawability of the yarn, necessitates a reduction in drawing speed, and yields a bulk yarn of excessive thermal shrinkage. The drawing of coarse denier conjugate filaments is often difficult to carry out at low

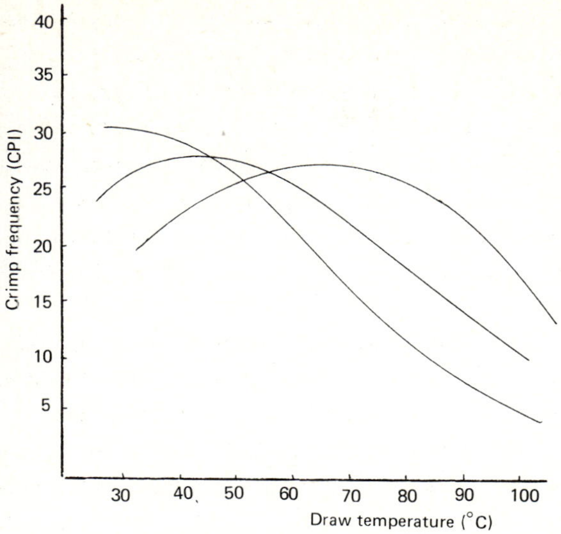

Fig. 30. Effect of Draw Temperature on Crimp Potential for Three Differently Spun Conjugate Fibers

temperature, which renders the production of carpet yarns problematic. In view of these considerations, a drawing temperature of 60 to 90°C is generally employed. To achieve drawing in this temperature range or higher, however, there must be a large difference in birefringence, molecular weight, polydispersity, and degree of cross orientation of the conjugate components (ref.46).

As is the case with drawing temperature, the structures of the two components in the fiber do not respond in similar manner to draw ratio. The differential in shrinkage potential of the two components, and hence the bulk potential of the conjugate structure, increases with draw ratio, reaches a maximum at about 2.0 to 2.5 draw ratio, and then falls off (see Fig. 31).

Because of the high shrinkage potential of one of the two components, the effective draw ratio of conjugate fibers is lower than the applied draw ratio: most of the stretching is recovered when the fiber is allowed to relax. The general level of birefringence in the fiber is not greatly increased by drawing process but the difference between the birefringence of the two components is reduced.

Heat Relaxation: Probably the most critical and the slowest step in the production of conjugate fibers is the heat relaxing step to develop crimp and stabilize the crimp configuration. To achieve maximum degree of bulk, the relaxation has to be carried out practically under no tension. This makes the handling of yarn bundle or filament tow extremely difficult. As the stress relaxation in the filament is inherently a slow process, filaments have to be treated in a hot environment for a prolonged period of time to develop optimum degree of relaxation

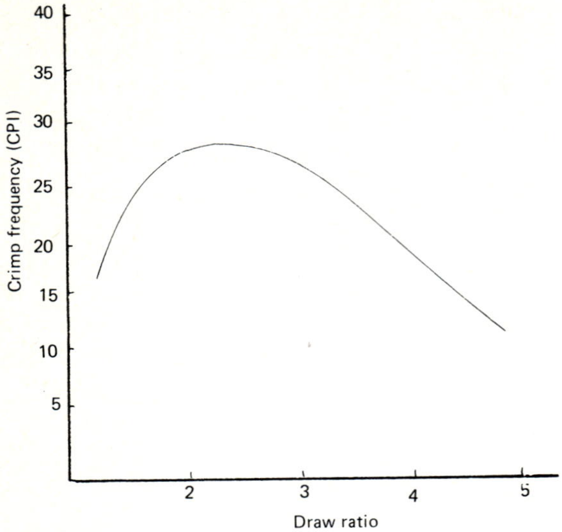

Fig. 31. Effect of Draw Ratio on Crimp Potential of a Conjugate Fiber

and crimp formation.

It is in the heat relaxation step that the potential of the drawn conjugate filaments is converted to real crimp and it is the processing conditions of this step that determine the quality of the finished product. Working with a given drawn conjugate filament yarn or tow, it is generally possible to achieve products of varied crimp frequency and crimp contraction by adjusting the relaxation condition. Higher tension on the yarn or tow decreases the crimp index of the fiber and higher temperature and longer times of treatment tend to increase it. Quality control is maintained by adjusting these three process parameters in the relaxation step.

Several different ways and means have been proposed to achieve bulk development with adequate economies. Figure 32, for example, depicts an arrangement for heat relaxation of continuous filament yarn in a heated gaseous medium. These methods are suitable for heat relaxation of relatively fine denier yarns but are not economical for heavy denier carpet yarns. A coil-type heat relaxing box, depicted in Fig. 33, is more suitable for economical processing speeds. As air is a poor heat transfer medium, the employment of steam is advisable for heat relaxing heavy-denier yarns at commercially feasible processing speeds. The treatment of drawn yarn with steam aspirating jets is another useful way of crimping the filament bundle continuously. In these processes, the yarn is led to aspirating steam jets which deposit it on a conveyor in a random and relaxed fashion.

Heat relaxation can be carried out, of course, as batch process. The drawn filament yarn is wound in skeins, tumbled in hot air, or autoclaved at desired temperature for required treatment times. The batch process suffers from poor process economy but the product obtained is quite stable. Moreover, yarns of relatively low crimp potential can also be converted into reasonably well-bulked products through this process.

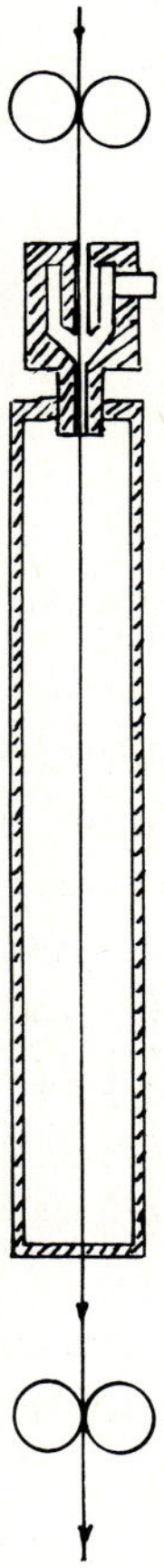

Fig. 32. Crimp Development of Conjugate Filament Yarn by Passage Through a Gaseous Medium (ref.64)

As is the case with multifilament yarns, there are also various possibilities for heat relaxation of filament tow. One such arrangement is shown in Fig. 34. The number of crimps needed for staple fiber is generally low as compared with continuous filament yarns. A considerable amount of tension can, therefore, be tolerated in heat relaxation of conjugate fibers for staple. Some relaxation is achieved on tow drawing machine and some in fabric finishing.

As mentioned earlier, spinning, drawing and heat relaxation of polypropylene

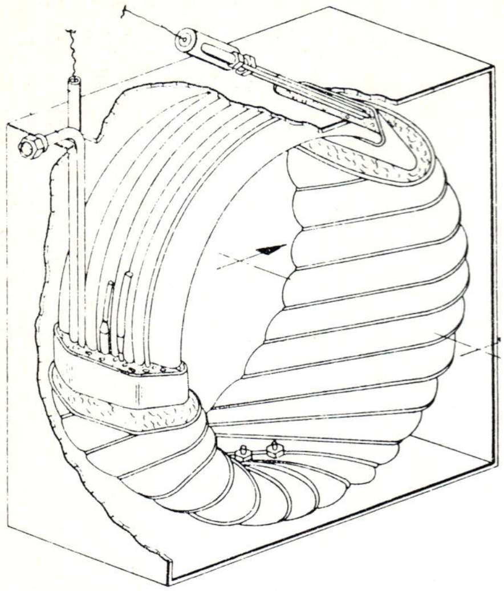

Fig. 33. A coil-Type Heater for Crimp Development of Filament Yarn (ref.65)

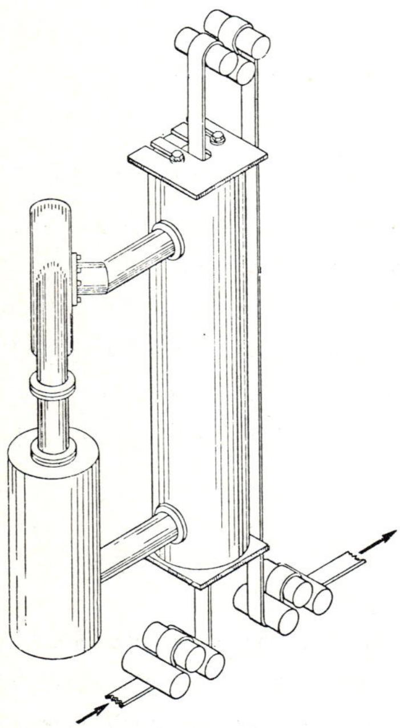

Fig. 34. A Tow Relaxing Arrangement for Conjugate Fibers (ref.66)

conjugate yarns can be undertaken in a one-step, continuous operation. These processes are particularly appropriate for the production of bulked continuous filament yarns for carpet, upholstery, and knitted goods. One such arrangement

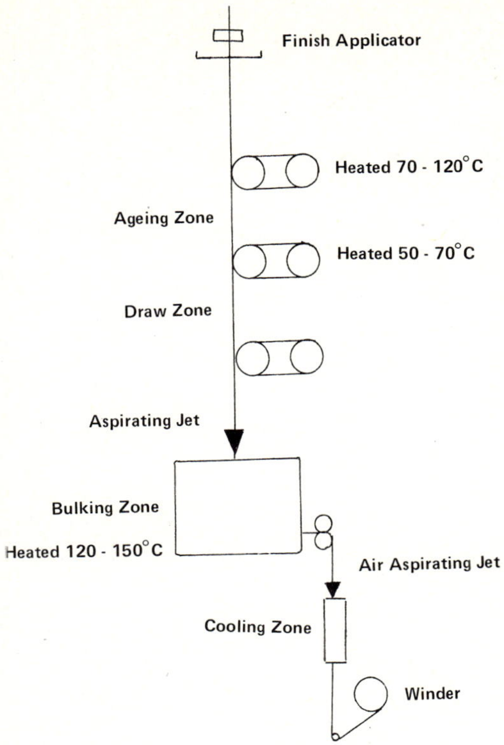

Fig. 35. A Schematic Diagram for Continuous Production of Conjugate Filament Yarns

is shown in Fig. 35. Similarly, Fourne (ref.36) has described an arrangement (Fig. 36) for the continuous production of conjugate staple fiber. The yarn is spin-drawn and cut into staple lengths in a continuous operation. The cut fibers are then air transported to a cyclone separator and heat-set in a J-tube from where the product is transported to a baling station. The crimping characteristics can be controlled by adjusting the temperature of drawing and of hot air.

VII. CONCLUDING REMARKS

The production of conjugate fibers requires a high degree of technological sophistication. This applies not only to the design of spinning, drawing, and relaxing equipment, but also to the ability of controlling the process parameters which determine the ultimate properties of the fiber. The spinning of conjugate filaments is more capital intensive than the spinning of homofilaments. The drawing and relaxing techniques are rather slow in speeds which tend to increase the labor cost of production. These factors have combined, with the result that

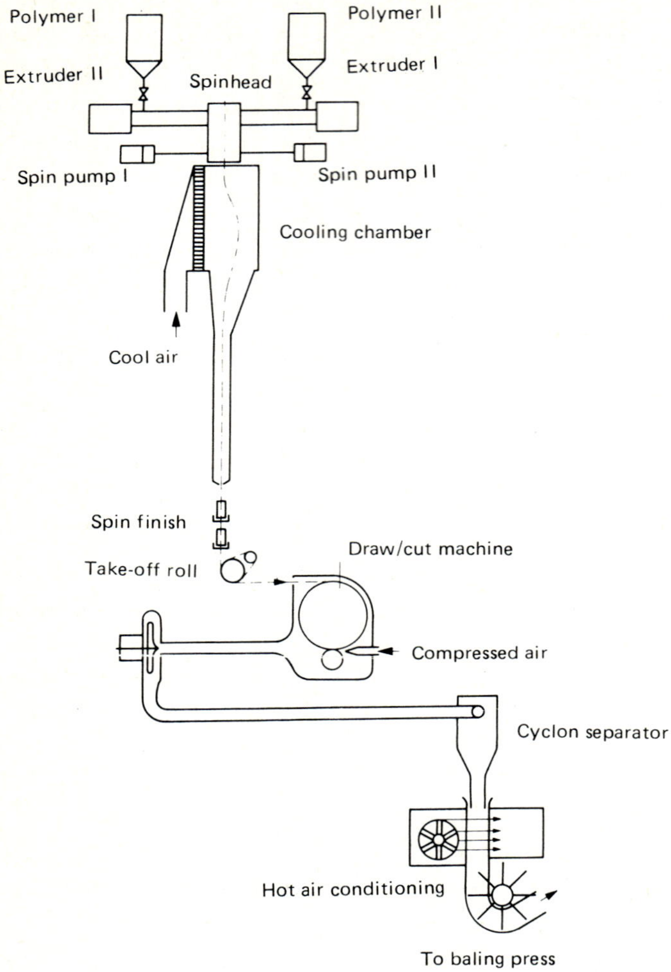

Fig. 36. A Schematic Diagram for Continuous Production of Staple Fiber (ref. 67)

only a few conjugate fibers have so far been commercialized. The few that have been produced include the "latent crimp" acrylic fibers from DuPont, Monsanto, American Cyanamide, and others; Cantrece® Monofil nylon from DuPont; Heterofil® from ICI; and Monvelle® nylon/spandex fiber from Monsanto. However, because of the aesthetic and property advantages which the conjugate fibers promise to offer, active development efforts are continuing in the synthetic fiber industry throughout the world.

Polypropylene has a special advantage in the production of conjugate or "self-crimping" fibers because it is relatively easy to create a conjugate structure in the filament without resorting to complicated chemical or physical modifications for the creation of a differentially shrinkable conjugate system. It is expected, therefore, that polypropylene conjugate fibers will form a major share of the "third generation" of engineered fibers. The experience of Canadian Celanese, which for several years marketed a self-crimping polypropylene carpet yarn, the developmental work of an acid-dyeable polypropylene conjugate yarn by Hercules for apparel goods, and the initial efforts of Mitsubishi in commercialization of conjugate pigmented polypropylene yarns for upholstery and knitted goods, have shown that the era of high-quality, specialty BCF yarns from polypropylene has indeed begun. The proliferation of nonwoven fabrics and ever-increasing demand on the quality of these products would certainly give an incentive for polypropylene conjugate fibers to be used in these or similar products. Carpet and knitting yarn spun from a mixture of conjugate and homofil staple fibers is another area where the commercial developments can easily be realized. A vast market lies in nonwovens: ICI has shown special advantage in properties and process details with the use of conjugate fibers. In fact, the use of conjugate or sheath-core fibers in such nonwovens is so unique that a new term, *melded fabrics*, has to be coined for these products. Polypropylene, in conjunction with polyethylene or other lower melting polymer, certainly has a large potential for this market.

VIII. REFERENCES

1. S. Timoshenko, J. Opt. Soc. Am., II(1925)23.
2. U.S. Patent 2,443,711, Am. Viscose, 1948.
3. U.S. Patent 2,517,694, Am. Viscose, 1950.
4. U.S. Patent 2,515,834, DuPont, 1950.
5. W.A. Sisson and F.F. Morehead, Text. Res. J., 23(1953)152.
6. Belg. Patent 604,988, B.N.S., 1960.
7. W.E. Fitzgerald and G.B. Hughey, Am. Dyestuff Rep., 55(1966)191.
8. R. Jefferies. Ciba-Geiav Review. I. 1974.
9. R. Jefferies, *Bicomponent Fibers,* Merrow, 1976.
10. British Patents 809,273; 1,137,027; 1,057,579; U.S. Patents 2,931,091; 3,366,722; 3,577,498; 3,233,023; 3,050,821.
11. German Patent 1,238,610, Hoechst, 1967; British Patent 964,406, ICI, 1964; Japan Patents 20,770/66, Toyobo; 21,577/66 Toyobo.
12. G.A. Ermilova et al, Soviet Plastics, (1965)36.
13. U.S. Patent 3,233,023, ICI, 1966.
14. U.S. Patent 2,931,091, DuPont, 1958; British Patents 979,083, Soc. Viscose Suisse, 1965; 1,121,314, Mitsubishi, 1968; 1,123,952, Asahi Kasei, 1968; French Patents 1,479,758, Asahi Kasei, 1967; 1,478,705, Mitsubishi, 1966; Belg. Patent 679,797, Mitsubishi, 1966.
15. U.S. Patent 3,663,376, Asahi Kasei, 1972; French Patent 1,479,758, Asahi Kasei, 1966.
16. U.S. Patents 3,233,023, ICI, 1966; 3,577,498, Toyobo, 1971; 3,050,821, Hoechst, 1969; 3,480,709, Hoechst, 1969; 3,366,722, Can. Celanese, 1968; British Patent 964,406, ICI, 1964; Japan Patents 20/770/66, Toyobo; 21,577/66, Toyobo.
17. British Patents 1,029,119, Rhodiaceta, 1966; 1,030,523, Monsanto, 1966; 1,088,592, Lansil, 1967; U.S. Patents 3,308,503, Text. Chem. Res., 1967; 3,295,552, Monsanto, 1967.
18. British Patent 1,038,670, Asahi Kasei, 1966.
19. W.E. Fitzgerald, Text. Res. J., 37(1967)447.
20. U.S. Patent 3,295,552, Monsanto, 1967.
21. W.E. Fitzgerald and J.P. Knudsen, Text. Res. J., 37(1967)447.
22. British Patent 1,087,823, Hoechst, 1967.
23. U.S. Patent 3,600,271, Toray, 1971; British Patents 1,096,249, Soc. Vis. Suisse, 1967; 1,058,448, Asahi Kasei, 1967; 1,151,272, Kanegafuchi Bo, 1969.
24. Japan Patent 4739/67, Toyo Rayon.
25. British Patent 1,050,042, Rhodiaceta, 1966; French Patent 1,393,572, Rhodiaceta, 1965; Netherland Patent 7,001,187, Mitsubishi, 1970.
26. Japan Patent 4737/67, Mitsubishi.
27. Fourne, Chemiefasern, 5(1968)535.
28. U.S. Patent 3,366,722, to Canadian Celanese, 1968.
29. U.S. Patents 3,399,259, ICI, 1968; 3,533,904, Hercules, 1970; British Patents 1,094,933, ICI, 1969; 1,183,410, ICI, 1970.
30. U.S. Patent 3,509,013, Hercules, 1970; British Patent 1,129,399, Hercules, 1968.
31. British Patent 1,129,399, Hercules, 1968; U.S. Patent 3,207,737; 3,207,739, Shell. (Also see Chapter 4.)
32. U.S. Patent 3,509,013, Hercules, 1970; British Patent 1,129,399, Hercules, 1968.
33. British Patents 1,134,924, Asahi Kasei, 1968; 1,171,274, Hercules, 1970.
34. U.S. Patent 3,505,164, Hercules, 1970; British Patent 1,134,924, Asahi Kasei, 1968.
35. British Patent 1,121,314, Mitsubishi, 1968.
36. Fourne, Chemiefasern, 535, 5/1968.
37. U.S. Patent 3,480,709, Hoechst, 1969.

38. U.S. Patent 3,050,821, Hoechst, 1969.
39. British Patent 1,151,272, Kanegafuchi Bo, 1969.
40. U.S. Patent 3,488,251, ICI, 1970.
41. U.S. Patent 3,511,824, Johnson & Johnson, 1970.
42. U.S. Patent 3,659,989, Kanegafuchi, 1972.
43. U.S. Patent 3,655,859, Hercules, 1972.
44. U.S. Patent 3,176,342.
45. British Patent 953,607.
46. British Patent 809,273.
47. British Patent 954,271, Japan Exlan.
48. British Patent 1,004,251, American Cyanamide.
49. British Patent 1,106,849, Japan Exlan.
50. British Patent 1,107,117, Japan Exlan.
51. British Patent 1,060,337, Japan Exlan.
52. British Patent 1,048,370, Kanegatachi.
53. U.S. Patent 3,408,277, Rhodiaceta.
54. Geleiji, J. Polym. Sci., 42(1973)713.
55. British Patent 979,083, Soc. Viscose Suisse.
56. British Patent 950,429, DuPont.
57. British Patent 944,424, DuPont.
58. British Patent 954,174, Japan Exlan.
59. U.S. Patent 3,245,113, American Cyanamide.
60. British Patent 1,019,052; U.S. Patent 3,308,503 to Text. Chem. Res. Co.
61. U.S. Patent 2,295,552, to Monsanto.
62. British Patent 1,045,047, to Mitsubishi Rayon.
63. British Patent 805,033, to DuPont.
64. U.S. Patent 3,458,390, to Kanegafuchi.
65. U.S. Patent 3,606,689, to Toyoke.
66. U.S. Patent 3,439,391, to Deering Milikan.
67. U.S. Patent 3,492,689, to American Cyanamide.

IX. BIBLIOGRAPHY

IX.1. Books

1. Bulk, Stretch and Texture, Papers, Annual Conf. Textile Inst. 1966.
2. R. Jeffries, Bicomponent Fibers, Merrow, Watford, Eng., 1971.
3. C. Placek, Multicomponent Fibers, Noyes Data Corp., Park Ridge, N. J., U.S.A., 1971.

IX.2. Journal Literature

1. M. Stratmann, Two-Component Fibers, Z. Ges. Textil. Ind., 63(1961)249-252.
2. F. Fourne, Two-Component Bulked Yarns, Chemiefasern, 14(1964)728-731.
3. W.E. Fitzgerald and J.P. Knudson, Self-Textured Fibers: Texture Development and Mechanical Properties, Bulk, Stretch and Texture, Papers Annual Conf. Textile Inst., 51(1966)134-164.
4. W.E. Fitzgerald and J.P. Knudson, Mixed-Stream Spinning of Bicomponent Fibers, Textile Research J., 37(1967)447-453.
5. F. Fourne, Bicomponent Yarns, Chemiefasern, 18(1968)353-359.
6. Heterofil Bicomponent Fibers, Rubber and Plastics Age, 49(1968)794-795.
7. R.A. Buckley, Development of Bicomponent Fibers, Chem. Eng. Progr., 65(1969)41-45.
8. P.A. Koch, Bicomponent Fibers, Z. Ges. Textil. Ind., 72(1970)253-256.
9. A.R. Freedman, High Bulk Yarns from Composite Films, Modern Plastics, 47(1970) 108-110.
10. E. Burd, Components of Conjugate Fibers, Fibre Chem., 3(1971)560.

11. A. El-Shiekh, Mechanics of Bicomponent Fibers. I. Theoretical Analysis, Textile Res. J., 41(1971)281-294. II. Experimental Investigations, Text. Res. J., 41(1971)916-922.
12. S. Backer, On the Mechanics of Bicomponent Fibers, Text. Res. J., 41(1971) 1008-1011.
13. R.K. Gupta, On the Mechanics of Bicomponent Fibers - Reply, Text. Res. J., 41(1971)1011-1012.
14. Heterofil Staple and Filament, Text. Mfr., 98(1971)20.
15. V. Krapotkin, Relative Position of the Polymers in Conjugate Fibers, Fiber Chem., 3(1971)335-336.
16. J.H. Ballman, Bicomponent Flow of Polymer Melts in a Tube, Polym. Prep. ACS, Div. Polym. Chem. 1972.
17. G. Ronca, Contribution to the Theory of Fiber Crimping, Faserforsch. u. Textiltech., 23(1972)55-60.
18. J.H. Ballman, Stratified Bicomponent Flow of Polymer Melts in a Tube, Appl. Polym. Symp., 20(1972)175-189.
19. R.H. Brand, Three Dimensional Geometry of Crimp., Text. Res. J., 43(1973) 544-555.
20. F. Geleiji, Bicomponent Fibers on Polypropylene Basis, J. Polym. Sci., Symp., 42(1973)713-716.
21. F. Fourne, Bicomponent Yarn Texturizing, Chemiefasern, 24/76(1974)832-834.
22. P.A. Koch, Nature of Bicomponent Fibers, Ciba-Geigy Rev., 1(1974)2-11.
23. R. Jeffries, Manufacture of Bicomponent Fibers, Ciba-Geigy Rev., 1(1974)12-17.
24. A.S. Abhiraman, Mechanics of Crimp Formation in a Bicomponent Fiber Model, Text. Res. J., 44(1974)319-323.
25. A. Ya. Malkin, Bicomponent Flow of Molten Polymers in Annular Dies of Extruders, J. Appl. Polym. Sci., 19(1975)375-390.
26. B. Bupta, Theory of Self-Crimping Bicomponent Filaments, Text. Res. J., 45(1975)338-349.
27. J.H. Southern Stratified Bicomponent Flow of Polymer Melts in a Tube, J. Polym. Sci., Phys. Ed., 13(1975)863-869.
28. Anon., Avoidance of Interfilament Sticking During Quenching of Bicomponent Filaments, Research Disclosure 148(1976)7.
29. Anon., Chisso Offering Sheath-Core Bicomponent Fiber, Nonwoven Report, 75(1977)
30. Tatsuki Matsuo, Polypropylene Fibers Crimped by Asymmetric Quenching, J. Text. Mach. Soc. Japan, 23(1977)2, 29-34.

IX.3. U.S. Patents

U.S. 2,987,797	DuPont	1961
U.S. 2,989,798	DuPont	1961
U.S. 3,038,240	DuPont	1962
U.S. 3,093,444	DuPont	1963
U.S. 3,117,362	DuPont	1964
U.S. 3,117,906	DuPont	1968
U.S. 3,118,011	DuPont	1965
U.S. 3,124,926	DuPont	1965
U.S. 3,174,184	Monsanto	1965
U.S. 3,176,346	Monsanto	1965
U.S. 3,182,106	Am. Cyanamide	1965
U.S. 3,209,402	Celanese	1965
U.S. 3,233,023	ICI	1966
U.S. 3,244,785	DuPont	1966
U.S. 3,245,113	Am. Cyanamide	1966
U.S. 3,262,257	DuPont	1967
U.S. 3,295,552	Monsanto	1967
U.S. 3,296,681	Shell	1967
U.S. 3,303,548	Monsanto	1967

U.S. 3,308,503	Text. Chem. Res.	1967
U.S. 3,315,021	Snia Viscosa	1967
U.S. 3,316,336	Dow Chem.	1967
U.S. 3,318,987	Monsanto	1967
U.S. 3,320,633	DuPont	1967
U.S. 3,332,110	Courtaulds	1967
U.S. 3,350,741	Toho	1967
U.S. 3,358,322	Monsanto	1967
U.S. 3,359,359	DuPont	1967
U.S. 3,362,149	Bancroft	1968
U.S. 3,375,548	Mitsubishi	1968
U.S. 3,381,074	DuPont	1968
U.S. 3,383,850	Ratti	1968
U.S. 3,387,327	Monsanto	1968
U.S. 3,397,426	Japan Exlan	1968
U.S. 3,399,259	ICI	1968
U.S. 3,401,516	Monsanto	1968
U.S. 3,406,427	Japan Exlan	1968
U.S. 3,408,277	Rhodiaceta	1968
U.S. 3,413,683	Am. Cyanamide	1968
U.S. 3,421,181	DuPont	1969
U.S. 3,424,834	Chemcell	1969
U.S. 3,425,091	Kanegafuchi	1969
U.S. 3,425,107	Kanegafuchi	1969
U.S. 3,438,193	Mitsubishi	1969
U.S. 3,438,439	Mitsubishi	1969
U.S. 3,446,005	Mitsubishi	1969
U.S. 3,446,703	DuPont	1969
U.S. 3,447,296	Monsanto	1969
U.S. 3,453,688	Monsanto	1969
U.S. 3,453,689	DuPont	1969
U.S. 3,453,689	DuPont	1969
U.S. 3,453,817	DuPlan	1969
U.S. 3,457,338	Dow Chemical	1969
U.S. 3,457,341	DuPont	1969
U.S. 3,457,610	Monsanto	1969
U.S. 3,458,900	Toyo Rayon	1969
U.S. 3,459,846	Kanegafuchi	1969
U.S. 3,458,390	Kanegafuchi	1969
U.S. 3,469,279	Brit. Nylon Spin.	1969
U.S. 3,458,615	DuPont	1969
U.S. 3,471,605	Snia Viscosa	1969
U.S. 3,480,709	Hoechst	1969
U.S. 3,480,996	Kanegafuchi	1969
U.S. 3,492,689	Am. Cyanamide	1970
U.S. 3,533,904	Hercules	1970
U.S. 3,538,544	DuPont	1970
U.S. 3,488,251	ICI	1970
U.S. 3,491,178	Mitsubishi	1970
U.S. 3,492,692	Japan Exlan	1970
U.S. 3,501,805	Am. Cyanamide	1970
U.S. 3,497,585	Monsanto	1970
U.S. 3,505,164	Hercules	1970
U.S. 3,500,498	Asahi	1970
U.S. 3,509,013	Hercules	1970
U.S. 3,536,802	Kanegafuchi	1970
U.S. 3,533,905	Hercules	1970
U.S. 3,553,774	Bemberg	1971
U.S. 3,540,077	Japan Exlan	1970

U.S. 3,551,279	Kanegafuchi	1970
U.S. 3,559,237	Am. Enka	1971
U.S. 3,560,604	Nat. Plast. Prod.	1971
U.S. 3,568,249	Matsui	1971
U.S. 3,570,235	Toray	1971
U.S. 3,577,213	Celanese	1971
U.S. 3,577,498	Toyo	1971
U.S. 3,584,339	Chisso	1971
U.S. 3,585,685	FMC	1971
U.S. 3,594,459	Celanese	1971
U.S. 3,596,459	Teijin	1971
U.S. 3,600,271	Toray	1971
U.S. 3,601,846	Monsanto	1971
U.S. 3,608,148	Am. Enka	1971
U.S. 3,613,170	Am. Cyanamide	1971
U.S. 3,618,166	Kanegafuchi	1971
U.S. 3,655,859	Hercules	1972
U.S. 3,657,062	Chisso	1972
U.S. 3,659,988	Phillips Petrol.	1972
U.S. 3,659,989	Kanegafuchi	1972
U.S. 3,663,352	Monsanto	1972
U.S. 3,663,376	Asahi	1972
U.S. 3,672,802	Kanegafuchi	1972
U.S. 3,677,880	Asahi	1972
U.S. 3,681,189	Kanegafuchi	1972
U.S. 3,681,910	DuPont	1972
U.S. 3,683,611	Rhodiaceta	1972
U.S. 3,761,348	Monsanto	1973
U.S. 3,861,843	Celanese	1973
U.S. 3,720,576	Asahi	1973
U.S. 3,724,038	Mitsubishi	1973
U.S. 3,725,192	Kanegafuchi	1973
U.S. 3,760,046	Avisun	1973
U.S. 3,761,348	Monsanto	1973
U.S. 3,787,162	ICI	1974
U.S. 3,785,919	DuPont	1974
U.S. 3,805,343	FII	1974
U.S. 3,807,917	Japan Exlan	1974
U.S. 3,831,231	FII	1974
U.S. 3,831,368	Uniroyal	1974
U.S. 3,843,098	Eastman	1974
U.S. 3,844,103	Teijin	1974
U.S. 3,849,044	Kanegafuchi	1974
U.S. 3,860,679	Fiber Ind.	1975
U.S. 3,860,685	Japan Exlan	1975
U.S. 3,861,843	Celanese	1975
U.S. 3,924,045	Toray Ind.	1976
U.S. 3,992,420	Bayer	1976
U.S. 3,930,772	ICI	1975
U.S. 3,940,543	Mitsubishi	1976
U.S. 3,995,420	Bayer	1976
U.S. 3,979,496	Schwartz	1976
U.S. 4,020,139	DuPont	1977
U.S. 4,019,844	Toray Ind.	1977
U.S. 4,052,146	Monsanto	1977

IX.4. British Patents

B.P. 896,955	DuPont	1962
B.P. 930,291	Courtaulds	1963
B.P. 930,629	Soc. Viscose Suisse	1963
B.P. 954,274	Japan Exlan	1964
B.P. 964,406	ICI	1964
B.P. 968,363	DuPont	1964
B.P. 972,932	Brit. Nylon Spin.	1964
B.P. 979,083	Soc. Viscose Suisse	1965
B.P. 981,003	NV Onderzuekingsinst Res.	1965
B.P. 984,919	Brit. Nylon Spin.	1965
B.P. 985,717	Monsanto	1965
B.P. 1,002,683	Mitsubishi	1965
B.P. 1,004,251	Am. Cyanamide	1965
B.P. 1,016,100	Chemcell	1966
B.P. 1,016,677	Brit. Nylon Spin.	1966
B.P. 1,029,119	Soc. Rhodiaceta	1966
B.P. 1,030,005	Am. Cyanamide	1966
B.P. 1,030,523	Monsanto	1966
B.P. 1,041,904	Brit. Nylon Spin.	1966
B.P. 1,045,047	Mitsubishi	1966
B.P. 1,050,042	Soc. Rhodiaceta	1966
B.P. 1,045,990	Kanegafuchi	1966
B.P. 1,048,370	Kanegafuchi	1966
B.P. 1,050,191	Toyo Rayon	1966
B.P. 1,053,916	Monsanto	1966
B.P. 1,057,579	ICI	1967
B.P. 1,058,448	Asahi	1967
B.P. 1,060,337	Japan Exlan	1967
B.P. 1,061,692	W. C. Heraeus	1967
B.P. 1,070,257	Chemcell	1967
B.P. 1,074,681	Chemstrand	1967
B.P. 1,079,790	DuPont	1967
B.P. 1,083,008	Kanegafuchi	1967
B.P. 1,085,057	Courtaulds	1967
B.P. 1,087,823	Hoechst	1967
B.P. 1,088,386	Monsanto	1967
B.P. 1,088,592	Lansil	1967
B.P. 1,091,367	ICI	1967
B.P. 1,094,688	Snia Viscosa	1967
B.P. 1,095,166	ICI	1967
B.P. 1,095,752	Courtaulds	1967
B.P. 1,097,003	Courtaulds	1967
B.P. 1,100,430	ICI	1968
B.P. 1,101,452	Onderzoekingsinst	1968
B.P. 1,103,528	Barmag	1968
B.P. 1,105,307	ICI	1968
B.P. 1,105,575	Asahi	1968
B.P. 1,106,849	Japan Exlan	1968
B.P. 1,107,117	Japan Exlan	1968
B.P. 1,112,358	Snia Viscosa	1968
B.P. 1,120,241	Kanegafuchi	1968
B.P. 1,121,314	Mitsubishi	1968
B.P. 1,123,952	Asahi	1968

B.P. 1,129,399	Hercules	1968
B.P. 1,130,065	Snia Viscosa	1968
B.P. 1,133,040	Toyo Rayon	1968
B.P. 1,134,924	Asahi	1968
B.P. 1,136,473	Chemcell	1968
B.P. 1,137,027	Toyo Boscki	1968
B.P. 1,137,028	Toyo Boscki	1968
B.P. 1,139,653	Kanegafuchi	1969
B.P. 1,148,773	Kanegafuchi	1969
B.P. 1,145,818	Asahi	1969
B.P. 1,150,587	Asahi	1969
B.P. 1,150,399	Inventa	1969
B.P. 1,151,272	Snia Viscosa	1969
B.P. 1,151,303	VEB Chemiefaser. Schwerza	1969
B.P. 1,167,390	Snia Viscosa	1969
B.P. 1,167,109	Chem. Lenzing.	1969
B.P. 1,167,182	AKZU	1969
B.P. 1,169,106	Teijin	1969
B.P. 1,171,274	Hercules	1969
B.P. 1,171,600	Bemberg	1969
B.P. 1,171,325	Mitsubishi	1969
B.P. 1,171,843	Toyo Rayon	1969
B.P. 1,172,045	Asahi	1969
B.P. 1,172,011	ICI	1969
B.P. 1,173,817	Asahi	1969
B.P. 1,174,869	Kanegafuchi	1969
B.P. 1,175,756	Toyo Boscki	1969
B.P. 1,179,598	Snam Progetti	1970
B.P. 1,179,615	ICI	1970
B.P. 1,181,013	AKZU	1970
B.P. 1,181,483	Japan Exlan	1970
B.P. 1,183,410	ICI	1970
B.P. 1,189,553	Snia Viscosa	1970
B.P. 1,192,270	Kanegafuchi	1970
B.P. 1,193,272	VEB Chem. Schwarza	1970
B.P. 1,194,406	Inventa	1970
B.P. 1,196,275	VEB Chem. Schwarza	1970
B.P. 1,198,035	Asahi	1970
B.P. 1,20-,134	Asahi	1970
B.P. 1,204,339	VEB Chem. Schwarza	1970
B.P. 1,205,331	Mitsubishi	1970
B.P. 1,205,959	Klinger	1970
B.P. 1,207,408	Allied	1970
B.P. 1,212,413	USSR	1970
B.P. 1,215,315	AKZU	1970
B.P. 1,217,226	Klinger	1970
B.P. 1,219,503	VEB Chem. Schwarza	1971
B.P. 1,229,842	Kanegafuchi	1971
B.P. 1,230,195	Toray Ind.	1971
B.P. 1,230,991	Rasmussen	1971
B.P. 1,232,282	Smith and Nephew	1971
B.P. 1,233,585	Snam Progetti	1971
B.P. 1,248,293	Kanegafuchi	1971
B.P. 1,259,034	Montecatini	1972
B.P. 1,262,030	Monsanto	1972
B.P. 1,286,416	Monsanto	1972
B.P. 1,287,945	Asahi	1972
B.P. 1,323,834	Mitsubishi	1973
B.P. 1,340,142	ICI	1973

B.P. 1,367,427	Monsanto	1974
B.P. 1,370,216	AKZU	
B.P. 1,381,865	ICI	1975
B.P. 1,402,360	Barmag	1975
B.P. 1,406,335	Rohn Poulenc	1975
B.P. 1,408,094	Hoechst	1975
B.P. 1,411,000	Kuraray	1975
B.P. 1,415,943	Asahi	1975
B.P. 1,416,322	Platt	1975
B.P. 1,444,319	Barmag	1976
B.P. 1,425,705	ICI	1976
B.P. 1,448,302	Nat. Spinning	1976
B.P. 1,465,038	James Mackie	1977
B.P. 1,473,835	James Mackie	1977

Chapter XII FIBERS FROM FILM

I. INTRODUCTION

Production of textiles from film is not altogether new: since the earliest recorded history, natural fibers have been produced from film-like materials. By mechanical working or chemical treatment, leaves from plants such as the palm and sheets of tree bark have been converted into fibrous products which were then spun into textile yarns. The production of fibers from disintegration of highly drawn polymer films has been known since the 1930s (ref.1). Very little progress was made, however, until the development of crystalline polymers, particularly high-density polyethylene and polypropylene. These two polymers have been the basis of nearly all subsequent developments in fiber from drawn film.

Fibers and yarns from synthetic film signify a whole new area of development in the technology of manmade fibers. The starting material is a sheet of polymeric film, in contrast to the discrete filaments emerging from individual spinnerette holes. Originally, the development of fibers from film represented an attempt to circumvent Montecatini's basic patents relating to the production of polypropylene fibers through spinnerette technology. The reduced manufacturing costs and small fixed capital required for the manufacture of film yarns gave an additional impetus to the development of these products. A comparatively low level of technological sophistication needed for the production of tape and fibrillated yarns helped to proliferate small manufacturing units, first in Western Europe and then in the United States and elsewhere. An active interest on the part of machinery manufacturers and polymer suppliers no doubt reinforced this trend.

In spite of some enthusiastic claims made in the early stages of their development, (ref.2), fibers from film were never meant to compete with fibers from spinnerette in their traditional usage. The two technologies are to a large extent complimentary. Film yarns and tapes are used exclusively in the high end of the denier range, which spinnerette fibers have been unable to penetrate to any great extent - largely for reasons of cost. For the time being, film yarns seem to have a clear edge in low-cost industrial applications, while in more sophisticated areas the spinnerette fibers are nonreplaceable. In the middle ground there are some areas of potential applications for film yarn alongside spinnerette fiber, to make the prospect of the next few years quite interesting.

Polypropylene and polyethylene were the two contending materials for the fiber-from-film market. Polypropylene has clearly emerged as the winner. Indeed, the showing of polypropylene in this field has been so strong that, in layman's language, "polypropylene yarns" has become almost synonymous with coarse, fibrillated film strands or thin, narrow tapes.

The technology (and the market) of film yarns is divided fairly clearly into three definite areas:

(1) Oriented tapes

(2) Split or fibrillated film yarns

(3) Fabric-like products based on stretched-film technology.

Each group comprises products with specific characteristics, although many products may defy any clearcut classification and may well lie in a borderline area.

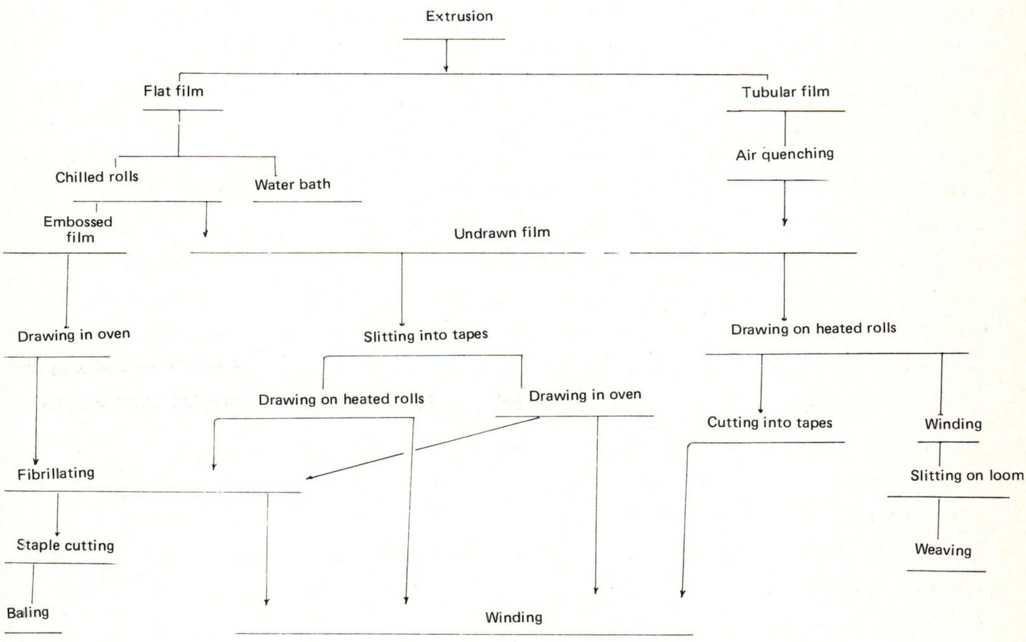

Fig. 1. Flow-Sheet Diagram for the Production of Fibers from Film

Figure 1 shows various steps leading to the formation of oriented tapes and fibrillated film yarns. A variety of machinery is available commercially to accomplish these processes. Table I lists some selected firms which offer the necessary machinery, most often as complete lines, for the production of polypropylene tape and film yarns.

A large variety of processes is known for the manufacture of fibers from film.

All of these processes share two steps in common: the production of film (or film strips), and uniaxially drawing of film or film strips. Therefore, before we proceed to discuss specific products and processes, a few comments on the general theme of film extrusion and drawing are in order.

TABLE I

Some Selected Suppliers of Machinery for the Production of Fibers From Film

ARCT, Rue Cuvier, F-42300 Roanne, France
Barmag Barmer Meschinenfabrik A. G., Remscheid-Lennep, West Germany
Beloit Corp., P. O. Box 350, Beloit, Wisc. 53511, U.S.A.
Black Clawson Co., Rt. 3W, Fulton, N. Y. 13069, U.S.A.
Bruckner-Trockenbau, P. O. Box 124, 725 Leonberg, West Germany
Bouligny Co., P. O. Box 10158, Charlotte, N. C. 28237, U.S.A.
Covema, na Podgoea 4, Milan, Italy
Husky Manufacturing Co., 201 E. Morehead St., Reidsville, N. C. 27320 U.S.A.
L. Dolci SpA, Cusanano Milanino, via Zucchi 18, Italy
Chemiefaser Lenzing, Lenzing, Austria
Japan Steel Works Ltd., Tokyo, Japan
Kawasaki Heavy Industries Ltd., 16-1 Nokamochi-Dori 2 Chome Ikuta-ku Kube, Japan
Modern Plastics Machinery Corp., Clifton, N. J. 07015, U.S.A.
Leesona/Johnson, Warwick, R. I. 02887, U.S.A.
Paul Pakleppa & Co., Munich, Germany
Plasticiser Ltd., Dringlington, Bradford, England
Rheinstahl Henschel A. G. Kassel, Germany
Reifenhauser K. G., 5210 Troisdord 15, Germany
Samafor, 60-62 Rue Emile Zola, 93 La Courneuve, France
Toray Engineering Co., 5-2 3-Chome Nakanoshima Kito-ku, Osaka, Japan
Toshiba Machine Co., 2-11 4-Chome Ginza Chuo-ku, Tokyo, Japan
Melville Plastics Engg, Kirkcaldy, Fife, Scotland
Emil Blaschke, 7057 Endersbach, West Germany
Utita, Via Moscova, Milano, Italy

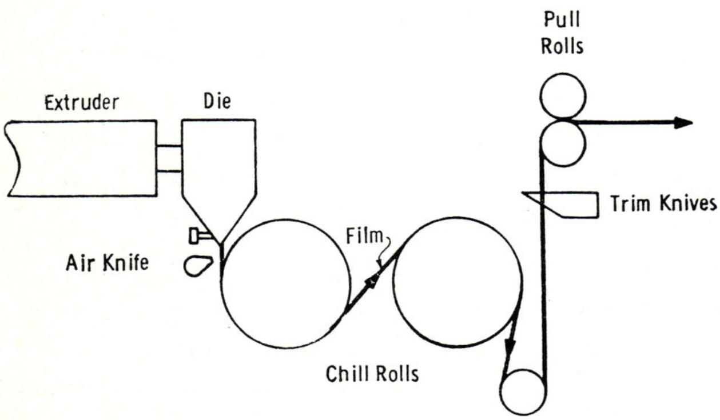

Fig. 2. Chill Roll Film Casting

Film Formation: Three main systems for film production are currently in use. The first two are based on extruding a sheet of film through a flat die and then either cooling it over chilled rolls (Fig. 2) or by means of a water quenching bath (Fig. 3). The third process consists of extruding a ring of molten material through a die, blowing through the center, and expanding the molten polymer tube so that the molten material is inflated and stretched like a balloon to the desired film thickness. Following cooling by a circular curtain of air, this balloon is collapsed and led away from the formation zone (Fig. 4). All three methods were originated by the plastics industry. These techniques are now well established (ref.49).

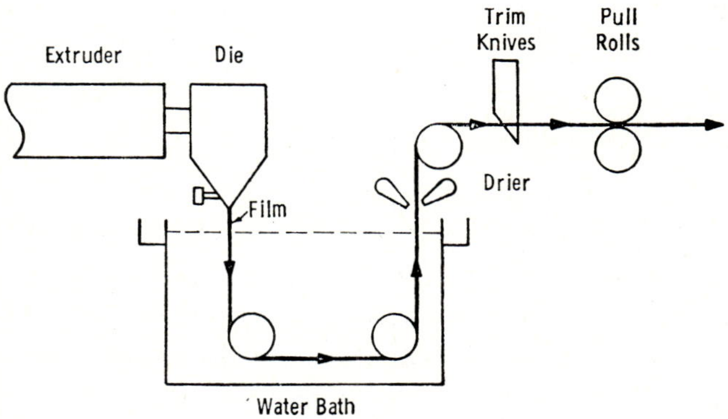

Fig. 3. Water Bath Film Casting

Water-bath quenching has the advantage of lower capital cost and ease of maintenance. Since both sides of the film are cooled simultaneously, more uniform cooling of the melt results. Water-bath quenching produces a film more easily stretched and with higher tenacity at a given stretch ratio than the film produced by either of the two other methods (ref.3). However, the process has several major disadvantages also. These disadvantages are poor gauge control, corrugation or distortion of the film due to water turbulence, and occasional water carryover into the orientation section.

In the chilled-roll casting system the die is similar to that used in the water-bath method, but the polymer sheet is extruded directly onto the polished surface of a revolving roll. The roll is cooled by the internal circulation of a liquid, and the cooling medium never comes into direct contact with the polymer film. Because the film does not come in contact with water, it is perfectly dry. The film is also usually free from cockles. Once the equipment is installed, the operation is relatively trouble free. This method gives excellent gauge control

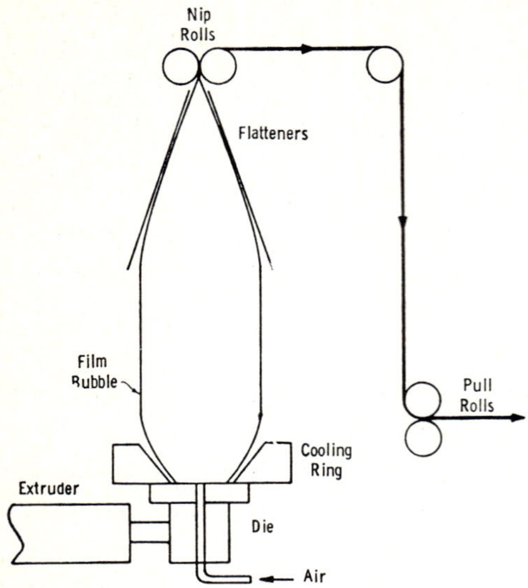

Fig. 4. Blown Film Production

and enables the production of very thin films at high extrusion rates. The disadvantages are comparatively high initial equipment cost and possible damage to the polished surface of the rolls in operation.

In the water-casting method a thick bead is formed at the edge of the film, while in the roll-casting method the film edges are not smooth. In both cases the edges have to be trimmed and treated as waste, resulting in *slit-off waste* - generally in the range of from 5 to 7 percent. The blow-film method has an economic advantage in that trim is minimized and the double layer of film means that twice as much poundage can be processed in a production unit of a given width. However, air cooling of melt is less efficient than the other methods because the cooling ring requires special design and is tricky to operate. The annular die is also more difficult to adjust than a longitudinal slot, and gauge control can be a problem. The gauge uniformity can be significantly improved by using "Rotatruder", as is used in the Chemiefaser-Lenzig process. In this arrangement the die and the cooling equipment are coupled with the barrel, the whole unit revolving through 360° every 20 minutes.

Blown-film method is almost universally used for the production of polyethylene film. However, for polypropylene the extrusion speeds have to be reduced because of the low melt strength and difficulties in cooling. Moreover, the film so produced is difficult to orient due to the development of spherulytic structure. To overcome these problems new processes have been recently developed in which extrusion is done downwards and the cooling is accomplished with chilled water. The film produced under these processing conditions has several desirable characteristics in addition to a substantial increase in production rate.

For any particular group of film products one of the fundamental decisions to be made is the choice between flat film made from a straight die and blown film made from an annular die. The main advantage of flat film is that higher extruder outputs can be obtained and higher working temperatures are common. Using the modern chilled roll, which is rapidly displacing water quenching, take-off speeds of unstretched polypropylene film can be in excess of 30 meters a minute. The more controlled cooling of the melt gives a finer grain structure which responds better to subsequent orientation than that produced with the conventional blown film technique. A flat-film setup is adaptable for a broader range of thicknesses, and is essential if very thick baler-twine type tapes or packaging bands are desired. Flat film is also capable of closer thickness control as a result of better facility for die-lip adjustment and ease of fitting film gauges. Flat-film processes for slit-film production are generally continuous and are most commonly used for polypropylene.

Blown-film units, although superficially more impressive, generally involve lower capital investment. If adjusted properly, the actual film production stage is less susceptible to film thickness variations, although, if present, such variations can give trouble later.

Probably the most interesting feature of blown film is the transverse stretch involved in the inflation of the film bubble. The inflation of the melt bubble gives a biaxial orientation which is never completely eliminated during subsequent unidirectional stretching. The end product has reduced splitting characteristics, usually reduced strength, and softer feel than does cast film. The reduced splitting tendency is desirable in weaving tapes but is undesirable in random splittable products. Blown films lead to more controlled splitting when a mechanical fibrillator of the pin-roller type slitter is used for the production of fine mesh film-fiber networks. However, where extreme strength and long splitting are needed, such as for twine purposes, the straight die remains the obvious choice. For certain processes, such as randomly fibrillated yarns produced by foamed melts, polymer mixtures, or by incorporating other fibrillating agents in the melt, there is no alternative but to use straight die arrangement.

Isotactic polypropylene can crystallize in form of large crystalline regions (spherulites) or minute crystals. In the later case one speaks of a *smectic* or amorphous structure. When the extruded polymer emerges from the die and is given time to cool slowly, as it is the case with blown tubular film, the crystals forming during solidification have sufficient time to aggregate into large spherulytic structures. The resulting film is rigid, opaque, and shows in polarized light the coarse gyrations of its large spherulites. In the case of extrusion of a flat film and quenching by either water immersion or on cooled chill rolls, the film is cooled much more rapidly and crystalline regions have no time to grow. The resulting film is more flexible, transparent, and reveals in polarized light minute gyrations typical for the microcrystalline, *smectic* state.

The differences in fine structure caused by the various ways of quenching can be demonstrated from the x-ray wide-angle diffraction patterns depicted in Fig. 5. The polypropylene film quenched at 11°C shows very diffuse x-ray wide-angle reflections, indicating the existence of relatively small ordered regions. The film solidified on a chill roll of 55°C shows more crystallinity. Similarly, the air-cooled blown film shows many more distinct diffractions, typical for large ordered regions.

The degree of order achieved in the extrusion step determines stretchability, and thus the achievement of good mechanical properties. It is reflected in the response of the system to the applied stretch for achieving molecular and crystallite orientation. As seen from Fig. 6, film solidified at 11°C on a chill roll shows a somewhat faster response to the orienting effect of stretch. The more rigid network of the large ordered regions in an air-cooled blown film is more difficult to orient, as shown by the lower orientation values at equal stretch ratios. This results in a somewhat faster development of tenacity on stretching in water-quenched film or film tapes, compared with air-cooled materials (Fig. 7). For these reasons flat die extrusion and water quenching is considered best suited for the production of thicker films and film tapes ($>$ 100 µm), such as used for baler twine or ropes.

Chill-roll casting has the advantage that the polymer sheet quenched on water-cooled drums never comes in contact with water. The chief disadvantages, however, are: (a) the risk of entrapping air pockets between film and drum, resulting in nonuniformities due to varying heat transfer - this problem can be overcome to some extent by using suction or air blowing jets ("vacuum or air knives"); (b) the need for edge trimming; and (c) structural nonuniformity, since only one surface of the film is cooled at a time. Furthermore, chill-roll casting equipment is more expensive than either water-quenching or air-blowing extrusion equipment.

The thickness of the film prior to stretching has to be adjusted in relation to the dimensions of the final film-tape product and to end-use requirements. It has a very marked effect on the economics of production and on the tape properties. The thinner the film, the lower the output of a given production line, the higher the stretchability (and the resulting tenacity of the tape), and the higher the tendency for fibrillation in further processing and use.

The restrictions that thickness places on line speed and stretch ratio are much more pronounced with polyethylene than is the case with polypropylene.

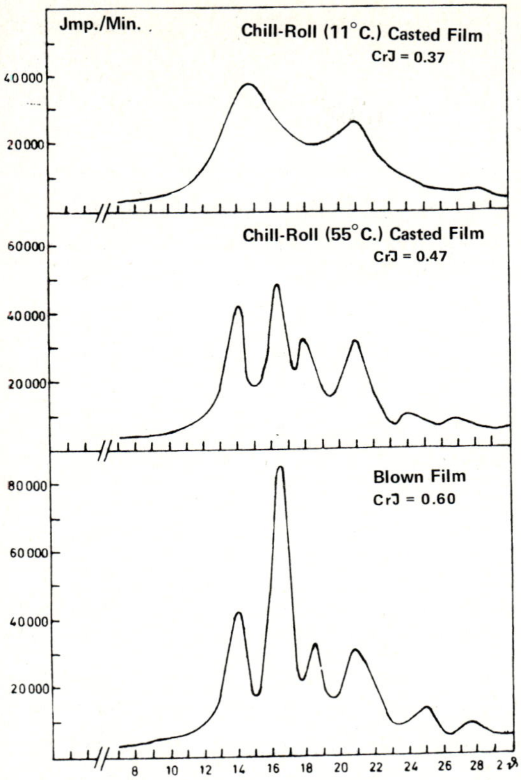

Fig. 5. X-Ray Wide-Angle Diffractograms of Polypropylene Films Solidified at Various Temperatures (ref. 66)

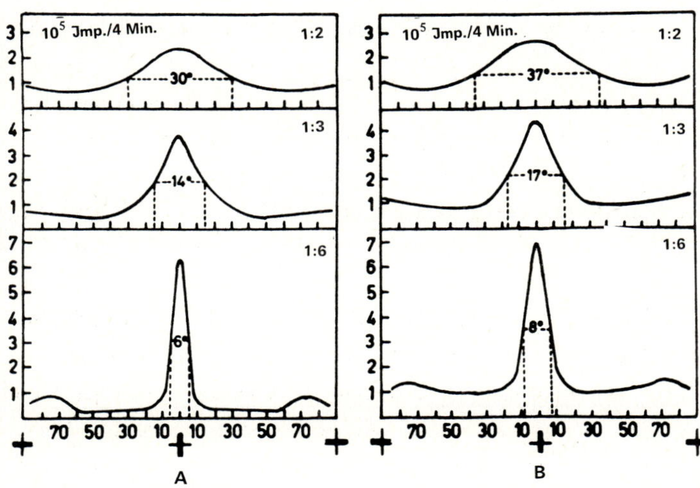

Fig. 6. Effect of Various Stretch Ratios on the Development of Crystallite Orientation (A) in Chill-Roll Cast (11°C) and (B) in Blown Polypropylene Film (Azimuthal Scan Through the X-Ray Diffraction at 2θ = 14°) (ref.66)

TABLE II

Differences Between Chill-Roll-Casting and Blowing Techniques With Respect to Maximum Stretch, Achievable Tenacities and Elongation, and Fibrillation Tendency of Polypropylene Film (m.f.i. 1.7 g/10 min; Extrusion Temperature 240°/200°C, Resp.; Stretch Temperature 135°C) (ref. 67)

Processing Method	Stretch Ratio	Maximum Stretch	Tenacity (g/dtex)	Elongation (%)	Fibrillation Tendency (Relative)
Chill-Roll Casting	1.6		3.4	33	2.8
	1.8		5.3	22	5.0
	1.10		6.1	16	7.0
		1:10			
Blown	1.6		2.7	54	2.2
	1.8		3.8	27	3.8
	1.10		5.8	19	6.7
		1.11			

TABLE III

Influence of Film Manufacturing Technique and Thickness on Polyethylene (m.f.i. 0.4 g/10 min; Density: 0.962 g/ml) Tape Processing (ref. 67)

Processing Method	Film Thickness (μm)	Maximum Inlet Speed for Even Stretch (m/min)	Maximum Stretch Ratio
Blowing	50	20	8.5.1
	75	13	10.1
	100	8	11.1
Chill-Roll Cast	50	20	7.1
	75	13	8.1
	100	8	10.1
Water Quenching	100	20	7.1

TABLE IV

Straight Twine Conversion Efficiencies of Polypropylene Fibrillating Tape and of Polyamide or Polyester Nonfibrillating Tapes (ref. 66)

Polymer Type	% Conversion Efficiencies
Polypropylene	90-105
Polyamide	30-60
Polyester	30-60

Normal line speeds of up to 30 m/min and stretch ratios in the order of 8.1 to 10.1 can be applied with polypropylene film of 1200 μm down to 40 μm. When polyethylene is used, however, increasing film thickness necessitates a reduction in line speed in order to obtain an evenly stretched tape or to avoid tape breakage in the hot-air stretching ovens. Some data illustrating this comparison are summarized in Tables II and III.

As already mentioned, decreasing film thickness increases fibrillation tendency. This is the case with both polyethylene and polypropylene. However, polypropylene tends much more towards fibrillation. This is an advantage when tapes are used in the manufacture of ropes, twines, cords, or for making fibrillated yarns. It can be a disadvantage, however, in film-tape fabrics such as those used for packaging purposes or as primary carpet backing. The tenacity of fibrillated polypropylene tape remains unchanged on twisting, as seen from the data in Table IV. In comparison tapes from polymers with lower fibrillation tendency do show large decreases on twisting.

While polyethylene is relatively stable against heat and light, polypropylene is relatively susceptible to oxidative degradation. The suppliers generally sell various grades, differing in their degree of stability. Table V gives an example of difference in heat stability of some commercial grades of polypropylene. When a given grade of polypropylene is extruded into film, the molecular chain length drops and the melt flow index increases with increasing processing temperature. Figure 8 shows some examples how differently stabilized polypropylenes change their melt flow index with increasing extrusion temperature.

TABLE V

Heat Stability of 1000 Denier Weaving Tape Made from Different Grades of Polypropylene Polymer (ref. 68)

Polymer Type	Time for Tenacity to Fall to 50% at 120°C (hours)
Packaging-Twine Grade	120
General-Purpose Grade	320
Extra-Heat-Endurance Grade	900
Very-Highly-Stabilized Grade	1500

The degradative changes in molecular weight of polypropylene during extrusion become evident in two major ways: (1) film-tape and fibrillated yarn tenacities will fall off as the chain molecules are shortened; and (2) owing to the decrease in melt viscosity with progressing degradation, less melt orientation is achieved during extrusion, which also results in reduction of product tenacity.

These effects can be recognized in measuring tensile strength of the unstretched

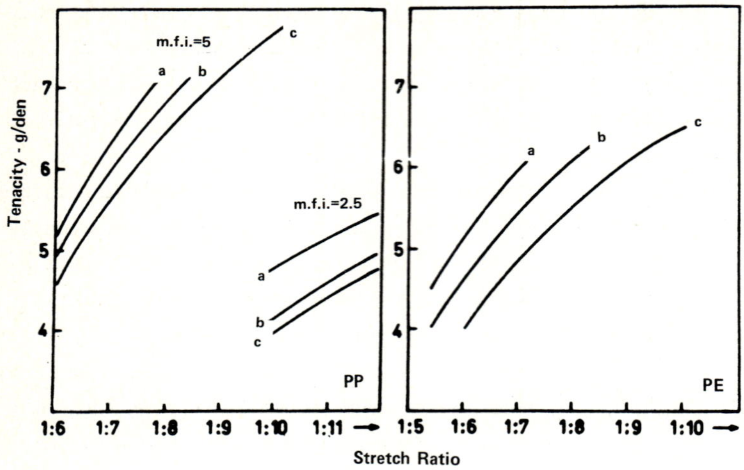

Fig. 7. Effect of Film Manufacturing Technique on Tenacity Development in Polypropylene (Left-Hand Side) and Polyethylene (Right-Hand Side) Film Tapes: A is Water Quenched; B is Chill-Roll Cast; C is Blown (ref. 66)

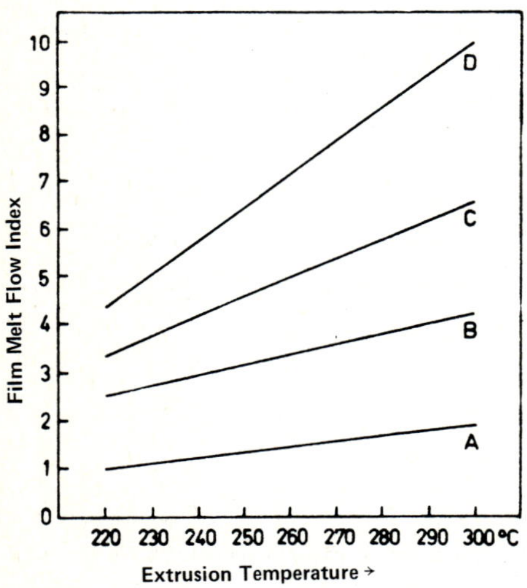

Fig. 8. Effect of Extrusion Temperature on the Film Melt Flow Index (MFI) for Polypropylene Polymers of Different Melt Stability (ref. 68)

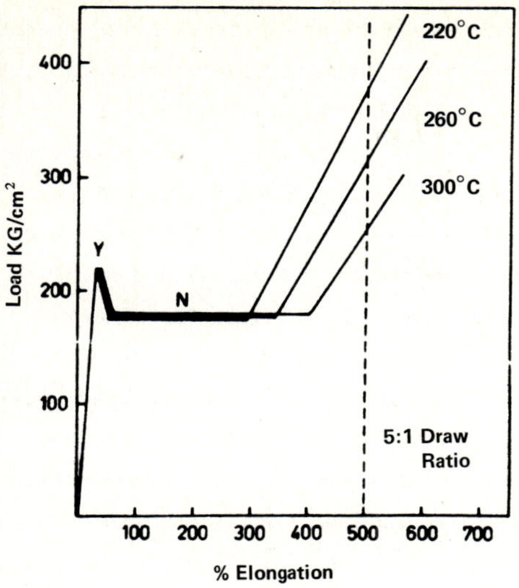

Fig. 9. Effect of Extrusion Temperature on the Tensile Strength of Undrawn Film and its Response to Drawing (Y is the Yield Point; N the Necking)

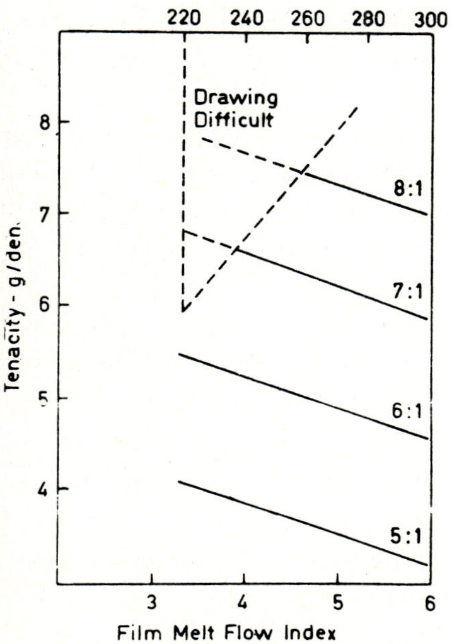

Fig. 10. Relationship Between Tenacity and Film Melt Flow Index at Different Draw Ratios With Respect to Extrusion Temperature (Scale at the Top) (ref. 68)

film and observing its drawing behavior with respect to the extrusion temperature. An example of such measurements is shown in Fig.9. As seen from these data, the film extruded at the lower temperature shows a shorter necking state (represented by the length of the horizontal portion N), responding earlier to the orienting force of drawing. This indicates higher levels of residual melt orientation. As the extrusion temperature is increased, more necking and a later response to the drawing occur, giving lower strength at a given draw ratio. The general effect of a decrease in molecular weight (or an increase in melt flow index) with increasing extrusion temperature, is summarized in Fig. 10. These data illustrate that, as extrusion temperature and melt flow index increase, the tenacity obtainable at a given draw ratio decreases. They furthermore indicate that films extruded at low extrusion temperatures are more difficult to stretch at high draw ratios due to their higher levels of residual melt orientation.

In order to increase the production capacity of given film lines, new and more sophisticated techniques and equipment have been developed in recent years allowing an increase in film speeds from approximately 30-50 to 100 ft/min and more. As film extrusion speed is increased, the level of melt orientation generally increases. This should raise the level of tenacity achievable at a given drawn ratio. However, as shown before, with increasing orientation, the stretchability at higher draw ratios decreases, resulting in a marked lowering of tenacity. This is shown in Fig. 11.

Drawing: Uniaxial stretching is the second most important step in the production of tapes and fibrillated yarns. Three different techniques are in use: (a) long-distance stretching in infrared, hot-air, or stream-heated ovens (see Fig. 12); (b) long-distance stretching on hot plates (see Fig. 13); and (c) short-gap stretching on hot rollers (see Fig. 14).

Drawing through heated ovens has probably been the most popular process. The film or slit undrawn tapes are fed through a set of godets into a long oven. A pull is provided by a second set of rolls which move at 4 to 11 times the speed of the feeding set of godets, depending on the draw ratio desired. A second oven is often used to heat-set the tapes or, in the case of fibrillated yarns, to further draw the film strips.

Most commonly, infrared-heated ovens with air circulation are preferred. Experience has shown that in such ovens a uniform temperature distribution can only be achieved with an even air distribution at air speeds not less than 20 m/sec. Normally, 2.5-4.0 m long ovens are used. The length has to be in relation with the film speed and the desired stretch ratio. In heated oven or steam-chamber stretching, the thickness and the width of tapes are reduced approximately by the square root of the stretch ratio. The width shrinkage is somewhat lower in hot-plate drawing.

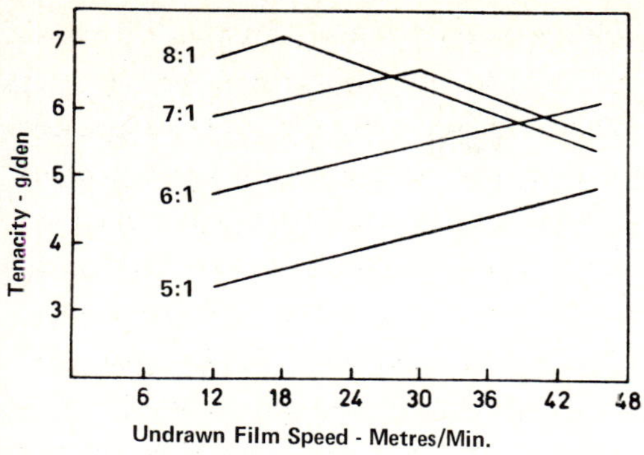

Fig. 11. Effect of Undrawn Film Speed on Tenacity at Various Draw Ratios (ref. 68)

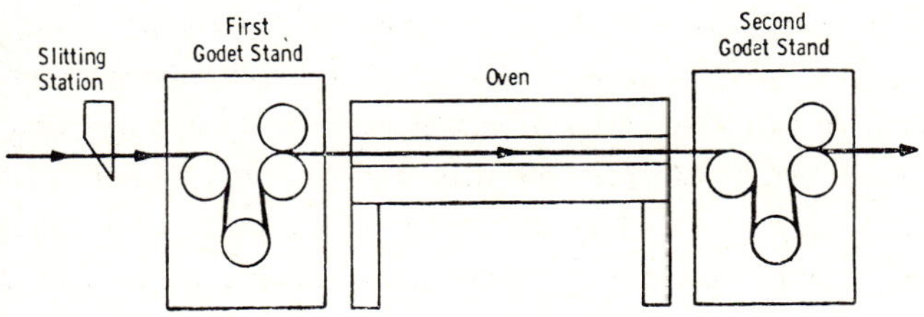

Fig. 12. Oven Orientation of Tapes

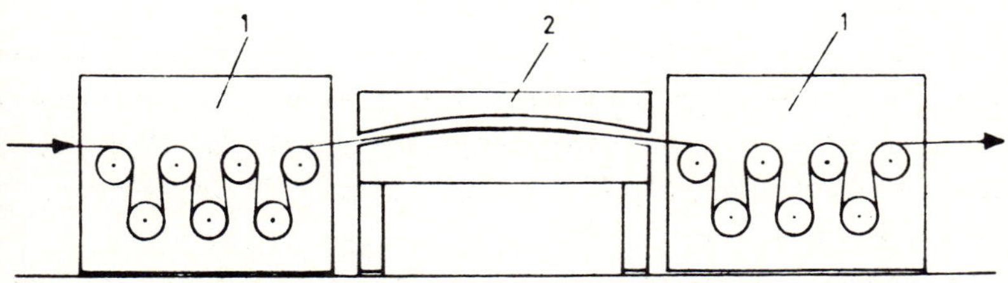

Fig. 13. Scheme of a Long-Distance Hot-Plate Stretching Apparatus: 1 is the Septettes (for Incoming Undrawn Film at Lower Inlet Speed on the Left-Hand Side; for Outgoing Stretched Film at Higher Take-off Speed on the Right-Hand Side); 2 is the Heated Plate for Long-Distance Stretching.

Oven drawing equipment is not too costly, and is relatively simple to operate. However, oven orientation has several significant limitations. An oven is relatively an inefficient heat transfer device and a long distance is required to raise the film to draw temperatures. Over this whole distance the film or strands are unsupported. This leads to a considerable degree of handling difficulties. Oven orientation imposes limitation on the size and production capacity of the line. There is a practical limit to the width of the oven, dictated by convenience in handling the strands through thread-up stage.

Drawing systems with hot plates are better in respect of more efficient heat transfer. However, there is the risk of temperature variations across the width of the hot plate, and use of high temperature is limited by film sticking.

Orientation by heated rolls requires a greater initial equipment investment than does oven orientation and is probably somewhat more complicated to operate. However, it does have major advantages in both product quality control and in the practical maximum size of a single production unit. Heated rolls are usually used in orientation of film in full width rather than in strips of film. Figure 14 depicts schematically one stretching line using heated rolls. The film is passed over two heated rolls where it is heated first on one side and then on the other. The heated film passes to a cold roll operating at a higher speed, and the drawing takes place in the small gap between the hot and the cold rolls. There is no need for the extensive length of unsupported film, which is a characteristic of oven orientation. With roll orientation, film approaching eight feet in width has been successfully oriented at high speeds with a gauge variation across the film of only $\pm$ 1 x 10^{-4} inches (ref.4).

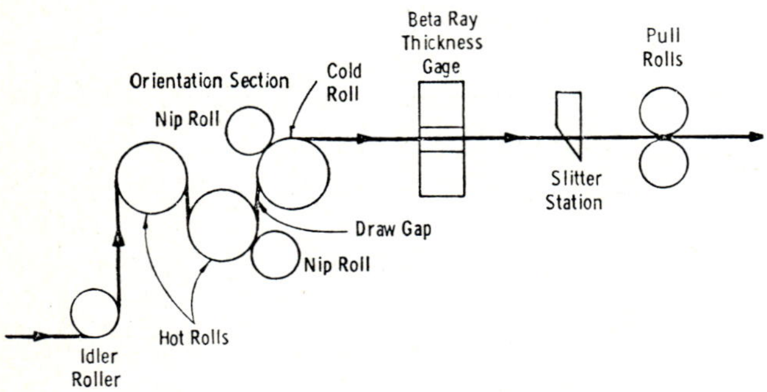

Fig. 14. Roll Orientation of Film

For roller drawing systems, constant heat distribution over the full width is very important. The orientation takes place in a gap of a few mm to about 30 mm. Necking is therefore reduced to a minimum. The film or tape width is more or less maintained while the thickness is linearly reduced with the stretch ratio. For this reason films or tapes drawn with short-gap stretching show more transverse orientation and thus somewhat lower splittability.

Short-gap stretching on heated rolls is advantageous in the production of film or tapes with reduced fibrillation tendency, such as in the manufacture of tapes for primary carpet backing or for bags and sacks. This technique is preferably used in the manufacturing of warp sheet material to be slit on the loom.

The mechanism of orientation on drawing and the processing conditions - properties relationship, have been discussed elsewhere in this book in connection

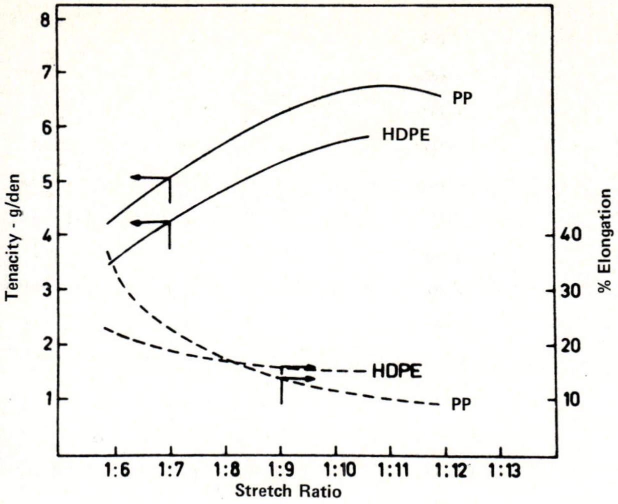

Fig. 15. Effect of Stretch Ratio on Tenacity and Elongation of Polypropylene and High-Density Polyethylene Film-Tape (ref. 67)

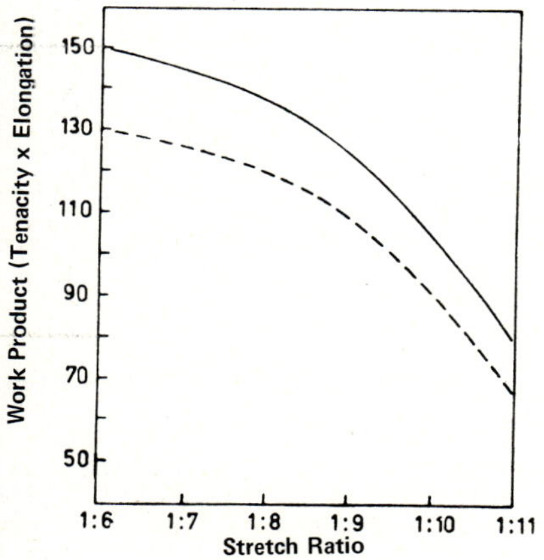

Fig. 16. Effect of Stretch Ratio on the Work Product (Tenacity x Elongation) of Weaving Tape Made From Polypropylene: _____, Flat Die Film; ---, Blown Film (ref. 66)

with multifilament or monofilament yarns. The same mechanisms and relationships apply to the orientation of film fibers or tapes. Here we would discuss only a few additional examples.

Increasing the draw ratio increases the tenacity and decreases the extensibility, as demonstrated in Fig. 15. Similar data have also been reported by other workers (refs. 57,68,60).

The effect of the stretch ratio on the work potential of film tapes, represented by the product of tenacity and elongation, is also interesting. One example is shown in Fig. 16. In general, the maximum stretch ratio that can be applied to a given film is controlled by the nature of the film-forming polymer, by the quality of the film, and by the drawing temperature. The degree of stretching, however, is mainly determined by the end-use requirements of the film, the film tape, or the film fibers. In manufacturing film tapes for weaving applications, low stretch ratios, in the range of 1.6 to 1.8 are preferable in order to keep the fibrillation tendency on the low side. High stretch ratios, in the range of 1:11 are used in processing film tapes for baler twine or rope applications where fibrillation is desired for reasons of bulkiness and high stretch conversion.

The drawing and orientation for achieving maximum tenacities have to be carried out at elevated temperatures, approaching the softening point of the polymer as a limit. For stretching high-density polyethylene film, temperatures between 110 and 130°C are used. For stretching of polypropylene film, a much wider temperature range, from approximately 120 to 170°C, can be applied, generally permitting high line speeds. With increasing line speed, higher drawing temperatures have to be used.

The drawing temperature has a marked effect on a number of important properties, such as strength and elongation, dimensional stability, and fibrillation tendency. Several authors have presented evidence for the effect of drawing temperature on tenacity. An example for such data is given in Fig. 17. The data in the figure demonstrate that for polypropylene an increase in drawing temperature above 140°C causes a significant drop in tenacity. This drop is due in part to thermal degradation and in part to the greater difficulty to achieve high levels of molecular orientation at the higher temperature. Furthermore, the data indicate that at lower drawing temperatures it is difficult to stretch film tapes which were extruded at low extrusion temperature (Fig. 17). Drawing at higher temperature, on the other hand, has certain advantages with respect to achieving lower heat shrinkage and better dimensional stability. At higher drawing temperature the system is in a more plastic state and shows shorter relaxation times, and the extension, therefore, does not cause as much internal stress. Some data on the effect of drawing temperature on heat shrinkage are summarized in Fig. 18.

Other factors influencing the effectiveness of drawing are the film thickness and the film speed. The thicker the undrawn film and higher the stretch ratio, higher is the tenacity or lower is the elongation obtainable from a given polymer (Fig. 19). These effects are observed for both polypropylene and high-density polyethylene. However, in the case of polypropylene, original film thickness does not cause any serious limitations on line speed or applicable stretch ratio, as is the case with high-density polyethylene. Normal line speeds

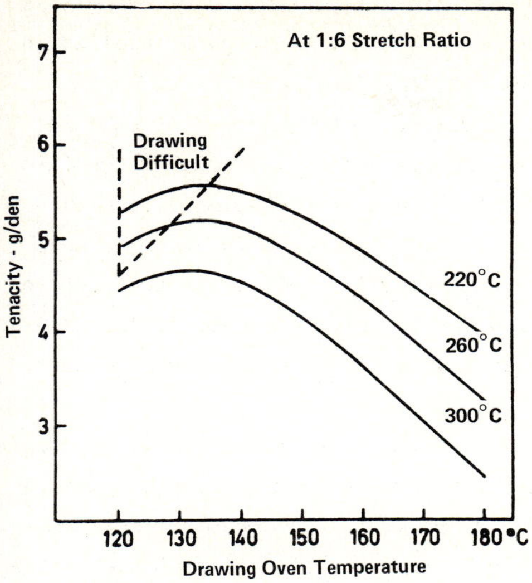

Fig. 17. Effect of Drawing Oven Temperature and of Extrusion Temperature at a Given Draw Ratio of 1.6 on the Tenacity of Polypropylene Film Extruded at Different Temperatures (ref. 66)

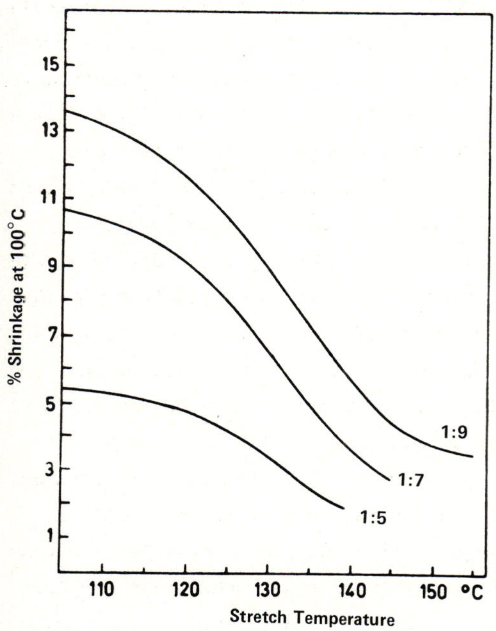

Fig. 18. Effect of Stretching Temperature on the Drawn Film Shrinkage at 100°C (After Heat Setting at 125°C With 15% Length Release) (ref. 60)

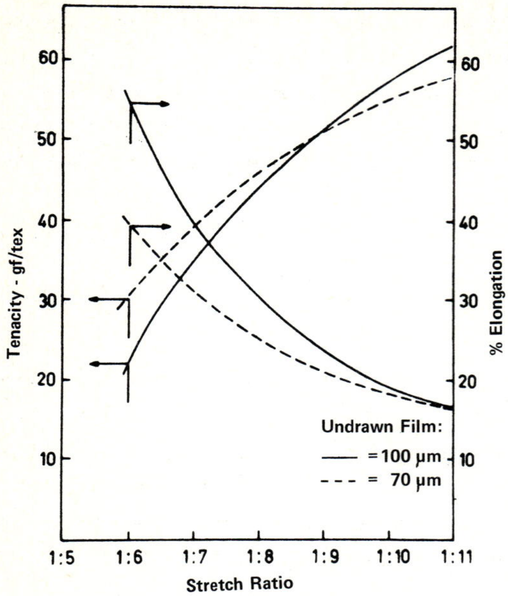

Fig. 19. Influence of Film Thickness and Stretch Ratio on the Development of Tenacity and Elongation in Polypropylene Weaving Tape (ref. 60)

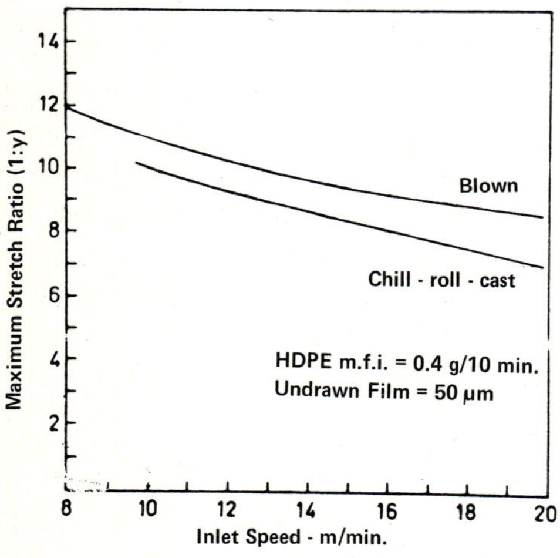

Fig. 20. Influence of Inlet Film Speed and Film Production Technique on Stretchability of High-Density Polyethylene Film (ref. 67)

of 25-30 m/min and stretch ratios in the order of 1:8 to 1:11 can be obtained with polypropylene film with undrawn thickness ranging from about 40 to 1200 μm. For high-density polyethylene, increasing film thickness calls for reduced line speeds in order to achieve uniform film or tape stretching and, at the same time, avoiding tape breakage. The maximum applicable stretch at a given film thickness decreases with increasing line speed, as illustrated by Fig. 20.

The higher the molecular weight of polypropylene polymer (or the lower the melt index), the lower the line speed at which best strength development in drawing is observed. Increasing extrusion temperature helps somewhat to overcome the adverse effect of high extrusion rates.

II. TAPE YARNS

Tape yarns are basically monofilament yarns of rectangular or low profile dumbbell cross section. These types of yarns can be produced by melt-spinning process similar to that used for round cross-section monofilaments, as briefly discussed in Chapter VIII. The extruded tapes are continuously drawn through hot oven or over heated rolls and wound onto suitable packages. This type of process offers no economic advantage over methods commonly employed for the production of monofilament yarns. A more economical method for tape yarn production is through slitting of cast or blown film. The heart of the process is the mode of drawing the film and slitting it into tapes of suitable width. Most of the processes in current use are variations of these two themes.

When polypropylene film or film strips are drawn, stiffness and strength are enhanced in the drawn direction, and generally both these properties tend to increase with increasing draw ratio. At the same time, however, the film becomes progressively easier to tear in the drawn direction until a point is reached when the slightest mishandling will cause the film to split or fibrillate. At lower draw ratios splitting is less marked, and it is possible to produce a narrow, drawn tape of high strength which is capable of being woven without much difficulty.

Two basic processes for the production of tape yarns have evolved:

(1) Slitting the film and drawing the tapes in heated oven;

(2) Drawing the film on heated rolls and cutting the drawn film into suitable tapes.

Slitting film into tapes is generally done with the help of bars equipped with razor blades mounted at the desired distances (see Fig. 21); between 15 and 35 mm for rope production and between 1 and 10 mm for textile uses. Circular cutting knives are also used. They have the advantage of enabling one to change the cutting edge without stopping the production. Razor blade life is still somewhat a problem in cutting film which contains titanium dioxide or inorganic pigments. The harder a pigment or additive, the greater the wear on the slitting knives or on fibrillating devices. Blunt knives or fibrillating tools result in tapes with ragged edges or nonuniform fibrillation, causing breakages during drawing and lowering the tensile strength. For quality control it is necessary therefore to replace cutting knives or fibrillation tools as soon as there are signs of wear. All pigments or dulling additives are knife blunting, but organic pigments are generally less damaging than inorganic additives. It is always of advantage to use the minimum additive concentration necessary to achieve the desired effect.

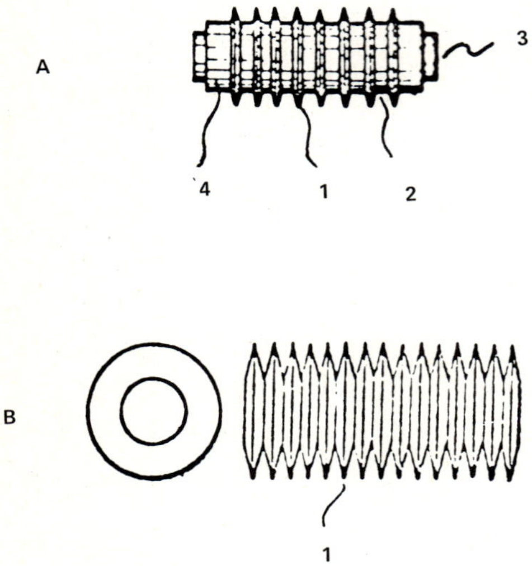

Fig. 21. Schemes of Rotary Film Tape (A) or Film-Fiber (B) Cutters: 1 Are the Circular Knives; 2 the Distance Plates; 3 the Axis with Screw Threads for Holding Screws on Both Ends; 4 the Holding Screws

Figure 22 is a diagrammatic representation of a conventional draw line based on the work sequence outlined in process 1 mentioned above. The cast or blown film is slit into tapes of appropriate width, then threaded in line through a set of godets operating at essentially the extrusion speed. The slitting may be combined with edge trimming to remove the bead, as mentioned previously. The tapes pass through a heated oven and on the exit end are threaded through a second set of godets, which operate at a speed sufficiently high to give the desired stretch ratio.

The drawing of each individual tape tends to occur at the neck, where both the thickness and width of the tape are reduced in a fashion that is controlled by the tape temperature and the overall draw ratio. In the course of necking down, the tapes shrink in from the edges and their final cross section becomes somewhat low profile dumbbell shaped. For weaving tapes the bead along the edge can be an advantage, since it tends to seal the edge and prevent fibrillation and fraying in the loom.

To decrease free shrinkage of the tapes and to prevent their fibrillation in

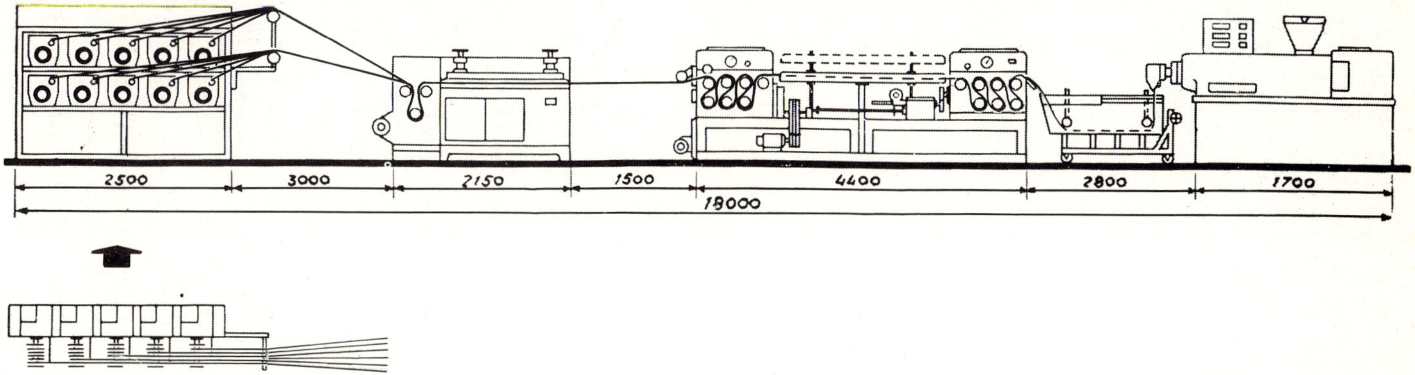

Fig. 22. Tape Yarn Production, Using Heated Ovens; Design Covema

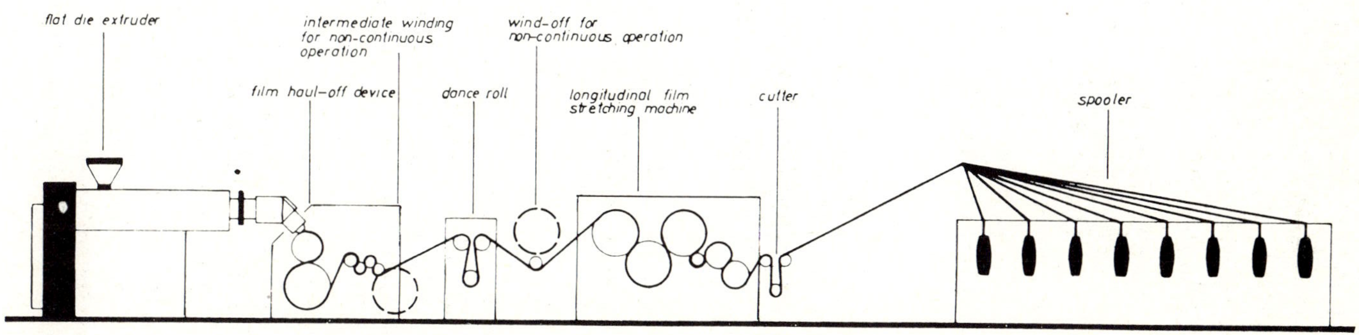

Fig. 23. Tape Yarn Production, Using Chilled Roll Casting and Heated Rolls Drawing; Design Buckner-Trockenbau

weaving, often an annealing treatment is given. This can be easily done between the second and third sets of godets, utilizing a second heated oven in between or, more preferably, a set of heated godets between the first and the third set of godets.

The layout of a conventional-type plant comprises an extruder, blown-film or flat die, film cooling equipment, slitter, stretching oven and rolls, relaxation oven and rolls, and cross-wound bobbin wind-up unit. These units are usually claimed to produce tape yarns at maximum speeds which can be handled easily by the operators in rethreading. In actual practice, however, annealing or heat setting speed is the limitation.

Other factors limiting production rates are tape thickness - a function of end-use requirements - and film width (and, therefore, the number of tapes). The latter limit is imposed by cantilever-type godet rolls whose lengths are limited by bearing loads and ease of tape threading. Most lines on the market are designed for narrower widths, between 600 and 800mm (23.6 and 31.5 in) since, apart from cost considerations, wider film units result in greater operating complexity.

Larger outputs would apparently result by using the blown-film extrusion method instead of the cast film. This is, however, not the case. Although the blown-film technique produces two separate layers of film (with theoretically twice the output the speed at which polypropylene can be blown is only about half of what is possibl by casting. Nevertheless, the two-deck system recently adopted by Covema, Italy, i which the two layers of tape are stretched in separate, double-deck ovens, is interesting.

Figure 23 is a schematic diagram of the process of drawing film on heated rolls, then slitting it into tapes. The advantage of this process is that it is not neces ary to handle a multiplicity of ends during thread-up. A single production unit ca therefore, be wider than an oven orientation line. In fact, it is not necessary to cut the oriented film into individual tapes of definite size directly on the orientation line. The spool of the stretched film can be transferred directly to weaving machine where it can be used as an instantaneous warp by slitting the film *in situ* on the loom.

One of the major disadvantages of the post slitting process is that the slitter blades have to be positioned close together, and there may be problems in obtaining high quality strands. Furthermore, post-slit tape yarns are generally prone to fraying and random fibrillation at their edges. These problems can, to some extent be avoided by maintaining sharp razor blades in the slitter, and employing a well-annealed film for the production of tape yarns. A high molecular-weight polymer film drawn to a moderate draw ratio (i.e., far below the maximum draw ratio) at considerably high temperature and subsequently annealed at 140°C, is less prone to these defects than is highly stretched film and low mw polymer film.

As mentioned previously, slitting of film into tape yarn can also be done on the loom and knitting machine instead of the draw unit. This technique has several features that recommend it to the weaver and the knitter. These include:
(1) An increased length of material on the supply beam can be accommodated;
(2) The tapes enter the weaving machine in flat position;
(3) The process is appreciably shortened, as individual package winding, crealing, and subsequent beaming are eliminated.

Loom slitters are available as part of the conversion line from Bruckner-Trocknerbau, and Kamp and Automatik, in Germany, Paul Packleppa & Co. in Switzerland, and Chemiefaser Lenzing AG in Austria, who all claim patent rights on the process. A slit-knit process has also been developed recently (ref.6). If desired, a part of the production can be slit into tapes on the draw line, taken up on individual spools, and used as filling yarns in conjunction with the *in situ* slit-warp yarn. Although the concept of converting film into tapes on the loom sounds simple, the techniques may have certain inherent disadvantages. Two important factors for trouble-free splitting are the straight positions of the blades and even tension across the film, and both need very careful adjustment.

The high width/thickness ratio of tape may be regarded as one of its advantages for weaving of flat goods but it also mitigates against its wider use. Manufacturers are often conservative in their acceptance of new materials, and tend to mistrust tape because it does not have the familiar circular cross section. "Giving" is another problem which is inherent in flat-tape fabrics. For this reason, some German and Japanese manufacturers produce a very wide, thin tape which folds readily into a superficially conventional form. This type of tape yarn is sometimes referred to as "bulked" tape yarn. Apart from the apparent high bulk and better frictional characteristics of the fabrics, folded tapes have the advantage of being easier to convert into fabrics with no need for special precaution for keeping them flat. Moreover, the folded tapes have a wide leeway in tape production as some accidental splitting can be tolerated without undue harm. The folded tape yarns, however, generally have lower strength than the flat tapes: tenacities of 3.5 to 4.0 g/den are common. These tapes can be woven into fabrics with a weight replacement ratio of 1 oz of polypropylene to 4 oz of jute. The flat, nonfolded tapes of 6 to 7 g tenacity may have a replacement ratio per unit area of cloth between 6 and 7.

Tape-yarn properties are dependent on several material and processing parameters. In general, the same kinds of relationships exist between tape yarn properties and processing parameters as those discussed in Chapter V in connection with multifilament spinning and in Chapter IX in connection with the preparation of high-tenacity yarns. In summary, it can be stated that:
(1) High viscosity polymers yield stronger tape yarns with less tendency to

fibrillate than does the product made with low molecular-weight polymers.

(2) Copolymers of propylene with 1 to 3 percent ethylene have been especially noted for their lower tendency to fibrillation, although tenacities are somewhat lower (ref.7).

(3) Rapid quenching, i.e., low temperature chilled rolls, water bath, or chilled air, assist drawing and permit higher tenacity products.

(4) Narrower and thicker tapes show higher tenacities than wider and thinner tapes of the same denier.

(5) Slower rates of film stretching give tapes of more uniform properties and yield fabrics of higher strength.

(6) An effective drawing temperature of 110° to 120°C seems to be the optimum for the attainment of high tenacities and lower break elongation.

(7) Annealing reduces the tendency of tape splitting and increases its elasticity. A slight decrease in tenacity can, however, be expected with annealing. A temperature of 120° to 130°C and a relaxation of 5 to 10 percent is often adequate for effective annealing at moderate rate of production.

(8) Longitudinal strength of tapes increases with increasing draw ratio, attains a maximum at a certain draw ratio, and then decreases on further drawing. The optimum draw ratio is a function of molecular weight and polydispersity of the polymer, rate of extrusion, temperature of extrusion and cooling, draw speed, finish, and the temperature of draw rolls.

As mentioned before, the transverse strength and the tendency to slitting decreases with increasing draw ratio. A maximum is seldom observed with regard to this property. A threshold draw ratio is often observed, as indicated in Fig. 24. The transverse strength decreases slowly at first, but at a certain draw ratio - depending on various material and processing parameters - it drops rapidly. For useful tape yarns, the optimum draw ratio should lie just below this inflection point.

For those applications in which uniaxially stretched polypropylene film tape or film fibers are making inroads, such as in tape fabrics laminated or coated with polymer films or coatings used for packaging applications, or in polypropylene tape fabrics for carpet backing, high dimensional stability during heat processing is an important requirement. The heat shrinkage in cases of conventional textile applications is measured by treatments in boiling water for 10 min, and, in cases of later heat-curing or extrusion coating applications, by treatments at 130°C for 30 min. The shrinkage in both treatments should not exceed about 1-2% when a film-tape or fiber product is required for the above-named applications.

To minimize heat shrinkage of stretched polypropylene films, film tapes, or film fibers, such products have to be heat set or annealed in separate heat treatments, allowing for a certain amount of relaxation. These heat-setting

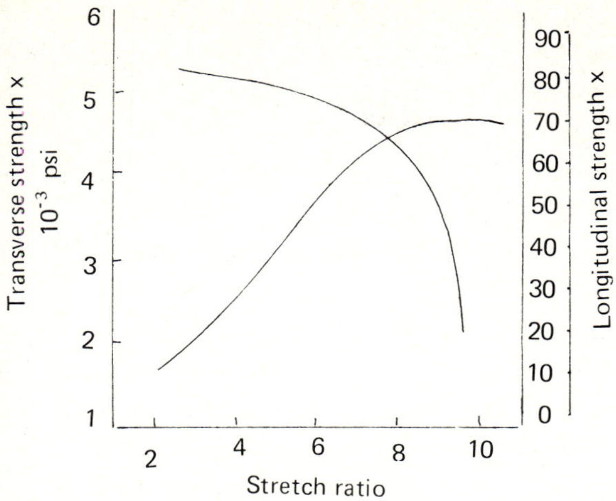

Fig. 24. Dependence of Tape Strength on Draw Ratio

or annealing treatments can be performed either in infrared, hot-air, or steam-heated ovens, similar to the ones used for long-distance drawing or on modified heated multiple-roller equipment, such as used in short-gap stretching operations. The out-going speed of the film product is reduced versus the in-going speed in speed in the ratio of the desired relaxation. Relaxations of more than 15-25% are not advisable, since higher relaxations are related to more drastic strength losses and tension loosening, causing entanglements on rollers or guides.

The heat shrinkage of stretched and non-annealed polypropylene film products is mainly a function of polymer M.F.I. line speed, stretch ratio, and stretch temperature. The shrinkage of non-annealed products is remarkably high and increases extensively with decreasing M.F.I. of the resin and with increasing line speed, especially when the draw ratio is also increased (Fig. 25). The effect of increasing draw ratio on the heat shrinkage at different temperatures is demonstrated by data summarized in Fig. 26. At exposures to low temperatures up to 110°C there is practically no increase in shrinkage with draw ratio, but there is quite a marked increase in shrinkage with increasing draw ratio for exposures to shrink temperatures over 100°C.

Recently, it has been shown (ref.69) that stretching polypropylene film extruded at high temperatures to stretch ratios higher than 1:12 results in a remarkable lowering of heat shrinkage of non-annealed film or film products, affording a lower relaxation ratio for achieving good dimensional stability. The influence of the stretching temperature has been discussed already (Fig. 18). High temperatures in the drawing operation result in higher dimensional stability of the non-annealed film product.

The effect of annealing on the reduction of heat shrinkage of polypropylene tapes stretched originally 1:9 is illustrated by data shown in Fig. 27. The data show that the heat shrinkage can be drastically reduced (to under 2%) by such heat-setting treatments when sufficient relaxation is allowed. The application of air- or infrared-heated ovens affords somewhat higher temperatures at relaxations of 20-25% than does the application of heated roller equipment, namely 160-180°C. The higher temperature in the heated ovens affords better control to avoid rupture by melting and sticking.

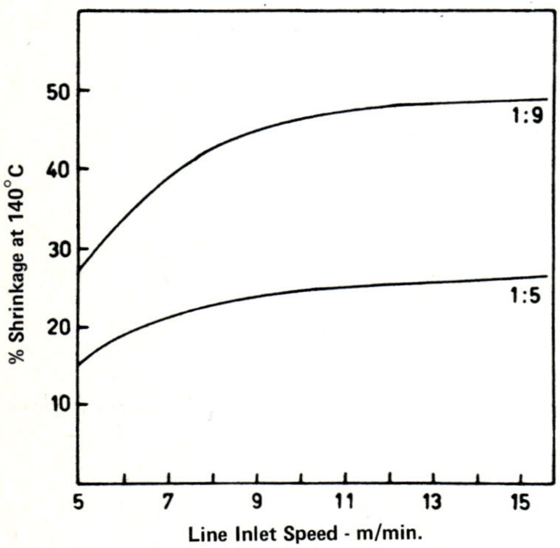

Fig. 25. Influence of Stretch Ratio and Inlet Speed on Shrinkage at 140°C (Ref.57)

III. SPLIT-FILM YARNS

Polypropylene film, when drawn beyond a certain draw ratio, loses its strength across the draw direction and begins to split into numerous fiber-like, interconnected micro-tapes of rectangular cross section. This property of uniaxially oriented polypropylene film is the basis of novel and relatively inexpensive yarns, known as split yarns, split-film yarns, fibrillated film yarns, or simply film yarns. The production of split-film yarns embodies the same basic steps of film production, quenching, and drawing discussed for the production of tape yarns. However, the conditions of film formation and drawing are considerably different because various steps are needed to fibrillate or split the tape into yarns of desired quality. Unlike tape yarns, split yarns are round in consistency and are more pliable.

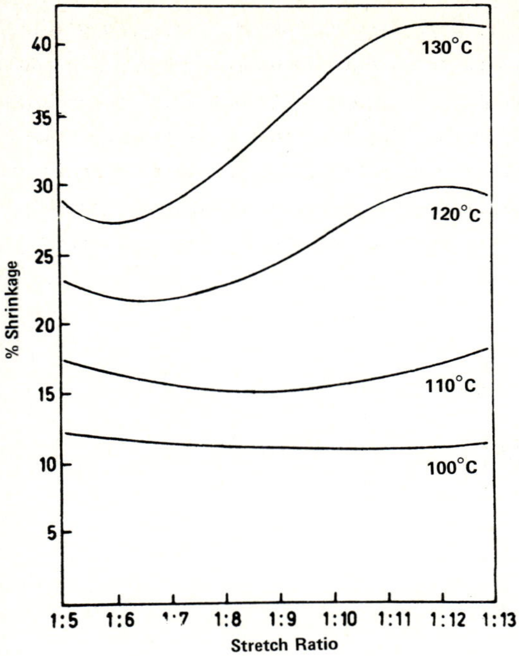

Fig. 26. Effect of Stretch Ratio on the Heat Shrinkage of Polypropylene Weaving Tape at Varying Shrink Temperatures (Blown Film; Undrawn Film Thickness 100 µm; No Heat Setting Treatment) (Ref.60)

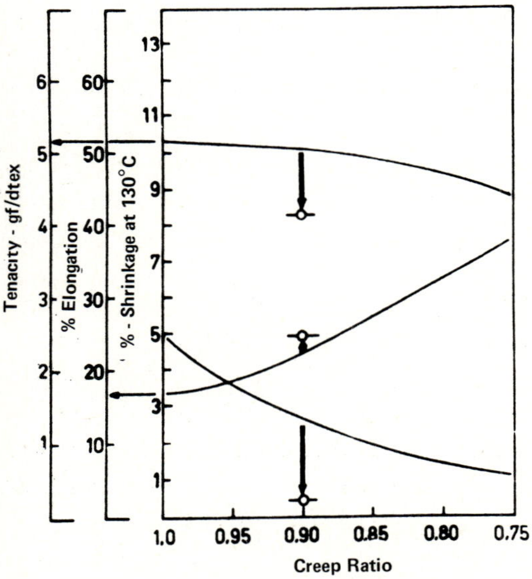

Fig. 27. Effect of Creep Ratio (Shrink Ratio) in Heat Setting on Tenacity, Elongation, and Heat Shrinkage at 130°C of Polypropylene Weaving Tape (Flat Die Extruded Film; 90 µm; Chill Roll at 25°C; Stretch Ratio 1.9; Stretch Temperature 105°C at Hot Plate; Annealing Temperature 160°C on Heated Rollers; Arrows Indicate Property Changes When Extruded With Hot Chill Roll and Treated at Higher Stretching Temperature) (Ref.66)

Figure 28 presents an overview of split-film yarn production. The heart of the process is, of course, the splitting or fibrillating step itself. Spontaneous splitting of highly oriented polypropylene film is generally not enough for the desired degree of fibrillation. The tendency of splitting or fibrillating of the film is augmented by performing work on the film tape. During recent years a large number of processes for fibrillation of film has been developed. A partial list of patents issued in this area during the last two decades is given in the Bibliography. These processes can be broadly classified as follows:

(1) Mechanical fibrillation
 (a) Random mechanical fibrillation
 (b) Controlled mechanical fibrillation
(2) Chemical-mechanical fibrillation
 (a) Polymer interphase fibrillation
 (b) Fibrillation through foaming agents

III.1. Random Mechanical Fibrillation

Random mechanical fibrillation, or random mechanical splitting, is defined as a process in which the film or narrow tapes derived from film is fibrillated mechanically to produce a web or net-like structure in a random pattern consisting of fibrils of random width and length. Highly drawn polypropylene film or tape will split into fine, interconnected microtapes on simple twisting or mechanical working. In this respect, polypropylene surpasses all other polymers, including polyethylene.

Although a large volume of literature has appeared in the last few years on the mechanical and commercial aspects of split-yarns production, few systematic studies on the splitting characteristics of polypropylene film and its dependence on material and process parameters have been undertaken (refs. 8,9,10). A large volume of knowledge about the splitting characteristics of highly oriented polypropylene tape yarns and their dependence on various material and process parameters is, however, available from empirical studies reported in the patent literature. The following may be noted:

(1) Lower molecular-weight polymers with narrow polydispersity yield easily splittable film.

(2) Slower quenching rate, high draw ratio, and high draw speed, favor splittability.

(3) Cast film splits more easily than the blown film.

Various methods have been used to facilitate spontaneous fibrillation of drawn polypropylene tapes.

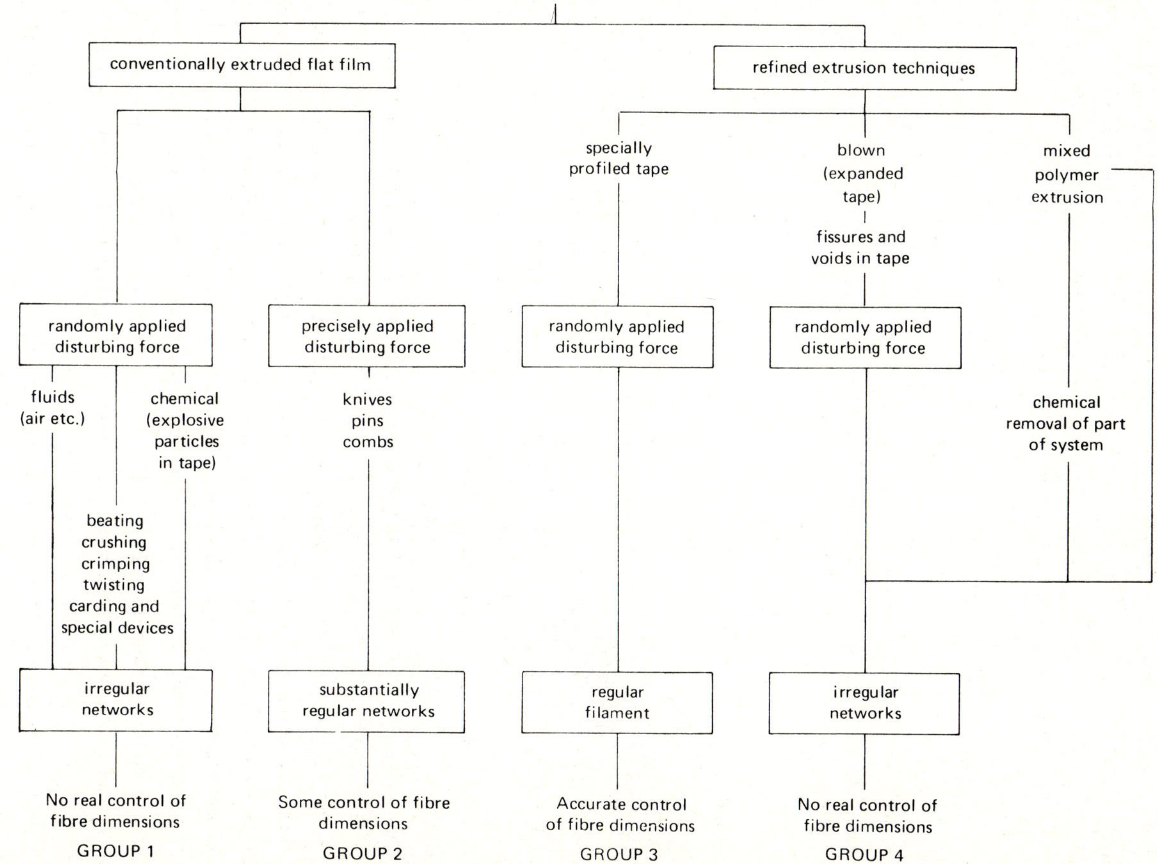

Fig. 28. Methods of Fibrillation (ref. 37)

Purely mechanical fibrillation can be achieved by applying shearing forces acting on a highly stretched film or film-tape perpendicular, or at a given angle, to the moving direction. Such shearing forces can be produced by brushing action using stationary or rotating brushes of various designs, mainly carrying steel needles or bristles. Similar action can be exerted on uniaxially stretched film by rubbing devices. Rubbing action on a uniaxially stretched film can be exercised by a pair of rubber-coated rollers, as shown in Fig. 29. At least one of the rolls is able to oscillate, as indicated in the figure. The oscillation movement in a direction transverse to the film movement and to the molecular orientation applies a transverse stress, causing the film-to-fiber fibrillation. The same effect can be achieved by devices consisting of two plates having rough surfaces and pressed together by spring action (ref.71). One of the plates is stationary, and the other is moved back and forth perpendicular to the film moving between them.

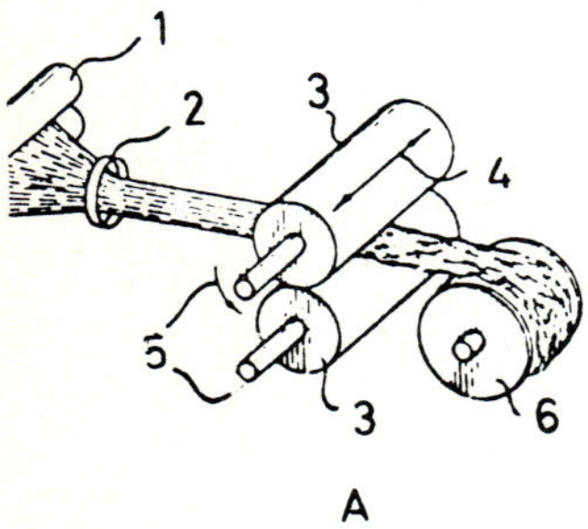

Fig. 29. Oscillatable Rollers; 1 is the Roller Assembly; 2 the Eyelet; 3 the Pair of Rollers; 4 the Oscillatable Roll; 5 the Counterrotation; 6 the Take-Up Package (ref. 70)

Stress action without actual rubbing action can also cause film-to-fiber fibrillation. Rasmussen (ref.72) has described a double-rubber-belt arrangement. Guide rolls exert multiple lengthwise action and relieve stresses on the film moving between the belts. In addition, compression zones are created by pressure members acting from both sides on the moving belt pair, as indicated in Fig. 30. A similar technique of stress fibrillation is applied in a process described by Soko Co. (ref.73). Here, fibrillation is achieved in a second stretching step between pairs of rollers or belts, some of which may have rough surfaces and rotate at different speeds. Pairs of conical rolls are also used for the same purpose. Quite often the effect of the rubbing and stress action is accelerated by introducing punctures or incisions into the film prior to the fibrillation step using rollers with needles or knives, as previously mentioned.

Stress action initiating uncontrolled mechanical fibrillation can also be

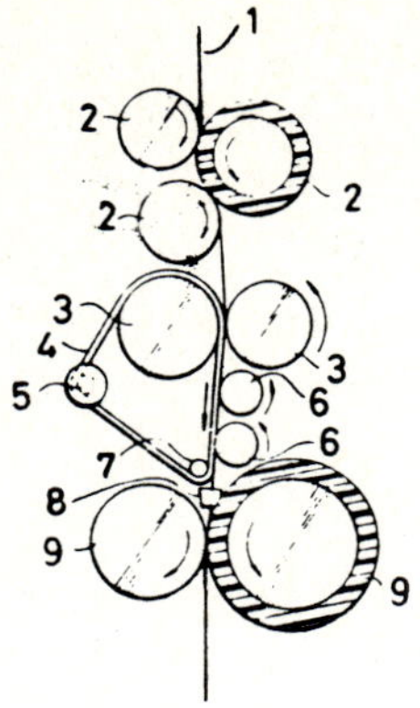

Fig. 30. Rubber Belt Stretch Device; 1 is the Film or Film Tape; 2 the Holding Roller Assembly; 3 the Middle Roller Assembly; 4 the Rubber Belt; 5 the Guide Roll; 6 the Pressure Rolls; 7 the Guide Rod; 8 the Collecting Funnel; 9 the Stretch Roller Assembly; Rotation Speeds: 9 > 3 > 2 (ref.72)

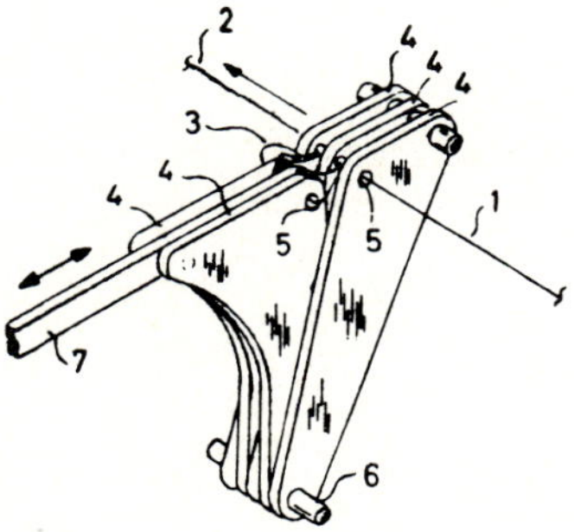

Fig. 31. Bending Plate Fibrillator; 1 is the Incoming Film Tape; 2 the Fibrillated Tape; 3 the Bended Film Tape; 4 the Bending Plates (Two on the Left-Hand Side are Movable); 5 the Tape Leading Holes; 6 the Bearing Rod; 7 the Transmission Rod (ref.74)

produced by bending operations. Devices that perform film-tape fibrillation by bending action have been described, for example, by Allied Chemical Corp. (ref.74) and by Rheinstahl Henschel AG (ref.75). The device claimed by Allied Chemical Corp. acts as follows: A film tape with dimensions yielding the desired yarn number is drawn through the holes in a set of three, five, or more plates, which are alternately stationary and movable in a parallel fashion. The back-and-forth movement of the movable plates causes the film tape running through the device to constantly bend 180°, as indicated in Fig. 31. The fibrillation tool described by Rheinstahl Henschel AG first leads the uniaxially stretched film tape through a funnel, in which the tape is folded together, then leads it over a running roll having a truncated conical running groove. The folded tape is forced into the conical groove and wound approx. 270°. Coming off this roll, the already splitting tape is drawn in a sharp angle over the incoming tape and lead to a sidewise displaced second conical groove roller. By this multiple bending, and the additional rubbing action working on the tape running through the device under tension, the tape fibrillates readily. The principle of this tool is shown in Fig. 32.

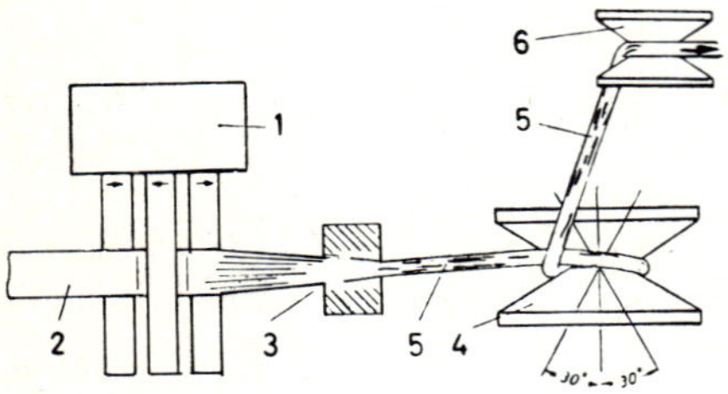

Fig. 32. Bending Roll Fibrillator. 1 is the Stretching Trio; 2 the Film Tape; 3 the Collecting Funnel; 4 the Roll with Conical Guide Notch; 5 the Folded Tape (Being Fibrillated Between 4 and 6); 6 the Sidewise Displace 2nd Roll with Conical Guide Notch. (ref.75)

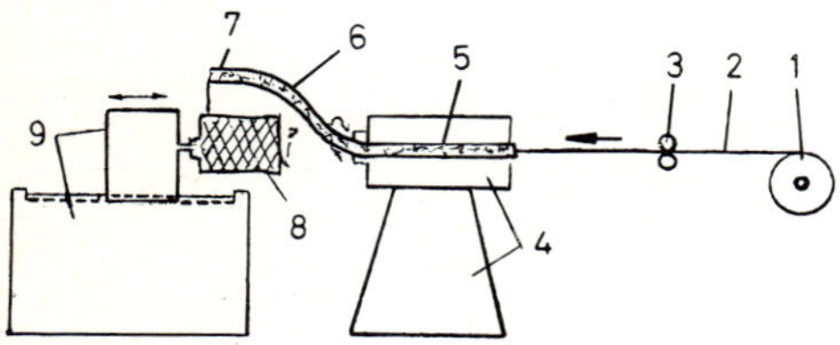

Fig. 33. Scheme of a Special False Twist Apparatus for Twist Fibrillation: 1 is the Film-Tape Package; 2 the Film Tape; 3 the Guide and Tension Rolls; 4 the Mount, Drive and Heater for False-Twister; 5 the Twist Tube Entrance Side; 6 the Bended Tube Section; 7 the Twist Tube Exit Side (Parallel to 4); 8 the Take-Up Package; 9 the Yarn Winder. (ref.77)

Another technique to initiate uncontrolled film or film-tape fibrillation is based on twisting action. For this purpose, normal ring-spinning frames or false-twisting machines of special design can be used (ref. 12,76). Shell International has described a special device based on a false-twist principle which can be used successfully for uncontrolled mechanical film-tape fibrillation. The design of this special device is shown in Fig. 33. The uniaxially oriented film cut into tapes of desired dimensions is transported into a rotating bended tube having a length of approximately 20 cm and an inner diameter of approximately 12 mm. The rotation can be controlled such that the film tape is twisted with 30-50 turns/m. The inlet speed of the film tape is limited to about 100-200 m/min. Via rotation in the tube and twisting action on the tape fibrillation into a fibrous network

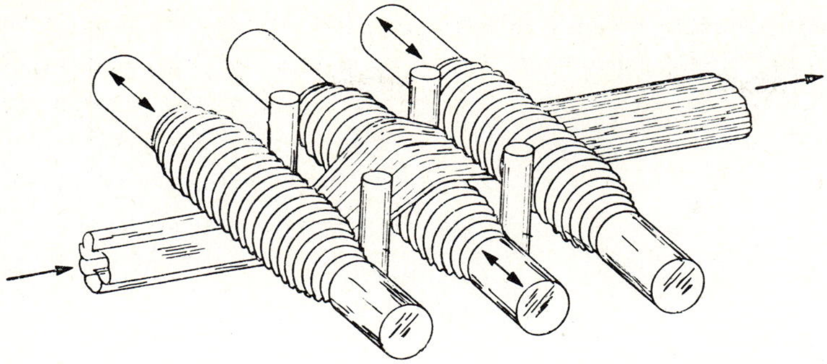

Fig. 34. Random Fibrillation Technique by False-Twisting and Rubbing Action (ref.50)

Figure 34 depicts a means of fibrillating uniaxially oriented tapes by means of a combined action of false-twisting and rubbing between a number of corrugated rolls. A simple method, but applicable only to easily splittable film, is to pass the stretched film or tape in pleated condition over a groove at an acute angle under high tension (ref.13). This method of fibrillation can be combined with the cutting of the drawn film into individual tapes, as shown in Fig. 35. Several other methods utilize friction of opposite rotating surfaces to induce random fibrillation in the drawn film or tape.

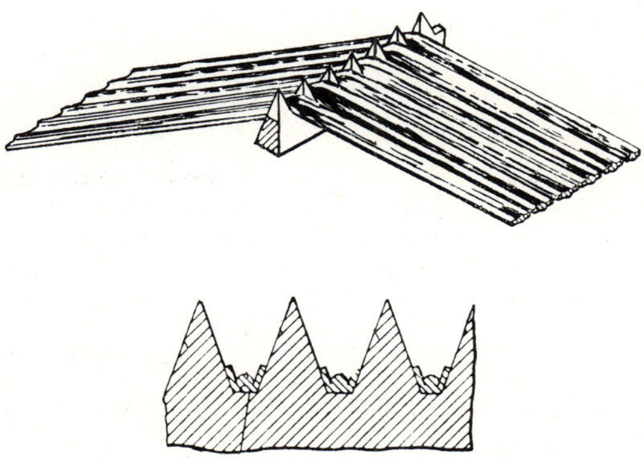

Fig. 35. Fibrillation Through Friction and Point-Stressing (ref.51)

A few fibrillation processes are based on the action of high-velocity turbulant fluids on the highly oriented tape (ref.15). E.I. du Pont de Nemours (ref.78), for example, has described an air-jet device for splitting film tapes. In this device (see Fig. 36), the oriented tape of desired dimensions passing through the jet is subjected to 5-80 psi air or gas stream having a velocity near that of the sound. Preferably, the gas passageway should be inclined slightly forward along

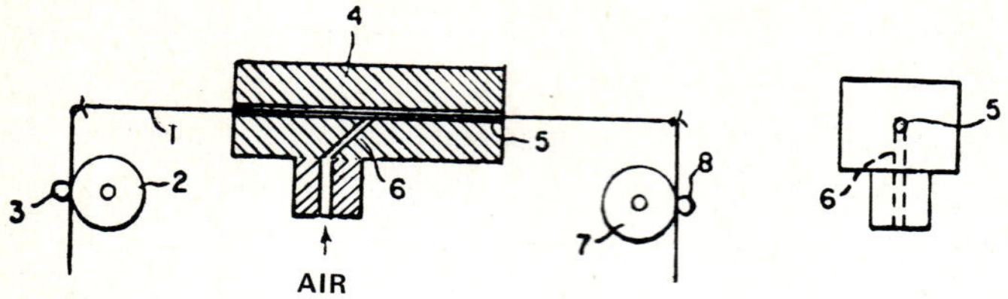

Fig. 36. Scheme of a Jet Device for Film-Tape Fibrillation: 1 is the Film Tape; 2,3 the Feed Rolls; 4 the Jet Device; 5 the Cylindrical Yarn Passage; 6 the Gas Entrance Way; 7,8 the Take-Off Rolls (ref.78)

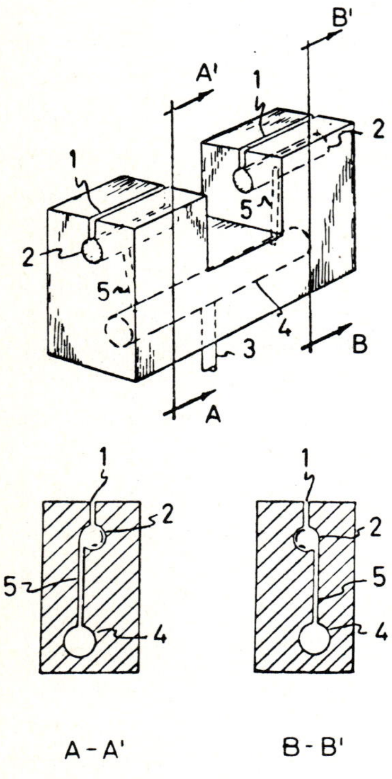

Fig. 37. Scheme of a Jet Twist Fibrillator: 1 is the Film-Tape Slots; 2 the Film Tape Passage Ways; 3 the Gas Entrance Way; 4 the Gas Manifold Box; 5 the Gas Passage Ways (ref.79)

the line of tape movement. In some instances, when increased interlacing is desired, the gas channel entrance can be tangent to the tape passageway to provide more twisting action and torque.

Another air-jetting device for tape fibrillation was claimed by Fiber Industries Inc. (ref.79). In their process, the fibrillatable tape is subjected to action of at least two fluid-twisting steps so that the direction of twist imparted to the tape is completely reversed by adjacent twisting actions. Figure 37 shows a perspective view of this fluid-twister and sections to illustrate fluid flow in both twisting steps with respect to the tape movement.

As the fibrillatable tape is being subjected to the action of at least two such fluid twisting means, it is maintained under tension of about 0.05-0.20 g/denier to insure good fibrillation. This tension can be maintained either by pulling the tape with multiple rollers or with the winding apparatus. The air or gas supplied to the fluid-twisting positions should be under a pressure of about 10-250 psig. With this process, about 5-300 fibrils/tape will be produced in the fibrillation of tape with a thickness of 10-25 μm.

Depending on the efficiency of the fluid jet used, this method of splitting produces comparatively finer fibrils than those discussed before. One characteristic of such fibrillated yarns is their high bulk and woolly appearance. Fluid jet fibrillation method, however, produces numerous broken fibrils which do not contribute to the yarn strength until highly twisted.

III.2. Controlled Mechanical Fibrillation

Randomly fibrillated yarns, described in the previous section, are composed of networks of individual fibrils which show considerable variation in length and denier. The cross-sectional profile of the fibers in such yarns remains that of a flat tape, thus giving relatively coarse and harsh yarns. More recent developments of the mechanical fibrillation are based on controlled fibrillation to carefully designed products of predetermined length and thickness of the fibrils. The end product is a much more regular fibrous network with respect to size of lattice, denier, tensile properties, and bulk of the yarn. This type of yarn is more pliable and softer in hand than the randomly split yarns described previously. For the production of controlled, mechanically fibrillated yarns, essentially non-fibrillating film strips are used. This removes limitations on material choice, film forming, and drawing conditions. Basically three methods are being used to produce film yarns of controlled fibrillation:

(1) Mechanical fibrillation with "fibrillators"

(2) Profiled-die yarns

(3) Embossed roll yarns

Mechanical Fibrillation with "Fibrillators". The process which has probably received the greatest attention and popularity is the one in which the oriented film or tape is passed over a rotating fibrillator, which can best be described as a tube of 2 to 6 inch diameter, covered with precisely positioned teeth, pins, or wires. The fibrillator can be an independent unit or it can be incorporated into the second or third set of draw godets. The teeth, pins, or grooves penetrate into and cut the film to a regular web of interconnected fine fibers.

To explain the cutting action of a fibrillator, imagine a roll covered with a number of axially spaced pins. If the linear speed of the roll equals the linear speed of the film, only holes will appear in the film. If either speed is varied, however, the pins will cut a slit into the film or the film will pull a slit over a pin. Either way the desired effect is achieved. The factors which affect network geometry (ref.16) are the following:

(1) Linear speed of the film
(2) Linear speed of the fibrillator
(3) Contact length of film on fibrillator
(4) Distance between adjacent cuts or point contacts
(5) Distance between cutting edges taken circumferentially
(6) Height of cutting edge from roller surface.

From these factors it is possible to devise a formula for the manufacture of any type of network yarns (ref.17).

The manufacturers of fibrillators based on needle or sharp-edge action have developed a variety of such tools. In general, the basic rule that applies is that the combination of small diameter and short contact length between film and pinroll gives strong irregular networks with random small meshes consisting of rather coarse interlinking fiber segments, while the application of pinrolls with large diameters at long contact lengths with the film results in weak fibrillates having an irregularly long mesh network. The mesh length (L), the degree of film slitting (DF), and the average denier of the resulting fiber segments are controlled by the contact length (a) between film and fibrillating roller, the film or tape speed (vf), the surface speed of the roller (vr), and the pin density per unit area (Nm). The mathematical treatment, making some simplifying assumptions, gives the following relations (refs.37,80,81):

mesh length: $L = a(1 - vf/vr)$

degree of fibrillation: $DF = a(vr/vf - 1)Nm$

mean fiber denier = denier of unit width of film /DF

(where N is the number of needles or cutting points per unit length of fibrillator width and m is the number of needles or cutting points per unit length of fibrillator circumference).

By the use of these formulas one can estimate the average expected fiber denier from the data of the uniaxially stretched film, the dimensions and characteristics of the fibrillating roller and the running conditions, such as film speed and roller speed, as demonstrated on two examples in Table VI using typical conditions.

TABLE VI

Examples of Needle Roll Film Fibrillation Conditions and Resulting Fiber Fitness (ref.80)

	Shell-LIRA-Process MK-I	Shell-LIRA-Process MK-II
Tape Characteristics		
Tape Width (before Stretch)	2 cm	2 cm
Tape Thickness (ditto)	100 µm	100 µm
Stretch Ratio	1.8	1.8
Tape Width (after Stretch)	0.7 cm	0.7 cm
Tape Thickness (ditto)	35 µm	35 µm
Tape Titer (ditto)	2200 dtex	2200 dtex
Tape Titer per Unit Film Width	3150 dtex	3150 dtex
Processing Conditions		
Line Speed (Vf)	120 m/min	120 m/min
Fibrillator roll speed (Vr)	360 m/min	360 m/min
Fibrillation Ratio (FR)	3.0	3.0
Fibrillator Diameter	12.5 cm	29.0 cm
Contact Length	2.5 cm	15.0 cm
N Needles per cm of Needle Bar	13	13
m Bars per cm of Roll Circumference	1.1	1.4

Resulting Film-Fiber Titer

For MK-I-Process:

$$\frac{3150}{2.5.(3-1).13.1.1} = 44 \text{ dtex}$$

For MK-II-Process:

$$\frac{3150}{15.0.(3-1).13.1.4} = 6 \text{ dtex}$$

In practice fibrous products produced by needle or edge roller slitting have a somewhat lower denier (approximately 30 instead of calc. 40), due to self-splitting, which often cannot completely be avoided. Also, the fact that the film sometimes will not be penetrated by all the needles plays a role. The polymer constituting the film and possible additives acting as splitting activating nuclei, such as pitments and heat or light stabilizers, may effect the practical result with respect to actually achieved fiber or fiber segment fineness. Proper polymer choice and the use of suitable polymers or polymer blends in needle or edge roller slitting allows us to control slit-fiber fineness.

In most of the fibrillating processes, penetration of the points into the film is obtained by applying tension to the film on either side of the fibrillation tool. The tension required is high, and ripping or tearing of the film can be a problem. As the teeth dull, there develops a tendency to introduce skips in the pattern. It has been found helpful to use a press roll to force the film into the points (refs.18,19). The pressure roll must have a substantially uniform surface and it must be resilient enough to permit the points to penetrate its surface. Good results have also been obtained with pressure rolls covered with stiff, short bristles (ref.18). Methods have also been described in which pressure is applied on the film by water jets (ref.20).

In some processes - for example, one developed by Toshiba (ref.21) - only the cutting action of the roll has been used and no slitting takes place. In such cases, the cutting points take the form of elongated edges arranged in any geometrical pattern around the rolls. Often multiple cutter rolls are used which also act as pressure devices.

The mechanical details of the process of slitting and fibrillation are numerous and varied. Three representative arrangements can be mentioned here:

(a) A continuous process where individual slit tapes are fibrillated by their own pinned rollers (e.g., plasticizers)

(b) A continuous process where the tapes are fibrillated by one single roller of the width of the machine (e.g., Henschel)

(c) A discontinuous process wherein the tape formation is separated from fibrillation process (e.g., Mackie).

A number of tools have been described for performing controlled film-to-fiber slitting by defined cutting methods. These yarns resemble normal, coarse, multifilament yarns. The first process for film conversion into more or less continuously separated filaments using controlled slitting was described by Dow Chemical Co. (ref.82). The cutting apparatus consists, according to their suggestion, of a pair of rolls, one grooved parallel about its periphery, the other a resilient backing roll. The grooves of the grooved roll are separated from one another by sharp ridges capable of scoring or slitting the film. The distance of the sharp ridges are in the range of 1 mm. The scheme of the grooved rolls as given in the corresponding patent is outlined in Fig. 38.

About ten years later, Mirsky (ref.83) claimed a process in which separation of a stretched polymer film into continuous strips using various techniques is achieved. Among the means described are stationary, rotary, or oscillatory blades, having distances of 2 mm or less.

Shortly thereafter, Hercules Inc. (ref.84) was granted a patent describing a process in which a uniaxially stretched film, which was striated in the extrusion or by scoring thereafter, is passed under tension over a stationary block, carrying a series of serrations along its width. The distance from the sharp tip of one serration to the tip of an adjacent serration is made substantially equal to the distances of the ribs in the striated film. The serrations of the block penetrate the thinner webs connecting the ribs, thus separating the film into continuous filaments. The inventors claim that this type of film-to-fiber separation also permits the introduction of latent crimp properties.

A somewhat similar process was developed independently by Chemiefaser Lenzing AG (ref.85). In this process, an unstretched, or only partially stretched, film is drawn under stretching tension over a heated bar carrying saw-tooth-like cutting tools arranged transverse to the feed direction of the polymer film. The sharp edges of the cutting tool are spaced less than 0.5 mm from each other. The heated cutting tool serves several purposes in this process: (a) heating, (b) film-to-fiber separation, (c) creating the braking effect for generating the stretching force, (d) defining the neck-in point, and (e) edge bending which introduces latent crimp properties. The continuously separated filament drawn through the cutting tool with line speeds exceeding the feed speed by the desired stretch ratio can be produced with fineness range down to approximately 6-10 denier. A scheme of the separating tool used in this process is shown in Fig. 39.

A process claimed simultaneously by Shell Oil Co. (ref.86) uses a similar separating tool mounted on a rotating body.

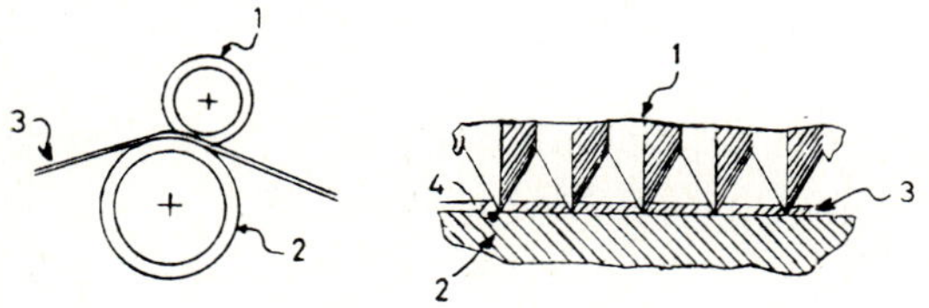

Fig. 38. Roller Cutter Assembly: 1 is the Roll Carrying Closely Spaced Parallel Narrow Circular Ridges; 2 the Resilient Backing Roll; 3 the Film or Film Tape; 4 the Sharp Ridges (ref.82)

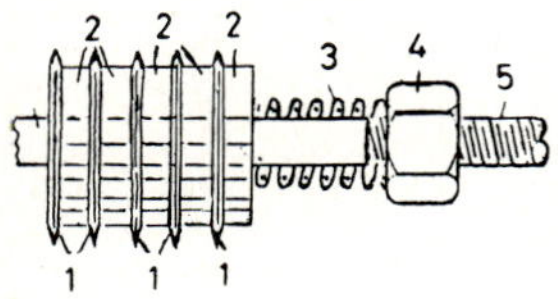

Fig. 39. Circular Film Cutter: 1 the Circular Cutter Blades; 2 the Distance Pieces; 3 the Compression Spring; 4 the Threaded Nut; 5 the Mounting Bar Carrying Screw Thread (ref.86)

Profiled-Die Yarns. Improvements in the uniformity of the cross-sectional profile and in the tensile characteristics within and between the filaments have been achieved by the extrusion of thermoplastic melt through a flat film die of which one or both lips have a fixed geometrical profile (ref.22). The shape of the profile may be rectangular, triangular, or circular. The profiled melt, with the ridges connected by thin webs of film, may be extruded to full width but a long, continuous slot without some form of reinforcement is not practical because very often it warps. Therefore, the extrusion is done in the form of tapes with discrete numbers of filaments.

It is generally found that spontaneous subdivision of the individual filaments does not occur during stretching and must be assisted by a mechanical process such as twisting, flexing, and compression, or by treatment with compressed air.

With a suitable selection of polymer, die profile, and processing conditions, it is possible to produce yarns with different tensile characteristics and denier varying from 7 to 70. This type of yarn, especially when subdivided completely into individual filaments, more than superficially resembles a multifilament-type yarn produced by the spinnerette, and approaches it in regularity of cross-sectional shape. The uniformity which can be achieved considerably extends the usefulness of these yarns, especially if fine-denier filaments are produced. At present, yarn made from 20 to 25 denier filaments is the most common. The main usage of such polypropylene yarns appears to be in better-quality cordage, packaging and protective materials, and in industrial textiles.

One process, known as "Barfilex", has been commercially developed by Barmag-Barmer of Germany (see Fig.40). The Barfilex method replaces the usual film extrusion stage with triple straight dies, each with castellated outlet profiles, giving altogether 105 separate tapes (3 x 35) each with 25 thick ribs. In the orientation stage each tape is split into 25 separate filaments. Thus, Barfilex yarns consist of virtually continuous filaments without the network characteristics of mechanically fibrillated yarns. The lower limit of Barfilex polypropylene deniers is about 7, and the upper limit about 70. The pin roller system leads to a bulkier fabric than does the Barfilex system. A bicomponent system is available for the production of bulked yarns through Barfilex process.

Compared with the pin-roller system, equipment cost is much higher for the Barfilex system. Barfilex process shares with fiber spinning a dependence on a high degree of polymer purity and the use of specially fine and easily dispersed pigments to avoid die blocking. These factors generally increase feed stock price.

Film-emboss Process. A comparatively new process of controlled fibrillation comprises the embossing of extruded flat film with a profiled pressure roll. Embossing can be undertaken while the film is still fluid or after it has

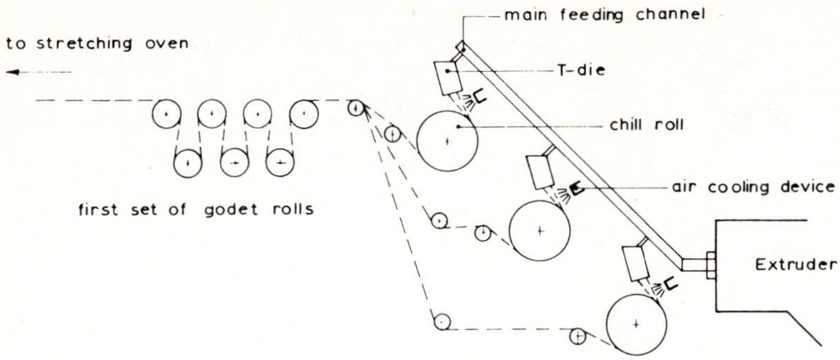

Fig. 40. Barfilex Process of Barmag-Barmer Maschinenfabrik

solidified. On drawing, this embossed film is converted into quasi-multifilament yarns of unconventional cross section. The melt-embossing method has been popularized by Smith and Nephew, who hold the patent rights to the technique (ref.23).

The processing equipment for the production of grooved sheet-film fibers through the embossed roller technique uses a conventional screw extruder fitted with a flat coat-hanger type die. The hot melt is extruded into a pressure nip formed by a plain roll and an embossing roll. The embossing roll is engraved with fine circumferential V grooves, similar in profile to a screw thread of between 75 and 400m in pitch. Cooling of the melt takes place while it is in contact with the cool roller, and an accurate mirror image of the embossed roller surface is produced on the sheet. The grooved sheet is next subdivided into tapes of the required denier and is finally stretched and heat-set by passage through hot ovens as is the case with a conventional polypropylene tape line. Where a complete fibrillation of the tapes is required, a simple mechanical beating system may be necessary; this is usually sufficient and involves no serious loss of tenacity.

Although the profiled-die and melt-emboss systems at the present stage of development produce basically similar types of yarns, the melt-embossing process has a number of advantages which make it more versatile and capable of producing a more uniform yarn than the profiled-die system. Some of these advantages are the following (ref.24):

(1) With a given pitch and profile of the embossing roller, it is possible to alter, within quite wide limits, the shape of the filament cross section.

(2) The number of filaments in the yarn can be simply and quickly altered by change of the filament selection bar.

(3) There is virtual freedom from blockage of the fine extrusion die which leads to variability in denier and breakage of the filaments in the profiled-die extrusion system.

Work of Shell Laboratories and others (refs.25,26) on cold forging of polyolefins into various end products pointed out the possibility of continuously forging an embossed pattern on an extruded film or sheet. The cold-embossed film route to fiber can be carried out either as in-line or as a separate operation. The in-line approach will probably be favored, although for economic reasons there may be a case for separating it from the extrusion stage (ref.27).

III.3. Polymer Interphase Fibrillation

The polymer interphase fibrillation methods of producing yarns from film depend on introducing discontinuities in the film by incorporating a foreign polymeric or oligomeric material in the polymer. On cold drawing, these discontinuities function as stress concentration points and initiate subsequent mechanical fibrillation. Polymer interphase fibrillation methods are geared to produce low-denier spun-look yarns for textiles and decorative purposes. Because of the presence of discontinuities and weak points in the extrudate, only casting methods of film formation come in question; the blow-film method is not applicable. Moreover, because of its easy splittability even at low draw ratios, the cast film has to be split into tapes of appropriate length before drawing. The fibrillation is random, the fibril denier is generally low, the fibril length is short, and they are extensively interconnected with each other throughout the yarn. In principle these yarns do not differ structurally from pliable monofilaments discussed in Chapter IX. The essential feature of the added polymer is its incompatability with polypropylene and its ability to cold draw to form fibers. Every conceivable combination has been mentioned but in view of availability and cost, linear polyesters, nylons, and polystyrene are the most suitable. Polyethylene and its copolymers are generally not suitable because of their partial compatability with polypropylene.

There are several different mechanical means to form a composite cast film. The simple extrusion of polymer blend would produce a film wherein the additive particles are randomly dispersed, their average size being determined by their relative compatability with polypropylene, viscosity differential, temperature of extrusion, type of extruder screw, and screw back pressure. Several mixing devices are possible: they have been discussed in Chapter VIII.

Rasmussen (refs.28,29) has pioneered a polymer interphase fibrillation method by creating a controlled microanatomy of the film or film strips by means of a lamellar extrusion technique. Two streams of different polymers are dispersed with each other in a mixing head just prior to the die exit. The dispersion follows a lamellar-shaped pattern and is in part carried out mechanically by means of two rotating elements. A series of layered ribbons is produced in which the added component is interspersed at regular intervals, sandwiched by polypropylene layers. These ribbons spontaneously fibrillate in prescribed fashion on mechanical working.

III.4. Fibrillation through Foaming Agents

As early as 1960, it was realized that a three-dimensional array of interconnected thin films could be drawn and that this, after fibrillation, would give rise to a three-dimensional network of very fine fibers, resembling a yarn. The concept was demonstrated with foamed polystyrene, which, after being drawn and fibrillated, produced yarn of surprisingly soft handle and good tenacity (ref.30). This work was extended to a wider range of polymers and considerably refined (ref.31). The polymer is mixed with a blowing or foaming agent and extruded through dies of suitable shape to give foam strips or wide film. By careful choice of suitable conditions, a foam of the required structure and necessary draw characteristics may be produced continuously at high production rates. A variety of types of foamed strand is available through the control of die opening, the nature and concentration of the blowing agent, and the extrusion conditions which influence the linear density, and the cell structure of the foam product.

The foam strand, or tapes of suitable quality, is fed into a heated drawing oven where it is subjected to a controlled draw. Here, foam of about 0.02 to 0.05 g/cm^3 density, is converted into a soft silvery ribbon at low draw or into a highly lustrous pearlescent strand of density about 0.4 to 0.5 g/cm^3 at a high draw ratio. The linear production rate of drawn strand naturally depends on the feed rate and draw ratio.

A foamed tape or film that has been drawn more than about four-fold can be readily broken down into a three-dimensional network of fine, interconnected fiber elements that contain essentially no free ends. This disintegration is achieved by doing work on the strand which can be effected by mechanical flexing, rubbing, beating, and twisting, and by various opening processes discussed in Section III.1. The linear density of the yarn is very simply controlled by changing the size of the foam strand without affecting the fineness of individual fiber elements, although this may be changed if required. Yarns ranging in linear density from 200 to 2000 denier can readily be produced.

A polymer foam consists of a three-dimensional assembly of minute interconnected films. When drawing takes place, the individual films of foam become oriented and will fibrillate at a high draw ratio.

The network yarns, prepared by the drawing of foamed polypropylene film strips, resemble those prepared from a polyblend system, discussed in the previous section. They contain no discrete fibers as do continuous filament or staple yarns, but resemble continuous filament yarns in their negligible number of free fiber ends. Visually, they resemble most closely a low-twist staple-fiber yarn, but they differ from the latter in the high tenacity they show at zero twist.

Depending on the type and concentration of blowing agent, melt viscosity of the polymer, and the operating conditions in extrusion, polypropylene sheet of "controlled" foam structure can be prepared. Structures of 0.23 to 0.35 g/cm^3 density have been prepared experimentally (ref.32). However, because of the difficulties in drawing and the resulting low tenacities of the yarn, a lower limit of foam density is in the range of 0.6 g/cm^3 for 0.05 mm thickness of the tape.

Drawn-foam strands are understandably weaker than mechanically fibrillated drawn film. A useful range of tenacity is markedly dependent on foam structure, which also introduces a range of interesting new characteristics to the product. The attractive lustre of the drawn strand arises from total reflection of light from the multitude of oriented films. The low density, high compressibility, and improved thermal insulation of the strand impart a texture to woven products that gives them a greater consumer appeal than is the case with split film.

A large variety of foaming agents has been mentioned in patent literature, ranging from volatile liquids - propane, butane, pentane, and halogenated compounds to inert gases such as CO_2 and N_2 (ref.33). Numerous chemical blowing agents, which emit CO_2, CO, and N_2 by chemical decomposition at elevated temperature, have been used. As examples of these compounds are ammonium carbonate, oxalic acid, diethylazodiformate, dinitrosopentamethylene tetramine, azodicarbonamide, azoisobutyronitrite, derivatives of hydrazides, barium azodicarboscylate (ref.34). Some compounds part with their water of hydration at extrusion temperature. These compounds, several of which are available commercially, can also be used as blowing agents for polypropylene. Such compounds include aluminum hydrate, aluminum sulphate, sodium carbonate and potassium alum (ref.35).

IV. MISCELLANEOUS PRODUCTS FROM FIBRILLATED FILM TECHNOLOGY

IV.1. Textured Fibrillated Yarns

Polypropylene fibers or continuous filament yarns made via film splitting or slitting, generally have a smooth, somewhat distasteful waxy hand. In order to impart a more textile touch to such products, they are sometimes subjected to texturizing or bulking processes, giving them a helical or zig-zag configuration.

These yarns can be textured through stuffer box or gear box crimper but the products so obtained are not very attractive for most of the potential uses. Knife-edge treatments have proved more attractive. In some cases, the knife-edge technique has been incorporated into the design of stationary cutting tool. This imparts latent crimp properties to the film fibers in continuous fashion during their production (refs.84,85,87,88,90,91,92).

A latent crimping behavior is also introduced into the filaments during a one-sided heat treatment accompanying the slitting operation. A definite conjugate structure is thereby obtained, as demonstrated microscopically with coarse model filaments produced similarly. Light-scattering and infrared dichroism studies using the frustrated multiple internal total reflection technique, and x-ray wide-angle diffraction studies also show their bicomponent nature (Table VII). All results indicate that the side exposed to short period heat treatments, while running over the edge of a slitting tool carrying blades without teeth, showed fine-structure characteristics different from that of the opposite side. The heat-treated side showed lower orientation and better crystallinity.

Besides these crimping techniques, chemical crimping methods have also received attention during the last decade. The bicomponent concept, namely the coextrusion of at least two polymers having good adhesion to one another, has found application in the processing of film fibers. A number of processes using this concept have been developed (refs.93-103). The two components can be either the same polymer, differing, however, in molecular weight or molecular weight distribution, or different polymers having different melting points or different swelling characteristics in solvents, but having reasonable adhesion. This method imparts a more stable and higher degree of crimp than the mechanical method. It also provides a one-step process for the manufacture of crimped film fibers.

The manufacture of conjugate film fibers has its origin in the well-known principle of coextruding two polymer components through specially constructed die heads to form a laminated film. The process has recently become feasible because of the commercial availability of reliable equipment for laminated film extrusion, primarily developed for the packaging industry. The material and processing parameters for the manufacture of conjugate film in general correspond to those discussed in Chapter XI for the production of conjugate multifilament yarns and staple. Polypropylene/low-density polyethylene combination has proved quite satisfactory for most of the uses at relatively low material cost. If 100 percent polypropylene film fibers are desired, an MFI differential (high viscosity/low viscosity) or polydispersity differential (degraded/undegraded) can be used.

The laminated film so produced is cut into tapes, stretched below $80^{\circ}C$ and then mechanically fibrillated, relaxed at $140^{\circ}C$ under minimal tension, and wound up. If the production of staple is the aim, the extruded laminated film is drawn, cut into small length, carded, and heat treated in swirling hot air or steam.

Considerable amounts of R&D efforts are being presently expended to impart a random, sinosoidal, gentle crimp to controlled fibrillated film yarns for the production of synthetic turf. Stuffer-box and gear crimpers produce sharp, zig-zag crimp while the edge crimping and conjugate spinning methods make a product which disadvantageously changes its character in use. The technique of steam or hot air jet texturizing has been reported to be very suitable for this purpose (ref.104). This method has been outlined in some details in Chapter IX in connection with multifilament yarns.

IV.2. Staple from Film

From the early days of fiber-from-film technology, numerous workers have pursued the aim of shredding oriented film into staple in which the individual fiber elements are as fine as conventional textile fiber. The technical feasibility of this is long established but most research has been hampered by unfavorable costs, the special treatments that secure fine splitting, the need for over-elaborate equipment, and the low speed of operation. The efforts, however, continue and noticeable progress has already been made in streamlining the process and simplifying the process machinery. The Linen Industry Research Association of Northern Ireland (LIRA), in cooperation with Shell N.V. of Delft, has done extensive work on this aspect of film-fiber production (refs.36,37,38).

TABLE VII

IR Absorption of the Two Layers of Polypropylene Film having Undergone a Heat Treatment on a Heated Bar Simulating the Conditions of the Chemiefaser Lenzing AG "Juvlen (R)" Slit-Fiber Process (FMTR Technique with Nonpolarized and Polarized Radiation) (ref.66)

	Tactivity (%)	Order Index J_{1455}/J_{840}	Orientation by Dichroitic Ratio 1168 cm^{-1}	995 cm^{-1}	840 cm^{-1}
Polypropylene Film					
Side Touching the Heated Bar	70	1.21	1.40	2.07	2.48
Side Away From the Heated Bar	75	1.08	1.55	2.27	3.02
Ratio Heated to Nonheated Side	0.93	1.12	0.90	0.91	0.82

Two different methods can be used to convert a fibrillated film tow into staple. The tow can be stretch-broken and converted directly into top or it can be cut and carded. Carding is carried out first on a garnet and then on conventional wool cards, but flax cards are also satisfactory. During carding, branched fibers or networks of fibers will tend to break down to single fibers and, because these fibrillated materials are sensitive to abrasion, single fibers may tend to shorten. It is desirable to keep the proportion of short fibers to a minimum since short fibers may lead to the production of irregular spun yarns or uneven nonwoven structure. It has been found that if the film web is stapled by stretch-breaking techniques rather than by cutting, significant fiber and yarn improvements result (ref.38). Because of the special characteristics of the fibrillated web, the stretch-breaking operation is simplified, consuming less power and producing a more level sliver than might be expected from a knowledge of the breaking of normal continuous filament tows.

IV.3. Network Fabrics

Closely related to the production of fibers from film is the formation of a unique type of net-like structures (see Fig.41). The process involves the steps shown in Fig.42. The extruded sheet is drawn to low draw ratios without inducing much fibrillation, embossed with the help of heated rollers and stretched successively in two perpendicular directions. Stretching of the embossed film (A) results in cracking parallel to the direction of the first draw, with cracks being halted by the thick bosses (B). On drawing in transverse direction the structures open up, yielding an open fabric which has the superficial appearance of a textile. By using hexagonal bosses, three different directions are available for the primary extension. The shape of the boss and the number of the embossed points determine the size of the structure; thickness of the original film determines the plyability of the sheet. Very fine net-like sheets have been produced commercially by adjusting these parameters. They are being used as base material for laminated fabrics, decorative drapery, skirm for needle punch nonwovens, and as secondary backing for carpets.

V. PROPERTIES AND APPLICATIONS

There are three considerations to be taken into account when judging the future development and prospects of film tape and film fibers with respect to their competition with conventional fiber manufacturing by spinnerette spinning:

(a) the specific characteristics and properties of film tape and film fibers; (b) the application areas open to film tape and film fibers; and (c) the competability of film-tape and film-fiber production methods with respect to technology and costs.

Fig. 41. Netlike Structure from Stretched Film

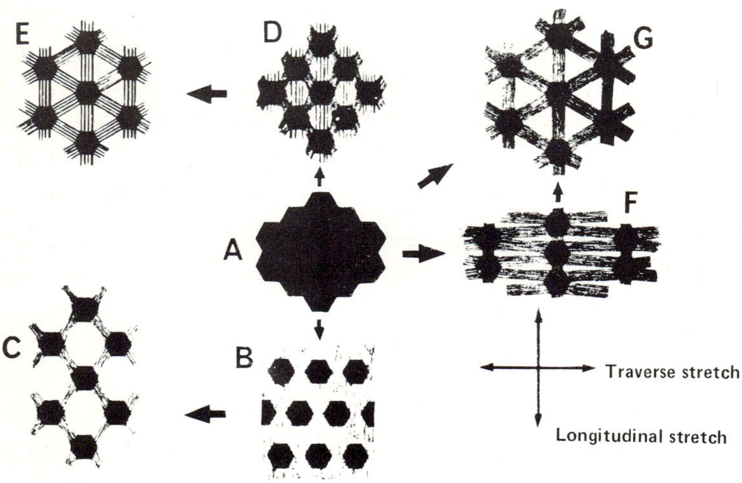

Fig. 42. Processing Steps to Produce Net-like Structures from Film (ref.14)

V.1. Properties

In the manufacture of film-tape or film-fiber from thermoplastic polymers, fundamentally the same flexibility exists as in the manufacture of conventional fiber, allowing one to "engineer" the properties of the respective products. This applies to tensile strength, elongation at break, work potential, tendency to further fibrillation, and dimensional stability of the film tapes or film fibers depending on the extrusion conditions, the draw ratio, the drawing conditions, and the proper heat setting and relaxation.

In general, the production processes for film-tape or film-fiber products allow us to achieve mechanical properties similar to conventional fiber manufacturing. This is illustrated by the data summarized in Table VIII. Also, the strength of twisted film-tape and of split- or slit-film-fiber yarns is by no means inferior to conventionally produced continuous filament or staple fiber yarns (ref.14).

The most apparent and basic difference between spinnerette fibers and film fibers is the cross-sectional characteristic. Spinnerette fibers are, in the majority of cases, circular or multilobal, while film fibers are rectangular or, in cases of bicomponent composition, sometimes "banana" shaped. In addition, film fibers are generally less regular in cross-sectional dimension, thus having a wider denier distribution than conventionally produced man-made fibers (Fig.43 and 44). In cases of pin-roll fibrillation, film-fiber products often have a more or less expressed network structure and comparatively more fiber ends. The properties which will be primarily influenced by these differing geometric characteristics are the hand, the resilience, the bulk, the pilling, and the friction of the fiber.

Nott (ref.2) published some interesting considerations concerning the effect of the cross-sectional difference of circular and rectangular fibers with respect to softness or hand. He came to the conclusion that, for equal softness, the thickness (t) and width (w) of film fibers should correspond to (d is the diameter of circular fiber of same denier):

$$t = 0.866d; \quad w = 1.047t$$

If the width of film fibers differs significantly from the thickness, the hand of a textile product will be caused primarily by the alignment of the rectangular fibers along their larger face. Table IX, taken from Nott's paper, gives the variations of film-fiber dimensions having equal denier with circular

TABLE VIII

Comparison of Spinnerette Spun Polypropylene Fibers and of Polyolefin Fibers Made by Film-to-Fiber Technology (ref.27)

Processing Technology	Polymer	Single Filament Titer (dtex)	Tenacity (gf/dtex)	Elongation (%)
Spinnerette Spun multifilament	Polypropylene m.f.i. 3.5	5	7.5	18[a]
Mechanically fibrillated	Polypropylene m.f.i. 3.5	5-45	1.5-4.5	11
Profiled Film (Barfilex)	Polypropylene m.f.i. 3.5	16	5.2	12
Roll Embossing (Shell Res.)	Polypropylene m.f.i. 3.5	15	6.7	16
Roll Embossing (Shell Res.)	80% Polypropylene 20% High-impact Polystyrene	23	5.7	19
Stretch Cutting (Chemiefaser Lezing AG)	Polypropylene m.f.i. 3.5	25	5.4	21[b]
Roll Embossing (Shell Res.)	H.D. Polyethylene m.f.i. 0.4 Density: 0.955	23	6.5	21

[a]Depending on fibrillator type and degree of twist.

[b]Heat setting without crimp.

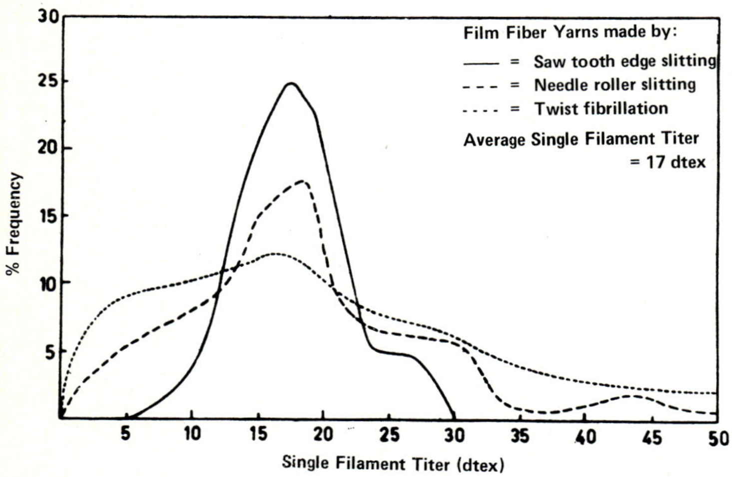

Fig. 43. Distribution of the Single Filament Titer for Film-Fiber Yarns Made by Different Film-to-Fiber Separation Techniques (ref.66)

fibers for various film thicknesses, and indicates the dimensions for film fibrils of equal softness with corresponding conventional fibers.

Film fibers made by uncontrolled mechanical fibrillation, have a relatively pronounced distribution width. They show a remarkably harsh handle. For film fibers made by controlled mechanical fibrillation or slitting, much narrower width distributions are obtained. However, since the practically achievable distances of the cutting edges of slitting tools, or the distances of the tops of embossing rollers, are today in the range 100-200 μm, they show a textile hand in agreement with Nott's rule.

This limitation in achieving deniers below 10 determines the appropriate place for film fibers with respect to their end uses. In Fig. 45, an attempt is made to define this place. The main areas where film tape and film fibers are of interest, and have their potential, are the fields of nonwovens, technical fabrics, heavy upholstery fabrics, carpet and floor covering manufacturing, and the manufacturing of cords, twines, ropes, tows, and nets.

Another difference of film fibers in comparison to conventional spinnerette fibers is their wider denier distribution. This is particularly the case for film fibers made by uncontrolled mechanical or chemo-mechanical fibrillation.

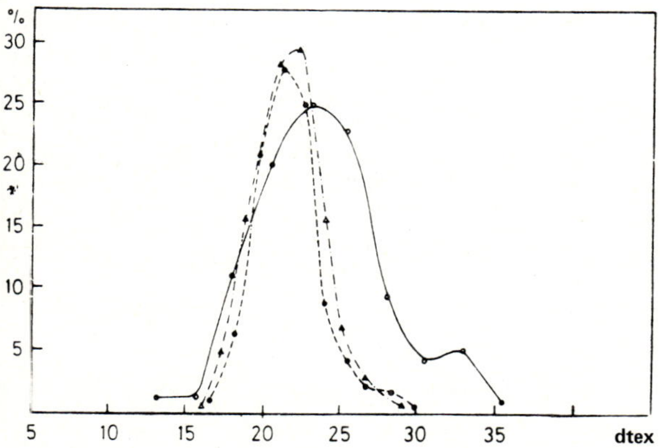

Fig. 44. Comparison of the Single Filament Titer Distribution of a Slit-Film Polypropylene Carpet Yarn with Those of Conventionally Processed Continuous Filament Polyamide and Polypropylene Carpet Yarns: _______, Polypropylene Slit-Film Yarn (24.9 dtex); -------, Spinnerette Spun Polypropylene Yarn (20.8 dtex); ---, Spinnerette Spun Polyamide Yarn (21.4 dtex) (ref.66)

The film fibers made with processes applying controlled film-to-fiber separation techniques in many cases approach the good uniformity in denier distribution characteristic for spinnerette spun man-made fibers, as can be seen in Fig. 44 (cf. also Fig. 43).

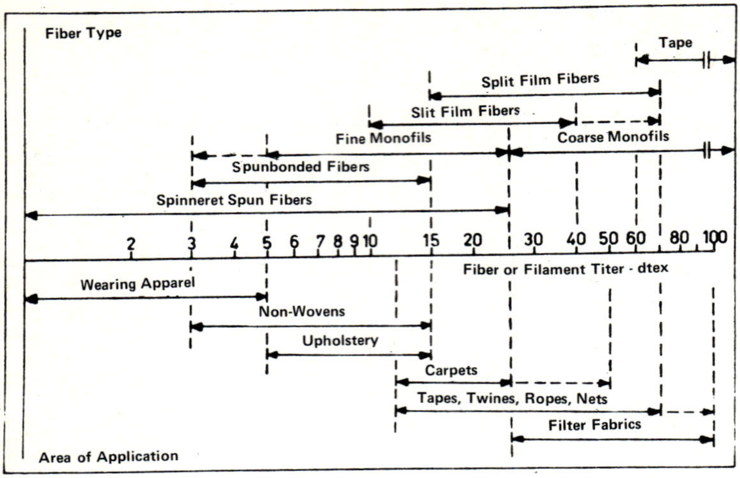

Fig. 45. Fiber Type and Area of Application in Relation to Their Titer and Method of Manufacturing (ref.105)

From theoretical considerations, film fibers of rectangular cross section pack to higher bulk density. They, therefore, should tend to give less cover than the fibers with circular cross section. This tendency should further be enhanced by the wider distribution in fibril denier. In practice, however the random arrangement of the fibrils in yarns, the remainders of network structures, and the presence of fibrillated ends hinder the formation of optimum bulk density of the yarn. Furthermore, we know from practical experience that spinnerette spun fibers of high "filament ratio" (dimensional anisotropy) have a better covering power than fibers with circular cross sections.

With respect to appearance, film fibers in many cases are designated as being "shiny" or "lustrous". In this regard, they are similar in appearance to spinnerette fibers of multilobal cross section. The incorporation of small amounts of dulling agents, such as titanium dioxide, does not work satisfactorily.

Film fibers generally have a markedly higher fiber-to-fiber friction than the circular spinnerette fibers. This results in ropes or twines with a much higher tenacity yield. However, the greater friction also gives rise to more shedding in production, more heat development on friction points in processing, and continued fibrillation in use, particularly when high draw ratios have been used to enhance film-to-fiber separation.

V.2. Applications

Within the past several years polypropylene fibers from film have been offered

TABLE IX

Variation of Polypropylene Fibril Dimensions with Denier (ref.2)

Multifil Diameter (μm)	Denier	Cross-Sectional Area (mm^2)	Width of Fibrils for Film Thicknesses (μm) 90 (μm)	70 (μm)	45 (μm)	33 (μm)	22.5 (μm)	16.5 (μm)	11 (μm)
12.5	1	0.123×10^{-3}	1.37	1.75	2.75	3.7	5.5	7.5	11^x
17.7	2	0.25×10^{-3}	2.75	3.5	5.5	7.5	11	15^x	22.5
25	4	0.5×10^{-3}	5.5	7	11	15	22^x	30	45
35.4	8	1.0×10^{-3}	11	14	22	30^x	44	60	90
50	16	2.0×10^{-3}	22	28	44^x	60	88	120	180
75	36.5	4.5×10^{-3}	50	64^x	100	135	200	270	400
100	64	7.87×10^{-3}	87.5^x	113	175	240	350	400	720

Note: x Fibrils of equal softness to multifils of given denier.

as alternatives of coarse natural fibers. These consist of the bast or semihard fibers such as jute, and the hard leaf fibers such as sisal and manila hemp.

With the sole exception of cotton, more jute is consumed in the world today than any other textile fiber. Approximately 70 percent of all jute is used in the form of woven bags and sacks for the packaging of commodities such as coffee, grain, sugar, flour, wool, etc., the remainder being consumed for tufted carpet backing, industrial fabrics, twine, packing, and a variety of products where low cost is more important than strength or durability.

The world production of sisal and manila hemp is just under one million tons per year. Since the fiber from each plant exhibits different physical characteristics, sisal fiber is generally qualified by the country or district of origin. The chief end-use application for sisal is agricultural twines and cordage. Manila hemp is the strongest of all vegetable fibers and is used for the manufacture of heavy duty ropes and cordage for marine, industrial, and farm uses.

Low cost coupled with adequate performance has led to the widespread acceptance and use of jute, sisal, and manila. These fibers, however, have few other assets since they are degraded by sunlight, rotted by moisture and microorganisms, and are prone to chemical attack. These natural fibers have also been subjected to severe price fluctuations over the years. As the growth of jute and other coarse fibers is concentrated in geographically small areas of the world, any political upheaval or adverse weather conditions in these areas disrupt the supply and upset the cost structure. Even under the normal conditions, there is doubt whether sufficient jute will be available in the next decade to cope with potential demands. Since jute, sisal, and manila are grown in developing countries, the price of these fibers must increase as the standard of living increases in those areas. Moreover, the population pressure in these areas would make more demand for rice production, which can only be obtained by cutting down the vegetable fiber acreage.

The United States, which has no large jute, hemp, and manila spinning industry, has felt these shortcomings and other potential problems probably more than Britain or Europe. Its huge committment to tufting, which requires large quantities of prewoven backing, has spurred efforts to find a domestic replacement for jute and other hard or semihard fibers. Since jute, sisal, and manila are low-cost materials, nylon and polyester fibers are at a distinct economic disadvantage as replacements. The choice of polymer for a low-cost synthetic hard or semihard fiber is, therefore, reduced to polypropylene and polyethylene, both of which are rapidly penetrating the traditional markets of jute, sisal, and manila and in some cases replacing cotton. Polypropylene has several definite advantages over polyethylene in not only in properties but also in price. It is therefore emerging as the dominant fiber for jute, sisal, and manila replacement.

The main advantages of film yarns, which have led to their rapid development into many outlets, are their ease and cheapness of manufacture. Film yarns are cheaper to manufacture than other types of synthetic-fiber yarn because of their higher through-puts in terms of poundage, they can be made with simpler and cheaper manufacturing machinery, and less labor is required to run the process. These reasons, plus the relative ease and cheapness of processing in many outlets, lead to a price or price/performance advantage over existing products.

Ropes and twine, a market previously dominated by sisal, was the first significant application for polypropylene fiber. The following years saw the introduction of woven film tape sacks, a jute replacement application. This was followed in 1967 by a penetration of the jute carpet-backing market. In 1968 mechanically fibrillated twine was introduced providing a soft fiber application, effectively penetrating the cotton market.

Table X summarizes the various present and potential applications of film yarns. The main areas of market penetration can be basically grouped under three headings: jute replacement, sisal and manila replacement, and soft fiber applications corresponding to film tape, coarse fibrillated film, and fine fibrillated film, respectively. The following discussion will follow these broad lines. Polypropylene is a low-price raw material and to benefit from this advantage, it is necessary that priority be given to those uses in which the raw-material cost represents a considerable share of the total manufacturing cost. It is, therefore, not pertinent here to discuss film yarns that can be used for woven stockings or knitted outerwear: in this discussion emphasis is placed on large-scale uses such as sacks and bags, carpet backing, twine and ropes, nonwovens, etc.

Twine, Cordage, Ropes. For many years twine has been the cheapest and most effective way of performing various packaging functions. These include binding the contents of a package together, securing the protective cover, forming a carrying

TABLE X

Some Applications of Polypropylene Film Yarns

Flat Tape

1. Primary backing for tufted carpets
2. Scrim for needle-punch carpets
3. Sacks for commodities transport and storage
4. Wool packs
5. Coated fabrics for industrial applications
6. Auto seatcovers
7. Wall coverings and decorative fabrics
8. Fabrics for soil conditioning, such as those required for building roads and runways on shifting soil
9. Temporary runways and roads for military uses
10. Cover fabrics for construction industry and temporary storage areas
11. Sandbags for military, construction industry, flood control, etc.
12. Outdoor furniture webbing
13. Travel baggage
14. Post office bags
15. Ladies handbags, hats, shoe uppers, etc.
16. Book covers

Coarse Fibrillated Yarns

1. Package cordage
2. Agricultural twines
3. Ropes
4. Fishing nets
5. Mats
6. Sport nets
7. Laundry bags
8. Industrial fabrics
9. Sacks for potatoes and onions
10. Chain yarns for woven carpets
11. Coated fabrics for industrial uses
12. Fashion articles such as shoe uppers, handbags, decorative fabrics, hats, table covers, etc.
13. Bast substitute for handicrafts

Fine Fibrillated Yarns

1. Cotton yarn replacement for industrial fabrics
2. Chain yarns for woven carpets
3. Upholstery fabrics
4. Filter fabrics
5. Knitted casual wear
6. Embroidery
7. Carrier yarn for knitted pile fabrics

Staple from Film

1. Nonwovens, carpets, and filter material
2. Cement reinforcement
3. Jute/polypropylene carpet face yarns

handle, bundling rigid articles such as pipes, affixing labels, and forming part of a security seal. Twines have been traditionally made from natural fibers - cotton, soft hemp, jute, and flax for the lighter twines and sisal for the heavier ones. The most important agricultural twines are baler and binder twines. Binder twine, the coarser of the two, was probably the first major utilization of a textile product based on film. The choice of polymer has been almost exclusively polypropylene which has advantages over polyethylene with respect to easy splittability, greater bulk, and higher strength.

Film-based twines are superior in many properties to their natural-fiber equivalents, which they are replacing rapidly. Commercially produced polypropylene film twines have a tenacity 4.5 g/den, compared with 2.5 for flax, 2 for sisal and soft hemp, 1.5 for jute, and 0.9 for cotton. Since most twines are joined by means of a knot, the knot efficiency will determine the strength of a twine loop. The knot efficiency of the lighter cotton, hemp, jute, sisal, flax, and polypropylene film twines is about 70 percent of the twine tenacity. This falls steadily with increasing weight to 45 percent for the heaviest sisal and 55 percent for the heaviest polypropylene twine (ref.40). Well-fibrillated polypropylene film is superior in knot efficiency to poorly fibrillated tapes. Elongation is important in packaging and binding, particularly when automatic tying machines are used. Polypropylene twines and straps have an elongation of about 12 to 15 percent when manufactured for optimum tensile strength. Natural twines of low twist have an elongation of about 10 percent but higher twists increase this figure.

The conversion efficiency of fibrillated polypropylene film strength from initial yarn to twine is extraordinarily high - about 100 percent. Table XI compares the straight twine and knot efficiencies of polypropylene film yarn with those of monofilament and multifilament yarns. The reason for the significantly higher conversion efficiency for the fibrillated-tape materials is not entirely clear, but it appears to be due to the fact that the tape tenacity is artificially low, probably owing to inhomogenities in the drawing process.

Polypropylene-film twines have proved satisfactory at $40^{\circ}C$ during baling in Australia on one hand and at $-20^{\circ}C$ in cold stores on the other. Since polypropylene does not absorb moisture, film-twine spools do not swell and lock the twine even though it is soaked with water. This is particularly important in agricultural operations. In abrasion and fusion resistance, film twines are inferior to their natural-fiber equivalents, particularly if replacement is made on equivalent breaking strength. Abrasion resistance can be improved by the use of polypropylene-polyethylene co-polymers, by use of air-cooled instead of water-quenched film, or by surface-emulsion treatments. But all these methods raise problems, the most common of which is a drop in knot efficiency. Because of their relatively

TABLE XI

Strength Conversion to Twine from Polypropylene Split Film and Multifilament and Polyethylene Monofilament Yarns (ref.41)

	Yarn Tenacity	3-Ply Twine Tenacity	Twine Conversion	Knot Tenacity	Knot Conversion
	g/den	g/den	%	g/den	%
Tape	4.6-4.8	4.7-4.9	102	2.9-3.1	64
Multifilament	8.0-8.5	6.0-6.2	74	3.0-3.3	38
Monofilament	5.5-6.0	3.4-3.6	61	2.6-2.8	47

TABLE XII

Typical Physical Properties of Ropes (Rope Circumference 7/8 Inch) (ref.52)

Rope material	Weight (15/100 ft)	Breaking strength (lb)	Breaking strength weight per ft	Performance ratio to Manila
Nylon monofilament	2.16	2250	10,400	3.04
Polypropylene split film	1.5	1500	10,000	2.93
Polyethylene monofilament	1.57	1200	7,180	2.09
Manila	2.4	820	3,420	1.00
Sisal	2.4	730	3,040	0.89

TABLE XIII

Typical Properties of Polypropylene Ropes (Circumference 3 Inches) (ref.43)

Properties	Nufil rope (split film)	Ulstron rope (Multifilament)	Sisal rope
Breaking load, (Tons)	8-9	7.5	4.0
Breaking extensional (%)	16-22	30-45	10-15
Weight (kg/100 meters)	26.8	29	42.5

low melting points, care must be taken with film twines to prevent the generation of excessive heat, for example, in the eye of a slipknot. Polypropylene-film twines are very resistant to rot, which is particularly important in agricultural twines, since they can be in service for several years.

Ropes manufactured from polypropylene split-film yarns have extremely high strength, while being less extensible and stiffer than ropes made from extruded fibers. The kink resistance is especially high. Rope fabricated from polypropylene-film yarns closely simulate the performance of sisal and manila ropes. A comparison of typical physical properties of ropes is shown in Table XII.

A similar comparison is given in Table XIII.

Split-film ropes are most suitable for outlets where stretch resistance is preferred, for example fishing mounting ropes, mooring and hauling ropes, and cargo slings. The low melting point of polypropylene in comparison with nylon and polyester has been cited as a drawback in high-speed usage. However, the low melting point of polypropylene is largely offset by its high specific heat. Improvements have been sought by properly lubricating the rope which considerably decreases the tendency of fusion on high-speed mooring on the winch. The use of internal lubricants such as hydrocarbon oils, greases, and microcrystalline waxes in 2 to 5 percent concentration, have also been recommended.

Abrasion-resistance data are notoriously unreliable. However, it has been variously shown that for a given rope compactness and exciting load, abrasion resistance depends upon the inherent resistance of the component yarns. Low-friction materials with high resistance to abrasion perform well, and vice versa (ref.44). In laboratory tests unlubricated polypropylene film ropes show a poorer performance than other synthetic materials and polypropylene multifilaments. The relative abrasion resistance of different ropes has been shown to have little relation to the wear life of bottom trawls. Here, the low abrasion resistance of polypropylene split-film twines is partly compensated by their rigidity, which makes them ride more clear of abrasive obstructions than do completely flexible twines.

Woven Fabrics for Jute Replacement Markets. Complementing the intrusion of fibrillating film into the sisal cordage and related trade, the jute weaving trade has received much attention by firms exploiting the high strength, covering power, and durability of coherent narrow tapes of polypropylene. Paradoxically, the efforts to produce highly splittable tapes have proved undesirable in jute replacement fabrics and, indeed, weaving processes have necessitated the use of nonfibrillating narrow tapes. Tape yarns are increasingly being used in packaging bags, sacks, and bale wrappers. The world market for jute used for these outlets is estimated at over 2,000 million pounds and represents a tantalizing prospect as an outlet for polypropylene tape. Polypropylene tapes also provide base fabrics in open woven form, primary carpet backing, scrims, underlays for needle felts, and similar products. Horticultural uses include vegetable bags, greenhouse shades, etc. Their use in Britain in post office bags has proved to be a successful innovation. Similarly their use in sand bags for construction industry, landscaping, dike building, and military use is now well established.

Fabrics can be manufactured in a variety of ways - by weaving, by knitting of all kinds, and by interlacing. For tapes, weaving is a primary method of conversion into fabrics; knitting and raschel techniques can be used for fibrillated yarns. There are many types and makes of looms available to the fabric manufacturer. The selection of the most suitable will depend to a large extent on the fabric to be woven. For example, a circular loom is most suitable for the manufacture of bags, and a wide loom with rapier-weft insertion is the best for tufted-carpet backing. Whatever loom is selected, there are several problems associated with the weaving of oriented slit-film yarn that are common to them all. Among the foremost of these is the twisting and fibrillating of tape in the process of weaving. These difficulties have called for important advances in winding and beaming.

Packaging Applications: The commonest use of polypropylene slit yarn is in packaging application. The main uses in this field are packing bags for textile materials and fabrics, agricultural products, industrial products, and other materials in general use. Jute sacks have been traditionally used for packaging of these commodities. The consumption of jute sacks is large. According to some estimates, the annual world need is approximately 10,000 million sacks, of which more than 3,500 million are new sacks produced annually, the rest being used sacks still in circulation. The world demand for packaging materials is growing. Jute sacks are not taking their share of this growth because, in essence, they do not meet the economies and technical requirements of the market. Bulk handling techniques, paper and film sacks, and - more recently - woven-tape polyolefin sacks, have been taking up the expansion in this market as well as making inroads into a variety of other segments traditionally serviced by jute. There are several reasons for this situation.

Table XIV shows a comparison of the properties of typical jute and polypropylene-tape fabrics. Table XV qualitatively compares the properties of jute and polypropylene fabric when each is used as packaging material.

In some European countries raschel-knit fabrics are used to bag vegetables and fruit, which require an open-mesh structure to ensure adequate ventilation. In the United States lenoweave fabrics have been developed for this application, owing primarily to the large broadloom capacity. Initially there was some difficulty with lenoweave fabrics from yarn slippage, but this has been overcome by the use of special coatings, and a considerable quantity of lenoweave fabrics was produced in 1970-1975 to bag carrots, cabbages, onions, and citrus fruits.

Polypropylene tape-yarn fabrics intended as a replacement for woolpacks have been successfully evaluated in Australia. Over the years, the jute pack has fairly adequately fulfilled the demands of the handling system. However, the trade has been conscious of wool contamination with loose jute fibers. The textile machinery used in worsted manufacture is designed and set for the mechanical processing of wool fibers, and any contaminating fibers from the jute packing will be processed as wool. Jute has no affinity for wool dyes and consequently, in many classes of manufacture (light pastel fabrics, for example), these fibers

TABLE XIV

Comparison of Properties of Weaving Polypropylene Tape and Jute

Property	Jute	Polypropylene
Yarn:		
Density, g/cm^3	1.45	0.91
Denier	1,000-1,200	750-1050
Tenacity, g/d	1.1 - 1.2	5.0 - 6.0
Elongation @ break %	1-5	15-20
Abrasion resistence (cycles to failure)	400-800	16,590
Fabric:		
Length (inch)	43	43
Width (inch)	28	28
Weight (oz)	35	7.4
Drop strength (Filled sacks cropped by increment of 1 ft.)	7	13-16
Coefficient of friction	0.75	0.69

TABLE XV

Comparison of the Properties of Cloth Woven from Polypropylene and Jute (ref.53)

Property	Polypropylene cloth	Jute cloth
Economics	Good	Variable
Availability	Potentially good	Generally good
Physical strength	Good	Good
Siftproof and cloth stability	Generally good	Good
Flame-resistance	Good	Smoulders
Weather-resistance	Generally good	Fair
Rot-resistance	Good	Poor
Coefficient of friction	Satisfactory	Good
Ease of handling	Good	Good
Toxicity and contamination from package	Good	Some limitations
Reuse	Potentially good	Good
Aesthetic appearance	Very good	Limited

show as visible flaws in the final product. In the early development stages of polypropylene woolpacks wraps severe problems were encountered from bales distorting under the internal pressure of their contents and the external pressure of other bales in the stack, which finally would cause the stack to topple. This necessitated changes in the fabric construction to induce greater fabric stability. Later constructions appeared to have eliminated this problem. Today the use of polypropylene woolpack wraps is mandatory in Australia and New Zealand.

<u>Carpet Backing</u>: Traditionally, carpet backings have been made from jute, whether from yarns incorporated into woven carpets or from prewoven jute fabrics in the case of tufted carpets. The reasons for the wide acceptance of jute as a backing fabric are pertinent to any discussion on alternative backings, defining as they do the proven requirements. The success of jute in this use is based firstly on low cost. This makes possible the economic production of relatively heavy fabrics which are strong and durable, have adequate resistance to stretch, and possess both body and flexibility. In addition to those properties the bond between jute and elastomers used for backing is high, thus ensuring good tuft or loop anchorage.

Jute withstood competition from manmade fibers almost completely until the 1960s. The substitution of jute, initially by polypropylene tapes in woven fabrics and now also by polypropylene spunbonded nonwovens, has been motivated by a variety of reasons, among which economy of cost is the most prominent. While it has been said that dissatisfaction with jute is not a major motive in its replacement, jute has a few shortcomings. Some of these are inherent, while others are associated with woven construction. Among the former defects are low resistance to attack by mildew and rot, poor dimensional stability after wetting or at high humidity, and the presence of lignins which can adversely affect the color and light-fastness of pile yarns in piece-dyed goods. Defects associated with woven construction include needle deflection caused by yarn irregularity, needle damage to weft yarns, bowing, and width variation within rolls. The increasing popularity of shampooing carpets means that rotting of the backing yarns can occur through over-wetting. The wetting of jute can also be accompanied by buckling and dimensional changes. Lignins released from the fiber or spinning oils from yarns can affect piece dyeing. Jute fabrics are often not very regular, and may suffer from bowing and other defects. However, the salient advantages mentioned previously far outweigh these latter defects.

The synthetic backings are much lighter than the jute fabrics generally used for tufted carpets. This is inevitable since raw material cost has a major influence on ultimate cost. Jute yarns are, at most, about one-third the cost of any form of synthetic fiber so, if the latter is to become economically viable, compensating reduction in the weight of synthetic backings is essential. This means that carpets with synthetic backings will be approximately 6 to 8 oz a sq yd lighter than corresponding jute backed carpets.

Briefly, jute yarns of 6,000 den to 12,000 den equivalent, composed of fibers of 1.45 g/cm^3 density, are generally replaced by polypropylene tapes of 750 to 1,100 den with density of 0.90 g/cm^3. Its lower density and tape form (e.g., cross section of 3 mm x 0.05 mm) combine to give polypropylene extremely high covering power, while its much higher breaking extension is associated with increased toughness. For these and other reasons, it becomes satisfactory to work at primary backing weights of 3 oz/sq yd, substituting for 10 to 11 oz per sq yd of jute, at which polypropylene has a slight advantage in cost. Common constructions are 10 warp tapes per inch by 10 to 13 weft tapes per inch, all of 1,000/1,100 denier, and 24 warp tapes per inch of 500 denier by 11 weft tapes per inch of 1,100 denier.

Flat tapes provide the maximum cover for a given weight of material, and these backing fabrics average only 200 mm in thickness. At $3\frac{1}{4}$ oz/yd^2 they are cheaper than equivalent jute backings of 9 to 12 oz/yd^2. Reduced thickness is an obvious disadvantage in terms of lateral tuft support, but it does permit maximum utilization of the pile yarn at the surface of the carpet, i.e., less pile yarn is buried in the backing.

The primary backing of polypropylene is advantageous in the dyeing process. The jute carpet backing passing through the dye liquor picks up a considerable weight of water, all of which has to be driven off. The polypropylene backing, on the other hand, will not absorb any liquor. This represents substantial savings in the cost of carpet drying.

Polypropylene film fabrics have attained only limited success against jute in the shag carpet area because of dyeability problems, and against nonwovens in the fine-gauge area because of needle deflection. One method which will probably enable film fabrics to compete more successfully in both these areas is to needle-punch into them another fiber which is dyeable. This not only produces a dyeable backing but reduces needle deflection and at the same time eliminates edge-raveling. In these composite structures, the weight of the nonwoven component varies between 2 and 6 oz per sq yd and the denier ranges between 3 and 60, depending upon the manufacturer's preference and upon whether the fiber was drawn or undrawn. So far, the only fibers used have been nylon 6 and nylon 66, but other fibers will almost certainly be tried in the future.

Because of the extra material and processing, the price of the composite backing is considerably higher than that of film fabrics alone, and this may ultimately limit its market. But compared with jute there are considerable cost savings to the tufted-carpet manufacturer. These include the elimination of the second dye cycle required for jute backings and the reduction in dye liquor needed in the first cycle as the synthetic backing absorbs far less dye than does jute. There may also be some savings compared with nonwoven backings, as the aesthetics of the composite backing may enable the manufacturer to reduce the amount of face yarn in certain carpet constructions.

All primary backings are damaged to some extent by the penetration of the needles, and this results in reduced strength of the fabric after tufting. This loss in strength can be substantial, but lubrication has been found to prevent the loss to a large extent, and it is now common practice to lubricate the fabric before tufting so that the strength after tufting is adequate for subsequent processing in dyeing and back-coating.

Another effect of lubrication is to prevent the tufting needles from getting unduly hot. Moreover, the stitch lines run straighter in the carpet with lubricated backing, the height of the pile appears to be more uniform, and knots and joints in the pile yarn pass through the backing more easily. In addition, the noise level of the tufting machine when operating is reduced, and it is an interesting point that noise is usually a good criterion of ease of tufting.

After dyeing, the carpet has to be dried, and here temperatures usually rise. So long as water is being driven off, the temperature of the carpet may never exceed 105°C, but when all the water has gone the temperature of the fabric will inevitably rise towards the temperature of the oven with increasing risk of shrinkage. The risk of excessive shrinkage also exists in latexing operation. The oven temperatures used in latex curing process may exceed 150°C, and at this temperature some contraction of polypropylene is inevitable. Where it is known that the carpet will be back-coated on a stenter and that the stenter will control the passage of the carpet through the entire length of the oven, shrinkage can be controlled. If the stenter operates over only a part of the passage, adequate heat-setting of the backing material is still necessary. Most weaving tapes are heat-set to some extent, and although there is still a tendency to contract on reheating, this contraction is rarely more than perhaps 3 percent at 130°C. In attempts to reduce the temperature to which carpets must be subjected during the back-coating operation, efforts are being made to provide lattices which can be dried at lower oven temperatures.

The difficulty of making a good bond between latex and polypropylene means that if the carpet is laid directly on the floor, some of the latex may rub off and form a loose powder underneath the carpet. This can be overcome by a secondary backing, and indeed a secondary backing of a lightweight jute fabric may be necessary in order to improve the "handle" and to bring back the weight of the finished carpet to the levels normally found with carpets containing a jute primary backing. This secondary backing is at present of great interest for replacement by polypropylene film yarns. Since the nonslip characteristic of secondary backing is important, a simple woven fabric, made of tape yarns, would not do the job. Similarly, the fibrillated film yarns could not be used because of cost considerations. Three distinct approaches are currently being taken:

(1) The use of tape yarn fabrics, similar to the primary backing fabrics, which are abraided on one or the other sides to make them rough for better adhesion to the primary backing.

(2) Woven fabrics of light weights in which the warp is of narrow film tape and filling is spun yarns of polypropylene staple.

(3) Woven fabrics with narrow film tape warp and air-textured, fuzzy, continuous filament yarns as filling.

Miscellaneous Applications: Film fabrics are competing with burlap as biased-sewn tubing to package textile roll goods. The polypropylene fabric price is below that of burlap, there are savings in freight costs because of lower tare weight, and colored fabric is available for brand identification.

A surprisingly large market for film fabrics is nursery squares for balling roots of shrubs. It was previously thought that this market would be closed to polypropylene, which does not rot when buried in the ground as burlap does. But rotting is not required so long as the weave is wide enough to let the roots grow through the fabric. Indeed, the high price of jute nursery squares is partly due to copper-naphthanate impregnation to prevent rotting before planting.

Cotton bale covers are expected to provide another outlet for polypropylene woven fabrics. A cotton bale cover of film fabrics was first developed in 1965, but could not find application because of the practice of gross weight trading. Since film fabric is only one-sixth the weight of burlap, for a 12-lb burlap bale cover the cotton grower would have had to include 10 lbs more cotton. In 1971, however, the United States cotton industry switched to net weight trading.

Other markets in which plain-weave film fabrics are now established include the interior linings of upholstered furniture, swimming pool covers, and grass catchers attached to lawn mowers.

An interesting application of film-yarn fabrics is in the construction of foundations for dykes in the Dutch "Delta Project". Large areas of these fabrics are being used in place of the more conventional structures of plaited reeds in this large-scale land-reclamation project. The fabrics are laid on rafts, loaded with stones and sand, and sunk, to form the dyke foundations.

The staple fiber from film is eminently suited to the manufacture of nonwoven fabrics because, in comparison with spun fibers, its special cross section and rough surfaces give a high coefficient of friction between the film fibers. By needle punching, felts can be obtained which are slightly stronger than those made from comparable spun fibers. In addition, the fibrillated fiber felts show a much better mechanical stability. If required, various scrims can be used to improve the mechanical properties further.

These felts are suitable for uses such as cushioning, sound insulation, filtration, and dust collection. They may be used in polishing discs or papermaker felt tips. Their use as backing materials for floor or wall coverings may be particularly interesting, as they have an advantage over cheap natural fibers in being insensitive to moisture, rot, and mildew.

Staple from film can be spun into yarns for normal textile operations such as weaving and knitting. They can also be blended with other natural fibers and spun into yarns without appreciable difficulties. Fibrillated film staple/wool blends have been used as carpet pile yarns in England but their widespread use has not so far materialized. Extensive work has been reported on film staple/jute blend yarns for woven carpets. A 30 percent addition of polypropylene staple with jute contributes a 200 to 400 percent improvement in wear life of floor coverings over those made from 100 percent jute pile.

Considerable work has been done in the past attempting to reinforce concrete with fibers. This use requires a high modulus reinforcing material such as steel or glass fibers. Organic polymeric fibers such as polypropylene are unsuitable for concrete reinforcing. Polypropylene fibers of short length are, however, finding increasing use in two sectors of the construction industry in conjunction with concrete:

(1) The production of highly porous, air-entrained concrete mix with improved thermal insulation and better resistance to frost and weathering,

(2) Thixotropic concrete mixes for "face molded" articles such as sculptured panels, and "moldless forming", such as the operation of river bank reinforcement.

In concrete technology, air entrainment is a well-established method of producing concretes with varied low densities. However, a maximum of about 10 percent by volume of air can usually be tolerated in the mix before segregation of the constituents begins to occur with settling of the heavier particles (ref.48). If about 0.1 percent by weight of short, inert fibers are dispersed in an air-entrained mix, segregation does not occur until much higher percentages of air are reached - up to 30 to 40 percent by volume - even with coarse aggregates of 20 mm maximum size (ref.48). Also, separation of water on the surface of the concrete mix, a phenomenon known as bleeding, is suppressed.

Applications in which the use of fibrous polypropylene-reinforced concrete would save weight and material costs are said to include: cladding and roofing panels, posts and fences, pipe sections, covers and lids, boat hulls, paving slabs, flooring, and roofing screeds. The greatly improved impact strength suggests sea-defense applications, in which waves might damage conventional materials, and construction uses in which heavy vehicles or similar equipment are used.

Chemical action between polypropylene and the concrete is virtually nil. This is an advantage in long-term aging. When a fiber-reinforced concrete slab is damaged, the fibers extend to bridge the cracked section easily. The item retains its shape and integrity, and this allows time for a safe repair or replacement if necessary.

An unexpected bonus of fibers containing concrete mixes is that they show thixotropic properties - stiff when stationary but flowing freely when vibrated. Concrete mixes to which air-entraining agents (such as saponifiable resins or stearates) and short polypropylene fibers have been added have unique handling characteristics: they have little tendency to creep and are self-supporting. A unique use of this property has been made in the reinforcing of difficult-to-reach river banks, steep slopes, and bunkers, where the use of a mold is either not possible for reasons of accessibility or not practical due to emergency situations. The fiber containing air-entrained concrete can be sprayed directly on the soil surface where, in contrast to normal concrete mix, it remains in position without flowing.

These polypropylene fiber/air/concrete mixes are also widely used for the manufacture of sculptured concrete panels or other articles of construction. Mixtures of concrete with 0.1 wt percent polypropylene fiber of 15 to 20 mm length and more than 10 percent air entrainment are suitable for such uses. The concrete can be molded in the plastic state by conventional methods. The thixotropic properties of the mix permit the easy working of very complicated patterns which will not deteriorate while the concrete is left to harden. The conventional moldings of concrete are done face down in the molds and cannot be inspected until the concrete is hardened. Molding of polypropylene fiber containing air-entrained concrete is much simpler and more flexible as the sculpturing is done face up. There should be no rejects since several attempts at a pattern are possible.

The volume consumption of polypropylene fiber in concrete mixes is relatively insignificant at this time but this use does present an attractive opportunity for large-scale consumption of polypropylene fiber. The proliferation of prefabricate houses should be a boost for this use.

Several interesting offshoots of polypropylene tape or fibrillated yarn have appeared. The similarity in texture of stretched polypropylene film yarns to that of natural products of the raffia type has caused interest in the craft industries, particularly in southern Europe, the West Indies and South East Asia.

Fashion goods such as sunhats, handbags, beach shoes, table mats, and wall coverings have been made by weaving and plaiting techniques. The supply of the natural products is not always satisfactory and there are problems of coloration that is overcome by use of colored polypropylene tape materials.

An interesting and potentially large-scale use of polypropylene tape or partially fibrillated yarns is for artificial seaweed, where tufts of film yarns resembling natural seaweed are anchored to the sea bottom. These tufts alter the ebb and flow velocities of incoming waves, which results in the trapping of offshore sand particles in the seaweed and subsequent reduction of beach erosion.

V.3. Comparative Cost Structure

Spinnerette and film-fiber production are both based on the same technological elements, namely extrusion through a die, cooling and quenching, orientation, heat setting, and collecting. From a basic point of view, there are no additional steps in either the spinnerette process or in the film-fiber process which could be considered specific only for one or the other of these techniques.

A number of investigators have tried to make comparisons of production costs between conventional fiber spinning and film-fiber production(ref.106). However interesting these comparisons may seem, the exercise is rather pointless. First of all, the nature of the two types of yarns are completely different and their markets generally do not overlap. Secondly, the short-spin processes, with which film-yarn processes are often compared are most suitable for the production of staple fiber, needing an additional process of staple spinning for conversion of fibers into finished yarns.

After reviewing the industry as a whole, however, a few intuitive conclusions can be drawn. The production costs for film tape and fibrillated yarns for use in carpet backing, bags, heavy ropes and agricultural twine, are definitely lower than the comparable multifilament yarns produced via spinnerette technology. This differential widens as the denier of the yarn increases. On the other hand, the yarns used in a multitude of other industrial uses have a definite advantage for conventional spinning, both in quality and production cost. There are of course some grey areas where one or the other process (and the product) may be preferable. One such use is 300-400 denier yarn used in sliver knitting. Here the production cost of a fibrillated yarn becomes quite high but this yarn can be used without a twist. Conversely, the production cost of a multifilament yarn would be somewhat lower(especially, via spin-draw route) but the yarn will have to be twisted for use.

VI. REFERENCES

1. British Patent 479,202 (1936), US Patent 2,185,789 (1936), I.G. Farbenfabriken.
2. R.E. Nott, Conf. on Textiles from Film, II, Manchester, Plastic Institute, 1971.
3. L.H. Gouw and W.H. Skoroszewski, Plastics and Polymers, Manchester, 3(1968).
4. Darragh and Johnson, Mod. Text., 49(1968)24.
5. Anon., Melliand Textillerichte, 50(1969)148.
6. H. Peuker, Z. ges. Textilind., 70(1968)774; 71(1969)10, 82.
7. US Patent 3,461,198.
8. N.P. Novikov and V.I. Selikhova, Mech. Polym., 3(1967)95.
9. A. Peterlin, J. Mat. Sci., 6(1971)490.
10. P. Van der Zee and D.T. Pals, Mod. Text., 76(1972).
11. US Patent 3,395,525 (1968), Phillips Petroleum.
12. US Patent 3,382,663 (1968), Shell International.
13. US Patent 3,372,222 (1968), Dow Chemical Co.
14. Condit and Johnson, Mod. Text. Mag., 55(1969).
15. US Patents 3,214,899; 3,242,035; 3,177,557; 3,242,035; 3,695,025.
16. R. Fletcher, Conf. on Textiles from Film, II, Manchester, Plastic Institute, 1971.
17. British Patent 1,073,741.
18. P.C. Condit and G.B. Johnson, Mod. Text. Mag., 53(1969).
19. South African Patent 671,672; French Patent 1,518,108; British Patent 1,128,274.
20. US Patent 3,460,416 (1969), Phillips Petroleum.
21. US Patent 3,427,912 (1969), Toshiba.
22. British Patents 1,035,571; 1,176,357; US Patents 3,164,948; 3,273,771; Japan Patent 16,450/65.
23. British Patent 914,489 (1963), Smith and Nephew.
24. O. Dow, A.G. Patchell, and Z.T. Piskozub, Conf. on Textiles from Film, II, Manchester, Plastic Institute, 1971.
25. D.H. Latham, Plastics and Polymers, 38(1970).
26. P.M. Coffman, PINTEC Conf., 1970.
27. G.J. Nichols, Conf. on Textiles from Film, II, Manchester, Plastic Institute, 1971.
28. US Patent 3,499,822, O.B.R. Rasmussen.
29. O.B.R. Rasmussen, Plastics and Polymers, Manchester, 3(1968).
30. P. Volan, Spinning in Seventies, Merrow Publishing Co., Manchester, 1970.
31. P. Volan and P.D. Changani, Ann. Conf. Text. Inst., Manchester, 1968.
32. H. Peuker, A. ges. Textslindustrie, 70(1968)774.
33. British Patent 1,123,538 (1968), Shell.
34. British Patent 1,188,968 (1970), Mitsubishi; 1,132,166 (1968) Monsanto.
35. H. Mehta, Conf. on Textiles from Film, II, Manchester, Plastic Institute, 1971.
36. R. Fletcher, ibid.
37. M. McMeekin, ibid.
38. McMeekin, Plastics and Polymers, Manchester, 3(1968).
39. News Release, Fabric Research Lab., 1970.
40. R.I. Taylor, Plastics and Polymers, 3(1968).
41. W.H. Broatch, Plastics and Polymers, 3(1968).
42. Coarse Fibers from PP Stretched Tape, Shell Chem. Co.
43. J.E. Ford, Man-Made Textiles, 26(1966).
44. M.R. Parsey, Plastics and Polymers, 3(1968).
45. I.H. Thomas, ibid.
46. B.W. Overton, ibid.
47. H.T.H. de Monte, Text. Inst. Ind., 116(1971).
48. J.A.G. Thomas, Composites, 3(1972)71.

49. O.J. Sweeting, Science and Technology of Polymer Films, Interscience, N.Y., 1968.
50. US Patent 3,395,525 (1968), Phillips Petroleum.
51. US Patent 3,372,222 (1968), Dow Chem.
52. Shell Chem. Trade Publication, 1968.
53. B.W. Overton, Plastics and Polymers, 3(1968).
54. K. Burggraf, Kunststoffe, 57(1967)78.
55. K. Okabe and T. Chino, Plast. Aage, 13(1967)39.
56. Anon., British Plastic, 41(1968)165.
57. L.H. Gouw, Conf. Text. from Film, Manchester, 1968.
58. P.T. Slack, Conf. Text. from Film, Manchester, 1968.
59. P.C. Condit, Modern Text. Mag., 50(1969)23.
60. D. Moorwessel, Kunststoffe, 59(1969)205, 539.
61. G. Schuur, Text. Prax. Int., 24(1969)65.
62. J.E. Ford, Text. from Film, Shirley Seminar, Manchester, 1970.
63. J. Balsler, Conf. Text. from Film, II, Manchester, 1971.
64. D.C. Hossak, Conf. Text. from Film, II, Manchester, 1971.
65. R. McKehagger, Conf. PP in Text., York, 1975.
66. H. Kraessig, J. Polym. Sci., Macromol. Rev. 12(1977)321-410.
67. W.H. Skoroszewski, Conf. Text. from Film, II, Manchester, 1971.
68. M.E. Evans, Conf. Text. from Film, II, Manchester, 1971.
69. M.L. Cooper, Conf. Polypropylene Fibers in Textiles, York, 1975.
70. US Patent 3,596,816 (1971), Phillips Petroleum.
71. US Patent 3,003,304 (1961), O.B. Rasmussen.
72. Austrian Patent 251,751 (1967), O.B. Rasmussen.
73. German Patent 2,021,986, Soko Co., 1971; Swiss Patent 509,434, Soko, 1971; Swiss Patent 534,232, Soko Co., 1973.
74. German Patents 1,807,145; 1,807,143; Allied Chemical Corp., 1969; US Patent 3,539,663, Allied Chemical Corp., 1970.
75. British Patent 1,238,120, Rheinstahl Henschel, 1971.
76. German Patent 1,207,543, Dow Chemical Co., 1965; French Patent 1,426,288, Sociate Rhodiaceta, 1965; British Patent 1,073,741, Plasticisers Ltd., 1967.
77. Austrian Patent 282,799, Shell International, 1969.
78. US Patent 3,242,035, DuPont, 1966.
79. US Patent 3,695,025, Fiber Industries, 1972.
80. W.H.J. Badrian, Text. Production, 8(1971)27.
81. J.H.L. Choufoer, Conf. Polypropylene in Textiles, York, 1975.
82. US Patent 2,728,950, Dow Chemical, 1956.
83. British Patent 1,067,514, Mirsky, 1967.
84. US Patent 3,470,285, Hercules Inc., 1969.
85. British Patent 1,287,874, Chemiefaser Lenzing, 1970.
86. US Patent 3,542,909; 3,500,517, Shell Oil, 1970.
87. J. Harms, International Man-Made Fiber Conf., Dornbirn, Aust., 1971.
88. J. Harms, Lenzinger Ber., 32(1971)81.
89. J. Harms and H. Kraessig, Melliand Textilber. Int., 54(1973)439.
90. J. Harms, Chemiefasern/Text. Ind., 23(1973)845,979.
91. H. Kraessig, Faserforsch. Textiltech., 26(1975)135.
92. H. Kraessig, Conf. Polypropylene Fibers in Textiles, York, 1975.
93. Netherlands Patent 66-02,387(1966); British Patent 1,230,991 (1971); British Patent 1,243,512 (1971), O.B. Rasmussen.
94. US Patent 3,539,663 (1970); German Patent 1,807,143 (1969); British Patent 1,260,957 (1972); British Patent 1,308,998 (1973) to Allied Chemical Corp.
95. German Patent 1,921,678 (1969), Eastman Kodak.
96. A.R. Freedman, Mod. Plast., 47(1970)109.
97. French Patent 2,030,319 (1970); German Patent 2,031,686 (1971); British Patent 1,232,282 (1971), Smith and Nephew.
98. H. Mehta, Conf. Fibers from Film, II, Manchester, 1971.
99. Japan Patent 71-36,890 (1971), Mitsubishi Rayon.
100. US Patent 3,582,418 (1971), Shell Oil Co.
101. British Patent 1,260,837 (1972), Shell International Research.

102. W. Berger, Faserforsch. Textiltech., 24(1974)36.
103. W. Berger, Faserforsch. Textiltech., 25(1975)140.
104. International Fibers Inc., Personal Communication.
105. F. Hensen, Intercarpet Baden b. Wien, Austria, 1974.
106. R.E. Holt and G. Sarwar, Conf. Polypropylene Fibers in Textiles, York, 1975.

VII. BIBLIOGRAPHY

VII.1. Books and Symposia

1. Textile from Film, I, (Symp.) Plastics Inst., London, 1968.
2. Textile from Film, II, (Symp.) Plastics Inst., London, 1971.
3. Tape Yarns, D.C. Hossack, Merrow, Watford Hertz, England, 1971.
4. Polypropylene Fibers in Textiles, (Symp.), Plastics Inst., London, 1975.

VII.2. Journal Literature

1. H.J. Zingel, Split Fibers Technology, Spinner Weber Textilveredlang, 83(1965)361-362.
2. Battle for Twine, Mod. Plastics, 42(1965)92-93,216.
3. J.E. Ford, Thermoplastic Split Fibers, Brit. Plastics, 38(1965)157-159.
4. How Olefin Film and Fiber Aid Conservation, Mod. Plastics, 43(1965)112-113.
5. Polypropylene Film as a Sisal and Jute Replacement, Brit. Plastics, 39(1966)38-39.
6. M. Nozawa, Japanese Split Fiber of Regular Geometric Form, Brit. Plastics, 39(1966)642-643.
7. S. Schwenkedel, Polyolefin Tape - a New Basic Material for Textile Sheeting, Text. Prax., 22(1967)507-510.
8. F. Hensen, Monoaxially Stretched Film Tapes, Melliand Textilber, 48(1967)379-381.
9. Extruding Polypropylene Tape, Man-Made Textiles, 44(1967)41-42.
10. Polypropylene Seaweek Fights Coastal Erosion Problem, Man-Made Textiles, 44(1967)101.
11. K. Burggraf, Monoaxially Stretched Tapes from Plastic Film, Kunststoffe, 57(1967)78-84.
12. S. Schwenkedel, Textiles from Polyolefin Ribbons, Z. ges. Text. Ind., 69(1967)835-838.
13. F. Hensen, Production and Processing of Monoaxially Stretched Film Tapes, Textil-Praxis, 22(1967)329-331,410-413.
14. Tapes and Fibers - a New Industry, Part I. Olefins as Replacement for Jute and Sisal; Part II. Equipment for the Production of Oriented Polypropylene Tapes, Brit. Plast., 4(1967)92-101.
15. Range of Equipment for Producing Polyolefin Tapes, Textile Recorder, 85(1968)68.
16. What is New in Slit Film Textile Systems, Mod. Plast., 45(1968)122-124,128.
17. G. Klust, Yarns Made From Polypropylene Film Slivers, Text. Praxis, 23(1968)361-363.
18. Film Yarn and Film Fibers, Plastics, 33(1968)161-171.
19. R. Huschka, Production and Properties of Textiles from Film Yarns, Deut. Textiltech., 18(1968)137-141,210-216.
20. B.D. Berkman, Yarns from Plastic Film: Slit and Split Polypropylene Yarns Make Rapid Progress, Mod. Text., 49(1968)28-29.
21. P. Lennox-Kerr, Textiles from Film: Fast Growing Technique, Text. Ind., 132(1968)69-77,115,118.
22. D.C. Hossack, Polypropylene Tape: Production and End Uses, Text. Inst. and Ind., 6(1968)274-276.
23. J.E. Ford, Textiles from Film, A Review, Chem. and Ind., 42(1968)1417-1422.

24. D.J. Brittain, The Properties of Polypropylene Tape and Yarns, Text. Inst. and Ind., 6(1968)271-273.
25. Tape Fiber Direct from Polymer Gains, Textile Month, (1969)115-118.
26. A.R. Freedman, Coarse Fibers from Polypropylene Stretched Tape - Polymer and Processing Variables, SPE Am. Tech. Conf., 27(1969)488-493.
27. J.E. Ford, Development in Film-Fiber Machinery, Textile Month, (1969)89,92, 94,97.
28. New Machinery for Film Yarn Production and Processing, Z. ges. Textilind., 71(1969)555-556.
29. G. Schurr, Yarns and Fibers from Fibrillated Film: A Review, Textile Month, (1969)45-46.
30. P. Lennox-Kerr, Split Film Gets a Second Look, Text. Ind., 133(1969)60-66.
31. D. Moorwessel, New Polypropylene for the Coarse Textile Sector. I. General, Kunststoffe, 59(1969)205-210.
32. Machinery, Products and Processes for the Production and Processing of Films and Tape Yarns from Polypropylene and Polyethylene, Z. ges. Textilind, 71(1969)198-202.
33. H. Peuker, Filaments from Films: Production, Processing, Uses and Some Development Possibilities of Tape Yarns and Split Fiber Based on Polyolefins, Z. ges. Textilind., 70(1969)774-782,867-874,10-20.
34. New Processes for the Production of Monoaxially Stretched Film and Film Tapes Yarns, Melliand Text., 50(1969)148-150.
35. Fibrillated Polypropylene for Concrete Reinforcement, Rubber and Plastics Age, 50(1969)44.
36. G. Schurr, Yarns and Fibers from Fibrillated Film: A Review, Textile Month, (1969)65-69.
37. A.S. Baumgartner, Processing Parameters and Their Effects on Polypropylene Fibrillated Products, Can. Text. Seminar, 12(1970)116-119.
38. R.J. Nightingale, Machinery for Film Tape and Yarn Production: The State of the Art, Brit. Plastics, 43(1970)121-125.
39. G.O. Thomas, New Process for Filament Fiber Production, Chem. Processing, 17(1971)22,25.
40. G.J. Nichols, Newer Routes to Fiber from Film, II, Plast. Polym., 40(1972)84-88.
41. H. Kralssig, Production Methods, Properties and Applications of Film Yarn and Fibers, Deut. Textiltech., 29(1972)641-646.
42. How to Make Your Own Polypropylene Yarn, Mod. Text., 53(1972)42,44.
43. R. Eller, Orient Filled Tape to Stretch Your Dollar, SPEJ, 28(1972)54-58.
44. J.E. Ford, Future Applications for Film Textiles, Plast. Polym., 40(1972) 129-135.
45. M. Velluire, New Developments in the Field of Synthetic Fibers by Film Fibrillation, Melliand Textilber. Intern., 1(1972)109.
46. A.G. Patchell, Newer Routes to Fiber from Film, Part I and II, Plastics and Polymers, 40(1972)80-88.
47. Development Trends in Split Fibers: Products, Techniques, and Machinery at K '71, Chemiefasern, 3(1972)239-242.
48. Film Fibers: Recent Developments in Polymers and Processes, Textile Inst. and Ind., 10(1972)170-173.
49. H. Balk, Production of Slit Film Fibers and Yarns, Chemiefasern, 3(1972)236-239.
50. J. Harms, Threads and Fibers from Film, Melliand Textilber. Int., 2(1973) 335-339.
51. W. Berger, Fibrillation of Films Made of Polyblends, Textiltech., 24(1974)36-41.
52. Fasciated Yarn Combines the Best Characteristics of Spun and Filaments, Textile World, (1974)124,134,136.
53. J. Colign, Split-Film Non-Wovens, Textil Praxis Intern. 30(1975)1201.
54. W. Berger, Split Fibers from Polymer Blends, IV. Investigations on the Flow Properties of Polymer Melts in Extrusion, Faserforsch. u. Textiltech., 26(1975)340-347.

54a. W.Berger, Physical-Chemical Basis of Fibrillation, Lenzingerber. 48(1978)121.

55. J.J.V. Colijn, Split Film Nonwovens, Textil-Praxis Intern., 30(1975)1192, 1201,1202.
56. Split Film Preparation, Fiber and Fabrics, 128(1975)8-9.
57. H. Kraessig, Film to Fiber Technology, J. Polym. Sci., Macromol Rev., 12(1977)321-410.

VII.3. U.S. Patents

3,061,401	Soc. Viscose Suisse	1962
3,081,519	DuPont	1963
3,112,160	Phillips Petrol.	1963
3,199,284	Ernest Scragg	1965
3,242,035	DuPont	1966
3,249,668	Ernest Scragg	1966
3,264,384	Soc. Viscose Suisse	1966
3,273,202	Bayer	1966
3,273,329	Ernest Scragg	1966
3,290,207	DuPont	1966
3,293,293	Eastman Kodak	1966
3,331,719	Branson	1967
3,323,978	Phillips Petrol.	1967
3,332,228	Chevron	1967
3,332,828	Monsanto	1967
3,346,916	Monsanto	1967
3,361,849	Hercules	1968
3,372,221	Dow Chem.	1968
3,372,222	Dow Chem.	1968
3,395,525	Phillips Petrol.	1968
3,395,825	Phillips Petrol.	1968
3,398,441	Asahi Kasei	1968
3,398,220	Parker Pace	1968
3,400,041	Phillips Petrol.	1968
3,400,193	Dow Chem.	1968
3,401,517	Phillips Petrol.	1968
3,402,548	Eastman Kodak	1968
3,416,714	Phillips Petrol.	1968
3,416,772	Phillips Petrol.	1968
3,422,616	Soc. Rhodiaceta	1969
3,423,888	Phillips Petrol.	1969
3,427,912	Toshiba Mach.	1969
3,427,654	Rasmussen	1969
3,444,683	Barmag	1969
3,457,609	Allied Chem.	1969
3,460,337	Hercules	1969
3,460,416	Phillips Petrol.	1969
3,461,662	Eastman Kodak	1969
3,470,285	Hercules	1969
3,470,594	Hercules	1969
3,470,685	Hercules	1969
3,470,594	Hercules	1969
3,470,685	Hercules	1969
3,473,206	Phillips Petrol.	1969
3,474,611	Mitsubishi	1969
3,486,674	Monsanto	1969
3,488,941	Teijin	1970
3,490,663	Phillips Petrol.	1970
3,491,419	Klinger	1970
3,491,928	Phillips Petrol.	1970

3,494,522	Hercules	1970
3,495,752	Hercules	1970
3,496,259	Chevron	1970
3,496,260	Chevron	1970
3,500,517	Shell Oil	1970
3,500,627	Hercules	1970
3,501,565	Johnson and Johnson	1970
3,505,162	Rasmussen	1970
3,511,901	Phillips Petrol.	1970
3,515,325	Johnson and Johnson	1970
3,526,349	Ikegai Tekko	1970
3,526,569	Celanese	1970
3,527,393	Chori	1970
3,539,663	Allied Chemical	1970
3,542,267	Mitsubishi	1970
3,553,069	Rasmussen	1971
3,562,369	Celanese	1971
3,565,308	Plasticizers	1971
3,566,735	Phillips Petrol.	1971
3,574,804	W.R. Grace	1971
3,576,931	Celanese	1971
3,577,724	Phillips Petrol.	1971
3,594,495	Celanese	1971
3,596,816	Phillips Petrol.	1971
3,595,454	Johnson and Johnson	1971
3,606,114	Hercules	1971
3,606,115	Mitsubishi	1971
3,608,024	Polymer Processing	1971
3,608,025	A. Mirsky	1971
3,608,042	Rasmussen	1971
3,611,699	Eastman Kodak	1971
3,630,114	DuPont	1971
3,639,573	Avisun	1972
3,640,921	Grunzweig and Hartman	1972
3,641,760	Celanese	1972
3,645,085	Celanese	1972
3,645,299	Chemiefasern Lenzing	1972
3,656,672	Johnson and Johnson	1972
3,676,243	Chemiefasern Lenzing	1972
3,679,111	Monsanto	1972
3,693,850	Hercules	1972
3,693,851	Polymer Processing	1972
3,726,079	Hercules	1973
3,726,955	Phillips Petrol.	1973
3,727,392	FII	1973
3,739,053	Polymer Processing	1973
3,756,484	Chevron Research	1973
3,786,706	Marshall and Williams	1974
3,787,261	VEB Textiltombinat	1974
3,796,778	Am. Cyanamide	1974
3,801,252	Ernest Scragg	1974
3,830,604	Gullifiber	1974
3,835,513	Stanley	1974
3,853,977	Kamebo	1974
3,883,936	R.K. Stanley	1975
3,927,957	Chevron	1975
3,954,361	Beloit	1976
3,985,600	Bathurst	1976
3,985,933	Shell Oil	1976

3,950,476, Akzona, 1976; 4,065,538, Fibron, 1977; 4,102,969, ITF, 1978

4,129,632, Chevron, 1978; 4,133,088, Hikobe, 1979; 4,134,951, Smith & Nephew, 1979; 4,140,509, Saint Gobain, 1979.

VII.4. British Patents

864,488	Soc. Viscose Suisse	1961
941,723	Rasmussen	1963
960,067	Dow Chemical	1964
977,535	Soc. Viscose Suisse	1964
982,114	FMC Corp.	1965
1,010,206	Ernest Scragg	1965
1,014,778	Ernest Scragg	1965
1,035,657	Courtaulds	1966
1,037,765	Rasmussen	1966
1,040,663	Plasticizers	1966
1,067,089	DuPont	1967
1,062,936	Rasmussen	1967
1,063,227	J.M. Arnold	1967
1,063,772	Eastman Kodak	1967
1,067,704	Soc. Rhodiaceta	1967
1,067,089	Ryottaro Ichikawa	1967
1,073,741	Plasticizers	1967
1,078,732	Rasmussen	1967
1,080,762	ICI	1967
1,080,970	J.M. Arnold	1967
1,081,774	Chevron	1967
1,084,390	Polymer Processing Res. Inst.	1967
1,087,036	Ube-Nitto	1967
1,089,044	Hercules	1967
1,103,114	Rasmussen	1968
1,103,141	Rasmussen	1968
1,104,694	Bakelite	1968
1,104,758	ICI	1968
1,106,908	Soc. Normande	1968
1,109,750	Barmag	1968
1,123,538	Shell Int.	1968
1,124,109	W.R. Grace	1968
1,126,759	Mitsui	1968
1,127,900	Asahi Kasei	1968
1,128,274	Chevron	1968
1,129,377	J.C. Lewis	1968
1,131,282	Monsanto	1968
1,131,438	Phillips Petrol.	1968
1,132,574	United Gas	1968
1,132,641	Polymer Processing	1968
1,139,165	Steiffert	1969
1,141,367	Rasmussen	1969
1,145,982	Cotton, Silk, and Man-Made Fibers	1969
1,149,631	Plasticizers	1969
1,150,234	Phillips Petrol.	1969
1,150,740	Dunlop	1969
1,151,076	Chori	1969
1,155,120	Cigarette Components	1969
1,155,961	Rasmussen	1969
1,156,321	Brit. Cellophane	1969
1,157,695	Monsanto	1969
1,159,216	Shell Int.	1969
1,165,571	Monsanto	1969

1,165,934	Celanese	1969
1,166,687	Monsanto	1969
1,167,631	Shell Int.	1969
1,191,167	Monsanto	1969
1,171,219	ICI	1969
1,171,962	ICI	1969
1,176,357	Barmag	1970
1,180,545	Kalle	1970
1,181,249	Shell Int.	1970
1,181,250	Shell Int.	1970
1,181,256	Polymer Processing	1970
1,188,968	Mitsui Petrochem	1970
1,190,087	Phillips Petrol.	1970
1,190,639	Celanese	1970
1,197,950	Monsanto	1970
1,198,943	Mitsubishi	1970
1,199,284	Courtaulds	1970
1,199,724	Rasmussen	1970
1,202,347	Shell Int.	1970
1,206,180	Monsanto	1970
1,207,412	Phillips Petrol.	1970
1,207,733	Mitsubishi Rayon	1970
1,210,644	Hercules	1970
1,212,682	Rasmussen	1970
1,213,269	Chevron Research	1970
1,214,543	Ernest Scragg	1970
1,217,498	Dunlop	1970
1,221,781	Rasmussen	1971
1,225,574	DuPont	1971
1,225,824	Eastman Kodak	1971
1,228,026	ICI	1971
1,229,217	Plasticizers	1971
1,231,866	Celanese	1971
1,233,713	Mitsubishi Rayon	1971
1,234,782	Courtaulds	1971
1,235,121	ICI	1971
1,235,122	ICI	1971
1,235,146	BASF	1971
1,236,097	AKZU	1971
1,237,360	DuPont	1971
1,238,120	Reinstahl Henschel	1971
1,239,910	Rasmussen	1971
1,240,661	Rasmussen	1971
1,242,346	Celanese	1971
1,243,512	Rasmussen	1971
1,243,668	AKZU	1971
1,244,860	Phillips Petrol.	1971
1,244,982	Mekie and Sons	1971
1,245,583	TMM	1971
1,245,670	Chevron Research	1971
1,246,052	Bemberg	1971
1,250,855	Polymer Processing	1971
1,253,036	ICI	1971
1,258,393	Rasmussen	1971
1,258,401	Ernest Scragg	1971
1,258,743	Rasmussen	1971
1,259,285	Ernest Scragg	1972
1,260,837	Shell Int.	1972
1,260,838	Shell Int.	1972

1,260,957	Allied Chemical	1972
1,261,397	Rasmussen	1972
1,262,693	Allied Chemical	1972
1,266,125	Kanegafuchi	1972
1,267,031	Smith and Nephew	1972
1,267,298	Hercules	1972
1,269,823	Monsanto	1972
1,275,866	Chemiefasern Lenzing	1972
1,287,874	Chemiefasern Lenzing	1972
1,305,653	BASF	1973
1,308,998	Allied Chemical	1973
1,326,772	Chemiefasern Lenzing	1973
1,327,459	Chevron Research	1973
1,427,281	Hercules	1976
1,427,282	Hercules	1976
1,464,475	ICI	1977
1,475,571	Barmag	1977
1,495,067	ICI	1977
1,497,540	Shell Int.	1978
1,513,820	Leesona	1978
1,522,952	Lamberg	1978
1,524,251	Stewart	1978
1,529,488	Flemings	1978

Chapter XIII FIBER PROPERTIES AND APPLICATION

I. INTRODUCTION

The discussion in the foregoing chapters of this book bears evidence to the variability of fiber properties which can be obtained by variations in material parameters and processing conditions. Commercial polypropylene fibers are available that differ widely from one manufacturer to the other. In addition, the quest for dyeable and fire retardant fibers has resulted in the production of fibers that differ from other polypropylene products not only in physical characteristics but also in several chemical properties. This situation often limits the usefulness of information about polypropylene fiber as a class, and a point is reached at which the particular product must be considered individually if information of value is to be conveyed. There are, however, some areas of interest where the differences in various products are not significant. This applies to various chemical and biological properties of unmodified polypropylene fibers. There are also some areas of interest, especially the viscoelastic behavior of polypropylene fibers, where the subject can be advantageously dealt with in general terms.

Cook (ref.1) has surveyed the physicomechanical properties of polypropylene fibers commercially available in the United States and Europe, and Galanti and Mantell (ref.2) have compiled extensive data on the properties of polypropylene fibers and compared them with those of established fibers. A review, mostly based on Japanese-manufactured polypropylene fibers, is available in the proceedings of an ICAFE seminar (ref.3). The following account of the properties and applications of polypropylene fiber is an attempt to consolidate this information.

The properties of a fiber may be grouped into the following categories:
(1) Inherent properties of the fiber material
(2) Geometric form of the individual fibers
(3) Physical and mechanical properties of the individual fibers
(4) Characteristics of bulk fibers.

Table I lists in more detail the individual properties that can be distinguished in each category.

The inherent physical properties of the fiber material, in combination with the geometrical form determine most of the physical properties of the individual fibers. The latter characteristics, in combination with some characteristics of

TABLE I

Classification of Fiber Properties

Inherent Properties of Fiber Material: Chemical	Inherent Properties of Fiber Material: Physical	Geometrical Form of the Individual Fibers	Physical Properties of the Individual Fibers	Characteristics of Bulk Fibers
Dye affinity and retention	Specific gravity	Fiber length	Tensile load-elongation time relationship	Compressibility
			Breaking load	Recovery from compression
Resistance to	Stress-strain-time behavior	Fiber crimp		
acids	Moisture regain	contraction	Elongation at break	Apparent density
alkalies	Water absorbency	amplitude	Elastic recovery	
salts	Swelling	frequency	Stiffness	Degree of opening
hydrocarbons		Cross section	Compliance	
organic compounds		area	Toughness	
		shape	Resilience	Orientation (Entanglement)
	Resistivity			
Resistance to	Dielectric constant		Complex mechanical properties	Bulk fiber friction
photo-chemical degradation	Dielectric strength	Moment of inertia	Knot strength	
			Loop strength	
			Shear strength	Foreign matter content
biological attack			Strength-Twist Relationship	
moths & other insects	Optical properties	Slenderness ratio	Resistance to and recovery from uncrimping	
	Thermal Properties		Interfiber friction	Electric properties
	Specific heat			
	Thermal stability		Luster	
	Softening			
	Melting			
	Decomposition			
	Burning rate			

the bulk fibers, are responsible for the yarn-forming characteristics. In other words, the translation of the inherent physical properties of the fiber material into those of the bulk fibers is profoundly affected by the geometrical form of the individual fibers. In the more organized forms of fiber assemblies such as sliver, roving, and yarn, additional geometrical factors are introduced, and by the interlacing of the yarns in a fabric the mechanism becomes still more complicated. In spite of the many form factors, the character of the individual fibers will determine to a large extent the behavior of fabrics.

II. INHERENT PROPERTIES OF FIBER MATERIAL

II.1. Chemical Resistance

The chemical resistance of polypropylene is exceptionally noteworthy. Polypropylene fibers are very resistant to acids and alkalies at room temperature (Table II). The substances that attack polypropylene are highly oxidizing agents, such as hydroperoxides, fuming nitric acid, halogens, and chlorosulphonic acids.

TABLE II

Resistance of Polypropylene to Acids and Alkalies

Acids	Concentration	Temperature	Time (Hr.)	Strength Retention
Sulphuric Acid	98%	20°C	120	100%
Nitric Acid	40%	20°C	120	90%
Formic Acid	75%	20°C	96	100%
Alkalies				
Caustic Soda	35%	20°C	216	96%
Caustic Potash	40%	20°C	96	90%
Ammonia	15%	20°C	216	95%

Other chemicals that adversely affect polypropylene are organic solvents such as chlorinated compounds, aromatic hydrocarbons, and higher aliphatic hydrocarbons, which, even at room temperature, produce swelling and softening on prolonged contact. At elevated temperatures polypropylene is soluble in these organic solvents (Table III). Because of the chemical affinity of polypropylene for aliphatic hydrocarbons, the mineral oils commonly used as finishes and lubricating oils in processing other synthetic and natural fibers are not

TABLE III

Effect of Organic Solvents on Polypropylene Yarn

Solvent	Residual strength (%)			
	Immersion for 24 weeks at			
	30°C (86°F)	70°C(158°F)	8 hours in boiling solvent	
Benzene	90	90	95	(80°C)
Toulene	90	85	0	(110°C)
Xylene	90	85	dissolves	(136°C)
Methylene chloride	100	-	100	(40°C)
Chloroform	100	-	95	(61°C)
Trichloroethylene	95	90	60	(87°C)
Perchloroethylene	80	80	dissolves	(121°C)
Carbon tetrachloride	90	85	85	(77°C)
1,1′,2,2′-Tetrachloroethane	-	-	dissolves	(146°C)
Acetone	100	-	100	(56°C)
Methanol	100	-	100	(65°C)
Ethanol	100	100	100	(78°C)
Isopropanol	-	-	100	(82°C)
n-Butanol	-	-	90	(116°C)
Ethyl acetate	100	-	100	(77°C)
Amyl acetate	-	-	85	(119°C)
Light petroleum (b.p.80-100°C)	100	100	100	(87°C)
White spirit	95	95	dissolves	(167°C)
Dimethylformamide	100	100	high shrinkage	(152°C)
Pyridine	100	90	90	(115°C)

employed polypropylene fibers. Alcoholic solvents generally have little or no effect on polypropylene fibers. The silicon oils and greases also have no effect. Thus, as far as chemical resistance is concerned, polypropylene fiber surpasses all other synthetic fibers, including nylon and polyester.

II.2. Resistance to Insects and Micro-organisms

Polypropylene fiber will not support the growth of mildew or fungi, nor can it be digested by insects and related pests. Polypropylene fiber is, therefore, not liable to attack unless it becomes a barrier beyond which the insect must pass to reach an object. In this case, the insect may cut through the fabric without digesting it. Examples have been found where polypropylene ropes and cordage have been damaged by rats. It appears that rats sharpen their teeth on such objects. A few scattered examples have also been noted where rats inflicted damage to polypropylene sacks filled with edibles. No such damage by rats or other pests have been reported in field testing of sand bags.

TABLE IV

Relative Moisture Absorption at 65% RH

Fiber	Moisture Uptake
Polypropylene	0.04%
Nylon	4.5%
Polyester	0.4%
Acrylics	1.6%
Acetate	6.5%
Rayon	13.0%
Cotton	8.0%
Wool	16.0%

II.3. Moisture Absorption

Polypropylene is a paraffin hydrocarbon and, consequently, it does not absorb water. The moisture regain of polypropylene fibers is insignificantly small in comparison with other natural and synthetic fibers (Table IV). Because of this inertness to moisture, the physical properties of wet and dry polypropylene fibers are essentially identical. This is, however, true only in the case of unmodified polypropylene fibers. The acid dyeable fibers, such as Meraklon D, do absorb some moisture when immersed in water or kept in humid atmosphere. A moisture absorption of 1 to 3 percent has been noticed in relative humidity of 50 to 80 percent. The residual spin and process finishes also affect the moisture regain of polypropylene fibers.

II.4. Specific Gravity

The specific gravity of polypropylene fibers varies with the degree of crystallinity, draw ratio, draw temperature, degree of annealing, and the presence of additives. It has been previously noted that the specific gravity of polypropylene fibers increases with the degree of crystallinity. Amorphous polypropylene has a specific gravity of 0.85; commercial, well-annealed fibers are in the range of 0.90 to 0.91, and highly crystalline fibers may reach 0.92 to 0.94. High-shrinkage yarns, produced by drawing at considerably low temperatures, have very low specific gravity, sometimes reaching 0.65. The specific gravity of polypropylene fibers, therefore, range between 0.70 and 0.92 (the most commonly quoted figure is 0.91). Even considering the highest-quoted value of specific gravity, polypropylene is the lightest of all commercial textile fibers (Table V).

TABLE V

Relative Specific Gravity of Various Fibers (ref.6)

Fiber	Specific Gravity
Polypropylene	0.90-0.92
Cotton	1.50-1.55
Wool	1.30-1.32
Viscose rayon	1.52
Acetate	1.32
Nylon 6, 66	1.14
Acrylic	1.14-1.17
Polyester	1.38-1.39
Dynel	1.30
Saran	1.72
Glass fiber	2.54

The specific gravity of fibers determines the weight of the finished goods. From the standpoint of comfort and ease of handling, a strong, lightweight fiber is most desirable. Furthermore, a low-density fiber affords appreciable cost savings in the extra yield per pound of material. Polypropylene fiber enjoys these advantages to the maximum. The significance of this lowest specific gravity fiber is realized by the greater surface and volume of the fiber. On the basis of filament thickness, for example, a 15 denier polypropylene is equivalent to 19 denier nylon, 19 denier acrylic, 22 denier polyester, and 25 denier viscose fiber. On a volume basis polypropylene fiber has a 26 percent advantage over nylon and the acrylics, 54 percent over polyester and 65 percent over viscose rayon.

II.5. Soiling and Staining

Polypropylene fiber has very low propensity to soiling and staining. This is because polypropylene fibers show little tendency to accumulate electrostatic charge and do not react chemically with the substances encountered in general use. The soiling and staining which does occur is rather superficial. The stain is held in the interstices of the fabric by capillary attraction, and is readily removed by laundering or shampooing. In fact, stain resistance or easy cleanability is one of the major advantages of polypropylene over other synthetic fibers that makes it an ideal fiber for home furnishing. This property has been extensively capitalized upon as a marketing point by major polypropylene fiber manufacturers. In comparison with nylon and other common synthetic fibers, however, polypropylene is somewhat more easily soiled by oil-bearing soils. Its removal is also a trifle more difficult.

The above remarks on soiling, staining, and cleanability apply to the unmodified polypropylene fiber; the behavior of the acid-dyeable fiber is somewhat different. In general, dry soiling properties are not markedly affected but the wet soiling resistance is somewhat decreased.

II.6. Electric Properties

Table VI summarizes the electrical properties of various fibers. The data show that polypropylene compares favorably in electrical characteristics with other competitive materials. It can be considered as a very good electric insulator with exceptionally low power loss even at radio frequencies. Dielectric constant of polypropylene remains unchanged from 50 cycles per second to 5×10^7 cycles per second for a given polymer formulation. It also remains virtually independent of temperature, from room temperature to $100^{\circ}C$. Because polypropylene does not absorb appreciable amounts of moisture, the good electrical properties remain virtually unaffected by relative humidities of the atmosphere.

III. GEOMETRIC FORM OF INDIVIDUAL FIBERS

Several properties of the fiber are a function of geometric form - length, denier, fiber cross section, and crimp configuration - of individual fibers.

TABLE VI

Electric Properties of Various Fibers (ref.3)

Fiber	*Dielectric Constant at 65% RH*		*Volume Resistivity (SL CM)*	Dielectric Strength (KV/mm)
	1 KC/s	*100 KC/S*		
Cotton	18.0	6.0	10^7-10^{12}	5-8
Vicose Rayon	8.4	5.3	10^{11}-10^{13}	25-50
Acetate	3.5	3.3	10^{11}-10^{13}	20-35
Wool	5.5	4.6	--	--
Nylon	3.7	2.9	10^{14}-10^{15}	20-35
Polyester	2.3	2.3	10^{19}	18-25
Polypropylene	1.7	1.7	7 10^{19}	25-45

For example, thermal insulation and resilience are affected by crimp configuration; fiber-to-fiber friction is affected by fiber length, crimp configuration, and fiber cross section; luster and soiling properties are markedly affected by fiber cross section and denier per filament; etc. These interrelations have been briefly mentioned elsewhere in this book and need not be repeated here. For more detailed informations, excellent books and monographs are available in the textile literature.

IV. PHYSICAL AND MECHANICAL PROPERTIES OF INDIVIDUAL FIBERS

IV.1. Stress-Strain Behavior

Polypropylene fibers can be prepared with a wide range of viscoelastic properties, depending on the molecular parameters, degree of stabilization, nature of additives, processing conditions, after treatments, and the like. In view of these facts, only a general treatment of the viscoelastic properties of polypropylene fiber is possible: for specific information of various commercial products, the manufacturer's data sheets should be consulted.

The important data on stress-strain behavior or viscoelastic properties of any synthetic fiber generally include rupture stress (tenacity), rupture elongation (elongation at break), modulus of elasticity, yield-strain, and work of rupture (toughness). The most striking features of the mechanical properties of synthetic fibers are: first, their considerable variation dependent on the mode of deformation; second, the complex relationships between stress, strain, and time for a given mode of deformation; and third, the remarkable changes in all these features which occur with change in temperature. All these remarks apply to polypropylene fibers as well.

Tensile strength, elongation at break, and toughness, measured at room temperature and 50 to 60 percent extension rate, are the most important mechanical properties of a fiber. Engineering tensile strength (lb/in^2) of polypropylene is related to textile measure (g/den) as follows:

Tensile strength (lb/in^2) = 12,800 x 0.91 (specific gravity) X tenacity (g/den).

Commercial polypropylene fibers are available in the range of 2-4 g/den for carpeting, 3-6 g/den for textiles, and 6-9 g/den for industrial purposes. In general, the strength and elongation of polypropylene fiber are interchangeable properties. It has been noted in Chapter IX that from a given spun yarn, fibers of different tenacity and elongation can be prepared and that the product (tenacity x elongation) generally remains constant. Commercial polypropylene fibers are available in the elongation range of 15 to 20 percent for industrial products, 20 to 30 percent for textile applications, and 30 to 100 percent for BCF yarns.

The toughness of a fiber is a measure of the ability of the material to absorb energy. This energy absorption is quantitatively determined by the area of the stress-strain curve. Polypropylene fiber ranks well with the toughest fibers of cotton, nylon, and polyester; these tough fibers have energy absorption values of about 0.5 to 0.7 g-cm/den-cm. Toughness of polypropylene fibers can further be improved by utilizing high molecular-weight polymers and annealing under tension.

In order to fully characterize the tensile properties of a material, the stress-strain relationships must be determined at both static and high-speed loading rates. Figure 1 depicts the stress-strain behavior at different strain rates. Table VII presents comparative data on the tenacity, elongation, and energy absorption at various strain rates for nylon (high and regular tenacity), Fortisan ^{R} (high-tenacity cellulose acetate), fiberglass, and polypropylene. The table shows that polypropylene exhibits the qualities for an outstanding, all-purpose fiber. Over the wide range of strain rates tested, polypropylene exhibits excellent energy-absorption values, even exceeding those of regular-tenacity nylon. The breaking tenacity of polypropylene is slightly lower than Fortisan but comparable to that of nylon. The data in Table VII and Fig. 1 show that the tenacity of fibers increases and rupture elongation decreases with increasing strain rates. These changes result from the fact that at the lower speeds of loading the viscoelasticity of the fiber permits partial deformation rather than immediate failure. At the ballistic speeds of loadings, the creep elongation is negligible, resulting in lower fiber extensibility.

The ratio of the change in stress to the change in strain within the elastic limits of a material is defined as Young's modulus. The modulus of fibers is

TABLE VII

Tenacities, Elongations, and Rupture Densities For Various Fibers (ref.2)

Material	Rate of strain, %/min	Elongation at break, %	Tenacity at break, %	Rupture energy density, g-cm/den-m
High-tenacity nylon[a]	1	16.7	6.28	41.3
	10	17.6	6.76	50.1
	100	16.1	6.97	46.2
	300,000	14.7	7.57	43.8
	400,000	--	--	32.7
Fortisan[b]	1	5.4	6.33	17.6
	10	5.4	6.80	19.3
	100	5.3	7.04	20.4
	120,000	5.2	9.10	25.0
	350,000	--	--	18.0
Fiberglas[c]	1	2.8	4.73	7.6
	10	2.5	5.58	7.5
	100	2.4	6.07	7.7
	60,000	1.8	6.12	5.1
	200,000	--	--	5.1
Regular-tenacity nylon[d]	10	27.4	7.65	94
	60	27.1	6.66	93
	200	27.0	6.65	88
Polypropylene[e]	20	38.0	5.7	105
	240	32.2	6.0	98
	200,000	28.0	6.9	92
	370,000	27.0	8.0	94
	500,000	28.4	8.4	95

a High-tenacity nylon - Du Pont, type 300, 1085 den total.
b Fortisan-high-tenacity deacetylated cellulose acetate yarn, 787 den total.
c Fiberglas-type 150-1/0, 204 filament, 323 den.
d Regular-tenacity nylon-Nylon 66, 840 den, type 300 bright yarn.
e Isotactic polypropylene - 230-den, 70 filament yarn.

based on an elongation at a rate equal to the initial slope of the stress-strain curve and is represented in terms of g/den. Polypropylene fibers widely differ in their elastic moduli, ranging from 120 g/den for well-annealed monofilaments to 25 g/den for carpet staple and BCF yarns. For textile fibers a mean value of 50 to 60 g/den is common.

The fiber resists tensile impact under instantaneous stresses of high intensity, as when the fiber is subjected to a sudden blow or struck with a ballistic trajectory. This impact resistance is important in several industrial uses of polypropylene fiber, such as mooring ropes and fabrics for antiballistic personnel armor. In contrast to normal tensile strength, impact strength is dependent on the denier of the fiber and is relatively insensitive to temperature over a wide range, as shown in Fig. 2.

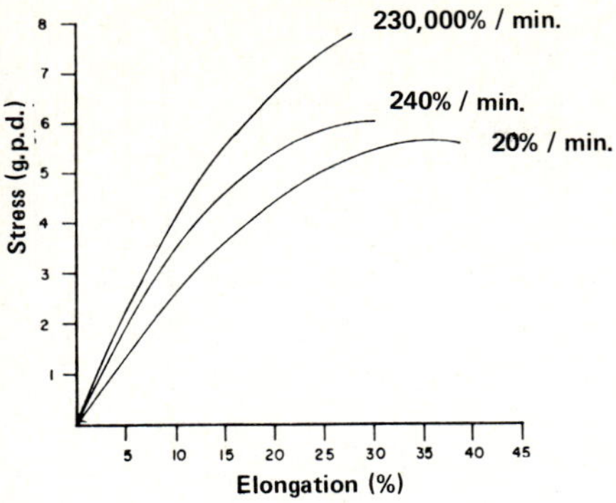

Fig. 1. Stress Strain Behavior of a Highly Drawn Isotactic Polypropylene Yarn at Different Strain Rates (ref.1)

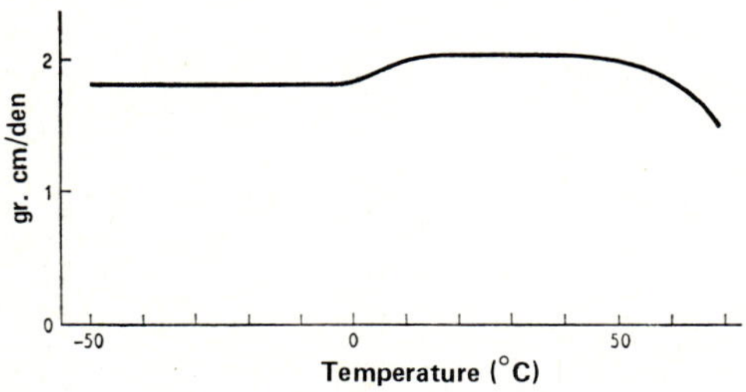

Fig. 2. Effect of Temperature on Impact Strength of Polypropylene Yarn (ref.1)

IV.2. Creep and Elastic Recovery

The above discussion on the stress-strain relationships of fibers is based on the application of a stress which ruptures the specimen. In practice, however, fibers are seldom subjected to breaking loads, but rather to cyclic stresses which are small relative to the rupture-load values. Consequently, important properties of fibers not evaluated by stress-strain diagrams are the elongation-recovery characteristics upon exposure to repeated stresses considerably smaller than the breaking load. These characteristics dictate the extent of wrinkle resistance, dimensional stability, and abrasion resistance exhibited by the fiber. On removal of a stress which has produced an elongation below the rupture value of the specimen, a recovery toward the original fiber dimension will result, the extent depending upon the inherent properties of the material under test.

In consonance with the general behavior of viscoelastic materials, the recovery of polypropylene fiber does not take place instantaneously; it is rather a time-dependent phenomenon. Cappuccio et al (ref.12) developed an empirical relationship expressing the stress relaxation of polypropylene at 25°C, which is:

$$y/y_1 = 1 - A \log t$$

Here, y_1 is the stress after 1 sec, y is the stress at time t, t is the time in seconds, and A is a factor dependent upon fiber crystallinity and orientation. The elastic recovery properties decrease as the value of A increases, i.e., the decrease of the stress with time is accelerated as A becomes larger.

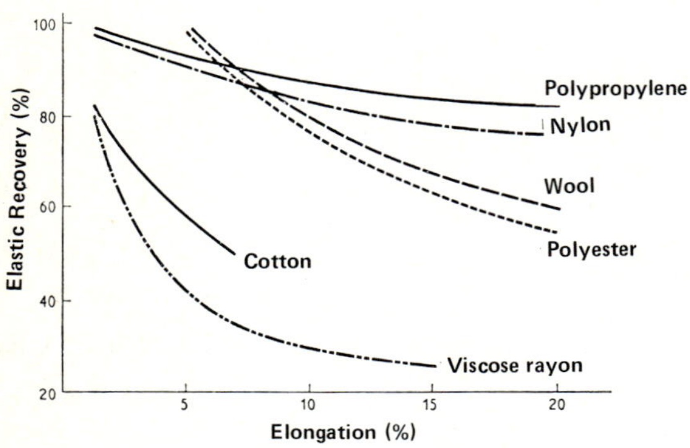

Fig. 3. Elastic Recovery of Various Fibers (ref.1)

The comparative data on recovery properties of polypropylene (ref.8) show that polypropylene fibers exhibit a high degree of immediate elastic recovery - equivalent to that of nylon and better than polyester - and is fairly constant up to 10 percent elongation (Fig.3). Polypropylene fiber can ultimately recover from almost all original deformation up to extensions as high as 8 percent. Above 8 percent there appears to be a significant amount of permanent set or creep. Here again it compares well with nylon and has advantages over polyester fiber. In comparison with the recovery properties for short stress times and low strain rates, however, the recovery of polypropylene for long stress times is slower than polyester recovery.

The fiber recovery from bending is an important property determining to a large extent the creasing and wrinkling characteristics of a fabric and its appearance in use. Guthrie (ref.9) has recently discussed the bending recovery behavior of polypropylene along with other natural and synthetic fibers. The bending recovery of polypropylene filament was shown to be generally poorer than nylon (Fig.4). Higher bending recoveries occurred after a heat-setting treatment. Fibrillated film is, as a rule, inferior to filament fiber in this respect.

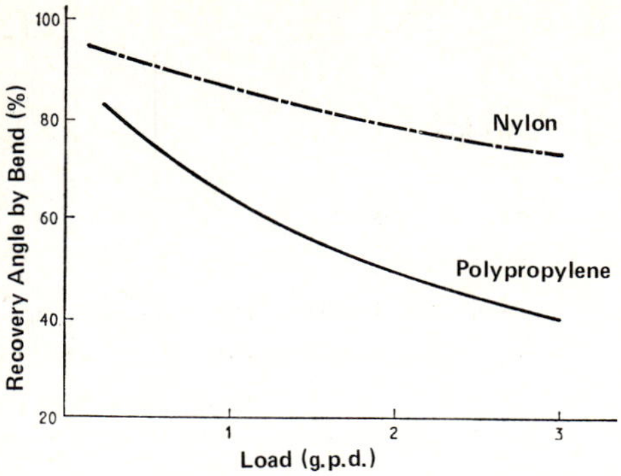

Fig. 4. Bending Recovery of Polypropylene and Nylon (ref.1)

IV.3. Knot and Loop Strength

The loop strength is the tensile load required to rupture two interlaced fibers or yarns. The looping resistance is the ratio of this breaking load of the looped fibers to twice the breaking load of the fiber alone multiplied by 100. The knot strength is the tensile load required to rupture a fiber or yarn looped into a simple overhand knot; the knotting resistance is the ratio between the breaking load of a fiber with the knot multiplied by 100 to the breaking load of the unknotted fiber. Loop and knot strength is important in several industrial applications - ropes, package and fish net twines, and sewing thread.

Upon elongation, brittle fibers quickly develop concentrated stresses which cannot be dissipated, and tensile failure results. Therefore, the brittle fibers, which have a high modulus of elasticity and high strength properties, possess poor loop and knot strengths. In contrast, more extensible fibers with relatively much higher rupture elongations than the brittle fibers are characterized by high loop and knot strengths; this is because the elastic fibers can stretch under stress rather than fail immediately. For textile denier (2 to 7 dpf), polypropylene yarns show a 80 to 85 percent loop strength and about 95 percent knot strength.

IV.4. Frictional Characteristics

The coefficient of friction of synthetic fibers is important in the choice of processing equipment and in optimization of processing conditions. It is also important in use of the product as some of the wear properties are strongly dependent on fiber friction. Two kinds of friction are involved:

(1) fiber-to-metal friction

(2) fiber-to-fiber friction

A distinction is further made between "static coefficient of friction" and "dynamic coefficient of friction".

Polypropylene fiber has a relatively high coefficient of friction against smooth surfaces, particularly metal or porcelain. The coefficient of friction against a matte guide is much less than against a polished guide, owing to the smaller actual area of contact between the yarn and matte surface. This factor is important in the use of polypropylene as mooring ropes on winches. The polished or painted posts tend to generate heat and fuse the upper surface of the rope at high speeds of mooring.

The fiber-to-fiber friction is important in flexing and abrasion resistance of the product. High coefficient of friction is generally associated with low flexing resistance and shorter life of such products as cargo slings and mooring ropes. Figure 5 compares the interfilament coefficient of friction of polypropylene fibers with that of nylon and polyester. The coefficient of friction of polypropylene between the filaments is very high. This is probably due to plastic deformation of the filament skin. The lubrication of the filaments with suitable finishes helps to improve this aspect of product performance.

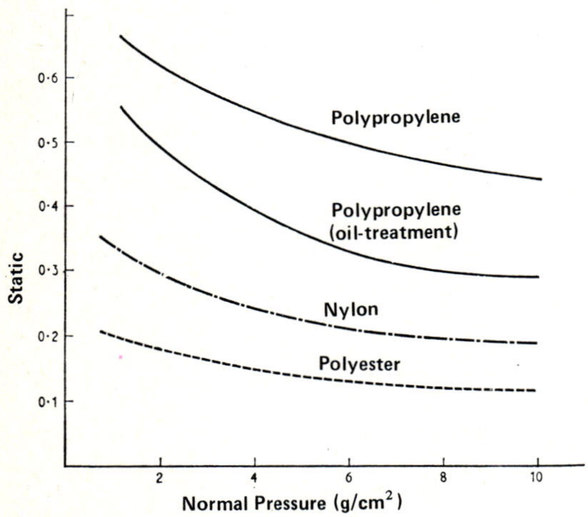

Fig. 5. Fiber-to-Fiber Friction of Different Fibers (ref.1)

IV.5. Abrasion Resistance

The abrasion resistance of fibers is important in applications such as twine, fish nets, and marine ropes as well as in carpets and upholstery fabrics. Polypropylene fibers are known to have good abrasion-resistance properties. Since the abrasion resistance of a fiber is related to the energy absorption, and since the deformation and recovery characteristics of a fiber are determined by the interaction of various material characteristics and fiber processing conditions, the abrasion resistance of polypropylene fiber is not a single value characteristic. In common with other physical and mechanical properties of the fiber, the abrasion characteristics are markedly influenced by molecular parameters of the polymer, crystallinity, fiber orientation, processing conditions, the nature and amount of additives in the polymer, filament cross section, and any finishing oils used in the production of the fiber.

Extensive studies have been done (ref.10,11,12) to characterize the various factors influencing the abrasion resistance of polypropylene fibers and to elucidate the mechanism of abrasion. These works show that abrasion resistance of polypropylene fibers increases with increasing molecular weight of the polymer and increasing crystallinity of the fiber. A large improvement in abrasion resistance is observed on annealing the fiber, especially the undrawn fibers. In addition, spin orientation of the fiber seems to have a significant effect on abrasion behavior. It appears that the yarns of lower spin birefringence gives drawn yarns of higher abrasion resistance than those obtained from fibers of higher spin birefringence (ref.10). Surface and frictional characteristics are undoubtedly very important since the presence of lubricants on the fiber surface can extend the life of the fibers many fold. Abrasion resistance varies with applied load and is also affected by filament size - abrasion resistance increases as the filament denier decreases.

V. THERMAL PROPERTIES

V.1. Boiling Water or Thermal Shrinkage

Hot water shrinkage denotes the decrease in linear length of the fiber when subjected to boiling water treatment for a specified period of time under relaxed conditions. It is also called boiling water shrinkage and is expressed in percent of the original length. The hot water shrinkage of polypropylene fiber is purely a function of heat relaxation and is in no way connected with the effect of moisture on the polymer, as is the case with nylon and other synthetic fibers. The same effect is, therefore, attained when the fiber is relaxed under steam or hot air. Boiling water shrinkage is important in processing (scouring, dyeing) as well as in use of the fabric (washing). Similarly, hot air shrinkage is important in finishing (latexing of carpets and upholstery fabrics) as well as in use.

As has been shown in Chapter IX, polypropylene fibers of varied shrinkage values can be easily prepared. For general purpose filament and staple fibers, attempts are made to keep the free shrinkage in boiling water as low as possible whereas for BCF yarns, some boiling water shrinkage is a desirable property. High-shrinkage polypropylene fibers are speciality fibers for various end uses.

Commercially available general purpose staple fibers and filament yarns have a shrinkage range of 2 to 5 percent in boiling water, comparable to that of polyester and nylon fibers. Recently some well-stabilized yarns have been introduced that have boiling water shrinkage values less than 1.5 percent. They are specially meant for woven carpet backing and coated fabrics.

V.2. Thermal Insulation

The thermal insulation characteristics of a fabric are determined largely by the amount of air entrapped in the fabric and the thermal conductivity of the fiber. While the amount of air entrapped in the fabric depends on the fiber crimp and yarn and fabric construction, the thermal conductivity of the fiber is purely a material characteristic. Thermal conductivity of polypropylene fiber, as shown in Table VIII, is lower than any of the other fibers. In this respect polypropylene is the "warmest" fiber in commercial use - a distinction primarily held by wool.

TABLE VIII

Thermal Conductivity of Various Fibers Relative to Air (ref.2)

Material	Heat conductivity, relative to air
Air	1.0
Polypropylene	6.0
Wool	6.4
Acetate	8.6
Viscose	11.0
Cotton	17.0
P.V.C.	6.4

VI. CHARACTERISTICS OF BULK FIBERS

VI.1. Resilience

Resilience of a fibrous structure is its ability to recover from compressional loading. This property is of utmost importance for polypropylene fiber in its use in carpets and as fiberfill for stuffing. The ability to recover from compressional loading depends on yarn and fabric construction, crimp configuration, and crimp type (regular or irregular, two dimensional or three dimensional) as well as on the inherent mechanical properties of the individual fiber. There is general agreement that resilient fiber must exhibit high crystalline orientation and high fraction of a-axis oriented crystallites. Very highly drawn fibers are, however, low in resilience. This is probably due to the incipient fibrilation which may occur at high draw ratios. Resilience is also markedly influenced by heat setting. Figure 6 shows the relative resilience of polypropylene fibers as a function of draw ratio and heat setting.

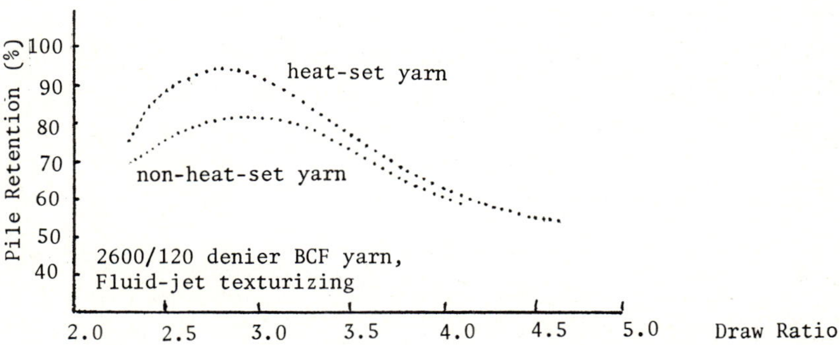

Fig. 6. Pile Height Retention of test Carpets under Cyclic Compression. (Relative values after a fixed number of compression-relaxation cycles)

Commercial polypropylene fibers widely differ in resilience characteristics. In general, polypropylene is considered intermediate between polyester and nylon. Specially prepared fibers for carpet end use or for fiberfill material even surpass nylon fiber in resilience and may approach wool. Table IX gives comparative values of resilience for various fiber as measured by "cylinder-wafer" method. Under cyclic compression and relaxation, the polypropylene fiber performs very well. When held under load for extended periods of time, it tends to take a set and is slow in recovery. In suitable construction, polypropylene carpets have been shown to be as resilient as wool.

TABLE IX

Comparative Values of Resilience as Measured by Cylinder-Wafer Method (ref.1)

Fiber	Resilience (%)
Wool	285
Polypropylene Type "SR" (15 den)	280
Acrylic fiber	280
Polyamide fiber	265
Cellulose acetate	40
Viscose rayon	35

VI.2. Static Loading

When two dissimilar surfaces are rubbed together, there will be some degree of charge separation; potential differences of several thousand volts may be produced between the objects. The magnitude of the potential developed is a function of the nature of the surfaces involved and their electric resistivities. Static loading is important for textile materials in apparel as well as in home furnishing use.

TABLE X

Comparison of Walk and Friction Tests at 50% RH, Volts[1] (ref.13)

Carpet	Rubber Sole		Rubber Sole		Leather Sole, Rubber Heel		Composition	
	Walk[2]	Friction[3]	Walk	Friction	Walk	Friction	Walk	Friction
Brunsmet	-520	900	-387	350	- 60	700	-220	670
Polypropylene	89	-645	55	-290	171	-655	113	-745
Polypropylene	- 49	-760	-202	-405	228	-900	268	-865
Nylon	-1395	1350	-1660	430	-2380	1500	-2190	1350
Wool	-59	1500	-200	970	-109	930	-298	935
Acrylic	119	400	99	350	21	105	57	80
Polyester	-720	1400	-952	605	-1450	1415	-1600	1335
Nylon	-916	1000	-1130	450	-1980	1150	-1820	1150

1) The negative values reported are caused from voltage being measured from the carpet in the friction tests, and from the shoe material in the walk trials.
2) Average of five walking tests in all cases except the Brunsmet, for which only one sample was available.
3) Average of machine tests on two samples of each carpet.

Carpeting is an area of special interest for polypropylene fiber. As static loading of a carpet is affected not only by its fiber content but also by its weight, construction, relative humidity, rubbing material, etc., a straightforward comparison of various fibers is a difficult task. Ralston (ref.13) has conducted an extensive series of testing of polypropylene fiber for its static loading properties and compared them with those of other fibers. Table X presents comparative data on various fibers. It is apparent from this table that polypropylene is considerably less prone to static loading than nylon and polyester. A further fact which is apparent from this table is that the electrostatic potential on polypropylene has an opposite charge to that on nylon and polyester. This opens up the possibility of mixing polypropylene fiber with nylon or polyester fiber to produce a static-free product. The positive potential produced on polypropylene fibers would be neutralized by the negative potential produced on nylon or polyester fibers (ref.14). Table XI demonstrates this fact.

TABLE XI

Comparative Static Values of Nylon and PP Shag Carpets at 10% RH

100% Herculon	4300 - 9500 volts
100% Nylon	3700 - 13,000 volts
50% Herculon/50% Nylon	150 - 650 volts

VII. IDENTIFICATION OF POLYPROPYLENE FIBERS

Due to the availability of several types of polypropylene fibers in recent years and because of several possible finishing processes and texturizing techniques available for each type, it is almost impossible to recognize polypropylene fiber by visual observation or by the touch of the product alone. The typical "waxy" hand, thought to be a characteristic of polypropylene in its early days of development, no longer serves the purpose because of the various noncircular cross sections of the newer products and because of the novel texturizing techniques which alter the "waxy" hand of polypropylene to the extent that it is no longer recognizable by touch. Most of the polypropylene BCF yarns available to the textile and carpet trade are very similar to nylon and polyester BCF yarns in texture and appearance. Similarly, the polypropylene staple of modified cross sections can be easily confused with acrylic, nylon, and polyester fibers if judged by touch and appearance alone. The "pigmented look" of polypropylene fibers is now also a thing of the past as newer organic pigments are capable of imparting to the fiber a "dyed" look. The usual dye-staining tests scheme, commonly used to identify natural and synthetic fibers, fail on polypropylene because of the nonstaining nature of unmodified polypropylene fibers and because of the availability of various dyeable fibers with staining characteristics common to other fibers. Although the very low density is a distinguished property of polypropylene fibers, this property cannot form a basis for identification because of the availability of low-density hollow fibers of nylon and polyester. Experience has shown that density measurement cannot be used even for solid fibers. This is due to the presence of air bubbles on the fiber surface, slow wetting, and inclusion of air in the fiber bundle.

The following remarks (ref.16-19) may be helpful in identification and separation of polypropylene fiber.

Melting Point Determination: The melting point of polypropylene fibers is between 150°C to 170°C for most of the moderately drawn and textured products. Some of the highly drawn and annealed samples may show a melting point in the range of 180 to 185°C. Thus, the fact that polypropylene fiber melts under 200°C makes it readily recognizable in comparison with other synthetic fibers - except polyethylene fibers, which melt at lower temperatures than polypropylene. The melting range can be precisely determined with the use of a hot-stage polarizing microscope. Melting point is the temperature at which fiber birefringence disappears. By working with a strong transmission light and crossed polarization, subjective factors can be avoided.

Shrinking and Burning Tests: If polypropylene fibers are exposed to a flame, they shrink away from it. The shrinking process continues even after the fibers are lit. This strong shrinking behavior of fiber prior to melting is very characteristic of polypropylene fibers. Other synthetic fibers melt without too much shrinkage or shrink without subsequent melting.

Polypropylene fibers melt, burn, and drip like a candle. The burning occurs slowly with light-blue flame and yellow tip. The soot buildup is not noticeable. The smell is that of a burning paraffin wax and the residue is not of any characteristic color. The modified polypropylene fibers may burn with the development of some characteristic smell and soot buildup may also occur. The "flame retardant" fibers, for example, may smell like burning PVC and smoke as they burn. The "tagged" polypropylene fiber may burn with a characteristic color of the flame. For example, polypropylene fibers tagged with strontium stearate may develop a crimson-tipped flame.

Dye Test: Unmodified polypropylene fibers are not dyeable with common dyestuffs. This can, therefore, be a basis of differentiating them from other fibers (except, of course, from polyethylene). There are some dyeable polypropylene fibers available in the market but they are not yet in widespread use. Where such fibers are at hand, the dye and staining tests fail and cannot be relied upon.

Swelling and Solution Tests: Probably a more positive and better suited test than those aforementioned is the swelling and dissolution behavior of polypropylene fibers. A hot-stage microscope, fitted with a concave shallow glass pan, is suitable for these tests. Several possibilities of swelling and dissolving agents are available; the most important and specific to polypropylene are paraffin oil, tetralin, decalin, and glycerine.

In working with paraffin oils, mild shrinkage of fibers occurs at 85°C. At about 90°C the first signs of solution are noticed. The fibers break into small pieces or the fiber ends begin to dissolve with the formation of tapered ends. At 125° to 130°C no residual fiber is left. The other common fibers remain unaffected in paraffin oils.

Tetralin (tetrahydronaphthalene) and decalin (decahydronaphthalene) are strong swelling and high-temperature dissolving agents for polypropylene. The dissolution begins at 70°C and ends at 90°C in tetralin. In decalin the dissolution process starts at 80°C and ends at 120°C. By slowly warming the reaction pan, the polypropylene fibers are seen splitting in the axial direction. PVC is the only other common fiber which is affected by these solvents, but the swelling and dissolution process is very slow in this case.

Glycerin and silicone oils have only a shrinking effect on polypropylene while other fibers are strongly swelled, are converted into polymer beads, or change in color.

For quantitative separation of polypropylene fibers from other fibers in a blend, the common technique of dissolution can be applied. The most suitable solvent for this purpose is tetralin or a mixture of paraffin oils and xylene. The filtration should be accomplished at elevated temperature, preferably between 100° to 130°C. Another way of quantitatively separating the fibers in a blend is by first roughly separating by floatation method and then quantitatively dissolving out the blended fibers from the polypropylene-rich portion of the float. Suitable solvents for most synthetic and natural fibers are known and described in textile testing manuals.

VIII. INDUSTRIAL APPLICATIONS

Polypropylene fiber is making rapid inroads into a wide range of industrial uses. Originally intended for use in carpet pile and backing areas, polypropylene fiber has found a home in a broad spectrum of industrial products where its physical and mechanical properties have special functional values. A partial list of these uses is given in Table XII. The properties which make polypropylene fiber a leading contender for industrial applications are its high strength, high toughness, and high chemical resistance. With the availability of the polymer at reasonable cost and the proliferation of small-scale plants for converting the polymer into fibers, a rapid growth of polypropylene fiber in industrial applications is in the offing. A few areas of use are discussed below in brief.

TABLE XII

Some Industrial Uses of Polypropylene Fiber

Staple	Continuous Multifilament Yarns
Asbestos carrier	Industrial sewing threads
Coated and protective fabrics	Wrap fabrics and bags
Filter cloth	Coated and protective fabrics
Papermaker's felt	Conveyor belts
Other felts	Cordage, twine, ropes
Medical and surgical	Filter cloth
Nonwoven fabrics	Fishnets
Paper reinforcement	Braids and trimming
Plastic reinforcement	Laundry and dye bags
Shoe fabrics and other canvass	Medical and surgical
Carpet backing (nonwoven)	Paper reinforcement
Sliver knit fabrics	Filament winding
Cement reinforcement	(Plastics applications)
Miscellaneous	Webbing
	Window lace
	Separate cloth (for rubber)
	Cigarette filter tips
	Bristles
	Artificial turf
	Miscellaneous

VIII.1. Ropes, Cordage, and Twines

The principal requirements of a fiber to be used in making ropes for maritime, agricultural and industrial use include the following:

(1) high strength, wet or dry

(2) resistance to repeated loading and flexing

(3) resistance to abrasion

(4) minimum water absorption

(5) lightness

(6) resistance to weathering, micro-organism and other degradative agents.

Polypropylene fibers satisfy all these requirements to a high degree. Consequently, polypropylene staple, multifilament yarns, and fibrillated film yarns have found an important market in the field of rope and cordage making.

A notable success in the use of synthetic-polymer fibers in ropes has been polypropylene staple fiber. This is provided in a length and linear density to match the natural fibers it is designed to replace, namely, manila hemp and sisal. The fiber can, therefore, be processed directly on conventional rope-making machinery. Imperial Chemical Industries has expended considerable efforts in popularizing this product in the United Kingdom (ref.6). It is used in most marine applications, including mooring and tow lines and fishing ropes.

TABLE XIII

Test Results For Ropes

	Minimum Breaking Load(tonf)	Extension (%)	Weight (lb/100 ft)	Yarn Tenacity (gf/den)	(gf/tex)
Spunstron	7.5	26-35	17.5	6.5-7.0	58.5-63.0
U201 Ulstron	7.5	31-41	20.3	8.0	72.0
Nufil	7.5	15-25	17.5	4.5	40.5
Type H Terylene	9.0	23-33	31.0	8.0	72.0
Nylon 266	11.8	36-46	25.2	8.0	72.0
Polyethylene monofil	6.0	27-37	19.3	5.5-6.0	49.5-54.0
Sisal	4.0	10-20	28.6	2.4	21.6
Polypropylene monofil	7.5	25-30	17.5	7.5	69.5

Ropes are also made from polypropylene multifilament yarns and from fibrillated-polypropylene film. As discussed in Chapter XII, the fibrillated-tape yarns are somewhat cheaper than yarns of equivalent linear density based on staple fiber, and considerably cheaper than those of equivalent multifilament yarns. Since fewer twisting operations are required in their manufacture, the finished products based on tape yarns are often considerably cheaper than equivalent products based on monofilaments, multifilaments, or staple-fiber yarns. Table XIII compares the mechanical properties of ropes made from polypropylene staple (Spunstron®), multifilament yarns (Ulstron®), fibrillated film (Nufil®) and monofil, with those of polyester fiber (Terylene®), nylon, polyethylene, and sisal. The breaking strength and resistance to extension per unit weight is very favorable for all the four kinds of polypropylene ropes. These ropes may differ among each other in kink resistance, abrasion resistance, and flexural fatigue. In determining their use for any specific job comparative price plays a major role.

The commonest causes of failure in ropes are frictional melting and abrasion. For this reason, it is important that all surfaces over which the ropes move be kept clean and smooth. In particular, it has been found with marine mooring ropes that synthetic fiber materials are more prone to fusion on winches with painted surfaces, since the heat generated during surges is prevented by the paint from being dissipated into the winch barrel. The frictional forces that come into play in using ropes can be divided into two different classes: (a) where the force on the rope is substantially independent of the material friction, and (b) where the force depends upon the material. In cases of category (a), lubrication of the material can improve the performance and the life of the rope. In category (b), lubrication is of no help in improving the wear resistance of the rope.

The end uses of ropes in which a high degree of flexing and dry abrasion are involved are typified by endless slings for cargo loading and unloading. In a controlled trial conducted by ICI, it was seen that Ulstron® polypropylene (multifilament yarns) and Terylene® polyester multifilament ropes performed better than the harsher Nufil® fibrillated film ropes, which, in turn, outperformed the coarser Spunstron® staple spun ropes. It was also noted that lubrication helped to improve the performance of the Nufil® and Spunstron® ropes by reducing the abrasion. The nylon ropes in this trial failed prematurely owing to work-hardening, an indication that these ropes should not be used for this type of application (ref.6).

The end uses, where the frictional force depends upon the material of friction, is typified by mooring ropes used on conventional winch systems. A field test conducted by ICI on a Shell tanker showed a considerably lower potential life for the fibrillated film rope than for the staple spun and multifilament polypropylene ropes. Even on a cost/life basis, the fibrillated film showed an unfavorable performance (ref.6).

A several-year long study of $5\frac{1}{2}$ inch three-strand laid tow ropes (ref.6) on 70-ton motor tugs showed that neither polyethylene nor nylon ropes can be used successfully because of work-hardening and creep. Judged solely on a cost basis, the fibrillated film and staple spun ropes appear to be the best. Overall, the staple spun ropes have proved to be economical, easy to handle, and free from work-hardening, and this type is now widely used for tow ropes (ref.6).

A recent development in the use of synthetic fibers in ropes is "parafil" ropes. Parafil is basically a class of rope construction based on parallel continuous filament yarns encased in thermoplastic sheaths with suitable end fittings. Some parafil ropes have a strength-per-unit area of cord yarn more than twice that of conventional ropes. These ropes are said to be already competing with steel in fixed-rope outlets, such as aerial-stay and control ropes. Several types of parafil rope are available, and others are being developed.

Parafil ropes are being used for sag-free buoy moorings, and their use in applications such as drilling-rig moorings is being actively considered. Parafils are also being evaluated for yacht stays, cooling-tower stay ropes, sewer ropes, control ropes, and so on. One disadvantage of these types is that they have insufficient flexibility and abrasion-resistance to be used satisfactorily as working ropes or in applications where they rub over metal surfaces.

Baler twine represents a large potential market for polypropylene fiber. Agricultural baler twine is currently made from sisal. Previous attempts to use synthetic material have generally failed on economic grounds, and on poor knot stability. Efforts are being made to replace sisal with polypropylene split fiber yarns in this area. As indicated in Chapter XII, polypropylene split fiber has good knot stability, low weight, good regularity, and high strength. The future of polypropylene fiber for baler twine will rest largely on the relative price position of sisal and polypropylene yarns. Only split film can come under consideration.

Sisal twines are commonly used for the heavier sizes of packaging twines. These tend to be rough on the hands, and their irregularity causes trouble on automatic tying machines. Synthetic twines of various types, including polypropylene multifilament yarns, have been tried in the past as alternatives of sisal, but these have not as a rule proved satisfactory. They are smooth and regular, but are often too smooth and slippery to knot easily and quickly by hand, too low in bulk, and possess insufficient resistance to stretch.

Single-ply twist twines made from polypropylene slit film have the properties desirable in packaging twines and are trouble free when used in automatic tying machines. Packaging twine made from polypropylene coarse staple is even better as it combines high strength, regularity of product, and extensibility with good bulk and high coefficient of friction. The staple spun twines are, however, expensive and generally do not compete with sisal twines on efficiency/cost basis.

VIII.2. Fishing Gear

The important properties of yarns to be used in fishnets, trawls, and lines are essentially similar to those listed for ropes. In the past, cotton and manila and a few other natural fibers have been used for making fishing gear. Their shortcomings were such, however, that synthetic fibers made rapid progress in this field, and nylon in particular has been able to take over a large share of fishing gear market since World War II. In 1961, polypropylene was introduced into this field, and it has made rapid progress in direct competition with nylon and polyester fibers. By 1964, for example, the production of polypropylene nets in Great Britain exceeded the overall total of nets produced from all other synthetic fibers (ref.1). The reasons for the ready acceptance of polypropylene in the production of fishing gear are to be found in its unique combination of properties rather than in its low cost.

Polypropylene fiber, in both the continuous filament and the film-yarn forms, is presently being used for the manufacture of fishing nets and twines. Monofilament yarns may also be used with advantage for certain applications. The three fiber forms differ in their extension at break, their initial resistance to stretch, their buoyancy, and their flexibility. Fishnets made from monofilament have properties similar to those of the equivalent multifilament nets but with a higher resistance to abrasion and usually a stiffer hand. The fishnets made from film yarns generally have higher bulk and are less resistant to abrasion than those made from multifilaments and monofilament yarns.

The properties of interest in a fishnet-twine material are its straight breaking load and extension, and its breaking load when in the configuration of a conventional netting knot. A fiber with a good straight breaking strength does not necessarily have an equally good knot strength. Continuous-filament polypropylene yarn has good straight breaking strength as well as knot strength and compares well with nylon. Fibrillated-film polypropylene has a straight strength inferior to that of nylon or continuous-filament polypropylene yarn but it gives a comparable performance in the knot.

During fishing, twines are subjected to fairly small loads for most of the time. During the landing of full nets, however, and under trawling and snagging conditions, much greater loads are experienced, and it is important that a twine does not retain any permanent stretch after these loads are removed. Permanent creep is negligible for polypropylene fibers, and, in addition, their immediate recovery from loads of the order of two-thirds of the breaking load is high compared with that of the natural fibers and polyethylene fiber.

Knot slippage results in meshes varying in size, which greatly reduces the effectiveness of the net and can also lead to offenses against the minimum-mesh-size laws. This problem is peculiar to monofilament and multifilament twines but does not occur to the same extent with fibrillated film twines. For most purposes, nets made from continuous-filament polypropylene fiber have to have their knots stabilized by net-stretching or dipping, or both, in a knot-bonding medium. Nets from fibrillated-film polypropylene fiber, however, because of the high surface friction of the netting twines, do not normally require more than net-stretching to ensure adequate knot stability.

Fishing trials with nets made from polypropylene fiber have been carried out in important fishing grounds throughout the world. The nets have shown good knot stability and possess fishing power similar to nylon nets (ref.1). In some cases nets made from polypropylene have shown superior fishing power. This may be due in part to the extent to which the fisherman is able to make use of the buoyancy of polypropylene to increase the catching power of his net, or to the different extensibility of the two netting materials.

Polypropylene can absorb more work than any other textile fiber. For this reason, gill nets made from it are tougher than nylon gill nets of equal weight, and they are less likely to be broken by struggling, enmeshed fish. Measurement of the wet mesh strength of cod gill nets fished in Norway have confirmed that nets made from polypropylene will retain their strength well during their working life.

Bottom trawls made of polypropylene fiber have been shown to be more efficient in fish catching efficiency than those of manila trawls. This is probably due to the reduced drag and, consequently, less power consumption for a given speed of pull in the case of polypropylene trawls.

VIII.3. Safety Belts

Because of the legislative requirements in the United States and a growing awareness for safety in driving, an increasing amount of industrial textiles will be used for safety belts in the future. Both polyester fiber and nylon are used in the belts but polypropylene high-tenacity filament yarns have a good potential for this market. The worldwide production of safety belts now exceeds 100 million units per annum.

The safety belt is considered as basically an energy-absorbing device that, in the event of an accident, keeps the load on the wearer's body within survivable limits. The degree of energy absorption of the webbing is determined not only by the fabric construction, but also to a great extent by the finishing conditions employed. Polypropylene fibers have a high capacity of absorbing mechanical energy on impact loading. This property can be further improved by specially engineering a tough fiber (Chapter IX) and properly annealing the yarn in the webbing.

VIII.4. Sewing Thread

High strength, excellent chemical resistance, versatility, and low cost have contributed greatly to the acceptance of polypropylene fiber for sewing thread used in multiwall bag and other industrial sewing applications. A typical sewing thread is about 1,000 denier multifilament yarn with 1.5 turn/inch twist and 5.0 to 7.0 g/d tenacity. This sewing thread runs satisfactorily with 0.067-, 0.073-, and 0.080-inch gauge needles. Because of its low melting point, polypropylene sewing thread is not suitable for high-speed applications. By applying suitable finishes, the sewing speed can be improved but to minimize the heat buildup at the eye of the needle a sewing thread made of staple polypropylene is more appropriate. Such sewing threads are indeed extensively used in Japan.

VIII.5. Laundry Nets

Polypropylene continuous-multifilament yarns are used with advantage in the production of laundry nets. Tests carried out with polypropylene laundry nets have shown that they compare in the following ways with nets made from nylon (ref.1):

(a) Polypropylene nets shrink only half as much as nylon nets,
(b) Polypropylene nets last twice as long as those made from nylon,
(c) Polypropylene nets have a better resistance to bleach than nylon nets.

VIII.6. Filter Fabrics

The outstanding chemical resistance of polypropylene yarns is of great importance in the field of filter fabrics. In uses where filter fabrics are not subjected to temperature above 125°C, polypropylene has many advantages over competitive fibers, and may be considered as a replacement for fluorocarbon fibers. Both woven or nonwoven types of fabrics are being used for filtration. In addition to wet filtration, polypropylene fiber is finding increasing uses in air filtration. Here again, nonwovens are as important as woven fabrics. Sprayed polypropylene fibers and mats have been used for ultra filtration of suspensions in water as well as in air (Chapter IX).

VIII.7. Coated Fabrics

Polyethylene film, ethylene/vinylacetate copolymers, or plasticized PVC is applied to polypropylene fabrics by a melt-coating technique, providing a fabric which may be made into protective clothing, vent tubing, etc. For example, a 4-inch diameter tube with a weight of only 0.9 oz per linear foot is made from this coated fabric and used as mine vent tubing. Polypropylene fabrics coated with bituminus or coal tar have been used as pond lining. Other coated fabrics are used as protective clothing and temporary shelters.

VIII.8. Cable Covering

The good mechanical properties, excellent dielectric properties, brilliant colors, and low price have enabled polypropylene fiber to enter the field of cable covering and related applications. Monofilaments and thin tapes are predominantly used but the paperlike structures, such as those prepared from superfine fibers, spun bonded fabrics, and wet-laid nonwovens are making rapid inroads.

VIII.9. Artificial Turf

One development in industrial textiles that could lead to a large usage of polypropylene in the future is that of artificial turf for sports stadia. Manmade "grass" was originally developed for use in grassless areas such as city playgrounds, where it not only improved the appearance but helped to convert unused space into additional play areas. Astro-turf® of Monsanto and Tartan Turf® of 3M Company have made a notable success in this field. Both of these products are based on nylon.

The American Biltrite Company has developed Uni-Turf®, which has a ribbon-like fiber pile of polypropylene. Monsanto also markets a less rugged variety of artificial turf called Chemgrass®, involving the use of polypropylene, for highway central reservations, service stations, and other landscaped areas. The 3M Company is said to be developing a similar material. One shortcoming of polypropylene for this use is its rather low resistance to direct sunlight. The advent of more effective and less expensive UV stabilizers is sure to help this product.

VIII.10. Paper Reinforcement

The engineering of polypropylene fiber and synthetic pulp for the production of paper has been briefly discussed in Chapter IX. Polypropylene fiber has a great potential as a reinforcing component for kraft stock. Experimental work by Hercules has shown that an increase in tear resistance from 2 to 10 fold can be achieved by using 20 percent polypropylene staple of 15 denier and $1\frac{1}{2}$-inch length. Shorter lengths are recommended for smaller denier fibers. It has been variously experienced that fibrillated film fiber is more suitable for pulp reinforcement than the spinnerette spun fiber. Its flat cross section imparts to the fiber a laying characteristic which is similar to that of cellulose fibrils. It has further been experienced that polypropylene fiber needs a good wetting agent to achieve an adequate dispersion in the pulp.

VIII.11. Papermaker's Felts

Some of the largest and most expensive pieces of industrial textile are used on the press sections of paper machines. These may be up to 320 feet long and 400 inches wide. These felts are used for mechanically pressing water from the sheet of paper stock by passage through a nip. In the past they were all-wool felts; synthetic fibers were introduced after the second World War to improve wear-resistance. In the early 1960s, the batt-on-base felt was introduced by needling a batt of fibers onto a relatively low-weight base. The result is a permeable cloth with no surface weave structure to mark the paper. Half of all press felts are now of this type, a high proportion containing more than 50 percent of synthetic fibers.

Polypropylene fiber is finding extensive use in this application. Its strong points, in comparison with other synthetic fibers, are high chemical resistance and low moisture absorption. The first factor prolongs the useful life of the felt while the second factor contributes to the economy of drying.

VIII.12. Conveyor Belting

Polypropylene multifilament yarn combines many of the desirable properties of a conveyor belting yarn. These include high strength, low cost, high modulus, high extension at break, high work of rupture, and low sensitivity to moisture. Polypropylene fiber, however, suffers from two shortcomings as belting material. Firstly, it is extremely difficult to bond to rubber or PVC. Secondly, it melts at about 165°C and shrinks below this temperature if heated in an unrestrained condition.

The problem of adhesion with rubber and PVC has limited the application of polypropylene fiber in belting to blend fabrics containing cellulose staple fiber. This is best achieved by core spinning with 15 to 20 percent of cotton, which gives adhesion values of the same order as 100 percent cotton ducks. In certain cases, a warp of polypropylene yarn can be combined with the weft of cotton or rayon. The problem of excessive shrinkage and melting of polypropylene fiber at gelling and curing temperatures has limited the covering material to low-temperature curing rubbers and excludes the use of PVC which gels adequately above 150°C. Due to these limitations, polypropylene is not expected to carve any significant share of conveyor belting market for itself, except in the food industry.

IX. HOME FURNISHING APPLICATIONS

Home furnishing is by far the most important sector of textile market for polypropylene multifilament yarn and staple. By virtue of its high bulk, good stain resistance, good abrasion resistance, resilience, and low static buildup, polypropylene fiber seems just natural for carpets and upholstery fabrics. Its light weight, good thermal insulation, and washability without felting makes polypropylene a very suitable fiber for blankets and drapery. This combination of properties makes polypropylene fiber also suitable for wall coverings, bed spreads, slip covers, etc. A partial list of current uses of polypropylene fiber in home furnishing is given in Table XIV. A few of these applications are discussed in the following pages.

TABLE XIV

Use of Polypropylene in Home Furnishing

1. Carpet face fiber (tufted, woven, needle punch)
2. Backing for tufted carpets
3. Chain warps in woven carpets
4. Blankets
5. Upholstery fabrics
6. Underlays
7. Bedspreads
8. Drapery, slipcovers
9. Wallcoverings
10. Fiberfill

IX.1. Carpet and Rugs

Floor covering is the single most important application of polypropylene fiber today. Table XV lists some important performance characteristics required in a carpet. The rapid growth of polypropylene as carpet fiber has been due to its excellent properties, attractive colors, and comparatively low price. Polypropylene products have emphasized low soiling, stain resistance, easy cleanability, reduced static buildup, and resilience, durability, and fast colors

TABLE XV

Important Carpet Performance Characteristics

1. Wear
2. Scuffing footprint resiliency
3. Furniture mark recovery
4. Traffic pattern recovery
5. Soiling
6. Cleaning
7. Static generation
8. Fuzzing and pilling
9. Biological inertness
10. Flammability

TABLE XVI

Wear Resistance of Carpets (ref.5)

Fiber	Pile Wt. Oz/Sq Yd	Weight Loss Per 1000 Cycles	Classification
Polypropylene	16-22	22-50	Excellent
Nylon	16-22	20-40	Excellent
Wool	16-22	400-450	Good
Viscose	28-32	1000-1500	Poor

achieved through pigmentation. Polypropylene carpets and needle-punch products have proved specially suitable in high traffic areas where other synthetic fibers did not perform as well.

Wear resistance is one of the most important properties of a carpet. It has been shown in laboratory tests as well as in actual use that polypropylene carpets compare very well in wear resistance with nylon carpets of equal weight. They are definitely superior to wool, viscose, and acrylic fiber in this respect. Table XVI shows tabor abrasion classification of different carpets.

High abrasion resistance of polypropylene fiber has been used in enhancing the wear resistance of carpets made by predominantly natural fibers. For example, a significant increase in wear resistance results from blending polypropylene with wool. A four-fold improvement in wear resistance of jute pile carpets has been noted by blending as little as 30 percent polypropylene fiber. A similar improvement has been noticed in cotton/polypropylene mattings and woven floor coverings, popular in Latin American and Asian countries. Canadian Celanese has made ingenuous use of the high abrasion resistance of polypropylene fiber by blending it with their poorly wearing triacetate fiber. The blend not only increases the wear resistance of the carpets many fold, but it also imparts to the blend yarn a certain degree of heat settability.

Carpet resiliency is a complicated subject with many different facets. The term generally refers to the matting and distortion of the pile in use. Three varieties of resiliency are of primary interest for a carpet used in the home. These are footprint resiliency, furniture resiliency, and traffic pattern resiliency.

Footprint resiliency is important because it is undesirable for a carpet to show individual footprints when a person walks on it. Cotton carpets are notoriously poor in this respect, but even nylon plush carpets, wool velvets, and viscose carpets of all types show scuff marks and footprints. Tests prove that carpet made from polypropylene is very resistant to footprints or other single, quick-acting deformations. In laboratory tests it is found that filament nylon and polypropylene are better than other conventional fibers. Both show about 95 percent recovery and no visual effect from simulated footprints and scuffing.

The second type of resiliency, furniture resiliency, refers to the deep, matted indentations from furniture that is left in one position for a long period of time. All types of carpets exhibit this distortion to some degree, and the measure of its severity is the appearance of the mark one or two days after a light brushing. Most wool carpets and a number of viscose carpets will not recover from such furniture marks even a week after the load has been removed and the carpet vacuumed. Nylon is superior in this type of recovery. Polypropylene also has good furniture resilience although not quite as good as nylon. After the load is removed and the pile is lightly brushed, the mark effectively disappears from the polypropylene carpet after one or two days.

Traffic pattern resiliency is the recovery from matting and compression caused by repeated loading. This type of matting is noticed as a loss in pile height in those areas where there is frequent walking. Such matting is particularly objectionable if it cannot be corrected by a vacuum cleaning.

In actual walkout tests it is found that bulked continuous-filament nylon and good-quality, high-twist wools have the best traffic pattern resiliency. Viscose pile carpets are particularly poor in recovery from traffic pattern matting. In the middle range are found the acrylics, many spun nylons, and some wool carpets. Polypropylene is also in this middle range for traffic pattern resiliency.

Resistance to soiling and ease of cleaning is another important property of carpet fiber. Generalizations are difficult to make in this area and a comparison of fibers for soiling and ease of cleaning is not possible. This is due to several facts: lab results generally do not corroborate with actual field tests, fiber cross section plays an important part, fiber finish makes an overwhelming difference, carpet construction is an important variable, and the type of soiling and the mode of cleaning confuse the experimental results. In actual use, polypropylene has been found to soil less than nylon of equivalent fiber cross section. A number of basic characteristics contribute to the low soiling noted for polypropylene carpets. Among these characteristics are: the low moisture absorbency which resists the penetration of stains and the low surface tension of polypropylene which inhibits adhesion of foreign material.

When it comes to cleaning, polypropylene fiber has some clearcut advantages over all other carpet fibers. Vacuuming and brushing is more effective in removing loose soil than is the case with most carpets. A normally soiled carpet can be shampooed or washed with good soil removal and excellent retention of texture and appearance, in marked contrast to the unpleasant effects often encountered with viscose and wool carpets.

Accumulation of static charge is commonly cited as a problem with carpets of synthetic fibers. It is not unusual for a person to receive an unpleasant electrical shock after walking across a nylon carpet and then touching a metal object which is grounded. This tendency to induce a static charge is particularly common during winter months when the humidity is low. It is recognized that this effect can be alleviated by the use of various fiber finishes; but such finishes usually have the disadvantage of being only temporary in their effect and often causing rapid soiling of the carpet. As Table X shows, polypropylene permits the production of a carpet with a minimum of this unpleasant static effect without the use of any special finishes on the fiber.

At present, almost as much polypropylene fiber is being consumed in needle-punch carpet as in tufted and woven carpets. Needled carpets now amount to at least 5 percent of the total floor covering in the United States. The same properties which make polypropylene a desirable carpet fiber in tufted carpet are also responsible for its wide popularity in needle-punch carpets.

A small quantity of high-pile polypropylene sliver knit fabrics are being used for throw rugs for bath rooms and living areas in homes. As Table XVII shows, polypropylene commands a distinct advantage over acetate and acrylic fabrics in heat retention and matting characteristics, and fares well in abrasion resistance.

TABLE XVII

Comparison of Properties of Sliver Knit Pile Fabrics

	Polypropylene	Acetate	Acrylic
Heat retention (cooling time in sec)	122	79	98
Matting, dry 500 cycles	20%	23%	53%
1000 cycles	23%	23%	53%
Matting, wet 1000 cycles	10%	52%	-
Flat abrasion (cycles to destruction)	8,567	942	1,998

IX.2. Carpet Backing

Carpet backing for tufted carpets is another important application of polypropylene yarns. Flat-film yarns are used almost exclusively to weave fabrics for primary backing. This usage has been discussed in Chapter XII. Increasing yardage of polypropylene spun-bonded fabrics are being used as primary backing for tufted carpets. This subject has been discussed in Chapter IX.

Recently, attention is being paid by the fiber producers and the weaving mills to the development of secondary backing. A lightweight, high-covering fabric is required for this use. The requirements on strength and uniformity are not as stringent as those for primary backing. The use of fibrillated film yarns as filling and nonfibrillated tape yarns as warp are being examined for the development of secondary backing material. Lightweight needle-punch scrims are also being evaluated. Another route is the development of brushed, raised, nonwoven, light-weight fabrics. The brushing or raising is needed for good adhesion of the secondary backing to the latexed primary backing.

IX.3. Upholstery

The production of upholstery fabrics from polypropylene fibers is an application of rapidly growing importance. The growth of this sector of polypropylene market has been the fastest in the United States during the last 5 years. In this field, the properties of polypropylene fiber offer many advantages over competitive fibers. The low specific gravity results in excellent covering power, high abrasion resistance, negligible water absorption, resistance to microorganisms and insects. Easy cleanability is another characteristic that makes polypropylene an ideal fiber for upholstery fabrics. The built-in stain resistance of polypropylene fiber renders the wet finishing of the fabric unnecessary, which results in considerable savings in production.

Upholstery fabrics are commonly woven from 100 percent polypropylene filament yarns in the United States, but staple yarns are more common in Europe. In the United States the warp is generally 2800 to 3000 denier BCF yarn or 200 to 500 continuous-filament yarns. The filling is generally air-textured yarns (Tastan® type), while the warp is BCF or staple spun yarns. The fabrics made of 100 percent BCF yarns or a combination of CF yarn, air textured yarns and BCF yarns, are free from pilling problem. The use of staple spun yarns in velvet fabrics is very attractive and is gaining importance in the market. Polypropylene CF yarns have been used in the production of freeze' fabrics although the current fashion trend is away from this style. Sliver knit pile fabrics for upholstery are not yet in common use, but polypropylene fiber certainly has a potential in this field. Considerable designing and marketing attention is currently being paid to knit upholstery based on medium-denier, false-twist, textured polypropylene yarns.

Polypropylene upholstery fabrics are generally pigment-colored, but with the availability of acid-dyeable fibers, printed fabrics are also being introduced to the market in limited quantities. The dyeable fabrics, however, do not claim the same degree of stain resistance and easy cleanability as the pigmented fabrics. Ni-dyeable polypropylene fabrics have been introduced for printed applications but because of their somewhat dull colors, the acceptance of these fabrics has not been widespread.

IX.4. Blankets

Polypropylene fiber has proved particularly suited to the production of blankets, and manufacture of blankets of different types, weights, and sizes is increasing rapidly. The yarns used for conventionally raised blankets are spun on the woolen system with as low a twist as practicable. The warp consists of fine-denier continuous-filament yarns or very narrow flat tapes. Recommended constructions for twill and plain-weave blankets are given in Table XVIII

TABLE XVIII

Typical Construction For 100 Percent Polypropylene Fiber Blanket (ref.1)

Weave	Yarn Count (YSW)		Yarn Twist (tpi)		Loom Set (threads/ in)	Reed Width (in)	Finished Width (in)
	Warp	Weft	Warp	Weft			
$2/_2$ Broken twill	1/20	1/10	10"s"	4.0-5.0"s"	16x24	98	70
plain	1/20	1/8.5	10"s"	3.5-4.5"s"	16x20	98	70

Conventional-type polypropylene fiber blankets may be raised using normal equipment. Polypropylene fiber blankets tend to contract in width during raising to a greater extent than woolen or other blankets. Cellular blankets may be left with a plain finish, or raised lightly to brush the surface. A fiber finish is generally applied to the fabric before raising. This can be done during scouring operation.

Polypropylene blankets are characterized by high thermal insulation power, moth and insect resistance, easy launderability, low shrinkage, and light weight. These properties make polypropylene blankets suitable for household as well as for military use.

IX.5. Wallcoverings, Bed Spreads, etc.

Tufted wallcovering fabrics are growing in importance in the United States. These fabrics not only have a decorative value but also serve as insulation material for sound and heat. The use of polypropylene fiber in this sector is promising because of the fast colors of the pigmented fiber, lightness, and easy cleanability. Staple spun yarns as well as bulked continuous-filament yarns are suitable for these fabrics. Nonwoven fabrics have not been tried because of the lack of printability of these fabrics. However, as the art of printing and dyeing of Ni-dyeable and acid-dyeable polypropylene fiber advances, the nonwovens may replace tufted or woven fabrics.

IX.6. Fiberfill

Polypropylene fiber has not gained any significant share of the fiberfill market in the United States because of the abundantly available polyester staple in this country. However, polypropylene staple plays an important role as fiberfill material in Japan and Europe. Unannealed, stuffer-box, crimped staple has not proved as successful as well-annealed fiber of high crimp frequency. The advent of conjugate fiber is expected to strengthen this market for polypropylene. The resilience of normal polypropylene staple is not very high when used as padding or filling material, but by proper engineering (see Chapter IX) a product can be obtained that is competitive with polyester and superior to nylon fiberfill.

X. APPAREL

Apparel is still the most tantalizing market for any textile fiber. Polypropylene fiber has failed so far to penetrate this sector of the market because of its low melting point (offering difficulties in ironing), excessive wrinkling, and most of all, lack of dyeability. With the advent of acid-dyeable fibers, polypropylene fiber is gathering strength to enter into selected areas of apparel use. Coincidental with the introduction of dyeable fibers is the development of false-twist textured polypropylene yarns which is helping polypropylene to claim a share in knitwear goods. Unlike the United States and Western Europe, in Japan the use of polypropylene fiber in apparel is much advanced. In fact, about 80 percent of all filament and 35 percent of staple is consumed in this market in Japan (ref.3). The areas of penetration by polypropylene fiber are listed in Table XIX, and a few selected uses are discussed below.

TABLE XIX

Selected Uses of Polypropylene Fibers in Apparel

1.	Knitwear (conjugate fibers, false-twist textured yarns)
2.	Underwear and hosiery (spun, textured, conjugate)
3.	Pile fabrics (long staple)
4.	Handknitting yarns (conjugate, "popcorn", BCF plied yarns)
5.	Work clothing (staple spun)
6.	Blend fabrics (rayon, cotton, wool)

X.1. Knitwear

Polypropylene fiber is making steady progress in the knitwear field. It offers many attractive properties in this application, including an excellent hand, loft, dimensional stability, easy care, and all-round wearing qualities. All the standard texturizing process may be used on polypropylene yarns, including false-twisting, stuffer box crimping, edge crimping, and other systems. Stuffer box textured yarns are somewhat "lazy" and are only suitable for low-bulk fabrics. False-twist texturizing imparts a more permanent and lively crimp to polypropylene. These yarns are beginning to find increasing use in knitwear and hosiery. In addition to textured, continuous-filament yarns, blended yarns of high-shrinkage polypropylene staple with rayon, cotton, or wool are being used for knitwear production. These items represent quite a significant sector of polypropylene consumption in Japan. Sweater fabrics made of 50/50 PP/wool blends is an important end-use not only in Japan but also in Italy. On finishing such fabrics, polypropylene fiber shrinks and the blend yarn is bulked in "popcorn" fashion. Cotton/PP blend fabrics are used for underwear. They need not have any loft and bulk as the outerwear fabrics do.

Standard finishing equipment may be used to finish polypropylene goods. Unless a special hand is required, steaming is usually adequate. Silicone softners are particularly useful for improving the hand of polypropylene knits.

X.2. Pile Fabrics

The manufacture of knitted fur pile fabrics represents one of the swiftest growing branches of the knitted yard goods field. Most knitted fur pile fabrics are produced from acrylic fibers. Recently, however, there have appeared on the market pile fabric goods made of polypropylene staple fiber. They are primarily used for boot lining, coat lining, toys, cushions, upholstered furniture, coat material, children's outerwear, etc. Long pile fabrics are used as area rugs and seat covers. The finishing is extremely important and in general consists of brushing. Silicone finishes significantly improve the hand and other tactile properties.

X.3. Hand Knitting Yarns

Due to its light weight, stain resistance, thermal insulation, and easy washability, polypropylene is ideally suited for handknitting yarns. It is surprising, however, how little effort has been put into the exploration of this end use. Recently, Canadian Celanese Company and Montecatini have pioneered the use of polypropylene fiber in the production of hand knitting yarns. While Canadian Celanese has concentrated on bulked continuous-filament yarns, Montecatini's efforts have been directed to the use of staple in combination with wool. A third promising approach is to spin a mixture of high-shrinkage/low-shrinkage polypropylene fibers, ply and heat-relax in skin form. The use of bulked continuous-filament yarns has the advantage of better economy and complete freedom from pilling. The aesthetics are, however, different from the normal knitting yarns. The yarn prepared from wool and polypropylene staple has good loft and aesthetics but generally suffers from excessive pilling. The "popcorn" yarns give the aesthetics of acrylic hand knitting yarns, are low in pilling, and command all those properties that are the characteristics of polypropylene fiber.

X.4. Protective Clothings

Due to its high chemical resistance and easy washability, polypropylene fiber is well suited to the production of protective clothing for the chemical industry. Due to its high energy-absorbing capacity, polypropylene fiber has been used for the production of antiballistic personnel armor for military and police forces. Woven cloth as well as nonwovens have been tested with results comparable to those of high-tenacity nylon fibers.

X.5. Polypropylene/Cotton and Polypropylene/Rayon Blend Fabrics

PP/cotton and PP/rayon blend fabrics are common in Japan. A 30/70 mix has proven to be the best. Polypropylene fiber imparts to the fabric good abrasion resistance, dimensional stability, reduced moisture absorption, easy drying characteristics, and light weight. Underwear, sports wear, work clothes, pajamas, coat cloth, and children's clothes are important uses of these blend fabrics. Jersey fabric, crepe, flannel, and denim weave have proved specially suitable for these uses.

XI. CIVIL ENGINEERING APPLICATIONS

The use of fibers and textiles in civil engineering applications has been recognised from the earliest recorded history. However, until the advent of strong and durable man-made fibers, this potential could not be fully exploited. Polypropylene has proven to be the most attractive fiber owing its relatively low cost and inert n-ture. The textiles used in civil engineering applications are often called *Geotextiles*. Geotextiles are used within, beneath, or on top of the solid materials derived from the earth's crust. They provide lower construction costs through savings in time or material, and make possible new approaches to design and construction.

At present four main functions of geotextiles are recognised, namely, filtration, separation, reinforcement, and drainage. In practical applications geotextiles invariably perform more than one of these functions simultaneously.

The separation function for geotextiles is one of the most widely recognised area in civil engineering. Many examples are known in which temporary construction roads have been formed by placing a permeable fabric membrane on cohesive subgrades which have been covered with layers of roadstone. The fabric separates the roadstone layers from the subgrade, preventing puncturing of the stone into the soil beneath. This provides savings in the total quantity of stone required. On many sites where access roads over soft ground are required, geotextiles make viable an otherwise difficult proposition. In such cases fabric, in addition to separation, acts as a mild reinforcement, mobilizing in-plane tensile stress to bridge localized soft areas. Fabrics have also been used beneath embarkments, separating the earth fill from the virgin subgrade. There is some evidence that differential settlement of embarkment may be reduced by placing fabric beneath it.

Another widespread application of geotextiles is within sub-surface drains. They are used to prevent the seepage of ground water or surface water into or beneath civil engineering structures, or to locally lower the ground water table. During construction such drains may be seen along the sides of highway pavements, at the crest of cuttings, and on soil slopes.

Sub-surface *French* drains are frequently formed by placing a perforated drainage pipe at the base of a trench 0.5m - 0.6m wide, and back filling with a graded granular material. This material is selected so that it will drain more rapidly than the surrounding *in situ* soil, and that it will prevent the ingress of fines from the soil, i.e., it acts as a filter and a hydraulic conductor. The cost of such drains can be reduced by dispensing with the relatively expensive graded material. This is achieved by lining the trench with a geotextile, and backfilling with a cheap free-draining stone which does not require careful selection. A further cost saving is also achieved in that no drainage pipe is needed.

Several other applications of geotextiles in cicil engineering field are being examined in Western Europe and the U.S.A. However, it will take considerable time to assess these applications.

Cement paste, mortar or concrete, although capable of great strength in compression, are weak in tension, have low strain to failure, and are generally brittle. Fibers have therefore been introduced to overcome some of these weaknesses, and to develop products which otherwise would not be possible. Whilst asbestos fibers have been used successfully for some time to produce thin cement board products with satisfactory flexural strength, their impact strength is still low. The alternatives, steel, glass and cellulose or pulp fibers, have several disadvantages including difficulty in handling, sensitivity to the highly alkaline or moist environment and price. Polypropylene fibers have good chemical resistance, are unaffected by moisture, have high tensile strength and have low specific gravity, and can be easily handled. These advantages have given the industry an incentive to try the use of polypropylene fibers in cement reinforcement. A vast amount of experimental work is going on at present and potential looks quite bright.

XII. REFERENCES

1. G. Cook, Handbook of Polyolefin Fibers, Merrow, Manchester, 1968.
2. A.V. Galanti and C.L. Mantel; Polypropylene Fibers and Film, Plenum Press, New York, 1965.
3. ICAFE Seminar on Man-Made Fiber Industry in Asia and the Far East, Tokyo, 1966.
4. R.G. Stoll, Synthetic Fibers, Fairchild Publishers, Inc., New York, 1953.
5. M.B. Truslow, Mod. Text. Mag., (1963)94.
6. R.S. Goy and J.A. Jenkins, Textile Progress, (1970)2.
7. R.C. Laible and H.M. Morgan, J. Appl. Polym. Sci., 6(1962)269.
8. D. Pollar and McDougall, Text. Res. J., 34(1964)117.
9. J.C. Guthrie, J. Text. Inst., 62(1971)6151.
10. S.E. Ross and H.W. Wolf, J. Appl. Polym. Sci., 10(1966)1557.
11. K. Kamide, Sen-i Gakkaishi, 23(1967)16.
12. V. Cappucio, et al, Chim. Ind., (Milan), 44(1962)463.
13. R.H. Ralston, Text. Ind., May, 1969.
14. Flooring, (1970)76.
15. Hercules Data
16. H. Appelt and G. Appelt, Deutsch Text. Tech., 17(1967)169.
17. M. Stratman, Polyolefin Faserststoffe, Faserforschumgsberichte des Landes NRW, Germany.
18. M. Stratman, Chemiefasern, 14(1964)484.
19. M. Stratman, Z. Ges. Textilind, (1964)746,850,948.
20. Textile Organone 1972.

XIII. BIBLIOGRAPHY

XIII.1. Books

1. G. Cook, Handbook of Polyolefin Fibers, Merrow, Manchester, 1968.
2. A,V, Galanti, Polypropylene Fibers and Film, Plenum Press, New York, 1965.

XIII.2. Journal Literature

1. ICI, Outlets for Ulstron Polypropylene Fiber, Skinner's Silk and Rayon Record, (1962)36, 330, 332.
2. M. Gralinski, New Solvent for Polypropylene Fiber in the Analysis of Fiber Mixtures, Text. Res. J., (1962)32, 170.
3. G.B. Southward, Longshoremen Find Polypropylene is the Rope to Fit the Job, Daily News Record, 218(1963)25, 30.
4. R.W. Moncrieff, Man-Made Fibers for Industrial Protection, Text. Weekly, 63(1963)924-925.
5. W.R. O'Donnel, Polypropylene and Its Uses, Am. Dyestuff Report, 54(1965)109-111.
6. P. Van Oost, Polypropylene Fiber, Chemiefasern, 16(1966)480-484.
7. J.E. Ford, Properties and Uses of Polypropylene and Polynosic Fibers, Text. Inst. Ind., 4(1966)135-138.
8. Hercules, Inc., Identification of Polyolefin Fiber, Hercules, Inc., Wilmington, Del., Bull. FTS-18, 1966.
9. K. Brockhaus, The Importance of Man-Made Fibers in Furnishing, Chemiefasern, 16(1966)385-386.

10. J.D.Couvoisier, Textiles, Cordage and Hooping: Three New Applications for drawn Polyolefins, Plast. Mod. Elst., 19(1967)68-74.
11. Anon., Polypropylene: Apparel Next?, Chem.Eng., (1969)76.
12. F.Riedel, Possibilities of Applications for Synthetic Fibers in the Field of Carpet Manufacturing, Chemiefasern, 21(1971)951,954-956.
13. G.Pamm, Technical Advances in the Development of Synthetic Fibers for Carpets, Z.Ges.Textilind., 72(1970)98-104.
14. M.Wandel, Modern Carpet Fibers - Problems and Possibilities, Chemiefasern, 22/74(1972)527-530.
15. C.East, The Soiling behavior of Man-made Fibers, J.Text.Inst., 64(1973)273-278.
16. Anon., Man-made Fibers used in Carpet Manufacturing, Mod.Text., 55(1974)35.
17. M.Pechocova, Stress-Temperature-Time Relationaship of Polypropylene Filament, Faserforschung u. Textiltech., 27(1976)27-31.
18. D.J.Hannant, Polypropylene Fibers in Concrete, Mortor and Cement, in Fibre Cements & Fibre Concretes, John Wiley, New York, 1978.
19. D.J.Hannant, Polypropylene Fibre in Cement Based Materials, Composite, 9(1978)83.
20. B.Ladeur, Factors Affecting the Structure of woven Polypropylene Fabrics for Carpet Backings, Plst. Rubber Inst., London, Conf. Polyp. Fibres in Textiles, York, Sept. 26-28, 1979.
21. N.Cormack, The Use of Polypropylene in Tufted Carpets, Plast. Rubber Inst., London, Conf. Polyp. Fibres & Textiles, York, Sept. 26-28, 1979.
22. B.K.McGuiness, Synthetic Turf, Plast. Rubber Inst., London, Conf. Polyp. Fibres & Textiles, York, Sept. 26-28, 1979.
23. A.L.Bell, The use of Polypropylene Textiles in Civil Engineering, Plast. Rubber Inst., London, Conf. Polyp. Fibres & Textiles, York, Sept. 26-28, 1979.
24. J.P.Giroud, Ten Years of Nonwoven Geotextiles, EDANA Index, 78 Congress on Nonwoven Fabrics, Amsterdam, 1978.
25. L.Belloni, Road Embarkments on Compressible Soils constructed with the Aid of Synthetic Fabrics, Proc. Int. Conf. Fabs. in Geotechnics, Paris, 1977.
26. W.Wolman, Reinforecement with Fabrics, a New Technique to Improve the stability Embarkments on Weak Soils, Proc. Int.Conf.Fabs. in Geotechnics, Paris, 1977.
27. A.L.Bell, Construction and Analysis of a Fabric Reinforced low Embarkment, Proc.Int.Conf.Fabs. in Geotechnics, Paris, 1977.
28. A. McGown, The Properties of Nonwovens important in Public works Applications, EDANA Index, 78 Congress on Nonwovens, Amsterdam, 1978.
29. D.G.Ellis, The Use of Polypropylene as a Reinforcement Medium in Various Substrates, Plast. Rubber Inst., London, Conf. Polyp. Fibres & Textiles, York, Sept. 26-28, 1979.
30. N.J.Dave, Polypropylene Fibre Reinforced Cement, Int.J.Cement Composites, 1(1979)
31. D.J.Hannant, Polypropylene Film in Cement Based Materials, Composite, 4(1978) 83-88.
32. G.Sawar, The Properties and Potential of Polypropylene Fibres in Domestic Textiles and Apparel, Plast. Rubber Inst., London, Conf.Polyp.Fibres & Textiles, York, Sept. 26-28, 1979.
33. B.Piller, Application of Polypropylene Filament Yarn in Knitted Fabrics, Plast. Rubber Inst., Conf. Polyp. Fibres & Textiles, York, Sept. 26-28, 1979.
34. J.Suominen, The Use of Polypropylene Fibre in Nonwoven Applications, Plast. Rubber Inst., London, Conf.Polyp.Fibres & Textiles, York, Sept. 26-28, 1979.

Chapter XIV ECONOMICS AND MARKET SURVEY

I. GROWTH OF POLYPROPYLENE FIBER

Commercial production of polypropylene fiber was started in early 1960 when small quantities of Italian-made staple and monofilament were made available to the European market. Soon followed supplies from the United States, United Kingdom, and Japan, yet, even as late as 1963, the total world production amounted to only 80 million pounds per year, of which the United States' share was probably not more than 35 million pounds. Most of this early production was monofilament to replace polyethylene and Saran in ropes, cordage, and outdoor furniture webbings; the United States consumption of textile-grade polypropylene fibers reached only 10 million pounds that year. Henceforth, the world polypropylene fiber production grew at an average annual rate of about 40 percent up to 1968 compared to a combined average annual growth rate of about 20 percent for nylon, polyester, and acrylic fibers. During 1969-1970, polypropylene fiber production declined to a yearly growth of about 10 percent but so did the production of other synthetic fibers. The United States took a major share of this recession, recording only a 5 percent growth rate for polypropylene fiber. Recovery started in late 1971 and continued through 1973, recording an estimated growth rate of 30 percent in the United States and about 25 percent on a worldwide basis. During the 1974-1975 recession, polypropylene fiber experienced a sharp drop in production, although not as severe as that experienced by nylon. Recovery started in late 1975, and in 1976 polypropylene fiber was again on its growth curve. Confident in today's market and in firm prices, some producers are forecasting an average growth of 20 percent for polypropylene fiber through the 1980s, leading to a world output of 2 billion pounds. More conservative estimates put the growth rate at 15 percent, the developing countries experiencing a more rapid growth than the industrially developed countries.

The growth of polypropylene fiber did not occur uniformly around the world. The growth rate has been the highest in the United States and Western Europe, while countries like Japan and Canada experienced only a modest, but steady, rate of growth.

The first commercial production of polyolefine yarn and textile monofilament in the United States was started in 1958. Reeves Brothers, Inc., of Spartanburg, S. C., and Dawbarn Brothers, Inc., of Waynesboro, Va., were the first producers of polypropylene monofilament and tape yarns. Their combined output of olefin fiber in 1958 totaled 2,600,000 pounds, but the records do not indicate how much of this production was polypropylene. The Reeves Brothers plant at Spartanburg was purchased in 1964 by Alamo Polymer Corporation and is currently being operated by Phillips Fibers Corporation, a subsidiary of Phillips Petroleum Company. The Dawbarn plant at Waynesboro is now operated by Wayntex Inc.

Reeves was probably the first producer of polypropylene staple in the United States, but the output of staple was nominal through 1961. Later on, some quantities of coarse staple became available from Canadian Celanese Company in the early 1960s and from Industrial Rayon Company. In the mid-1960s Canadian Celanese Company was acquired by Celanese Corporation of America. It is still an important producer of polypropylene fiber in North America. Industrial Rayon was purchased by Hercules, Inc., during the same time. At present Hercules is the biggest polypropylene fiber producer in the United States.

Polypropylene film fiber began to appear on the United States scene in 1965. The earliest producers were American Manufacturing Company, Chevron Chemical Company, International Harvester Company, Patchoque Plymouth Company, and Tennessee Eastman Company. Their collective output in 1966 amounted to some 16 million pounds.

Textile denier, textured polypropylene multifilament yarn was pioneered by Canadian Celanese Company in 1965, in an attempt to use it in underwear garments, socks, children's outerwear, filter cloths, etc. These efforts were not successful because of the deficiency in the stabilization system and the increasing pressure from nylon yarns. It was in late 1960s or early 1970s that Vectra Company (then a subsidiary of Exxon Chemicals and now a subsidiary of Chevron) introduced low-denier polypropylene filament yarns to the textile trade for upholstery fabrics. This lead was quickly followed by Hercules Inc., and other producers of polypropylene fibers. By 1975, the use of Taslan air-textured polypropylene yarns for upholstery fabrics became one of the most important segments of the home furnishing industry.

In the production of polypropylene BCF yarns for carpets, Canadian Celanese was again a pioneer. In 1963, a "self-crimping" (conjugate structure) multifilament yarn was introduced to the tufted carpet trade. The carpets competed very well in aesthetics, performance, and price with heavyweight nylon BCF yarn carpets for a number of years, until the very low prices of nylon BCF yarns forced it out of the market in the early 1970s. In 1965-1966, Hercules made its entry into the production of polypropylene BCF yarns for carpets via the stuffer-box crimping route and dominated the market until the early 1970s when the application of steam jet texturizing process gave it even a higher advantage in the market. The extensive current use of multiple-ply, pigmented polypropylene for contract carpets owes its origin to the pioneering efforts of Hercules, Inc. in the mid-1960s. Polypropylene BCF yarns were later introduced to the textile trade for upholstery fabrics by Vectra and others.

TABLE I

U.S. Polypropylene Fiber Production 1965–1980 (million of pounds)

Year	Yarn and Mono*	Staple	Total
1965	61	12	73
1966	97	35	132
1967	114	46	160
1968	204	54	258
1969	200	63	263
1970	198	59	257
1971	260	57	317
1972	345	66	411
1973	422	63	485
1974	460	65	525
1975	439	52	491
1976	511	58	569
1977	554	78	632
1978	597	92	689
1979	639	119	758
1980	633	112	745

*Multifilament yarn, monofilaments, spunbonded and film yarns
(Source: Textile Organon)

From these modest beginnings, polypropylene fiber usage in the United States grew to 800 million pounds in 1981 and its future prospects may be reflected in current production capacity already in excess of a billion pounds. Table I shows the United States output of polypropylene fiber over the past decade in million of pounds. Table II gives the growth rate on a worldwide basis for recent years.

The producers of polypropylene fiber and yarn are divided into three distinct categories. There are those who are primarily engaged in the production of film yarns and monofils, there are those who produce spunbonded fabrics, and there are those who are the producers of multifilament yarns and staple. The production of film yarns and monofils is fragmented and a substantial portion of the production is captive for internal use. There are only eight polypropylene spunbonded fabric producers in the U.S.A. and only a few throughout the world. The production of multifilament yarns and staple is also concentrated in a few hands. However, as several new producers are entering the field, this preponderence of large producers is expected to decrease considerably.

A review of the overall growth of polypropylene fiber in the last ten years shows that it has generally kept pace with the annual growth of all other synthetic fibers (Table III). In view of the numerous synthetic fibers of comparable

TABLE II
World Olefin Fiber Production - 1974-1980

	(million lbs)						
	1974	1975	1976	1977	1978	1979	1980
Western Europe	571	604	741	816	902	933	926
Eastern Europe	46	55	79	99	103	117	121
U.S.A.	527	496	575	632	692	758	745
Other Americas	138	154	177	223	239	246	245
Japan	198	178	190	187	95	104	95
Other-Africa, Asia, & Oceania	88	79	81	118	123	128	119

Source: Textile Organon, July, 1981

and interchangeable properties already established in the textile market, and in view of the past attempts to commercialize various new fibers such as polyvinyl alcohol, polyvinyl chloride, casein-based fibers, etc., the acceptance of polypropylene fiber has been considerable. However, development of polypropylene fiber did not proceed as rapidly as optimistic forecasts had predicted in the late 1950s and early 1960s. The reasons for this comparatively slower rate of development than that expected on the basis of its potentially cheap raw material base and good physical properties were partly strategic, partly economic, and partly technical.

Strategic Factors: With the exception of Montecatini, ICI, and Celanese, none of the early polypropylene manufacturers had any background in the art of fiber manufacturing and marketing. The major polypropylene fiber manufacturers, such as Hercules, Phillips, Amoco, etc., all had to acquire the necessary know-how for fiber manufacturing as well as the knowledge in the textile field. This severly hindered development work. The companies who had the know-how for fiber manufacturing and were textile oriented, and who could afford the capital required for development of a new fiber, were already deeply entrenched in nylon and had recently invested huge sums of capital for the development of polyester

TABLE III

U. S. Fiber-Consumption Growth

Annual average rate of growth (%)

	1957-1962	1962-1967	1967-1972	1972-1977	1977-1982	1982-1987
Acrylics	14.0	21.3	7.5	1.5	1.9	1.6
Nylon	13.8	13.2	13.2	2.6	3.5	3.4
Olefins	17.7	33.3	18.0	15.3	7.5	7.5
Polyester	13.5	36.8	28.8	7.6	6.8	4.5
Subtotal	13.9	20.6	18.2	5.4	5.4	4.0
Rayons	1.5	3.4	(1.1)	(9.5)	(2.0)	(2.0)
Textile Glass	12.6	11.5	13.4	6.8	4.3	3.7
Cotton	0.8	1.0	(2.7)	(3.8)	(3.5)	(4.0)
Silk	--	--	--	--	--	--
Wool	3.1	(5.8)	(5.2)	(11.3)	(5.5)	--
Subtotal	1.0	0.4	(2.9)	(4.1)	(3.5)	(3.8)
Total	2.6	5.0	5.4	0.9	2.7	2.2

fiber. They were reluctant to embark upon the introduction of another new fiber with comparable and interchangeable properties. As polypropylene fiber was considered a competition with nylon and polyester fibers in most of their uses, these companies had little incentive to develop it as a general-purpose textile fiber. Polypropylene was largely regarded as "poor man's nylon" and all development efforts were directed to the production and popularizing of coarse staple and monofilament or tape for non-critical industrial uses. It was mainly due to the continued efforts of Canadian Celanese Company, Hercules, Inc., Enjay Fibers, Reeve Brothers, and Vectra in North America, and several companies in Japan, who did not have any vested interests in polyester and nylon, that polypropylene fiber began to shed the image of a cheap industrial fiber and gradually established itself as a commodity fiber for carpet and textile industry.

Economic Factors: In the early stages of polypropylene fiber development, too much emphasis was placed upon the prospective cost of the polymer. These early forecasts proved to be unrealistic. In the United States, where the development of polypropylene was pushed ahead with great enthusiasm, the cost of polymer was maintained initially at 42 to 45 cents per pound. By 1964, the price of polypropylene dropped to 24 to 32 cents per pound, and it was not until the late 1960s that the price of polymer dropped below 20 cents per pound. Meantime, however, the prices of established synthetic fibers also dropped, and a slump appeared in the synthetic market in 1969-1970. In 1971-1972 the price of general-purpose polypropylene dropped further and reached a low level of 15 cents per pound. This happened without a sharp reduction in nylon price. The price for polypropylene started rising rapidly but so were the raw material costs for nylon and polyester fiber. It is estimated that the price for polypropylene will continue to rise but it will be considerably lower than that for nylon and polyester in the foreseeable future. This situation has given polypropylene fiber a substantial cost advantage over nylon and polyester fibers, with which polypropylene fiber is destined to compete in several applications.

Technical Aspects; Polypropylene fiber arrived on the market at a time when other synthetics were thoroughly established. The new fiber offered a range of properties different from those of other fibers, but these differences were not as dramatic as those which distinguished early synthetics from natural fibers and rayons. Polypropylene had to fight for its market share against established synthetics, offering advantages that were more marginal than novel and dramatic. For the great majority of textile uses, polypropylene has been competing with rival fibers on the basis of relative performance. This increasingly slowed down the growth of polypropylene fiber and denied it a place among the large-volume textile fibers.

Polypropylene fiber has many attractive properties to offer the textile trade. It has a built-in stain-resistance property and has the basic advantage of low cost which puts it economically in a class by itself. However, these technical and economical advantages have to be weighed against some technical problems and end-use shortcomings. Among the technical problems are polymer's sensitivity to heat, high viscosity of the melt, and difficulty of dyeing. Among the end-use shortcomings are low stability to light degradation and the low melting point of the fiber. Some of these problems - heat and light stability, for instance - have already been overcome through development of effecient stabilizer systems; some problems, such as dyeability, will be resolved in time; but the low melting temperature will remain with us as an inherent property of the polymer.

II. MARKET SURVEY

The markets for polypropylene fiber can conveniently be discussed in terms of the following four distinct types of products:

(1) Monofilament
(2) Multifilament yarns and staple
(3) Tapes and fibrillated film yarns
(4) Spun-bonded fabrics and related products

The term monofilament is usually applied to filaments with cross-sectional diameter of about 0.006 inch or larger. Multifilament yarns form the broad category of textile fibers; they include untextured straight filament yarns as well as textured or bulked continuous-filament yarns, having a cross-sectional area per filament of less than 0.0015 inch. This category also includes staple fiber and tow. The former is basically multifilament yarn cut into 1-inch to 8-inch lengths; tow is a large bundle of crimped multifilament yarn - generally in the range of 50,000 to one million denier - from which staple is cut. Tapes are essentially monofilaments with a rectangular cross section. Fibrillated film yarns are interconnected fibrous structures produced through the fibrillation of thin tapes. Spun-bonded fabrics are nonwoven fibrous structures produced in the form of flat fabrics directly from the melt in one continuous operation. The traditional nonwoven market is excluded from this category.

The equipment needed for the production of monofilament, tapes, and fibrillated yarns is relatively simple and inexpensive. It is possible to produce these products on a small scale. The manufacture of multifilament textile yarns and staple, on the other hand, requires expensive equipment and much more sophisticated manufacturing technology than is required for monofilament, tape and fibrillated yarns. Spun-bonded fabrics and related products are largely based on the economy of mass-production, employ heavy outlays in capital expenditure, and make use of sophisticated production and control equipment.

As a general rule, the markets of monofilament, tape, and fibrillated film yarns do not overlap with those of multifilament yarns and staple fiber. The manufacturers are generally divided into two groups: most manufacturers produce either monofilament, tape and fibrillated yarns or multifilament and staple. In 1981 there were probably not more than 25 manufacturers in the world who produced multifilament yarns and staple but there were more than 600 plants which produced monofilament, tape, or film yarns. There were only six producers engaged in the production of more or less all products. The markets for these products are generally well defined although there might exist some gray areas between uses.

The markets of spun-bonded fabrics overlap some of the markets of staple fiber and tape yarns. These products are relatively new and are manufactured by a third small group of manufacturers who generally produce neither polypropylene multifilament yarns and staple nor monofilament, tape and film yarns.

TABLE IV

U. S. Polypropylene Fiber Consumption (MM Lb/Year)

	1965	1968	1971	1974	1977	1980	1983 (est)
Continuous multifilament yarns	24	43	75	91	106	120	210
Staple	11	60	70	90	76	110	200
Nonwovens (spun-bonded fabrics)	--	15	35	46	60	140	160
Monofilament	31	50	80	100	120	170	220
Tape and film fibers	--	110	110	187	140	300	350
Total	--	280	325	504	602	840	1140

Table IV shows domestic shipments by United States polypropylene fiber producers since 1965 in terms of above mentioned categories of products. Table V summarizes the various applications of these major categories of polypropylene fiber. Table VI shows a breakdown of polypropylene fiber shipment for 1965-78 according to trade or end use and Table VII shows a similar breakdown for 1980-81. Fig. 1 shows an en-use breakdown for the UK.

TABLE V

Some current and Prospective Applications for Polypropylene Fiber

Fiber type	
Monofilament	Bristles, cords, ropes, fishing nets, furniture webbing.
Tape	Sacks, bags, primary carpet backing, bale wraps (wool packs, cotton and staple bales), furniture webbing, light-duty tarpaulins, fishing nets (leno weave), synthetic grass.
Slit and split film	Twine, cords, ropes, matting, concrete reinforcement, woven carpet-backing yarn, wallcoverings, upholstery.
Staple	Blankets, spun yarns for upholstery and carpets, knitting yarns, some apparel, needle-punch carpets, felts, fiber-fill, nonwovens of all sorts except spun-bonded fabrics.
Continuous-filament yarns	Ropes, cordage, carpet face yarn, upholstery, apparel, heavy-duty tarpaulins, coated fabrics, synthetics grass.
Spun-bonded fabrics	Carpet primary backing, electric insulation, battery separators, coated fabrics.

Monofilament: The monofilament was the first large market in which polypropylene made rapid progress, mostly at the expense of Saran and polyethylene monofilament. Within three years of its introduction to the United States market, polypropylene monofilament had virtually replaced Saran in outdoor furniture webbing and in car seat covers. Polypropylene monofilament also made a particular impact in the rope and cordage market, initially displacing nylon and polyethylene and then manila and jute.

The current United States production of polypropylene monofilament is in the region of 120 MM lbs per year (Table IV). The manufacture of cordage, twine, and related products, such as fishing nets and wire rope cores, accounts for almost half of the monofilament consumption. Agricultural and filtration fabrics are the second largest market segment. Other significant uses are in wide fabrics including monofilament warp for heavy duty fabrics, car seat covers, swimming pool covers, laundry bags, and industrial fabrics. The remaining outlets for monofilament include bristles for street sweeping rolls and brooms, kitchen scouring pads, nets, electrical cable fillers, and scrims. The major future market will be in the area of soil erosion and stabilization fabrics which are competing successfully with nonwoven fabrics. Synthetic bristles for the industrial brush market, while filling an important need, can be expected to have a slower growth

TABLE VI

U. S. Domestic Shipments of Polypropylene Fiber by Trade or End Use (million lbs)

	1965	1966	1967	1968	1969	1970	1971	1972	1973	1974	1975	1976	1977	1978
Knit Goods - Total	3	4	3	3	3	2	2	3	5	5	5	5	7	9
Carpets & Rugs														
Face - yarn	18	37	31	44	37	32	37	36	40	42	39	39	34	42
- monofil + film	+	+	1	3	3	4	6	9	16	18	21	27	34	37
- staple + tow	+	1	1	3	5	6	7	7	7	Included with backing----				
Backing - film fiber						36	49	66	78	78	89	109	122	137
- mono + spun-bonded)	5	11	16	42	66	28	44	57	71	65[f]	68[f]	69[f]	75	77
- staple	1	3	2	9	13	6	15	18	20	16	10	9	11	16
Needle-punched - staple	7	23	31	32	34	33	24	30	29	26	20	19	28	30
TOTAL CARPETS & RUGS	31	75	82	133	158	145	182	223	261	245	239	272	304	339
Other Broad Woven & Spun-bonded Goods														
Upholstery & slipcovers	4	4	5	6	12	24	30	34	45	46	55	47	57	54
Bags - sand	+	+	12	44	7	+	3	4	3	+)	42	47	44	43
- other	1	1	2	3	5	7	11	22	40	58)	--	--	--	--
Other[a]	4	4	4	5	8	7	9	15	25	30	41	57	--	--
TOTAL OTHER BROAD WOVEN, ETC.	9	9	23	58	32	38	53	75	113	134	133	146	163	159
Narrow Woven - Total[b]	7	7	8	9	7	9	10	11	11	10	8	10	10	11
Rope & Cordage[c]														
Rope	10	11	12	18	20	17	21	25	26	30	34	40	38	39
Cordage under 3/16"														
Tying twine)							15	16	21	19	18	18	19	21
Harvest twine)	3	13	19	18	18	21	9	10	15	23	30	13	16	22
TOTAL ROPE & CORDAGE	13	24	31	36	38	38	45	51	62	72	82	71	73	82
Nonwovens[d] - staple	Included in Other Uses Below -------------------									7	18	22	31	29
All other uses[e] - total	3	3	4	5	6	9	9	9	8	8	6	9	31	39
TOTAL DOMESTIC SHIPMENTS	66	122	151	244	244	241	301	372	460	481	486	534	619	668

(Data from Textile Economics Bureau)

Footnotes to Table VI

a. *Includes grille cloth, filtration (wet and dry), tarps and pool covers, agricultural cloth, rubber liner, bale wrap, marine fabric, protective covers, tent and luggage liners, florist nursery squares, brattice cloth, and substrates. Also includes knit poundages for 1975-1978.*

b. *Chiefly outdoor furniture webs.*

c. *Includes braided, twisted, and filler + core.*

d. *Wet and dry laid and needled batts only.*

e. *Includes fishlines, nets and netting twine;pot cleaners and pads; vacuum cleaner hose covering; novelty and other braids; book binding and similar scrims, etc., as well as unknown end uses.*

f. *Includes face yarns, 1974-1978.*

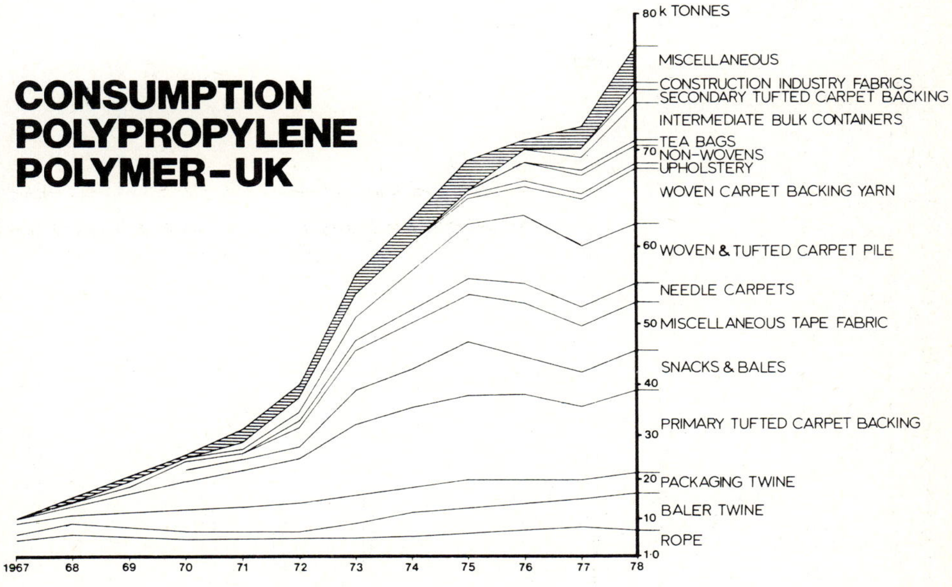

Fig. 1. Fibre Applications of Polypropylene in U.K. (Source: Shell Chemicals UK Ltd.)

TABLE VII

U.S. Olefin Yarn, Monofilament[a] & Film Fiber Shipments by Trade or End Use

	1978	1979
APPAREL		
Sliver Knit & Other	9.2	9.2
HOME FURNISHINGS		
Carpets & Rugs		
Face Yarn-Filament Yarn & Film Fiber	78.9	85.3
Primary & Secondary Backing	213.7	220.7
Drapery, Upholstery, & Other	45.1	39.9
Total Home Furnishings	337.7	345.9
INDUSTRIAL & OTHER CONSUMER		
Narrow Woven		
Outdoor Furniture Webs	6.7	7.5
Other Narrow Woven[b]	4.4	4.3
Rubber Industry	0.8	1.6
Medical, Surgical & Sanitary	19.0	23.8
Auto Seat Slipcovers	0.4	0.4
Filtration, Wet & Dry	2.3	2.5
Sewing Thread, Fishline & Nets	1.0	2.8
Rope & Cordage		
Rope-Braided	9.7	9.5
-Twisted	26.5	28.1
-Filler & Core	3.1	3.0
Tying Twine-Monofil	6.8	8.0
-Film Fiber	14.5	16.1
Harvest Twine-Mono & Film	21.8	30.1
Bags & Bagging		
Close Woven, Leno Mesh, & Knit	43.2	51.9
Bale Wrap	16.7	18.1
Tarps, Pool Covers, Etc.	0.5	2.6
Civil Engineering Fabrics[c] & Agricultural Cloth[d]	6.0	16.6
Cable Binding[e]	4.1	4.2
Macrame & Other Craft Yarns	18.5	18.5
Substrate Fabrics, Construction Fabrics, & Other Industrial & Consumer	26.8	19.5
Total Industrial & Other Consumer	232.8	269.1
DOMESTIC SHIPMENTS		
Filament Yarn	113.3	107.7
Monofilaments	183.6	209.2
Film Fiber	282.8	307.3
Total Domestic Shipments	579.7	624.2
EXPORT	15.2	21.1
TOTAL SHIPMENTS	594.9	645.3

a. Monofilament includes spunbonded polypropylene.
b. Other Narrow Fabric is mainly weatherstripping and auto windlace & trim.
c. Civil Engineering Fabrics includes soil erosion and stabilization cloth, roadbed and railbed stabilization fabric, drainage applications, etc.
d. Agricultural Cloth includes poultry and shade cloth. (Prior to 1979 also included soil erosion and stabilization cloth.)
e. Cable Binding includes colored identification fiber.

rate than in the past. Fig. 2 depicts the growth pattern of monofilament markets in the United States.

Multifilament: The multifilament market includes both continuous filament yarns (straight and textured) and staple. Straight filament is produced in a range of deniers from 60 to over 2,500. As mentioned earlier, shortly after its introduction as a commercial product, polypropylene fiber made an easy and rapid penetration into the fiber industry as monofilament. However, penetration into the much larger market of multifilament textile yarns did not occur until 1966. Multifilament yarns are now the largest dollar volume segment of the polypropylene fiber market and have had the largest growth during the last five years. The principle United States manufacturers of multifilament products are Hercules, Cheveron, Phillips, and Amoco. In Europe the dominant producer is Neofil of Italy, which is a joint venture of Hercules and Montedison. In Japan, Mitsubishi, Toyobo, Toyo Rayon, Toa-bo, Daiwan Bo, and Chisson, are important producers of polypropylene multifilament yarns and staple.

Polypropylene continuous multifilament yarns are used for industrial fabrics, sewing thread, and cordage. Polypropylene's unique chemical and stain resistance make it particularly valuable for such industrial uses as filter cloths, air receivers, gas filter fabrics, bagging, etc. But by far the largest and the fastest growing use of multifilament yarns is in carpets and upholstery fabrics. In 1981 about 100 million lbs of polypropylene filament yarns were used in tufted carpets in the United States alone. Similarly, upholstery uses for polypropylene filament yarns have grown from one million lbs in 1961 to 50 million lbs in 1981.

The polypropylene filament yarns used for carpets are in the form of bulked and textured yarns produced through several texturizing processes available to the synthetic fiber industry. The multifilament yarns used in upholstery are generally textured but some yarn is also used in untextured form. Consumer acceptance of polypropylene carpets was prompted by the fact that these carpets are easy to clean. They are practically free of static build-up, have good durability, and are colorfast to light. To upholstery markets, polypropylene fibers bring aesthetics (color and hand), performance (abrasion resistance, weatherability, and light fastness), cleanability, and economics.

The penetration of the apparel field by polypropylene multifilament yarns has, to date, remained limited in scope because of technical as well as economic reasons. The hesitation in introducing polypropylene fibers into apparel fabrics does not mean that the fiber has a doubtful future in this field. On the contrary, it is here that the future of polypropylene fiber must lie, once the dyeing and other developmental difficulties have been overcome. In Japan, for example, about half of the total production of multifilament yarns and staple is being used in wearing apparel and stockings. With the recent attempts at

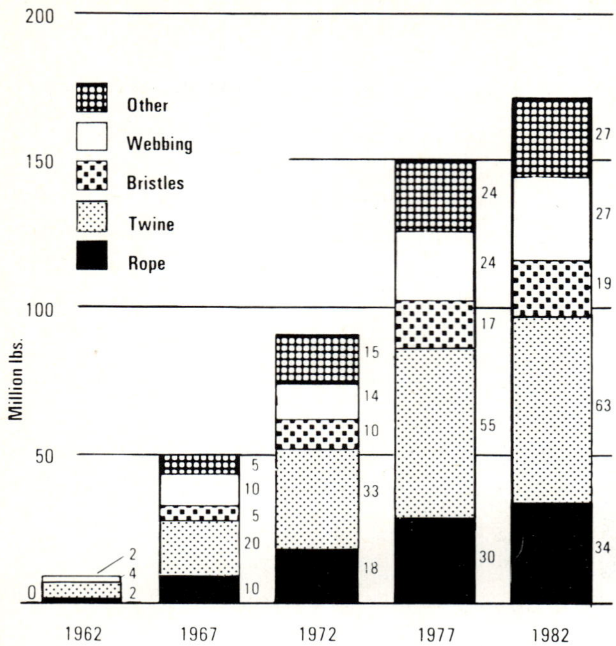

Fig. 2. Demand for U.S.-Produced Olefin Monofilament (ref.1)

commercialization of textured filament yarns by Mitsubishi, Toyobo, Hercules, and others, and the development of acid-dyeable fibers, the prospects for penetration of polypropylene filament yarns into the broad field of woven and knitted apparel goods seems promising. In fact, the U.S. market is already showing a hightened interest in polypropylene underwear and aesthetic garments. The wicking property of polypropylene fabrics is being sited as a selling point. The United States filament yarns market may increase to about 150 million lbs. by 1982 (Fig. 3). The main growth will occur in carpet and upholstery area.

Until recently, essentially all polypropylene staple sold in the United States was 15 to 20 denier pigmented staple used for needle-punched carpets. Most of this staple was highly light stabilized so that it could be used in outdoor carpets. In 1981 about 35 million lbs. of polypropylene staple was consumed in this market. The needle-punched carpet market has considerably declined in recent years in the United States but light weight wet or dry laid nonwoven fabrics for diaper lining, disposable garments, filter cloth, tea bags, etc., is believed to be an expansion area for polypropylene staple fiber. Geotextiles is another area of expansion. It is estimated that the United States staple market will grow to 142 million lbs. by 1985 (Table VIII). In Table IX is shown some additional areas where one feels that polypropylene staple has good growth potential in nonwovens.

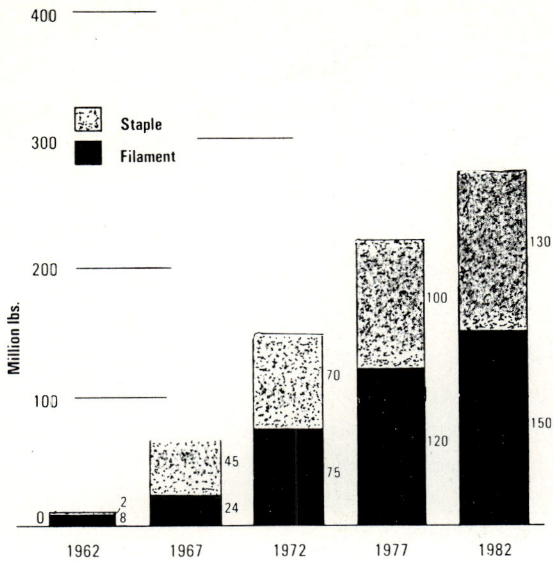

Fig. 3. Demand for U.S.-Produced Polypropylene Multifilament Yarns (Ref.1)

TABLE VIII

Major Uses for Polypropylene Staple in Dry-Laid Nonwovens (1979-1985)

End Use	1979 MM Lbs.	1985 MM Lbs.
Needle-punched Carpet	33	42
Geotextiles	10	40
Furniture Construction	7	10
Vinyl Coating Substrate	7	12
Other Uses	3	38
Total	60	142

Diaper cover stock holds very good potential for polypropylene staple. A number of companies now have latex binder systems that provide good adhesion properties with polypropylene at safe curing temperatures. Another approach to the use of polypropylene in diaper cover stock is to use a bi-component fiber, such as the Chisso ES material, which has one component with a lower melting point than the polypropylene component. Another approach is to calender the 100% nonwoven fabric, using patterned engravings on the calender, to provide an adequate number of binding points without appreciably stiffening the fabric.

TABLE IX

Additional Growing Uses for Polypropylene Staple in Nonwovens

Diaper Cover Stock
Filtration
Auto Seating and Trunk Liners
Oil-Spill Sorbents
Wet-Laid Applications

Filtration, particularly wet filtration, represents product opportunities for polypropylene because of the good chemical resistance of the fiber. This is another application where a self-bonding fiber system can be advantageous.

Greater use of polypropylene nonwovens is under consideration by the automobile companies because of favorable economics, high strength and the low specific gravity for seating and trunk lining uses. Staple nonwovens in these applications also compete with spunbonded products.

Some needle-punched polypropylene fabrics made of waste fibers have found use in oil spill cleanup products. There are opportunities to develop special products for oil and chemical spill cleanups, using polypropylene alone or in combination with other materials such as urethanes.

In the USA (as well as in western Europe) the use of polypropylene staple in carpet manufacturing rarely extends beyond its inclusion in a few down market constructions like low level loops, serolbs, and velours. Only 7 million lbs. of staple were used as carpet face yarn in the USA in 1980. The situation in the U.K. is, however, different. Here, polypropylene staple was taken at first as a replacement for an equally volume-priced man-made fiber, viscose rayon. Beyond these substitution periods, the consumption of polypropylene staple fiber has climbed steadily in a unique manner until today it is found in every type of construction from shags to berbers.

One of the reasons for this spread of interest is the legacy of a spinning industry that tufting inherited from the U.K.'s one-time vast concentration on Wilton and Axminster weaving. Some of the tufting pioneers were offshoots of those same weavers and naturally had access to yarn supplied by spinning mills within vertically integrated organizations. Later came greater availability as the jute spinners of Dundee diversified into semiworsted carpet yarns.

Most of the U.K.'s polypropylene staple goes to tufting, with a smaller proportion going to needlefelt. Polypropylene-wool and polypropylene-acrylic blends are more common than 100% polypropylene yarns.

Another promising area of use for polypropylene staple is in the woven and narrow gauge tufted upholstery fabrics. In 1981 about 20 million pounds of polypropylene staple was used in the U.S. upholstery market. This level of penetration is rather low considering a total market of 300 million lbs. staple. The same situation exists in the U.K. and Western Europe as well. The non-availability of good quality staple in fine deniers and attractive colors has mainly contributed to this poor showing so far. However, the picture is rapidly changing, especially in the United States.

In general, staple processes are much less complex because they require less rigid uniformity and rather laxed quality control than do continuous-filament yarns. Consequently, polypropylene staple sells for one-half the price of continuous filament with the same denier per filament. This factor, coupled with the lower wages involved in yarn preparation, is expected to greatly spur the use of polypropylene staple in developing countries of Asia and Latin America.

Tape and Fibrillated Film Yarns: Tape and fibrillated film yarns include all those products that are prepared by extrusion of film with subsequent slitting, splitting, or fibrillating. Tapes are practically all used in woven fabrics. Carpet backing is by far the largest volume use in the United States, bagging is important in Western Europe, and wool pack covers are finding extensive use in Australia. Its share in tarpaulin and coated fabrics and synthetic grass is small but slowly growing. Fig. 4 shows the pattern of growth for tape and fibrillated yarn in the United States market. Currently over 300 million lbs. of polypropylene tape is being woven into fabrics in the United States. Of this yardage about three-fourths is estimated to be used in carpet backing and the rest for industrial fabrics and bagging. The U.S. tape yarn market is expected to grow to about 350 million lbs. by 1985.

The growth of fibrillated film has been rather slow in the United States: it has a limited number of uses, rather deficient properties, complicated technology, and an expectedly high production cost. About 100 million lbs. of fibrillated film was sold in the United States in 1981. Of this quantity, probably 35 million lbs. were used for artificial turf and the balance for ropes, cordage, and secondary carpet backing. This volume is expected to grow.

The world market for tape and fibrillated yarn is estimated to be about 4 billion lbs. in 1981, of which probably 80 percent is tape and 20 percent fibrillated film. This market is expected to grow more in developing countries and Japan than in the United States and Western Europe.

In general, it can be said that flat-film yarns are being used for many jute applications, such as woven sacks and carpet-backings, whereas the fibrillated material is replacing sisal and manila hemp in such applications as rope, twine, and matting. This, however, is an oversimplification. In the first place, the volume of manufactured jute goods, on a world basis, has just about doubled over the past twelve years or so, and for some applications, particularly packaging, there is a shortage of natural fiber. The potential for film yarns,

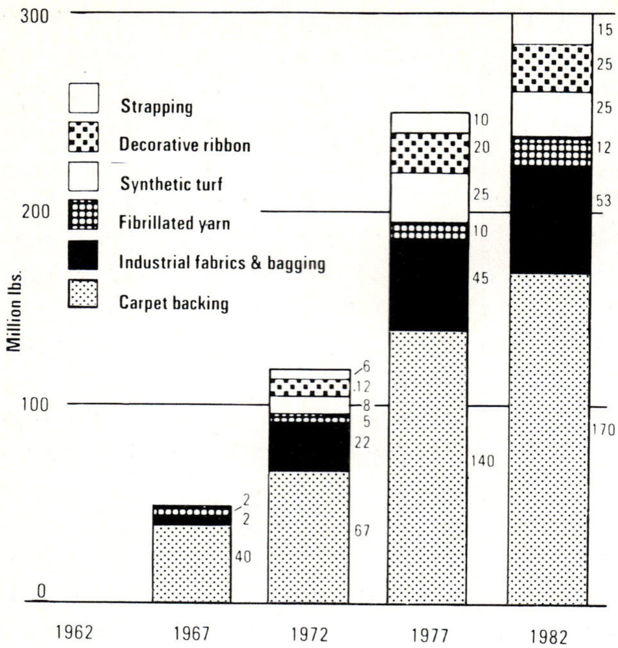

Fig. 4. Demand for U.S.-Produced Olefin Fiber (All Types) (ref.1)

therefore, can be seen to be much greater than that represented, for example, by the replacement in some market segments of jute sacks by woven-film sacks. Also, it is likely that sooner or later many applications for film yarns will be developed that will in part replace materials other than jute and sisal. According to a conservative estimate, the world market for these products may develop to over 4.5 billion lbs. in 1985.

TABLE X

Major Uses for Spunbonded Polypropylene (MM Lbs.)

End Use	1980	1985
Primary Carpet Backing	30	50
Furniture Construction	17	22
Geotextiles	9	24
Bed Components and Bedding	7	8
Other Uses	18	44
Total	81	148

Spun-bonded Fabrics and Related Products: About 100 million lbs. of polypropylene was used in 1981 in the United States for the production of spun-bonded fabrics. Table X shows some of the major uses for spun-bonded polypropylene products. By far the largest outlet is in carpet backings, followed by furniture construction. Increasing but still very small quantities of nonwovens are used as scrim, fabric liners, and bagging. A potentially large market is the secondary backing for carpets and soil stabilization fabrics for civil engineering applications.

Du Pont's "Typar" spun-bonded polypropylene is the preferred primary backing for printed carpeting because of its superior dimensional stability over woven ribbon yarn products. "Typar" is also preferred in carpets for use with fine gauge tufting equipment, where it provides a more precise pattern. The spun-bonded product causes less needle deflection than occurs when using woven ribbon yarn backings.

Furniture construction uses have become important for spun-bonded polypropylene products. Products used by the furniture industry are Du Pont's "Typar", Kimberly-Clark's "Accord" and Crown Zellerbach's "Celestra".

The three suppliers of spun-bonded polypropylene for geotextile use have all taken somewhat different approaches to producing the product. Du Pont's product is based on the technology that was used to develop "Typar" as a primary carpet backing. Celanese Fibers Marketing Company's "Mirafi 140S" is based on technology that was originally developed by I.C.I. in England with their "Heterofil" approach. Crown Zellerbach's "Celestra" manufacturing process is based on Lurgi technology, which they have modified. "Celestra" also incorporates meedlepunching into the process. Spun-bondeds have been used in a wide range of geotextile applications, from temporary road building for logging and mineral exploration to use in the Alaskan pipeline and for railroads.

Spun-bonded polypropylenes are used in dust covers, in coil spring pockets and spring insulators for mattress and box spring manufacturing for beds. Lightweight spun-bondeds have found extensive use as cover material for "Pinsonic" quilted mattress pads and for quilt backings for popular-priced bedding items.

In Table XI, we show additional uses for spundondeds with good growth potential.

TABLE XI

Additional Growing Uses for Polypropylene Spun-Bondeds

Auto Seating and Trunk Liners
Diaper Cover Stock
Hospital/Medical Uses
Specialized Packaging
Upholstery Fabric Backing

Several upholstery fabric manufacturers are now having a light-weight spun-bonded polypropylene (o.5 oz./sq. yd. or lower) applied to their fabrics to improve the dimensional stability and to lock in yarns. This technique also provides a very attractive appearance for the back and lends opacity to the overall fabric.

The markets for spun-bonded fabrics will grow considerably in Western Europe and to some extent in the United States. Here, an estimated 150 million lbs. of polypropylene will be used for the production of spun-bonded fabrics and related products in 1982 (Table XII).

TABLE XII

U. S. Consumption of Spun-Bonded Fabrics (ref. 1)

	(MILLION LBS.)
1962	0
1967	10
1972	60
1977	100
1982 (est)	150

The markets for polypropylene fiber can also be analyzed in terms of the major area of applications. These are:

(1) Apparel
(2) Home furnishing
(3) Ropes, cordage, and twine
(4) Bags and wraps
(5) Geotextiles
(6) Other industrial uses
(7) Speciality products

II.1. Apparel: As already mentioned, polypropylene fiber has not been able to penetrate apparel market to any significant extent so far. It was partly due to its non-dyeing characteristics and to its low melting temperature. But mostly it was due to the lack of marketing interest on the part of polypropylene fiber suppliers and to the extreme competiveness of polyester fibers. Nevertheless, about 10 million lbs. of polypropylene continuous filament yarns were used in 1981 for sliver knit backing in the USA and probably 1 million lbs. of staple for miscellaneous apparel uses. Interest and technical activities have been recently stimulated by development of ladies panty product. In addition, there has been real commercial progress in staple spun polypropylene use in thermal underwear, socks and atheletic outerwear.

II.2. Home Furnishing: Home furnishing is the one single most important application of polypropylene fiber today. About 60 percent of the total production is being presently consumed for the manufacture of carpets, upholstery fabrics, mattress ticking, bedspreads, and other decorative fabrics. In addition to these direct uses, polypropylene fabrics are finding increased uses in several indirect ways; the primary uses of them being bed components and bedding, furniture construction, upholstery fabric backing, and filling.

II.2.1. Floor Coverings: Floor coverings is the largest component of polypropylene consumption in home furnishing area. This market can be loosely discussed in terms of the following major groupings:

(i) Face yarns for tufted carpets
(ii) Primary backup for tufted carpets
(iii) Secondary backing for tufted carpets
(iv) Needle-punch carpets
(v) Face yarns for woven and knitted carpets
(vi) Chain and yarns for woven carpets

To better understand the U.S. carpet market, and by reference, that of Western Europe, and to more easily see why specific fibers are used, the market can be segmented into five basic individual markets. Table XIII is a summation of these five markets and their relative size.

TABLE XIII

U. S. Carpet Industry - Five Market Segments

Household	56.5%
Institutional Contract	18.0%
Low-End Contract	16.0%
Outdoor	4.5%
Transport (automotive)	5.0%
	100.0%

(a) Household: The household, or residential market, is a more fashion oriented market than any of the other segments. This emphasis on fashion is clearly shown by the fact that more preference is given to color and style when purchasing carpet than price and durability. For this reason nylon fibers have attained a major share of this market segment. Because of the large size of this market and the wide preference of styles, other fibers, particulary polypropylene, are used when price is a prime consideration.

(b) Institutional contract: The institutional contract market is generally a performance oreinted market where durability, ease of maintenance, soil hiding, and anti-static protection are prime considerations in selecting carpet. Historically, wool and acrylics have dominated this market. However, in recent years there has been a shift to the synthetic fibers, particularly BCF nylon. BCF nylon does have the majority share of this market, mainly because of its soil hiding and performance characteristics. In recent years BCF polypropylene has been used more in these areas in the fine gauge level loop pile tufted carpet

constructions. These pigmented BCF polypropylene products have offered advantages of inherently better light fastness which could not be obtained with piece dyed nylons.

Buildings which utilize this type of carpeting include office buildings, schools, hospitals, banks, churches, auditoriums, municipal buildings, shopping centers, retail stores, etc. Because of the nature of these buildings, carpets going into these types of buildings are generally sold through architects, specificers, government agencies, or building contractors.

(c) Low End Contract: The most important consideration in selecting carpet for this market segment is price and apparent value. Long term performance in many cases is not a prime consideration since this is a very cost sensitive market with the primary uses being in mobile homes, mass produced lower cost housing, lower priced apartments, retail stores, and motels. Because of its good bulking characteristics, BCF nylon has a large portion of this market. Additionally, polypropylene is used extensively because of its lower price in relation to BCF nylon.

(d) Outdoor Carpets: The outdoor market is comprised almost exclusively of the tufted artificial grass type carpets which utilize ribbon polypropylene yarns. A small portion of this market, the athletic sports fields, utilize nylon products. However, this is a diminishing market which has almost completely disappeared. The decorative grass area has virtually all of the market share. The most important criteria for polypropylene yarns in this area is light fastness performance properties. Most fibers used in this area are guaranteed by the fiber company for a minimum of three years, and in many cases, five and even seven years against ultra-violet light degradation.

(e) Transport Market: The majority of this market is automotive carpets which is dominated by a relatively small number of consumers; especially, General Motors, Ford, and Chrysler. In this market, price and strict performance specifications are the prime criteria for carpet. The automobile manufacturers continually demand greater performance factors such as wear, color fastness, etc., at a minimum of price increase. There is more performance testing in this market than any other of the USA carpet market segments. Currently, this market uses mostly staple nylon tufted products which give the performance criteria required by the automotive manufacturers at low weight constructions in the desired cut pile plush styling. Auto industry is now showing increasing interest in polypropylene carpets.

Table XIV summarizes the data for the consumption of various fibers as carpet face yarn in the United States. As is evident from this table, the share of polypropylene as carpet face yarn has been steadily increasing over the last 10 years while that of several other fibers has been decreasing. At present polypropylene commands about 6 percent share of the market. This share is ex-

TABLE XIV

U.S.A. Carpet Industry Face Yarns Consumed ($\overline{M}$ of Pounds)

FIBER	1971	1972	1973	1974	1975	1976	1977	1978
Cotton	13.5	12.0	10.9	3.6	3.0	3.0	1.7	1.0
Rayon-Acetate	73.8	75.4	54.2	23.3	19.1	12.0	7.5	5.7
Wool	74.3	68.8	40.0	10.1	15.0	13.2	13.5	14.1
Nylon: Filament	348.8	480.9	579.1	567.5	495.5	539.5	624.2	683.7
Nylon: Spun	210.3	337.5	417.2	370.4	371.8	501.5	590.0	591.8
Acrylic/Modacrylic	150.4	147.1	140.1	127.3	93.6	84.0	77.2	69.0
Polyester	184.7	185.3	187.0	158.2	150.2	185.5	184.3	184.3
Polypropylene	53.7	66.8	83.7	69.8	66.0	57.6	73.0	92.2
All Others	7.6	5.0	3.8	3.1	2.6	2.7	7.4	8.4
TOTAL	1,117.1	1,378.8	1,516.0	1,333.3	1,216.8	1,399.0	1,578.8	1,650.2

pected to grow to about 10 percent by 1985, claiming the second position after nylon which will continue to be the dominant fiber in this market. This forecast takes into account the rapidly increasing cost of nylon raw material, decreasing weights of carpets, and increasing innovations in polypropylene yarns.

TABLE XV

Worldwide Production of Textile Floorcoverings in Percentage

	1974	1975	1976	1977	1978	1980
Tufting	73.0	73.0	73.0	74.0	74.0	75.0
Needled	14.0	13.5	13.5	12.5	12.0	11.0
Woven	10.0	10.0	10.0	9.5	9.0	10.5
Others	3.0	3.5	3.5	4.0	5.0	3.5

i) Face yarns for tufted carpets: Table XVI is a summary of the world carpet production in percentages broken down according to method of production. Table XIV presents similar data for the U.S. market. As can be seen from these tables, tufted carpets represent by far the major portion of the U.S.A. carpet market as well as on the worldwide basis. Tufted carpets are expected to maintain this dominate share of the market, while the wovens and other methods of manufacturing continue to be used for specialized products for specific market segments. These areas will probably continue to show slight percentage decreases of the total share. Total face yarn shipments for tufted carpets (all fibers) over the last five years were as shown in Table XVII.

As can be seen, 1980 and 1981 were very depressed years. This lag in carpets is attributed to the poor economy, most particularly the continued drop in housing starts and the very high interest rates.

Prior to 1977, the average consumption of fiber for tufted carpets increased about 90 million pounds per year, representing as average growth rate of about five percent per year as of 1979. Under the current economic conditions, it is difficult to project the future for the carpet business, but it is believed that since carpet is a deferrable, high ticket item, a strong recovery will develop when economic conditions are more settled. In its favor is the fact that no other product threatens the use of carpet as the dominant floor covering product.

Based on the above, it is projected that the consumption of face yarns for tufted carpets will grow to almost 2,500 million pounds by 1985. If anything, this is conservative considering the 1979 consumption was almost 2,000 million pounds. In 1980, BCF carpet yarn accounted for 50% of all the face fiber, spun yarns accounting for the other half. This surge in BCF percentage was probably due to the fact that residential sales were very depressed and the use of spun

TABLE XVI

U.S.A. Shipments of Carpets and Rugs (M of Square Yards)

	1971	1972	1973	1974	1975	1976	1977	1978*
Grand Total	755.1	943.0	1,025.4	939.1	834.0	939.3	1,024.6	1,128.0
Woven - Total	26.2	26.7	24.7	23.2	19.0	18.1	15.7	18.7
Axminster	3.8	3.3	3.2	2.8	2.3	2.2	1.3	1.5
Wilton	2.7	2.8	2.6	1.9	1.3	.9	.7	.8
Velvet	17.1	18.3	16.6	15.5	12.3	11.6	11.9	14.1
Other Woven	2.5	2.3	2.2	3.0	3.2	3.4	1.9	2.3
Tufted - Total	680.9	856.9	944.3	866.0	783.7	890.3	977.1	1,074.7
Scatter rugs and sets (rugs 6' x 9' and smaller)	80.4	78.2	83.2	71.0	64.3	66.0	76.0	81.0
Roll goods and rugs larger than 6' x 9'	558.6	734.3	809.4	756.4	676.7	762.2	834.8	931.6
Auto and Aero	42.0	44.4	51.6	38.6	42.7	62.1	66.4	59.9
Other Total	48.0	59.3	56.4	50.0	31.4	31.0	31.7	34.6
Knitted	5.2	7.4	8.1	9.3	7.7	9.6	9.3	7.4
Needlepunch	40.0	48.6	44.4	37.1	20.1	17.1	16.8	21.5
Braided, Hooked and Other N.E.C.	2.8	3.4	3.9	3.4	3.6	4.3	5.6	5.7

*Estimated

TABLE XVII

U.S. Shipments of Face Yarns for Tufted Carpets

	MILLION LBS.
1977	1,706
1978	1,762
1979	1,949
1980	1,617
1981	1,750

yarns for Saxony construction was sharply reduced. As is generally known, BCF nylon and polypropylene yarn dominate the level loop type of carpet used extensively by the commercial/contract segment of the market.

Normally BCF accounts for 40 to 42% of the yarns used in carpet and in this area, nylon and polypropylene are in direct competition, particularly in level loop construction. Nylon is more versatile since it can be tak dyed, space dyed, printed or in some cases piece dyed. Polypropylene is cheaper and the difference in price has been widening in favor of polypropylene.

TABLE XVIII

U.S. Polypropylene Fiber Consumption

Carpet and Rug Face Fibers (M of Pounds)

YEAR	BCF	RIBBON	STAPLE	TOTAL
1971	26.7	3	24.0	53.7
1972	27.0	7	32.8	66.8
1973	32.2	12	39.5	83.7
1974	24.0	16	29.8	69.8
1975	18.7	22	25.2	65.9
1976	11.2	28	16.8	57.6
1977	19.3	38	15.7	73.0
1978	24.2	46	19.8	90.0
1979	26.4	48	17.6	92.0
1980	29.9	52	16.1	98.0

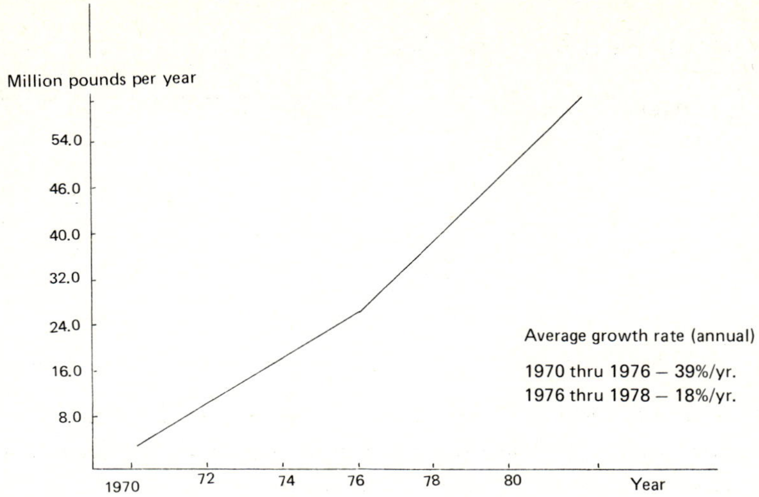

Fig. 5. Total U.S. Consumption of Polypropylene Yarn for Grass Carpeting (ref. 11)

Table XVIII gives the consumption of polypropylene carpet face yarns in the USA. In the early 1970s, polypropylene BCF yarns experienced very good growth in the United States. However, it decreased somewhat during the mid 1970s. This was mainly due to the fact that at that time polypropylene was generally offered only in solid colors and its styling capabilities were limited. During 1976-1977, the polypropylene fiber manufacturers made concerted efforts to improve the styling capabilities of polypropylene by offering space-dyed-look, heather look, and spun-look yarns. Because of those developments, polypropylene BCF yarn has begun increasing its market share again, and is expected to continue this growth on into the early 1980s. At present about 30 million pounds of polypropylene BCF yarn is being consumed in the U.S. carpet trade. By 1985, this market should consume about 60 million pounds.

In Western Europe and the U.K., the situation is somewhat different. Here, polypropylene staple is being widely used in tufted and woven carpets, mostly in blends with wool and acrylics (Table XIX). The U.K. carpet industry has been particularly very inventive in this regard.

Presently, polypropylene's share in the carpet market seems to have been stabilized at about 6 percent in the United States and 10 to 20 percent in Western Europe (Table XX). The reasons for a comparatively restrained growth for polypropylene face yarns in carpets in competition to nylons are many. First, the price of nylon BCF carpet yarns and staples declined considerably during 1970-1971, leaving little incentive for tufters to use polypropylene. Second, polypropylene carpet yarns remained confined to the low end of the market, thus giving it a poor image in the consumer's eye. Third, polypropylene carpet yarns

TABLE XIX

Production of Tufted Carpets from Polypropylene Staple (Million sq.m.)

Western Europe	1979	1981	1983
West Germany	3.5	4.2	4.5
Netherlands	1.1	1.3	1.4
Belgium	4.8	6.0	6.8
France	0.8	0.7	0.7
Italy	0.8	1.1	1.2
Great Britian	10.0	12.0	13.0
Denmark	3.3	2.5	2.3
Switzerland	0.3	0.3	0.3
Austria	0.5	0.6	0.7
Spain	0.5	1.0	1.2
Scandinavia and Others	1.5	2.0	2.2
Total Western Europe	27.1	31.7	34.3
U.S.A.	3.0	2.8	3.1

TABLE XX

Share of Polypropylene as Face Fiber in Tufted Carpets (in percentage)

	1975	1976	1977	1978	1979
Germany	4	5	6	6	7
Netherlands	6	9	11	10	10
Belgium	6	7	8	9	10
France	10	8	10	9	9
Italy	18	20	22	25	28
Great Britain	2	3	4	11	14
Denmark	10	12	12	12	11
Sweden	7	9	10	11	11
Switzerland	4	5	5	4	5
Austria	8	9	10	10	10
United States	5	5	6	6	6

were marketed by the major producers as cheap yarns, thus by implication excluding their use in quality carpets. In Europe, several small producers tried to introduce fibrillated yarns to the carpet trade as carpet face fiber. The performance and aesthetics of these carpets were very poor, contributing to a bad image for polypropylene fiber. It appears that the damage done to the image of polypropylene carpet is permanent in the United States. Until nylon prices rise considerably higher than polypropylene fiber prices, the chance of regaining a greater share of the carpet market is remote for polypropylene. In Europe, however, there is still some enthusiasm for polypropylene carpeting, and this may lead to its substantial growth in the carpet sector.

One special market for tufted polypropylene yarns is in synthetic turf, in the United States this market has shown the most dynamic growth for polypropylene, growing from approximately 3 million pounds of fiber in 1971 to 50 million pounds in 1980. Most of the growth in synthetic grass carpets has occurred at the expense of needle-punch indoor/outdoor carpets. Carpet manufacturers and fiber producers are predicting a growth rate of 18 to 25 percent a year in the United States and probably 15 percent in Western Europe through 1980s. Fig. 5 presents the total United States consumption of fibrillated yarns to make grass carpeting since 1970. Projecting these figures through 1980s, the annual growth rate is estimated at 24 percent. It is reasonable to say that, the consumption of polypropylene yarn for grass carpeting will reach 72 million pounds by 1985.

From the traditional and predictable residential applications - - terraces, balconies, around swimming pools - - the product has moved in many directions. One important category is in recreational playing surfaces. A tough, sturdy product in various face weights and pile constructions is required. In the commercial market, grass is going on public walkways, highway medians, public areas in restaurants, hotels, motels, condominiums, country clubs, and as landscaping around office buildings. Low maintenance cost and permanent green are among the reasons for replacing natural grass with synthetic turf. Grass has moved from summer resorts to ski and winter resorts and to such places as the decks of cruise ships.

ii) Primary backing for tufted carpets: On a world wide basis, about 80 percent of all tufted carpets have a polypropylene primary backing -- 80 to 85 percent woven and 15 to 20 percent non-wovens. The share of tufted carpets with polypropylene woven and non-woven primary backing is presently approaching the 90 percent mark in the United States and Western Europe. It will increase possibly to 95 percent by 1985. Table XXI and XXII show the estimated growth of woven polypropylene will increase while in others the share of non-woven polypropylene will show a gain. In Western Europe woven fabrics are more commonly used but in the United States a considerable market has been captured by non-woven fabrics (35 million lbs. in 1980). However, the use of woven fabrics is

TABLE XXI

Production of Tufted Carpets with Polypropylene Primary Backing (million sq.m.)

Western Europe	1979	1981	1983
West Germany	76.0	72.0	78.0
Netherlands	33.2	35.0	36.5
Belgium	90.0	87.0	90.0
France	19.0	19.5	22.0
Italy	8.0	10.0	12.5
Great Britian	100.0	101.0	108.0
Denmark	19.9	21.0	22.5
Switzerland	3.0	3.0	3.5
Austria	8.6	8.8	9.5
Spain	6.5	5.1	7.0
Others	18.7	21.2	23.5
Total Western Europe	380.9	383.6	413.0
U.S.A.	603.0	658.0	721.0

TABLE XXII

Production of Tufted Carpets with Non-Woven Polypropylene Primary Backing (million sq.m)

Western Europe	1979	1981	1983
West Germany	27.0	30.0	32.0
Netherlands	3.6	4.3	4.7
Belgium	14.0	10.9	22.0
France	3.0	4.0	4.7
Italy	2.0	2.5	2.8
Great Britian	18.0	22.0	23.7
Others	5.3	6.5	7.4
Total Western Europe	72.9	88.3	97.3
U.S.A.	72.0	77.0	81.0

still by far the most common. Table XXIII shows the total consumption of polypropylene in primary backing.

iii) Secondary backing for tufted carpets: The market potential for polyprolene fiber in secondary backing is presently quite interesting. This is the largest growth market for polypropylene in textile floor coverings within the

TABLE XXIII

Consumption of Polypropylene in Carpet Backing (million lbs.)

	1967	1970	1973	1974	1975	1976	1978	1980
United States	3.5	22.5	123.5	131.5	120.0	135.0	177.0	220.0
Western Europe	1.0	13.5	54.5	64.5	55.0	78.0	91.0	110.0

TABLE XXIV

Production of Tufted Carpets with Woven Polypropylene Secondary Backing (million sq.m.)

Western Europe	1979	1981	1983
West Germany	2.0	7.0	13.5
Netherlands	0.8	3.7	7.5
Belgium	1.5	4.5	9.2
France	0.5	1.2	2.3
Italy	0.1	0.5	1.3
Great Britian	14.0	10.9	27.4
Spain	2.2	1.9	2.6
Others	1.2	3.5	8.0
Total Western Europe	22.3	41.3	71.8
U.S.A.	151.0	395.0	461.0

next decade. In order to analyze this potential market, one has to differentiate between the developments in Western Europe, Great Britain, and the United States (Table XXIII).

Until recently, more than 98 percent tufted carpets had foam as secondary backing in Western Europe. Only lately, a certain switch towards fabric secondary backings can be noted. It is expected that by 1983 about 10 percent of tufted carpets will have secondary backing, most of it woven from polypropylene.

In Great Britain, the share of foam backed carpets has been traditionally lower than that in the Western European continent. Jute had historically a total market. Lately, however, woven polypropylene secondary backing gained significantly. In the United States consumer historically demanded a textile or jute like secondary backing. Foam has never been really a favorite item. Jute still dominates the secondary backing market and is expected to maintain this dominant position, at least during the next five year period.

In the secondary backing area, jute has not been affected by the man-made products as greatly as in the primary backing area due to several factors:

--Man-made backing producers did not actively attempt to replace jute until 1973,

TABLE XXV

Consumption of Secondary Backing in the United States (million sq. yards)

Jute	648	63	592	63	509	61	563	60	573	56	575	51	545	46
Foam/Sponge	265	26	225	24	192	23	197	21	195	19	203	18	296	25
Precoat Only	92	9	94	10	92	11	131	14	174	17	226	20	178	15
Woven Man-Made	10	1	19	2	25	3	38	4	82	8	124	11	166	14
Non-Woven Man-Made	10	1	9	1	16	2	9	1	0	0	0	0	0	0
TOTALS	1,025	100%	939	100%	834	100%	938	100%	1,024	100%	1,128	100%	1,185	100%

partly due to the low prices of jute and partly due to technical difficulties in producing a comparative fabric.

--The technical problem of uniform quality is not critical in the secondary backing market as it is with primary backings.

--The installer and ultimate consumer are more familiar with and accustomed to the appearance of jute.

--There is a tradition in using jute secondary. Jute has an image of quality with the ultimate consumer who is very difficult and expensive to reach and convince to change. (In the case of primary backing, the consumer in most cases was not aware that a new product was being substituted for jute.)

--Jute suppliers have maintained relatively low prices and carpet manufacturers have switched to light weight jute materials in order to maintain price advantages over the man-made secondary backings.

Table XXV shows the secondary backing market for tufted carpets in the United States for the years of 1973 through 1980. Since its introduction in 1973, the woven polypropylene secondary backings have shown a steady growth pattern. It is expected that these backings will continue to take away from jute's share of the secondary backing market. However, it will be at a much slower rate than what happened in the primary market.

The non-woven polypropylene secondary backings were used in the early 1970s on a limited basis but were quickly phased out of the market due to a lack of consumer acceptance as well as technical problems experienced by the carpet manufacturers.

At the present time the largest producer of synthetic secondary backing is the Patchogue-Plymouth Division of Amoco Fabrics. The product offered by Amoco utilizes a ribbon polypropylene warp yarn and a spun polypropylene filling yarn. The only other producer in the United States market is Wayne-Tex, Inc. (formerly Thiokol), which offer a material made from 100 percent fibrillated ribbon polypropylene. Based on the current volume levels of 1981, this represents approximately 15 million pounds of ribbon polypropylene yarn and 25 million pounds of staple fiber. Table XXVI summarizes the growth for each of these fiber forms in the primary and secondary backing.

iv) Needle-punched carpets: Polypropylene staple found extensive use in needle-punched nonwoven carpeting in the mid to late 1960s. The fiber's high strength and abrasion resistance, coupled with pigments with good light fastness and the appropriate light stabilizers, have enabled this type of carpeting to become a low cost utilitarian floor covering. Much of the needle-punched carpeting is sold to the "do-it-yourself householder" who will install it in his finished basement, family room or screened porch. The product is also sold in square form with a pressure-sensitive backing for simplified installation purposes. The 1980 market for polypropylene staple used in needle-punched carpets was

TABLE XXVI

Consumption of Polypropylene Fiber for Primary and Secondary Backing (million lbs.)

Year	Non-Woven Primary (Staple)	Woven Primary (Ribbon)	Woven Secondary Ribbon	Woven Secondary Staple	Total
1973	47.5	103.8	--	--	151.3
1974	39.0	117.4	--	--	156.3
1975	29.3	111.6	1.5	2.3	144.7
1976	26.3	129.1	2.9	4.5	162.8
1977	25.0	139.0	6.1	9.2	179.3
1978	22.5	169.2	10.2	15.3	217.2
1979	20.0	176.3	11.7	17.6	225.6
1980	19.5	190.0	13.6	20.4	243.5

estimeated as 33 million lbs. The growth in this sector will be slower. By 1985 about 42 million lbs. of polypropylene staple is expected to be used in this market.

v) Face yarns for woven carpets: Some quantities of polypropylene BCF yarn are being used in Italy, Belgium and the U.S. and Canada in the manufacture of Wilton carpets. Small quantities of staple, mixed with wool and nylon, are also being used in the U.K.Statistics are not available for this market. However, this market size should not be more than 10 million lbs/year in total.

vi) Backing for woven carpets: An estimated quantity of 3 to 5 million lbs/year of polypropylene yarns are being used as chain and stuffer yarns in the manufacture of woven carpets in the United States. About 5 to 8 million lbs/year are probably being used in the United Kingdom and Western Europe. This is relatively a small market, the quality specifications are rigid and the change over from jute and cotton is rather slow. Therefore, only a modest growth can be anticipated in this market for polypropylene yarns, which are generally fibrillated, twisted, heat-set yarns.

II.2.2. Upholstery Fabrics: Historically, the upholstery market has been one of the more successful marketing areas for polypropylene filament fiber. In this area it has been readily accepted by the consumer as being a good quality fiber with excellent cleaning and anti-soil characteristics. This is mainly due to the marketing efforts of Hercules Inc. As with carpets, polypropylene in upholstery market has gone through a product evolution, including improvements in performance and aesthetics. Better texturizing systems, different fiber cross sections, new colors, changes in luster, and additional deniers. These developments have contributed heavily to the success of polypropylene in upholstery segment.

TABLE XXVII

U.S. Consumption of Polypropylene in Upholstery Trade (million lbs.)

Year	Multifilament	Staple	Total
1971	22.7	5.9	28.6
1972	29.6	4.1	33.7
1973	36.9	6.0	42.9
1974	40.5	4.3	44.8
1975	49.3	5.6	54.9
1976	41.0	3.6	44.3
1977	52.0	5.5	57.5
1978	57.0	6.0	63.0
1979	58.0	6.0	64.0
1980	62.0	6.5	68.5

Table XXVII summarizes polypropylene's growth in upholstery market since 1971. As can be seen from this data, multifilament polypropylene yarns have enjoyed the most of success in this market. Less than 600,000 pounds of polypropylene filament yarn was used in the United States for upholstery in 1967 when it was first introduced by Vectra. By 1981 this use exceeded 70 million pounds. This gives a fairly impressive growth rate of 43 percent per year, compounded through 1975, and 25 percent per year, compounded through 1981.

There are no reliable statistics published giving the total upholstery market, but it is believed that polypropylene has about 30 percent of the total market in the United States. Because of the availability of several new yarns, polypropylene fiber is now moving into fabric constructions where it was not used before, and it is expected that the polypropylene share of the market will grow substantially in the coming years.

The entry of polypropylene into the upholstery market was initially a case where properties dominated and aesthetics were secondary. The fabrics were functional but rather plain. However, stain resistance, ease of cleaning, and abrasion resistance made polypropylene an ideal fiber for upholstery. No less significant was the low density, which really comes into play in flat-woven goods such as upholstery, with the resultant favorable economics. With time, however, new fiber variations were introduced. The introduction of BCF and air-textured (Taslan) yarns was a great leap forward. The recently introduced false-twist textured yarns for woven upholstery and low denier staple fibers for plush and velour fabrics has made polypropylene an all-purpose fiber in the upholstery area.

It is estimated that polypropylene fiber will dominate in this end use, and that by 1985, polypropylene fiber will have captured 50 percent of the total upholstery market.

The merchandising of polypropylene in upholstery fabrics has been a key factor in its growth and acceptance by the furniture manufacturers and upholstery fabric producers. While Hercules has been the leader in merchandising activity, the presence of Vectra Company (now Cheveron) and other smaller producers has enabled healthy competition to prevail and has accelerated the growth of the total market for polypropylene.

Multifilament yarns have so far dominated in upholstery area. The use of polypropylene staple fiber in upholstery has been much smaller. There are few spinners that can handle pigmented fibers. Those that do, generally process only rayon and do not want to risk contamination of the mill with another fiber. However, the use of staple spun polypropylene yarns in plush and velvet fabrics has been growing rapidly in recent years.

Table XXVIII gives an estimated breakdown of the fabric types used for upholstery. Table XXIX gives the estimates of market share of major fibers in the United States upholstery market. It seems that polypropylene and polyester fibers will increase their share at the expense of acetate, rayon, cotton, and wool. Polypropylene is expected to gain both in filament and staple spun fabrics.

TABLE XXVIII

Estimated Use of Uncoated Upholstery Fabric Types (USA)

	Percent of all fabric shipped	
Type Fabric	1975	1978
Velvets	33	28
Flat-woven jacquard	24	27
Novelty types	17	17
Prints	15	18
Plain	9	7
Miscellaneous	2	3

II.2.3. Mattress ticking, bedspreads, curtains, draperies, and casements: The mattress ticking business uses about 70 million pounds of fiber per year. Typical woven fabrics are dobby and Jacquards with combinations of rayon, acetate, and polyester. Cotton's total share of the mattress ticking business has fallen below 20%.

The only polypropylene used in mattress tickings in the USA is in the form of spun-bondeds from Kimberly-Clark. The only potential for polypropylene multifilament yarns is in 80 denier or finer yarns used in warp kints.

TABLE XXIX

U.S. Fiber Shipment for Uncoated Upholstery

	1974		1977	
Staple Fibers	% of Total	Millions of lbs	% of Total	Millions of lbs
Rayon	27.5%	75.6	24.2%	63.4
Cotton	19.1	52.5	19.5	51.1
Nylon	10.4	28.6	9.7	25.4
Polypropylene	2.5	6.9	3.5	9.2
Acrylics	0.9	2.5	1.6	4.2
Wool	0.8	2.2	0.8	2.1
Polyester	1.0	2.8	1.4	2.7
Total Staple	62.2%	171.1	60.7%	158.1
Filament Fibers				
Acetate	0.4%	1.4	0.4%	1.1
Rayon	12.5	34.4	8.5	22.3
Nylon	10.3	28.3	10.7	28.0
Polypropylene	13.1	36.0	16.2	42.4
Polyester	1.5	____	3.5	____
Total filament	37.8%	104.2	39.3%	103.0
Total fibers	100.0%	275.3	100.0%	261.1

Bedspreads consume about 140 million pounds of fiber per year. About 50% of this market is still cotton. Many of the fabrics are printed and are coordinated with curtains. Yarn-dyed fabrics are a relatively small percentage of the business. It is felt that any potential for polypropylene multifilament yarn in bedspreads would be very limited. Any potential use would require colored yarns in the upholstery denier ranges.

Draperies are hung on either side of a window but can usually be drawn across the window where they block vision and much or all of the light. Draperies are made with pinch pleats and this is sometimes regarded as the important distinction between curtains and draperies.

Curtains are ordinarily drawn across the window, although they can be pulled back or tied back at the sides. They are typically made with rod pockets or rings rather than pinch pleats. The fabrics used usually admit light, yet permit some degree of vision. Cafes, shorty drapes and panels are all included in the curtain category.

Casements are a cross between curtains and draperies. They replace one or both. The fabrics used permit some passage of light. Vision may be partly or wholly obscured. The fabrics employed include some sheers, but many are made with coarse filling yarns.

About 65 MM pounds of all fibers were used for curtains, draperies and casements in 1981 in the United States. Because of its low resistence one sees no opportunity for polypropylene in curtains. There may be some potential for draperies and casements but so far none has been realized.

II.3. Ropes, cordage, and twine: Table XXX summarizes the data on the consumption of polypropylene yarns (fibrillated film, monofilament and multifilament) for twine, cordage, and ropes. In 1978 Western Europe consumed almost twice the amount of polypropylene that the United States consumed for this market. This has been a rapidly growing area of consumption for polypropylene yarns but the future growth rate seems limited.

TABLE XXX

Consumption of Polypropylene Yarns for Cordage and Ropes

	(million lbs.)				
	1967	1970	1973	1976	1979
United States	17.5	23.5	31.5	40.1	40.6
Western Europe	17.3	34.8	71.0	89.0	92.0

Several factors have contributed to this marked penetration:

One, manila hemp is not as much of a competitor as it once was. The fiber, once purchased solely as a cordage product, is being used more and more as a pulp material and, in some special rope applications such as mooring lines, for naval operations.

Also, polypropylene rope has a number of advantages over its hard fiber competitors:

- o Superior Strength -- 50 percent stronger than sisal or manila ropes on a size-for-size basis.
- o Long Lasting -- Rot and mildew proof.
- o Floats on Water -- Absorbs very little water and makes handling large hawsers easier.
- o High Visibility -- Pigmented colors allow bright colors or color combinations.
- o Easy to Handle -- Clean and smooth to handle; easy to splice and thermally seal cut ends.
- o Economical -- Comparable in price to manila and less expensive than nylon or polyester.

With these factors taken together it is not difficult to see why polypropylene is gaining volume from sisal, rayon, polyester and nylon and at the same time losing volume to synthetic strappings. This loss in volume to strappings is particularly evident in the industrial tying-twine segment. There is no specific data on the volume of strappings consumed in this market segment. Some industry sources place this volume in the 30 million pounds per year for the United States, others as high as 90 million pounds per year. This variance in strappings volume makes it difficult to get a complete picture of polypropylene's impact on this market but the inroads made against sisal demonstrate polypropylene's capabilities here.

Agricultural twine market is one of the major areas of interest for polypropylene. Harvest twine is used in the bailing of hay and straw and as such its main markets are the temperate zones, predominately of North America and Europe, but also of Australia, Japan and South Africa. Baler twine largely superseded binder twine in the major countries on the introduction of the pick-up baler in the 1940s, although the latter is still used in some areas, notably Poland. Hay baling for fodder accounts for the great majority of the totals requirement in North America, but straw baling (both wheat and barley) is relatively important in some areas of Western Europe.

According to the latest available statistics the total world demand for agricultural twines, both baler and binder, is now in the range of 380,000 to 405,000 tons. The size of each market in any season naturally varies according to the stock carry-over situation and, most important, the weather conditions prevailing, but normal demand, the approximate share currently held by sisal and polypropylene can be estimated as indicated in Table XXVII.

From Table XXXI it can be seen that polypropylene twine has an overall share of about 32 percent of the total market but within the territorial lands there are considerable variations. An earlier F.A.O. estimate was that polypropylene twine would have had an average 40 to 50 percent share of the market by 1978, but competitive, and at times even uneconomic, sisal twine prices over the past few years have probably been largely responsible for preventing this.

In the United States and Canada polypropylene twine is more popular and enjoys a premium. Polypropylene twine has also made important inroads into the market previously held by wire baling -- especially in California, where large 150 lb. bales are produced for long-distance transportation.

In the United Kingdom while the heavier twine traditionally used has assisted polypropylene twine vis-a-vis the weight/runnage factor, it is probably true to say that there has been more active promotion of the synthetic twine than elsewhere in Western Europe. Polypropylene's share of the market in France (easily the largest European user) has increased considerably over the past five years but now appears to have stabilized. Poland is by far the largest market in Eastern Europe and sisal is used practically exclusively. On the other hand, in

TABLE XXXI

Relative Share of Polypropylene and Sisal in Agricultural Twine (tons)

Region	Total Market	Sisal	Polypropylene
USA	110 - 115,000	82 - 87,000	28,000
Canada	32 - 35,000	20 - 22,000	12 - 13,000
E.E.C.	160 - 165,000	105 - 110,000	55,000
Other Western Europe	18 - 20,000	12 - 14,000	6,000
Other Developed Countries	25 - 30,000	14 - 17,000	11 - 13,000
Eastern Europe	35 - 40,000	25 - 28,000	10 - 12,000
Total	380 - 405,000	258 - 278,000	122 - 127,000

smaller markets of Hungary and Czechoslovakia, polypropylene is dominant. The overall world market for agricultural twines has fallen considerably over the past five or six years due mainly to the introduction of new harvesting methods using less twine, or in some cases, dispensing with twine altogether.

In Europe the most significant impact has been made by various silage methods of fodder preservation -- the great advantage being reduction in the farm labor force. The twine market, particularly in North America, has also suffered from the introduction of chopper-blower-stacker system requiring no twine, and the "big-round" or "jumbo" balers using about half the twine normally required for baling a given weight of hay. Recently, some publicity has been given to "All crop" methods of harvesting but it is too early for these to have made any real impact. The data presented in Table XXXII for EEC can serve as an indication of the declining demand.

It will be seen from Table XXXII that over this period the overall market has decreased by approximately 700,000 tons (over 30 percent) athough the market for polypropylene twine has increased substantially over the same period. Polypropylene twine has been relatively more successful in capturing the straw baling market than the hay baling market. One estimate is that within EEC straw baling now represents 40 percent and hay baling 60 percent, whereas five years ago the percentage were respectively 25 percent and 75 percent. This is largely the result of hay baling being lost to silage, and any increase in straw baling over the period has not compensated for this.

The evidence would seem to point to agricultural twine requirements falling still further over the next few years in the main areas of North America and Europe -- despite an increasing animal population. It seems likely that the pace of the fall will be slower than over the past few years and it is signifi-

TABLE XXXII

Consumption of Agricultural Twine in EEC (tons)

Year	Sisal	Polypropylene	Total
1972/73	212,000	12,000	224,000
1973/74	194,000	22,000	216,000
1974/75	133,000	30,000	163,000
1975/76	140,000	39,000	179,000
1976/77	119,000	44,000	163,000
1977/78	110,000	45,000	155,000

cant that sales of conventional balers and round balers continue in most countries at a fairly steady level. The overall deficit in the market may be made up in part at least by increased demand in Eastern Europe (and perhaps the USSR where wire baling dominates), and other areas.

In conclusion, assuming that the price relationship between sisal and polypropylene remains about unchanged, then it seems likely that polypropylene can make further modest inroads into the overall market over the next few years, after which a period of stability is foreseen. The scope in this respect would, however, be relatively small and no doubt in some cases of dubious commercial viability. Should the price relatively change substantially, then naturally a different picture could emerge and the results of recent market movements in both polymer and sisal must, therefore, be awaited.

It is more difficult to forecast a growth rate in the harvest twine segment because of the more pronounced price and supply fluctuations. But even here, some experts think that within the next 10 to 15 years polypropylene should capture 60 to 70 percent of this market. There will no doubt be years of big gains, no gains, and even some losses before this conversion occurs. But the overall superiority of polypropylene fiber over competing natural fibers, as well as its availability, will ensure growth in all segments of the cordage industry.

There are five basic product groups within the industry: (1) rope, with a variety of applications; (2) tying twine, used primarily for packaging; and (3) baler or harvest twine, used exclusively for baling; (4) fishline and nets; and (5) macrame and other craft yarns. Table XXXIII gives a breakdown of this market for the United States.

A quick look at the rope and cordage industry as a whole reveals a competition among polyolefin fibers (95 percent of which is polypropylene), nylon, polyester, rayon, natural fibers, and in some cases, wire. In 1979, 645 million

TABLE XXXIII

U.S. Polyolefin Fiber Shipments

	(million lbs)					
	1972	1973	1974	1975	1976	1977
Total	312.3	395.4	426.3	436.1	482.0	520.0
Cordage	53.4	63.6	71.5	82.0	70.8	84.0
% of Total	17	16	17	19	15	16

pounds of polyolefin fiber were shipped in the United States. Of that amount 116 million pounds were used in the cordage industry - 18 percent of the total. The cordage industry has been fairly flat in its efforts to increase this overall consumption but it is expected that this market will not consume more than 20 percent of the total polypropylene fiber production.

No segment of the cordage industry currently demonstrates the potential of polypropylene more than the rope market, which accounts for the largest portion of United States polyolefin shipment with the cordage industry. In 1968, the United States shipment totaled 12 million pounds. Since that time they have more than tripled as the fiber made inroads against manila, the competing natural fiber, and other synthetics like nylon (Table XXXIII). Here again, Western Europe consumed almost twice the amount of the United States (Table XXXIIIa).

TABLE XXXIIIa

U.S. Rope Shipments (ref.5)

	(million lbs.)					
	1972	1973	1974	1975	1976	1980
Polyolefin	27.2	27.8	32.2	34.0	40.1	45.0
Nylon	23.4	29.1	21.6	20.4	22.5	-
Polyester	8.6	7.3	7.9	5.9	7.0	-
Rayon	6.3	5.3	6.5	4.0	4.0	-
Hard Fiber	52.0	52.4	49.4	40.6	44.6	-

A special segment in cordage and twine market is that of macrame and craft yarns. About 20 million pounds of polypropylene are being consumed in this market in the United States. Most of this market segment is dominated by BCF yarns in the United States. Substantial quanitites of staple spun yarns are also involved, particularly in Mexico, Canada, and the U.K. where there seems to be practically no growth for BCF yarns in this market, staple spun yarns may see a modest growth.

II.4. Bags and Wraps: Table XXXIC shows the growth of polypropylene use in the manufacture of bags and balewrap cloth. Here again, Western Europe consumed more polypropylene for this use than did the United States. Table XXXIV shows the total size of the United States industrial bag market and polypropylene's share of it. Woven polypropylene's share of the market grew continuously until 1976 when the entire market, which had been in a state of decline for several years, suddenly reversed itself. Woven polypropylene peaked in 1974 at 220 MM yd, then suffered two successive years of decline before recovering in 1978. This was due in large part to a reversal in prices that gave burlap an advantage for the first time in several years.

TABLE XXXIV

Consumption of Polypropylene for the Manufacture of Bags and Balewrap Cloth

	1967	1970	1973	1974	1975	1976	1977	1978
United States	6.5	4.5	20.5	25.0	24.4	23.7	24.2	25.0
Western Europe	2.5	17.0	53.5	61.0	60.0	61.5	64.2	66.0

TABLE XXXV

Growth of Bags Market in the United States (ref. 9)

Year	Total market	Woven PP	% PP
1970	840 MM yd^2	30 MM yd^2	4
1971	775 "	110 "	14
1972	700 "	140 "	20
1973	650 "	155 "	24
1974	620 "	220 "	35
1975	445 "	195 "	44
1976	510 "	190 "	37
1977	490 "	195 "	40
1978	480 "	200 "	42
1979	470 "	215 "	46
1980	460 "	230 "	50
1981 (est.)	450 "	240 "	53

Currently about 80 million pounds of polypropylene is being consumed for bags and bale wraps. It is felt that during the next five years there will be a definite improvement in polypropylene's share of the industrial bag market (Table XXXV). Such growth will not be automatic, however, as much research in new areas will be necessary to accomplish it. For example, the development of minibulk bags to offset, in part, the trend to bulk handling, is important. Such bags, capable of holding 1,000 to 3,000 lbs, can eliminate 10 to 60 small bags and yet allow the user the advantage of a bag. Tubed bags that are heat sealed with an adhesive and eliminate the need for sewing are also important in future growth, as are insect repellent-treated fabrics that will reduce or eliminate bug infestation in food stuffs, such as flour while it is in transit or storage.

One bag area which is often overlooked is that of open-mesh, or leno as it is popularly known. Products normally packed in leno bags are oranges, grapefruits, lettuce, cabbage, onions - anything that needs plenty of fresh air to remain fresh. The leno market is in the neighborhood of 90 million square yards yearly in the United States alone. Before woven polypropylene was introduced it was dominated by twisted paper, but since 1971 paper's share has been steadily decreasing. Now it is well below 50 percent. As with the aforementioned industrial bags, price was the big factor in the changeover. This is a relatively stable market and one does not see it growing by more than 10 to 15 million square yards in the next 5 years. During that time, paper's share will continue to fall to extinction.

Another interesting new development in packaging is bale wraps for cotton and wool. The annual United States cotton crop runs 8 to 12 million bales depending upon weather, prices, etc. As each bale uses about 8 square yards of wrapping, this market presents a potential of 65 to 100 million square yards yearly in the United States.

The use of polypropylene woven fabrics for cotton bale packaging started in the 1973-1974 season in the United States. During the same period the packaging of wool with woven polypropylene fabrics was allowed in New Zealand and Australia. While polypropylene fabrics have taken almost all of the wool packaging market, it has penetrated the cotton packaging up to 60 percent. It is estimated that this market will grow rapidly and that by 1985 almost all of the United States cotton will be baled with polypropylene fabrics.

A third bale-covering market that is almost as big as the cotton market is that for synthetic staple fibers. For years the staple industry used corrugated boxes to ship its product. In the past few years, however, the high cost of the boxes and the additional freight rates involved in shipping have forced the industry to start changing to flexible bale packaging. It is estimated that 15 to 18 million square yards of woven polypropylene is already being used in the United States market, a market which could consume some 75 million square yards at present if all staple were shipped in flexible packaging. Basically, a plain woven fabric with a latex-type coating is used. For exports or domestic shipments where moisture resistance is desirable, a thin film is laminated to the fabric.

In total, the three major United States markets for bale wraps mentioned above, consumed approximately 20 million pounds of polypropylene in 1981. Woven polypropylene had more than 40 percent of the total market. By 1985 polypropylene could have more than 60 percent of the total market. In pounds, this converts to 50 to 60 million pounds. If we add in 10 percent of the multiwall paper market, the figure could jump to 100 MM pounds of polypropylene consumed in bags and industrial fabrics in the United States by 1985.

II.5. Geotextiles and agricultural applications: Geotextiles is perhaps the fastest growing area for polypropylene. Both woven and non-woven fabrics are presently used. These fabrics provide a combination of reinforcement separation, drainage, and filtration. Some of their uses are listed in Table XXXVI. In agricultural fabrics are mainly included poultry and shade cloth.

Geotechnical applications of textile products is relatively a new area of activities. Starting just a few years ago, the market increased to about 30 million pounds of fibers in the United States alone. Estimates for 1985 range from 200-300 million square yards or about 60-90 million pounds of fibers. Polyester, nylon and polypropylene fibers are competing for this market. Presently, the polypropylene share of the market is about 50% of the total market. This share is definitely going to increase, probably to the extent of 90% by 1985. The increased total market and an improved share for polypropylene fiber in the total market makes it a fascinating growth market for polypropylene woven and non-woven fabrics. In 1981 about 20 million pounds of polypropylene fiber was used for this application. The estimates for 1985 are between 50 to 80 million pounds, of which about 25 million pounds will be non-wovens and the balance will be woven fabrics produced from tape, fibrillated film yarn and monofilament in varying combinations. The use of filament yarns for this application will be limited.

Europe has been the leader in geotextiles. A partial explanation is that European construction firms and government agencies recognized that building materials could be used more economically when geotextiles were properly applied. Building materials like gravel and some types of sand are in relatively short supply in Europe and are more expensive than in the United States. Energy costs for extraction and transportation expenses for these products have been much higher in Europe. The escalation of energy costs and the growing scarcity of some building materials should be a strong motivating force to use more geotextile materials in the United States.

II.6. Other Industrial Uses: Apart from the above mentioned major areas of market, i.e. home furnishing, cordage and twine, bags and wraps and geotextiles there are some other markets in which polypropylene fiber is playing an important and growing role. These markets are small but, taken togeher, form a sizeable sector of total consumption of polypropylene fiber. This category covers a miscellany of application, which can be grouped as follows:

A. Industrial Applications

Automotive uses (mainly rubber liner)

Filtration

Narrow fabrics and webbing

Cable binding

TABLE XXXVI

Some Applications for Geotextiles

Road Stabilization	Drainage For:
Access Roads To:	Highways
Building Sites	Tennis Courts
Drilling Sites	Septic Systems
Logging Areas	Driveways
Farm Areas	Parking Lots
Grazing Areas	Sport Facilities
Mining Sites	Airport Runways
Materials Storage Sites	
Power Lines	

Area Stabilization	Paved Surface Reinforcement Of:
Feed Lots	Highways
Race Tracks	Driveways
Golf Bunkers	Parking Lots
Tennis Courts	Tennis Courts
Parking Lots	Airport Runways
Materials Storage Areas	
Baseball Fields	
Football Fields	
Soccer Fields	
Pipelines	

Erosion Control For:
Canals
Beaches
Dams
Jetties
Levees

Brushes and bristles

Strapping

Sewing thread

Coating substrates

Vacuum cleaner hose

Traps and pool covers

Wind screens

Nursery squares

B. Hospital, Medical and Disposable Products

Surgical packaging

Hospital uniforms

Incontinent pads

Disposable diapers

Industrial uniforms and coveralls

Tea bags

Sanitary towels

C. Reinforcement Applications

Cement reinforcement and thixotropic products

Paper reinforcement

Because of the fragmented nature of this market, reliable statistics are not readily available in terms of the above mentioned application categories. However, some idea of the extent of this market can be obtained from Table XXXVII and the following stray remarks about the U.S. market.

TABLE XXXVII

U.S. Polypropylene Consumption for Industrial Uses

	(millions lbs)			
	1977	1978	1979	1980
Narrow fabrics		11.1	11.8	
Rubber industry		0.8	1.6	
Medical and Sanitary		19.0	23.8	
Auto seat strip covers		0.4	0.4	
Filtration		2.3	2.5	
Sewing thread		0.5	1.4	
Tarps, pool covers, etc.		0.5	2.6	
Cable binding		4.1	4.2	
Substrate fabrics, etc.		5.0	5.0	

The use of polypropylene *sewing thread* is very well restricted to bag manufacturing. It may also be used to some extent by the ultimate bag user for closing the bags but this is not believed to be the general case. It is believed that the current U.S. consumption of polypropylene thread for bag manufacturing amounts to 3/4 to 1 million pounds/year. Some significant amounts may also be used for bag closing by the ultimate users but this does not seem to be a major market. In addition, nominal amounts are used for seaming bale wraps and joining industrial and agricultural polypropylene fabrics. There is no apparent reason to expect a big increase in the use of polypropylene sewing thread. The estimated U.S. consumption in 1985 could be 2 to 3 million pounds.

Narrow fabrics are divided into four general classifications:

Trimmings - including ribbons and labels

Tapes - elastic and nonelastic (weight less than 15 oz. per sq. yd.)

Webbing (weight more than 15 oz. per sq. yd.)

Braided Fabrics

In general, polypropylene is not an important factor in trimmings or tapes (the first two classifications) at the present time. In the case of trimmings for apparel use, there is the concern about the effect of ironing temperatures. Polypropylene has been used to some extent for non-apparel uses such as fringes, curtain trimming and cords for home furnishings products.

Elastic tape manufacturers in general use yarn covered with cotton or polyester but serious development work with polypropylene is being undertaken by a few U.S. manufacturers for bra and girdle elastics. If successful, this could lead to a major new market for polypropylene.

The *webbing* area is divided into several categories:

Apparel/Sporting Goods - straps for back packs, life preservers, harness, camping

Household/Auto - car top, bike racks, luggage, outdoor furniture, etc.

Safety Belts - auto and other

Other - industrial including military, slings, tie downs, etc.

Polypropylene is an important yarn in all of the above except for auto safety belts where polyester yarn is used almost exlusively, having better physical properties for this particular application. Polypropylene has essentially replaced cotton and nylon for general strapping use and is taking over the narrow strapping area. It is significant that Elizabeth now extrudes all of their polypropylene requirements and Southern Weaving is going in that direction. There are other areas in webbing such as high performance slings and tie downs where nylon still dominates. Polypropylene monofilament for outdoor furniture webbing is the single most important sector in narrow fabrics.

The current U.S. consumption of polypropylene filament in narrow fabrics is estimated at 8 million pounds per year. The total market exclusive of auto seat belts is estimated at 40 to 45 million pounds per year. Since polypropylene has proven to be the best fiber for most webbing and strapping uses, a very good growth rate can be projected. The consumption could reach 18 million pounds in 1985.

A special case of narrow fabrics is the narrow woven *pile fabrics* for home, auto and industrial weathersealing around windows, doors, and specialty closures. The current U.S. consumption is under 2 million pounds. The expected U.S. market for 1985 is about 3 to 4 million pounds.

In 1980, an estimated quantity of 12 million pounds of polypropylene was used in filtration in the U.S. About 7 million pounds were used in staple form,

3 million pounds in multifilament form and about 2 million pounds in monofilament.

There is a trend toward finer denier yarns and greater use of monofilament yarn. There is also a trend toward wider fabrics, since much of the new filter equipment requires 72" or wider fabric. By 1985 the estimated market for multifilament polypropylene could reach 5 million pounds.

II.7. Speciality Products: This sector of the market mainly comprizes the following items:

a) Micro fibers

b) Cement reinforcement

c) Synpulp

d) Fibrillated network nonwovens

Mineral wool products made from slag date as far back as 1840 and were actually microdenier products. In the mid-1940s, Owens-Corning developed the Flame Attenuation Process for making microdenier glass fibers. The commercialization of microdenier products from organic polymers has been a long process.

The basis for some of our present-day microdenier technology is rooted in work done by Van A. Wente at the Naval Research Laboratory in about 1951. Wente's work was discontinued in the late 1950s and lay dormant until Esso Research (now Exxon) picked up where Wente left off and developed "Melt Blowing" technology for making fibers generally in the 2 to 4 micron range, using polypropylene and other thermoplastic fibers.

Exxon has licensed the technology to a number of firms now making commercial products using microdenier technology. Riegel Products was the first merchant source of microdenier materials from polypropylene and polyester. In the 2 to 4 micron range, polypropylene fibers have low strength, and high strength. They produce microporous structures when calendered. These fabrics are characterized by very high surface areas.

Because of the high cost of making microdenier nonwovens, their use is confined to applications where the high-performance requirements of these materials is justified. In Table XXXVIII is shown some of the major uses for microdenier products.

Since some microdenier polypropylene products sell as high as ten times the base resin price on a per pound basis, large poundages of resin are not consumed. Table XXXIX gives a best estimate of the consumption of polypropylene in these products and the projected usage.

Inmont, who purchased the PNC process technology for fibrillated network nonwovens several years ago, is the only known practitioner of this technology in the United States. In Europe, the technology is used to make diaper cover stock and for covers for incontinent pads. These markets are also under investigation in the United States. Other uses for PNC polypropylene products are for protective covers for fiber glass products and for protective networks for wallboard.

TABLE XXXVIII

Major Uses for Polypropylene Microdenier Products

Industrial
Battery Separators
Oil Spill Cleanup
Filtration Air and Liquid
Face Masks
Consumer
Diaper Cover Stock & Sanitary Napkin Cover
Vacuum Cleaner Bags
Hospital/Medical
Surgical Face Masks
Surgical Instrument Wrap
Apparel
Home Furnishings
Limited-Use Table Cloth

TABLE XXXIX

Estimated Consumption of Polypropylene in Microdenier Nonwovens

	1979-1985
1979	8-10 Million Pounds
1985	15-20 Million Pounds

The statistics for synpulp and reinforcement fibers are not available. Both of these markets are not yet stabilized and may take several years to develop. The total impact is expected to be high.

III. LONGTERM FUTURE PROSPECTS

As important as the immediate future growth of polypropylene fiber is its long-term future. Prospects for polypropylene fiber depend on several interrelated factors, the most important of which are the growth of synthetic fibers in general compared with natural fibers, the relative availability of raw material for polypropylene vis-a-vis nylon and polyester, the availability of technical know-how for fiber production, and the innovations in the use of polypropylene fibers in textile, home furnishing, and industrial uses.

Although synthetic fibers were developed and produced commercially in the 1930's, it was not until the 1950's that production exceeded 100,000 tons per annum. World production of fibers in 1978 was 28 million tons, of which synthetics represented almost 30 percent of the total. The prices of synthetic fiber were initially high, but with increasing volume of production these have fallen steadily over a number of years and the synthetics have become competitive with natural fibers, particularly with cotton and wool and lately with jute and sisal, on cost/performance basis.

The base chemicals required for fiber production are derived primarily from naphta, which is in turn obtained from crude oil distillation. In the United States, natural gas is also extensively used as a source of chemical feedstocks. Through the 1950s and 1960s the prices of naphtha and natural gas rose little, which meant a low-cost feedstock for synthetic fibers. This situation continued until the oil crises of 1973. The escalation of crude oil prices also reflected in the increased cost of naphtha. As a result, in Western Europe, propylene cost increased from 72 units to 190 with a reflected price increase in polypropylene polymer from 450 to 750 units per metric ton. In a similar manner benzene (the raw material for nylon) increased from 62 to 335 units per metric ton over the same period.

In the production of fibers the effect of oil prices is lessened at each subsequent stage of processing, as is the effect of energy input. The base chemicals used account for 15 to 20 percent of the final cost of the staple fiber and energy cost about 20 to 25 percent. Taken overall, the increase due to the energy crisis probably did not exceed 20 percent over the last decade. The oil crisis also contributed to the increased cost of producing cotton, in particular the increased cost of fertilizers and pesticides and power for machinery operation and transportation. For example, the cost of fertilizer is estimated to have risen by more than 160 percent from 1972 to 1978. Thus, the impact of the oil crisis has been more adverse to cotton than to synthetic fibers. With any further increase in oil prices, natural fibers will suffer more than the synthetics. This is in addition to the steadily growing population pressure for food production which is taking away agricultural land to produce wheat instead of cotton and rice instead of jute. Similar factors are contributing to a shrinkage of pasture land needed for wool production.

Cotton, wool, jute, and sisal are subject to climatic variations; the yield varies from year to year, which can give rise to shortages or surpluses causing considerable price fluctuation. Natural fibers are often grown in regions far from the market place for the end product. Transport and handling costs thus greatly exceed the equivalent costs for the synthetic fibers. Due to the fluctuation in natural fiber production it is necessary to hold substantial stocks, again contributing to cost. Labor and machinery costs are significant due to the labor-intensive nature of natural fiber production, even though wages are normally low by comparison with those in the synthetic fiber industry. Conversion costs, i.e., those incurred in converting the raw fiber into a condition suitable for textile processing, are additional. Synthetic fiber (staple) cost should therefore be compared with the cost of the converted natural fiber.

In view of the above considerations, it is safe to say that synthetic fibers will continue to gain at the expense of natural fibers. The overall price index for synthetic fibers will no doubt continue to rise, but mainly due to inflation; over the longterm the rise in price for synthetic fibers will be less than the price rise for natural fibers.

The growth of synthetic fibers during the rest of this decade and through the 1980s is estimated at 4 to 7 percent annually. Of this growth, polypropylene fiber is expected to claim a major share. According to various industrial estimates, a growth rate of 10 to 12 percent is predicted for the United States or about 15 percent on a global basis. The current share of polypropylene fiber in noncellulosic synthetic market is about 10 percent on a worldwide basis. Western Europe ranks first in this respect (12 percent) followed by the United States (9 percent), and Japan (7 percent). With production gains of 10 to 15 percent yearly, the market share of polypropylene fiber could approach 12 percent for the United States by mid-1980. A 12 percent figure may be reached on the basis of total world production.

The reasons for these optimistic forecasts for polypropylene fibers are many. (1) An increase in relative cost and price pressure for jute and sisal will make polypropylene coarse yarns still more competitive with these natural fibers. (2) Expiration of basic patents for polypropylene production will attract several new producers, with the accompanying competition, technological innovations, novel uses, and generally increased "visability" of polypropylene fiber, are all bound to help it penetrate the traditional and nontraditional markets.

(3) As the technology of fiber production improves and becomes widespread, more and more fine-grade polypropylene staple will be available which will compete directly with cotton and wool. Similarly, the acquisition of improved continuous-filament yarn manufacturing technology by smaller producers and the easy availability of texturizing techniques and hardware will put pressure on nylon CF and BCF yarns.

(4) The continued trend of higher prices for the raw materials of nylon and polyester in comparison with polypropylene will eventually create a sizeable gap between the prices of polypropylene fibers and those of the "big three" (nylon, polyester and acrylic fibers).

Figure 6 shows that a number of principal synthetic fibers, with which polypropylene fiber competes, can be traced back to naphtha as the common feedstock.

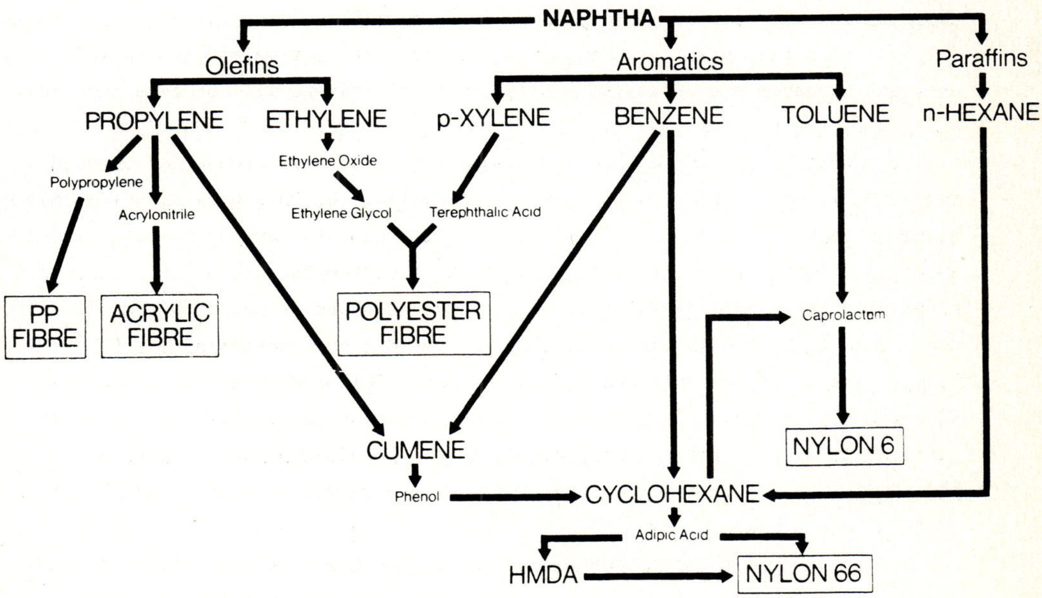

Fig. 6. Flow Chart of Production of Synthetic Fibers

In cracking of naphtha the prime purpose is to produce ethylene and it is therefore surprising that ethylene is the main cost bearer of all the product streams. On the other hand the price of propylene is usually about 70 percent of that of ethylene reflecting the relatively lower demand of propylene for other products (e.g. acetone, acrylonitrile, and propylene oxide) and the potential availability of propylene from other refinery operations at perhaps lower cost. The strong demand for the aromatics fraction for upgrading gasoline, particularly when lead free gasoline is required, has caused the price for these aromatic products to increase relatively more sharply than the olefin feedstocks. It is of course conceivable that alternative methods and additives may be developed commercially for gasoline in due course, reducing the demand for aromatic fraction, but one would anticipate any effect of such a development only becoming significant after a number of years and therefore not influencing the situation until towards the end of the next decade.

As mentioned earlier, polypropylene fiber has a distinct advantage over other synthetic fibers in terms of related energy costs as shown in Table XL, which is based on data published by Woodhead. Unfortunately there are no reliable figures for jute and sisal, which have been principal natural fibers with which polypropylene fiber has been competing up to now, but they would probably be of the same order as cotton. However, it is also important to bear in mind the comparative technical equivalence of materials. For example, in certain carpet constructions it might be necessary to offset some particular performance limitations of polypropylene, such as resillience, for instance, by using more of it in comparison with, say, nylon thereby reducing the apparent price/performance advantage enjoyed by polypropylene. The reverse is, however, also true, especially when compared with jute, sisal, cotton, cellulosics, and some extent with acrylics and wool.

Figure 7 shows the price movements that have occurred over the past five years for polypropylene staple fiber as well as for those other synthetic staple fibers, such as nylon and acrylic, with which it is chiefly competing. In view of what has just been said, one would expect this relative price pattern to continue for the foreseeable future although there may of course be periodic fluctuations reflecting temporary changes in the supply/demand picture of any one polymer.

From the viewpoint of long range future prospects, the big growth potential for polypropylene fiber is in Eastern Europe, the Soviet Union, China, and the developing countries of Asia and Latin America. The production of polypropylene fiber in Eastern Europe and the Soviet Union has just started. Because of the unavailability of adequate technology, a lag of a few years is expected before polypropylene fiber production can really pick up in these countries. As the basic patents of Montecatini on polypropylene and polypropylene fibers run out, a great upsurge in polypropylene fiber manufacturing activities is anticipated

TABLE XL

Comparative Energy Contents of Fibers

	Polyester	Polyamide	Polyacrylics	Polypropylene	Cellulosics	Cotton	Wool
Energy Content (Tons oil Eq./ton)	3.9	5.5	6.9	2.8	2.7	0.7	1.7
Specific gravity	1.38	1.14	1.17	0.91	1.52	1.54	1.32
Relative energy content per unit volume	2.1	2.5	3.2	1.0	1.6	0.4	0.7

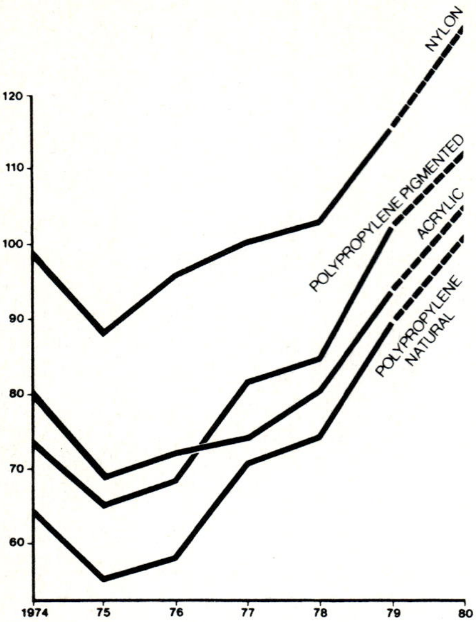

Fig. 7. Indexed prices of synthetic staple fibers

in the late 1970s and early 1980s. Due to a relatively low standard of living, the market constraints which are operative in Western Europe and North America are not likely to play a role as a braking factor for polypropylene growth in socialist countries in the foreseeable future.

The rapidly increasing population and the steadily rising standard of living in developing countries are the two important factors which will influence the demand for synthetic fibers in general and polypropylene fibers in particular in these countries. The increasing demands for food will result in increasing pressure on the land which will be given over more and more to food production. Natural fibers will have to struggle against heavy odds for the growing space they need. Thus, the cotton fields will be needed for wheat production and jute acreage for paddies. As the less developed regions of the world increase their living standard, the need for textile fibers will increase at a faster rate than the population growth. At the present time, polypropylene is the only fiber developed or theoretically feasible that can combine serviceability with economy and availability. It is safe, therefore, to predict that the next decade or two will see a sizeable growth in polypropylene fiber in less developed countries of the world.

All of the developing countries suffer from a chronic shortage of capital, especially in the form of foreign exchange for the purchase of machinery and equipment. Because of the amenability of polypropylene fiber to production on a small scale and because of the recently available "short-spin" processes for the production of polypropylene fibers, such an investment becomes within reach of these countries. The short-spin routes of manufacturing fibers and yarns are small in capacity and low in capital expenditure. They are slow in production rate and are rather labor intensive. These factors are, however, of less importance to developing economies than to labor-expensive industrial societies.

For the next few years, until OPEC production of polymer is initiated, the fiber producers of developing countries will be relying mainly on polymer produced by industrially developed countries. On a regional basis the cost of producing fiber in Europe, OPEC developing countries and non-OPEC developing countries, will be about equal. (Fiber cost comparison is given in Table XLI.)

After OPEC-produced polymer becomes available in the 1980s, the regional cost of fiber will be lower for developing countries than in Europe or the United States. Their regional advantage derives from assumed lower polymer cost, as seen in Table XLII. This cost comparison is based on the assumption that propylene prices to OPEC polymer producer will be two-thirds those prevailing in Europe and the United States. Another factor for lower fiber prices in the OPEC region will be the cheap rates of electricity.

Would the developing countries be able to export any sizeable quantity of polypropylene fibers to Western Europe and the United States? There is little possibility for this happening. Developing countries which are remote from developed markets and have no duty treaties with the industrially developed countries will not be able to export to the developed markets on a profitable full-cost basis because of transportation cost and duty penalties. Even the OPEC fiber producers, whose manufacturing costs are lower than the non-OPEC developing countries; will not be able to compete in Europe and the United States because of transportation and duty costs. Aside from cost considerations, many industrial end uses require close collaboration between fiber producer and processor which is difficult if a developing country is remotely situated from his market. For example, the apparently "simple" process of weaving carpet backing from polypropylene film fiber now necessitates close cooperation between producer and processor due to the stringent quality standards imposed by the tufters - especially, but not exclusively, for printed carpets. Furthermore, most processors are reluctant to rely on long lines of supply which are vulnerable to all kinds of interruption, whether it be political, meteorological, or strategic in origin.

For the above reasons, it is difficult to see more than a limited amount of "spot" business being open to exports from the developing countries and then only at very low prices which may not be remunerative even on the basis of marginal costs.

TABLE XLI

Fiber Cost Comparison (ref. 4)

	Relative units of cost					
"Short-spin" staple 2 lines-1,200 T/a each 15 shifts/week	(Europe)	1975-1980 Developing countries (OPEC)	(Non-OPEC)	(Europe)	1980-1985 Developing countries (OPEC)	(Non-OPEC)
Polymer (S.P. granules)		=	=		-	-
Other materials		=	=			
Manpower		-	-			
Power		-	=		Assume relative costs	
Packaging		+	+		of all other elements	
Gen. exp.		+	+		remain the same	
Depreciation (5 years S.L.)		+	+			
Transp. (regional)		=	=			
Marketing exp. (regional)		=	=			
Cost of capital		-	-			
Cost of sales (regional) (1)	100	98	98	100	90	94
Profit	10	10	10	10	9	10
Selling price (regional)	110	108	108	110	99	104
Add transp. to MDC markets (Europe)	-	10	10	-	10	10
Duty 10%		12	12	-	11	11
Available price MDC area (Europe)	110	130	130	110	120	125

(1) MDC (Europe) is Base = 100

TABLE XLII

Polymer Cost Comparison (ref. 4)

			Relative units of cost	
			Developing countries	
Capacity 60,000 t/y 1 line		(Europe)	(OPEC)	(Non-OPEC)
Propylene			-	=
Other raw material			+	+
Total raw material			-	+
Manpower	(1)		-	-
Maintenance material			+	+
Power			-	+
Packaging			=	=
Gen. exp.			-	-
Depreciation			+	+
Transportation (regional market)			+	+
Marketing (regional)			-	-
Cost of capital			-	=
Total cost of sales	(3)	100	77	99
Profit		10	8	10
Selling price		110	85	109
Granulation cost	(2)	29	29	29
Selling price (granules)		139	114	138
(regional market)		--	--	--

(1) Includes operating, maintenance and laboratory labor; (2) includes stabiliser, transformation, packaging in sacks and profit; (3) MDC (Europe) is Base = 100.

Therefore, it is important that the fiber production facilities in developing countries be carefully matched to the size of their captive market. This situation gives rise to the opportunity to apply what is termed a "total marketing" concept, that is, a marketing service in which the needs of the total polypropylene fiber industry within each interested developing country would be studied from A to Z, to evaluate local needs and opportunities, define fiber production requirements, assist with plant procurement, erection, and commissioning, define down-stream transformation industries, establish local applied research programs, local and overseas marketing consultancy (pricing) etc.

This type of "total marketing" approach is feasible in the case of polypropylene fiber because of the relatively limited number of end-use variations. This contrasts with the conventional textile fibers (polyester, acrylic, and nylon) where the availability of an enormous variety of processing techniques, in combination with an infinite variety of taste at the consumer level, have created the opportunity for a seemingly endless number of end-use variations for which it would be very difficult to establish a balanced industry plan. The present difficulties in the textile industries of developing countries derive partly from this difficulty.

WORLD POLYPROPYLENE FIBER PRODUCERS

Australia

Patchogue Plymouth, Division of Amoco Chemicals Pty. Ltd. Liverpool, N.S.W.,(film)
Sarlon Australia, Punchbowl, (monofil and film)
Vilene Australia, Minto, N.S.W., (spunbonded)
The following companies produce olefin fiber from film:
A. Abrahams & Sons Ptd. Ltd., Clayton, Victoria (Koala, Polytex)
Broons Plasweave Pty. Ltd., Altona, Victoria, (Polyfabric)
Cordex Industries Pty, Ltd., Gladesville, N.S.W.
Geo. Kinnear & Sons Pty. Ltd., Footscray, Victoria, (Emu)
Miller Ropes, Twine & Textiles, Geelong West, Victoria
H. J. Penno, Adelaide, South Australia
Tapex Pty. Ltd., Sydney, N.S.W.

Austria

Chemie Linz A.G., Linz, (Yarns, staple and spunbonded)
Chemiefaser Lenzing A.G., Lenzing, (film)
M. Teuffelberger, Mech. Hanf. Und Drahtseilerei, Wels TWEW (film)

Belgium

N.V. Neofil, Tessenderlo, (filament yarn and staple)
Thiokol-Ten Cate Belgium, Overpelt (Limberg), (monofil)
The following companies produce olefin fiber from film:
N.V. Fibrillo, Zele
Ostend Stores & Rope Works S.A., Ostend
Poly-Texco, Desselgem
S.A. Firme Pardou-Duthoo, Roeselaere
S.A. Uco, Gent
Savicom, Zele
Luxilon Thread Industries
Polycarpets

Brazil

Brasil Viscose S/A, Jundiapeba (Sao Paulo), (staple)
Equipesca - Equipamentos De Pesca S.A., Campinas (film)
Ethuria S.A. - Industria De Fibras E Fios Sinteticos, Sao Paulo, (staple)
Manap - Manufatura Nacional De Plasticos, Osasco, (film)
Monofil - CIA. Industrial De Monofilamentos, Sao Paulo, (film)
Spumar S/A Industria E Comercio, Sao Paulo, (yarn, staple and film)
The following companies are known to produce olefin fiber from film:
Cacique De Embalagens S.A. - Industria E Comercio, Londrina, Parana
C.A.T.A. - CIA. Amazonia Textil De Aniagem S.A., Belem, Para
CIA. Fab. Yolanda S.A., Recife, Pernambuco
CIA. Jauense Industrial, Sao Paulo, Sao Paulo
CIA. Uniao Manufactoro De Tecidos, Rio de Janeiro, Rio de Janeiro
Commercial Importadora Cauduro, Santos, Sao Paulo
Dicoplast S.A. Industria E. Comercio De Plasticos, Presidente Prudente, Sao Paulo
Empla - Embalagens Plasticas S.A., Rio de Janeiro, Rio de Janeiro
Fibrasa - Ind. E Com. De Fios E Embalagens S.A., Espirito Santo

Fibrotex - Tecelagem De Fibras S.A., Sao Paulo
Fitesa/Olvebra - Fiacao Textile De Embalagens Plasticas S.A., Proto Alegre. Rio Grande do Sul
Ibis De Plasticos S.A., Compinas, Sao Paulo
Industria De Plasticos Zaniolo S.A., Curitiba, Parana
Industria Textil Tsuzuki Ltda., Suzano, Sao Paulo
M. Lepper & Cia. Ltda., Joinville, Santa Catarina
Oscar Bergrin & Cia. Ltda., Americana, Sao Paulo
Pigmaplast - Industria E Comercio De Plasticos Ltda., Sao Paulo, Sao Paulo
Plasticos Do Parana Ltda., Londrina, Parana
Plasticos Linil S.A., Joao Pessoa, Paraiba
Plastral - Industria E Comercop Ltda., Sao Paulo, Sao Paulo
Plastilira Industria De Embalagens Ltda., Campinas, Sao Paulo
Polimas - Ind. De Plasticos Textil, Santa Catarina
Sacoplast/Planosa - Plasticos Do Nordeste S.A., Natal, Rio Grande do Norte
Saronn Embalagens Plasticas Ltda., Sao Paulo, Sao Paulo
Textiplast, Joinville, Santa Catarine
Tecplast S.A., Brasilia, D.F.
Tomce - Industria E Comercio De Plasticos Ltda., Sao Paulo, Sao Paulo
Uniao Textil Quim. Brasileira S.A., Tiete, Sao Paulo

Canada

Calko Mills Ltd., Riviere du Loup, Quebec, (monofil)
Canada Western Cordage Co., Ltd., New Westminister, B.C. (monofil)
Celanese Canada Limited, St. Jean, Quebec, (yarn and staple)
Doon Twines Ltd., Kitchener, Ontario, (yarn, monofil and staple)
Du Pont of Canada Limited, North Bay, Ontario, (film)
Fiberworld, Ltd., Hawkesbury, Ontario, (film)
General Wire & Cable Company Ltd., (subsidiary of Belden Corp.), Cobourg, Ontario, (monofil)
Guelph Twines Ltd., Guelph, Ontario, (monofil and film)
Novastrand, St. John, New Brunswick, (monofil)
Patchogue Plymouth, Division of Amoco Canada Petroleum Co., Ltd., Hawkesburg, Ontario, (film)
Poli-Twine Corp. Ltd., Belleville, Ont. & Saskatoon, Saskatchewan, (monofil)
Polymer International (N.S.) Ltd., Truro, Nova Scotia, (film)
Schlegel Company Canada Ltd., Oakville, Ontario, (monofil)
Strudex, Division of Kraus Carpet Mills Ltd., Waterloo, Ontario (staple and yarn)
Thiokol Fibres Canada Ltd., Brantford, Ontario, (fiber, monofil)
York Bag Company, Toronto, Ontario, (film - captive)

Chile

Tejidos Plasticos Interplast S.A., Santiago, (film)

China (Taiwan)

Hsieh Chin Fishing Net Company, Ltd., Kaohsiung, Hsieh-Chin (monofil)
Pao Tsung Tai Industrial Company, Ltd., Tainan, (film)
San Yang Industrial Company, Ltd., Nei Hu Hsiang, (film)

Columbia

The following companies are known to produce olefin fiber from film:
Compania Nacional De Empaques S.A., Medellin
Plasticos De Magdelena Ltda, Santa Marta
Polymer De Colombia S.A., Cartagena

Czechoslovakia

Chemicke Zavody Juraja Dimitrova, N.P., Bratislava, (yarn and staple)
Chemosvit, N.P., Svit Pod Tatrami, (staple)

Denmark

Esbjerg Tovvaerksfabrik A/S, Esbjerg, Westfex, (monofil and film)
Jakob Holm Varde, Varde, Ihs, (staple and film)
O-B Rasmussen Ingeniorfirma, Birkerod (pilot plant), (yarn)
The following companies produce olefin film fiber:
F.D.B.'s Fabriker, Viby Jylland
Randers Rebslaaeri A/S, Randers
R.F.W. Plast-Industri A/S, Vejle
Roblon A/S, Fredrikshavn

El Salvador

Filamentos Y Perfiles, S.A. (Filpersa), San Salvador, (monofil)
Salvaplastic, S.A., San Salvador, (monofil)

Finland

The following companies produce olefin fibers from film:
Okarla Oy, Piispanristi
Piippo Oy, Lempaala

France

The following companies produce olefin film fiber:
Amman, Bataille (Ayello Group)
Bemis
Bihr
C.E.P.P. (Weille Group)
C.I.C.O.R.
La Cellofibre
Ets. Carmichael
Filatures & Tissages De Saulxures
Lainiere De Roubaix
Rigot-Stalars & Vandesmet
Saint Greres
Sein & Lys
Sommer

Germany (Federal Republic of)

Chemische Werke Huls A.G., Marl, Kreis Recklinghausen, Westphalia, (monofil)
Dynamit Nobel A.G., Troisdorf, Bez Koln, (monofil and tow)
Ernst Jaeger, Dusseldorf-Reisholz, (staple and film)
Patchogue Plymouth, Division of Amoco Deutschland GmbH, Gronau, (film)
Reinhold, Selbitz/Ofr., (staple)
Spinnerei Bruckmuhl/Schaeffler, Bruckmuhl, (staple)
Steen, Hamburg, (staple)
The following companies produce olefin film fiber:
J.F. Adolff, Backnang
P. Baumhuter, Batenhorst
Bremer Woll-Kaemmerei, Bremen-Blumenthal
Hfi Hanfwerke Fussen-Immenstadt, Fussen
Hanfspinnerei Westfalia, Minden

Kusga, Mainz
Nordhessische Haarwarenfabrik, Felsberg
Odenwald-Chemie, Schoenau
Polyunion Kunstoffwerk, Mesum
Rhein-Main-Plastik, Michelstadt
J. Schilgen, Emsdetten
Spohn & Knoll, Freiburg
Textilwerke Todtnau, Todtnau
Dr. Karl Wetekam & Co., Melsungen
Ewald Wolf, Weissenburg

Greece

The following companies produce olefin film fiber:
Egli-Zinnober, Argos/Penoponnes
Elvisak A.G., Athens
Niva, Moschaton
C. Vellis Heirs, Piraeus

Guatemala

Cordeleria La Rapida, Guatemala, (monofil)

Honduras

Polymer, Subsidiary of United Brands, San Pedro Sula, (film)

Hungary

Kender Juta Textillipari V., Budapest, (film)
Kenderfono Es Szovoipari V. Szeged, Szeged, (film)
Magyar Viscosagyar, Nyergesujfalu, (staple)

India

Noemar Limited, Baroda, Panchmahal (Gujurat), (staple)

Iran

Several plants under planning - all film yarns

Ireland (Eire)

Aberclare Ltd., (formerly J. & L. F. Goodbody Ltd.), Clara, Offaly, Aberclare, (staple)
Goulding Plastics, Waterford, Waterford, (film)
Irish Polysacks Ltd., Blarney, Cork, (film)
Irish Ropes Ltd., Newbridge, Kildare, (film)

Italy

I.T.S. (Aerfil), Parabiago (Lombardia), (yarn)
Montefibre S.P.A., Terni (Umbria), (yarn and staple and monofil)
Neofil S.P.A., Terni (Umbria), (yarn)
The following companies produce olefin film fiber:
Emilio De Magistris, Milan
Linificio E Canapificio Nazionale, Milan
Montedison Film, Milan
Nutini, Montemurlo

S.E.A.L. S.P.A., Milan
S.I.R.P.A., S Dona di Piave

Japan

Asahi-Dow Ltd., Suzuka, (monofil)
Chisso Polypro Fiber Co., Ltd., Moriyama, (staple and film)
Daiwa Spinning Co., Ltd., Harima, (staple)
Mitsubishi Rayon Co., Ltd., Toyohashi, (yarn, monofil and staple)
Mpc-Spunbonded Inc., Yokkaichi, (spunbonded)
Taisei Rayon Co., Ltd., Mito, (monofil)
Toa Wool Spinning & Weaving Co., Ltd., Tomari, (staple)
Toray Industries, Inc., Shiga, (yarn and staple)
Toyobo Co., Ltd., Tsuruga, (staple)
Ube-Nitto Kasei Co., Ltd., Fukuyama & Koriyama, (yarn, monofil and staple)
The following companies produce olefin film fiber:
Daido Kasei, Osaka
Fujimoto Sangyo, Osaka
Hagihara, Kurashikl
Okamoto Chemical, Ehime
Sakai Kasei, Osaka
Sanso Jushi, Yokkaichi
Santo Chemical, Aichi
Sekisui Chemical, Amagasoki & Nara
Showa Polymer, Ibaragi
Toyama, Chiba
Wako Spinning Co., Gifu
Yamajin Textile, Fukui

Korea (Republic of)

Jeil Synthetic Fibers Industrial Company, Yung Dung Po, (yarn and staple)
Korea Hapsum Company, Ltd., Shihung, (staple)
Kum Sung Synthetic Fiber Co., Ltd., Seoul, (staple)

Luxembourg

Du Pont Fibres S.A., Contern-Hesperange, (spunbonded)

Malaysia

Folex Industries Bhd., Malacca, (film)
Westmalex Plastic Industries Sdn. Bhd., Malacca, (film)

Mexico

Fibrasomni, S.A., Txtapalapa (D.F.), (yarn and monofil)
Industrias Polifil, S.A., Lardizabal (Tlaxcala), (monofil, filament yarn and staple)
Vischisa, Chihuahua, (staple)
The following companies produce olefin film fiber:
Hilos Y Cordeles, S.A., Tlalnepantla, Mexico
Plasticel, S.A., Monterrey, Nuevo Leon (Plasticel)
Polisac, S.A. (affiliate of Nylon de Mexico, S.A.), Guadalajara, Jalisco (Polisac)

Netherlands

Kunstzijdespinnerij Nyma N.V., Nijmegen, (monofil)
N.V. Lankhorst Touwfabrieken, Sneek, (film)

Thiokol-Ten Cate N.V., Nijverdal, (monofil)
The following companies produce olefin film fiber:
N.V. Plastic-Extrusiebedrijf Verdelta, Vlaardingen
N.V. Verenigde Touwfabrieken (United Ropeworks), Rotterdam
Von Schuppen, Veenendaal

Nicaragua

Manufacturera Centro Americana S.A. (Macen) Managua, (film)

Norway

A/S Platon, Avd. Tekstil, Notodden, (film)
Norfil-Norpack A/S, Stavern, ner Larvik, (yarn, monofil and film)
Stal Og Tau, Tonsberg, (film)

Pakistan

Polypropylene Products Ltd., Karachi, (film)

Phillippines

Industrial Textile Manufacturing Co., (Intemcop) Mandaluyong, Rizal, (flim)

Poland

Chemitex-Wiskord, Szczecin-Zydowce, Staple)

Portugal

The following companies produce olefin film fiber:
Derfil, Maia-Douro
Cotesi, Grijo-Gaia
Companhia Uniao Fabril (C.U.F.), Lisbon
Eurofil, Lisbon

South Africa

Ropes and Mattings Limited, Jacobs (Natal), (film)

Spain

Paular, S.A., Puertollano (Ciudad Real) Istrakin, (monofil and film)
The following companies produce olefin film fiber:
Beniplast, Benijanim
Hilcosa, Barcelona
Intermas, Cardedeu
Juan Miro, Granollers
Sacofiplas, Valladolid
Saites, Sabadell
Augustin Vila, Tarrasa

Sweden

Sonesson Plast Ab, Malmo, Laveten, (monofil)

Switzerland

Arova A.G., Schaffhausen (Schaffhausen), (film)

Societe De La Viscose Suisse, Emmenbruecke (Luzern), (film)

Trinidad

Polymer (Caribbean) Ltd., Port-of-Spain, (film)

United Kingdom

Anglia Fibres Ltd., Hadleigh (Suffole), (monofil)
British Celanese Ltd., Coventry (Warwickshire), (monofil and film)
F. Drake (Fibre) Limited, Colcar, Huddersfield (Yorkshire), (staple)
Filamex Ltd. (Subsidiary of Ernst Scragg & Sons (Holdings) Ltd., Winsford (Cheshire), (yarn)
Film Fibres Division, Belfast Ropes Group (formerly Belfast Ropework Co.) Bluebell, (monofil, staple and film)
Filtrona Textile Products Limited, Dundee (Angus, Scotland), (film and staple)
James Mackie & Sons Limited, Belfast (N. Ireland), (staple)
Marlow Plastics Ltd., Hailsham (Sussex), (staple and film)
Patchogue Plymouth Division, Amoco (U.K.) Limited, Consett (Durham), (film)
Plasticisers Ltd., Drighlington (Yorkshire), (staple, monofil and film)
Thiokol Chemicals Limited, Forfar (Angus, Scotland), (yarn, monofil, ribbon and film)
The following companies produce olefin fiber from film:
Bridon Fibres & Plastics Ltd., Gateshead & Sunderland (Co. Durham) & Wallsend (Northumberland), Broplene, Red Star, Viking
Dawson Ltd., Oldham (Lancashire)
Fairburn Lawson Packing Ltd., Leeds (Yorkshire)
Polytape Ltd., Dundee (Angus, Scotland), Propatape, Propafil
Rothes Plastics Ltd., Glenrothes (Fife, Scotland)
Smith & Nephew Polyfabrik Ltd., Newhy, near Rochdale (Lancashire), Polyweave
Tay Textiles, Dundee (Angus, Scotland)

United States of America

American Manufacturing Company, Inc., Honesdale, Pa., (monofil and film)
Amitech, Inc., Odenton, Md., (monofil)
Andray Products, Inc., East Orange, N. J., (film - captive)
Arlin Manufacturing Company, Lowell, Mass., (film)
Armstrong Cork Company, Charlotte, N. C., (film - captive)
Bemis Company, Inc., Talladega, Ala. & St. Louis, Mo., (film)
Berkley & Co., Inc. (formerly Industrial Wire & Plastics Co., Inc.), Spirit Lake, Iowa, (monofil - captive)
Blue Mountain Industries, Blue Mountain, Ala., (film)
Bridon Cordage Ind., Albert Lee, Mn.
Chase Bag Co., New Orleans, La.
Chevron Chemical Company, Fibers Division (Subsidiary of Standard Oil of California), Dayton, Tenn., (film and staple)
Concorde Fibers, Inc., Columbia, Md., (yarn)
The Cordage Group, Division of Columbian Rope Company, Auburn, N.Y., (monofil and film)
Crowe Rope Company, Warren, Me., (monofil)
Crown Zellerback Corporation, Nonwoven Fabrics Division, Camas, Wash., (spunbonded)
E.I. Du Pont De Nemours & Company (Inc.), Textile Fibers Dept., Old Hichory, TN, (spunbonded)
E-B Industries, Inc., Simsburg, Conn., (film - captive)
Exxon Chemical Company U.S.A., Fabricated Products Department Fabrics Division, Summerville, S.C., (monofil and film)
Polybale Twine Company Div., Tracy, Minn., (monofil)

Fibron Incorporated, Chattanooga, Tenn., (film)
General Fabrics & Fibers, LaGrange, Ga., (film)
Georgia Synthetics, Inc., Elberton, Ga., (film)
H L Industries, Inc., Glen Buwnie, Md., (yarn)
Hercules Incorporated, Polymers Department, Home Furnishings Division, Oxford, Ga. & Covington, Va., Herculon, (yarn, staple and tow)
The Hooven & Allison Company, Xenia, Ohio, (film)
International Harvester, New Orleans Plant, New Orleans, La., (film)
International Fibers, Inc., Reidsville, N.C., (yarn)
Jackson Rope Corporation (Subsidiary of Aspro, Inc.), Reading Pa., (monofil and film)
Keller Industries, Inc., Swainsboro, Ga., (monofil - captive)
Kimberly-Clark Corporation, Neenah, Wis., Kimkloth, (spunbonded)
Lambeth Corporation, New Bedord, Mass., Lambeth, (monofil)
Moultrie Textiles Division of Moultrie Cotton Mills, Inc., Moultrie, Ga., (film - captive)
Mountain Corp., Pilot Mountain, N.C., (monofil)
NSI, Inc., (formerly Nowers Synthetics, Inc.) Soperton, Ga., (film - captive)
Newton Filaments, Inc. (Subsidiary of Newton Line Company), Homer, N.Y., (monofil)
The Osterneck Company, Lumberton, N.C., (film - captive)
Patchogue Plymouth Company, Division of Amoco Fabrics Company, Bainbridge, Hazelhurst & Nashville, Ga., Patlon, (film)
Phillips Fibers Corporation, Spartanburg, S. C., (yarn, Staple, tow and fiberfill)
Plant Industries (Texas) Inc. (Subsidiary of Plant Industries, Inc.), Houston, Texas, (film)
Shuford Mills, Inc., Hickory, N.C., (monofil)
Sunshine Cordage Corporation (formerly Poncar Plastic Corp.), Miami, Fla.,(monofil)
Synthetic Industries, Chickamauga, Ga., (film - captive)
Tennessee Eastman Company, Division of Eastman Kodak Company, Kingsport, Tenn., (film)
Tennessee Fibers (Subsidiary of Synthetic Industries), Decatur, Tenn., (film-captive)
Wayn-Tex Corporation, Fibers Division, Waynesboro, Va., (filling yarn)
Tubbs Cordage Company, Orange, Cal, Loktite, (monofil)
Vectra Corporation (Subsidiary of Chevron Chemical Company), Odenton, Md., Vectra, (yarn and staple)
Wall Industries, Inc., Beverly, N.J., (monofil)
Waltrich Plastic Corporation, Tennent, N.J., (monofil)
Wellington Synthetic Fibres, Inc., Montair Division Pilot Mountain, N.C., Montrel, (monofil), Plastic Woven Products Division, Leesville, S.C., (monofil), Poly Fibers Division, Trussville, Ala., (monofil and film)
Western Textile Products, Inc. (Subsidiary of Chase Bag Co.), Orange, Cal., (monofil)
The following companies are also producing a variety of textile products from olefin:
Conwed Corp., Conwed Plastic Netting (embossed film for netting)
E. I. Du Pont De Nemours & Company (Inc.),(embossed film for netting)
Hercules Incorporated Polymers Division, (embossed film for netting)
Nalle Inc., (embossed film for netting)
Reigel Products Corporation, (melt blown)
The 3M Company, (spunbonded)

IV. REFERENCES

1. T.B. Teeple, Jr., Chemtech, April 1973.
2. U.S. Dept. of Commerce, Current Industrial Reports, Carpets and Rugs.
3. Bernhardt Ten Hovel, Symp. Polypropylene Fibers in Textiles, Manchester, 1974.
4. S.A. Mossman, Man-Made Textile in India, Oct. 1976.
5. Data from Textile Economics Bureau, New York.
6. R.D. Evans, paper presented at Symposium Polypropylene's 20th Year in Textiles, Sept. 13-14, 1977, New York.
7. R.M. Hendricks, paper presented at symposium, Polypropylene's 20th Year in Textiles, Sept. 13-14, 1977, New York.
8. E. Silosberg, paper presented at symposium, Polypropylene's 20th Year in Textiles, Sept. 13-14, 1977, New York.
9. J.W. Deas, paper presented at symposium, Polypropylene's 20th Year in Textiles, Sept. 13-14, 1977, New York.
10. Exxon Chemical Company, U. S. A.
11. C.W. Manzione, paper presented at symposium, Polypropylene's 20th Year in Textiles, Sept. 13-14, 1977, New York.

SUBJECT INDEX